Small Angles, Large Benefits...

The Unique Behavior of Tapered Piles

Front cover: This graphic was originally created by the writer several years ago as an animated GIF to illustrate the unique shaft-resistance mechanism of *wedging* that occurs with tapered driven piles. This resistance mechanism develops in two distinct stages that are synergistically cumulative with respect to the final load-bearing capacity of the pile:

➢ Stage I occurs during initial driving when the all-important radial stresses acting along the shaft-soil interface develop in excess of those that develop around a comparable constant-perimeter pile. Furthermore, these radial stresses are locked-in after driving to a much greater extent with a tapered pile compared to a constant-perimeter pile. This results in a beneficially stiffer pile-soil system response to externally applied axial-compressive loading.

➢ Stage II occurs under post-installation axial-compressive external loading when the radial interface stresses increase further as the result of pile settlement.

These two behavioral stages are nowadays well-modeled physically and mathematically by *cylindrical-cavity creation* and *cylindrical-cavity expansion* respectively.

The writer has long championed the recognition and acceptance of *shaft resistance due to wedging* as a third resistance mechanism that is unique to tapered piles. This resistance mechanism complements the traditional resistance mechanisms of *shaft resistance due to sliding friction* that applies to constant-perimeter piles and *toe resistance due to end bearing*. The load-bearing uniqueness of wedging, combined with the fact that relatively small taper angles (typically of the order of one degree of angle or less) produce outsized benefits in terms of increased pile resistance to axial-compressive loads, are reflected in the title of this monograph.

Rear cover: This photograph was taken in Summer 2011 during expansion of Terminal 4 at the John F. Kennedy International Airport (IATA/FAA LID: JFK, ICAO: KJFK, but referred to as JFKIA throughout this monograph) in New York City. Terminal 4 is the 21st-century replacement of the venerable International Arrivals Building (IAB) that was, in many ways, the defining structure at JFKIA for decades. The IAB hosted numerous national airlines from around the world whose presence at JFKIA was too limited to justify a standalone terminal but who collectively made JFKIA a truly international airport. On a more personal level, the IAB was, indirectly, responsible for the writer's introduction to tapered piles in Summer 1972 and was thus the genesis of this monograph almost 50 years later.

The tapered piles shown in this photo during an intermediate stage of installation are carrying on a tradition of tapered piles being the deep-foundation element (DFE) of choice at JFKIA dating back to the original airport construction in the 1940s. The original airport name was New York International Airport/Anderson Field but always just 'Idlewild" to New Yorkers.

The piles in this photo also nicely frame the JFKIA control tower, the third such tower in JFKIA history. At the time of its construction in 1992, it was the tallest airport control tower in the world, at 320 feet (97.5 metres) above ground surface and approximately twice the height of the 1950s-era tower that preceded it. [photo credit: John McCullough]

Small Angles, Large Benefits...

The Unique Behavior of Tapered Piles

A Scholarly Monograph
by

John S. Horvath, Ph.D., P.E., Life Member.ASCE

Charter Independent Individual Member - Deep Foundations Institute

Consulting Professional Engineer
d/b/a
John S. Horvath Consulting Engineer
Scarsdale, New York, U.S.A.

and

Professor of Civil Engineering (retired)
Manhattan College
School of Engineering
Civil and Environmental Engineering Department
Bronx, New York, U.S.A.

Published
by
John S. Horvath Consulting Engineer
Scarsdale, New York, U.S.A.

Small Angles, Large Benefits...The Unique Behavior of Tapered Piles
by John S. Horvath

This document is a scholarly monograph written solely for the purpose of dissemination of technical knowledge and information for consideration by licensed design professionals and academicians educated in the field of civil engineering. As such, it is not a textbook, design manual, or standard, and should not be used or referenced as such. Furthermore, no liability is expressed or implied as to the suitability or accuracy of any information presented herein for any specific application in engineered construction or academic research. Use of any of the methods presented herein remains solely the liability of the licensed design professional using these methods.

Published May 2020
by
John S. Horvath Consulting Engineer
Scarsdale, New York, U.S.A.
linkedin.com/in/jshce
researchgate.net/profile/John_Horvath3

Library of Congress Control Number: 2020903881

ISBN-13: 978-1-7320953-2-8

Contents

CHAPTER 4 - HISTORICAL DEVELOPMENTS AND CURRENT TRENDS IN GEOMECHANICS AND STRUCTURAL MECHANICS RELATED TO TAPERED-PILE BEHAVIOR

CHAPTER 6 - CASE-HISTORY COMPARISON OF RESISTANCE-FORECASTING METHODOLOGIES

CHAPTER 9 - DENOUEMENT

APPENDIX D - MATHEMATICAL RELATIONSHIPS FOR MODELING SHEAR-MODULUS DEGRADATION

SUBAPPENDIX D1 - ENGINEERING VS. TRUE NORMAL STRESS AND NORMAL STRAIN

Preface

The writer Ella Winter is credited with the interrogative sentence *"Don't you know you can't go home again?"* that was the genesis of the title of a 1940 novel for which authorship was posthumously assigned to Thomas Wolfe. This monograph is my attempt to defy this admonition and go home again, at least professionally. It documents my return to literally the roots of my professional career as a civil engineer specializing in geotechnical and foundation engineering, specifically, the first days, weeks, and months of my full-time professional employment that began in June 1972.

As an entry-level civil engineer in the then-Soils and Foundations Division of the Port Authority of New York and New Jersey (PANYNJ[1]) Engineering Department, I was assigned to a major building project that was referred to within the PANYNJ organization by the acronym IAB-STRAP. This project involved the design of a 'structural' parking garage (the STRAP part of the acronym and never built at that time as it turned out) for the former International Arrivals Building (the IAB part of the acronym) at the John F. Kennedy International Airport (JFKIA) in New York City. The IAB was, for decades, the cornerstone and hub of passenger traffic at JFKIA as it served the myriad national airlines of other countries that did not have their own separate terminal at JFKIA. The IAB has since been replaced by what is simply called Terminal 4.

As the PANYNJ was wont to do in the latter decades of the 20th century as part of the planning and preliminary-design phases of a major project or building campaign at one of their facilities, they conducted what they called an 'indicator-pile' program. This was essentially a field-test program where various deep-foundation alternatives (typically driven piles although then-novel geotechnologies such as drilled shafts, a.k.a. bored piles/drilled piers/drilled caissons, were sometimes included in later years) would be installed and load tested (sometimes with internal force and displacement instrumentation) to establish realistically achievable values of axial-compressive pile resistance (capacity) to be used for production work. The overall purpose of these test programs was to define the most cost-effective, technically acceptable foundation alternative for a project and allow contractors to bid on the foundation work with less risk, greater confidence, and, presumably, increased cost competitiveness.

The PANYNJ was a very technically progressive, intellectually informed, and forward-looking facility owner in the latter decades of the 20th century and knew that construction technology and its concomitant costs were ever-changing. This meant that a deep-foundation alternative that was optimal in the past might not be so in the present. In addition (and this was the case at JFKIA by the late 1980s), although nature did not change, how geotechnical engineers interpreted nature changed over time. Specifically, in 1972 when the IAB-STRAP construction was being planned, seismic design did not enter the minds of foundation designers in the New York City metropolitan area. Less than 20 years later it did, and there were significant concerns at JFKIA over seismic-liquefaction potential in portions of what for over 40 years prior had been considered the bearing stratum for deep foundations. The lateral loads from direct shaking on structures was also a design issue.

In any event, it was for the purpose of updating the deep-foundation selection process that an indicator-pile program was conducted at JFKIA in Summer 1972 for the aforementioned IAB-STRAP project. Although timber 'friction/floating' piles had been the

---

[1] The original name of this bi-state, quasi-governmental agency was The Port of New York Authority (PONYA). The name was changed in 1972, the same year that I began employment there.

deep-foundation alternative of choice at JFKIA since construction of the airport began in the early 1940s (I presented and discussed the evolution of deep-foundation alternatives at JFKIA in detail in a 2014 white paper), the PANYNJ Engineering Department was, as noted previously, always conscious of and sensitive to the fact that construction technology and associated economics are dynamic and ever changing. Consequently, a decision was made to test several types of steel shell and pipe piles (both constant diameter and tapered) in addition to timber. The PANYNJ was also exploring the possibility of raising the maximum allowable axial-compressive service load on timber piles from that which had been used in the past both at JFKIA and throughout the New York City metropolitan area. This goal alone was considered somewhat controversial and progressive at the time.

In retrospect, what might otherwise have been a relatively mundane, long-forgotten test-pile program on a project that, as it turned out, never went forward into construction turned out to be a profound and lasting professional experience for me as I was fortunate to see a number of then-novel, cutting-edge geotechnologies used. One was the office use of wave-equation software (keep in mind that this was still the mainframe-computer era when access to computer software was still a novelty for most civil engineering practitioners) as part of the post-driving assessment of the IAB-STRAP test piles. Another noteworthy and memorable aspect was what I believe was the first use of on-pile dynamic measurements, then known as the *Case-Goble Method*, during pile driving anywhere in the New York City metropolitan area. This was the era when an engineer had to be hoisted to the top of a pile that was suspended in the pile-driving leads in order to attach hard-wired sensors to an analog field computer (no microcomputers/personal computers then!).

While either of these experiences would have been noteworthy at the time (I still remember that the measured pile-hammer energies, well below the manufacturer's rated energy in some cases, astounded everyone involved), my most lasting memory of the IAB-STRAP indicator-pile program was the inability to match axial-compressive pile resistances measured in static load tests on the naturally tapered timber piles with calculated resistances using the analytical methodologies that were in use at the time. These analytical tools included not only the then-prevalent 'pile-driving formulas' such as the *Wellington* (a.k.a. *Engineering News*) *Formula* that was found in the New York City Building Code and textbooks but also the results from the wave-equation analyses and dynamic measurements from both the on-site *Pile Driving Analyzer*® and *CAPWAP*® post-processing. The takeaway I got from this project was that there was something mysterious...almost magical...definitely unique... about tapered piles that caused them to have actual axial-compressive resistances that were often well in excess of what conventional analytical tools of that era indicated.

As an aside, although I had acquired both Bachelor and Master degrees in civil engineering (with the latter concentrating in geotechnical engineering) by the time I started working for the PANYNJ in 1972, I have no recollection of any particular discussion or emphasis of tapered piles in any of my academic coursework. This total unfamiliarity with tapered piles no doubt contributed to my perceptions about them as being something unusual and unique in the world of driven piling.

In any event, my interest in tapered piles and the technical enigma I perceived them to be that began back in Summer 1972 persisted in my professional psyche even after I left the PANYNJ in 1974 and my research interests and time were focused on other topics. The first of these topics was subgrade models for soil-structure interaction, the subject of my doctoral research as well as a monograph that I self-published in 2018. This was followed by cellular geosynthetics (geofoams and geocombs), the subject of two self-published monographs, one in 1995 and the other more recently, in 2018. On the cusp of the 21st century, my active involvement with tapered piles once again moved to the forefront as the result of some self-supported academic research and concomitant publication I did that was related to new-product development launched at JFKIA (the *TAPERTUBE®* pile) as well as my involvement with some local consulting projects that involved *Monotube* piles.

The highpoint of my academic research into tapered piles in the early years of the new millennium was the publication of a Manhattan College research report in 2002 that focused on the use of tapered piles at JFKIA. When I left academia in 2014, I realized that I had the time to devote more attention to the subject and subsequently authored a tetralogy of white papers between 2014 and 2016 that again focused on JFKIA and tapered piles. This most recent research also provided the opportunity for me to explore state-of-art concepts involving analytical methodologies that generate complete load-settlement curves as part of their outcome as opposed to just an estimate of ultimate pile resistance as has been the standard since the beginning of modern analytical foundation engineering in the 1950s.

It was during this most recent episode of research and writing that I came to the conclusion that the time was right to document my thoughts related to tapered piles in a single document and that a scholarly monograph was the appropriate publication vehicle for this. This document that you hold in your hands is the result. I can tell you that had someone told me at the outset that this effort would eventually fill 600-plus pages, I would not have believed it. I certainly did not start out with this or any other page-count in mind, it just happened.

It is important to note that, as with all three monographs I published prior to this one, I do not see this document as being the final word on the subject. Not by any means and more so on this topic, tapered piles, than any other topic that I have researched and written about in my professional career that now spans almost 50 years. As any researcher of technical subjects knows, whatever topic one is writing about is never static but is constantly producing new information and/or new interpretation of existing information and sometimes changing direction as a result. I am well aware of the fact that despite the human usage of tapered piles for at least 7000 years (after all, naturally tapered timber piles were the first deep-foundation element), there is still much research and observation in practice to be done to verify (and modify as necessary) existing analytical methods as well as to develop new analytical methodologies for use in practice. However, to borrow a line from the film "Wag the Dog" (which is actually a modified version of a quote attributed to General George S. Patton):

"A good plan today is better than a perfect plan tomorrow."

In this vein, I feel that it is better to write and publish a monograph documenting what we know about tapered piles at present rather than wait until some vague, uncertain time in the future when the knowledge-base of the profession on this subject may be greater. More importantly, this monograph will hopefully stimulate not only greater consideration and concomitant usage of tapered deep-foundation elements of all types but more and better research that will ultimately extend the knowledge-base for the benefit of all concerned. This was certainly one significant outcome of my first monograph on geofoams that was published in 1995. It resulted in global research into geofoams and their many functional applications that continues to the present.

As a final, but very important and essential, comment, the preparation of this monograph and the decades of my research that led up to it are totally in the 'labor-of-love' category. I have never received any financial support of any kind from any entity or person who is or was involved in the manufacture, distribution, sale, installation, or promotion of tapered piles. Any financial compensation I ever received that was related to tapered piles was solely for design-related consulting work on specific projects that happened to use tapered piles based on decisions made by others prior to my involvement as a consultant. Thus, for better or worse, all observations, opinions, etc. expressed in this monograph are strictly my own; of my own doing; and without any underlying or hidden commercial or financial motivation.

I emphasize these last points as recent history has taught us that there are numerous instances, some going back decades, where scientific researchers in academia and other nominally independent, non-commercial organizations published, often repeatedly, what amounted to work-for-hire on behalf of, and biased toward, some commercial enterprise. Furthermore, and more importantly, the financial connection in these publication efforts was not made known at the time of the initial publications.

As a consequence of this, the public has understandably come to look at published academic and scientific work with some concern about its objectivity. This skepticism is understandable and, in many cases, justified. Thus, I feel that it is necessary to be proactive and up front on this issue of potential conflict-of-interest. There is none in this case. Period.

That having been said, I am, however, very grateful to several individuals as well as business entities and organizations that have provided technical information to me over the course of the past 48 years that is relevant to the contents of this monograph. In each case, information was provided on a pro-bono basis and with no quid-pro-quo expectation in mind. These contributors include (listed alphabetically): ARUP; Mr. Terrence Carroll, P.E.; ConeTec; Mr. Jack Dougherty; Mr. Stanley 'Stan' Merjan, P.E.; Mr. Bruce Miller; Port Authority of New York and New Jersey; Dr. Mark Styler; Underpinning & Foundation Skanska; Mr. Donald L. 'Don' York, P.E.; and several owners of buildings that used tapered piles who have asked to remain anonymous. I am especially grateful and appreciative to Don York who mentored me during my work on the IAB-STRAP project in 1972-1973 and thus provided me with a 'firm foundation' for a decades-long professional interest that has culminated in this monograph.

John S. Horvath, Ph.D., P.E., Life Member.ASCE
Scarsdale, New York, U.S.A.
May 2020

Chapter 1

Introduction

___

1.1 PROLOGUE

1.1.1 Introduction and Overview

Some (many? most?) might wonder why someone would write a book about *tapered*[2] *piles*[3] in the first place and, if they did, how they could fill over 600 pages of such a book. The genesis of these questions is understandable as almost 50 years of professional experience in foundation engineering has taught the writer that a surprisingly large percentage of design professionals involved in foundation selection and design simply do not even consider tapered piles when evaluating deep-foundation alternatives for a project on which 'friction', a.k.a. 'floating', piles are called for. This is understandable as most undergraduate textbooks, at least in the U.S., give tapered piles passing mention at best so most instructors in academia, who nowadays tend to have little to no practical experience, do not discuss them. So, there tends to be a self-perpetuating cycle of ignorance when it comes to tapered piles.

The underlying reasons for this informational dearth about tapered piles appear to be many, varied, and complex but boil down to the fact that academic research, which is the cornerstone of the educational process, has systematically failed to consider, no less explore, the unique geomechanical behavior of tapered piles both during driving as well as under post-installation axial loading. As a result, with few exceptions, the entire gamut of field and office tools used for assessing the *resistance*[4] of deep-foundation elements (DFEs) does not distinguish between those that are tapered and those that have a depth-wise constant perimeter. It follows that because research has not properly investigated and identified the unique aspects of tapered piles that technology transfer (education) itself has not had the basic information to provide to generations of engineering faculty no less students and practitioners. Hence, the aforementioned self-perpetuating cycle of ignorance.

___

[2] The term *conical pile*, e.g. as used in Gregersen et al. (1973), is a much less common synonym for tapered pile.

[3] In this monograph and consistent with U.S. usage, the term 'pile' when used alone means a conventional preformed deep-foundation element (DFE) that is installed into the ground by impact driving. Thus, the term *driven pile* is redundant in this monograph. However, the term 'driven pile' does appear widely in the published literature, especially outside the U.S. where the term 'pile' typically has a much broader meaning and includes all types of DFEs, thus always requiring an adjective or modifier (bored, driven, jacked) to indicate the specific type of pile. That having been said, in some cases in this monograph the phrase '(driven) pile' is used where the writer feels that it is necessary to be abundantly clear and without ambiguity as to what is being discussed.

[4] The term 'resistance' when used alone in this monograph means the geotechnical support provided by the ground along the embedded portion of a DFE in response to axial loading. 'Resistance' is used in lieu of the terms *capacity* and *geotechnical capacity* that are often used in U.S. practice and, as will be seen in Chapter 2, have an entirely different meaning in this monograph. In some cases, the otherwise-redundant term *geotechnical resistance* is used in this monograph to clearly distinguish the geotechnical resistance from *structural resistance* of a DFE as a structural element. Note, however, that geotechnical resistance or structural resistance are not necessarily the ultimate (maximum) values unless clearly specified as such.

Because of this broad ignorance concerning tapered piles, it is necessary for the benefit of the reader that some background and context for the subject of this monograph be provided at the outset. A brief, introductory history of the key role that tapered piles have played in the history of humankind and foundation engineering up to circa 1970 is provided in the following section. This is followed by an assessment of tapered piles from circa 1970 to the present that is based on the writer's first-hand experiences in U.S. practice. Note that the information provided in this introductory chapter is intentionally relatively brief as a much more detailed treatment of these subjects is presented in subsequent chapters.

1.1.2 Early History

Although the terms[5] *civil engineer* and *civil engineering* only date to the 18th century CE[6], the practice of what we now consider to be civil engineering is, in many ways, intuitive, if not instinctive, for humans. For example, the use of piles for foundations, like the building of retaining walls for earth retention, is a construction activity that goes back much farther in time, long before the basic science used to develop civil engineering principles and theories was developed. There is archaeological evidence of multiple European lake-dwelling cultures that inhabited structures supported on wooden columns sunk into the ground as far back as 5000 BCE[7]. Thus, the *timber pile*[8] was born at least 7000 years ago.

These human endeavors were, no doubt, enabled by nature providing an abundance of trees from which timber piles were crafted with relatively modest effort. Thus, the fact that timber piles inherently have a non-uniform diameter along their entire length (what we nowadays refer to generically as *taper*) was simply a natural occurrence. However, as will be seen in this monograph, this naturally tapered geometry of timber piles appears to have had a profound influence on advances in pile technology thousands of years later.

After the first use of timber piles 7000-plus years ago, the subsequent millennia mainly saw the development of ever-more-efficient ways to install piles by impact driving, eventually moving from human to machine power. Technology related to the pile itself remained remarkably unchanged for almost 7000 years until the 19th century CE. This is when materials such as iron, steel, and Portland-cement concrete (PCC) began to emerge as alternatives that could be used for piles in lieu of wood.

It is of interest to note that this 19th-century evolution of new pile materials was separate from, but contemporaneous with, the first efforts (circa 1850) to quantify the post-driving ultimate resistance of piles based on measurements made during the pile-driving process. This is what nowadays is called as the *dynamic approach* for estimating the ultimate resistance of piles. It is interesting to note that a full century would pass before the *static approach* for estimating the ultimate resistance of DFEs in general, and piles in particular, based on geomechanics concepts and principles would begin to emerge in the 1950s.

As discussed in Chapter 5, there were two different physical models for pile-driving dynamics that were postulated throughout the latter half of the 19th century. The second and much more enduring model of *rigid-body impact* produced the so-called *pile-driving formulas*

---

[5] The key terminology used throughout this monograph is formally defined and discussed in Chapter 2, with additional terms defined as required in subsequent chapters and appendices as they are introduced. However, for clarity of presentation, certain terms are defined in advance in the current chapter.

[6] *Common Era*, used in this monograph as an alternative to *AD*.

[7] *Before Common Era*, used in this monograph as an alternative to *BC*.

[8] The terms *wooden pile* and *wood pile* are much less common than the term 'timber pile', at least in U.S. practice in the recent past and present.

or *dynamic formulas* (the latter is used in this monograph) as they were later called. Arguably, the best known of these is the *Wellington Formula*, a.k.a. *Engineering News Formula*. Despite the fact that the physical model and concomitant algebraic equation that forms the basis for all of these dynamic formulas was shown to be grossly incorrect when E. A. L. Smith[9] published his seminal paper on the one-dimensional (1-D[10]) wave equation in 1960[11], they still persist in routine practice to the present. This is, no doubt, because of their irresistible simplicity and concomitant ease of use in practice, with minimal education and training required of the end user. The latter fact is especially important to budget-driven organizations such as state departments of transportation or highways in the U.S. as it allows them to have a lower-wage technician perform pile-installation inspection on a project as opposed to a licensed professional engineer as is legally required in some jurisdictions.

It seems plausible that as the 19th century progressed, the synergy of the availability of new pile materials; the concomitant ability to create piles of arbitrary length as opposed to accepting timber piles in lengths dictated by nature; and the ability (however flawed) to estimate ultimate resistance by calculation using some simple arithmetic formula, the concept of load testing installed piles began to take on greater importance and frequency of use. This would have been significant at the time for several reasons, not the least of which would have been the ability to compare the ultimate resistance of piles of different types and geometries. The writer posits that this load-testing capability that evolved in the 19th century led to the realization that there was a benefit, at least in some ground conditions, to intentionally construct piles of these then-novel alternative materials such as steel and PCC with a taper that mimicked the natural taper of timber piles.

The basis of this opinion is the historical fact that beginning in 1896 and continuing for about a decade thereafter, several U.S. patents were filed for a variety of pile designs and installation concepts involving metals and PCC in various combinations[12]. The one common element that most of these patent filings contained was that the final installed pile had a taper that was both continuous and uniform, and thus broadly mimicked the natural shape of timber piles[13]. It thus seems likely that the perceived benefit of using a tapered, as opposed to a non-tapered (constant-perimeter), pile geometry was the result of comparing tapered vs. non-tapered piles in actual applications that were subjected to post-installation load testing.

The reason for this presumption by the writer is because it would have been just as easy, if not easier, to construct piles with a constant-perimeter had they proven to be as good

---

[9] Some earlier publications (which are rarely cited in civil engineering literature), such as Smith (1957), give his name as Edward A. Smith.

[10] The '1-D' notation as a compound adjective (modifier) of the term 'wave equation' as applied to piles and pile driving is generally omitted in the published literature. However, this notation is retained throughout this monograph to emphasize the fact that the wave transmission assumed in this model is one-dimensional. This emphasis is done for its relevance to the theme of this monograph as the inherent 1-D nature of the wave equation is a significant issue for tapered piles.

[11] The concept of the 1-D wave equation applied to pile driving was actually made approximately three decades earlier, by David Victor Isaacs in 1931 (Isaacs 1931). However, this fact seems to have been widely missed, lost, or ignored over time with the result that the vast majority of references incorrectly give Edward A. L. (E. A. L.) Smith sole credit for applying the 1-D wave equation to pile foundations.

[12] The patent applicant in each of these cases, Alfred A. Raymond, appears to have been involved in the contracting side of deep foundations, through the Raymond Concrete Pile Company, as opposed to being involved primarily in design or academic research. As will be seen, the fact that early research and development (R&D) of tapered piles in the U.S. was industry-driven is significant for its profound effect on subsequent technology transfer or lack thereof.

[13] As discussed in Chapter 3, while timber piles are continuously tapered, they are not uniformly tapered. This is simply an artifact of the way that trees grow.

or better than tapered piles. Such load tests would have provided the necessary 'ground-truth' of the actual geotechnical resistances as no known dynamic formula in use then or since explicitly differentiates between tapered and non-tapered pile geometries.

In summary, tapered piles have been used since the earliest recorded history of pile foundations that now extends back over 7000 years before present, although for most of that time the use of tapered piles was simply happenstance by virtue of the natural pile materials, i.e. the trunks of trees, readily available to humans. However, it appears that beginning with the appearance of manufactured piles more than 100 years ago there was explicit recognition of the benefit of intentionally using tapered piles as a more cost-effective alternative to non-tapered piles, at least for some ground conditions. This appreciation of the overall combined technical and economic benefits of tapered-pile geometry sparked a flurry of proprietary product R&D activity on the cusp of the 20th century and the decades immediately thereafter. This was also the timeframe when steel and PCC were entering the mainstream for use in engineered construction and thus significantly expanding the suite of structural materials civil engineers could work with when selecting and designing piles. As discussed in Chapter 3, the net result of this R&D led to the existence of several commercially viable versions of tapered piles by the early decades of the 20th century, at least in the U.S.

As a final comment in this broad overview of how tapered piles came to be, it is of interest to note that in the 1950s, a time when many different types of polymeric[14] materials were just beginning to move from the research laboratory to commercial production and use in both industrial and consumer products, a U.S. patent was obtained on behalf of the Raymond Concrete Pile Company for a tapered pile that replaced the steel component with a polymeric material. This demonstrates that the pile industry, in the U.S. at least, remained active with R&D related to tapered piles well into the 20th century.

However, it does not appear that polymeric tapered piles ever entered the commercial mainstream. When the writer entered practice in 1972, all tapered-pile product lines marketed by Raymond International (the corporate successor to the Raymond Concrete Pile Company) were either already gone from, or on their way out of, the commercial marketplace, at least in the U.S.

1.1.3 The Last 50 Years

1.1.3.1 Introduction and Overview

As noted in the Preface, the writer first encountered tapered piles in Summer 1972 at the John F. Kennedy International Airport (JFKIA) in New York City. To provide a basis for understanding the genesis, objectives, and scope of this monograph as presented later in this chapter, it is useful to continue the history of tapered piles from that time to the present based on the writer's first-hand perspective.

Note that all comments herein going back to circa 1970 are made based on subsequent knowledge acquired and insight developed throughout the writer's professional career to the present, not the writer's knowledge and understanding at the time events were occurring. Insight from hindsight has its benefits in this case. Note also that, for the most part,

---

[14] The term 'polymeric' is used in this monograph as the preferred term, as opposed to the colloquial term *plastic*, to indicate a specific category of materials that includes polystyrene, polyethylene, and many other materials of broadly similar chemistry, The term 'plastic' is reserved for use in this monograph in its traditional engineering-mechanics context as defining the behavioral state in which permanent displacement or deformation of a material occurs.

the comments made herein are limited to developments in the U.S. with which the writer has the greatest familiarity although some broad observations of a more-global nature are also made.

In the simplest of terms, there are two broad aspects that shaped the usage of tapered piles in recent decades:

- pile technology and

- resistance analysis.

Although these topics are addressed in detail in later chapters, brief overviews are presented here to provide the basis of the synthesis that follows.

1.1.3.2 Pile Technology

1.1.3.2.1 Generic Alternatives

By circa 1970, the available types of tapered piles in the U.S. fell into two distinct categories: generic and proprietary. On the one hand was the continued use of timber piles that were generic and could be driven by anyone who had access to even a simple drop hammer and the primitive hardware to use it. Thus, there were still railroads and other infrastructure owners with suitable equipment who were driving timber piles for their own structures (typically bridges and trestles in the case of railroads) in addition to construction contractors that specialized in or were otherwise capable of performing such work on a project-specific basis for buildings and other types of structures.

However, timber-pile usage throughout the U.S. was in decline in the latter decades of the 20th century. The limitations of both their geotechnical and structural resistances (which had always been dictated by the physical dimensions and material properties of the harvested trees over which there was relatively limited human control) had become significant considerations on many projects. When the cost of ancillary foundation components such as pile caps was figured in, there was an ever-increasing 'toolbox' of deep-foundation alternatives that consisted of steel and/or PCC that could provide a lower cost per unit resistance than timber piles on all but the smallest projects or for all but the most lightly loaded structures. In fact, the 1972 IAB-STRAP indicator-pile (test-pile) program at JFKIA that was noted in the Preface and with which the writer was involved is believed to have been the last such program at JFKIA in which the airport's operator, the Port Authority of New York and New Jersey (PANYNJ), considered using timber piles.

Another factor that contributed to the decline in the relative use of timber piles throughout the U.S. was their availability. Unlike other structural materials, PCC in particular, that can be manufactured or reasonably be made available almost anywhere in the world, trees that are suitable for use as timber piles and in a sustainable quantity adequate for ongoing commercial production, sale, and driving were increasingly limited to certain geographic regions within the U.S. toward the end of the 20th century. Thus, the economics of delivering timber piles in suitable quantity to project sites was often a significant factor in foundation engineering practice.

Finally, another issue that began to evolve in the latter decades of the 20th century in the U.S. and elsewhere involved environmental concerns. The focus in this case was on the inherent toxicity of creosote that had been used as the most common preservative treatment for timber piling for decades up to that point in time.

The only generic alternative to timber piles that existed in the latter decades of the 20th century was, and still is, a precast (and possibly prestressed) PCC pile cast locally in the region of usage although, at times, specific design details were patented and thus proprietary until the patent expired. As best as the writer has determined, PCC tapered piles have seen use in geographically diverse areas (Florida in the U.S, Norway, Slovakia, and the former Soviet Union are known to the writer) where the combination of regional subsurface conditions, foundation requirements, and economics presumably made these cost-effective alternatives. Such piles appear to be particularly attractive in areas where structural steel is relatively expensive and/or saltwater corrosion is a significant design consideration[15].

1.1.3.2.2 Proprietary Alternatives

With regard to proprietary types of tapered piles, three different product lines of commercially available composite (multi-component) steel-and-PCC tapered piles had been developed in the early decades of the 20th century and remained in widespread commercial use throughout the U.S. for much of the 20th century. Although patents on these piles had long since expired by the latter decades of the 20th century, these piles remained effectively proprietary in the marketplace as their use was tightly controlled to varying extents by the two companies that had developed them:

- The Raymond Concrete Pile Company was both the exclusive manufacturer and exclusive installer of the *Raymond Standard Pile* and *Raymond Step-Taper Pile*. The former was the first commercially viable non-timber tapered pile and a direct outcome of the intense period of U.S. patent development circa 1900 that was note earlier in this chapter. The latter was developed later in the 20th century.

- The Union Metal Company (later renamed Union Metal Corporation) was the exclusive manufacturer of the *Monotube* pile. However, unlike the aforementioned Raymond products, any pile-driving contractor could install *Monotube* piles although a proprietary drive-head was required on loan from Union Metal.

Although the basic patents on these three pile types had long since expired by the latter decades of the 20th century, each required both some special manufacturing equipment/process for the steel shell or pipe (cold-formed steel in the case of *Monotube* piles) as well as unique driving hardware (special-purpose mandrels or drive-heads) to install them. Thus, there were significant economic issues related to both the manufacture of these piles and their installation that presumably made them unattractive and cost-prohibitive for either widespread generic manufacturing or installation by a wide range of construction contractors. This assessment is supported by the fact that the *Monotube* pile was (is?)[16] produced in only one location (Canton, Ohio) which means that the pile components

---

[15] PCC DFEs in general have historically been viewed by many as being more robust than steel in open-water, especially saltwater, environments. The writer has been involved in projects where project stakeholders viewed PCC piles as being maintenance free, even in a saltwater environment. However, these views are, at best, overly simplistic and, at worst, simply incorrect, and, in the writer's experience, not universally viewed by all stakeholders. For example, the writer has been involved in the design of shallow-water marine structures such as piers for the U.S. Navy and their preference was always steel pipe piles with cathodic protection.

[16] As discussed subsequently, at the time this monograph was published in 2020, the business status of the company that produces the *Monotube* pile was uncertain.

must always be shipped from there to a project site. It is also obvious that these economic considerations also limited the use of these piles outside of the U.S. and even within the U.S.

While this tight control of market availability of their products may have worked to the business advantage of both Raymond and Union Metal for some period of time early in the 20th century when DFE choices were extremely limited by today's standards, by the latter decades of the 20th century this was no longer the case. By circa 1970, both Raymond product lines were for all practical purposes extinct in the marketplace. It appears that the need to use special mandrels and concomitant *shelling up* (the relatively ponderous process of getting the rigid mandrel into the thin-wall steel pile shell, usually using a *doodle hole*) simply made them economically unviable in view of other pile and, increasingly, non-pile DFE alternatives.

There is also indication that changing technical demands as a result of the great seismic 'enlightenment' of geotechnical and foundation engineers in the 1960s as the result of major earthquakes in Alaska and Japan (Niigata) in 1964 played a role in the demise of Raymond's tapered piles, at least in some regions and markets, as evidenced by this statement that was found on the Web in August 2018:

"Circa January 1970 I was employed by Raymond International, Inc., and held the office of Northern California District Manager. And, at this time I determined that due to the revised seismic section of the Uniform Building Code the traditional Raymond "Step Tapered" piles were no longer marketable. The new seismic requirements had in my opinion ended an era..."

- Frank Cavin - Founder, Foundation Piledriving Contractors (Oakley, CA)

While Union Metal's *Monotube* pile remained commercially available circa 1970 and well beyond into the 21st century, its use was, and still is, constrained by its being a unique, cold-formed-steel product that is manufactured in only one location in the world. In addition, when the *Monotube* constant-diameter extension (upper) sections are used with the tapered lower section, a special drive-head must be used to install the piles as noted previously. To avoid this as well as for greater overall economy, toward the end of the 20th century there was a trend that persists to the present to using generic hot-rolled steel pipe for the constant-diameter extension and only the *Monotube* cold-formed-steel tapered lower section (Brand 1997).

An additional technical consideration that evolved toward the end of the 20th century was that the cold-forming process used to manufacture both the tapered and constant-diameter components of *Monotube* piles places significant limits on the maximum wall thickness of the final products that can be produced. This constrains the axial-normal stresses that these piles can sustain both during and after installation. As a result, the post-installation resistance of a *Monotube* pile can be constrained by structural considerations, not geotechnical considerations as is typically the case with DFEs in general and piles in particular.

It appears that all these facts combined to impact the economic viability of the *Monotube* pile. After a number of corporate spin-offs and changes in ownership that began in 1984, the *Monotube* brand of piling was subsequently manufactured by a standalone company named the Monotube Pile Corporation. Ownership of the Monotube Pile Corporation changed hands several times after 1984 and at the time that this monograph was being finalized in early 2020 the ownership is once again in a state of flux that appears to be impacting the commercial availability of *Monotube* pile components.

In the writer's opinion, the U.S. marketplace constraints placed on a proprietary construction product such as the *Monotube* pile are well-illustrated by considering the Interstate highway program that began in the U.S. in the late 1950s. This program generated the demand for a tremendous amount of foundation piling in subsequent decades. A highway-industry publication authored by Peck (1958) in the early years of this drastically increased road-related R&D and construction in the U.S. noted:

> *"...it is obvious from an inspection of Figure...that taper has a beneficial influence on the capacity of piles in sand...it would appear reasonable to conclude that a taper of 1 percent or more is likely to increase the capacity of a pile, for a given length of embedment, between 1½ and 2½ times."*

Thus, the clear technical benefit of using tapered piles, at least as friction/floating piles in coarse-grained soil, was known to the highway community early in the Interstate highway building program. However, government contracting requirements in the U.S. typically require the use of generic materials and products whenever possible. Highway structures in the U.S. are no exception to this. Thus, the use of generic pile types, especially steel H-piles, was, and still is, favored on such projects, even in ground conditions where a tapered pile would clearly be a much more cost-effective alternative.

In any event, as the 20th century drew to a close, the *Monotube* pile was the only type of steel-and-PCC tapered pile that was still commercially available in the U.S. However, this changed on the cusp of the 21st century with the development and market introduction of the *TAPERTUBE®* pile in the U.S. Initially, this pile was intended to be simply a geometric clone of the *Monotube* pile that was much more commercially attractive for several reasons. To being with, only the tapered lower section of a *TAPERTUBE* pile, which is made of hot-rolled steel sheet so can be fabricated in thicknesses substantially greater than the cold-formed *Monotube*, is proprietary. From the beginning, the *TAPERTUBE* was intended to be used with generic hot-rolled steel pipe as the constant-diameter upper section or extension. Not only does this minimize material costs but it means that the pile can be driven without the need for a special drive-head. Furthermore, the *TAPERTUBE* pile does not require special machinery and tools to manufacture so fabrication of the proprietary tapered lower section can be outsourced to numerous steel fabricators throughout the U.S. and, presumably, outside of the U.S. if desired. Thus, the overall *TAPERTUBE* pile can be fabricated on a regional, if not local, basis which minimizes shipping costs.

As a consequence of the development of the *TAPERTUBE* pile, at the present time, and in the U.S. at least, the marketplace with respect to tapered steel piles is significantly different than at any time in the past. Since the beginning of the 21st century, there have been two commercially available products that have the same basic geometry and can thus function in the same basic manner although there are nuanced differences between the *Monotube* and *TAPERTUBE* that have been highlighted in this chapter and are discussed further in Chapter 3. Thus, although both the *Monotube* and *TAPERTUBE* piles have at least one component (the tapered lower section) that is proprietary, there is an overall competitive element that opens the door to their use even on government projects.

As a result of the profoundly changed commercial landscape for tapered piles not only within the U.S. but globally (given that the *TAPERTUBE* pile can, in principle, be manufactured anywhere in the world that has the capability to fabricate hot-rolled steel sheet), the need for analytical methodologies for tapered piles that can be used by any design professional has never been greater.

1.1.3.3 Resistance Forecasting

By circa 1970, the ability to forecast[17] the ultimate resistance of piles by calculation was in the process of significant advancement on several fronts. The dynamic approach, which dated back to the middle of the 19th century as noted earlier in this chapter, had been improved dramatically as both the 1-D wave equation as well as on-pile, real-time dynamic measurements of forces transmitted to a pile and pile velocities were transitioning from the world of academic research to engineering practice. These geotechnologies were advanced initially by the needs of the offshore oil and gas industry and later by the aforementioned terrestrial road construction of the U.S. Interstate highway system. Over time, these methods would only improve and become more widespread as microcomputer, microelectronics, and wireless communication capabilities developed and continue to develop in the present.

By circa 1970, the static approach for forecasting ultimate resistance of piles using what is now referred to as the *indirect* or *rational* method was also well into development for almost 20 years. The static approach had been essentially non-existent about 30 years earlier when Terzaghi (1943) wrote in his seminal English-language textbook (Page 137):

> *"Since the bearing capacity of piles cannot yet be computed on the basis of the results of soil tests performed in the laboratory we are still obliged either to estimate this value on the basis of local experience or else to determine it directly in the field by loading a test pile to the point of failure."*

In the five decades since circa 1970, resistance-forecasting methodologies based on the static approach have increased substantially, primarily using what is called the *direct* method. This is a direct result of the enormous advancements in in-situ testing, especially the cone penetrometer. More recently, it has even become relatively easy and routine to forecast an entire load-settlement curve for a DFE and not just the ultimate resistance. Advances in computer hardware and software have also played a major role in all of these developments.

Unfortunately, this broad-based advancement in resistance-forecasting methodologies for DFEs has offered mixed results for tapered piles. While the dynamic approach, both the 1-D wave equation and on-pile dynamic measurements, has routinely been used with tapered piles (as noted in the Preface, the writer was involved in the early use

---

[17] Foundation engineers have used a variety of terms over the years to define the expected performance of some design. The term 'prediction' is clearly the most commonly and universally used of these terms as can be seen in the many 'prediction symposia' that have been held over the years throughout the world. However, the writer feels that more attention should be paid to the terminology used as this aspect of foundation engineering is one that can interface with the non-engineering public at times. Specifically, foundation engineers should be sensitive to how the public interprets engineering terminology. The writer posits that the term 'prediction' conveys a degree of clairvoyant certainty that is not the actual intention as foundation engineers understand (or should understand) that there is always uncertainty in their 'predictions'. Consequently, the writer suggests that the term 'forecast' is better suited for use in foundation engineering. The public is familiar with this term from weather forecasting and should understand that a forecast conveys an estimated behavior that is reasoned and has scientific basis for high probability yet is subject to some uncertainty. Thus, the public should understand that a forecast allows for some leeway if actual outcomes deviate from those made initially. In summary, the writer feels that foundation engineering outcomes should be presented as forecasts, not predictions, and will thus use the term 'forecast' as opposed to 'prediction' throughout this monograph.

of both technologies in 1972), these methodologies often produce mixed results in the writer's experience. In part, this may be the result of the physical model used for both methodologies, i.e. a 1-D stress wave traveling through a constant-diameter elastic rod with ground resistance from the traditional resistance mechanisms of shaft sliding friction and toe end bearing. As will be discussed in Chapter 4, research conducted since the 1990s clearly indicates that tapered piles (and all tapered DFEs for that matter) have the additional shaft-resistance mechanism of *wedging* that is well-modeled using what is called *cavity mechanics*, specifically, that of a cylindrical cavity. Wedging is an inherently three-dimensional (3-D) physical mechanism so it should not come as a surprise that current versions of dynamic analyses, which are based on a 1-D physical model, do not properly model this wedging mechanism. Consequently, current dynamic resistance-forecasting methodologies based on 1-D wave transmission should not be expected to yield correct results for tapered piles.

With respect to the static approach, although both indirect and direct methods of analysis specifically for use with tapered piles have been developed and published going back to the early 1960s, for the most part they appear to have received limited attention in textbooks, design manuals, analytical software, practice, and academic research. As a result, in any practical context based on the metric of widespread acceptance and use, there has been no advancement in using the static approach to forecast the ultimate resistance of tapered piles since the early 1960s when R. L. Nordlund published his first paper on the subject.

It is relevant to note that Nordlund was professionally affiliated with the Raymond Concrete Pile Company at the time he wrote his seminal 1963 paper. This reinforces the observation made earlier in this chapter that, with few exceptions, all of the advances in tapered-pile technology dating back to the late 19th century and continuing to the present have been made by people and organizations associated with the deep-foundation industry, not academia or engineering practice. This includes the relatively recent *TAPERTUBE* pile (which is now about 20 years old) that also owes its origins to people involved in the pile-manufacturing and pile-installation business sectors.

1.1.4 Synthesis

The writer's assessment of tapered piles based on the synopsized histories of pile technology and resistance forecasting presented up to this point is that, for the most part, tapered piles are indeed an enigma in the overall history of deep foundations. On one hand, tapered piles constructed of steel and PCC were among the first types of non-timber piles developed on the cusp of the 20th century as humans moved on from a world in which the timber pile was the only type of pile in existence. However, the tapered piles that eventually saw commercial production beginning in the early decades of the 20th century were kept on close-hold as proprietary technology by the U.S. companies that developed, manufactured, and, in the case of Raymond, installed them. This may have been a good business strategy at one time in an age when foundation engineering was still mostly experience-based art with little science, and geotechnical engineering and engineers as we know them today simply did not exist[18].

---

[18] Well into the 1960s, in the U.S. at least, it was possible to obtain a Bachelor degree in civil engineering (at that time still considered all the formal professional education one needed to practice civil engineering for a lifetime) without having taken even one course in geotechnical or foundation engineering. Furthermore, when the writer began professional practice in 1972 there was a prevailing attitude among some long-time civil engineering practitioners that geotechnical engineering as a specialization in engineering practice was a 'solution in search of a problem' in that foundation design was simply a task that was adequately handled by any structural engineering specialist.

However, this business strategy turned into a handicap by the final decades of the 20th century as generic alternatives, first other pile types and then various drilled technologies, became available to satisfy a marketplace that increasingly demanded generic, not proprietary, DFEs. More importantly, contemporaneous with this was the active R&D and subsequent promulgation of analytical and measurement tools that allowed for the reliable forecasting of deep-foundation resistance in routine practice by all foundation engineering practitioners, not just a select few in industry hoarding their decades of experience. In simple terms, the net result of all these changes was that the stranglehold that Raymond and Union Metal had on tapered-pile technology for much of the 20th century turned on them and, in the end, strangled <u>them</u> into oblivion business-wise.

However, since the beginning of the 21st century there has been a significant paradigm shift with regard to tapered piles. As noted earlier in this chapter, for the first time there are now two brands of steel-pipe tapered piles, *Monotube* and *TAPERTUBE*, that are functionally identical, at least geotechnically. Furthermore, in each case only the lower tapered portion of the pile is proprietary as the constant-diameter extension is, or at least can be if desired, generic hot-rolled steel pipe so that the overall fabricated pile can be driven by any construction contractor with normal pile-driving equipment. This means that tapered piles constructed of steel pipe and filled with fluid PCC after driving are now more in the generic mainstream than at any time in the history of modern deep foundations (i.e. the post-19th century period when products other than timber piles were available). The only limitation to these tapered piles is their geographic availability which is more of an issue with the *Monotube* pile as it requires very specialized cold-forming machinery that at the present time is believed to exist in only one location. As the *TAPERTUBE* pile is constructed using generic hot-rolled steel plate, only a local steel fabricator capable of bending steel plate into the desired final shape of the tapered lower portion of the pile is required. This is significant as not only is the *TAPERTUBE* pile capable of being fabricated in more than one location within the continental U.S. at the present time, it is readily amenable for fabrication globally in the future.

When these facts are considered together with the continued regional availability of timber piles and locally produced tapered precast-PCC piles as well as recent developments of formed-in-place tapered DFEs that are created by drilling, it appears that, more than ever, there is a need for reliable, generic analytical methodologies for forecasting the load-settlement behavior of tapered DFEs in general and tapered piles in particular. These resistance-forecasting methodologies should be within the capability of all foundation engineering practitioners so that independent, generic designs of tapered piles can be performed for routine projects and on a routine basis worldwide.

1.2 OBJECTIVE AND SCOPE

The synthesis in the preceding section laid the groundwork for the argument made here by the writer that more than at any time in the past there are a sufficient number of choices in tapered piles of varying material composition that, taken as a group, can compete as generic, commodity products along with all the other deep-foundation alternatives in the marketplace and on a potentially global basis. However, in order for this to happen both practitioners and academicians need to become more aware of the tapered-DFE choices that are available and confident in the analytical methodologies for forecasting, as a minimum, ultimate resistance and, preferably, complete load-settlement curves of tapered DFEs. Note that academicians play a crucial role in this process because they not only need to educate future civil engineers about tapered DFEs to a much greater extent than at present but also conduct relevant research as significant improvements to existing resistance-forecasting methodologies that are still based on a circa-1960 geotechnical knowledge-base are needed.

This perceived timely informational need with regard to tapered-DFE technology combined with the writer's career-long interest in tapered (driven) piles was the genesis for this monograph. The primary objective of this monograph is to make a contribution toward addressing the significant knowledge gap concerning tapered (driven) piles that currently exists as was explained earlier in this chapter.

This monograph focuses on tapered (driven) piles as these are by far the most commonly and widely available type of tapered DFE available at the present time, at least in the U.S., and the writer is most familiar with them. Note, however, that some of the material presented in this monograph is applicable, in concept at least, to either DFEs in general or tapered DFEs that are formed-in-place using some drilling technique as some researchers have studied, e.g. Khan et al. (2008). However, the writer is unfamiliar with tapered non-pile DFEs and they do not appear to have been used to any significant extent in routine practice to date. This is not surprising as the ability to reliably and repeatedly create a stable tapered hole in the ground on a production basis as opposed to doing so in closely monitored research would seem to be difficult. Thus, little material is presented in this monograph with specific regard to tapered DFEs created by some drilling process other than to cite published work by others that address such DFEs. The writer leaves it to others to extend the concept of tapered DFEs to anything other than a (driven) pile.

To accomplish this educational goal related to tapered piles, this monograph is intended first and foremost to be a resource document for state-of-knowledge, practice-oriented technical information for the benefit of licensed design professionals. It focuses on practical, usable technical information, especially analytical methodologies, some of them published here for the first time, that to date have not entered the mainstream of geotechnical and foundation engineering textbooks and similar reference publications to any significant extent. Thus, it is intended that the material presented in this monograph will help to fill the informational void in engineering practice that was noted earlier in this chapter.

An important secondary objective of this monograph is its use as a resource document for academic education and research as proved to be the case for the writer's 1995 monograph on geofoam (Horvath 1995) that was well-received internationally. With regard to education, as noted earlier in this chapter, foundation engineering textbooks have historically contained relatively little information about tapered piles which means that little, if any, information on this subject has found its way into courses at colleges and universities.

In addition, there appears to have been relatively little academic research and concomitant scholarly publication devoted to tapered piles when compared to the overall body of research and publication that has been devoted to deep foundation in general[19]. In retrospect, this is not surprising as academic research, in the U.S. at least, tends to be

---

[19] The one notable exception to this is the University of Western Ontario (UWO) in London, ON, Canada. UWO was "corporately rebranded" (to quote their Wikipedia entry) as Western University in 2012 and apparently is now commonly referred to simply and colloquially as 'Western'. The use of 'UWO' is retained in this monograph where historically appropriate, i.e. when referring to occurrences prior to 2012. Since the late 1990s, there has been a relatively steady stream of research at UWO/Western under the overall direction of Prof. M. Hesham El Naggar, Ph.D., P.Eng. that has been devoted to tapered DFEs. However, most of this research has been devoted to alternative, non-traditional DFE materials, geometries, and installation methods. Thus, the tapered DFEs studied at UWO/Western are, for the most part, on the fringe of the focus of this monograph although they are noted throughout this monograph for the sake of completeness. However, most of the concepts related to tapered piles that are presented in this monograph are likely relevant to these tapered DFEs, possibly with some modification. This work is left to others in the future.

dominated by topics that receive funding from sources outside of academic institutions. It appears that the business entities that have had and, in some cases still have, a business interest in tapered piles felt that there was no benefit in supporting academic research into their products. Furthermore, other potential stakeholders in tapered-pile technology such as government agencies have been loath to fund work involving what they perceive to be an overall proprietary technology. Hopefully, publication of this monograph will be an incentive for these attitudes in both industry and government to change in the future (but the writer is not holding his breath for this to happen).

Because this monograph is intended to be as informational, informative, and, above all, objective as possible, names of specific persons and organizations as well as specific business entities and tradenames of their products are mentioned extensively throughout this monograph. This has been deemed by the writer to be necessary for the stated informational and educational purposes and overall objectivity of this document. ***However, in no way should this be interpreted as a comment or endorsement by the writer of the named person, organization, business, or product, nor any guarantee or warrantee, expressed or implied, that a specific product will perform as desired in a specific project application. Liability for any potential application discussed in this monograph rests solely with the licensed design professional(s) involved in the design and construction of that application*** (emphasis intentional).

The writer emphasizes these last points as recent history has taught us that there have been numerous instances going back decades where scientific researchers in academia and other nominally independent, non-commercial organizations published, often repeatedly, what amounted to work-for-hire on behalf of, and biased toward, some commercial enterprise. Furthermore, and more importantly, the financial connection in these publication efforts was not made known at the time. Therefore, the public has understandably come to look at published academic work with some concern about its objectivity. This skepticism is understandable and the writer thus feels it is necessary to be proactive and up front on this issue of potential conflict of interest. There is none in this case. Period.

1.3 OVERVIEW AND ORGANIZATION

The content of this monograph has been developed and organized assuming no prior knowledge by the reader of tapered-pile materials, products, or analytical methodologies although it is assumed that the reader has at least an undergraduate-level civil engineering education. Although the focus of this monograph is tapered piles loaded in axial compression, as noted above, some of the analytical methodologies presented are sufficiently general so that they could be applied to other types of tapered DFEs with either no or modest modification.

One of the changes that the writer has seen in foundation design in recent decades is the increased need to design for uplift and lateral loads on all types of foundations. This is a combination of increased seismic design in areas such as the East Coast of the U.S. where this was not done historically before the latter decades of the 20th century as well as more-efficient structural design of tall, slender buildings that results in uplift loads under some loading conditions becoming the rule rather than the exception. Thus, both uplift and lateral loads are addressed in this monograph, especially the former. This is because one of the reasons for not considering tapered piles that the writer has heard from time to time over the years is that they have little uplift resistance. There is a simplistic perception among some design professionals that somehow a tapered pile will lift out of its tapered 'hole in the ground' under a relatively small uplift force. This perception, which is perhaps intuitively appealing but simply incorrect, is likely abetted by the fact that there is a dearth of uplift load-test data for tapered piles and a complete lack of any analytical methodology for uplift loading. Therefore, some discussion of this topic is very much needed and is provided.

The following is a list and synopsis of the remaining chapters of this monograph:

- <u>Chapter 2</u> discusses fundamentals of the tapered-pile problem such as systems of units, terminology, and notation that are used throughout this monograph. This is necessary as geotechnical and foundation engineering are historically notorious for their lack of standardization with respect to both terminology and notation compared to other specializations in civil engineering. Thus, it is essential to have well- and clearly defined parameters to minimize the potential for errors or ambiguities in this regard given the global audience for which this monograph is intended.

- <u>Chapter 3</u> builds on and extends the summary of tapered-pile history that was presented earlier in this chapter. Technical information concerning pile geometries and pile materials is presented in detail.

- <u>Chapter 4</u> contains a detailed presentation and discussion of the many and varied fundamental, generic geotechnical issues that affect the physical mechanisms that go into the resistance of tapered DFEs in general to axial loads. The most significant part of this chapter is the discussion of shaft resistance due to wedging that is unique to tapered DFEs and constitutes a third resistance mechanism in addition to the traditional mechanisms of shaft resistance due to sliding friction and toe resistance due to end bearing. The overall goals of this chapter are to provide an understanding of the key geomechanical concepts in the current state of knowledge that control the behavior of tapered piles when subjected to axial loading in both compression and tension (uplift). This is to provide a theoretical basis for understanding the pluses and minuses of the resistance-forecasting methods presented in Chapters 5 and 6, and the logic behind the schema for an improved resistance-forecasting methodology that is presented in Chapter 7.

- <u>Chapter 5</u> contains a detailed presentation and discussion of analytical methodologies for forecasting the resistance of tapered piles as well as producing complete load-settlement curves for axial-compressive loading. Included are methods developed specifically for tapered piles that have appeared in the published literature as well as the writer's research (much of it never before published) illustrating how widely used methodologies for constant-perimeter DFEs can be extended to tapered piles. Because no existing resistance-forecasting methodology for tapered piles is ideal in its present form, this chapter concludes with a list of what the writer feels are desirable features and capabilities of an improved resistance-forecasting methodology for tapered piles that also includes generating a complete load-settlement curve in a single, one-step process.

- <u>Chapter 6</u> presents a limited numerical comparison of some of the resistance-forecasting methodologies presented in Chapter 5 using an actual tapered-pile case history as the ground-truth for this comparison. This is not intended to be a conclusive determination of the best or most-accurate analytical methodology but more to illustrate some of the practical issues to be kept in mind when using these methods in routine practice. Consequently, this chapter is expected to be of particular relevance and interest to design professionals as well as to academicians involved in teaching foundation engineering.

- <u>Chapter 7</u> uses the 'wish list' of desirable features for an improved, one-step resistance-forecasting methodology for tapered piles presented at the end of Chapter 5 to create the schema or framework for such a methodology. Consequently, this chapter is expected to be of particular relevance and interest to academicians who are contemplating conducting research into tapered piles.

- <u>Chapter 8</u> discusses the concept of quantifying taper benefit within the broader concept of *resistance efficiency* of DFEs in general. This discussion is central to project-specific foundation design in practice because it addresses the issue of how to compare deep-foundation alternatives on an apples-to-apples basis to identify the most efficient alternative that satisfies the technical needs and constraints of a project. This chapter is expected to be of particular relevance and interest to stakeholders involved in making project-specific decisions with respect to foundation selection.

- <u>Chapter 9</u> provides a concise denouement of the contents of this monograph. Because of the overall length of this monograph and the broad range of topics that are presented and discussed, this chapter serves as a concise summary of the writer's opinions on how to perform resistance-forecasting for tapered piles at the present time as well as what the research priorities are to advance the state of art. Thus, this chapter should be of interest to both design professionals and academicians.

As is common for scholarly published works such as master's theses, doctoral dissertations, research reports, and monographs, appendices are used to present the detailed treatment of narrow topics that support portions of the main body of text. This monograph is no exception to this and contains eight appendices and two subappendices:

- <u>Appendix A</u> uses case-history data from two instrumented tapered piles to illustrate how empirical model parameters used for some of the common, all-purpose resistance-forecasting methodologies in use at present, such as the β Method and UniCone Method, can easily be extended to include tapered DFEs.

- <u>Appendix B</u> explores concepts for an improved toe-bearing methodology that considers load-settlement behavior based on a hyperbolic model as opposed to the traditional approach to DFE toe bearing that considers the ultimate resistance only.

- <u>Appendix C</u> contains a detailed discussion of the three behavioral phases for the shaft-resistance mechanism of wedging based on cylindrical-cavity expansion that were postulated by Kodikara and Moore in their seminal 1993 paper on tapered DFEs. Also discussed are the subsequent simplifying assumptions made by El Naggar and Sakr. A thorough understanding of cylindrical-cavity mechanics for the wedging resistance mechanism is essential for understanding the behavior of tapered DFEs.

- <u>Appendix D</u> contains a detailed treatment of the subject of modulus degradation (reduction), with an emphasis on shear modulus. Shear-modulus degradation is an essential element of modern resistance-forecasting methodologies for all types of DFEs, not just tapered piles. Because the issue of shear modulus in general is important to the contents of the monograph, there are two subappendices to this appendix:

- o <u>Subappendix D1</u> discusses the issue of engineering vs. true strain, a subject that historically is not addressed in either civil engineering education or publications to the extent that it should and
- o <u>Subappendix D2</u> discusses shear-strain fundamentals.

- <u>Appendix E</u> contains an updated and revised version of previously published work by the writer to develop a new 3-D analytical model for axial and radial normal stresses and displacements of a composite (multi-component) DFE cross-section. A comparison to the simple 1-D model that is typically used in practice to apportion axial-normal stresses to a composite DFE cross-section is also included.

- <u>Appendix F</u> discusses the Sabry Model for forecasting the radial stresses acting along the interface between a DFE shaft and adjacent soil after DFE installation. The ability to forecast these stresses is an essential component of developing an improved resistance-forecasting methodology for tapered piles.

- <u>Appendix G</u> explores some of the modern alternatives for forecasting the ultimate toe resistance (bearing capacity) for all types of DFEs. These include Vesic's spherical-cavity expansion (SCE) solution and NTH Limit Plasticity Theory. The attraction of these alternatives to traditional solutions based on rigid-plastic models that all evolved from Terzaghi's original bearing-capacity solution is that they lend themselves to rational assessment using cone-penetrometer data.

- <u>Appendix H</u> presents a detailed assessment of the Holmen Island test-pile program conducted by the Norwegian Geotechnical Institute in 1969. This program was unique in that instrumented test piles that differed only in geometry (tapered vs. constant-perimeter) were driven in close proximity in coarse-grained soil conditions and then loaded to geotechnical failure in both compression and tension (uplift). Extensive assessments of the original results, which were relatively obscurely published in the proceedings of a 1973 international conference and thus are not widely known, were performed by the writer in order to abstract the most information possible that is relevant to the theme and goals of this monograph.

This monograph concludes with a section that contains a combined, integrated list of references cited throughout the monograph as well as selected additional references that provided a supplemental bibliography.

Chapter 2

Problem Fundamentals

2.1 INTRODUCTION AND OVERVIEW

Compared to other areas of specialization within civil engineering, geotechnical engineering, as well as its link to structural engineering, foundation engineering, have historically used a multiplicity of terminology and notation that seem to defy efforts at standardization. This has only become more problematic and confusing in recent decades as the ability to distribute information globally with relative ease via digital documents and the Web has increased dramatically. The result is that a civil engineer anywhere has relatively easy access to information generated everywhere. This makes the issue of terminology and notation more important than it has ever been.

The writer's personal experience is that the geographic location(s) of a person's professional education and practice tend to influence what a particular material property or parameter is called, labeled, and sometimes even how it is spelled. This means that the average civil engineer has access to information wherein the same property or parameter may be called a different name or assigned a different notation or spelled a different way depending on the author's predilections. This diversity can create needless confusion, even for those who have been in the profession for a relatively long period of time.

Unfortunately, nowhere is the lack of standardization more acute than with deep foundations, especially piles. Using the writer's experience within the U.S. as an example, variations in pile terminology occur within geographic regions and even with the particular type of pile, i.e. depending on the specific type of pile, the same parameter has different colloquial names. Examples of this are noted in this and subsequent chapters.

Similar comments can be made about variations in units. Although the *Système International d'Unités* (SI) version of metric units is supposed to be universal among civil engineers, the writer's experience is that there are national, regional, and application-specific variations that do not adhere to this (although in some cases the non-SI deviation is pragmatic and thus defensible).

A related problem occurs with parameters that have compound units. For example, density is always mass per unit volume but it can be expressed using different combinations of mass units (e.g. tonne vs. kilogram vs. gram) and length units (e.g. metre vs. centimetre) within the metric system. As a result, depending on an author's preference one sees soil density expressed in both kilograms per cubic metre (kg/m^3) and grams per cubic centimetre (g/cm^3 = g/cc), with the latter theoretically deprecated in SI. Again, the writer's experience is that there are distinct national and regional preferences for such things.

Therefore, before proceeding further it is essential to clearly define certain fundamentals such as systems of units, terminology, and notation that are used throughout this monograph. When deemed relevant, reasons will be given for choosing a particular term. When there are competing terms and/or notation, an attempt will be made to mention them for the sake of completeness and cross-referencing.

This chapter concludes with introductory comments on other topics that are used or referred to repeatedly in different contexts throughout this monograph. This is done to establish a basic understanding and vocabulary for these topics that are used consistently in the remainder of this monograph.

2.2 SYSTEMS OF UNITS

The primary system of units used in this monograph is the *Imperial* (a.k.a. *U.S Customary* or *English*) system, largely because most tapered piles that are currently in use are U.S. manufactured and exactly dimensioned in such units. In general, the particular Imperial units or combinations of units and their abbreviations that are used are those that the writer feels are most commonly associated with a particular material property or parameter in U.S. practice.

Consistent with the Imperial system of units, a decimal point (.) is used to indicate a fractional value of a number. Also, a comma (,) is used to separate groups of three integer values. This usage is highlighted as in some parts of the world a comma is used instead of a decimal point to indicate a fractional value.

The secondary system of units used throughout this monograph is the aforementioned SI version of metric units that is considered the preferred standard for civil engineering. This means that certain non-SI metric units that routinely appear in some geotechnical and foundation engineering publications, especially in some countries, such as the centimetre (cm) and what are supposed to be forces but are expressed in kilograms or the oxymoronic unit of kilogram-force (kilopond) instead of the correct Newtons are avoided.

This use of the SI system of metric units also means that the basic length unit is spelled 'metre', not 'meter'. This is an often-overlooked part of the SI system and is done not with a nod to British/Canadian English spelling preference but to avoid any confusion with the use of the word 'meter' to mean a measurement device. In addition, in keeping with SI guidelines the use of a decimal (to indicate a fractional value) and comma (to separate groups of three integers) are avoided to the greatest extent practicable although there are exceptions to be found throughout this monograph where the writer deemed them efficient to use.

There are a few instances in this monograph where the primary and secondary systems of units are switched to SI and Imperial respectively. These involve case histories where metric units were primary so it is logical to maintain that precedence.

2.3 DEFINITION OF FAILURE (LIMIT STATE)

The concept of 'failure' in a civil engineering context is best defined as the 'loss of function'. This means that whatever function (service or role) a structure or structure component was intended to provide is no longer being provided to an acceptable degree.

Within this definition, there are two distinctly different types of failure that can occur:

- *Serviceability* failure due to excessive displacement and/or deformation that renders a structure or structure component unable to continue to function as intended.

- *Collapse* or *ultimate* failure as the result of material rupture (in the case of solid materials) or, in the case of soils which do not exhibit classical material rupture, the development of some physical mechanism composed of one or more zones or surfaces along which the shear strength has been fully mobilized and large-scale, open-ended relative displacements are occurring.

Note that the term *Limit State* has for some time now been preferred to 'failure' in civil engineering usage, especially outside of the U.S. Thus, these two types of failure are referred to as the *Serviceability Limit State* (SLS) and *Ultimate Limit State* (ULS). This concise SLS/ULS terminology are used in this monograph.

2.4 PILE TERMINOLOGY AND NOTATION

2.4.1 General

As noted at the beginning of this chapter, foundation engineering stands out for its lack of standardization in terms of notation, terminology, and sometimes even spelling. Nowhere is this this lack of standardization more apparent than with deep foundations in general and (driven) piles in particular.

The basic rule of thumb adopted for this monograph with regard to notation, terminology, and spelling is to use consistent terminology for a material property or parameter based on what, in the writer's experience and opinion, is used colloquially and commonly in U.S. practice at the present time. The one notable exception to this rule is that the use of the force unit of *ton*, more correctly called a *short ton* and equivalent to 2,000 pounds (= 2 kips = 8900 N), is not used[20].

The reason that this deprecation of ton is emphasized here is that in the writer's experience, the use of ton as a force unit, either alone or as part of the stress unit of tons per square foot (tsf), is curiously persistent in U.S. foundation engineering practice to the present (at least in some regions such as the New York City metropolitan area where its usage is ubiquitous). Apparently, this continued usage is for historical reasons as 'ton' was in use as a foundation engineering force unit long before use of *kip* (= 1,000 pounds = 4450 N) occurred[21]. For example, one routinely continues to hear a colloquial phrase such as '100-ton pile' used when someone is referring to a pile that has an allowable (maximum-service) load of 100 tons (= 200 kips = 890 kN). In any event, given the intended international audience for this monograph, the writer feels that it is time for this terminological anachronism to be retired to the Foundation Engineering Hall of Fame (at least in this monograph) as it is out of step with Imperial-system force units used in current civil engineering research and practice.

However, even within this adopted rule of thumb of following current U.S. practice in terminology used in this monograph there is still substantial variation to contend with. For example, the lower end of an installed DFE is variously referred to as the (listed alphabetically): *base, bottom, end, foot, point, tip,* and *toe*. There are somewhat fewer terms that are used for the upper end: *butt, head, top.* Therefore, the writer has chosen to adopt (with one exception[22]) the recommendations of Fellenius (1999b) as summarized in Figure 2.1 (for the specific case of a (driven) pile to be consistent with the focus of this monograph). These recommendations appear to the writer to be reasonable in that they are at least among the terms used in U.S. practice plus have the additional benefit of being directly and logically translatable into other languages. So, for example, *head* and *toe,* which are certainly universal in their meaning in any language, are used in this monograph for the upper and lower end, respectively, of an installed DFE of any kind.

As an aside, the notational standard used in this monograph and as reflected in Figure 2.1 is to use capital letters for parameters that are forces and lowercase letters for parameter that are stresses (note that *unit shaft resistance, unit drag load,* and *unit toe resistance* are all stresses). There is no set standard for parameters that involve only length dimensions.

---

[20] Note that ton/short ton should not be confused with *tonne* (a.k.a. *metric ton*). A tonne has an equivalent mass of 1000 kilograms (kg) or 1 Megagram (Mg). Note further that 'tonne' is not used in this monograph either, but for the fundamental reason that it is not an SI unit for mass.

[21] 'Kip' is a contraction of the term *kilopound* and should not be confused with the arcane metric unit of *kilopond* (abbreviated *kp*) that is an alternative term for *kilogram-force* and equal to 9.81 Newtons.

[22] The notation used for pile diameter in this monograph is *d*, not *b* as recommended by Fellenius, as the notation *d* is near-universally used for diameter while *b* is equally associated with width.

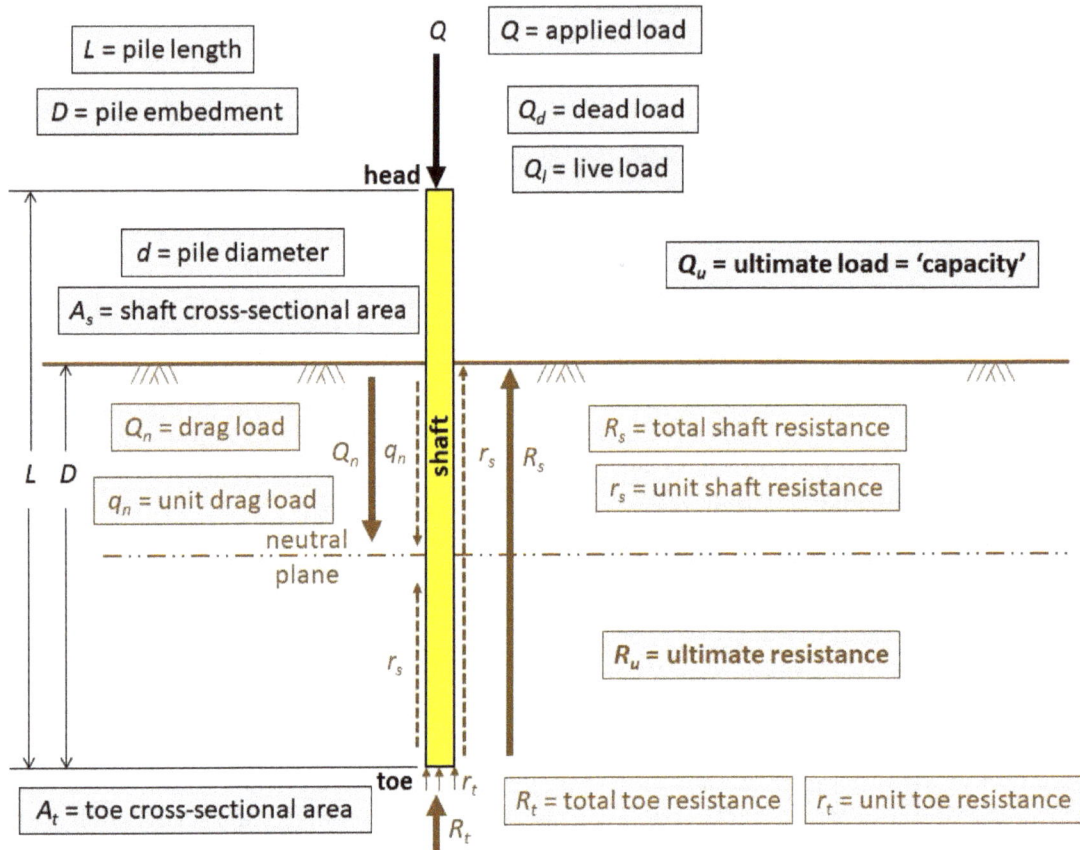

Figure 2.1. Deep-Foundation Terminology Used in Monograph.

In a similar vein, the terms *vertical* and *horizontal* are used when referring to the spatial orientation or direction of a force or stress as these terms are also unambiguously translatable. There are two exceptions to this, both involving the horizontal direction:

- When referring specifically to earth pressure[23] theory, the term *lateral* is used in lieu of horizontal to be consistent with colloquial English-language usage.

- In some contexts, *radial* is used in lieu of either horizontal or lateral when it is desirable to convey the sense of stress or displacement occurring simultaneously in 360° of horizontal direction, usually around some object such as a DFE shaft that has a circular cross-section.

---

[23] Note that the use of the term 'pressure' in this context is itself a technically incorrect, but common, colloquial usage. Technically, 'stress' should be used in this context as pressure is a subset of stress to be used only when fluids (gases or liquids) are involved. Furthermore, because pressure at some point within a fluid is omnidirectional by definition, the use of a directional adjective (vertical or horizontal) is redundant and thus unnecessary, However, given the historical view of stresses in the ground, especially in the horizontal direction, of being 'equivalent fluid pressures', the use of pressure in this context is, therefore, used in this monograph. Because earth pressures are not omnidirectional, the direction of earth pressures must always be stated explicitly.

Additional terms that are used frequently throughout this monograph are defined as follows (some were previously defined in Chapter 1 of necessity but are repeated here for the sake of completeness):

- _Pile_ when used alone in this monograph is the traditional U.S. usage to mean a preformed structural element (which may or may not reflect the final, complete installed element as in the case of a steel shell or pipe that is filled with fluid PCC after installation) that is driven at least to its final depth using an impact hammer of some type. Given the intended international audience for this monograph, this relatively narrow terminological usage of the term pile is very important to emphasize as in most countries other than the U.S. the term pile means any type of DFE and an adjective or modifier (bored, driven, jacked, etc.) is always required to distinguish the specific type of DFE. Thus, when deemed necessary for clarity or emphasis, the term '(driven) pile' is used in this monograph.

- _Resistance_ when used alone in this monograph refers to the geotechnical support forces provided to the embedded portion of a DFE shaft or the DFE toe by the ground in an orientation that is parallel to the longitudinal axis of the DFE. When deemed necessary for clarity, the term '(geotechnical) resistance' is used in this monograph to distinguish the geotechnical contribution from the _structural resistance_ (always so noted in this monograph) that is used for the support forces provided by the structural materials that comprise the DFE shaft. Note that (geotechnical) resistance is not necessarily the ULS value, R_u. It can be some intermediate value of mobilized support depending on the context so the term _ULS resistance_ is used when the ultimate value is specifically intended. Note also that the ULS (geotechnical) resistance, R_u, is almost always called 'capacity' in current U.S. practice. However, the term capacity in this monograph has a different meaning as defined in the following bulleted item.

- _Capacity_ as used in this monograph and illustrated in Figure 2.1 is the ultimate load, Q_u, that can be applied to the head of a DFE in axial loading (compression is shown in this figure but it applies to uplift as well). A subtle point is that capacity is usually governed by the (geotechnical) resistance as opposed to the structural resistance, and modern design codes are crafted to ensure this. However, it is possible when analyzing an existing DFE that structural resistance may govern Q_u for any number of reasons. Thus, both geotechnical and structural resistance should always be evaluated when analyzing an existing DFE as it is never known beforehand which of the two resistances may be the more critical for future performance of the DFE.

- _Load,_ whether used alone or with a modifier such as _externally applied load_, always means a force that is parallel to the longitudinal axis of the DFE that acts on a DFE in some fashion. Note that 'load' may apply to a force external to a DFE (usually at the head), within a DFE shaft, or along the DFE shaft-ground interface. Therefore, the distinction as to the source of the load (e.g. _externally applied load_) will be made in this monograph when it is important to be clear as to where the load is applied to the DFE.

- _Taper_ is the change in cross-sectional dimension (diameter/radius or width depending on the cross-sectional geometry of a DFE) in a direction parallel to the longitudinal axis of a DFE over at least a portion of its overall length. Unfortunately, there are multiple ways in which taper is numerically defined or expressed. Because this parameter is so central to this monograph, a separate section of this chapter is devoted to the subject of defining taper and taper-related parameters.

- *Tapered pile* refers to any pile that incorporates taper for at least a portion of its shaft. There are three subcategories of tapered pile:
  - *Continuously tapered* (the term *fully tapered* is also seen in the literature) from head to toe although the taper angle may be constant in magnitude (*uniformly tapered*, as is typical for manufactured piles) or variable in magnitude (*non-uniformly tapered*, as is typical for timber piles).
  - *Step-tapered*, in which the diameter-change occurs as a series of contiguous constant-diameter segments, each of a different diameter, so that the overall head-to-toe taper is really a virtual quantity.
  - *Partially tapered* (*semi-tapered* is an alternative term) where a constant-perimeter segment is connected to a continuously tapered segment (which may be either uniformly or non-uniformly tapered depending on the material used), typically the former above the latter although some model piles used in research testing have used the latter above the former as some researchers have concluded that this is theoretically more desirable.

  These sub-category terms are generally not used in U.S. practice (use outside of the U.S. appears to be somewhat more common) except in one instance where 'step-taper' was used as for decades as part of a specific product tradename that was also a registered trademark in the U.S. at one time. In this monograph, the term 'tapered pile' is used to mean any and all of these three sub-categories unless it is important in a particular case to distinguish between and among these sub-categories. In such cases, the specific sub-category term is used.

- *Taper benefit* is a term that to date has been used broadly in an ill-defined, qualitative manner to indicate the technical benefit, and the cost benefit that derives from this, of using a tapered DFE as opposed to a non-tapered DFE in the same ground conditions. Because the concept of taper benefit has potential use as a metric for allowing a relatively quick comparison of DFE alternatives, e.g. during the preliminary design phase of a project, an entire chapter (Chapter 8) is devoted to exploring this subject.

- *Forecast* is used in lieu of *prediction* when referring to the final results or outcome of some analytical methodology or process. As discussed in Chapter 1, while 'prediction' has a long history of informal and formal usage in geotechnical and foundation engineering, the writer has become increasing troubled by the fact that laypersons tend to view something labeled a prediction as implying a rather high degree of certainty or finality that has a mystical/psychic 'crystal-ball' element to it. On the other hand, thanks to weather forecasting in particular, the public has (for the most part) become educated to the fact that a forecast has a rational, scientific basis but one that allows for the vagaries and uncertainties of nature in the actual outcome. In short, there is (or at least should be) an unconscious allowance or forgiveness for some error in a forecast that does not exist with a prediction which (in the writer's opinion) has a more ironclad connotation of certainty associated with it. Thus, in the writer's opinion, foundation engineers would be much better served as a profession if their work-products were viewed as forecasts and not predictions.

2.4.2 Taper

Because taper is the feature characteristic of all piles that are the focus of this monograph, it is important to clearly define this parameter. Figure 2.2 depicts the key geometric parameters for the tapered portion of a pile with a circular cross-section. The parameters shown are defined as follows:

- d = diameter at any cross-section along the length of the tapered portion of the pile,

- $r = d/2$ = radius at any cross-section along the length of the tapered portion of the pile,

- x = increase in radius that occurs over some defined length y that is parallel to the centerline (longitudinal axis) of the pile, and

- α = angle (referred to as the *taper angle*) defined by x and y, $\alpha = \tan^{-1}(x/y)$.

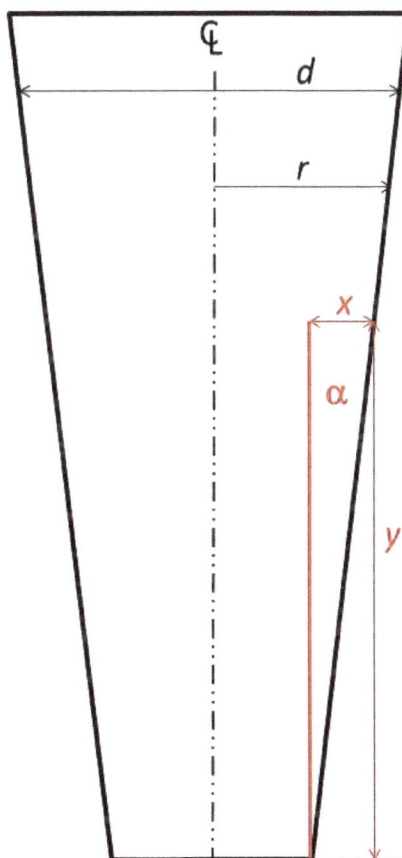

Figure 2.2. Generic Parameters Used to Define Taper.

To date, three distinctly different definitions of taper have been identified by the writer in the published literature (there may be more that the writer is not aware of), with variations or alternatives possible for each:

1. Manufacturers of both the *Monotube* and *TAPERTUBE* piles in the U.S. have historically expressed taper as the ratio of <u>diameter</u> change over some defined length, i.e. $2x/y$ in terms of the parameters shown in Figure 2.2. Both manufacturers use inconsistent

Imperial units for 2x and y, specifically, inches and foot respectively. Thus, the 2x dimension is always expressed as a fraction of one inch and is usually expressed to two decimal places of precision. The y dimension is always one foot. For example, a commonly used taper in both product lines is catalogued as "0.40 inches per foot".

2. A much less common variation of this uses consistent units for the 2x and y parameters and then expresses the result as a percent. So, the preceding 0.40 in/ft (= 0.40 in/12 in) example would be a 3.3% (= [0.40/12] x 100) taper.

3. The writer (Horvath 2002, 2014b, 2015, 2016a, 2016b) as well as Kodikara and Moore (1993) and Manandhar and Yasufuku (2011b, 2012, 2013) have all used Nordlund's (Nordlund 1963, 1979) definition of what is illustrated in Figure 2.2 and defined above as the 'taper angle', α, expressed in degrees of angle[24]. Using this definition, both the *Monotube* and *TAPERTUBE* piles typically have taper angles no greater than 1° (the 0.40 in/ft example given above in Taper Definition #1 is 0.95°). For comparison, timber piles typically have an average taper angle of the order of 0.25°. Note that these relatively small values of angle are the reason for noting this inherent attribute of tapered piles in the title used for this monograph. Note also that if one wanted to be consistent with SI protocol and eliminate usage of degrees of angle as well as fractional values, the unit of milliradians would seem most appropriate for taper angle. In this case, 0.95° = 17 mrad.

As noted above, each of these three definitions has alternative versions. For example, if it were desired to use Taper Definition #1 with SI units, in order to maintain the same two-digit precision and avoid the use of fractional values, which are eschewed in SI-unit protocol, the logical dimensions to use would be millimetres per metre. So, the above example of 0.40 in/ft would be 33 mm/m.

Care also needs to be exercised when using either the Taper Definition #1 or Taper Definition #2 as it is possible that some manufacturer or publication author might choose to express taper as the ratio of <u>radius</u> change per unit length (x/y) in which case the values would be exactly one-half of those obtained using the above definitions.

In a similar vein with respect to Taper Definition #3, taper angle is not always defined consistently. For example, Fellenius et al. (2000) defined taper angle using <u>diameter</u> change, i.e. as $\tan^{-1}(2x/y) = 2\alpha$, so the resulting angular values they quote are double those obtained using the Nordlund et alia definition. Based on the quotation cited in Chapter 1, it appears that Peck (1958) may have defined taper angle in the same way as Fellenius et al. (2000).

It appears that the rationale for this alternative definition may derive from viewing the tapered portion of a pile as defining an imaginary cone that comes to its apex/vertex some distance below its actual pile toe. In this visualization, the taper angle is the apex angle of the imaginary cone. Note that there is a history in geotechnical engineering of using the apex angle of a conical tip to define a tapered object. For example, the cone penetrometer

---

[24] Note that Nordlund used the notation ω, not α, for this angle, possibly because ω is often used to define the angle that the assumed planar inside face of a rigid earth-retaining structure makes with respect to the vertical (Horvath 2018a). Because Nordlund modeled the shaft-resistance mechanism of wedging using 2-D lateral earth pressure theory, pile taper was viewed as being physically and mathematically equivalent to the rotation of a rigid retaining wall into a mass of retained soil. Consequently, his notational usage would seem to follow from this. The notational change from ω to α for this monograph was made by the writer to be consistent with later published work by Kodikara and Moore as well as Manandhar and Yasufuku who used α together with a more-modern physical and mathematical 3-D model for wedging that is based on cylindrical-cavity mechanics.

(discussed at the end of this chapter) has historically been defined by the apex angle of its conical tip (60°).

In summary, taper can be defined as either a fractional length dimension, percentage, or angle, and based on either diameter or radius for a total of six potential definitions, four of which can be expressed using either Imperial or SI units. The resulting ten theoretical possibilities are illustrated in Table 2.1 for one taper (the aforementioned 0.40 in/ft that is highlighted in **bold green font**) that is commonly used in practice in both the *Monotube* and *TAPERTUBE* product lines. The published literature indicates that many of these theoretical combinations have been used at various times by manufacturers and authors in published works. Consequently, before using a document dealing with tapered DFEs, the definition of taper as used in that document should always be ascertained.

Table 2.1. Exemplar Summary of Taper Definitions and Their Variations.

Basis	Taper Definition		
	#1: length, in/ft (mm/m)	#2: percentage, %	#3: angle, ° (mrad)
diameter	0.40 (33)	3.3	1.9 (33)
radius	0.20 (17)	1.7	0.95 (17)

Taper Definition #3 based on taper angle, α, as defined in Figure 2.2 and with units of degrees of angle (the combination highlighted in **bold red font** in Table 2.1) is used exclusively throughout the remainder of this monograph. This definition has proven efficient to use by researchers going back almost 60 years to Nordlund's seminal 1963 paper as it avoids the built-in problems associated with a specific system of units that may be inconsistent as well (Taper Definition #1) and is more insightful than using percentage (Taper Definition #2).

Before closing the discussion of taper, there are additional comments to be made on this topic that are explored further in Chapter 3 when types of tapered piles are discussed in detail:

- As noted earlier in this chapter, taper is not necessarily uniform along the length of a continuously tapered pile. This is certainly true of timber piles by virtue of the way that trees, or at least the softwood species (Southern Yellow Pine) that is the most commonly used for timber piles in the U.S., grow. This is another reason why Taper Definition #3 based on taper angle is preferred as it is the easiest and least ambiguous definition to use in these cases as it readily allows for rational definition of taper at various places along the pile shaft.

- One of the prominent types of manufactured tapered piles in the U.S. (the Raymond *Step-Taper* that was mentioned in Chapter 1) that was used for many decades in the 20th century consisted of two or more constant-diameter segments of different diameters. As a result, the pile had an overall quasi-taper that was actually achieved in a series of constant-diameter segments or steps. Furthermore, because these segments did not necessarily have the same length (the change in diameter from segment to segment was, however, always the same), the quasi-taper angle of the final pile as installed could vary along its length, similar to the aforementioned variation in taper angle of timber piles. Again, such variable taper is most easily handled using Taper Definition #3.

- As noted earlier in this chapter, several researchers have explored and subsequently recommended use of partially tapered piles wherein the tapered portion is placed <u>above</u> the constant-diameter portion. This is the reverse of what has been done historically in practice with respect to partially tapered piles such as the *Monotube* and *TAPERTUBE*. Once again, Taper Definition #3 is readily usable with such a pile whereas the other two taper definitions are not.

2.5 SOIL- AND SYSTEM-PROPERTY TERMINOLOGY AND NOTATION

Compared to piles, the terminology and notation used for soil properties and what are referred to in this monograph as *system properties*, i.e. the properties for mechanical (stress-strain-time) behavior exhibited along the contact between two dissimilar materials such as a pile shaft and adjacent ground, are reasonably standardized, not just within the U.S. but on a global basis. Nevertheless, the first time that either a soil or system property is used in this monograph it and its notation are clearly defined. The same soil/system property terminology and notation are then used throughout the remainder of the monograph.

The one soil property that requires some elaboration at this point is the soil shear modulus, G. As will be seen in several later chapters as well as Appendix D, soil shear modulus figures very prominently nowadays in many state-of-art analytical methodologies for all types of DFEs, not just tapered piles. Shear modulus has replaced Young's modulus, E, as the preferred parameter to use for defining ground stiffness in many foundation engineering applications. This is because it has become much easier to measure shear modulus in routine practice using a *seismic piezocone* (sCPTu) sounding or other in-situ testing tool or methodology than to measure Young's modulus. Because most foundation engineering analyses can be formulated using either modulus, there is no reason to continue to use Young's modulus. Furthermore, there are some geotechnical and foundation engineering applications where shear modulus is actually conceptually the more correct modulus to use as it is more consistent with the soil behavior inherent in that application, i.e. a shearing mode vs. a pure-compression mode.

For reasons explained in later chapters and explored in detail in Appendix D, it has become common to reference the shear modulus to what is called its *small-strain value* that serves as a constant-magnitude, baseline reference value under the assumed linear-elastic conditions that occur under very small magnitudes of strain (how small is "very small" is discussed in Appendix D). The notation used in this monograph for the small-strain value of shear modulus is G_{max}.

This notation is highlighted here because the notation G_o is often used as an alternative notation to G_{max} and actually appears to be globally more common[25]. In fact, the writer used the G_o notation in published work related to tapered piles as recently as 2015 (Horvath 2015). However, beginning in 2016 the writer began using the G_{max} notation, solely and specifically to avoid a notational conflict as the G_o notation is used in a completely different context in a state-of-art analytical methodology related to deep foundations that

---

[25] Prof. Kenneth H. Stokoe of the University of Texas at Austin, a recognized expert on small-amplitude wave propagation in the ground, presented the 2017 F. E. Richart Lecture at the University of Michigan on 18 October 2017. This lecture, which is titled "The Increasing Role of Seismic Measurements in Geotechnical Engineering", was broadcast globally as a webinar by the International Society for Soil Mechanics and Geotechnical Engineering (ISSMGE). In this lecture/webinar, Prof. Stokoe commented that the use of the G_{max} notation appears at present to be more common in the U.S. while the G_o notation is more common in the rest of the world. This is consistent with the writer's perceptions based on a review of the published literature in recent years.

figured prominently in recent publications (Horvath 2016a, 2016b). Because this methodology also figures prominently in this monograph, the G_{max} notation has been retained in this monograph to prevent any confusion or ambiguity and thus has nothing to do with regional preferences.

For the sake of completeness, it is noted that the same concept of a small-strain modulus value applies to Young's modulus as well, with the notation of E_{max} or E_o used as desired. Although the use of Young's modulus for soil stiffness has been eclipsed by shear modulus, especially in deep-foundation applications, for historical reasons theoretical solutions that make use of Young's modulus are mentioned throughout this monograph. The notation of E_{max} as opposed to E_o is used for these as well.

2.6 SECANT VS. TANGENT PARAMETER INTERPRETATION

A universal challenge in geotechnical and foundation engineering analyses is that the solution to some theory or analytical process typically requires a constant value of some problem parameter that is inherently non-linear in nature. This typically occurs with soil stiffness (modulus) and soil strength. As will be seen, this situation arises numerous times throughout this monograph in various contexts for different analytical methodologies so a generic discussion of how to deal with this challenge is appropriate at this point.

There are two broad solution approaches that can be used to deal with this dilemma. Although they have been in use at least since the advent of the commercial digital computer in the 1960s and are thus widely known, they are summarized here to establish the terminology and notation used throughout this monograph. The common problem of shallow-foundation bearing capacity for a 'cohesionless' soil ($\phi > 0$, $c = 0$) will be used to illustrate this issue and its solution approaches as the writer has studied this particular application in some detail in the past (Horvath 2000a, 2000b, 2011).

Regardless of which specific bearing-capacity solution is used (Hansen, Meyerhof, Terzaghi, Vesic, etc.), the calculated results are extremely sensitive to the value of the soil friction angle, ϕ. This means that the single most important issue in a bearing-capacity assessment is selecting the appropriate value for ϕ.

Separate from, and in most cases more important than, the issue of using a *peak*, ϕ_{peak}, vs. *constant-volume* (a.k.a. *critical-state*), ϕ_{cv}, value of ϕ is selecting an 'operative' value of ϕ_{peak}. As can be seen in Figure 2.3, the Mohr-Coulomb failure envelope for ϕ_{peak} is always curved concave-downward whereas the failure envelope for ϕ_{cv} can reasonably be assumed to be a straight line. Thus, the challenge in this case is rationally selecting a single value of ϕ_{peak} for use with whichever solution (Hansen, etc.) is desired.

As illustrated in Figure 2.3, the key to rationally selecting a value of ϕ_{peak} begins by determining what is known as the *operative stress level*. In simple terms, this is the stress level that is applicable or relevant to a particular problem and application so cannot be generalized further.

In the case of shallow-foundation bearing capacity, the operative stress level involves the gross ultimate (ULS) bearing capacity, i.e. the final answer, so an iterative solution is required (details can be found in the writer's publications cited above). As it turns out, the iteration converges very rapidly so even manual calculation is reasonable although automating the process using some type of computer software will accelerate the process as well as allow for efficient repeat use in other project-specific applications.

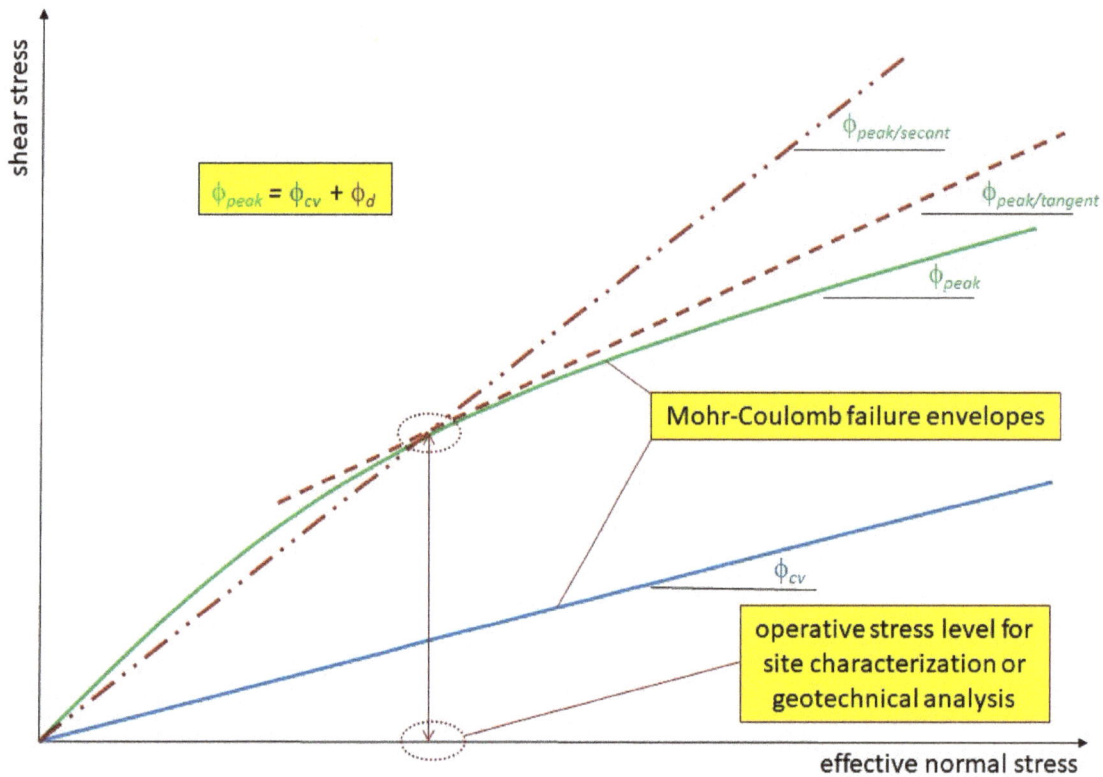

Figure 2.3. Exemplar of Secant vs. Tangent Linear Parameter Definitions for an Inherently Non-Linear Parameter.

Once the operational stress level has been determined, there are two concepts for estimating the corresponding operational value of ϕ_{peak}:

- a <u>secant</u> value, defined as $\phi_{peak/secant}$, that is the slope of a straight line from the origin to the peak Mohr-Coulomb failure envelope at the operational stress level and

- a <u>tangent</u> value, defined as $\phi_{peak/tangent}$, that is the slope of a straight line that just touches the point on the peak Mohr-Coulomb failure envelope corresponding to the operational stress level.

Both concepts are illustrated in Figure 2.3.

As can be seen, even with a qualitative plot such as this, the two values of ϕ_{peak} differ. However, in the limit as the magnitude of the operative stress level (the effective normal stress in Figure 2.3) increases, both values of ϕ_{peak} approach ϕ_{cv} but at a different rate. As can be seen in Figure 2.3, $\phi_{peak/tangent}$ will always approach ϕ_{cv} at a faster rate compared to $\phi_{peak/secant}$. This general trend always exists, i.e. the tangent value of a parameter will always approach some limiting value at a faster rate than the secant value of the same parameter.

The choice of which conceptual approach is better to use always varies with the specific analytical methodology and depends, to some degree at least, on the type of analysis as well as the desired outcome. For example, if only a one-shot, single-valued final answer is sought (as is the case with shallow-foundation bearing capacity), then the secant approach is generally preferred for its overall computational simplicity and efficiency, even if iteration is sometimes required as with bearing capacity.

On the other hand, if a sequential calculated outcome is desired, e.g. a complete load-settlement curve, then the tangent approach is usually necessary so that subsequent results can build on prior results in an additive, cumulative fashion. Essentially, the tangent approach approximates some curved relationship with a contiguous series of straight-line segments whereas the secant approach provides incorrect information concerning behavior intermediate to the final result. Obviously, the more and shorter the linear segments using the tangent approach, the closer the approximation to the actual curve.

However, there are exceptions to this general rule, As will be seen in this monograph, there are special cases where a sequential outcome such as a load-settlement curve can actually be reasonably obtained by repetitive application of the secant approach, with each repetition using a different parameter value that is estimated as part of the overall analytical methodology.

In conclusion, each situation where it is desired to approximate a non-linear parameter relationship with either a secant linear approximation or tangent linear approximation should be evaluated on its own merit, and the appropriate approach selected based on a careful assessment of the analytical process involved. There are numerous examples in this monograph where the secant vs. tangent approach choice needs to be made. For each one of them, the reasoning behind the selection is provided, even in cases that involve published work by others where the original author(s) did not explain the logic behind their decision.

2.7 CONE PENETROMETER

The category of in-situ testing devices referred to broadly as *penetrometers* have revolutionized site characterization and, as a result, many aspects of geotechnical, foundation, pavement, and geoenvironmental engineering research and practice. This trend is only expected to increase in the future as miniaturization of various types of measurement devices allows more and varied sensors to be incorporated within the bodies of penetrometers.

As a group, penetrometers are based on the common, core principle of performing a *test* or *sounding* by advancing a metal rod (which can be solid or hollow) with some type of shaped tip (generally conical) into the ground using a downward-acting axial force applied concentrically to the top of the rod. This force is delivered by either pushing (either manually or mechanically) or impact-driving in a manner similar to advancing the *Standard Split Spoon* in the *Standard Penetration Test* (SPT)[26]. As a minimum, in a penetrometer sounding the *tip resistance* to rod penetration, which is most commonly expressed as a stress but can be expressed as either a force or some arbitrary dimensionless number as in the case of the SPT *N*-value, is measured and recorded in some fashion.

Unfortunately, the writer has learned first-hand in recent years that there is conflicting and overlapping terminology used for different types of non-SPT penetrometers, some of which come in variants in terms of not only their physical dimensions but the physical parameters they measure. This is exacerbated by the fact that not all penetrometers have been formally recognized and standardized by ASTM International (formerly known as the American Society for Testing and Materials). In addition, different practice areas in civil engineering, e.g. foundation engineering and pavement engineering, and different

---

[26] The SPT is nowadays generally considered to be the oldest member of the penetrometer family.

geographical regions of the world have seemingly developed and named different penetrometers with apparently minimal or no knowledge, interaction, consideration, or technology transfer between or among the different groups.

Consequently, the net result is that there is some ambiguity in practice as to exactly what is meant by a 'cone penetrometer' or 'cone penetration test'. All too often, it depends on which professional is asked and where they are located geographically. Further complicating the terminology issue is the fact that the terms *direct push* and *direct-push system* seem to be used synonymously in at least some technical and geographical markets as an alternative term to *cone-penetration test*.

The cone-type penetrometer that appears to be most relevant for foundation engineering and deep-foundation purposes is the classical device with a 60° apex angle at the tip and a downward-projected tip area[27] that is typically either 10 cm² or 15 cm². Collectively, devices of this general design are referred to generically in this monograph as a *cone penetrometer*.

The cone penetrometer, often referred to in the past throughout the U.S. as the *Dutch cone* in deference to its development and evolution in The Netherlands (Horvath 2014b), figures prominently in the contents of this monograph. Equally significant are the technical differences between the numerous versions of this device used at different times, past and present, in foundation engineering research and practice. Thus, because of the need to clearly distinguish between and among the three versions of the cone penetrometer that figure most prominently into the subjects covered in this monograph, the acronyms used to refer to them throughout this monograph are, of necessity, very specific in their meaning and will be used only as such. Thus, in this monograph the acronym:

- **CPT** refers <u>only</u> to the second-generation cone penetrometer that measures two parameters:
  - *uncorrected tip resistance*, q_c, and
  - *sleeve friction*, f_s.

  Note that the earliest versions of the CPT measured these parameters mechanically while later versions used electronic load cells. Note also that in older English-language literature the CPT is sometimes referred to as the *Begemann Cone* in recognition of H. K. S. Begemann who, in the 1960s, was instrumental in developing the friction sleeve.

- **CPTu** refers to the third-generation cone penetrometer, what is colloquially called a *piezocone*, that also measures porewater pressures, u, in the now-standard location just behind the cone tip ('shoulder') that is referred to as the u_2 location. Earlier CPTu versions measured u on the face of the tip (u_1 location) and there are piezocones that also measure pore pressures just behind the friction sleeve (u_3 location) and even farther up the device (u_3 location). Only piezocones with u_2 data are referenced in this monograph.

- **sCPTu** refers to the fourth-generation cone penetrometer, what is colloquially called a *seismic piezocone* or *seismic cone*, that can also measure the shear-wave velocity, v_s, at operator-defined depth intervals. Note that the alternative notation for shear-wave velocity, V_s, appears to have become more common in published usage in recent years and is adopted for this monograph.

---

[27] The area unit of *square centimetre* (cm²) is retained here for its past historical and current colloquial usage in this context even though the use of centimetre is deprecated in the SI system of units. The reason that cm² is used in this context is solely because it produces a 'usable' value. The use of either square millimetre or square metre would be clumsy.

There are several additional comments with respect to cone penetrometers and their reference and usage in this monograph:

- The complete, generic term 'cone penetrometer' is used in this monograph when referring to all three of the above versions of this device in a broad, collective sense. Note that many (most?) publications use the acronym CPT when referring generically to all versions of the cone penetrometer which is <u>not</u> the case in this monograph as is emphasized here and above. The acronym CPT has a very precise and narrow meaning in this monograph.

- No acronym is defined in this monograph for the earliest (circa 1930s) first-generation version of the cone penetrometer that only measured tip resistance mechanically (as there was no friction sleeve there is no need to refer to this parameter as 'uncorrected tip resistance' as there was nothing to correct for). The reason for ignoring the first-generation cone penetrometer is that none of the material presented in this monograph is based on data obtained with this version that was the only type in existence until the friction sleeve was developed by Begemann in the 1960s.

- No acronyms are defined here for other, relatively recent versions of the cone penetrometer that have unique features that are typically intended for use in geoenvironmental applications. One example is the so-called *vision cone penetrometer* that has a video camera built into the device.

This page intentionally left blank.

Chapter 3

Pile Types

3.1 INTRODUCTION AND OVERVIEW

This chapter contains a detailed presentation and discussion of the evolution of tapered pile types, a topic that was only previewed briefly in Chapter 1 in order to provide background for the genesis of this monograph. The key technical characteristics of piling products that are currently available commercially are also addressed in considerable detail. There is a focus on the U.S. market where most of the historical development of tapered piles has occurred and with which the writer is personally very familiar. However, developments in other countries that are known to the writer are included to illustrate the fact that there is some global appreciation of the benefit of tapered piles.

It cannot be emphasized too strongly that the contents of this chapter are intended to be as objective, factual, complete, and overall informative as possible for the benefit of all stakeholders (owners, developers, design professionals, material suppliers, construction contractors) to make informed decisions on a project-specific basis. Nothing that is presented herein should be interpreted or construed as expressing the writer's preference or recommendation for one product over another. Only the stakeholders involved in a specific project are in a position to determine which product offers the optimal combination of technical performance and cost for that project.

3.2 WOOD

Wood piles, which are essentially the delimbed and debarked trunks of trees that have a suitable straightness, have long been referred to as *timber piles* in U.S. practice and will be referred to as such in this monograph[28]. It is beyond the scope of this monograph to trace the complete history of timber-pile usage which spans over 7000 years as noted in Chapter 1. Rather, just those aspects deemed relevant to this monograph are discussed here.

For the purposes of this monograph, only timber piles that are driven 'upside down' relative to how the original tree grew are considered. This means that the end of the timber pile with the greater circumference[29] is always considered to be the head of the pile. This may seem trivially obvious but driving timber piles the opposite way is known to have been done at times in the past so that the toe would have the largest diameter for maximum toe-bearing so it is necessary to be unambiguous in this regard.

For many decades in the U.S. and continuing to the present, timber piles used for foundation piling in both terrestrial as well as shallow-water marine applications such as piers and relieving platforms have generally been some species of softwood, primarily

---

[28] It is of interest to note that in his seminal paper on why and how foundation engineers and construction contractors in the U.S. transitioned from the almost exclusive use of timber piles to piles constructed of manufactured materials (steel, PCC), Gow (1917) used the term *wooden piles*. It is not known to the writer when colloquial usage in the U.S. transitioned to using 'timber piles'.

[29] While timber piles are nominally circular in cross-section, they are not perfectly so. Consequently, the preferred method for determining the nominal diameter at any point along a timber pile is to measure its circumference at that point and use that measurement to calculate a nominal diameter.

Southern Yellow Pine and, to a much lesser extent, Douglas Fir. The former dominates what is available in the U.S. currently according to the Timber Piling Council, the industry organization that oversees the promotion and technical support of timber pile usage in the U.S.

In the more distant past, it appears that other, regional softwood species were also used. For example, a number of 19th-century structures in Boston, Massachusetts that are still in use in the 21st century are known to be supported on spruce harvested regionally from New England forests (Zelada-Tumialan et al. 2013, 2014). Therefore, it would appear to be advisable that whenever dealing with an existing foundation supported on timber piles of relatively significant age, the species of the wood (which influences its mechanical properties such as strength) should not be assumed or taken for granted unless there is explicit prior knowledge in this regard.

Note that hardwood species such as oak have generally not been used for foundation piling in the U.S. (one exception is noted below) although hardwoods are sometimes specified for use in specific shallow-water marine applications such as breasting and fendering systems for vessels where impact resistance and concomitant physical durability are important design issues that historically have justified the greater cost of hardwood piles. Such applications are outside of the scope of this monograph and all further reference in this monograph to timber piles will imply use as foundation piling only.

The most significant change that has occurred over time with regard to the material properties of timber piles is pre-installation chemical treatment of the wood fiber. This is done for long-term preservation purposes using a pressure-injection process. Such piles are referred to in U.S. practice as *treated timber piles*.

Note that although the use of wood treatment in general for above-ground applications apparently dates back to the early 19th century (*Timber Pile Design & Construction Manual* 2016), with treatment of timber piles being done by later in the 19th century, the writer's first-hand experience is that *untreated timber piles* were used well into the middle of the 20th century, either alone or as part of *hybrid piles*[30] that are discussed later in this chapter. This continued use of untreated timber piles for many decades after treated timber piles were available in the marketplace was presumably for cost reasons. Thus, when assessing existing timber-pile installations for continued or modified future use, treatment should not be taken for granted based on pile age alone.

Initially and for many decades, *creosote* was the chemical of choice for treated timber piles and its use continues to the present although alternative treatments such as *chromated copper arsenate* (CCA) are now also used. Note that although CCA is now deprecated for use with residential timber in above-ground applications such as decks where long-term human contact can occur, it is still routinely used with timber piles.

As an alternative to the chemical treatment of timber piles, in some cases (primarily marine structures such as piers in saltwater environments where marine borers can be a significant design issue) certain naturally durable hardwood species such as Greenheart (*Chlorocardium rodiei*) are used without any chemical treatment. This is one of the rare exceptions to the otherwise exclusive use of softwoods for timber foundation piling.

Wolfe (1989) provides an objective, critical discussion of the mechanical (stress-strain-time) properties of timber piles as currently supplied in the U.S. Included are some important issues not normally addressed in most references dealing with timber-pile properties, e.g. the effect of chemical treatment on timber-pile material properties.

---

[30] 'Hybrid piles' are generally called *composite piles* in the published literature. However, for reasons explained later in this chapter, the term 'hybrid piles' is used in this monograph and the term 'composite piles' is reserved for other use.

As noted above, the conventional reasoning for using any kind of chemical treatment of timber piles is to prevent various naturally occurring degradation mechanisms such as dry rot or marine borers that can occur within the vadose zone of soil in terrestrial applications as well as in open air and open water in marine applications. Conventional reasoning has long held that timber piling that is both embedded in soil and permanently submerged in water is immune to natural degradation mechanisms indefinitely. This explains why untreated timber piles continued to be used in some hybrid-pile applications into middle of the 20th century.

However, recent research in Europe has found that in the very long term (time-in-ground of the order of 100 years or more) there can be degradation of untreated timber piling even within the saturated zone of soil (Zelada-Tumialan et al. 2013, 2014). This is significant as it would seem to call into question the above-stated, long-held position that permanently saturated wood embedded in soil has a more-or-less indefinite lifespan. The root cause of the degradation found in this recent research appears to be naturally occurring microorganisms in the groundwater, and is not related to any human-induced activity in the form of ground or groundwater contamination.

While all technical issues are important to consider when selecting a type of DFE, the most relevant aspect of timber piles for the purposes of this monograph is their continuous natural taper from one end of the pile to the other that is simply an artifact of how trees grow. Unfortunately, it is not possible to generalize on the taper angle of timber piles for several reasons:

- Taper angle is generally non-uniform for a pile, typically being least near the head (the former base of the tree) and increasing toward the toe (the former top of the tree).

- The overall average taper angle of a pile as determined over its entire pre-installation length will vary depending on the wood species and length of the pile.

- The relative variation in taper angle, i.e. the value of taper angle of some portion compared to the overall average, also depends on the wood species and length of the pile.

- Average taper angle varies from one pile to the next, even in the same group of nominally identical piles used on a given project that all meet some construction specification, e.g. piles all of the same class per ASTM Standard D25.

To illustrate these points with specific examples for which the writer has data, for creosoted Southern Yellow Pine piles used at JFKIA between approximately 1970 and 1990 that were in the range of 50 to 60 feet (15 to 18 m) long prior to driving, the writer found that the average taper angle, α, based on the entire pre-installation pile length typically varied between approximately 0.2° and 0.3°[31]. For a given pile, the maximum variation in taper angle along the pre-installation pile length was typically ±⅓ of the average. So, for example, a pile with an average taper angle = 0.21° might have a taper angle $\cong$ 0.14° near the head and $\cong$ 0.28° near the toe.

The importance of taper angle is that the magnitude of this angle is a significant input variable in several analytical methodologies for ULS resistance that are presented and discussed in Chapter 5. Therefore, whenever timber piles are used it is not sufficient to

---

[31] As a comparison, it is noted that Gregersen et al. (1973) indicated that timber piles used in Norway up to the time of their paper had a taper angle toward the upper end of this range.

measure the 'true' head[32] and toe diameters alone and use these values to determine an overall average taper angle. As a minimum, additional measurements should be made at the quarter-points so that taper-angle calculations can be made for quarter-length segments of the pile. Timber piles installed for research purposes warrant even more-closely spaced diameter measurements so that taper variations along the length of the pile can be determined even more precisely.

As for the future of timber piles, it is likely that their overall use will continue to decline, especially for terrestrial applications and even in areas such as the Eastern U.S. where they have historically enjoyed considerable use even to the present due to ready access to sources in the Southeastern U.S. There are several reasons for this but high on the list is that there is an increasing variety of alternative DFEs that are more cost-effective to use, especially in terrestrial applications and even for small projects involving lightly loaded structures that have long been the primary market for timber piles. For example, during the rebuilding and replacement of single-family houses along coastal New Jersey after Superstorm Sandy in October 2012, there was use of conventional treated timber piles as would be expected but also newer alternatives such as helical piles[33]. The latter can be installed much more easily and quickly compared to any type of (driven) pile, especially in confined environments.

That having been said, even if timber-pile usage were to end suddenly at some point, because timber piles were used so extensively, even exclusively, in the past it is likely that foundation engineers worldwide will continue to encounter them well into the future when assessing older, especially historical, structures. This fact combined with the increasing desire to reuse existing foundations to the greatest extent practicable rather than replace them completely means that knowledge about timber-pile characteristics and behavioral issues, especially with regard to durability, as well as how to assess their structural and geotechnical resistances remains essential for foundation engineers, present and future.

3.3 STEEL

3.3.1 Introduction

As will be seen in the following sections, there is now well over a century of experience with tapered-pile alternatives to timber piles. In the U.S. at least, the overwhelming number of these alternatives are what are classified as 'steel' piles for the purposes of this monograph although such piles typically contain a substantial amount of PCC relative to the total pile volume once installed. Therefore, in presenting and discussing tapered-pile alternatives to timber piles, steel piles are discussed first because of their historical prominence and dominance that has continued to the present.

---

[32] Typical timber-pile specifications in the U.S. such as the widely used ASTM D25 define the head (which is almost always referred to as the *butt* in U.S. practice) diameter as the nominal diameter 3 feet (900 mm) in from the true head of the pile. The genesis of this curious (to the writer at least) requirement is not known to the writer and there is no obvious technical reason for it. However, an educated guess is that this is a concession to the timber industry to allow for greater dimensional variability at the butt end of a given pile while still fitting into a given standardized class (category) of pile size (A, B, C). For taper-angle calculation purposes, measurements should always be made and recorded at the true head of a pile.

[33] At least some of the evidence for this comes from episodes of the well-known TV show "This Old House" that featured post-Sandy rebuilding during Season 34/Episodes 1-8 that first aired in Fall 2013. Videos of episodes that include deep-foundation installation can be found on the Web.

3.3.2 Background

A review of U.S. patent applications and other early publications (*Bulletin No. 113* 1909, *Concrete* 1910, Gow 1917) indicates that the years around the cusp of the 20th century were the 'Golden Years' of early R&D into developing alternatives to timber piles that were essentially all PCC. This corresponded to the ascendancy of PCC in engineered construction that accelerated as the 19th century progressed. Numerous variations in piling alternatives were explored during this timeframe including:

- pile geometries that were of constant perimeter, continuously tapered, or some combination of the two;

- various cross-sectional geometries from square to approximately circular (multi-sided) to truly circular;

- either unreinforced or lightly reinforced with steel but not prestressed[34]; and

- either precast[35] and cast in place (CIP)[36].

In retrospect, it is not surprising that the first pile developed in the U.S., reportedly circa 1890 (*Bulletin No. 113* 1909, *Concrete* 1910), that used PCC as the predominant material in the finished pile was both CIP and continuously tapered. Furthermore, many of the U.S. patents that followed in the early years of the 20th century were for nominally PCC (mostly CIP) piles with a continuously tapered geometry. As noted by Gow (1917, page 144):

> *"It was natural that the earlier adaptations of concrete to piling should follow in form and methods the previous experience with wooden piles."*

Furthermore, the CIP-PCC piling alternative in general appears to have gotten both most of the R&D attention as well as commercial development in the early years of the 20th century. This appears to be the result of its being a logical evolution from caisson-type deep foundations (which were largely hand dug in the 19th century) that were filled with PCC after excavation and problems with damage during driving that was routinely encountered when using early precast-PCC piles. The latter issue was noted by Gow (1917) and based on what we know now from 1-D wave mechanics and pile driving is not surprising given that the precast piles of that era were not prestressed.

The tapered CIP-PCC piles of that early era required a steel casing to be installed into the ground first. However, the steel component was relatively thin and not relied on for providing any long-term structural resistance. Rather, the steel was intended to act only as temporary formwork for pouring an unreinforced PCC core and then essentially abandoned

---

[34] Although the concept of *prestressed concrete* evolved toward the end of the 19th century and contemporaneous with the greater use of PCC in engineered construction, prestressing technology was apparently either not sufficiently advanced for use with foundation piling or not deemed necessary to use with foundation piling until well into the 20th century.

[35] Early publications (*Bulletin No. 113* 1909, *Concrete* 1910) use the term *cast-and-driven* in lieu of 'precast' when applied to piles.

[36] Early publications (Gow 1917) also use the term *built in place* in lieu of 'cast in place'.

in place. The resulting pile was viewed essentially as a CIP-PCC pile and patent applications, etc. of that early timeframe refer to the finished pile as such. In fact, archived product literature that the writer found while performing background research for this monograph indicated that these piles were marketed as PCC piles, with no mention of steel.

However, for the purposes of this monograph it is more appropriate to consider such CIP-PCC piles that used a relatively thin, sacrificial steel casing as steel piles which are discussed in this section. There are two reasons for this.

First, the experience in the U.S. throughout much of the 20th century was that such CIP-PCC piles did not compete in the marketplace with 'true' PCC piles that were precast (and, by later in the 20th century, prestressed to minimize installation-damage issues) and thus did not require a temporary steel casing for installation. Rather, such CIP-PCC piles competed in the marketplace with other versions of tapered steel piles that had much more substantial steel casings that were assumed to contribute to post-installation, long-term structural resistance. 'True' tapered PCC piles, the vast majority of which are precast, are discussed in a separate section later in this chapter.

The second reason is more recent and that is the emergence of tapered CIP-PCC DFEs that do not require use of any casing as part of the construction process. While these new tapered DFEs are drilled-in-place and thus are not piles in the context of the monograph, it is better, in the writer's opinion, to treat the traditional CIP-PCC tapered piles that used a steel casing as a nominally steel pile to avoid any confusion.

3.3.3 Overview

The type of steel casing used to construct a tapered steel pile provides a convenient way to distinguish between the two types of such piles. Specially, the following discussion of tapered steel piles is divided into sections dealing with what are referred to in this monograph as *thin-wall steel shell piles* (the commercial successors to the early CIP-PCC piles discussed above) and *thick-wall steel pipe piles*[37]. The shell vs. pipe distinction made in this monograph is based on the ability or not to install the pile by direct driving on the head of the steel component that is installed in the ground alone and then filled with fluid PCC (both shell and pipe piles are driven with a closed toe). Specifically, shell piles cannot be driven directly and thus require a driving aid of some kind whereas pipe piles can be driven directly.

Note, however, that this terminology as adopted here is not always applied consistently in U.S. practice as the terms 'shell' and 'pipe' are often used synonymously and thus interchangeably. However, the distinction is very significant with regard to tapered steel piles so is adhered to in this monograph.

Note also that although shell and pipe piles are both referred to as steel piles, once installed they always have a *composite*[38] cross-section as the interior (core) of the shell or pipe is filled with fluid PCC after installation. In the case of the thin-wall shell pile, the steel shell is typically considered to act only as formwork for the PCC and the PCC core is assumed to carry all the structural load as was noted previously. On the other hand, with the thick-wall pipe pile the steel cross-sectional area, perhaps reduced by a corrosion allowance, is

---

[37] Outside of the U.S. (e.g. Poulos 1987), steel pipe piles are sometimes referred to as *steel tube piles*.

[38] The term 'composite' in this context refers to the multi-component nature of the cross-section of the final installed pile that has an outer cylinder of steel and an inner solid core of PCC. This should not be confused with the term *composite pile* as commonly used in published literature which generally means two distinctly different types of piles stacked one on top of the other to create a single DFE. To avoid confusion, as noted earlier in this chapter such one-above-the-other piles are called *hybrid piles* in this monograph and are discussed in a separate section later in this chapter.

relatively substantial so is considered along with the PCC core when calculating post-installation, long-term structural resistance.

In both cases, the PCC core is generally unreinforced except perhaps locally near the head of the pile for structural connection to the pile cap. Nowadays, with seismic design common in many areas, the PCC core may be reinforced to a greater depth to provide additional structural strength and stiffness against lateral loading.

As noted above, the distinction between 'shell' and 'pipe' relates solely to how the steel component is installed into the ground. The thin-wall shell piles that were in commercial use in the U.S. for many decades and are discussed subsequently consisted of relatively thin corrugated steel pipe similar to that used for drainage purposes (the corrugations provide some nominal flexural stiffness for handling) that cannot withstand direct driving and thus requires the use of a *mandrel*[39]. This is a temporary steel insert (it can be expandable or solid, the former version appears to have been developed first) placed within the steel shell prior to driving (much like one puts a foot into a sock) in a process referred to colloquially as *shelling up*. During driving, the pile hammer only impacts the mandrel and the mandrel essentially drags the engaged pile shell into the ground with it. After driving the pile to the desired depth, the mandrel is withdrawn and the steel shell filled with fluid PCC.

One of the significant constructability considerations with any mandrel-driven pile, tapered or constant diameter, is the aforementioned need for shelling up. The process can become ponderous as pile lengths increase. This is because it becomes impractical to raise the pile hammer with attached mandrel high enough in the pile-driving leads to be able to insert the mandrel into pile shell that is suspended beneath it in the leads. In such cases, what is colloquially called a *doodle hole*[40] in the U.S. must be used. This is usually a constant-diameter, thick-wall steel pipe pile with a closed toe and an inside diameter somewhat greater than the outside diameter of the production-pile shells that is driven beforehand into the ground to a depth of the same order as the planned length of the production-pile shells. A production-pile shell is first lowered into the doodle hole and the mandrel inserted into it. The shelled-up mandrel is then lifted out of the doodle hole and the production pile driven.

As can be appreciated, the need for shelling up, especially when a doodle hole is required, seriously impacts pile-installation productivity and overall pile costs. As will be seen, this is part of the reason why tapered piles of the thin-wall shell type disappeared from U.S. practice several decades ago.

On the other hand, thick-wall pipe piles consist of a pipe section of sufficient thickness that can withstand direct driving although a special drive-head for the pile-driving hammer is required for some products. The pipe section may be composed of cold-formed or hot-rolled steel, and be either smooth or vertically fluted on the exterior depending on the particular product used.

3.3.4 Thin-Wall Shell

It appears that in the aforementioned Golden Years of U.S. pile-technology development around the cusp of the 20th century there was a succession of concepts that were explored for creating a steel shell; advancing it into the ground; and then filling it with fluid PCC, with each concept being patented in turn. No doubt constructability and economics ultimately decided which of these concepts was ultimately commercially viable in the long

---

[39] In publications in the early years of the 20th century such as Gow (1917) when such piles represented the state of art at the time, the mandrel was also referred to as a *core*.

[40] Also colloquially known in the U.S. (although less commonly in the writer's experience) by equally colorful alternative names such as *dummy hole*, *rat hole*, and *make-up pile*,

term. Rarely is a product or concept, even something relatively simple, a commercial success in its initial version. An excellent illustration of this is the mundane, every-day paperclip that had a surprising number of conceptual forms before evolving to what has been used now for many decades (Petroski 1992).

Most, if not all, of this development and patent work was undertaken by Alfred A. Raymond, apparently on behalf of the Raymond Concrete Pile Company to which each of A. A. Raymond's patent was assigned. At one time, Raymond (the company, as it was simply and widely referred to colloquially in practice in the 20th century and is hereinafter referred to in this monograph) was a dominant market force and innovative leader in the field of deep foundations, especially piles. Therefore, it is no surprise that A. A. Raymond's early patents evolved into two products that defined Raymond's pile business for much of the 20th century.

References of the era such as *Bulletin No. 113* (1909) and *Concrete* (1910) indicate that A. A. Raymond began his research into piles circa 1890 as an outgrowth of his work with caisson-type deep foundations that were largely hand-excavated at the time. These references indicate that A. A. Raymond was issued the first of several pile-related patents in 1896, and that the first piles derived from these patents were installed in 1901.

The commercial pile that eventually evolved in the early decades of the 20th century that most closely resembles these early patents was known as the *Raymond Standard*[41,42] pile. It consisted of a continuously tapered steel shell with circumferential corrugations that mimicked a timber pile in overall geometry but with a greater taper angle (0.95°), i.e. three to four times that of a typical timber pile. The use of a one-piece corrugated steel shell must have been an evolutionary improvement to the overall concept as A. A. Raymond's early patents clearly show a segmental shell composed of nominally flat sheets that had to be assembled in the field into a conically shaped element prior to driving.

Chellis (1961) includes tables of the standard sizes in which the *Raymond Standard* pile was available. Toe diameters were either 8 or 10.8 inches (203 or 274 mm) and head diameters varied from 9 to 23 inches (229 to 584 mm). Because the taper angle was always fixed at 0.95°[43], this severely limited potential pile lengths. The maximum possible length was 37.5 feet (11.4 m) which undoubtedly placed a real constraint on potential applications.

The second and more-common Raymond pile that also evolved commercially early in the 20th century was the *Raymond Step-Taper*[44] pile. It consisted of an assemblage of two or more constant-diameter, circumferentially corrugated shell segments of different diameter that were called 'steps'. A flat steel plate that Raymond called a 'boot' was used to close off the toe. The final assemblage of steps had the overall appearance of a tapered pile, hence the tradename.

---

[41] It appears that the *Raymond Standard* pile may be the commercial outcome of U.S. Patent No. 985,549 that was filed on 21 June 1907 by, and granted on 28 February 1911 to, Alfred A. Raymond who assigned the patent to the Raymond Concrete Pile Company of Chicago, IL.

[42] A search of the U.S. Patent and Trademark Office (USPTO) Trademark Electronic Search System (TESS) indicates that this pile name was never a U.S.-registered trademark (®) although it may have been used as an unregistered trademark (TM).

[43] How and why this taper angle was chosen is not known to the writer at this time although it clearly differed substantially from typical timber-pile taper angles as noted previously so was clearly not simply a replication of timber piles. Regardless of how this taper angle came to be, as will be seen later in this chapter, this taper angle is noteworthy as it set the benchmark taper angle used for tapered steel piles in the U.S. for the next 100 years and to the present.

[44] A search of the U.S. Patent and Trademark Office (USPTO) Trademark Electronic Search System (TESS) indicates that this pile name was once a U.S.-registered trademark (®). However, this mark has been legally 'dead' since the early 2000s.

The diameter difference between adjacent steps was fixed at 1 inch (25 mm) but otherwise there were numerous standard component diameters and lengths from which to choose (Chellis 1961):

- Toe diameters ranged from 8 to 14 inches (203 to 356 mm).

- Step diameters ranged from 9 to 17 inches (229 to 432 mm).

- Step lengths ranged from 4 to 24 feet (1219 to 7315 mm) although Chellis (1961) indicated that 8 and 12 feet (2438 and 3658 mm) were most common.

Clearly, there was a very large number of potential component combinations that could be created in practice.

Note that there was no fixed or standard configuration for this type of pile. A project-specific design of step diameters and lengths would always be developed, presumably based on Raymond's experience combined with a site-specific assessment of subsurface conditions and foundation loading. In theory at least, very long piles could be configured (certainly much longer than for the *Raymond Standard* pile) although the aforementioned constructability issues involving shelling-up were always a consideration in practice.

It is of interest to note that the quasi-taper angle that could be achieved with the *Raymond Step-Taper* pile was relatively modest, varying between 0.10° and 0.60° depending on the chosen step lengths. For what were reportedly (Chellis 1961) the most commonly used step lengths of 8 and 12 feet (2438 and 3658 mm), the quasi-taper angles ranged from 0.20° to 0.30°, i.e. exactly the same as the range for timber piles noted previously and well short of the 0.95° taper angle of the *Raymond Standard* pile. The primary benefit, then, of the *Raymond Step-Taper* pile was not so much its taper but the fact that piles of relatively substantial length could be created. The down side, of course, was that shelling-up issues increased with increasing pile length.

Ultimately, the biggest negative with both the *Raymond Standard* and *Raymond Step-Taper* piles was that, with few exceptions, to the end of their commercial lives only Raymond would install them. Even after any patent protection for these piles had expired, these piles required both special manufacturing facilities for the corrugated shells and special mandrels for installation (the *Raymond Step-Taper* pile required a unique solid-steel stepped mandrel). This undoubtedly deterred others from developing the capability to manufacture and install these piles. As a result, this significantly limited opportunities for these piles to be used and also linked their usage to a single business entity (Raymond). As the fortunes of this business entity declined during the latter decades of the 20th century, these piles essentially disappeared from the marketplace.

The writer is not aware of any other proprietary steel shell-type taper piles or attempts to market a generic steel shell-type tapered pile, at least in the U.S. The need to use a mandrel with any type of steel-shell pile, tapered or constant diameter, was and is a significant constraint to any business entity contemplating commercial production of a shell-type pile (the commercial use of constant-diameter steel shell piles and associated mandrels is discussed later in this chapter).

Although these two proprietary Raymond piles disappeared from the marketplace decades ago (the last use of the *Raymond Step-Taper* pile that the writer is personally aware of was at JFKIA in Summer 1972 and was one of the very rare cases where Raymond allowed another pile-driving contractor to install their pile as it was a one-off test pile), and the Raymond Concrete Pile Company along with them, a successor company, Raymond International, still exists but does not appear to be involved in the installation of deep foundations. It is doubtful that the specific concepts on which these Raymond piles were based would ever be worthwhile to resurrect given all the other DFE alternatives available to the present-day foundation engineer and contractor. However, these Raymond piles can be found supporting countless existing structures not only in the U.S. but in other countries as well. Consequently, foundation engineers now and into the foreseeable future will undoubtedly encounter them from time to time and thus need to be familiar with them.

3.3.5 Thick-Wall Pipe

3.3.5.1 Monotube

During the same timeframe of the early decades of the 20th century in which Raymond (the company) was introducing its thin-wall steel shell piles to the marketplace, an alternative steel tapered pile that consisted of a thick-wall steel pipe was emerging in the U.S. What is interesting is that this new entry into the foundation-piling marketplace initially occurred as an afterthought and the result of extenuating circumstances rather than intentional R&D combined with astute business planning as in the case of Raymond's tapered piles. Yet this 'happy accident', as it were, turned out to outlive Raymond's tapered piles by many decades and into the 21st century.

The company behind this alternative tapered-pile effort was the Union Metal Company (some references, including U.S. patents, list it as the Union Metal Manufacturing Company although it is clear that the name was later changed to Union Metal Corporation) in Canton, Ohio. Then, as now[45], the primary Union Metal product line is streetscape lighting fixtures (light poles). Union Metal, which was founded in 1906, is reportedly the first such company in the U.S. that was devoted solely to this type of product.

Of relevance to this monograph is that a late-1920s addition to the Union Metal streetscape-lighting product line was characterized by a tapered, cold-formed, seamless steel tube with a nominal circular cross-section that had visually distinctive vertical fluting on the exterior. Light poles with this distinctive fluting became a signature architectural feature of the Union Metal product line at the time and continues to the present in their *Nostalgia Series* product line.

Also of relevance is that it appears that the tradename *Monotube* was applied initially to only light poles of this general design as the U.S. patent[46] covering this invention of a tapered, vertically fluted, steel pipe makes no mention of foundation piling as one of its potential uses. Also, there is no indication that in its first decades of existence that Union Metal manufactured anything other than light poles. However, in later years, Edmund W. Riemenschneider (who was apparently employed by Union Metal beginning in 1926) and others subsequently filed and received several U.S. patents specifically related to foundation piling that noted the vertically fluted design (among other details). All of these patents were assigned by Riemenschneider et alia to Union Metal.

Available information suggests that Union Metal got into the foundation-piling business circa 1930, most likely as a way to bolster their product sales at the onset of the Great Depression. In essence, Union Metal simply turned their *Monotube*-brand light poles

---

[45] As discussed subsequently, during the timeframe in which this monograph was written this business went through a cycle of short-term closure in late 2017; sale; and subsequent re-opening in mid-2018 under new ownership and name, Union Metal Industries Corporation.

[46] U.S. Patent No. 1,821,850 filed on 24 September 1929 by, and granted on 1 September 1931 to, Edmund W. Riemenschneider who assigned the patent to the Union Metal Manufacturing Company of Canton, OH.

upside down and sold them as a line of foundation piling called the *Monotube*[47] pile, simply repurposing the existing tradename of their line of light poles. However, unlike Raymond, Union Metal did not then or ever get into the pile-installation business but focused on manufacturing only. This meant that any pile-driving contractor could purchase and install *Monotube* piles. This remains true to the present, or at least until very recently[48]. However, a special drive-head is required to install *Monotube* piles whenever the proprietary *Monotube Type N* constant-diameter extension is used. This drive-head is provided on a loan basis.

As an aside, one important consequence of the business model followed by Union Metal as opposed to that followed by Raymond is that it is the end users of the *Monotube* pile who are ultimately responsible for its performance in terms of determining both the structural and geotechnical resistances. This compares to Raymond who, as both the manufacturer and installer of their piles, would essentially guarantee some minimum specified load-bearing performance of the overall foundation system as is often done by specialty deep-foundation contractors nowadays. Details such as how many piles, pile dimensions, etc. were required to provide this foundation performance were essentially solely Raymond's problem to deal with.

The basic *Monotube* pile consists of a tapered lower section (available in three standard tapers as discussed below) with closed toe that is connected to a constant-diameter upper section that Monotube called an 'extension'. Both sections are made of cold-formed steel and have the distinctive vertical fluting that has always been the visual hallmark of the *Monotube* pile. In fact, in some applications such as highway bridges (more so in the past than the present), the constant-diameter extension is continued above ground all the way up to the bridge superstructure to serve as part of a bridge pier as well as an architectural element of the final overall structure.

Because special machinery is required to cold-form the steel into the distinctive fluted shape, the *Monotube* product line has remained remarkably static over time:

- The tapered sections come in three standard tapers called 'types' that are designated *Type F*, *Type J*, and *Type Y* with taper angles, α, = 0.33°, 0.57°, and 0.95° respectively. Note that the *Type Y* has the same taper angle as the *Raymond Standard* pile. This may have been an intentional design choice given that the *Monotube* pile evolved circa 1931, not long after commercial introduction of the *Raymond Standard* pile.

- The toe is always a nominal 8 inches (200 mm) in diameter and has a hemispherical shape that is another signature detail of the *Monotube* piles. However, it appears that this hemispherical toe-design detail may have been inspired by the same A. A. Raymond U.S.

---

[47] A search of the U.S. Patent and Trademark Office (USPTO) Trademark Electronic Search System (TESS) during the course of research related to preparation of this monograph indicated that this pile name was once a U.S.-registered trademark (®) that had been in commercial use since April 1931 but which has been legally 'dead' for its original use since the early 2000s. However, in March 2018 this mark was reregistered by Magnum Piering, Inc. of Cincinnati, OH. This company is involved primarily with helical piles. Another check of the USPTO TESS during finalization of this monograph in early 2020 indicated that Magnum Piering had abandoned this trademark in January 2019 after holding it for less than one year. This means that at the time this monograph was published in 2020 the Monotube name is no longer a U.S.-registered trademark related to foundation piling, However, for the sake of completeness, it is noted that there are other commercial products, some with 'live' registered trademarks, that use 'Monotube' in the product name but none has anything to do with deep foundations.

[48] As discussed subsequently, it appears that the *Monotube* pile is unavailable commercially at the time that this monograph was published in 2020.

patent (No. 985,549) issued in 1911, approximately 20 years before Union Metal started marketing their light poles as foundation piles, that appears to have led to the *Raymond Standard* pile. Figure 3.1 is a figure that accompanied this patent filing. Although the steel shell shown in this patent reflects an early conceptual iteration of what ultimately became the *Raymond Standard* pile (which used a one-piece shell with circumferential corrugations, not the segmental shell of multiple smooth sheets as shown in this patent), the hemispherical toe detail is nevertheless quite apparent.

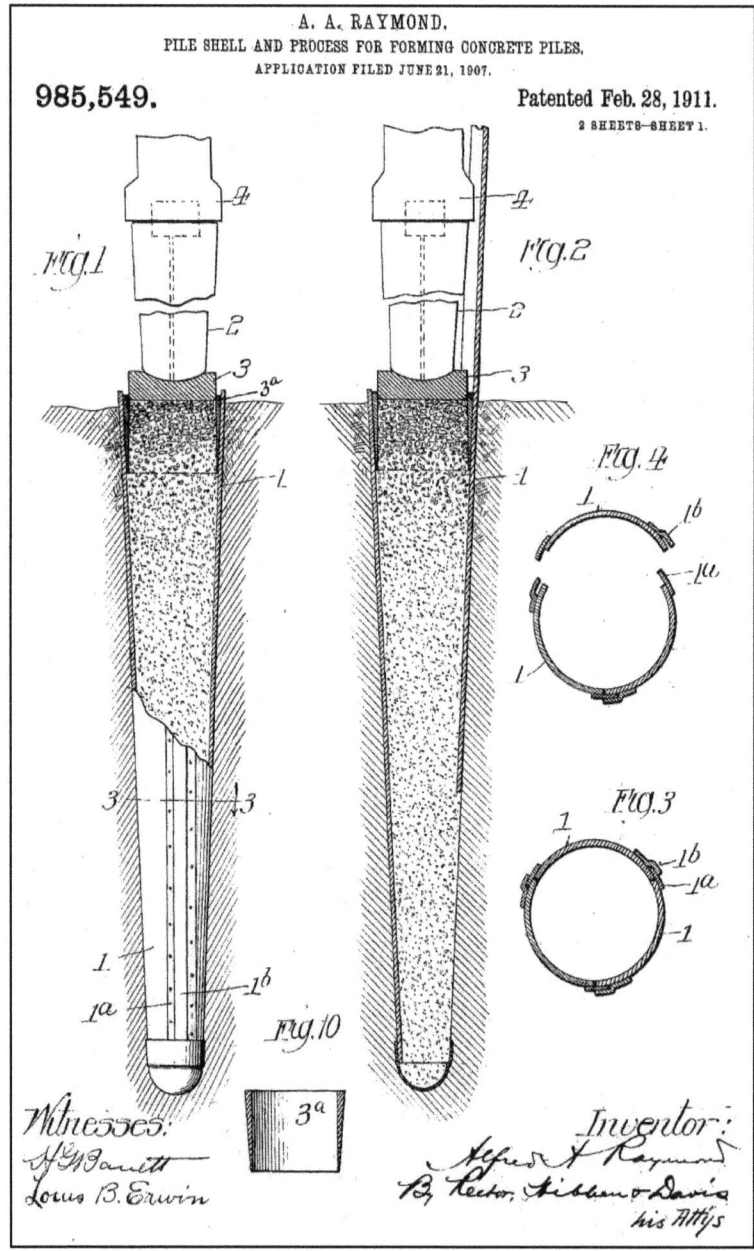

Figure 3.1. Excerpt from U.S. Patent No. 985,549 Issued 28 February 1911.

- The diameter at the top of the tapered section and concomitant diameter of the constant-diameter extension vary from 12 to 18 inches (305 to 457 mm).

- There are several wall thicknesses (called *gauge thickness* in *Monotube* product literature) for both the tapered section and constant-diameter extension but the cold-forming process has always placed an upper limit on what can be produced for this line of piling: 3 gauge, which has a thickness of 0.2391 inches[49] = 6073 μm.

At the time this monograph was being finalized for publication in early 2020, it appears that the commercial availability of the *Monotube* pile is in a state of flux. Up until circa 2017, the *Monotube* pile was still being manufactured and actively marketed, at least within the U.S., although since the 1980s this was no longer done under the auspices of Union Metal. Since 1984, the *Monotube* line of piling had been marketed by a standalone company called the Monotube Pile Corporation (more recently listed as Monotube LLC in some places), but still located in Canton, Ohio. In reality, the spinoff of Monotube Pile Corporation/Monotube LLC[50] from Union Metal Corporation may simply have been a legal separation of business units (Union Metal continued to manufacture and market only light poles after 1984) within the same organizational umbrella as a review of online information shows Monotube and Union Metal essentially at the same physical location but with different official street addresses.

However, it appears that more-substantive changes occurred beginning in 2017. To begin with, Union Metal essentially ceased operations at the end of 2017 as noted previously. Its assets were sold in 2018 to a new buyer (American Industrial Acquisition Corporation of New York, NY) who renamed the company Union Metal Industries Corporation (UMIC). It appears that by mid-2018 UMIC had resumed light-pole manufacturing operations at the original Canton, OH location but with a much-reduced workforce and sales territory that was no longer national in scope. As 2018 progressed, it became clear to the writer that Monotube disappeared from the commercial marketplace (no website, etc.) although there are indications that the *Monotube* tradename and trademark and possibly the pile-manufacturing assets[51] were acquired circa 2018 by a new owner, Magnum Piering. as noted earlier, Magnum Piering abandoned the *Monotube* trademark less than a year later, in early 2019. At the time that this monograph was being finalized for publication in early 2020, it appears that the *Monotube* pile has disappeared from the marketplace. Whether this is temporary or permanent remains to be seen.

In any event, although whatever patent(s) that may have existed originally for *Monotube* piles have undoubtedly long since expired, no other manufacturer in the U.S. or elsewhere is known to have produced a generic duplicate. Likely the cost of acquiring the special manufacturing equipment needed to cold-form the fluted pile sections as well as

---

[49] This four-digit precision is actually used in *Monotube* product literature. From a practical manufacturing perspective, it seems questionable whether such manufacturing tolerances actually exist in reality.

[50] Hereinafter in this monograph simply referred to as 'Monotube' (without italicization). When the specific pile product is being referred to, the name is italicized (*Monotube*).

[51] As was explained to the writer while conducting background research for this monograph during a confidential information-gathering meeting in 2016 with someone who has been involved in the manufacturing of deep-foundation products for several decades, the essential business assets related to the *Monotube* pile are primarily the machinery used to cold-form steel into the proprietary product shapes. In principle, any business that owns this or similar equipment could produce tapered steel piles that were clones of the *Monotube* or even extend the product line to manufacture tapered sections with tapers other that which has been available for many decades now.

producing the special drive-heads necessary to install these piles whenever the proprietary *Monotube Type N* extension is used has been a deterrent. This is conceptually identical to the deterrents to manufacturing clones of the former Raymond piles that were discussed previously.

The most significant technical change that has occurred over time with regard to the *Monotube* pile is that use of the proprietary *Type N* extension has been increasingly replaced on projects by generic hot-rolled steel pipe, usually spiral-welded as opposed to seamless for economic reasons. This change appears to have originated with foundation engineers and pile-driving contractors, not Monotube itself (Brand 1997). This is significant to note as this means that the burden of properly designing and executing the resulting connection between proprietary and generic components of the same pile rests not with Monotube but with the design professionals and construction contractors of a project. In general, this connection detail between the tapered and constant-diameter sections of a tapered steel pipe pile is not trivial as is discussed in the following section.

In any event, the reasons for this change in the constant-diameter section used with *Monotube* piles likely involve a combination of factors including cost, constructability (no special drive-head is required), and technical (material-stress issues, especially under driving). This last factor sets the stage for the discussion of a competitive brand of thick-wall tapered steel pipe pile that emerged on the cusp of the 21st century.

3.3.5.2 TAPERTUBE

In the final decades of the 20th century, foundation engineers and pile-driving contractors in the Northeastern U.S. in general, and New York City in particular, pushed the design envelope for *Monotube* piles to the point that by the end of the century the capacity (Q_u in Figure 2.1) was routinely being limited by the ULS structural resistance, not the ULS (geotechnical) resistance. Specifically, on a project at JFKIA in New York City it reportedly became increasingly difficult to consistently achieve the desired capacity, Q_u, of 600 kips (2670 kN) minimum per pile, even using the heaviest-gauge *Monotube* sections available.

This situation led directly to the commercial development of the *TAPERTUBE*® pile as an alternative (Horvath 2003b, 2003c; Horvath and Trochalides 2004; Horvath et al. 2004a, 2004b). Initially, the *TAPERTUBE* was intended to simply be a more-structurally-robust clone of the *Monotube* pile so that pile design would always be governed by geotechnical, not structural, ULS resistance as is generally the case with friction/floating piles. This goal was achieved by using hot-rolled steel for all sections of the *TAPERTUBE* pile which removed the wall-thickness manufacturing limitations of the cold-formed *Monotube* pile. Thus, the *TAPERTUBE* pile is identical to the *Monotube* pile in terms of its overall geometry and components (i.e. it always consists of a tapered, closed-toe lower section and constant-diameter upper section or extension) and is assumed for analytical purposes to have the same geotechnical behavior[52]. It is important to note that this observation is not meant to trivialize or deprecate the manufacturing innovation reflected in the *TAPERTUBE* pile but simply emphasize the fact is that there is nothing fundamentally different about it geotechnically

---

[52] The writer is not aware of any published, objective research that demonstrates conclusively one way or the other whether or not the vertical fluting that is a signature physical detail of *Monotube* piles plays any role in geotechnical resistance as has reportedly been suggested from time to time over the years. As can be seen from the preceding discussion of *Monotube* piles, available information indicates that this vertical fluting was never an intentional design detail with respect to foundation piling but rather an architectural artifact of repurposing Union Metal's *Monotube*-brand light poles for foundation piling.

compared to the *Monotube* pile. Both the *Monotube* and *TAPERTUBE* piles are analyzed in exactly the same manner.

That the *TAPTERTUBE* pile design concept was successful in its intended goal is borne out by the fact that within a relatively short time after its introduction and use at JFKIA circa 2000, ULS resistances were achieved that were well beyond the aforementioned 600 kips (2670 kN) minimum-per-pile goal that had become problematic to achieve with *Monotube* piles. Specifically, from the very beginning of the *TAPERTUBE* pile's commercial use circa 2000, ULS resistances in excess of 800 kips (3560 kN) were routinely achieved and ULS resistances approaching 1,000 kips (4450 kN) were measured in some cases. In all cases, the limiting design factor was geotechnical, not structural. In the years since the introduction of this pile at JFKIA, even greater ULS resistances have reportedly been achieved on projects at other locations within the New York City metropolitan area.

Although the *TAPERTUBE* pile broadly mimics the *Monotube* pile in terms of its overall appearance and geotechnical performance, the similarity ends there in several important respects. To begin with, the proprietary component of the *TAPERTUBE* pile is and has only ever been the overall tapered lower section. However, even the shaft portion of that component is made from generic hot-rolled steel sheet that is bent using generic steel-fabrication machinery (not specialized cold-forming machinery as in the case of *Monotube* piles) into a contiguous series of flat surfaces to approximate the overall shape of a circle in cross-section. These physical features can be seen in Figure 3.2 that shows a stockpile of *TAPERTUBE* piles at a construction site awaiting installation. The multiple flat faces of the shaft portions of the tapered sections (all aligned with their toe closures toward the right-hand side of the photo) are quite evident in this figure.

Figure 3.2. Geometry of Proprietary *TAPERTUBE* Tapered Section [photo credit: Underpinning and Foundation].

A proprietary component of the tapered section is a steel toe-closure piece that is welded to the formed shaft portion of the tapered section to complete the overall tapered section and form a watertight seal at the toe. As can be seen in Figure 3.3, the shape of this toe closure is much more conical and with a rounded point compared to the hemispherical toe-closure detail of the *Monotube* pile.

Figure 3.3. Proprietary Toe-Closure Component of *TAPERTUBE* Tapered Section [photo credit: Underpinning and Foundation].

Note also in Figure 3.3 that the maximum diameter of the conical toe-closure piece of the *TAPERTUBE* tapered section is slightly larger than the diameter of the bottom of the shaft portion to which it is attached. This is unavoidable as the toe-closure piece is truly circular in cross-section whereas the tapered shaft portion only approximates a circle in cross-section and, in reality, consists of a series of connected flat sections.

From the beginning of its commercial use circa 2000, the constant-diameter upper section (extension) of the *TAPERTUBE* pile has been generic steel pipe made of hot-rolled steel. Spiral-welded pipe, as opposed to seamless pipe, is generally used for economy. This is the same type of pipe that is used for other types of piles such as generic constant-diameter steel pipe piles with either a closed or open toe.

Figure 3.4 shows the 'production' connection detail that is used between the proprietary tapered section and generic constant-diameter section to create a complete *TAPERTUBE* pile that ultimately evolved and is used to the present. However, this connection detail, while deceptively simple in appearance, apparently proved tricky to perfect during the development and initial field testing of the *TAPERTUBE* pile circa 2000. There are some useful lessons to be learned from this experience that are applicable to other types of piles, especially when generic hot-rolled pipe is used in lieu of the proprietary *Monotube Type N* extension with *Monotube* piles.

The patent[53] for the TAPERTUBE pile showed several connection details that were each very different from that shown in Figure 3.4. However, the common element of all of them was that the connection area itself was <u>larger</u> in diameter than either of the two components (lower and upper sections) that were being connected.

---

[53] U.S. Patent No. 6,468,003 B2 that was initially filed on 27 May 1998 by, and granted on 22 October 2002 to, Stanley Merjan and John Dougherty. Note that John 'Jack' Dougherty is the owner of DFP Foundation Products, LLC that is noted subsequently.

Figure 3.4. Connection Detail between Proprietary *TAPERTUBE* Tapered Section and Generic Constant-Diameter Extension [photo credit: Underpinning and Foundation].

The "preferred" connection detail shown in this patent is shown in Figure 3.5 and is reportedly the one actually used on the very first *TAPERTUBE* piles installed at JFKIA circa 2000 (what was effectively alpha-testing for these piles). This is what Horvath and Trochalides (2004) arbitrarily called a Type I connection detail, solely for convenient reference purposes in their paper. Note that this is not, and has never been, an official designation for the *TAPERTUBE* pile.

As can be seen in Figure 3.5, the two pile sections (tapered and constant diameter) were connected using a proprietary steel ring that fit <u>inside</u> the top of the tapered lower section but <u>over</u> the bottom of the constant-diameter upper section (extension). In the process, this connecting ring produced a localized bulge in the overall pile diameter of approximately one-half inch (12 mm). While this may seem modest, the post-installation effects on the measured geotechnical resistance in conventional static load tests were not.

Post-installation static load testing of piles installed with this so-called Type I connection detail suggested that this localized bulge in pile diameter caused a not-insignificant reduction in ULS shaft resistance along the constant-diameter upper section. This was most likely due to the annular space created as the bulged section was driven into the ground, and the concomitant reduction in post-installation lateral earth pressures acting along the upper, constant-diameter portion of the pile shaft. Note that while the ULS resistance of a *TAPERTUBE* pile (or *Monotube* pile) always derives mainly from the lower tapered section, the ULS resistance from the upper constant-diameter section is, in general, not insignificant so this was clearly an issue that had to be addressed by those who held the legal rights to the *TAPERTUBE* pile.

50

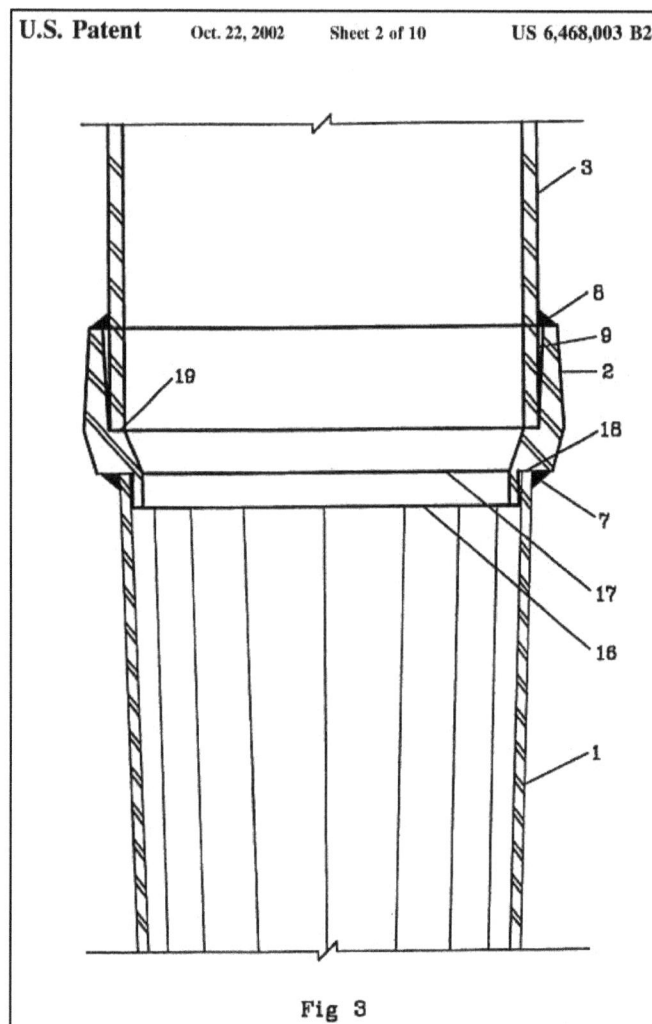

Fig 3

Figure 3.5. Original Connection Detail between Proprietary *TAPERTUBE* Tapered Section and Generic Constant-Diameter Section (Extension) Shown in U.S. Patent No. 6,468,003 B2.

In very short order while work at JFKIA was progressing, the developers of the *TAPERTUBE* pile switched to the connection details shown in Figure 3.4 and what Horvath and Trochalides (2004) called the Type II connection detail (again, this is <u>not</u> an official designation used by the manufacturer of the *TAPERTUBE* pile). This eliminated the troublesome diameter bulge by eliminating the proprietary ring connection. The upper end of the lower tapered section was bent during fabrication so that it fits <u>inside</u> the generic spiral-welded pipe that constitutes the upper constant-diameter section of the overall pile. A weld around the connection results in a watertight connection with an outer diameter that is the same as that of the upper section. This revised connection has proven to be successful in almost two decades of use to date.

It is important to emphasize that the creation of this connection between sections of a *TAPERTUBE* pile is nowadays typically done by a local steel fabricator who forms the proprietary tapered shaft; attaches the proprietary toe-closure piece to the bottom of the tapered shaft; and attaches at least a short stub of the constant-diameter extension to the completed tapered section (more on this fabrication process subsequently). Thus, the critical welds at the toe and transition between tapered and non-tapered sections that are required in order to create a watertight pile are done under controlled conditions, not in the field.

Some additional comments about the generic constant-diameter pipe extension used with both the *TAPERTUBE* and *Monotube* proprietary tapered sections are in order. For decades, only seamless pipe was typically specified and used for any type of foundation piling, at least in the U.S. Spiral-welded pipe, while less expensive, had a poor reputation (well-deserved in some cases) for being prone to splitting along the weld seam during pile driving. The writer is personally familiar with construction specifications used in the New York City metropolitan area in the latter decades of the 20th century that outright forbade the use of spiral-welded pipe for foundation piling for this reason.

However, more-recent experience has shown that the manufacturing quality of spiral-welded pipe has improved significantly to the point that the manufacturing welds do not split even under stresses that cause yielding of the overall pipe section (Buckland 2005, Horvath 2016a). This was independently and informally confirmed to the writer while conducting background research for this monograph during a confidential informational meeting in 2016 with a person who had been involved with the manufacturing of deep-foundation products in the U.S. for decades.

This improved behavior of the current generation of spiral-welded pipe is illustrated in Figure 3.6 that shows what had been the head of the constant-diameter upper section of a *TAPERTUBE* pile during driving. Note that the former head of the pile in this figure was first flame-cut with a torch from the installed portion of the pile <u>after</u> driving was completed then flame-cut again into the two pieces shown to better illustrate the damage to the pile head.

Figure 3.6. Installation Damage to Head of *TAPERTUBE* Pile Constant-Diameter Extension [photo credit: Thomas Quinn].

Figure 3.6 illustrates a situation where the head of the pile 'rolled' (yielded) during exceptionally hard driving that was apparently exacerbated by imperfect axial alignment between the head of the pile and the helmet of the pile-driving hammer (note that this was f_y = 50 ksi = 345 MPa steel). Despite the yielding of the parent material, the factory-welded seams of the spiral-welded pipe did not rupture.

Returning now to the discussion of differences between the *Monotube* and *TAPERTUBE* piles, as with the *Monotube* pile, the entity that controls the business aspects of the *TAPERTUBE* pile has changed hands over the approximately two decades that this pile has been in existence to the present although to date there has been critical continuity of the principals who were involved in the initial development of the pile.

However, the *TAPERTUBE* pile has always had a somewhat complex business structure behind its production in that the *TAPERTUBE* pile is and has always been more of a proprietary <u>concept</u> with a sole source to contact for purchasing information but no fixed manufacturing location like the *Monotube* pile. As discussed subsequently, there are important reasons for stakeholders to understand this business structure which may change in the relatively near future as patent protection of the *TAPERTUBE* pile expires. In addition, the inevitability of time takes its toll as always. Stanley 'Stan' Merjan, P.E., one of the two original *TAPERTUBE* patent holders, died in early 2019 and there will undoubtedly be other changes in principals involved with this pile.

The current business entity that controls the *TAPERTUBE* pile, DFP Foundation Products, LLC (DFP) of Franklin Lakes, NJ, only handles marketing and sales. DFP is not directly involved in manufacturing as Monotube and Union Metal before it were and, thus, DFP has no substantial brick-and-mortar location (their office is a private residence) and no substantial manufacturing assets. Actual forming of the proprietary *TAPERTUBE* tapered shaft from generic hot-rolled steel sheet; making and attaching the proprietary toe closure; attaching the completed tapered section to a relatively short length of constant-diameter pipe to create the tapered sub-assembly; and supplying additional lengths of constant-diameter steel pipe to be added to the tapered sub-assembly at a project site, are all always outsourced by DFP to local steel fabricators. On occasion, DFP has had geotechnical analysis and design support via outsourcing to a geotechnical consulting firm but this has generally been for its own internal benefit in pricing and bidding potential jobs, not for project-design purposes of record or guaranteeing the load-bearing performance of their piles as Raymond did in its heyday. Consequently, as with *Monotube* piles, end users (which nowadays may be a construction contractor who has to 'guarantee' a pile of a specific load-bearing capability, ULS or otherwise, to a project owner) are on their own with regard to a project-specific design involving *TAPERTUBE* piles.

To completely understand the current business structure employed in producing *TAPERTUBE* piles, imagine a typical project where the *TAPERTUBE* pile is to be used. The basic specifications for the piles (lengths, etc.) would be determined by a design professional working for either the project owner or construction contractor depending on how the project was structured with respect to design and construction responsibilities. The owner of the *TAPERTUBE* product rights (currently DFP) would be contacted by the pile-installation contractor involved in the project for a price quote and to ultimately place the order. DFP would then contract with any number of steel fabricators located throughout the U.S. who have the capability to bend generic hot-rolled steel sheet of the desired thickness into the proprietary *TAPERTUBE* tapered shaft. DFP also separately controls the outsourced production of the proprietary steel toe-closure piece. The steel fabricator would then mate the tapered shaft that they created to the toe-closure piece supplied through DFP and then to a relatively short section of generic hot-rolled steel pipe (the length varies depending on project requirements based on shipping and on-site storage and handling) and ship the completed sub-assembly to the project site along with additional sections of constant-diameter steel pipe for final on-site assembly and installation by a pile-driving contractor as shown on the back cover of this monograph.

Typical examples of the basic tapered sub-assemblies (proprietary tapered lower section that is factory-connected to a relatively short stub of generic constant-diameter pipe) that would be supplied to a project site by a steel fabricator working on behalf of DFP are shown in Figures 3.7 and 3.8. Note that even though these examples are from two projects at the same location (JFKIA) and only one year apart (2012 and 2011 respectively) that the length of constant-diameter pipe that is included with the tapered sub-assembly created by the steel fabricator varies noticeably. Specifically, the length of the constant-diameter section in Figure 3.7 is considerably shorter than that in Figure 3.8. This could have been due to a variety of shipping and handling factors (the specific project site shown in Figure 3.7 was much smaller and spatially restricted compared to the specific project site shown in Figure 3.8) as well as the expected final installed pile lengths and perhaps even the preferences of the pile-driving contractor. Most pile-driving specifications place a variety of limitations on the constant-diameter extensions that are added in the field, e.g. minimum length, minimum distance between the head of the pile and field splice during driving, etc.

Figure 3.7. Fabricated Tapered Sub-Assemblies of *TAPERTUBE* Piles at JFKIA in 2012 [photo credit: Thomas Quinn].

Figure 3.8. Fabricated Tapered Sub-Assemblies of *TAPERTUBE* Piles at JFKIA in 2011 [photo credit: John McCullough].

54

With regard to the current manufacturing business structure for *TAPERTUBE* piles, this has both positive and negative aspects in the writer's opinion. On the one hand, in principle it allows for manufacturing to be sited close to a project site which minimizes shipping costs and should make these piles competitive in more situations compared to the *Monotube* pile that was always shipped from one location (Canton, OH). On the other hand, it raises issues concerning *Manufacturing Quality Control* (MQC) and the resulting consistency of the final product because manufacturing know-how is not concentrated in one location.

It is important to emphasize that these MQC issues are not unique to the *TAPERTUBE* pile and there is nothing inherently 'bad' about localized, outsourced manufacturing. In fact, the vast majority of materials and products used in engineered construction are produced in this fashion. However, it is incumbent upon design professionals to know such details concerning product manufacturing so that they can ask the right questions concerning MQC and have the appropriate *Manufacturing Quality Assurance* (MQA) protocols in place to check and verify manufactured quality. The writer has extensive first-hand experience with this with respect to manufacturing block-molded expanded polystyrene (EPS-block) for geofoam applications.

3.3.6 Closing Observations and Comments

Since the beginning of the 21st century, the commercial marketplace with regard to tapered piles that utilize a steel shell or pipe as an essential component has been more competitive than at any time in history. Until recently, there were two commercially available tapered sections (there is no fundamental reason why the current market absence of *Monotube* piles could not be reversed quickly if desired), either of which can be mated with a generic constant-diameter upper section or extension to produce a complete pile ready for installation using any type of impact pile-driving hammer[54]. Furthermore, for the first time in history there is a steel tapered pile (the *TAPERTUBE*) that has the potential to be producible and thus marketable globally by virtue of the fact that it utilizes (for the most part) generic components that can presumably be fabricated at countless locations not only in within the U.S. but around the world.

It is significant to note that the relative ease with which the *TAPERTUBE* can be manufactured compared to the *Monotube*, which requires specialized fabrication tooling and machinery to produce cold-formed steel, has allowed the standard sizes for the *TAPERTUBE* product line to be significantly expanded within a relatively short period of time since the pile was commercially introduced approximately 20 years ago. By comparison, the *Monotube* product line remained virtually unchanged for the better part of a century.

The earliest (circa 2000) *TAPERTUBE* piles were manufactured to compete on a specific project with *Monotube* piles with a *Type Y* taper angle of 0.95° and nominal toe diameter of 8 inches (200 mm). However, the current *TAPERTUBE* product line lists toe diameters ranging from 8 to 14 inches (203 to 356 mm) and constant-diameter extension sections ranging from 12.75 to 24 inches (324 to 610 mm) in diameter. Taper angles range from 0.44° to 1.59°. Interestingly, the α = 1.59° taper angle actually matches a taper angle used on at least one prototype pile during initial product alpha-testing with the Type I connection detail circa 2000 (this is the Type Ia pile discussed by Horvath and Trochalides (2004)).

---

[54] At least one relatively recent project at JFKIA is known to have used a vibratory hammer for partial installation of *TAPERTUBE* piles. However, the final driving was performed with a traditional impact hammer.

In any event, because of the marketplace competition between the *Monotube* and *TAPERTUBE* piles, in the U.S. at least, there is no reason why foundation engineers could not design and specify a generic thick-wall steel tapered pile, even on projects where non-proprietary product usage is emphasized, and let market forces determine the specific brand used for the tapered portion of the pile. Thus, to reinforce a point made in the opening chapter of this monograph, more than ever there is a well-defined need to improve the analytical methodologies for forecasting the resistance of tapered piles as well as making this information more widely available than in the past or at the present.

3.4 PORTLAND-CEMENT CONCRETE

3.4.1 Introduction

As discussed in the preceding sections dealing with steel piles, the earliest tapered steel piles that were invented in the late 19th century to early 20th century and that culminated in the commercial development of the *Raymond Standard* and *Raymond Step-Taper* piles viewed the final pile as a CIP-PCC pile. The steel shells of these piles had no post-installation function and acted only as temporary, engineered formwork for the fluid PCC. Nevertheless, because these Raymond piles required the installation of a steel shell, for the organizational purposes of this monograph they were treated as steel piles.

Thus, the presentation in this monograph on tapered PCC piles is limited to a discussion of PCC piles that are constructed without a steel shell although they typically have internal steel reinforcement. Note that only precast-PCC piles are discussed in the following sections as tapered CIP-PCC piles constructed without a steel shell involve drilling, not driving, as part of the installation process. Consequently, they are not piles in the context defined for this monograph. However, for the sake of completeness, CIP-PCC drilled DFEs are briefly discussed in a separate section toward the end of this chapter.

One caveat concerning the following presentation of tapered precast-PCC piles is that it is almost certainly incomplete to an unknown extent. Overall, this is due to the inherently fragmented, regional business structure of manufacturing precast-PCC piling in general that has always existed and continues to exist to the present, at least in the U.S. This has resulted in incomplete knowledge-transfer about specific products beyond a local audience, especially when it involves crossing national boundaries and continents.

In addition, it is clear that many types of tapered precast-PCC piles have fallen by the technological wayside over time. During the Golden Age of R&D related to precast-PCC piles in general that occurred around the cusp of the 20th century, there were countless concepts for precast-PCC piles that were developed as discussed in Gow (1917). Some of these concepts were patented in the U.S. and elsewhere and some were not. Most of these concepts were developed by, or assigned to, a specific pile-installation contractor and never achieved a critical mass of widespread commercial acceptance and use beyond the person or organization that developed the pile. This was most likely due to various inherent complexities in their manufacture and/or installation as it is well-known that more than 90% of all U.S. patents never become commercially successful. Clever ideas, while patentable, much more often than not do not translate into a product that is viable to scale-up, manufacture, and distribute on a commercial, profit-making basis. It appears that the aforementioned *Raymond Standard* and *Raymond Step-Taper* piles were exceptions to this rule, largely because of the geographic dominance of Raymond for much of the 20th century.

The point being made here is that there are countless concepts for tapered precast-PCC piles that have long ago fallen by the wayside and to attempt to document all of them in this monograph would have had little value (in the writer's opinion) for the effort involved. Therefore, the approach taken here is to present and discuss a representative sampling of tapered precast-PCC pile concepts that have appeared over the last approximately 100 years. There are, no doubt, more examples that could be found with further historical research, especially outside the U.S.

3.4.2 Precast

3.4.2.1 Background

Several U.S. patents for tapered precast-PCC piles were issued around the same time as those for tapered steel piles in the early 1900s. Thus, tapered precast-PCC piles have been used in the U.S. for over a century and for about as long as tapered steel piles. However, the use of tapered precast-PCC piles, in the U.S. at least, has been substantially less than that of tapered steel piles. What use of tapered precast-PCC piles has occurred is not widely documented in the published literature and thus relatively unknown to most foundation engineers, including the writer. This is likely due to a combination of factors, with the lack of widespread product recognition and use being a major one.

To begin with, tapered precast-PCC piles are not produced in standard shapes and sizes by a national manufacturer or distributor who would advertise and otherwise promote their use on a national basis, at least in the U.S. although this may be true in smaller countries (e.g. the *Brynildsen* pile discussed by Gregersen at al. (1973)). What use of tapered precast-PCC piles is known to have occurred appears to be relatively localized and regional, limited to those areas where a combination of subsurface conditions, foundation loading, and cost factors have favored the local innovation, production, and use of such piles.

Consequently, because of the scattered nature of the tapered precast-PCC pile market compared to that for tapered steel piles, the presentation in this monograph is limited to the presentation and discussion of several distinct geometries of tapered precast-PCC piles that were identified by the writer while conducting background research for this monograph. This presentation is done to illustrate the breadth of product development rather than its depth.

3.4.2.2 Overview

The vast majority of tapered precast-PCC piles produced to date have had conventional geometries similar to that of timber piles and tapered steel piles, i.e. continuously tapered and partially tapered. However, the ability to cast PCC into unconventional shapes has allowed some geometric innovation not seen with other pile materials.

Discussed first are piles with a conventional geometry. Both continuously tapered piles similar to timber piles and the *Raymond Standard* pile and partially tapered piles similar to the *Monotube* and *TAPERTUBE* steel-pipe piles have been produced (there has been no known production of a step-tapered precast-PCC pile). This is followed by presentation of piles that have an unconventional geometry, specifically, in the shape of a bulb that is similar to the CIP-PCC *pressure-injected footing* (PIF) a.k.a. *Franki* pile.

Not surprisingly, precast-PCC piles with a continuous taper are the oldest type of tapered precast-PCC piles as they date from circa 1900 when pile developments using steel and PCC frequently mimicked the geometry of timber piles as has been noted earlier in this chapter. However, all continuously tapered piles, regardless of their material makeup, suffer from inherent length limitations as was noted earlier with the *Raymond Standard* pile.

However, this limitation has not hampered their use for over 100 years to the present in applications where a relatively short DFE is all that is required.

Partially tapered piles do not have the same length constraints as was seen previously with the *Monotube* and *TAPERTUBE* piles. However, extensive use of partially tapered precast-PCC piles does not appear to have been widespread to date, possibly because handling long precast-PCC piles in general without causing physical damage, especially if they are not prestressed, can be problematic. This fact may have limited the wider use of relatively long tapered PCC-piles.

Of course, length issues are not problematic if a precast-PCC pile is installed in segments. There are any number of proprietary connection details that have been developed for use specifically with precast-PCC piles. However, for whatever reason the use of segmental precast-PCC piles has never caught on in the U.S. although such usage is apparently much more common outside of the U.S. The aforementioned *Brynildsen* pile that was used in Norway more than 50 years ago is but one example but an important one as it indicates that producing partially tapered precast-PCC piles of substantial length is certainly technically feasible. Whether or not it is cost effective is a separate issue.

3.4.2.3 Conventional Pile Geometries

3.4.2.3.1 Continuously Tapered

One of the oldest, if not the oldest, tapered precast-PCC piles appears to have been developed at the beginning of the 20th century, contemporaneous with the early R&D and concomitant patents by A. A. Raymond for tapered steel piles that were presented earlier in this chapter. This precast-PCC pile is referred to in the literature of the time (*Concrete* 1910) as the *Gilbreth* pile after its inventor, Frank Bunker Gilbreth. He was a mechanical engineer and referred to in a recent (2012) American Society of Mechanical Engineers (ASME) publication as the "Father of Management Engineering". Gilbreth was apparently an early user of PCC, having started his working career in the latter decades of the 19th century as a bricklayer. It appears that his professional forte was developing efficiencies in construction activities and processes.

The *Gilbreth* pile was apparently also referred to at the time as the *Corrugated* pile (Gow 1917), including by Gilbreth himself (Gilbreth 1906), in reference to the vertical corrugations or flutes that were cast into the pile exterior. This physical detail can be seen in Figure 3.9 that is an excerpt from Gilbreth's U.S. patent for this pile.

Note that the vertical corrugations of the *Gilbreth* pile are broadly similar to those of the *Monotube* pile. However, in the case of the *Gilbreth* pile they were specifically called out in the text of the patent as being intentional and beneficial both during and after pile installation (whether or not they actually are is a separate issue not addressed here). On the other hand, for the *Monotube* pile they were simply a consequence of Union Metal repurposing their decorative light poles for use as foundation piling.

As noted earlier in this chapter, there have been attempts over the years to assign or explain away a geotechnical benefit to the vertical fluting of *Monotube* piles but this appears to be, in the writer's opinion, an attempt to 'engineer' a benefit after the fact to something that has a non-technical genesis. Specifically, the *Monotube* flutes are there because they looked nice on light poles. Period.

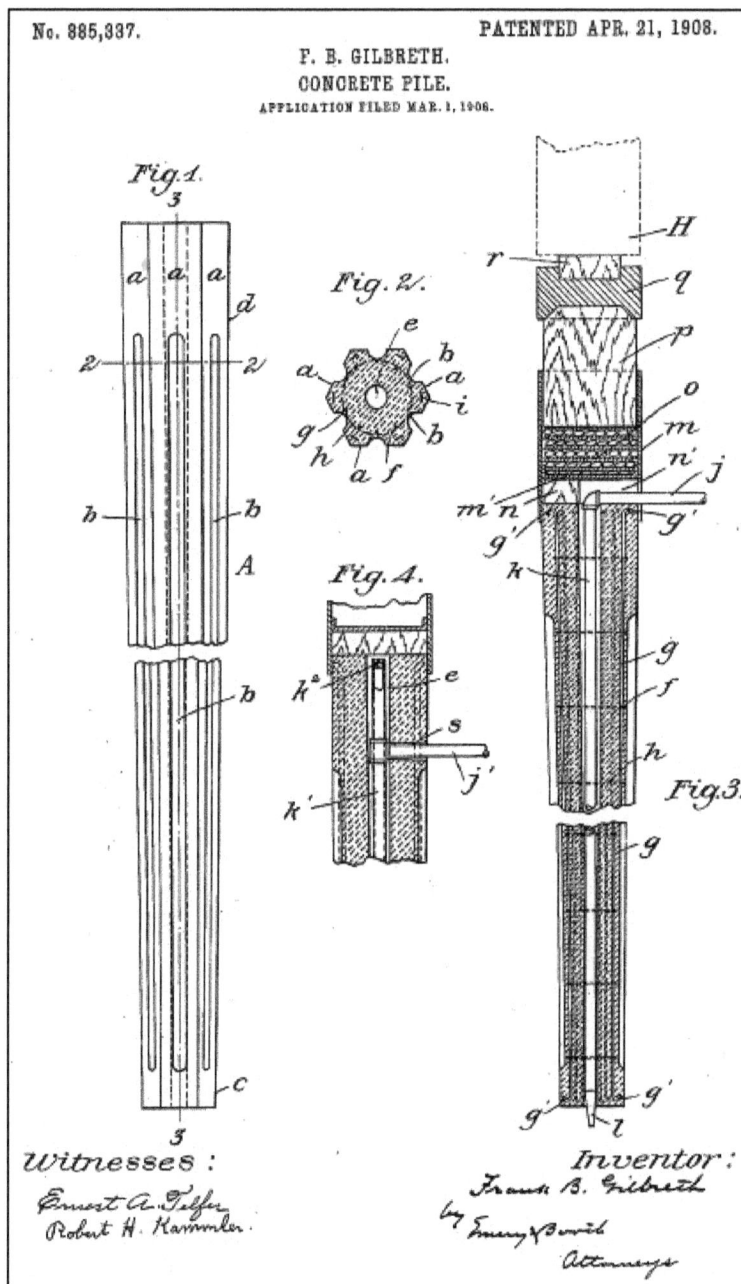

Figure 3.9. Details of the *Gilbreth* a.k.a. *Corrugated* Pile - U.S. Patent No. 885,337.

However, arguably the most notable feature of the *Gilbreth* pile is that it was intended from the start to be installed only by jetting using a jet pipe that was cast into the center of the pile as can be seen in Figure 3.9[55] (the installation process was covered by a separate U.S. patent, No. 885,520). Not only did jetting eliminate the issue of pile damage during impact

---

[55] Although the focus of this monograph is on piles installed by impact driving, an exception is made here for the *Gilbreth* pile in view of its place in the historical development and evolution of tapered DFEs and the fact that it appears to be the first tapered precast-PCC pile that saw production use, at least in the U.S.

driving (as noted earlier in this chapter, this was an inherent, systemic problem with early precast-PCC piles due to their lack of being prestressed) but it is clear from the basic pile patent (U.S. No. 885,337) that Gilbreth saw this installation methodology as a distinct benefit that was highlighted in the patent. In the patent text, he hypothesized the same ground-improvement mechanism that is nowadays recognized with vibrocompaction of coarse-grained soils, i.e. that the temporary liquefaction of soil particles (during pile installation in this case) would cause the soil particles to settle into a denser, confining pattern around the pile once the pile achieved its desired embedment and the jetting was terminated. Gilbreth also called out in the patent the cast-in corrugations on the pile face as being beneficial both during pile installation as well as afterward in the long term although the hypothesized mechanism for this was not given and what it might be is unclear to the writer.

There are indications that continuously tapered precast-PCC pile research, development, and commercial usage continued throughout the 20th century and continues into the 21st century. Several examples will be cited to support this observation.

To begin with, as part of a larger research study comparing tapered vs. constant-perimeter piles driven into coarse-grained soil, Gregersen et al. (1973)[56] compared the load-settlement behavior of two precast-PCC piles with a solid, nominally circular[57] cross-section and normal reinforcement. Each was 8 metres (26.2 ft) long. One had a nominal constant diameter of 280 millimetres (11.0 in) and the other was continuously tapered, with a diameter of 200 millimetres (7.87 in) at the toe and 280 millimetres (11.0 in) at the head. This works out to a taper angle, α, = 0.29° that Gregersen et alia noted was approximately that of a typical timber pile of that era (circa 1969) in Norway. Note that this taper angle is also consistent with the writer's first-hand experience with Southern Yellow Pine timber piles during the same early-1970s timeframe as discussed earlier in this chapter. In any event, it is of interest that Gregersen et alia referred to these piles as being more-or-less typical *Brynildsen[58]* piles, implying that they were a commercially available product in Norway at least at that time.

Continuously tapered precast-PCC piles with a square cross-section have also been used. Kodikara and Moore (1993) referenced such a pile that was used in the former Soviet Union although they did not provide details.

The writer has heard anecdotal information about such piles being used in the U.S. in south Florida but again, details have been lacking.

There is confirmed use of such piles in Slovakia in the early years of the 21st century. Several years ago, some unsolicited photos were sent to the writer via email[59] but no further information was provided as to the intended use of such piles, whether they were prestressed

---

[56] The case history presented in this paper was discussed in some detail by Fellenius (2002a). It appears that Fellenius incorrectly identified the pile-labeling nomenclature used in Gregersen et al. (1973). It is not known to the writer to what extent this misidentification may have affected interpretations and concomitant comments made by Fellenius.

[57] The reason why the piles discussed in Gregersen et al. (1973) were only nominally, and not exactly, circular in cross-section is explained in Appendix H. This appendix contains the writer's detailed reassessment and parsing of the piles in the Gregersen et alia paper.

[58] This pile name may be related to not so much the pile itself but the screw-connection detail used between precast-PCC segments of piles (segments used in the test program reported by Gregersen et alia were 4- and 8-metres (13.1- and 26.2-ft) long). This assumption is supported by the fact that U.S. Patent No. 4,157,230 titled "Joint for Pile Sections" was issued 5 June 1979 to Arne Tømt and Ivar Vamnes of Norway, and assigned to B. Brynildsen & Sonner A/A of Norway. Further research suggests that this company was a major PCC precaster in Norway, at least in the 1960s and 1970s.

[59] The writer did not record the name of the person who sent the email and photos, and regrets not being able to provide the proper photo credit for Figure 3.10.

or not, how common they are or were in Slovakia, etc. Figure 3.10 shows one of these Slovakian piles, apparently ready to be driven. It appears that at least the version shown in this and other photos is relatively short and sharply tapered.

Figure 3.10. Continuously Tapered Precast Portland-Cement Concrete Pile Used in Slovakia (Early 21st Century).

Such continuously tapered precast-PCC piles are not limited to a solid cross-section. Togliani (2010) discussed a pile used commercially in Italy (for how long and how extensively are unknown to the writer) that is essentially a tapered pipe with a closed toe that is centrifugally cast. The wall thickness of the pipe and type of reinforcement were not indicated. The specific example cited by Togliani was 12 metres (40 ft) long although there are implications in his paper that this was atypically long for such piles. The head and toe diameters were 420 millimetres (16.5 in) and 240 millimetres (9.5 in) respectively. This works out to a taper angle, α, = 0.43° which is somewhat greater than the range for timber piles with which the writer is familiar and toward the lower end of the range used for *Monotube* and *TAPERTUBE* piles.

3.4.2.3.2 Partially Tapered

An early-1970s U.S. patent filing (see Figure 3.11) depicts a partially tapered precast-PCC pile (the patent allows for the pile to be either normally reinforced or prestressed) that is square and solid in cross-section and with an overall appearance similar to a *Monotube* or *TAPERTUBE* pile, i.e. a tapered lower section and a constant-width upper section (extension). The patent was filed by a resident of Florida (William H. Medema of Fort Lauderdale) and assigned to a Florida company (Oolite Industries, Inc. of Miami). Whether this type of pile was ever used in Florida or elsewhere in actual production work is unknown.

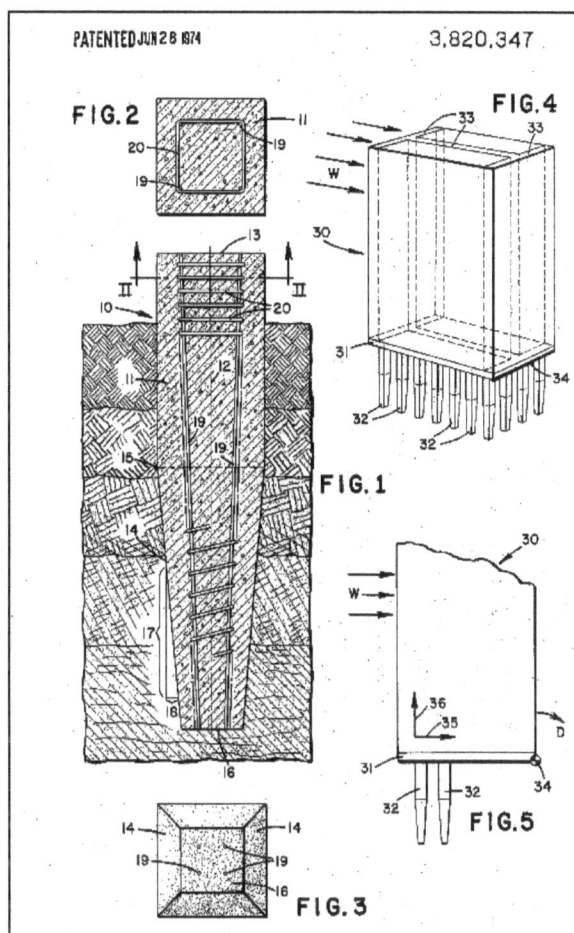

PATENTED JUN 28 1974   3,820,347

Figure 3.11. Details of the *Medema-Oolite Industries* Pile - U.S. Patent No. 3,820,347.

During the same early 1970s timeframe, the same Gregersen et al. (1973) paper noted previously contained comparative results between a nominally constant-diameter *Brynildsen* pile and a partially tapered *Brynildsen* pile. The latter pile had the typical arrangement of a tapered lower portion and nominally constant-diameter upper portion. In this case, both piles had a total length of 16 metres (52.5 ft), installed in two segments that were each 8-metre (26.2 ft) long. The taper angle, α, of the partially tapered pile was 0.29°.

3.4.2.3.3 Summary

Although the available information is fragmented and incomplete, it appears that in the over 100 years since tapered precast-PCC piles have been developed and used that they have enjoyed, and continue to enjoy, commercial success in local or regional applications that make them a cost-effective deep-foundation alternative. It is opined that with greater awareness of the technical benefits of tapered piles, combined with the availability of more-accurate resistance-forecasting methodologies for tapered piles, that the use of tapered precast-PCC piles could be even greater worldwide. This is because there are several physical parameters for such piles that can be changed and combined as desired to maximize their cost effectiveness. Specifically, tapered precast-PCC piles can be:

- continuously or partially tapered;

- circular, quasi-circular (multi-sided), or square in cross-section;

- solid or partially hollow in cross-section;

- normally reinforced or prestressed, with, of course, the usual choice of pre- or post-tensioning if prestressed.

It does not appear that all combinations of these variables have been tried to date which opens up the possibility of future developments in this area.

There is, however, a potentially significant issue concerning tapered precast-PCC piles that came to light during the research for this monograph. As discussed further in subsequent chapters, there is both anecdotal field experience as well as theoretical considerations that indicate that cross-sectional geometry may play a not-insignificant role in the shaft resistance of tapered piles. Specifically, all other things being equal, it appears that tapered piles with a square cross-section do not develop the same ULS unit shaft resistance as tapered piles with a circular cross-section. If correct, this is a significant paradigm shift compared to constant-perimeter piles where it is routinely assumed that the cross-sectional geometry of a pile does not affect the ULS unit shaft resistance, $r_{s(ult)}$.

The explanation for this apparent behavior of tapered piles appears to relate to the relatively complex mechanism for developing shaft resistance that is unique to tapered piles. However, at the present time, the differences between square and circular cross-sections for tapered piles is very poorly and incompletely understood. Consequently, additional research is required to better understand this issue so that it can be quantified.

Nevertheless, the issue is raised here in view of the fact that tapered precast-PCC piles with both circular and square cross-sections have been used for at least several decades now. Therefore, before future developmental commitments are made to tapered precast-PCC pile technology, it is desirable to first investigate the issue of pile cross-sectional geometry thoroughly.

3.4.2.4 Unconventional Pile Geometries

For the sake of completeness, there is one more tapered precast-PCC pile that is worth mentioning, if only for its uniqueness and thus worthy inclusion in the historical record even though it does not appear that this pile is in current production and use. This pile is the *Tapered Pile Tip* pile or simply *TPT* pile as it was colloquially and universally known when it was a commercially viable pile in the Northeastern U.S. during the latter decades of the 20th century. It is of interest to note that by the time this pile was developed, it appears that the practice of naming piles after the inventor[60] that had existed in earlier decades of the 20th century had been dispensed with.

Figure 3.12 shows an excerpt from the U.S. patent for the *TPT* pile. Note that although the patent was granted in 1979, the initial patent filing occurred almost a decade earlier,

---

[60] The late Stanley 'Stan' Merjan, P.E. in this case who was a prolific inventor of modern tapered-pile technologies that are in use to the present. As noted previously, he was the co-inventor and subsequent co-patent holder of the *TAPERTUBE* pile that was developed three decades after the *TPT* pile. A review of the legal record indicates that Stan Merjan was named as the inventor or co-inventor of at least 12 U.S. patents related to deep foundations.

1970. The writer knows from first-hand experience that the pile covered by this patent was in commercial production and project use in New York City by at least 1972.

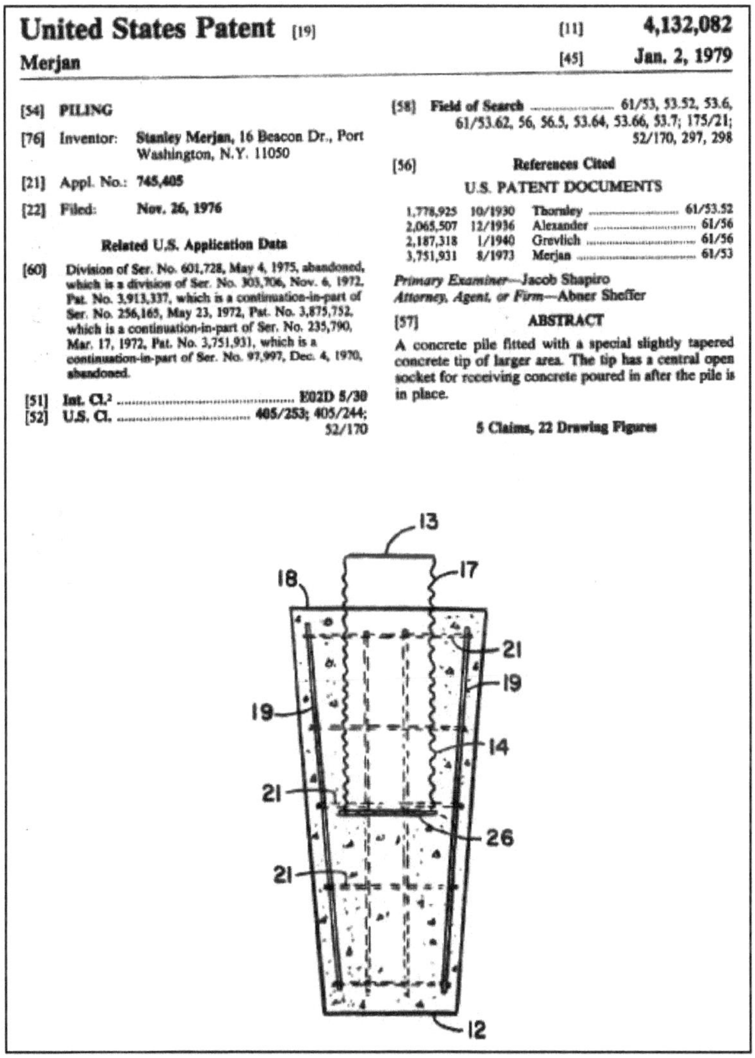

United States Patent [19] [11] **4,132,082**

Merjan [45] **Jan. 2, 1979**

[54] **PILING**

[76] Inventor: Stanley Merjan, 16 Beacon Dr., Port Washington, N.Y. 11050

[21] Appl. No.: 745,405

[22] Filed: Nov. 26, 1976

Related U.S. Application Data

[60] Division of Ser. No. 601,728, May 4, 1975, abandoned, which is a division of Ser. No. 303,706, Nov. 6, 1972, Pat. No. 3,913,337, which is a continuation-in-part of Ser. No. 256,165, May 23, 1972, Pat. No. 3,875,752, which is a continuation-in-part of Ser. No. 235,790, Mar. 17, 1972, Pat. No. 3,751,931), which is a continuation-in-part of Ser. No. 97,997, Dec. 4, 1970, abandoned.

[51] Int. Cl.² E02D 5/30
[52] U.S. Cl. 405/253; 405/244; 52/170

[58] Field of Search 61/53, 53.52, 53.6, 61/53.62, 56, 56.5, 53.64, 53.66, 53.7; 175/21; 52/170, 297, 298

[56] **References Cited**

U.S. PATENT DOCUMENTS

1,778,925	10/1930	Thornley	61/53.52
2,065,507	12/1936	Alexander	61/56
2,187,318	1/1940	Grevlich	61/56
3,751,931	8/1973	Merjan	61/53

Primary Examiner—Jacob Shapiro
Attorney, Agent, or Firm—Abner Sheffer

[57] **ABSTRACT**

A concrete pile fitted with a special slightly tapered concrete tip of larger area. The tip has a central open socket for receiving concrete poured in after the pile is in place.

5 Claims, 22 Drawing Figures

Figure 3.12. Primary Component of the *Tapered Pile Tip* (*TPT*) Pile - U.S. Patent No. 4,132,082.

The primary component of the *TPT* pile is the tapered precast-PCC base shown in Figure 3.12. It consists of a normally reinforced, precast-PCC element with the overall geometry of a truncated right-circular cone (typical dimensions are given subsequently) although the patent did allow for and illustrated a variant with a square cross-section.

Regardless of the overall cross-sectional geometry, there is a relatively short right-circular cylindrical cavity that is cast into the upper portion of this base using a section of either corrugated, thin-wall steel shell (as shown Figure 3.12) or thick-wall steel pipe, both of constant diameter, as the form for this cavity during casting of the PCC base. As also shown in Figure 3.12, the stub of steel shell or pipe that is cast into the base to create the cavity is left to protrude a short distance above the top of the precast-PCC base. The exposed stub of steel shell or pipe is to be used as the connection to the extension that is used to drive the base into the ground. At least three different types of extensions were covered by the patent and actually used in practice as discussed subsequently.

A logical question at this point would be why would someone invent a pile with this unconventional geometry. There was a very simple and good commercial reasons for this, broadly similar to the motivation for the same person (Stan Merjan) to co-invent the *TAPERTUBE* pile some 30 years later: to 'build a better mousetrap' compared to an existing pile in the marketplace.

In any event, the writer recalls that at one point circa the 1970s the *TPT* pile was catalogued as being available in at least 10 standard sizes that varied both in height and diameters of the tapered precast-PCC base. However, for order-of-magnitude reference purposes it is useful to refer to the patent that gives typical dimensions of the tapered precast-PCC base as being 60 inches (1524 mm) high with top and bottom diameters of 30 and 24 inches (762 and 610 mm) respectively. This works out to a taper angle, α, = 2.9° which is substantially greater than most of the tapered piles discussed earlier in this chapter. The diameter of the circular cavity cast into the top of the base was given as typically being 12 or 14 inches (305 or 356 mm) and was governed largely by structural considerations.

Installation of the *TPT* pile varied somewhat depending on the type of extension that was used with the tapered precast-PCC base. The writer recalls that initial commercial applications of the *TPT* pile used what can be interpreted as the "preferred embodiment" of the patent as shown in Figure 3.12, i.e. using a constant-diameter, thin-wall steel shell that was filled with fluid PCC after installation. Installation using this embodiment is essentially the same as driving a *Cobi pneumatic-mandrel*[61] pile (or simply *Cobi* pile as it is referred to colloquially in practice[62]).

As was discussed earlier in this chapter, thin-wall steel shell piles such as the two Raymond tapered piles always require use of a mandrel. The Cobi pneumatically expansible-contractible mandrel had been developed in the latter half of the 1950s[63] as a way to install a pile that is essentially a constant-diameter version of the Raymond tapered piles, i.e. a pile that once installed is basically a CIP-PCC pile as the shell is only short-term formwork for fluid PCC and has no meaningful contribution to permanent, long-term structural resistance.

Given the constructability issues involved with shelling up any type of mandrel-driven pile as was discussed earlier in this chapter, an alternative embodiment of the *TPT* pile allowed for use of a thick-wall steel pipe extension that removed the need for a mandrel for driving and also provided for additional structural resistance of the pile shaft once installed and the pipe filled with fluid PCC.

---

[61] In the writer's personal experience in the New York City metropolitan area where use of this pile was once not uncommon, the term 'expandable' was/is used colloquially and exclusively in lieu of the term 'pneumatic' with respect to the *Cobi* mandrel.

[62] In the writer's personal experience in the New York City metropolitan area where use of this pile was once not uncommon, the *Cobi* pile has, in the past, also been referred to colloquially as the *Cobi Helcor* (sic) pile. HEL-COR® is actually the registered trademark of a specific brand/product line of corrugated metal pipe (CMP). At the present time, this trademark is owned by Contech® Construction Products Inc. and the product is sold by Contech Engineered Solutions. Note that CMP is a generic, commodity construction product and there are other manufacturers and suppliers in addition to Contech. Presumably, a *Cobi* pile could be installed with any brand of CMP, not necessarily Contech HEL-COR, as the unique element of the *Cobi* pile is the pneumatic/expandable mandrel, not the specific brand of CMP used with it. Therefore, use of the term *Cobi Helcor/HEL-COR* for this type of pile should be deprecated unless that specific brand of CMP is installed using the Cobi pneumatic/expandable mandrel and it is desired to make that distinction.

[63] U.S. patent No. 2,911,795 issued 10 November 1959 to Walter H. Cobi, no assignee indicated.

As discussed subsequently, the primary intended use of the *TPT* pile was for applications requiring substantial geotechnical resistance from the completed pile. However, the patent also envisaged a lower-resistance version of the *TPT* pile for applications such as supporting sewerage and other settlement-sensitive utility lines wherein a treated timber pile could be used as an alternative to the steel shell or pipe extension. The bottom[64] of the timber-pile extension was simply set into the short section of thin-wall steel shell that was cast into the top of PCC base component. No physical connection between the timber-pile extension and precast-PCC base is believed to have been made. The overall pile was then installed by normal driving on the head of the timber-pile extension.

A significant technical consequence of the *TPT* pile that ultimately contributed to its commercial demise is related to the fact that because the diameter of the top of the tapered precast-PCC base is always significantly greater than that of the steel shell or pipe or timber pile extension, an annulus is created as the precast-PCC base is driven into the ground. That annulus has to be filled in during driving. This is typically done with sand that is imported to the project site for this purpose. As a minimum, this means that the shaft resistance of the completed pile is essentially nil and that all resistance has to come from the precast-PCC base. Furthermore, apparently early on there were concerns about uplift resistance of the finished pile which, during the 1970s timeframe, was solely the result of either wind or uplift-water-pressure loading on the completed structure supported on these piles (seismic design was not a consideration at that time in the regional marketplace in which the *TPT* pile was being sold). This uplift issue resulted in Stan Merjan securing an additional, follow-on patent[65] that called for filling the annulus with fluid PCC as the pile was being driven.

The writer is unaware of any actual projects on which this later embellishment was ever done with production piles. Having seen *TPT* piles being driven first-hand, the writer has to question as to how successfully this could have been done on a consistent, reliable basis, especially once a pile toe penetrates below the groundwater level (which, as an overall rule, tends to be relatively shallow in the New York City metropolitan area where this pile was developed and initially marketed). This is because the annulus created during driving tends to collapse in an unpredictable, uncontrollable manner to some degree as the pile is being driven, especially below the water table in coarse-grained soil.

Turning now to the practice-oriented aspects of the *TPT* pile, although neither of Merjan's patents specified an assignee, in reality the patents were for all intents and purposes assigned to Underpinning & Foundation (U&F), a New York City-based foundation contractor whose corporate history dates back to 1897 and that has been a subsidiary of the multinational contractor, Skanska, for several decades now. U&F was also Merjan's employer at the time and continued to be until just a few years before Merjan's death in early 2019. Thus, initially the *TPT* pile was not only proprietary technology in and of itself but closely held in terms of its commercial application by the contractor that was its de-facto patent assignee. This was likely due to its initially targeted market application for a very specific combination of subsurface and market conditions that were, at the time, present in New York City and surrounding areas. That is not to say that the *TPT* pile was not viable or could not be revived for a broader market. It is simply a statement of fact for the circa-1970 timeframe when the *TPT* pile was initially developed and used commercially.

The *TPT* pile was originally developed to meet a very specific market niche that existed in the New York City metropolitan area circa 1970. A very common subsurface condition in this region consists of a surficial stratum of non-engineered fill that can have highly variable origin and composition. The fill is often underlain by a Holocene Epoch

---

[64] Toe, had the timber pile been driven independently.
[65] U.S patent No. 4,293,242 issued on 6 October 1981 to Stanley Merjan, no assignee indicated.

stratum of organic clay and peat followed by a Pleistocene Epoch stratum of relatively loose sand. At the time, a viable foundation alternative for structures with relatively large foundation loads was a *bulb pile*, known locally as either a *Franki* pile or by its generic name, *pressure-injected footing* (PIF), that was founded within the uppermost portion of the Pleistocene sand. Typically, this might be of the order of 30 feet (10 m) below the ground surface at a site. The soil densification that accompanied the installation of a PIF resulted in relatively substantial (for the time) ULS resistances of the order of 500 kips (2200 kN) being achievable. This would translate into a pile with an allowable (service-load) capacity of 250 kips (1100 kN). This is well in excess (by factors of approximately two and four respectively) of what was routinely being sought at the time for *Monotube* and timber piles that represented competitive alternatives for these subsurface conditions.

The problem was that constructing a PIF was a relatively slow, labor-intensive process. The *TPT* pile was intended to essentially be a precast PIF that could develop comparable ULS resistances but with much greater piles-per-workday productivity from a pile-driving crew and, therefore, at a much lower cost per unit of resistance.

Initially, the *TPT* pile proved to be very successful in its intended goal and was used on several projects within the New York City metropolitan area where the soil conditions for which it performed best existed. The fact that the *TPT* pile was ultimately marketed in several standard sizes added to the pile's market success as it could be used cost effectively for supporting utility lines in addition to the building loads for which it was initially intended. However, there were several reasons why the *TPT* pile ultimately disappeared from the market.

To begin with, the *TPT* pile was a proprietary product that was produced locally and was never marketed or licensed to a larger geographical region no less nationally within the U.S. Furthermore, its niche for cost-effective application was a very specific type of subsurface condition that, unfortunately, is highly susceptible to seismic liquefaction in regions where this is a reasonable possibility. By the late 1980s, foundation engineers in the New York City metropolitan area were increasingly designing with liquefaction in mind which required piles with substantially deeper penetrations into the Pleistocene-sand bearing stratum in order to bypass potentially liquefiable zones that were, unfortunately, precisely at the depths in which both PIFs and *TPT* piles would typically be founded. While the *TPT* pile could, in principle, be installed to any depth (the writer personally visited one early project in 1972 where the piles were of the order of 100 feet (30 m) in length in order to satisfy an arcane provision of the New York City building code), they were not, as a rule, cost-competitive at greater depths. Finally, there was the issue of the annulus created by pile installation and the labor-intensive need to backfill it as pile driving progressed. The loose fill placed within this annulus would have be highly problematic with regard to seismic-induced lateral loads in particular.

Although the *TPT* pile was not marketed as a tapered pile per se, the writer knows from personal experience that the *TPT* designer, Stan Merjan, was clearly aware of the benefit of taper by virtue of his extensive professional experience with both timber and *Monotube* piles that his employer, U&F, installed with regularity throughout the greater New York City metropolitan area for many decades. This is why taper was explicitly incorporated into the geometry of the *TPT* base element and why Stan Merjan devoted his energies to developing the *TAPERTUBE* pile some 30 years after he developed the *TPT* pile.

3.5 ALTERNATIVE MATERIALS

3.5.1 Overview

For the sake of completeness of the historical record, mention needs to be made of the fact that academic researchers, at least, have investigated the use of materials other than traditional piles materials (wood, steel, PCC) for tapered DFEs. To date, most of this research appears to have be conducted since the beginning of the 21st century at UWO/Western University in Canada under the overall direction of Prof. M. Hesham El Naggar. This includes:

- tapered piles consisting of a shell of *fiber-reinforced polymer* (FRP) as presented in Sakr et al. (2004a, 2004b, 2005, 2007) and

- *helical tapered piles*[66] constructed of *spun-cast ductile iron* (SCDI) pipe (Fahmy 2015; Fahmy and El Naggar 2016a, 2016b, 2017a, 2017b, 2017c).

3.5.2 Polymeric Materials

The use of polymeric materials in general for piles is not new, especially for use in a marine environment that is notoriously harsh on materials used for engineered construction. The FRP used by Sakr et alia is just a particular composite material of this general type and its use in a terrestrial application was atypical, but understandable, given the experimental nature of Sakr et alia's work.

Typically, the polymeric material for a pile is in the form of a closed-toe shell that functions solely as formwork for post-installation placement of CIP-PCC. The CIP-PCC core provides all of the long-term structural capacity of the finished pile.

The practical problem is that polymeric materials as a group are relatively weak compared to conventional piles materials and cannot tolerate direct driving with any significant force, if at all. Consequently, piles composed of polymeric materials require alternative, sometimes unconventional, installation techniques.

This was the case with the aforementioned tapered FRP piles studied by Sakr et alia. They used an unconventional installation methodology that involved driving directly on the pile toe. In this case, the toe essentially dragged the FRP shaft into the ground behind it. This process is broadly similar to the use of a mandrel with steel shell piles, both tapered and constant diameter, that was discussed earlier.

3.5.3 Ductile Iron

Ductile iron is a class of materials that are the evolutionary successors to cast iron. The reformulated material chemistries of ductile iron are an effort to overcome the primary mechanical (stress-strain) shortcomings of cast iron, specifically, brittleness and low tensile strength relative to compressive strength. The primary civil engineering application of

---

[66] It appears that even the developers of this concept (Fahmy and El Naggar, doctoral student and faculty advisor respectively) cannot agree on the preferred term for this type of pile as their several publications on the subject interchange the order of 'tapered' and 'helical'. In the writer's opinion, the pile involved is essentially a tapered spun-cast ductile iron (SCDI) pipe pile that could be used without the helix. Therefore, since the helix is an optional embellishment of what is fundamentally a tapered pile, the term 'helical tapered SCDI pipe pile' will be used in this monograph.

ductile iron historically has been as bell-and-spigot jointed pipe for underground utility lines transporting aqueous-based liquids.

Ductile-iron pipe piles with a constant diameter are relatively new on the deep-foundations scene and appear to be essentially repurposed, redesigned (primarily at the connections) ductile-iron utility pipe that is installed by impact driving. The niche of ductile-iron pipe piles is as an alternative for use with relatively light loads in confined-installation environments with relatively compact installation equipment. Note that because these piles have a relatively substantial wall thickness, the ductile-iron pipe can be counted on for long-term structural resistance unlike the polymeric shells discussed above that depend entirely on the PCC core for structural resistance.

The aforementioned helical tapered SCDI pipe piles studied by Fahmy and El Naggar appear material-wise to be a niche application for, combined with an evolutionary manufacturing variation of, otherwise generic, run-of-the-mill ductile-iron pipe piles. First of all, it appears that these new helical-tapered piles are intended for use with relatively lightly loaded structures so that only one section of pipe is required. This inherently limits the possible pile lengths (the piles studied by Fahmy and El Naggar were 3.1 and 6.2 metres (approximately 10 and 20 ft respectively) long). Second, the tapered variations they studied (constant-perimeter versions were installed and tested for comparative purposes) were continuously tapered (something that is not normally done with ductile-iron pipe), with a taper angle of 0.46°. This is somewhat greater than timber piles and toward the low end of the range of tapered steel pipe piles in current commercial use in the U.S.

There were some significant geometric and installation aspects to the helical tapered SCDI pipe piles studied by Fahmy and El Naggar. These aspects are discussed later in this chapter as this section is limited to a pile-material discussion.

3.5.4 Comments

While tapered DFEs constructed of novel, alternative materials and possibly installed in novel ways may prove commercially viable in the future, the required extensive R&D for this does not appear to have been done to date. Therefore, they are neither mentioned nor considered further in this monograph.

However, a generic caution is raised here that anytime a novel material and/or installation methodology is used there is no guarantee that resistance-forecasting methodologies developed based on traditional pile materials and installation methods (impact driving on the head of the pile as assumed for the tapered piles that are the focus of this monograph) will apply to new piles and/or installation systems without further research and possible modification. Therefore, none of the material presented subsequently in this monograph should be construed as being applicable to alternative pile materials and installation methodologies without further research.

3.6 HYBRID

As noted earlier in this chapter, what are defined as 'hybrid piles' in this monograph are typically referred to as 'composite piles' in the published literature. The term 'hybrid' is used here as it more accurately describes what these piles are as well as avoids terminological conflict when using the term 'composite' that, in the writer's opinion, is better suited and used to describe a pile (or DFE in general for that matter) cross-section composed of more than one type of material such as the tapered steel shell and pipe piles with PCC core discussed earlier in this chapter.

In simple terms, a hybrid pile is a pile composed of two distinctly different components, one stacked directly on top of the other, where each component could be its own pile. Hybrid piles were popular in the U.S., at least, at a time when material was the more significant cost component of engineered construction compared to labor (in very broad terms, this was any time prior to the 1970s in the U.S.) so that designs that saved on material costs were often the most cost effective overall even if they were somewhat labor intensive. This situation reversed during the 1970s, which was an extended period of historically very high inflation rates in the U.S. due to episodes of petroleum shortages, and has remained that way to the present. Consequently, the focus of engineered construction in the U.S. for the last several decades has been on construction technologies that favor reducing labor (and, more recently, construction time) even if it means the use of more material.

As an aside, in recent years an additional consideration and concomitant complication has evolved that injects an additional (to material, labor, and time) consideration into decision-making in engineered construction, that of what might be called 'greenness'. This aspect is quantified by a combination of material or product sustainability/renewability and the overall carbon footprint of the manufacturing and construction process involved with that material or process. When combined with other evolving technologies with environmental implications such as 'energy foundations' that incorporate geothermal-retrieval systems into DFEs, these 'green' initiatives promise to alter the economics of deep-foundation alternatives further in the future.

In any event, with regard to hybrid piles, Chellis (1961) presents a good summary of the many variations that had been used in the U.S., at least, up to that point in time (circa 1960). There is probably very little new material with regard to development of hybrid piles subsequent to the circa-1960 cutoff date of this reference.

The relevance of hybrid piles to this monograph is that the most common type of hybrid pile in the U.S, at least, consisted of a lower portion that was a timber pile (often untreated to save even more money and considered technically acceptable at the time because the wood was embedded entirely in water-saturated soil) mated to an upper portion that was some type of CIP-PCC pile (usually using a constant-diameter, thin-wall steel shell). Note that considerable design and construction attention had to be paid to the splice detail between the two components, and there was, undoubtedly, not-insignificant on-site construction labor that had to be expended in order to effect this splice properly during the pile-driving process.

Because most hybrid piles had a component that was tapered, the overall installed pile was technically a partially tapered pile. Thus, hybrid piles are mentioned in this monograph so that foundation engineers who may be unfamiliar with such piles (they have not been widely used in the U.S., at least, in many decades and thus most foundation engineers currently in practice have not even been educated about them) are aware that they exist. As was noted earlier with regard to timber piles, although it is highly unlikely that hybrid piles would be used for new construction, it is certainly possible to encounter them when dealing with an older structure that may date from the first half of the 20th century.

Note that there is a situation that is sometimes encountered in current practice, especially with marine structures, that is visually related to hybrid piles that is noted here for the sake of completeness and to distinguish it from a 'true' hybrid pile. This is when some portion of an original pile (usually untreated softwood timber) just below and including the pile head has deteriorated over time, usually due to dry rot, and has been replaced with a piece of new material that is typically different from the original pile material. The replacement material is often simply a piece of chemically treated timber but it can be more elaborate such as replacing the entire pile cap with PCC. This repair technique is called *posting* and the repaired pile is referred to as a *posted pile*.

3.7 NON-DRIVEN ALTERNATIVES

3.7.1 Introduction and Overview

As established at the beginning of this monograph, the focus of this document is on impact-driven piles composed of wood, steel, and precast-PCC. This simply reflects the overwhelming historical usage to date with respect to both the types of tapered DFEs used in practice as well as the installation methodology that has been used. A consequence of this is that the vast majority of the resistance-forecasting R&D to date that has been done for tapered DFEs has focused on these variables.

That having been said, for historical completeness, this monograph includes at least mention of tapered DFEs that deviate from these norms. Alternative pile materials have already been discussed earlier in this chapter. The focus of the following sections is on installation methodologies other than impact driving.

The significance of installation methodology with respect to DFE resistance in general cannot be overemphasized for two important reasons. First and foremost is because it is now well-understood that the post-installation resistance of any type of DFE depends to a very significant extent on how it was installed or created in the ground. This understanding comes from one of the most significant geomechanics advances of the last several decades with respect to deep foundations which is an appreciation of the importance of the stress state in the ground both prior to and after DFE installation. This topic is discussed extensively in Chapter 4.

The second significant artifact of DFE installation methodology that was recognized at least as far back as the 1950s but seems to have been ignored in recent decades is the issue of *residual loads* from installation. Residual loads are axial forces locked-in to the shaft of a DFE due to the installation process, even when the applied axial load at the head of the DFE is zero. Note that the installation-induced *drag loads* (see Figure 2.1) that produce residual loads should not be confused or conflated with the drag loads that develop from settlement of the ground around a DFE relative to the head of the DFE (a completely unrelated geo-phenomenon) although the effect on a DFE is the same. The issue of residual loads is resurrected in this monograph and discussed at length in various chapters and appendices because available information indicates that residual loads are more significant for tapered (driven) piles than any other type of DFE.

Because of the combined influence of post-installation stress state and residual loads on DFE resistance, it cannot be emphasized too strongly that the resistance-forecasting methodologies discussed in Chapter 5 are based on knowledge gained from the installation of piles:

- that are constructed of traditional materials,

- have a traditional continuously or partially tapered geometry, and

- were installed completely by impact driving.

Any deviation from these criteria in terms of pile material, geometry, or installation methodology means that the resistance-forecasting methodologies established based on these criteria should not be used without due diligence to evaluate the effect of the deviation(s). This is particularly true when a deviation involves an installation methodology other than impact driving.

3.7.2 Jetting

As noted earlier in this chapter, what may have been the first tapered precast-PCC pile, the *Gilbreth* pile, was intended to be installed only by jetting. While it does not appear that this installation technique has ever been used with tapered piles other than the *Gilbreth* pile, it is possible that jetting might be used to partially install a tapered pile, with the final installation reverting to impact driving. In such a case, the portion of the pile shaft through the zone of jetting may have its resistance compromised.

3.7.3 Vibrating

Vibratory installation of foundation piling is not common, at least in the U.S. This is because resistance-forecasting of vibrated piles has proven elusive despite attempts from time to time over the years to develop a reliable forecasting methodology.

However, it is much more likely to use vibratory installation for a portion of the installation process then switch to a conventional impact hammer to drive a pile 'home'. This is apparently allowable with some construction specifications that only require some minimum pile embedment combined with some minimum driving criterion in terms of blows per some unit length for the final portion of pile installation. The writer has seen such a composite driving methodology used to install *TAPERTUBE* piles in the New York City metropolitan area just within the last several years.

Due diligence and good engineering practice suggest that such a composite installation strategy should be viewed with some caution unless and until research is done to evaluate the effect on pile resistance due to the vibratory portion of installation.

3.7.4 Cast-In-Place

Tapered DFEs consisting of CIP-PCC placed into an open hole in the ground have apparently been used in some countries, specifically the former Soviet Union (Kodikara and Moore 1993). This reference refers to such elements as "bored piles" so they would be called *drilled shafts* in U.S. terminology as it is assumed that they were constructed by placing fluid PCC under gravity as opposed to injecting grout under pressure. Other published work indicates that there has been some research devoted to tapered drilled shafts elsewhere, e.g. Khan et al. (2008) in Canada.

While there is nothing inherent in drilled-shaft technology that would preclude constructing tapered elements, this does not appear to be a globally common practice although it may have been done on a local or regional level. In the writer's opinion, it is not surprising that the use of tapered drilled shafts (as they will be referred to in this monograph to be consistent with U.S. terminology) has not been more widespread as part of the tremendous growth of drilled-shaft usage in general since the 1970s. This is when drilled shafts ceased being just a regional DFE alternative in the U.S. If nothing else, there is the obvious difficulty of creating a predictably tapered hole in the ground; keeping the geometry of this hole stable during the subsequent concreting phase; and then doing this repeatedly and reliably on a production basis on a real project as opposed to some research effort. One only has to look at the extensive history of trying to reliably create a conventional constant-diameter drilled shaft to see that creating a tapered drilled shaft on a repeatable basis would be problematic. There would also be the attendant difficulty of creating a tapered rebar cage and keeping it centered in the tapered hole during concreting.

3.7.5 Torqued-In-Place

Piles that have been torqued (screwed) into the ground have been used since at least the early 19th century with the development of the *screw pile* made of *cast iron* or *wrought iron*. They proved popular in shallow-water marine installations such as lighthouses underlain by relatively soft/loose bearing strata where conventional impact driving would have been difficult using the pile-installation technology available at that time.

In recent decades, the concept has been revived as what are now called *helical piles* and these have proven to be quite popular in terrestrial applications involving lightly loaded structures, including residential construction.

A very recent development (Fahmy 2015; Fahmy and El Naggar 2016a, 2016b, 2017a, 2017b, 2017c) along these lines is the helical tapered SCDI pipe pile that was noted earlier in this chapter for its non-traditional ductile-iron pile shaft. The helical tapered SCDI appears to be a novel extension of the more-common constant-diameter SCDI pile in two ways. One is that the shaft, which has a diameter that is substantially greater than today's helical piles and is much more reminiscent of a 19th-century screw pile, has a continuous taper to it, with taper angles, α, of the test piles reported in the literature ranging between 0.95° (that 'magic' value dating back a century or more to the *Raymond Standard* pile!) and 1.91°.

The second and more-unique aspect from an installation perspective is the inclusion of one helix that is attached to the exterior of the pile shaft just above the toe. This allows the entire pile to be torqued into the ground as opposed to driving the pile into the ground if it were just a tapered SCDI pile without the helix.

It appears that the reason for the inclusion of a helix with what amounts to an otherwise relatively generic spun-cast ductile-iron pipe pile (Fahmy and El Naggar tested constant-diameter versions, referred to here as a *helical SCDI pipe pile*) is primarily, if not solely, related to installation considerations. It appears that the helical tapered SCDI pipe pile is intended for use with relatively lightly loaded structures. The ability to install such piles by torqueing instead of impact driving allows for increased installation versatility and reduced installation costs, an important consideration when installation in potentially remote sites with difficult access is involved. As is well-known, the equipment required to torque a pile into the ground is relatively simple compared to impact driving when only relatively modest pile embedment and ULS resistance from the installed DFE is being sought.

3.8 EXPERIMENTAL AND PROPOSED ALTERNATIVE PILE GEOMETRIES

3.8.1 Introduction and Overview

With the exception of the unique bulb-shaped *TPT* pile, all of the known (to the writer) tapered piles that have been used commercially to-date have either been continuously tapered[67] (but not necessarily with a uniform taper angle) or partially tapered. The final section of this chapter considers two alternative tapered-pile geometries that have only been explored experimentally to date as part of academic research but, as a result, have been proposed or at least suggested for commercial use as alternatives to traditional tapered-pile geometries.

First discussed is an alternative partially tapered geometry that has received attention from more than one research group in recent years for its purported more-efficient

---

[67] Step-tapered piles such as the *Raymond Step-Taper* that are quasi continuously tapered are included in this category for the purposes of the present discussion.

development of resistance. The other alternative is a much more recent development of only the last few years and involves modifying the traditional geometry primarily for the purpose of installation considerations, not resistance.

3.8.2 Partially Tapered Alternative

Historically, partially tapered piles, which have existed at least since circa 1930 with the commercial availability of the *Monotube* pile, consist of a tapered lower section and constant-perimeter upper section or extension. Discussed here is an alternative partially tapered pile geometry where the lower section is the constant-perimeter component while the upper section is the tapered component. The genesis of such a pile geometry can be readily explained based on the writer's background research for this monograph over the last few years.

A number of published research studies into tapered piles were based, at least in part, on laboratory testing of small-scale model piles under 1-g conditions. Wei and El Naggar (1998) and Manandhar and Yasufuku (2013) are representative of such efforts. In general, the writer does not consider such 1-g testing of small-scale foundation models to be meaningful or useful due to the 'bowling-ball' effect[68] as well as the fact that model piles are generally installed in a test box in a manner that does not even remotely resemble real-world impact driving. This means that the stress state to which the model piles are subjected does not replicate the real-world stress state or residual loads of actual piles. Nevertheless, this does not mean that the results of such tests should be dismissed or ignored completely.

Based largely on measurements made in such tests, a number of researchers have come to the conclusion that, all things being equal, taper benefit is maximum under relatively low confining stresses in the ground. This implies that taper benefit is maximized with relatively shallow pile embedment. As a result, this has led some researchers to suggest that the optimal geometry of a partially tapered pile is to have the lower section be the constant-perimeter component while the upper section is the tapered component. Some, such as Manandhar and Yasufuku (2013), have gone so far as to perform laboratory tests on model piles with such a geometry.

The writer is not aware of any full-scale tapered piles with such a geometry that have been used in practice no less instrumented and tested under full-scale field conditions to allow a comparison to more-traditional tapered-pile geometries. Putting aside for a moment the issue of whether the manufacture and installation of such a pile geometry would be cost-effective compared to more-traditional tapered-pile geometries in particular and other types

---

[68] The writer credits this colloquial term to the late Prof. Stephen T. "Steve' Mikochik who was the writer's doctoral advisor at the then Polytechnic Institute of New York (currently the New York University Tandon School of Engineering). In an era prior to geotechnical centrifuge testing, when all laboratory testing of foundation models involved small-scale models and 1-g conditions but using normal (full-scale) coarse-grained soil, Mikochik likened the scale effects of such testing to a full-scale foundation element embedded in bowling balls. The image was intended to impress upon the student that such small-scale laboratory testing was inherently unrealistic because the relative sizes of the foundation element and soil particles did not scale up to the real world correctly. Consequently, the outcomes from such testing should always be viewed with some caution and skepticism. That the writer has remembered this analogy more than 40 years later is a testament to its effectiveness as an educational tool. It is of interest to note that, in reality, the analogy is not far from the truth, especially with respect to small-scale models of DFEs. The standard bowling ball is 8.5 inches (216 mm) in diameter and the model piles used by Manandhar and Yasufuku (2013) were approximately 12 inches (300 mm) in length. Thus, these model piles embedded in sand (as they were) would indeed scale up to a full-scale pile embedded in bowling balls.

of DFEs in general, it would appear reasonable to at least undertake research to investigate this alternative tapered-pile geometry in a controlled, scientific fashion. Such work could possibly be complemented and augmented by centrifuge testing and/or computer analyses that allowed for better approximation of full-scale confining stresses compared to the 1-*g* conditions that have been used for the most part to-date. Only after a meaningful assessment of the taper benefit of this alternative tapered-pile geometry to the traditional geometry of partially tapered piles is made can the necessary economic comparison be made considering manufacturing and installation costs.

3.8.3 Helical-Taper Hybrid

The helical tapered SCDI pipe pile (Fahmy 2015; Fahmy and El Naggar 2016a, 2016b, 2017a, 2017b, 2017c) has already been discussed in this chapter in the context of its use of spun-cast ductile iron as a non-traditional tapered-pile material as well as its non-traditional torqued-in-place installation methodology. This pile is discussed here again in the context that its addition of a single helix near the toe of the pile is the element that gives the pile its unique geometry compared to traditional tapered piles.

The addition of a single helix to an otherwise unremarkable spun-cast ductile-iron pipe pile is the structural element that in and of itself causes the resulting helical tapered SCDI pipe pile to have a unique geometry. Note that without the helix, the resulting pile would simply be either a constant-diameter ductile-iron pipe pile or a tapered ductile-iron pipe pile. The former has existed in the marketplace for several years now and is installed by impact driving. The latter is not known to have been produced and used commercially to date but would be a logical conceptual extension of a constant-diameter ductile-iron pile. It would also be installed using impact-driving. In any event, neither the constant-diameter nor continuously tapered geometries of these ductile-iron pipe piles would be unique.

Chapter 4

HISTORICAL DEVELOPMENTS AND CURRENT TRENDS IN GEOMECHANICS AND STRUCTURAL MECHANICS RELATED TO TAPERED-PILE BEHAVIOR

4.1 INTRODUCTION AND OVERVIEW

Since the first effort circa 1850 to forecast the post-installation ULS resistance of piles using an analytical methodology based on science and mathematics, there have been countless developments in geomechanics and structural mechanics that have continuously shaped and reshaped our understanding of what happens geotechnically and structurally when a pile is driven into the ground and then subjected to externally applied axial loading. These developments have, in turn, shaped and reshaped the analytical methodologies used to forecast the ULS resistance of tapered piles and DFEs in general.

As with any technological subject, the geomechanics and structural mechanics knowledge-base is constantly changing and evolving over time as the result of what the writer termed (back in 2001) the "trilogy of technology" (Horvath 2001) that has, at its root, fundamental research that sometimes turns out to be seminal and game-changing. Thus, this chapter is devoted to a presentation and discussion of historical developments as well as the writer's perception of current trends in both geomechanics and structural mechanics issues that are relevant, and in some cases game-changing, to tapered piles and DFEs in general.

The overall purpose of this chapter is to lay the theoretical groundwork for Chapter 5 that explores and critiques existing resistance-forecasting methodologies for tapered piles. In the writer's opinion, it is essential to first understand what should be incorporated into any modern analytical methodology for tapered piles before evaluating a methodology for potential use. The material presented in this chapter also lays the theoretical groundwork for a detailed schema for an improved resistance-forecasting methodology for tapered piles that is presented in Chapter 7.

As expected, most of this chapter is devoted to geomechanics issues that cover a wide range of topics, including resistance verification. This aspect of deep-foundation behavior is essential for evaluating the accuracy of resistance-forecasting methodologies and is, unfortunately, nowadays often trivialized and taken for granted. Issues related to structural mechanics and soil-structure interaction (SSI) between pile and ground that are also often overlooked are discussed at the end of this chapter. Note that very basic concepts are presented for some of the topics covered in this chapter, if only to establish common terminology and notation that is used throughout the remainder of this monograph.

4.2 GEOMECHANICS ISSUES

4.2.1 Soil Properties and Behavior

4.2.1.1 Introduction and Overview

One of the most significant changes affecting deep foundations in general that has occurred over time is the increased sophistication in the understanding of soil properties and soil behavior. This enhanced level of soil mechanics understanding has profoundly impacted both resistance-forecasting and resistance-verification methodologies for all types of deep foundations. Thus, it is useful to discuss the relevant advances in geomechanics that have made many of these forecasting methodologies possible.

It is important to note that this increase in theoretical understanding is strongly and synergistically linked to the dramatic changes in in-situ testing technologies during roughly the same timeframe of the last several decades. The changes related to in-situ testing have occurred along several parallel developmental tracks:

- Drastically increased understanding of traditional methodologies that date back to the earliest days of modern soil mechanics such as the Standard Penetration Test (SPT) and the parameter(s) measured therein, in this case the SPT N-value. In the case of the SPT, it has been the detailed understanding of how many aspects related to the performance of this test, such as the mechanical efficiency of the drive system, affect the measured N-value.

- Enhanced capabilities of traditional methodologies such as the cone penetrometer. From the 1930s to the present, this device has progressed from mechanical to electronic in terms of data acquisition and has progressively added to the suite of parameters measured (tip resistance, sleeve friction, pore pressure, etc.).

- The development of entirely new devices such as the *pressuremeter* (PMT) and *dilatometer* (DMT).

There are three general areas in which this evolutionary improvement in the state of soil mechanic knowledge has occurred:

1. stress-state in the ground with regard to horizontal/radial stress,

2. shear strength, and

3. stiffness (modulus).

Each is discussed in an introductory, overview manner in the following sections, with further discussion of specific aspects later in this chapter.

4.2.1.2 Stress State

The influence of horizontal stress in the ground, what is nowadays part of what is called the *stress state* in the ground, on deep-foundation resistance has been hypothesized since the beginning of what the writer calls 'modern' foundation engineering in the 1950s. The years following the Second World War marked the beginning of concentrated efforts to forecast the behavior of both shallow and deep foundations using a wide range of empirical and theoretical methodologies that all had the then-relatively-new technology of modern soil mechanics as their underlying basis. Note that modern soil mechanics was still somewhat of a novelty circa 1950 (as late as the 1960s, many undergraduate programs in civil engineering in the U.S. did not require even an introductory course in the subject) as it had its origins only a couple of decades earlier, in the 1920s and 1930s. By the late 1930s, soil mechanics was beginning to be recognized on a global basis but the onset of global war in 1939 temporarily suspended further global development until the late 1940s.

This broad understanding of the influence of horizontal stress on deep-foundation behavior has advanced in several specific areas since the 1950s:

- It is now well-established and broadly acknowledged that the depth-wise variation in horizontal effective stress, σ'_h, acting along a DFE shaft after installation can be defined in terms of the vertical effective overburden stress, σ'_{vo}, and a lateral earth pressure coefficient, K_h. The concept of a 'limiting vertical effective stress', which gives rise to a 'limiting horizontal effective stress', that was in vogue for a period of time in the latter decades of the 20th century (e.g. Poulos and Davis 1980) has long since been debunked by several researchers (Kulhawy 1984, Fellenius and Altaee 1995).

- It is now appreciated that horizontal stress influences <u>both</u> traditional resistance mechanisms of DFEs:
  - shaft resistance due to sliding friction and
  - toe resistance due to end bearing.
  Each of these resistance mechanisms is discussed separately in greater detail later in this chapter. This represents a profound and significant paradigm shift as historically it was assumed that toe resistance was a function of vertical effective stress.

- Horizontal stress plays a significant role in the recently defined third DFE resistance mechanism that the writer has termed *wedging* that is unique to the shaft resistance that develops along the tapered portion of a DFE in general and (driven) pile in particular.

- The coefficient of lateral earth pressure at rest, K_o, that exists prior to the installation of a DFE is now understood to play a significant role in deep-foundation resistance. Kulhawy (1984, 1991) showed that the coefficient of lateral earth pressure, K_h, acting along the shaft of a DFE <u>after</u> installation was a function of:
  - K_o <u>before</u> installation and
  - the specific type of DFE.
  DFE typology is a metric of the installation process that can either cause stress increase or stress relief in the ground around the installed DFE, with the overall net stress change related to installation of the DFE being the direct influencing factor. Kulhawy found it convenient to express the stress-related behavior due to DFE installation methodology using the dimensionless ratio K_h/K_o that was correlated to deep-foundation type and, in the case of piles, relative diameter. Unfortunately, all of the research to date to determine K_h/K_o ratios for different DFEs has focused solely on constant-perimeter DFEs. Therefore, at present there is a critical knowledge-gap on this topic for tapered DFEs in general and tapered piles in particular. Nevertheless, as will be seen, use of the K_h/K_o concept is a key and recurring concept throughout this monograph.

- One complication with dealing with K_o as part of an analytical methodology is that for overconsolidated soils there are different values for unloading and reloading. This understanding was an outcome of doctoral research by Prof. Raymond B. Seed in the early 1980s and is reflected in a very complex soil model he developed and used as part of subsequent published papers dealing with soil compaction (Duncan and Seed 1986, Duncan et al. 1991). Installation and subsequent structural loading of a DFE can involve both unloading and reloading depending on the specific situation. Unfortunately, despite the great advances made with estimating K_o using various in-situ testing devices, especially the cone penetrometer, the distinction between the unloading and reloading values of K_o is typically not made at the present time in the various empirical correlations that have been developed to relate data obtained in some in-situ test such as the cone penetrometer or DMT to K_o.

In summary, the significant advances in soil mechanics knowledge in recent decades have demonstrated that the pre-installation value of K_o is an essential soil property for understanding and estimating the resistance of all types of DFEs as it reflects the pre-installation stress state and starting point for the post-installation horizontal effective stress, σ'_h, acting both along the shaft and around the toe of a DFE. Advances in site characterization based on in-situ testing now allow estimates of K_o to be made in all types of soil on a routine basis although defining separate unloading and reloading values of this parameter is not yet routinely feasible.

4.2.1.3 Shear Strength

There have been equally significant advances in the understanding of the shear strength of soils, with the most fundamental being the appreciation that effective stress governs the shear strength of all soils under all loading conditions. This is true even if an alternative conceptual approach of interpreting shear strength, e.g. the common total-stress interpretation used with fine-grained soils under undrained conditions, is used. When this fundamental effective-stress behavior is viewed within the context of the Mohr-Coulomb failure criterion, this means that the *friction angle (angle of internal friction)*, ϕ, is the fundamental soil-strength property[69].

Of greatest relevance to resistance forecasting for deep foundations in general is the fact that it is now well-established that ϕ is never a constant in any application but varies as a function of what is called the *operative stress level* in a given application. This was shown incidentally in Chapter 2 as an exemplar of secant vs. tangent parameter assessment and is discussed more formally in detail later in this chapter. In addition, ϕ is also dependent on strain as it relates to the displacement necessary to mobilize shear strength.

The behavioral fact that ϕ is both stress- and strain-dependent is gradually being assimilated into practice and implemented into newer analytical methodologies for resistance forecasting. However, older analytical methodologies that are still widely promoted and used were based on the assumption of a fixed, single-valued ϕ for both shaft and toe ULS resistances. As a result, there is substantial mental inertia to overcome to get practitioners to embrace newer analytical methodologies that invariably require more work.

4.2.1.4 Stiffness

Historically, resistance forecasting for all types of DFEs was solely strength-based as design was based solely on calculations of the ULS total resistance with appropriate factors then applied in order build some level of 'safety' into the system. This was true for traditional designs based on the *Allowable Stress Design* (ASD[70]) concept with safety factors and even more so in the more recent *Load and Resistance Factor Design* (LRFD[71]) concept. There were some attempts early on in the 1950s to develop simple empirical relationships for the

---

[69] The alternative notation ϕ' is often used for the friction angle, especially when referring to fine-grained soils. This alternative notation is not used in this monograph.

[70] Alternatively called *Working Stress Design* (WSD).

[71] Alternatively called *Ultimate Strength Design* (USD).

settlement of a single DFE under 'working' (service) loads but these never caught on with practitioners or design/building codes. As a result, in most cases in practice, especially those involving terrestrial applications, the stiffness (modulus) of the ground around a DFE did not enter into the picture in any way as deep-foundation settlements under any load level were not explicitly considered and were always implied to be 'small', i.e. tolerable by the structure being supported, to the point of being insignificant.

Several attempts have been made over the years to develop analytical methods for forecasting load-settlement behavior, both for deep foundations in general as well as specific types of deep foundations, e.g. drilled shafts (bored piles). The more theoretically rigorous of these methods require site-specific, single-valued estimates of Young's modulus, E. However, none of the all-purpose load-settlement methodologies ever entered the mainstream of usage, at least in the U.S., for reasons discussed in detail in Chapter 5.

This situation is now changing. In fact, it has already changed profoundly over the last decade. This is due to a 'perfect-storm' convergence of concepts that are also discussed in detail in Chapter 5.

In brief, this paradigm shift has been largely due to the ability to relatively easily and routinely measure the site-specific small-strain shear modulus, G_{max}, using sCPTu soundings. This site-characterization capability, when combined with theoretical solutions reformulated to use shear modulus, G, instead of Young's modulus, E, and, more importantly, the introduction of relatively simple-to-use modulus-degradation (reduction) models, has resulted in the ability to forecast the load-settlement behavior of any type of DFE using nothing more than a spreadsheet calculation.

That having been said, at this point in time ability has far exceeded application with respect to routine load-settlement forecasting for DFEs in general and tapered piles in particular. This is due to the fact that, more than ever, design codes based on LRFD principles focus on strength as opposed to settlement.

4.2.2 Shaft-Resistance Mechanisms

4.2.2.1 Background and Overview

The single most significant geomechanics issue that has impacted the way in which the axial-compressive load-settlement behavior of tapered piles is analyzed has been the evolution of the hypothesized physical mechanism by which shaft resistance is mobilized as a tapered pile is both driven as well as loaded after installation. All the other geomechanics issues addressed in this chapter have just refined or fine-tuned this key, central issue. This is not surprising as identifying the physical mechanism by which any object reacts to and distributes applied loads is the first step toward creating a free-body diagram of forces and from this visualizing, postulating, and developing a physical behavioral model and concomitant mathematical solution. This process has been used time and time again for a wide variety of problems in both civil and mechanical engineering.

There were likely some broad, intuitive perceptions that date back to antiquity as to how piles in general develop resistance once installed in the ground. If nothing else, it is clear that the ancients understood the basic concept of a column as being a critical load-bearing component of a structure given the substantial use of columns as support elements for masonry structures of great size and cultural significance in many ancient cultures and civilizations. The same can be said about an arch and dome. It is, however, debatable as to whether in antiquity there was a perception that tapered piles developed resistance in some unique manner as tapered piles were the only 'game in town' in terms of DFEs.

The earliest published work with which the writer is familiar that specifically addressed pile resistance in general dates back to the middle of the 19th century. It is clear that as perceptions concerning pile resistance crystallized and were then formalized in published work throughout the latter half of the 19th century that they universally coalesced around the concept that post-installation pile resistance could be related to a physical model based on the physical mechanism of work done to the pile during driving. In simple, colloquial terms, if a certain amount of energy in the form of work is put into driving a pile into the ground, then that pile 'owes' energy in return afterward in the form of axial resistance. Note that a key implication of all of these simplistic work-based concepts for pile resistance is that pile geometry, specifically tapered vs. constant perimeter, plays no role.

These work-based conceptual and theoretical efforts culminated about 100 years later with the use of the 1-D wave equation that remains the current state of knowledge for the pile-driving process. In simple terms, the 1-D wave equation still makes use of the broad concept of relating pile-driving energy/work to pile resistance but in a manner that is theoretically and analytically much more sophisticated and closer to reality that the 19th century concepts. However, none of the 1-D wave-equation formulations that the writer has seen implemented to date properly account for the unique resistance mechanism of tapered piles and, therefore, do not, in the writer's opinion, come close to correctly and accurately modeling what occurs when driving tapered piles.

It was not until the evolution of modern foundation engineering in the 1950s (approximately 100 years after the work-related model for pile resistance initially appeared) that physical mechanisms and associated physical and mathematical models, theories, and solutions for what happens during the nominally static loading of a DFE post-installation began to be developed based on the framework of engineering mechanics combined with modern soil mechanics. It is relevant to note that these 'static' analytical methodologies based on soil mechanics did not replace the earlier work-/energy-based 'dynamic' methodologies. Rather, these two very different concepts for how static pile resistance can be forecast co-existed and continue to co-exist in the present. They are simply two different conceptual ways to view the same thing.

In the writer's opinion, in current foundation engineering practice these two conceptual approaches for resistance forecasting, static and dynamic, each play a crucial role and thus complement each other. However, in the writer's experience, at least some design professionals and academicians have developed a distinct bias of, or preference for, one approach over the other, almost to the exclusion of one approach. This is important to recognize when reviewing the published literature as published works naturally tend to reflect the preferences and biases of the author.

Whether viewed from a static or dynamic perspective, more than a century of research into how a pile behaves axially under both driving and post-driving conditions has coalesced around the two now-familiar traditional resistance mechanisms of *shaft resistance* and *toe resistance* shown in Figure 2.1. For reasons that will become clear shortly, the shaft-resistance mechanism depicted in Figure 2.1 is hereinafter in this monograph referred to as *shaft resistance due to sliding friction*.

A key precept put forth in this monograph is that the two traditional resistance mechanisms shown in Figure 2.1 do not adequately define the behavior of tapered piles. Rather, at third resistance mechanism must be recognized as well. However, in order to understand this third, unique behavioral mechanism of tapered piles and tapered DFEs in general that is presented and discussed subsequently, it is necessary to have a clear understanding of the two traditional resistance mechanisms, especially shaft resistance due to sliding friction. This discussion will also serve to present additional terminology and notation used throughout the remainder of this monograph.

To achieve these goals, first discussed in broad terms are the basic physical concepts for the mechanisms involved in DFE resistance in general and the generic equations that define these mechanisms. This is followed by a presentation of the historical evolution of specific hypotheses and research outcomes used to quantify the terms in these equations.

4.2.2.2 Basic Physical Concepts

For all types of DFEs, the mechanism of shaft resistance due to sliding friction has historically been assumed to be a straightforward 'textbook' application of the simple physical mechanism called *Coulomb dry friction*[72]. This is one of the most basic concepts in physics and engineering mechanics that defines the force system that develops between two materials (usually dissimilar) due to relative displacement (sliding) along an assumed zero-thickness material interface (planar in this case) under both non-displacing (static) and kinetic conditions[73]. A frictional resisting force[74], F, is assumed to develop (up to some finite limiting value, F_{max}) along and parallel to the planar interface in proportional response to resist an applied or driving force, T, that also acts along and parallel to the planar interface. Note that because both F and T are force vectors, in addition to a magnitude they each have a direction, with one always acting opposite the sense of the other in this case.

As an aside, the absolute directions of these two opposing forces, F and T, as applied to the free-body diagram of forces in an application are often important and deep foundations are no exception. With reference to Figure 2.1, the exterior of the DFE shaft is assumed to be the planar surface on which Coulomb dry friction develops. Whether the resisting force, F, acts upward and produces resistance or downward as drag load depends on the assumed sense of the relative displacement between the DFE and adjacent ground that will occur under the assumed nominally static conditions up to the ULS resistance.

Under axial-compressive loading, a DFE is generally assumed to move downward relative to the surrounding ground so that the developed frictional resistance acts upward providing what is called *positive shaft friction* that translates into *resistance due to sliding friction*. However, if the surrounding ground moves downward relative to the DFE <u>or</u> the DFE moves upward relative to the ground (as discussed subsequently, these are two very different phenomena that happen to produce the same outcome), the developed frictional resistance acts downward as what is called *negative shaft friction*. This produces a *drag load* on the DFE shaft that can have both DFE-settlement and structural-load impacts on the DFE.

Before proceeding further, it is essential to note that this classical assumption that the relative displacement between the shaft of any type of DFE with a depth-wise constant perimeter and surrounding ground occurs along a nominally zero-thickness shear surface defined by the exterior of the DFE shaft has been known to be incorrect for some time now (see the discussion in Appendix A). In reality, this relative displacement results from vertical shearing within a finite-thickness zone of soil adjacent to the shaft that can be visualized as a shear-band surrounding the shaft of the DFE.

---

[72] This term is somewhat of a misnomer as the interface along which this mechanism develops does not have to be physically dry.

[73] The static case is, in reality, only approximately and nominally non-displacing. Relative displacement between materials always occurs in reality in order to mobilize frictional resistance but it is neglected quantitatively in many basic problems in which Coulomb dry friction is assumed to develop. However, in geotechnical applications involving DFEs it should always be considered at least qualitatively for reasons discussed subsequently.

[74] If this force is distributed over the interface area, the resistance can be visualized as a shear stress.

For example, Fellenius (2018) noted (writer's comments highlighted in yellow):

"...the 'movement' [Fellenius is referring to the axial displacement of the DFE shaft relative to the adjacent ground.] is not a slippage, i.e., definite sliding of the pile element against a stable body of soil, but occurs as a shear deformation within some zone or band of soil next to the pile element. Therefore, the movement along the side of the element is the relative movement between the pile element surface and a [sic] outer boundary or shear zone, a somewhat undefined location, actually."

Because of the difficulty in defining the thickness of this shear-band as noted by Fellenius and dealing with this ill-defined shear-band analytically based on the current state of knowledge, the classical assumption and concomitant model of a nominally zero-thickness shear surface as defined above remains solidly entrenched in both routine practice as well as most academic research. Consequently, this classical interpretation of the traditional resistance mechanism of shaft resistance due to sliding friction is used throughout the remainder of this monograph solely as an analytical simplification that is known up front to be overly simplistic and inherently incorrect.

In any event, as noted above, the resisting force, F, is assumed to have a limiting value, F_{max}, under which static conditions can be maintained. Consequently, if the driving force, T, exceeds this limiting value there will be unlimited relative displacement parallel to the interface and kinetic conditions will develop.

It is important to note that a non-zero value of frictional resistance will still exist along the interface under kinetic conditions although it is allowed to have value that differs from the static condition. This is one of the key conceptual elements of Coulomb dry friction, that when 'failure' occurs along the interface, i.e. $T > F_{max}$, the frictional resistance does not drop to zero as when a solid material ruptures and loses all strength. Rather, there will always be some frictional resistance that is maintained along the interface for as long as kinematic conditions exist.

In the original formulation of Coulomb dry friction, F_{max} is postulated to be a function of the normal force, N, acting on the interface and the *coefficient of friction*, μ, that is a system property and thus a function of the two interface materials involved. In light of subsequent discussions, it is important to note that N is assumed to remain constant throughout all ranges of loading under static conditions and even if the interface transitions from static to kinetic behavior.

In geotechnical engineering, the coefficient of friction parameter, μ, has historically been terminologically and notationally replaced by its mathematical equivalent, the tangent of the *interface friction angle*, δ. These assumptions can be stated mathematically as:

$$F_{max} = \mu \cdot N = \tan \delta \cdot N \,. \tag{4.1}$$

Furthermore, in geotechnical applications the interface-resistance model expressed in Equation 4.1 was long ago extended using Mohr-Coulomb strength parameters to allow for the ULS resistance between two interface materials to have both a <u>stress-dependent</u> component based on δ (which is a carryover from the original Coulomb dry friction model) and a <u>stress-independent</u> component, C_a, that is usually called the *adhesion*[75]. Equation 4.1 is then replaced by:

$$F_{max} = C_a + (\tan \delta \cdot N) \, . \tag{4.2}$$

In keeping with the terminology adopted for this monograph and as illustrated in Figure 2.1, the overall parameter notation $R_{(ult)}$ is used for ULS resistance. The specific components of resistance and their subscript notation are discussed subsequently.

To summarize up to this point, the traditional shaft-resistance mechanism as applied to all DFEs with a depth-wise constant perimeter is assumed to be one of pure sliding friction. With reference to the notation shown in Figure 2.1, this component of shaft resistance is given the notation of R_{ss}, with $R_{ss(ult)}$ being the ULS value. It is necessary to make and emphasize this clear distinction for understanding the uniquely different shaft-resistance mechanism for the tapered portion of a DFE that is presented and discussed subsequently.

The component of shaft resistance from the tapered portion of a DFE is referred to in this monograph in a generic fashion as the *shaft resistance due to taper*, with the notation R_{st} and $R_{st(ult)}$. This is because the physical mechanism by which this resistance component is developed has undergone considerable conceptual evolution since analytical methodologies for forecasting DFE resistance based on the static approach began to appear in the 1950s. Therefore, it is pragmatic for the purposes of this monograph not to limit the terms used for R_{st} and $R_{st(ult)}$ to a specific physical mechanism. The subject of the resistance mechanism for the tapered portion of a DFE requires considerable discussion on its own later in this chapter.

The total shaft resistance, R_s, is the sum of R_{ss} and R_{st}. However, it is important to note that it is not a given that these two components of shaft resistance will each reach their ULS values at the same magnitude of displacement. The genesis of this caution is that the resistance mechanisms for these two shaft-resistance components are completely different. Thus, in principle, there is no theoretical reason why they should be mobilized at the same rate and it follows that there is no theoretical reason why these two components of shaft resistance should reach their ULS values at the same magnitude of displacement. In fact, there is compelling evidence that R_{st} does not reach a ULS value in any realistic application.

Note that the issue of whether the two shaft-resistance mechanisms mobilize and subsequently maximize at the same rate is conceptually identical to the well-known fact that the two traditional resistance mechanisms for a constant-perimeter DFE of shaft resistance due to sliding friction and toe end bearing are physically very different and thus mobilize resistance at very different rates. Consequently, simply summing these two traditional resistance mechanisms together to arrive at a ULS resistance, R_u, for the entire DFE for design purposes is now understood to be something that should be done only with a full understanding of the implied settlement magnitudes involved. In fact, Fellenius (2018) specifically noted this simple summation of shaft (due to sliding friction) and toe resistances is one of the "fallacies" of routine foundation engineering practice.

With regard to toe resistance, R_t and $R_{t(ult)}$, historically it has been assumed to be a single-valued result from a bearing-capacity type mechanism similar to shallow foundations that is based on some theory-of-plasticity solution. However, over the years there has been both evolutionary and revolutionary thinking about the toe-resistance mechanism with the result that the current state of knowledge is somewhat complex. Consequently, the subject of toe resistance warrants its own separate discussion later in this chapter.

---

[75] 'Adhesion' is typically seen in geotechnical literature using the notation c_a, which is a <u>stress</u> term. Because Equation 4.2 is expressed in terms of forces, the notation used for the adhesion <u>force</u> is capitalized to emphasize and highlight the conceptual difference.

Central to the resistance mechanisms of shaft resistance (in general) and toe resistance are the parameters of *unit shaft resistance*[76], r_s, and *unit toe resistance*, r_t, as shown in Figure 2.1. These parameters have stress units so when integrated over the shaft and toe areas respectively yield the aforementioned resistance forces, R_s and R_t. Note that as with the resistance <u>forces</u>, R, the unit resistance <u>stresses</u>, r, are not necessarily those that occur under ULS conditions. Depending on the context, they may be intermediate between zero and the ULS value. Consequently, when ULS values are intended, they are so noted.

Again, for clarity in subsequent discussions in this monograph there is a distinction made between the unit shaft resistance developed along the constant-perimeter and tapered portions of a DFE. The former is defined as the *unit shaft resistance due to sliding friction*, r_{ss}, and the latter as the *unit shaft resistance due to taper*, r_{st}.

As noted in the introductory overview in Chapter 1, from the circa-1950s beginning of the development of analytical methodologies for forecasting DFE resistance under post-installation static loading conditions there have been two distinctly different schools of thought as to how to evaluate r_s and r_t in practice (Niazi 2014). One is the *indirect method*[77]. It is a two-step procedure that first defines r_s and r_t using algebraic relationships based on fundamental soil-mechanics principles and fundamental soil properties for strength and stiffness. In the second step, some combination of application-specific in-situ and laboratory test results are used to assign values to the fundamental soil properties in these algebraic relationships which are then solved. The indirect method is explored in this chapter in broad terms and in much more specific terms in Chapter 5.

The other school of thought is called the *direct method*. It is a one-step methodology that uses measured values in some in-situ test to directly produce estimates of r_s and r_t based on empirical relationships that were presumably developed from a database of field measurements that included actual resistance measurements on DFEs. Although the direct method has existed contemporaneously with the indirect method since the 1950s, the direct method has seen explosive growth in recent decades, largely as a result of the dramatic increase in use of the cone penetrometer in all of its forms (CPT, CPTu, sCPTu). The direct method is noted in this chapter only for its historical context but is explored in considerable detail in Chapter 5.

As is often the case, there are pros and cons to both the indirect and direct methods. The writer's position in this monograph is to be as objective as possible about each approach and allow design professionals to make their own decisions in practice.

That having been said, the writer has found that the indirect method offers the advantage of more clearly illuminating all the factors that go into and influence the development of a component of resistance. On the other hand, the direct method is basically a 'black box' that simply produces results with little or no insight into the underlying theoretical behavior. More importantly, the user of a direct method generally has no knowledge about the breadth of the database that was used to develop a particular direct methodology. This is very important because a direct methodology can only be as good or useful as the database that was necessarily used to develop it. Thus, it is all too possible for an end user of a direct method to use the method in a completely inappropriate application that was not covered by the method's original database yet still get a forecast outcome that

---

[76] Note that r_s is the same as the parameter f_s that is commonly seen in the published literature and referred to by different names such as *unit shaft friction* and *unit skin friction*. To avoid confusion and ambiguity, in this monograph the notation f_s is used exclusively for the sleeve friction measured using a cone penetrometer (CPT, CPTu, or sCPTu).

[77] This is also called the *rational method* in the published literature. The term 'indirect method' is used in this monograph.

might be grossly incorrect. This is emphasized here as it is especially relevant for tapered piles.

The writer's opinion is that the indirect method offers a distinct advantage when new concepts are being explored as all the behavioral factors are clearly and explicitly identified. On the other hand, the direct approach tends to be useful for a mature technology where the underlying theory is reasonably understood and there is simply a desire to produce end results efficiently and quickly in routine practice.

Thus, as will be seen, the indirect approach is strongly favored in this monograph in order to more clearly elucidate all the behavioral factors and issues involved in the total resistance of tapered piles. That is not to say that a direct method or methods for tapered piles could not be developed and, indeed, some have (although they will be seen to produce results of poor or at least questionable accuracy). However, in the writer's opinion, the development of a reliable direct method or methods for use with tapered piles is well in the future and appropriate only after the necessary databases have been developed.

It is important to note that the definitional boundaries of what constitutes an indirect method have been extended over time to include methodologies that are combinations of soil mechanics principles and empiricisms. The empirical aspects of such extended indirect methodologies typically combine two or more fundamental soil and system properties together into one, usually dimensionless, variable. This variable is typically quantified in the published literature based on back-calculated load tests on actual DFEs and is often presented as a range of values from which the design professional must pick a single value.

The writer's logic for defining such hybrid analytical methodologies as indirect, as opposed to direct, methods is based on two facts:

- The overall methodology still retains theoretical elements which a direct method always lacks completely.

- The empirical component derives from load tests on actual DFEs, not in-situ tests on the ground component as is always the case with direct methods.

These characteristics are illustrated using an example later in this chapter and such extended, hybrid indirect methods are explicitly noted when presented and discussed in Chapter 5.

For the discussions of indirect and direct methods that follow, it is necessary to elaborate first and foremost on the unit shaft resistance, r_s, shown in Figure 2.1. This is essential for understanding the basis of the indirect method and is also useful for understanding what problem-parameters are incorporated into the myriad empirical relationships that collectively comprise the direct method.

Using an effective-stress formulation and neglecting adhesion, it is typically assumed that:

$$r_{s(ult)} = \tan \delta \cdot \sigma_h'$$
(4.3)

where δ = the interface friction angle along the soil-pile interface and σ_h' = horizontal effective stress[78] acting on, and perpendicular to, the DFE shaft. Note that this is essentially just a stress formulation of the force-based equation (4.1) that defines the Coulomb dry friction mechanism.

---

[78] For DFEs with a circular cross-section, this is sometimes alternatively referred to as the *radial effective stress*, with either the same notation or the notation σ_r' used. For the purposes of this monograph, σ_h' and σ_r' are interchangeable in this application.

Note also that the same basic definition presented in Equation 4.3 for $r_{s(ult)}$ can also be used separately for both $r_{ss(ult)}$ and $r_{st(ult)}$. However, beyond this point the definitions for $r_{ss(ult)}$ and $r_{st(ult)}$ diverge due to the very different ways in which the σ'_h term develops in each case. The traditional shaft-resistance mechanism due to sliding friction and the concomitant derivation of $r_{ss(ult)}$ are discussed first.

To being with, it is assumed that:

$$\sigma'_h = K_h \cdot \sigma'_v \qquad (4.4)$$

where K_h = a dimensionless lateral earth pressure coefficient and σ'_v = the vertical effective stress adjacent to the DFE shaft.

There can be further insightful breakdown of the K_h parameter but a detailed discussion of this is deferred to later in this chapter where the subject of the influence of horizontal stresses on DFE resistance (which was noted in general terms earlier in this chapter) is discussed in detail in a separate section. However, in simple terms for the purposes of the present introductory/overview discussion, it is assumed that for constant-perimeter DFEs that K_h only reflects the effect of installing the DFE in the ground. Thus, K_h is assumed to remain constant under post-installation axial load application.

Over the past several decades, there has been a back-and-forth debate about evaluating the σ'_v term. For a while in the latter decades of the 20th century, the concept of a 'critical depth' with a 'limiting vertical effective stress' was in vogue (Poulos and Davis 1980) but this concept has now been soundly debunked (Kulhawy 1984, Fellenius and Altaee 1995). Consequently, nowadays it is generally accepted that $\sigma'_v = \sigma'_{vo}$, i.e. the vertical effective overburden stress adjacent to the DFE shaft.

Combining Equations 4.3 and 4.4 together with the restriction that this is applicable to the mechanism of shaft resistance due to sliding friction only and thus the constant-perimeter portion (if any) only of a tapered pile yield:

$$r_{ss(ult)} = K_h \cdot \tan \delta \cdot \sigma'_{vo} . \qquad (4.5)$$

As an aside, with this generic equation (4.5) it is easy to see the underlying conceptual difference between resistance-calculation methodologies based on the indirect method vs. the direct method for resistance forecasting. With the indirect method, all of the parameters on the right-hand side of Equation 4.5 have to be evaluated independently and explicitly on an application-specific basis. The most challenging parameter to quantify is always K_h.

On the other hand, with the direct method the entire right-hand side of Equation 4.5 is replaced with some empirically derived algebraic relationship that involves, say, SPT N-values or one or more parameters that are either measured in (q_c, f_s, u_2, V_s) or derived from measured results of a cone-penetrometer sounding. Note, however, that the effects of each of the parameters on the right-hand side of Equation 4.5 must still be reflected in the empirical relationship used in a direct method which is why direct methods often have slightly different algebraic formulations depending on the specific type of DFE being analyzed as this will impact both the K_h and δ terms.

As a further digression, Equation 4.5 also presents an opportunity to illustrate how the definitional boundaries of what constitutes an indirect method have been extended over the years to include hybrid methodologies that have both theoretical and empirical components. Perhaps the most common of these hybrid indirect methods is the β *Method* that has long been championed by Dr. Bengt H. Fellenius, P.Eng. (Fellenius (2020) contains the most recent version) wherein the theoretical term $(K_h \cdot \tan \delta)$ is replaced by an empirical dimensionless parameter called β:

$$r_{ss(ult)} = K_h \cdot \tan \delta \cdot \sigma'_{vo} = \beta \cdot \sigma'_{vo} . \tag{4.6}$$

Because the vertical effective overburden stress still appears in the final result expressed in Equation 4.6, the writer considers the β Method to be an indirect method as opposed to a direct method. Furthermore, selection of the β value to use is left up to the design professional and is not dictated by some empirical correlation with, say, cone-penetrometer data.

Returning now to examination of Equation 4.5, the values for the δ parameter are most commonly reported the published literature as the dimensionless ratio δ/ϕ for a specific DFE material in contact with a general type of soil, e.g. 'sand on steel', where ϕ = the Mohr-Coulomb friction angle of the soil. Historically and to the present, these δ/ϕ values are based on laboratory tests performed using some type of direct-shear (shear-box) apparatus although alternative testing methods such as the ring-shear device are sometimes used when testing to relatively large strain levels is desired. Regardless of the specific type of laboratory test, an important test variable in addition to the two different materials involved is the normal stress applied to their interface as the test is conducted.

Unfortunately, use of a single-valued δ/ϕ ratio that ignores any variation due to normal-stress magnitude dates back to an earlier era in foundation engineering when it was at least implied, even if not stated explicitly, that a given soil had a singular, unique value of ϕ. Of course, it has long been recognized that this is incorrect and that a given soil can exhibit a range of ϕ values between some peak value, ϕ_{peak}, and a value associated with the constant-volume/critical state condition, ϕ_{cv} (ϕ_{cs}). This was illustrated in Figure 2.3 as the exemplar for defining secant vs. tangent parameter values.

Although ϕ_{cv} for a given soil can usually be assumed to be constant in magnitude, ϕ_{peak} never is. The relationship between ϕ_{peak} and ϕ_{cv} is typically expressed as:

$$\phi_{peak} = \phi_{cv} + \phi_d \tag{4.7}$$

where ϕ_d = the dilatancy angle, the value of which depends on the operative stress level. In this context, the operative stress level would be the normal effective stress applied to the material interface in the direct-shear or ring-shear test. This means that whenever δ/ϕ for some material-interface test is reported, details are required as to operative stress value at which the test was performed and whether the peak or constant-volume conditions are represented.

In an ideal world, the value of δ in Equation 4.5 should be replaced by:

$$\delta_{operativ} = \left(\frac{\delta}{\phi}\right)_{operative} \cdot \phi_{operative} \tag{4.8}$$

where the subscript 'operative' means the measured laboratory results for a specific normal stress that is the same as that assumed to be acting on the DFE-soil interface as well as a specific point along the stress-displacement curve (i.e. peak, constant volume, or somewhere in between). However, this is an unrealistic expectation as it would require an extensive suite of laboratory tests to be performed for every project as well as complicated analyses to make full use of the results. Given all the other variables and uncertainties involved in forecasting

DFE resistance, a simpler, more practical approach is required when calculating the ULS resistance of a DFE.

First of all, it is assumed that in routine practice δ/ϕ values from the published literature would be used. It is further assumed that these are end-of-test values after relatively large displacements have occurred so can be assumed to be values representative of the constant-volume condition. Now, it is well-established that the sliding-friction shaft resistance of deep foundations in general is fully mobilized at relatively small displacements (fraction of one inch or several millimetres). This is a displacement level at which only a small fraction of toe resistance has been mobilized. Because resistance calculations are typically made for the ULS condition when a much larger fraction of the toe resistance would be mobilized, it is reasonable to assume that a constant-volume condition would exist along a DFE shaft undergoing sliding friction by that point. Consequently, it is reasonable to assume that would be the operative stress state for sliding-friction shaft resistance in calculations.

Putting all of these concepts together, for the purposes of this monograph Equation 4.5 can be expanded on as follows to define the ULS unit shaft resistance due to sliding friction, $r_{ss(ult)}$:

$$r_{ss(ult)} = K_h \cdot \tan\delta \cdot \sigma'_{vo} = K_h \cdot \tan\left[\left(\frac{\delta}{\phi}\right)_{cv} \cdot \phi_{cv}\right] \cdot \sigma'_{vo} \ . \tag{4.9}$$

Considering next the unit shaft resistance due to taper, r_{st} (not assumed to be the ULS value for now), the most general form of the equation defining this parameter is:

$$r_{st} = \tan\delta \cdot \sigma'_h \ . \tag{4.10}$$

Note that this assumes that r_{st} has a vertical orientation to be consistent with the sense of Figure 2.1 even though the DFE-ground interface along the tapered portion of a DFE shaft is tilted slightly from the vertical. Historically, this inconsistency has generally been ignored as in calculations a tapered DFE is typically divided into artificial segments. Each tapered artificial segment is analyzed as though it were a right-circular cylinder with a constant diameter that is the average for the segment.

In principle, the σ'_h term in Equation 4.10 can only be expanded or quantified as was done previously for the unit shaft resistance due to sliding friction, r_{ss}, by defining an appropriate physical mechanism for what happens when the tapered portion of a DFE displaces downward relative to the surrounding soil as the result of an applied load. Unfortunately, there has not been a consensus for this over the last 60-plus years. How this has been handled historically and to the present is addressed in the following section.

4.2.2.3 History and Evolution of Modeling Shaft Resistance Due to Taper

The evolution of analytical methodologies for forecasting the resistance of tapered piles can be traced back to the late Prof. George Geoffrey 'Geoff' Meyerhof who is best remembered for his long and professionally prolific tenure at the Technical University of Nova Scotia. He was one of the pioneers in developing analytical methods for forecasting the static resistance of piles, especially in coarse-grained soil, beginning in the 1950s timeframe. A number of his early papers on the subject can be found in a 1982 compilation volume (Meyerhof 1982) issued on the occasion of his retirement from academia.

To provide the necessary background for understanding the geotechnical and foundation engineering 'landscape' in which Meyerhof's first contributions involving tapered piles appeared, it is useful to recall that the first American Society of Civil Engineers specialty conference devoted exclusively to what we would now call geotechnical engineering (the term did not even exist then) was not held until 1960 and even then it focused entirely on the theoretical aspects of shear strength. In addition, well into the 1960s, geotechnical engineering was still not a required subject in some North American undergraduate civil engineering programs. Furthermore, even in the early 1970s when the writer began his professional career there was still a prevailing opinion among some practitioners that foundation engineering was just a subset of structural engineering as it had been universally perceived since the 19th century and prior to the emergence of modern soil mechanics in the 1920s and 1930s. In this vein, these practitioners saw geotechnical engineering efforts to improve the science behind foundation design as a solution in search of a problem.

It is interesting to note that although Meyerhof was very theoretically oriented for some of his foundation-related research (his theory-of-plasticity bearing-capacity solution for shallow foundations is still routinely used to the present), his research related to piles focused on a more pragmatic, empirical approach to developing analytical methodologies for shaft resistance using the direct method. Specifically, he published empirical relationships for the ULS unit shaft resistance due to sliding friction, $r_{ss(ult)}$, that were correlated to both SPT N-values and cone-penetrometer sleeve friction, f_s, values.

As an aside with regard to the latter, it is of interest to note that Meyerhof was prescient and decades ahead of other North American researchers and practitioners alike in recognizing the benefit of using the cone penetrometer in geotechnical and foundation engineering. In fact, at that time (1950s), it was likely that a majority of geotechnical and foundation engineers in North America had simply never heard of the device. Meyerhof's circa-1950s use of sleeve friction in a direct-resistance methodology is all the more remarkable as the friction sleeve was only added to the cone after the Second World War and thus had only been in use even in The Netherlands for a relatively short period of time. Furthermore, sleeve friction was still being measured hydraulically in the 1950s as the electronic (Begemann) cone was not developed until the 1960s as was noted in Chapter 2.

In any event, Meyerhof's pile research is noted here because he clearly recognized that tapered piles behaved resistance-wise in a manner that distinguished them from piles with a constant perimeter in that he explicitly identified a unique ULS unit shaft resistance due to taper, $r_{st(ult)}$. However, he did not offer any hypothesis for a physical mechanism for this observed behavior and treated the 'taper benefit' in a very simplistic fashion as an empirical modification of the basic ULS unit shaft resistance due to sliding friction, $r_{ss(ult)}$.

Specifically, Meyerhof's empirical relationships for $r_{st(ult)}$ were the same as those for $r_{ss(ult)}$ but with a single-valued, 'one-size-fits-all' correction factor to increase the ULS shaft resistance for tapered piles above the base values for non-tapered piles. This may well have been adequate at the time as it is likely that most of the tapered piles in Meyerhof's database on which his empirical relationships were based were timber piles. Because Meyerhof worked primarily in Canada, it is unclear to what extent any of the Raymond steel shell and/or *Monotube* steel pipe piles were used on projects to which he had data access.

The first true breakthrough on the path to understanding the correct resistance mechanisms for tapered piles occurred in the early 1960s with the publication of R. L. Nordlund's seminal paper on the subject (Nordlund 1963)[79]. Unlike Meyerhof, Nordlund (an

[79] Nordlund's surname was misspelled "Norlund" in the ASCE journal paper as initially published, clearly a typographical error. In the writer's opinion, this has to go down as <u>the</u> most unfortunate typo in geotechnical and foundation engineering publication history. **[footnote continues on following page]**

employee of Raymond at the time) chose to develop a problem-solution methodology referred to herein as the *Nordlund Method* using the more theoretically rigorous indirect method. Figure 4.1, which was developed by the writer several years ago and adorns the cover of this monograph, is useful for visualizing the physical mechanism of *wedging* that was postulated by Nordlund (although he never used the term 'wedging') and subsequently used by other researchers (including the writer) to physically model shaft resistance due to taper.

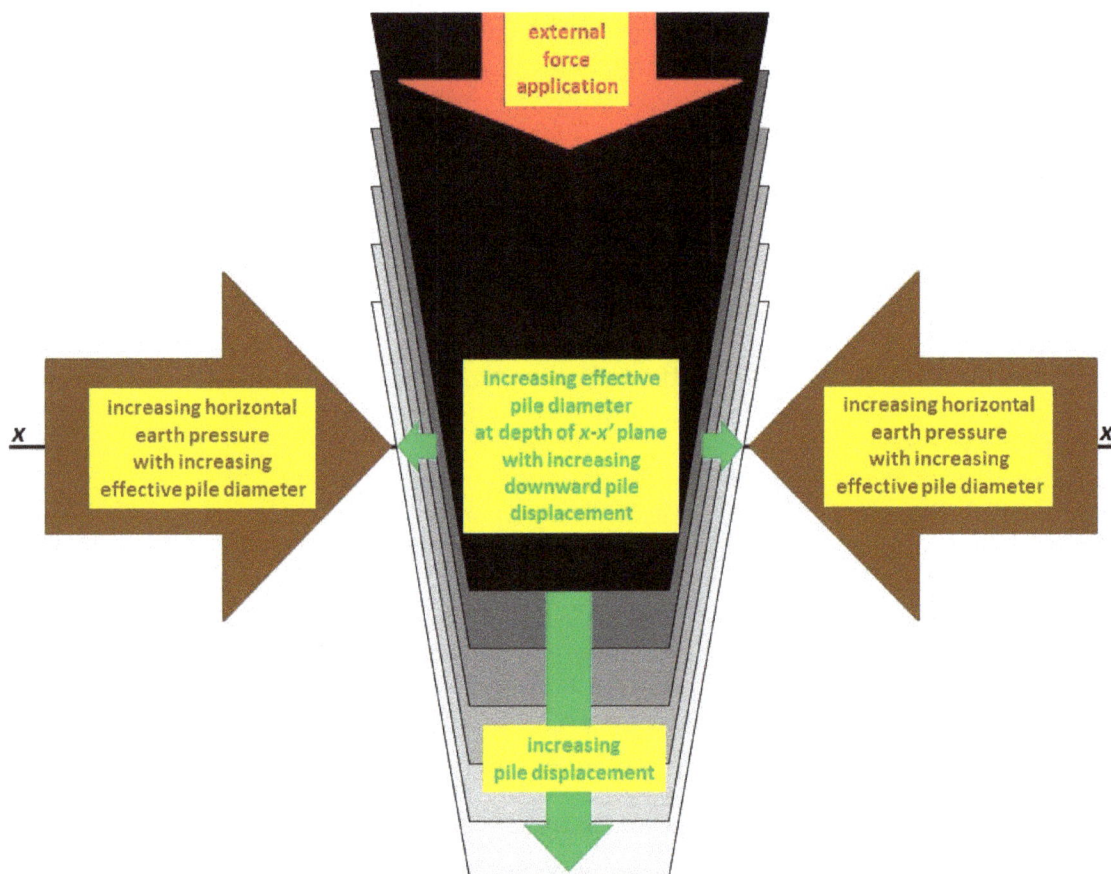

Figure 4.1. Writer's Visualization of the Wedging Mechanism for Tapered-Pile Behavior under Pile Driving and Post-Installation Externally Applied Axial-Compressive Loading.

Nordlund realized that the downward displacement of the tapered portion of a pile into the ground (depicted in Figure 4.1 by the shaded transition from black to light gray of the trapezoid that generically represents the tapered portion of a pile), whether during driving or under post-installation axial-compressive load application, was not the usual sliding-friction mechanism that developed with non-tapered piles with a depth-wise constant perimeter. Rather, the overall mechanism is first and foremost a continuous volumetric

---

This is because although the error was corrected in the errata/closure to the paper that was eventually published, the correction was clearly missed by the vast majority of later researchers and authors who cited his original paper and method. As a result, most subsequent publications by others have unfortunately perpetuated the misspelling of Nordlund's surname and refer to his resistance-forecasting methodology as the "Norlund Method".

expansion that is conceptually identical to driving a steel wedge into a piece of wood. This simple wedge mechanism has long been known as a very efficient way to geometrically turn a purely vertical force into one with a horizontal component based solely on vector statics of rigid bodies applied to the geometry of the wedge.

This wedging mechanism can readily be visualized by focusing on some arbitrary, spatially fixed horizontal plane such as defined by the line x-x' in Figure 4.1. Relative to this plane that is perpendicular to this figure, the diameter or width of the pile appears to displace radially (for a circular cross-section) or horizontally (for a square cross-section) into the ground and thus increase volumetrically with increasing pile penetration. This radial or horizontal displacement and concomitant volumetric expansion is accompanied by an increase in radial or horizontal earth pressures acting on the pile shaft.

To convert this physical mechanism into a mathematically solvable problem, Nordlund turned to basic soil mechanics and classical lateral earth pressure theory such as existed circa 1960 when he developed his analytical method. Nordlund envisaged the radial/horizontal displacement of the pile into the ground that is depicted in Figure 4.1 as being mechanistically equivalent to a rigid, planar surface of zero thickness, undefined height, and infinite length (perpendicular to the plane defined by Figure 4.1) rotating about its bottom (the toe of the pile in this case) and, as a result, displacing horizontally into a horizontally infinite soil mass in a passive lateral earth pressure mode. The assumed angle of rotation of this surface (which can be visualized as a rigid retaining wall) is the same as the taper angle, α, of the pile.

Of immediate note is that Nordlund used a two-dimensional (2-D) physical model and concomitant 2-D analytical problem and solution for what is very much a 3-D problem. Another significant implication of his assumptions is that the horizontal stress in the ground does not increase limitlessly with increasing taper angle but is delimited by the coefficient of passive earth pressure, K_p. Specifically, Nordlund used by the 'exact' solution published by Caquot and Kerisel in 1948 to evaluate K_p. In retrospect, this was actually quite advanced and insightful for the timeframe (circa 1960) as most would likely have used either Coulomb's solution or even Rankine's solution (in 2020, there are undoubtedly still those who would use either Coulomb's or Rankine's solution).

There are quite a few simplifying assumptions and empirical approximations that Nordlund had to make in order to turn this basic concept into a practice-oriented analytical methodology as was his goal. These are discussed in detail in Chapter 5 but for now it is important to note that at the top of the list is the fact that Nordlund limited his methodology to coarse-grained soils. From a practical perspective, this is not terribly important because to date tapered piles have primarily been used at sites where coarse-grained soils predominate.

For the purposes of the present discussion, it is only of interest to examine Nordlund's model in terms of Equation 4.10 which is the generic expression for the unit shaft resistance due to taper, r_{st}. Of greatest interest is how Nordlund defined σ'_h:

$$\sigma'_h = K_\delta \cdot \cos(\alpha + \delta) \cdot \sigma'_{vo} \tag{4.11}$$

where K_δ is an earth pressure coefficient that is <u>not</u> oriented horizontally and that varies in magnitude between the coefficient of <u>lateral</u> earth pressure at rest, K_o (which is vectorially inconsistent and which Nordlund assumed was <u>always</u> equal to 0.5), and K_p (which is vectorially consistent). The angle α is the pile taper angle and δ is nominally the pile-soil interface friction angle. Note, however, that Nordlund had a relatively complex and unique interpretation of δ (this is discussed in detail in Chapter 5) so the generic interpretation of δ presented earlier in this chapter does <u>not</u> apply in Equation 4.11.

Note also that in a given analysis, Nordlund assumed that K_δ is constant. In theory and as can be visualized using Figure 4.1, the lateral earth pressure acting along the shaft of a tapered pile should increase from K_o as the pile is driven and then increase further as the pile is externally loaded in axial compression and the pile settles. However, Nordlund did not explicitly consider post-installation settlement in his analytical methodology so implicitly lumped the outcomes of these two distinct stages of wedging together into one value.

With this definition of σ'_h, Equation 4.10 for the generic unit shaft resistance due to taper, r_{st}, becomes a ULS-based definition for $r_{st(ult)}$:

$$r_{st(ult)} = \tan \delta \cdot \sigma'_h = K_\delta \cdot \cos(\alpha + \delta) \cdot \tan \delta \cdot \sigma'_{vo} . \tag{4.12}$$

Nordlund assumed that α was negligibly small relative to δ so that Equation 4.12 could be simplified to:

$$r_{st(ult)} \cong K_\delta \cdot \cos \delta \cdot \tan \delta \cdot \sigma'_{vo} = K_\delta \cdot \sin \delta \cdot \sigma'_{vo} . \tag{4.13}$$

It might appear that by neglecting α in Equation 4.13 that Nordlund was neglecting taper effects. In reality, he was only neglecting the <u>effect</u> of the taper angle on the vector geometry and orientation of the ULS unit stresses acting along the shaft-soil interface. The effect of the taper angle was very much still part of the problem by virtue of the empiricisms and assumptions he made to come up with both the K_δ and δ values.

As is discussed in Chapter 5, the Nordlund Method as first presented in his 1963 paper[80] makes extensive use of a series of charts with relatively crude hand-drawn curves that require visual interpretation. That these charts were published in a relatively small-format journal[81] makes them all the more difficult to use. Of much greater importance is that the Nordlund Method requires that <u>all</u> soil properties be estimated based on <u>uncorrected</u> SPT N-values as appropriate for the circa-1960 timeframe combined with specific empirical correlations specified by Nordlund, facts that are typically left out of subsequent publications such as textbooks, design manuals, and handbooks that include the method.

As a final comment for the time being concerning the Nordlund Method, it is relevant to note that this method can be used for constant-perimeter piles, including the constant-perimeter portions of partially tapered piles. The taper angle, α, is simply taken to be zero and all the parameters in the methodology are evaluated accordingly. In fact, there are cases where design manuals intended explicitly for use with <u>non-tapered</u> piles use the Nordlund Method as the resistance-forecasting method of choice (NYSDOT 2015).

Moving on from the Nordlund Method, in order to produce an analytical methodology that was amenable to computer solution, circa 2000 the writer developed an analytical methodology for tapered piles (Horvath 2002) that was broadly based on Nordlund's approach for shaft resistance but in a highly modified and approximate form. The writer's *Modified Nordlund Method*, as it is referred to in this monograph, incorporates certain improvements such as using theoretically rigorous cone penetrometer-based site characterization to estimate the required soil properties as well as using a classical theory-of-plasticity bearing-capacity solution for toe resistance that included consideration of Vesic's soil compressibility (rigidity) factors as suggested by Kulhawy (1984, 1991).

---

[80] As discussed subsequently, Nordlund revised his method in 1979 but the basic, overall structure of the method remained unchanged.

[81] American Society of Civil Engineers (ASCE) journals of that era had a much smaller trim size for publication compared to their journals of recent decades.

The net results of the writer's modified version of the Nordlund Method are the following equations that replace Equations 4.11 and 4.13 respectively for the Nordlund Method:

$$\sigma_h' = K_h \cdot \sigma_{vo}' \tag{4.14}$$

$$r_{st(ult)} = \tan\delta \cdot \sigma_h' = K_h \cdot \tan\delta \cdot \sigma_{vo}' \tag{4.15}$$

where K_h = a lateral earth pressure coefficient that varies between the site-specific, pre-construction value of K_o and an empirically derived limiting value that is a function of the pile taper angle, α, and operative value of ϕ.

However, one of the most significant changes that the writer incorporated into the Modified Nordlund Method was to make use of Kulhawy's suggestions (discussed earlier in this chapter) that an accurate estimate of site-specific, pre-construction K_o values (as opposed to Nordlund's one-size-fits-all assumption of $K_o = 0.5$), combined with consideration of the effects of installing a particular type of DFE, are crucial to the overall analytical process. As noted previously, Kulhawy expressed the effects of installing any type of DFE as the dimensionless ratio K_h/K_o. This ratio conveniently reflects the discussion earlier in this chapter of the increased awareness of the importance of the stress state in the ground both before and after installation of a DFE.

With these suggestions in mind, Equation 4.14 is better expressed as:

$$\sigma_h' = K_h \cdot \sigma_{vo}' = \left(\frac{K_h}{K_o}\right) \cdot K_o \cdot \sigma_{vo}' . \tag{4.16}$$

As with Nordlund's original method, K_h and K_h/K_o in the Modified Nordlund Method are assumed to be independent of the magnitude of pile settlement and concomitant radial/horizontal displacement of the tapered portion of the pile into the soil.

As will be seen in Chapter 5, expressing the post-installation lateral earth pressure coefficient, K_h, acting on any type of DFE of any geometry, i.e. constant-perimeter or tapered, as $[(K_h/K_o) \cdot K_o]$ is a useful, behaviorally insightful enhancement to several other DFE resistance-forecasting methodologies, including the widely used β Method mentioned earlier in this chapter. This is because nowadays it is relatively straightforward to evaluate the site-specific, pre-construction profile of K_o for all types of soils based on a cone-penetrometer sounding. Furthermore, the K_h/K_o ratio can be developed experimentally for all types of DFEs based on instrumented static load tests as demonstrated decades ago by Kulhawy (1984, 1991). The general framework for enhancing DFE resistance forecasting based on using $[(K_h/K_o) \cdot K_o]$ in lieu of K_h alone is discussed in some detail in Chapter 5.

Nordlund's breakthrough insight into the unique behavior and concomitant resistance mechanism of tapered piles was constrained by the state of geomechanics knowledge in the years leading up to circa 1960 when Nordlund was presumably developing his concepts as an employee of Raymond. Thus, when viewed from the perspective of today's geotechnical knowledge-base there are four primary shortcomings in his work:

1. The analytical methodology can only be applied to piles bearing primarily in coarse-grained soil.

2. Classical lateral earth pressure theory only provides solutions to a 2-D plane-strain problem while the apparent radial/horizontal pile displacement relative to plane x-x' in

Figure 4.1 is clearly 3-D in nature. Nordlund acknowledged this fact in his original 1963 paper and stated that he made a correction for 3-D effects in some manner that he did not explain in his paper.

3. Classical lateral earth pressure theory is essentially an all-or-nothing analytical tool meaning that a soil mass must always be assumed to be in one of three earth-pressure states:
    a. active,
    b. at-rest, or
    c. passive.

    There is no theoretically rigorous way to calculate intermediate lateral earth pressure conditions. In addition, if the soil is assumed to be in either the active or passive state, there is no way to calculate (based on earth pressure theory alone) how much displacement is required to mobilize that state. Again, Nordlund acknowledged this fact in his original 1963 paper. He developed an empirical relationship for the variation in lateral earth pressure coefficient between K_o and K_p as a function of wall rotation (and, by implication, pile taper angle) based solely on large-scale retaining wall tests reported on by Terzaghi in the 1930s.

4. The additional wedging and concomitant taper benefit that clearly must occur incrementally during post-installation external load application (as illustrated qualitatively in Figure 4.1) is not considered explicitly as the Nordlund Method is a classical ULS-resistance-only forecasting methodology.

It is emphasized that these observations are not meant to deprecate or detract from the critical, seminal role played by Nordlund's work in the evolution of understanding how tapered piles develop resistance. If nothing else, he was the first to formally postulate the wedging mechanism shown in Figure 4.1 even though he never used that term. In fact, a review of English-language textbooks, handbooks, and design manuals suggests that even almost 60 years after it was first published, the Nordlund Method still defines the state of knowledge for tapered piles with very few exceptions. Nevertheless, these observations illustrate that Nordlund did the best he could with the relatively limited state of geomechanics knowledge at the time and that his passive earth pressure model is really just a starting point for a more accurate understanding of how tapered piles develop resistance.

Before proceeding further, it is important to note two facts. First, although Nordlund's 1963 paper is well-known, considerably less well-known is that Nordlund revised his method in 1979[82]. Unfortunately, it appears that Nordlund chose to publicize his revised methodology, which is hereinafter referred to as the *Revised Nordlund Method*, only via a set of participant notes distributed to attendees at a regional continuing-education short course (Nordlund 1979). It is fortuitous that in recent years these notes were republished in an accessible, public venue (NYSDOT 2015[83]) which is how the writer became aware of the Revised Nordlund Method.

That there are two versions of the Nordlund Method is important because, as discussed in detail in Chapter 5, although Nordlund did not change the basic structure of his methodology, he made changes to some of the charts and correlations used in the methodology and, more importantly, changed the specific way in which these charts and

---

[82] To wit, the writer did not become aware of this until 2016 as the result of conducting background research for this monograph.

[83] This reference also contains a copy of Nordlund (1963).

correlations were to be evaluated using SPT N-values. As it turns out, the cumulative effect of these deceptively subtle changes can have a relatively large impact on the calculated outcomes as Nordlund himself noted in Nordlund (1979).

The second and overall more significant fact to note is that although Nordlund's method, whether in its original or revised form or as modified by the writer, recognized the unique wedging effect of tapered piles, it did not define a unique shaft-resistance mechanism that reflected this uniqueness. Rather, the analytical algorithm that Nordlund incorporated into his methodology treats the shaft resistance along the tapered portion of a pile using the mechanism of shaft resistance due to sliding friction, albeit with an increase in sliding friction over and above that of an otherwise identical pile with a depth-wise constant perimeter.

Thus, in the end, Nordlund used the same approach as Meyerhof some years earlier, albeit with substantially more analytical sophistication as to how the relative increase in frictional resistance is determined. Taper benefit is reflected in Nordlund's methodology but in a manner that is conceptually inconsistent with the actual physical wedging mechanism. The significance of highlighting this inconsistency between reality and methodology will become clear shortly.

In the writer's opinion, the truly seminal publication to date with respect to defining the true physical nature of the wedging mechanism was published 30 years after Nordlund's original work and was based on research by Kodikara and Moore (1993). Kodikara[84] and Moore made use of the tremendous advances in geomechanics theory during that 30-year period between 1960 and 1990 to reinterpret the wedging mechanism depicted in Figure 4.1 as a true 3-D problem in *cylindrical-cavity expansion* (CCE). Rather than envisaging the apparent radial expansion of the pile relative to the spatially fixed plane *x-x'* shown in Figure 4.1 as a 2-D passive lateral earth pressure mechanism, it is visualized as a vertically oriented cylinder that is volumetrically expanding in diameter, similar to what occurs during the expansion phase of a PMT test.

Before proceeding further, there are several aspects of the CCE model used by Kodikara and Moore that need to be clarified and emphasized so that it is clearly understood what their model does and, more importantly, does not do in the context of tapered piles. This is because this model figures prominently in the remainder of this monograph so a clear understanding of the model is essential.

To begin with, the model used by Kodikara and Moore is a cavity-expansion model that, by definition, assumes as a starting point an initial, pre-existing cavity (with a concomitant cavity radius and initial radial pressure within the cavity) that is expanded further, with only this additional expansion replicated by the model. A CCE model is distinctly different from a *cylindrical-cavity creation* (CCX[85]) model that assumes there is no cavity initially, i.e. the initial cavity radius and cavity pressure are both zero, and that a cavity is created by some process such as the insertion of a cone penetrometer.

---

[84] In published work up to the present, the writer used only the name of the senior author of Kodikara and Moore (1993) when referring to the analytical methodology presented in that paper. This reflected the writer's understanding that the work was largely that of Kodikara while he was working as a protégé and research associate of Moore in Australia, under research funding secured by Moore. This understanding was based on the Acknowledgements section in Kodikara and Moore (1993) as well as letter correspondence initiated by the writer to both authors in 1993. However, the writer has chosen to use a broader recognition in this monograph to avoid any implication of slighting Moore's contributions, whatever they may have been, to the Kodikara and Moore 1993 paper. Consequently, the analytical methodology developed by Kodikara and Moore is referred to as the *Kodikara-Moore Method* in this monograph, not the *Kodikara Method* as in prior publications by the writer.
[85] As will be seen subsequently, there was a good reason why 'CCC' was not used as an abbreviation here, as would otherwise seem logical.

It is of interest to note that a common element of both cavity creation and cavity expansion is that at some finite magnitude of expansion a *limiting pressure* is reached in the radial direction so that further cavity expansion occurs with no increase in pressure. This is in consonance with the same conclusion that Nordlund reached decades earlier using a completely different theoretical basis. This suggests that there is a limit beyond which pile taper (or DFE taper in general) does not produce incremental benefits.

In the tapered-pile application, the initial, pre-existing cavity of a CCE-based model such as used by Kodikara and Moore is assumed to be created by installing the tapered DFE in the ground (the Kodikara-Moore Method allows for any type of tapered DFE, not just a (driven) pile). In essence, Kodikara and Moore simply 'wished' the DFE into the ground and gave that aspect of the problem no explicit theoretical consideration. The CCE model used by Kodikara and Moore only explicitly models the further cavity expansion caused by post-installation axial-compressive load application on the head of the DFE. As can be seen qualitatively in Figure 4.1, tapered-DFE settlement due to external load application causes radial expansion of the shaft-ground interface due to the simple geometry of the problem.

Thus, the first point of note with respect to the Kodikara-Moore Method is that the model used by Kodikara and Moore does <u>not</u> model changes to the stress state in the ground caused by installation of the DFE, tapered or otherwise. Rather, it only models changes to the stress state due to post-installation external axial loading on a DFE. As will be seen, it is incumbent upon the design professional using their analytical methodology to estimate or otherwise assume the post-installation stress state acting along the DFE shaft prior to external load application as this is the necessary initial condition and starting point for an analysis using the Kodikara-Moore Method.

As will also be seen, the fact that the Kodikara-Moore Method does not explicitly model the significant cavity-creation process caused by driving a tapered pile into the ground is a substantial drawback. Nevertheless, the fact that the Kodikara and Moore identified cavity mechanics as a much-improved (compared to Nordlund) 3-D physical model for the wedging that occurs with tapered piles was a substantial, seminal advancement in the state of knowledge and at least provides a direction for future research as outlined in Chapter 7.

The next point of note concerning the CCE solution used by Kodikara and Moore is that the problem of cavity expansion is traditionally formulated in geomechanics assuming that there is an increasing (from some known initial value) internal pressure within the pre-existing cavity that is causing the cavity expansion. The resulting radial displacements of the cavity wall are a consequence of this expansion. For example, this is precisely what happens when a pressuremeter (PMT) test is performed.

However, in the tapered-pile application the mechanism causing cavity expansion is completely reversed. The cavity expansion is caused by displacements of the cavity wall due to pile settlement and the increased radial pressures that develop (in this case envisaged as acting radially along the pile shaft at the pile-soil interface) become the consequence. Thus, in the manner in which CCE is used by Kodikara and Moore there is a complete reversal of cause and effect between the radial cavity pressure and radial cavity expansion. Fortunately, from a theoretical perspective it is the same problem no matter how the cause-effect is framed.

Setting aside for the time being the fact that the Kodikara-Moore Method does not address the CCX that occurs when a tapered pile is driven into the ground, the distinctly positive aspects of the CCE model used by Kodikara and Moore are that it overcomes the three primary shortcomings of Nordlund's passive earth pressure model enumerated above as it:

- can be applied to any type of soil, not just coarse-grained;

- captures the true three-dimensionality of the problem; and

- provides for an incremental, displacement-based increase in radial stresses acting on the pile shaft.

It is appropriate at this point to bring up a broader theoretical issue that extends beyond the specifics of the Kodikara-Moore Method. As noted earlier in this chapter, the writer has long held the position that the shaft-resistance mechanism for tapered piles is distinctly different from the traditional shaft-resistance mechanism due to sliding friction. Because of the wedging mechanism as illustrated in Figure 4.1, tapered-pile shaft resistance at any given depth develops under an ever-increasing radial stress acting on the shaft as axial-compressive load is applied whereas shaft resistance due to sliding friction occurs under an assumed constant radial stress. Note that this is true whether one views the wedging mechanism that develops with tapered piles as being the result of laterally rotating a rigid 2-D plane into the surrounding ground (as Nordlund did) or expanding a cylindrical cavity into the surrounding ground (as Kodikara and Moore did).

The point being made here is that the shaft resistance due to taper that was identified earlier in this chapter can now be seen as being caused by the physical mechanism of wedging and thus referred to as *shaft resistance due to wedging*. Note that earlier publications by the writer used the term *shaft resistance due to cylindrical-cavity expansion* but that is now seen to be somewhat incorrect as the wedging mechanism associated with tapered piles develops in two distinct, cumulative stages:

1. cavity <u>creation</u> (CCX) as the pile is driven and

2. further cavity <u>expansion</u> (CCE) as the pile is externally loaded in axial compression,

<u>not</u> just the one stage of cavity expansion during post-installation external load application as postulated by Kodikara and Moore in 1993.

Because the physical mechanism of wedging is distinctly different from the physical mechanism of sliding friction, there should be universal recognition that shaft resistance due to wedging represents a third, distinct resistance mechanism for DFEs. This is not an esoteric and unnecessary technicality, as it is possible to model wedging as a pseudo-sliding friction mechanism as demonstrated by Nordlund. This is because under the traditional shaft-resistance mechanism of sliding friction, the radial stresses acting along the shaft remain constant in magnitude. However, with the new shaft-resistance mechanism of wedging these radial stresses are constantly changing. Furthermore, because shaft resistance due to sliding friction and shaft resistance due to wedging are such completely different physical mechanisms, they will not be mobilized and 'max out' at the same rate so need to be treated as distinct and separate mechanisms.

One relevant example of this that is discussed further in Chapter 5 is the widely used 1-D wave equation and on-pile dynamic measurements that derive from it. These analytical methodologies are based on a physical model of a stress wave that propagates and causes pile displacement in one direction only combined with the traditional resistance mechanisms of shaft sliding friction and toe end bearing. As a result, it has been well-known since at least the early 1970s that these dynamic-analysis methods often do not adequately forecast the ULS total resistance of tapered piles. This is not surprising as these methodologies do not correctly model the axial response of a tapered pile as shown in Figure 4.1.

Returning now to specifics related to the Nordlund and Kodikara-Moore methods, an important element in this discussion is the primary problem variables used in each methodology. In addition to the obvious pile geometry, both methodologies require information about soil and pile-soil system properties.

First considered is the soil property of shear strength. Not surprisingly, both Nordlund and Kodikara and Moore used the traditional Mohr-Coulomb strength parameters of ϕ and c. Nordlund simplified things by assuming that $c = 0$ and limiting his analytical methodology to coarse-grained soil. However, the most important issue is that both Nordlund and Kodikara and Moore assumed that ϕ = constant for a given problem. As discussed earlier in this chapter, it is now well-established that this is never correct and at best is an approximation of reality.

The writer recognized this when developing the aforementioned analytical methodology that was based on modifications and enhancements of the Nordlund Method (Horvath 2002). The value of ϕ that was used for both shaft and toe resistances in the writer's methodology varied in accordance with Equation 4.7 and considered the effect of the operative stress level on the dilatancy component. This was done using Bolton's widely used empirical relationship that can be found in many references (the writer used Kulhawy and Mayne (1990)). Manandhar and Yasufuku (2011b, 2013) and Manandhar et al. (2013) used a similar approach with regard to ϕ in their slightly modified version of the Kodikara-Moore Method.

Related to ϕ by virtue of the commonly used δ/ϕ ratio is the pile-soil system property of the interface friction angle, δ. This has already been discussed in a generic sense earlier in this chapter. Also as noted earlier in this chapter, Nordlund viewed the δ/ϕ ratio in an unusual, atypical manner in his analytical methodology by making it dependent not only on pile-material type but also the pile volume per unit length of installed pile. His logic for this unique interpretation of the δ/ϕ ratio is discussed further in Chapter 5. The writer did not adopt Nordlund's perspective on this issue in the Modified Nordlund Method but did allow for δ to vary with the operative stress level by virtue of the aforementioned variation in ϕ.

The other soil property that plays a major role but only in the Kodikara-Moore Method is soil stiffness as characterized by the shear modulus, G. This is because of the use of cavity mechanics that always requires use of soil modulus.

In the writer's opinion, assuming a constant value for G in a given problem as both Kodikara and Moore and Manandhar and Yasufuku (in their research using the Kodikara-Moore Method) did is a much larger issue compared to assuming a constant value for ϕ. This is because the consideration of *modulus degradation* (reduction) as a function of either strain or stress as is more appropriate in a given applications is now understood to be a necessity, not a luxury, for correct problem solution in a wide variety of geotechnical applications that explicitly consider soil stiffness. This is illustrated in Chapter 5 and Appendix D. Because the Kodikara-Moore Method explicitly considers pile resistance as a function of displacement, it stands to reason that considering shear-modulus degradation is a necessity. Unfortunately, this has not been done to date with this method.

To complete the discussion of the historical evolution of resistance mechanisms for tapered piles, it is appropriate to discuss the published work of those who have chosen to follow a completely different path from the Nordlund-Kodikara wedge model and ignore the fact that loading a tapered pile results in a volumetric expansion as illustrated in Figure 4.1. This alternative behavioral model for tapered piles is shown in Figure 4.2 and is credited to Dr. Bengt H. Fellenius at some unknown date in the past (Fellenius (2019) is a recent reference that discusses the model). This model has also been used by Togliani (2010) who developed a direct method for forecasting the ULS resistance of tapered piles that is discussed in Chapter 5.

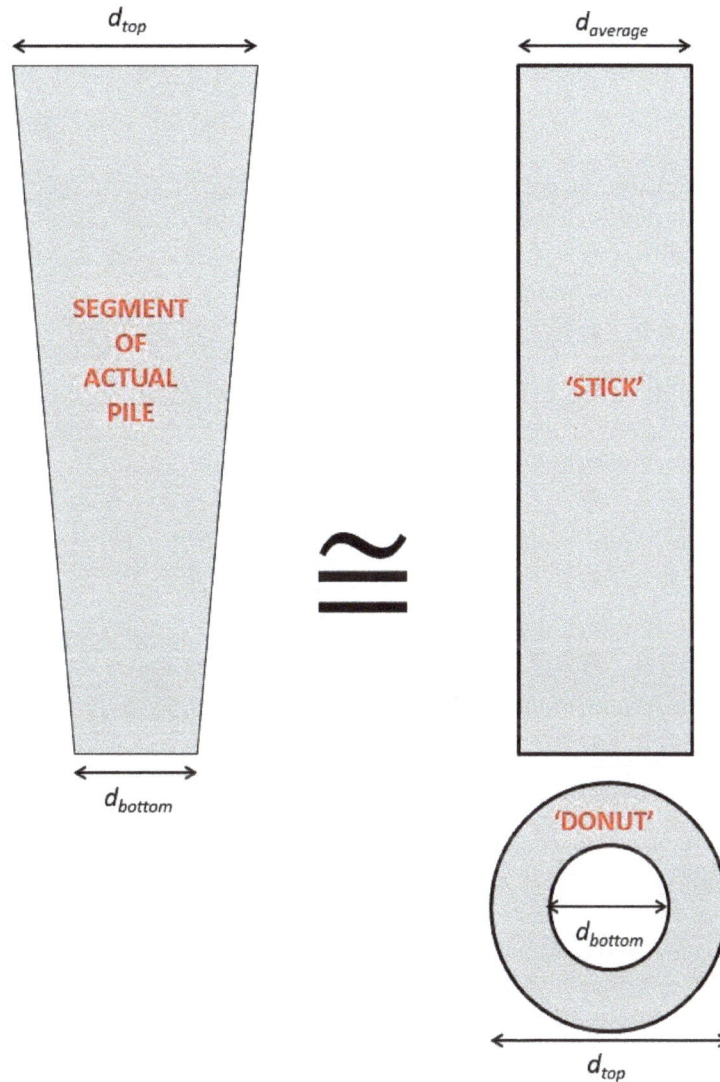

Figure 4.2. Visualization of the Fellenius 'Donut-on-a-Stick' Model for Tapered Piles.

The Fellenius model, which the writer refers to as the *donut-on-a-stick model* for reasons that will become clear, divides the tapered portion of a pile (or DFE in general) into an arbitrary number of artificial segments. Each artificial segment is then decomposed into two fictitious components as shown in Figure 4.2:

- A constant-diameter right-circular cylinder (the 'stick' component) that has the same length as the actual pile segment being analyzed and a diameter equal to the arithmetic mean (simple average) of the top and bottom diameters of the actual pile segment being analyzed. This component is assumed to develop resistance from the traditional mechanism of shaft resistance due to sliding friction.

- An annular area (the 'donut' component as Fellenius himself called it) is assumed analytically to be lying within a horizontal plane at the bottom of the constant-diameter cylinder ('stick') and defined by the dimensions shown in Figure 4.2. This component is assumed to develop resistance from the traditional mechanism of toe end bearing.

Note that any consideration of the actual physical mechanism of wedging is completely absent from this model.

These two fictitious components of each artificial segment of the pile are then quantified using whatever analytical methodology is desired for normal shaft resistance due to sliding friction and toe resistance due to end bearing without any explicit consideration, correction, or adjustment for taper.

There are several significant criticisms of the donut-on-a-stick model:

- While it is straightforward to work with and might be visually pleasing on some level, it does not even remotely capture the actual physical mechanism of wedging for tapered piles as illustrated in Figure 4.1.

- It forecasts a single-valued ULS resistance (identical in concept to the Nordlund Method) as opposed to a settlement-dependent resistance as with the Kodikara-Moore Method.

- Because it is just a physical model devoid of an accompanying resistance-forecasting methodology, it depends on a resistance-forecasting methodology to produce a numerical result. Because any traditional resistance-forecasting methodology, indirect or direct, can be used with this model, the number of possible outcomes is substantial, with no clear indication of relative accuracy.

- Most importantly in the writer's opinion, there does not appear to be the necessary substantive body of published work to support Fellenius' analytical model by showing that calculated results using this model together with any resistance-forecasting methodology agree reasonably well with field-measured results. The writer is aware of only one such comparison, in Togliani (2010). Togliani discussed exactly one case history that was analyzed using this model together with a resistance-forecasting methodology that he developed that is discussed in Chapter 5. However, a limited study of Togliani's resistance-forecasting methodology by the writer (Horvath 2016a) found considerable differences between forecast and measured results. Additional comparisons are presented later in this monograph.

In conclusion and summary, the current state of knowledge with regard to the correct shaft-resistance mechanism for tapered piles has been well-established since the early 1960s as the wedging mechanism shown in Figure 4.1. As it is distinctly different from the traditional resistance mechanisms of shaft resistance due to sliding friction and toe resistance due to end bearing, it deserves to be recognized as the third resistance mechanism for deep foundations in general as it would be the applicable mechanism for any tapered DFE regardless of the installation or construction method. Consequently, the current state of art with regard to resistance mechanisms for tapered deep foundations in general is to use an analytical methodology that is based on the wedging mechanism.

Currently available information indicates that the wedging mechanism was first identified in published work almost 60 years ago by Nordlund who modeled it approximately using the geomechanics tools available at the time. Consequently, the resulting analytical methodology was constrained by having the radial stresses acting on the tapered portion of a pile fixed in magnitude no matter how much the pile settled under load and volumetrically expanded into the surrounding ground as depicted in Figure 4.1.

Kodikara and Moore have since shown that a much more correct physical model to use for wedging is cavity mechanics assuming a cylindrical cavity. The shortcoming of their work is that they did not account for the cavity creation that occurs when a tapered pile is driven into the ground. They only considered the cavity expansion that occurs when the installed DFE (their method allows for non-piles) is loaded externally in axial compression. While there is always the possibility that some even-more-accurate physical and mathematical model might be proposed or developed in the future, what that might be is not obvious at the present time. Consequently, it appears logical to devote resources to extending, improving, and implementing the cylindrical-cavity model in both research and practice. Exploring this further is one of the major objectives of the remainder of this monograph.

With regard to Fellenius' donut-on-a-stick model, there are several important facts to note:

- It does not even come close to reflecting the true wedging resistance mechanism of tapered piles.

- There does not appear to be any substantive body of work that has been published to support its use as even an approximate analytical model.

- It is not a complete analytical methodology for forecasting the resistance of tapered piles as are both the Nordlund and Kodikara-Moore methods. Rather, it is just a simplistic geometrical tool to extend the application of traditional resistance-calculation methodologies that were developed for constant-perimeter DFEs, such as the Eslami-Fellenius *UniCone Method* (Eslami and Fellenius 1997) that was improved recently by Niazi (Niazi 2014, Niazi and Mayne 2015a), for use with tapered piles.

Nevertheless, in principle, Fellenius' donut-on-a-stick model could serve a useful role in routine practice as an interim, stand-in methodology but only if it can be shown to produce acceptably accurate results on a consistent basis when used with one or more traditional resistance-forecasting methodologies for constant-perimeter piles. But again, there does not appear to have ever been a comprehensive published study along these lines. Togliani's claim (Togliani 2010) that Fellenius' model produces accurate results when coupled with Togliani's direct-method empirical equations for shaft and toe resistance appears to be based on one case history so does not even come close to being such a study.

There is one final issue to raise before ending the discussion of tapered-pile resistance mechanisms and this involves the geometry of the pile cross-section. More specifically, the effect that cross-sectional geometry has on the wedging mechanism.

As was seen in Chapter 3, tapered piles with a square cross-section are quite common with precast-PCC piles. One of the most significant unresolved technical issues concerning tapered piles that requires fundamental research in the future is the subject of the effect of cross-sectional geometry on tapered pile behavior and resulting taper benefit.

With reference to Figure 4.1, it should be visually obvious that the cavity that forms around a tapered pile both when it is driven and when it is loaded in axial compression is significantly different for piles with circular and square cross-sections. While the more-common circular cross-section is that of an expanding right-circular cylinder, a square cross-section is more like driving four separate rectangular blocks of soil outward from the pile-soil interface. What happens physically in the four quadrants (one at each corner of the pile) between these rectangular blocks is not immediately obvious. What is obvious, in the writer's opinion, is that the overall geometry of the expanding soil mass(es) is likely markedly different between the two different cross-sectional geometries.

Because of its modeling approximations, the Nordlund Method (and all of its derivatives) actually applies to both cross-sectional geometries although the calculated results do not distinguish between them. On the other hand, the Kodikara-Moore Method is, strictly speaking, only applicable to circular piles. Presumably, the donut-on-a-stick model only applies to tapered piles with a circular cross-section as well.

The writer is not aware of any published study that specifically compares the performance of tapered piles with circular and square cross-sectional geometries. Nordlund made some comments in Nordlund (1979) that are worth quoting in their entirety (writer's comments highlighted in yellow):

> "Tests 38 and 39 [Nordlund is referring to two specific case histories cited in Nordlund (1979).] were made on square, tapered, precast concrete piles. No matter how I try I cannot get these test results to fit. The only way they will fit is if I ignore the taper and then the fit is quite good as shown in Figure 12 [This is actually a tabular comparison of forecast (by Nordlund using his methodology) vs. measured results.]. However, this solution offends my scientific nature! Why should the taper for a circular section be considered and that for a square section ignored? I suppose it is possible that when a square (or rectangular) tapered pile is driven into sand, there is a concentration of soil stresses and displacements at the corners, arching tin [sic] effect, and this could significantly reduce the pressure on the pile surface at points distant from the corners. In any event, I do not have a well defined [sic] explanation for the behaviors of Tests 38 and 39. However, if you are considering the use of such piles, the taper should be ignored. At least you will get reasonable answers, even if for the wrong reasons."

In the writer's opinion, Nordlund's suggestion to completely ignore taper effects on piles with a square cross-sectional geometry seems extreme and flies in the face of use of such piles in several diverse geographic locations where, presumably, such piles have been found to have benefit over piles with a constant perimeter. On the other hand, in view of Nordlund's extensive experience with tapered piles, his admonition should not be ignored completely either. It appears to the writer that a prudent, middle course of action would be to acknowledge that taper benefit is likely affected by cross-sectional geometry and that fundamental research is required to define the physical behavioral mechanism that occurs with square tapered piles so that an appropriate physical and mathematical model can be identified and developed into a usable solution.

4.2.2.4 Directional Reversal: Residual Loads, Downdrag, and Uplift Loading

4.2.2.4.1 Introduction and Overview

The discussion of shaft-resistance fundamentals has, up to this point, focused on axial-compressive applied loading for its dominance in practice. In this case, the unit stress from the soil acting along the shaft (which is always under compression), whether from sliding friction or wedging, always has an upward directional sense as shown in Figure 2.1. Furthermore, this unit stress is always a resistance that develops in response to the applied load thus would not develop if there were no externally applied axial-compressive load.

However, there are several scenarios where the unit stress on the shaft not only reverses direction 180° and acts downward but does so even in the absence of an externally applied axial load. With reference to Figure 2.1, these scenarios are as follows:

- **Drag load** due to pile installation. These unit stresses develop within the upper portion of the shaft when the shaft rebounds upward relative to the adjacent soil after hammer impact. These loads are what produce residual loads within a pile after installation.

- **Drag load** due to *downdrag* that is settlement of the ground relative to the pile shaft as the result of any number of causes such as consolidation.

- **Uplift loading**[86] on relatively slender buildings due to some extreme event such as wind or earthquake.

- **Uplift loading** on specialized structures such as electric transmission towers under certain design-load cases such as unbalanced transmission-line loading due to line breakage on one side of a tower.

There are many ways in which the downward-acting unit stresses caused by these three, diverse causal mechanisms differ from the normal upward-acting unit stresses. To begin with, the drag loads associated with residual loads and downdrag produce <u>compressive</u> forces in a pile while uplift loading produces <u>tensile</u> forces.

Depending on the particular situation, the reversed, downward-acting unit stress on the shaft can be either a load (in the cases of residual load from installation and downdrag) or a resistance (in the case of uplift loading). This is a conceptual paradigm shift from the usual case of an upward-acting unit stress acting on the shaft where it is always a resistance.

In addition, both residual load and downdrag are, in a sense, created by the ground and will exist even if the load at the head of the pile is zero. On the other hand, uplift loading is, obviously, external to the pile and will only exist when it is applied.

Finally, each of the three causal mechanisms is independent of the other. Consequently, any one or two or even all three could occur for the same pile at the same time.

There are theoretical reasons that are discussed subsequently why for the same pile-soil conditions the unit stress acting downward could differ depending on which of the above three causal mechanisms caused the unit stress to develop. The writer is not aware of any published work to date that explores this potential difference.

As with the normal unit shaft resistance, the reversed-direction unit stress needs to be modeled differently for a constant-perimeter vs. tapered shaft. Therefore, separate discussions are presented in the following two sections.

4.2.2.4.2 Constant-Perimeter Shaft

The overarching issue concerning the unit stress acting on a DFE with a constant-perimeter shaft is whether or not that stress is the same or different depending on whether it acts upward (the more-common condition) or downward (the less-common condition

---

[86] In general throughout this monograph, 'uplift' is used in lieu of 'tensile' or 'tension' for loads applied to the head of a pile in the upward direction unless there is some perceived need to emphasize the tensile force or stress created within a pile by such loading. This reflects colloquial usage in the U.S. for such loads.

being addressed here). Numerous published opinions and data claiming to support these opinions have been made over the years for each side of the same vs. different argument.

The writer has no independent data to support either position, nor an opinion one way or the other at this point in time. However, an objective observation can be made. In addition, Appendix H discusses observations made by others for a case history that involved both constant-perimeter and tapered piles.

As noted above, there is a subtle theoretical difference depending on whether a downward-acting unit stress is associated with residual loads or downdrag within the soil mass or uplift from an externally applied load. This difference is based on what the writer terms the *Poisson Effect*, a topic that is discussed in some detail toward the end of this chapter. For now, the Poisson Effect is defined simply as the radial-displacement outcome of Poisson's ratio of the material or materials that comprise the DFE shaft.

Drag loads always cause an increase in <u>compressive</u> forces and concomitant axial-normal stresses in a DFE shaft. Due to the Poisson Effect, this would cause the shaft diameter to <u>increase</u>. As discussed toward the end of this chapter, this can, in principle, cause an increase in the unit stress.

On the other hand, an externally applied uplift load will always cause <u>tensile</u> forces and concomitant axial-normal stresses in a DFE shaft. Due to the Poisson Effect, this would cause the shaft diameter to <u>decrease</u>. As discussed toward the end of this chapter, this can, in principle, cause a decrease in the unit stress.

In summary and conclusion, there is a theoretical argument that can be made that there is at least a potential for a unit stress that has a downward-acting sense to differ depending on how that unit-stress was generated and in both cases differ from the baseline unit stress that has an upward-acting sense.

4.2.2.4.3 Tapered Shaft

In principle, the same Poisson Effect issue that applies to a constant-perimeter shaft should apply to a tapered shaft. However, with a tapered shaft the increase or decrease in shaft diameter resulting from the Poisson Effect is likely small in comparison to what happens geometrically to a tapered shaft under an externally applied uplift load.

This geometric change can be understood by visualizing Figure 4.1 but in reverse, i.e. as the shaft moves <u>upward</u>, the effective shaft diameter at some imaginary fixed reference plane x-x' in the ground <u>decreases</u>. Thus, it should be readily apparent that a <u>decrease</u> in horizontal/radial stress at that imaginary reference plane and concomitant <u>decrease</u> in unit stress acting on the shaft should result.

As an aside, one of the comments that the writer has heard over the years with regard to tapered piles is that they are presumed to have minimal resistance to uplift loading. The imagined mechanism proffered in support of this position is that uplift loading would cause a tapered pile to simply lift up and out of some imagined tapered hole in the ground, i.e. the soil forming this imagined hole would just remain in its position as the pile moved up and out of it. It should be obvious that this imagined mechanism is simplistic to the point of being totally unrealistic and at odds with actual soil behavior.

However, the strongest arguments against this imagined behavior is the ground-truth from actual uplift tests on tapered piles. Such ground-truth is presented in Appendix H where the results of uplift static load tests on both continuously and partially tapered piles are presented and discussed.

Nevertheless, it is highly likely that the unit-stress acting along a tapered shaft is reduced, at least to some extent, from that which existed prior to the external application of an uplift load. Again, this is separate from any reduction due solely to the Poisson Effect. The question then becomes how to model this reduction for the purposes of developing a resistance-forecasting methodology.

It does not appear that any proposed solution to this problem has appeared in the published literature to date. Therefore, it seems reasonable, at least as an initial attempt, to model the uplift behavior of a tapered shaft as simply a reversal of the wedging mechanism that occurs under axial-compressive loading. The physical model for this would, therefore, be the reverse of CCE which is herein defined as *cylindrical-cavity contraction* (CCC).

4.2.3 Horizontal Stress Effects on Resistance

4.2.3.1 Background and Overview

As noted earlier in this chapter, since the beginning of modern foundation engineering in the 1950s, horizontal stresses acting along the assumed planar interface between the shaft of a DFE and adjacent ground have always been recognized as playing an explicit, important role in resistance-forecasting methodologies that are based on the indirect approach. This is reflected in the linear relationship (Equation 4.3) with unit shaft resistance, r_s, in the traditional two-resistance-mechanism model shown in Figure 2.1.

Also as noted previously, in recent decades the importance of horizontal stresses for all types of DFEs has broadened in two distinct ways. One is evolutionary and involves new insights into the traditional understanding of how horizontal stresses affect shaft resistance, including extension of these concepts to tapered piles and the wedging resistance mechanism. The other is revolutionary and involves a completely new perspective of toe resistance. Each of these is discussed in detail in the following sections. As will be seen, the overall conclusion drawn from this presentation is that it is now appreciated that horizontal stresses dominate the behavior of deep foundations of all types. This means that knowledge of horizontal stresses in the ground both before and after installation of a DFE is more important than ever.

4.2.3.2 Shaft Resistance

Long-term research at Cornell University into deep foundations (drilled shafts in particular) that was largely funded by the electric-power industry in the U.S.[87] and conducted in the latter decades of the 20th century under the overall direction of the late Prof. Fred H. Kulhawy showed that the lateral earth pressure coefficient, K_h, that defines conditions acting along the shaft of a DFE after installation per Equation 4.5 had two significant influences:

- the type of DFE and its installation methodology and

- the pre-installation coefficient of lateral earth pressure at-rest, K_o, in the ground.

---

[87] This point is noted as the primary deliverables of this funded research were in the form of research reports, the distribution of which was tightly controlled by the funding organization, the Electric Power Research Institute (EPRI). Relatively little of this work appeared in publications such as journal and conference papers that were more widely accessible to the public. Only in later years were some of the EPRI research reports made available without restriction to the general public. Consequently, the substantial state-of-art body of knowledge generated by this EPRI-funded research has, unfortunately, been slow to be disseminated to both practicing engineers and academicians alike.

The former hypothesized influence is not surprising. In the case of (driven) piles, soil is displaced and densified, and soil particles possibly crushed, as a result of driving[88] although it has long been appreciated (e.g. Meyerhof's published work from the 1950s) that there are noticeable differences between 'high-displacement' piles such as a closed-toe steel pipe and 'low-displacement' piles such as steel H-piles. With a drilled shaft, soil is excavated and there is concomitant stress relief in the vicinity of the shaft. Note that this stress relief was the primary motivation in recent decades for developing the concept of post-grouting drilled shafts, to restore and even increase the pre-installation soil stresses and densities in order to increase shaft resistance. Other, newer types of deep foundations that either involve various forms of injecting grout under pressure or rotating the pile into the ground as in the case of 19th-century *screw piles* and modern *helical piles* have their own unique effect on the horizontal stress state in the ground.

The latter hypothesized influence of the pre-installation K_o is, in retrospect, intuitive to the point of being obvious although Kulhawy clearly deserves considerable credit for not only formally calling attention to the concept but, more importantly, playing a significant role in advancing the site-characterization geotechnology necessary to make use of this concept. Historically, it had been practically impossible to estimate K_o, especially for coarse-grained soils, so recognizing the influence of this fundamental soil property on deep-foundation resistance was, at best, of academic interest only and of no use to practitioners. However, tremendous advances in site characterization based on various in-situ testing devices, especially the cone penetrometer, that occurred beginning in the latter decades of the 20th century and that continues to the present have now made estimating K_o relatively simple and straightforward on even the smallest of projects.

Kulhawy and Mayne's seminal research report (Kulhawy and Mayne 1990) that was produced during the aforementioned EPRI-funded research program at Cornell and largely launched Mayne's career as one of the world's leading experts in the area of site characterization, summarized state-of-art site-characterization knowledge in one applications-oriented document for the first time. As such, it was a landmark publication in the technology-transfer process of moving research into routine practice. This report was also an indispensable tool in the writer's initial published efforts (Horvath 1989, 1994, 2000a, 2000b, 2002, 2003a, 2004, 2011) to couple state-of-art site characterization and foundation analyses into one integrated, seamless analytical process. The information provided in Kulhawy and Mayne (1990) allowed the writer to develop a site-characterization algorithm[89] whereby even SPT N-values could be used to generate quasi-cone penetrometer q_c values and from this an estimate of K_o. Horvath (2002) illustrated how results obtained from both SPT N-values and a true CPT sounding compared for a tapered pile bearing almost completely in coarse-grained soil.

All of Kulhawy's concepts can easily be incorporated into the general equation (Equation 4.9) for unit shaft resistance due to sliding friction, r_{ss}, by using Kulhawy's suggestion (Kulhawy 1984, 1991) for representing installation effects using a dimensionless K_h/K_o ratio and assuming that a site-specific values of K_o would be determined as part of a project's site characterization. The result is the following expanded version of Equation 4.9:

---

[88] Note that partial pre-drilling or partial vibratory installation that are sometimes done as part of conventional pile driving can complicate this scenario.

[89] Horvath (2011) is the latest officially published version but it has been superseded many times over to the present by unpublished versions that the writer used for calculated results presented in Horvath (2014, 2015, 2016a, 2016b) as well as this monograph.

$$r_{ss(ult)} = K_h \cdot \tan\left[\left(\frac{\delta}{\phi}\right)_{cv} \cdot \phi_{cv}\right] \cdot \sigma'_{vo}$$
$$= \left[\left(\frac{K_h}{K_o}\right) \cdot K_o\right] \cdot \tan\left[\left(\frac{\delta}{\phi}\right)_{cv} \cdot \phi_{cv}\right] \cdot \sigma'_{vo} . \tag{4.17}$$

As further proof of the utility of Kulhawy's concepts, as noted earlier in this chapter in this monograph the category of indirect analytical methodologies includes those composed of hybrid combinations of theoretical and empirical problem variables. Arguably the best known of such hybrids is the β Method that is defined in Equation 4.6. Horvath and Trochalides (2004) showed that Kulhawy's concepts as applied to the generic theoretical equation for r_{ss} in Equation 4.17 can logically and easily be extended to hybrid methodologies such as the β Method. This is because the β parameter incorporates both deep-foundation typology and installation effects and, as such, is conceptually a combination of K_h and δ.

With this in mind, Equation 4.6 can be extended as follows:

$$r_{ss(ult)} = \beta \cdot \sigma'_{vo} = \left[\left(\frac{\beta}{K_o}\right) \cdot K_o\right] \cdot \sigma'_{vo} . \tag{4.18}$$

Furthermore, there are indications (Horvath and Trochalides 2004) that using an expanded visualization of the β Method as shown in Equation 4.18, i.e. replacing β with $[(\beta/K_o) \cdot K_o]$, provides improved correlation with measured resistances compared to using just β alone. This is especially true at relatively shallow depths where β values tend to exhibit the greatest variability for a given deep-foundation type and soil, likely because it is within shallower depths where K_o exhibits the greatest depth-wise variability on many sites and thus tends to produce relatively more variability and scatter in back-calculated β values.

Turning attention now to the unit shaft resistance along the tapered portion of a pile, r_{st}, the most general governing equation (Equation 4.10) for this parameter indicates that its value is dominated by σ'$_h$. Therefore, it is logical to assume that σ'$_h = K_h \cdot$ σ'$_{vo}$ as with r_{ss} which means that an equation identical in form to Equation 4.17 for r_{ss} would apply to r_{st} as well. However, the significant difference is that K_h and, by extension, K_h/K_o for r_{st} would be a function not only of installation effects and pile material but taper angle, α, and pile settlement under post-installation applied loads as well.

This latter issue is a significant paradigm shift as it indicates that the ULS unit shaft resistance is not always a fixed quantity determined by the stress state at the completion of DFE installation. Rather, for tapered piles at least the ULS unit shaft resistance is dependent on the magnitude of post-installation settlement under externally applied loads as well.

Unfortunately, Kulhawy did not make any recommendations for K_h/K_o values for tapered piles or any type of tapered DFE for that matter. Therefore, to the extent that it is desired to replace K_h in some tapered-pile solution for r_{st} with the more-general expression $[(K_h/K_o) \cdot K_o]$ in order to explicitly incorporate K_o effects, a way to evaluate K_h/K_o for tapered piles needs to be a part of the process. Note that this also opens the door for possibilities such as being able to correlate K_h/K_o with taper angle and/or other problem-specific parameters, including settlement under applied loads.

Considering first Nordlund's solution, in his problem formulation (Equation 4.11) he used an earth-pressure coefficient, K_δ (which is never aligned horizontally), in lieu of K_h. Thus, some analytical work is required to relate K_δ and K_h.

To deal with this fact, the generic definition of K_h (Equation 4.14) can be expanded using Kulhawy's concepts:

$$\sigma'_h = K_h \cdot \sigma'_{vo} = \left[\left(\frac{K_h}{K_o}\right) \cdot K_o\right] \cdot \sigma'_{vo} \qquad (4.19)$$

and inserted into Equation 4.12 that defines $r_{st(ult)}$ in terms of Nordlund's problem parameters:

$$r_{st(ult)} = \tan \delta \cdot \sigma'_h = \tan \delta \cdot \left[\left(\frac{K_h}{K_o}\right) \cdot K_o\right] \cdot \sigma'_{vo}$$
$$= K_\delta \cdot \cos(\alpha + \delta) \cdot \tan \delta \cdot \sigma'_{vo} . \qquad (4.20)$$

Rearranging and eliminating terms leaves:

$$\left(\frac{K_h}{K_o}\right) = \left(\frac{K_\delta}{K_o}\right) \cdot \cos(\alpha + \delta) . \qquad (4.21)$$

Noting that Nordlund assumed that α was small relative to δ and could thus be neglected in this specific equation produces the final desired result:

$$\left(\frac{K_h}{K_o}\right) = \left(\frac{K_\delta}{K_o}\right) \cdot \cos \delta . \qquad (4.22)$$

Note that this is a logical result as K_δ is always assumed to be oriented at an angle equal to $(\alpha + \delta) \cong \delta$ below the horizontal so the product $(K_\delta \cdot \cos \delta)$ is simply the horizontal component of K_δ with α neglected and thus conceptually equivalent to K_h.

In the writer's Modified Nordlund Method, one of the changes incorporated from the start when developing this methodology was to define the lateral earth pressure coefficient, K_h, as always acting horizontally. As a result, incorporating Kulhawy's concepts into this method is straightforward as was already shown in Equation 4.16.

4.2.3.3 Toe Resistance

The evolution of thinking with respect to toe-resistance mechanisms in general has gone through tremendous change over the years and is discussed in detail in its own section that follows. The discussion here is limited to insights learned in recent decades concerning how horizontal stresses in the ground factor into the broader discussion of toe-resistance mechanisms.

The broad geometrical and physical similarities between a cone penetrometer being pushed into the ground and a DFE subjected to axial-compressive loading have long been recognized. Thus, it is no surprise that from the earliest efforts to develop direct analytical methodologies for DFE resistance, e.g. Meyerhof's circa-1950s research that was noted earlier in this chapter, that correlations based on the basic parameters measured in a cone-penetrometer sounding (primarily q_c and f_s) were used to estimate deep-foundation resistance.

With this in mind, the outcomes of research performed in recent decades (e.g. Salgado et al. 1997, Salgado and Prezzi 2007) to understand the very complex stress state that develops around the tip of an advancing cone penetrometer are very enlightening for their obvious extension to deep foundations of all types under axial-compressive applied loading. Contrary to what intuition might suggest, it turns out that the uncorrected tip resistance, q_c, of an advancing cone penetrometer is governed directly by the <u>horizontal</u> effective overburden stress, σ'_{ho}, <u>not</u> the <u>vertical</u> effective overburden stress, σ'_{vo}. In turn, this means that K_o plays a significant role in determining the measured value of q_c.

Thus, the theoretical relationship between q_c and σ'_{ho} (= $K_o \cdot \sigma'_{vo}$) is the underlying reason why it has become possible to estimate K_o in routine practice for all types of soil as was noted earlier in this chapter. Research in recent decades has yielded empirical relationships between q_c and σ'_{ho} that have been upgraded and improved continuously over time although the ability to define separate loading and unloading values of K_o remains elusive. This evolution in developing empirical relationships for K_o has been reflected in the writer's aforementioned cone penetrometer-based site-characterization algorithm that remains a constant work in progress as discussed in Horvath (2003a, 2004, 2014b, 2015).

The most significant practical implication of this research from the perspective of deep foundations in general, and piles in particular, is that, all other things being equal, at the same value of σ'_{vo} the value of q_c will increase with increasing K_o. Thus, it stands to reason to expect that at a given value of σ'_{vo} the toe resistance of a DFE would be influenced primarily by K_o and, as a result, σ'_{ho}.

4.2.4 Toe-Resistance Mechanisms

4.2.4.1 Background

Historically and to the present, most types of DFEs were and are designed assuming that the ULS total resistance (R_u in Figure 2.1) is simply the sum of the ULS shaft and toe resistances. Central to this design approach is the assumption that there are unambiguous, well-defined, single values for each of these ULS resistance components.

Using traditional Allowable Stress Design, a safety factor is applied to the ULS total resistance to produce a reduced, maximum-allowable or design capacity that is matched against the estimated actual (working or service) loads. With the more modern Load and Resistance Factor Design, resistance (reduction) factors are applied to the ULS-resistance components that are then matched to factored (increased) service loads.

There are two key points being made here:

- DFE design for axial loading has always been and still is strength-based in most cases, with no routine consideration of settlements under any load level.

- Defining a single-valued ULS toe resistance is central to both the ASD and LRFD design methodologies. In fact, in many ways ULS resistance is more important than ever with LRFD given that ULS resistance is the theoretical basis of this design approach.

Given the clear significance in deep-foundation design of always being able to define a unique single value of ULS toe resistance, it is important to discuss in some detail the current methodologies for doing so analytically.

4.2.4.2 Traditional Bearing-Capacity Solutions

The classical problem of bearing capacity was one of the core concepts of modern soil mechanics when it emerged as a distinct body of knowledge within civil engineering in the 1920s. This is not surprising as the underlying theoretical basis of all traditional bearing-capacity solutions developed to the present is the so-called *Prandtl punch problem* from solid mechanics that was already well-established published science and thus undoubtedly well-known to Terzaghi and other early pioneers in soil mechanics.

It is relevant to the present discussion to recall that Prandtl's problem is a classical theory-of-plasticity problem that has nothing to do with soil. It assumes rigid-plastic 'mechanical' (stress-strain) behavior for a weightless sheet of metal that is brought to a localized state of material failure by a perfectly rigid circular punch that creates at least an indentation, if not a complete hole, in the sheet. The fact that the sheet is composed of metal is significant because this implies a material with inherent, stress-independent strength. In the context of the well-known Mohr-Coulomb failure hypothesis that is used for geomaterials, this means a 'true' solid with only cohesion and no friction. Thus, it is not surprising that generations of geotechnical researchers (Terzaghi was only the first) have had to make numerous modifications and additions to the basic Prandtl problem to bring it closer to solving problems involving geomaterials that are particulate materials, not true solids.

There are several specific aspects of the Prandtl punch problem, both in its original form and as extended by Terzaghi for use with shallow foundations (he had to add terms for soil weight and embedment effects to make it viable for geotechnical applications), that require highlight and emphasis as they are particularly relevant to various aspects of this monograph:

- The downward penetration of Prandtl's punch into the metal sheet creates what is now referred to as a *general-shear* failure mode (mechanism) that was appropriate for Prandtl's overall problem assumptions. In simple terms, this failure mode means that the assumed zero-thickness failure surface that defines the boundary between failed and non-failed solid material propagates continuously from beneath the centerline of the punch to the upper surface of the sheet. This is an often-unremarked implication of classical bearing-capacity theory, that the material involved is incompressible (the result of assuming rigid-plastic material behavior) and in order for the punch to penetrate downward into the material, the material beneath the punch has to 'flow' (displace) somewhere. According to the failure mechanism assumed by Prandtl and perpetuated by subsequent geotechnical researchers, that "somewhere" is first horizontally and then upward out of the way of the punch or foundation element as the case may be.

- There is always a well-defined, single-valued failure stress and, therefore, failure force. This stress magnitude is generally referred to as the *bearing capacity*.

- The single most important point for the purposes of the present discussion is that the extended solution for geotechnical applications that was developed by Terzaghi and, subsequently, many other researchers who made their own interpretation of the general problem and often added other effects (e.g. inclination of the applied load) assumes that only <u>vertical</u> overburden stresses influence the problem. This is because only vertical overburden stresses 'hold down' the failed material that wants to 'flow' outward and then upward from beneath the 'punch'. Horizontal stresses within the soil either before or at failure play no role in this assumed behavior.

The reason for emphasizing this last fact is that, as noted earlier in this chapter, it is now understood that <u>horizontal</u> (<u>not</u> vertical) effective stresses control toe bearing for all DFEs. This is explored further in the following section.

Although the bearing-capacity problem was developed initially for shallow-foundation applications, it did not take long for the same basic concepts and failure mechanism to be extended to deep foundations. Over time, it was recognized that the significant embedment of deep foundations compared to shallow foundations altered the assumed failure mechanism (e.g. it was no longer reasonable to assume that the general-shear failure surface propagated all the way to the ground surface and that 'failed' soil would 'flow' all the way to the surface) so bearing-capacity solutions specifically with deep-foundation applications in mind were developed. This typically involved postulating general-shear failure mechanisms that were fully contained within the soil around the toe of the DFE. This resulted in solution variables, especially the three *bearing-capacity factors*, having values that differed from those obtained for shallow-foundation solutions. However, the basic bearing-capacity concept and solution that has three distinct components ('soil strength', which was the holdover from the original Prandtl problem; soil weight; and embedment) remained the same.

As an aside, the coversheet for the version of Fellenius (2018) that Fellenius makes available on his website[90] is an insightful visual reminder of the significant range of theoretical results that have been generated over the last 100 years for bearing-capacity factors for use with deep foundations. Shown is a plot of the N_q factor for embedment (which dominates solutions for deep foundations) versus the ϕ angle of the soil. As is typical, such plots use a $\log_{10}$ scale for N_q so that the plotted range can be compressed. However, the range in results, which spans approximately two orders of magnitude (i.e. a factor of 100), is still overwhelmingly apparent. Given the linear relationship between N_q and bearing capacity (and thus the ULS toe resistance, $R_{t(ult)}$, shown in Figure 2.1), the enormous potential range in calculated outcomes is distressingly obvious.

In any event, arguably the most significant evolutionary development with regard to classical bearing-capacity theory for all types of foundation elements, shallow and deep, was the appreciation that grew over time that there were two other modes of failure that can develop in addition to the baseline general-shear failure observed in the original Prandtl Punch Problem:

- *local shear* and

- *punching shear.*

The key element of each is that the failure surface does not propagate all the way to the ground surface, even for shallow foundations. In fact, in the case of punching shear the failure surface does not even propagate much horizontally from beneath the footing or toe of the DFE.

Terzaghi had actually noted early on that a general-shear failure did not always develop with shallow foundations with loose/soft soil conditions and had suggested an approximate, empirical way to deal with this. However, it was not until the latter half of the 20th century that these alternative failure modes received formal research attention and, ultimately, theoretical solution. The late Prof. Aleksandar S. Vesic deserves credit for identifying *soil compressibility* (sometimes referred to using its complementary opposite, *soil rigidity*) as being the underlying causative factor for there being a range in failure modes in the bearing-capacity problem.

Simply stated, one of the significant holdovers from the original Prandtl problem is the assumption that the material being punched (i.e. the soil beneath a footing or toe of a DFE in foundation applications) has idealized rigid-plastic material behavior, i.e. there is no strain (displacement and/or deformation) up to the point of material failure. This turns out to be an essential requirement for development of general-shear failure pattern,

---

[90] http://www.fellenius.net/papers.html, last accessed 5 March 2020.

What Terzaghi had noted early on, and Vesic formalized decades later, was that while real soil will always strain (with concomitant settlement of the foundation element) to some degree prior to failure, in some cases (typically relatively dense/stiff soils under relatively low confining stresses such as in shallow-foundation applications) these strains and the settlement they produced could be ignored as a general-shear failure will develop. However, in other cases (looser/softer soils under low confining stresses in shallow-foundation applications or most soils under relatively high confining stresses as in deep-foundation applications), the soil strains cannot be ignored and either a local- or punching-shear failure mechanism will develop.

The way that Vesic addressed the issue of soil compressibility/rigidity was to introduce correction factors into the traditional bearing-capacity problem solution so that the calculated gross ultimate (ULS) bearing capacity value was reduced as necessary for soil-strain effects. Kulhawy wrote a seminal paper (Kulhawy 1984) to illustrate that considering soil-compressibility effects was absolutely essential for deep foundations.

Unfortunately, although Vesic's work related to soil compressibility/rigidity has been known for almost 50 years, it remains relatively unused in routine practice to the present time, at least in the U.S. The primary reason is that evaluating soil-compressibility effects has always been and still is complex and difficult to perform <u>accurately</u>. This is because soil stiffness in the form of the soil shear modulus, G, must be considered explicitly in the problem solution. Specifically, the algebraic expressions for evaluating Vesic's compressibility/rigidity factors require a single-valued G. However, it is now well understood that the correct, operative value of G to use in any geotechnical analysis is highly strain-dependent. This was noted earlier in this chapter and is explored in detail in Chapter 5 as it is relevant to several analytical methodologies for deep-foundation resistance. Furthermore, by definition, soil strains at failure tend toward infinity in the limit so it is ill-defined as to what value of G should be used in this case.

In summary and conclusion up to this point, indirect analytical methodologies for forecasting the ULS total resistance of all types of DFEs requires the use of some analytical methodology for forecasting the ULS toe resistance. Historically, this has been done by using a traditional bearing-capacity solution that has the simplistic Prandtl punch problem as its underlying theoretical basis.

Experience has shown that the current state of affairs with respect to using such traditional bearing-capacity solutions for calculating the ULS toe resistance of deep foundations is, in the writer's opinion, muddled. On the one hand, solutions for the baseline general-shear failure mode have often been found to overestimate the ULS toe resistance. On the other hand, solutions that take into account soil compressibility/rigidity using Vesic's correction factors as recommended by Kulhawy are problematic to use given the uncertainty in the operative value of soil shear modulus that should be used. In the writer's experience (Horvath 2002), the assessment of soil compressibility/rigidity effects on bearing capacity are very sensitive to the value of G used in the analysis.

This, then, opens the door to several alternative approaches:

- Use empirical correlations with in-situ test data, especially from cone penetrometers, to estimate the site-specific operative value of rigidity index, I_r, and reduced rigidity index, I_{rr}, that incorporate shear modulus (Mayne 2006a, 2019b).

- Use empirical values for the bearing-capacity factors that can be derived from a variety of approaches, especially instrumented load tests on actual DFEs (discussed in Chapter 5 in the context of specific resistance-forecasting methodologies).

- Use a direct methodology where the bearing-capacity factors are correlated to in-situ test results (also discussed in Chapter 5 in the context of specific analytical methodologies).

- Use a different theoretical concept for calculating the ULS toe resistance (this is discussed in the following section as well as Appendix G).

- Look at the broader picture as has long been suggested by Fellenius (2018) and question whether or not the concept of a well-defined, single-valued ULS toe resistance really exists (discussed later in this chapter).

4.2.4.3 Spherical-Cavity-Expansion (SCE) Theory

There have been attempts dating back to at least the 1960s to develop alternative theoretical models for assessing the ULS toe resistance of DFEs. The physical mechanism used most often to date for this purpose is *spherical-cavity expansion* (SCE). The basic assumption is that loading the toe of a DFE in axial compression can be visualized as expanding a spherical region of soil beneath the toe. The support for this hypothesis comes, in large part, from the observation of failure patterns that developed in model tests on piles that were conducted by Vesic and others many decades ago.

What is interesting is that the physical mechanism of SCE beneath the toe of a DFE implies that <u>horizontal</u>, not vertical, stresses play the more predominant role in the problem solution and, ultimately, the forecast value of deep-foundation ULS toe resistance. This is a marked departure from the traditional bearing-capacity solutions based on the Prandtl punch problem that were discussed previously.

Although the application of cavity-expansion theories (both spherical and cylindrical) in general to geotechnical applications apparently dates back to at least the 1940s, it appears that the first researcher to pursue this line of research with the specific application to deep-foundation toe resistance was Vesic. This work was apparently conducted during his extensive research into the general subject of bearing capacity in the 1960s and 1970s and resulted in several publications (Vesic 1972, 1975, 1977).

Vesic's work was followed approximately 20 years later by the work of Yu and Houlsby (1991) who investigated both spherical- and cylindrical-cavity expansion as geotechnical engineering tools. In fact, they used deep-foundation bearing capacity as a specific example of an application of their theoretical work.

Much more recently, after another two-decade gap, Manandhar and Yasufuku (2012, 2013) published results of research that have particular application to the theme of this monograph. They postulated (without any physical evidence as support) that the geometry of the mechanism of SCE (they used Vesic's original model) is influenced by the geometry of the tapered shaft of a tapered pile and developed a solution based on this hypothesis.

There are some important issues to be raised from the perspective of the utility of any of these SCE solutions in routine practice as a reasonable alternative to traditional bearing-capacity solutions:

- It is one thing to generate figures in an academic publication that show calculated results for some suite of assumed model parameters and quite another thing to show that the calculated results actually compare well with measured values in an actual application. The Yu and Houlsby paper is a superb example of doing the former adequately while largely ignoring the latter which, in the writer's opinion, calls the practical value of the presented methodology into question. Manandhar and Yasufuku's papers did go somewhat further in an attempt to correlate theoretical results with measured results. However, the latter were obtained using small-scale, 1-*g* physical models that had geometries that differed from any known commercially available tapered pile. In addition, these model piles were placed within the soil of the test chamber in a manner that did not replicate the installation of any real DFE. When taken together, the net result is that the relevance of the measured results to reality is highly questionable as has been elaborated on by the writer earlier in this monograph.

- Overall, the practical limitation of theoretical work involving SCE is that it suffers from the same problem discussed previously for Vesic's compressibility/rigidity factors for traditional bearing-capacity solutions. Specifically, use of cavity-expansion theories in general, both spherical and cylindrical, require use of single-valued soil properties such as shear modulus, G, and Mohr-Coulomb friction angle, ϕ. As has now been noted several times, both of these properties are never constant in any geotechnical application. The shear modulus, whether viewed on a tangent or secant basis as described generically in Chapter 2, decreases with increasing soil strain and the peak friction angle decreases due to decreasing dilatancy component under increasing stress. That having been said, recent developments involving empirical correlations with cone-penetrometer data (Mayne 2006a, 2019b) offer substantial promise to overcome these issues.

- Accounting for real-world behavioral effects such as shear-modulus degradation in any practical application is theoretically challenging, a fact acknowledged by Yu and Houlsby in the summary and conclusion section of their 1991 paper. As demonstrated by Cook (2010) in a research study involving geotechnical application of cylindrical-cavity expansion (CCE), properly considering effects such as modulus degradation and stress-dependent degradation of dilatancy angle results in a very complex analysis yet still does not guarantee good correlation between forecast and measured results.

The conclusion drawn here is that there is still a gap between theoretical development and practical application with regard to use of SCE to calculate ULS toe resistance for DFEs. However, this gap has closed substantially in recent years as the result of research related to cone penetrometers so at least after many decades of inactivity things are moving in the right direction. It is unclear how and when this gap might be closed but until it is it remains uncertain whether SCE theory is the clear-cut superior choice over traditional bearing-capacity solutions or the other alternatives enumerated at the end of the preceding section for this application.

In the interim, the writer has, in recent years, used Vesic's SCE solution as one of several methods for evaluating the ULS toe resistance of tapered piles. The writer's current strategy is to use several analytical methodologies for ULS toe resistance and then use engineering judgment based on a comparison of the forecast results.

4.2.4.4 Other Trends and Developments

From the preceding discussions of traditional bearing-capacity and spherical-cavity-expansion theories and concomitant solutions, it is clear that there are still issues and difficulties with using any type of theoretical approach for <u>accurately</u> assessing ULS toe resistance for any type of DFE. Unfortunately, making such an assessment is a necessary component of any indirect analytical methodology for resistance forecasting.

Consequently, it is no surprise that from the circa-1950s beginning of developing DFE resistance-forecasting methodologies based on the static approach that direct methods for toe resistance have been developed in parallel with, and as an alternative to, theoretical approaches for use with indirect methods. As noted earlier in this chapter, in the 1950s and 1960s Meyerhof developed and revised direct-method analytical methodologies for ULS toe resistance using both SPT N-values and cone-penetrometer q_c values. This trend toward using direct methods for estimating ULS toe resistance has only accelerated over time and recent decades in particular have seen a proliferation of new methodologies, primarily based on the cone penetrometer.

Recent research into direct methods for determining ULS toe resistance has actually brought to the forefront an issue that foundation engineers have undoubtedly been aware of since the first conventional static load tests were performed on deep foundations bearing entirely within soil (as opposed to end bearing on rock). Specifically, some (or many or most or even all, depending on whom is asked for an opinion) DFEs loaded in axial compression do not exhibit a well-defined 'failure' (ULS) load characterized by a more-or-less vertical post-yield 'plunge' of the load-settlement curve. Rather, the load-settlement curve displays a sharply curved zone of yielding that is followed by a relatively steep, but not vertical, slope in the load-settlement curve[91]. This is especially true of piles bearing in coarse-grained soil as shown in Figure 3 of Vesic (1977) and countless other publications.

The net result of this behavior is that the DFE develops some additional post-yield resistance, albeit while undergoing relatively large settlements. Due to practical limits in load-testing hardware, typically no more than several inches (tens of millimetres) at most of settlement is recorded before the test is terminated. Thus, it is not possible to say for how long this 'not quite failed' behavior continues.

Note that due to residual loads (noted earlier and discussed further later in this chapter), the reverse behavior can be observed in a static load test. A DFE may appear to exhibit the classic plunging failure but, in reality once residual loads have been taken into account, actually exhibits the above-described non-plunging behavior (Fellenius 2015).

As an aside, this observed behavior of there not being a well-defined plunging type failure should not be confused with load tests that are not carried to sufficiently large settlement magnitudes to define yield and post-yield behavior. In the writer's experience, this is a growing trend that only seems to have gotten worse in recent years as project stakeholders seem to focus only on doing the minimum required to reach and define some very conservative 'failure' metric such as the well-known Davisson Offset Limit criterion. This issue of defining a 'failure load' from load-settlement data is discussed in detail later in this chapter.

In any event, this observed load-test load-settlement behavior of an ill-defined or undefined failure load has had two distinctly different effects on the state of practice. On the one hand, it has resulted in some of the newer direct methodologies for toe resistance being linked to settlement in one way or another. For example, in some cases a methodology allows for different values of toe resistance to be forecast as a function of settlement, often expressed as the dimensionless ratio of settlement divided by the shaft diameter of the DFE. The simpler approach adopted by other analytical methodologies is to forecast only one value of toe resistance for some specified settlement magnitude or settlement/shaft-diameter ratio. Regardless of which approach is used, the point still remains that the traditional concept of there being a unique, unambiguous, single-valued ULS toe resistance is not valid in certain ground conditions, primarily coarse-grained soil.

---

[91] It is worth noting that these visual clues can be affected by something as simple as the plotting scale of load-test data. This is well-illustrated in Figure 4 in Vesic (1977).

The other outcome that has evolved from this observed load-test behavior of an undefined or ill-defined ULS load magnitude is rather radical in concept and to date appears to have been embraced only by Dr. Bengt H. Fellenius. However, given Fellenius' professional stature in the world of deep foundations in general and driven piles in particular over the past 50 years, his stated position is at least worthy of note. More importantly, there are implications of Fellenius' opinion that are worthy of consideration even if one does not embrace the totality of his position.

In simple terms, for some time now Fellenius has stated in the published record (e.g. Fellenius 1999a, 2016, 2018) that the concept of a bearing-capacity failure simply does not even exist in foundation engineering. Fellenius included both shallow (spread footing) and deep foundations in this absolute, broad-brushed stated opinion.

The only comment made here is that this broad statement is in conflict with research (Vesic 1977) that clearly shows that <u>some</u> type of failure pattern does develop beneath the toe of a pile, at least in 1-g model tests where failure patterns can be observed (Fellenius is not known to have performed any similar physical tests). Taking into consideration the issues with 1-g testing of small-scale models noted earlier in this monograph, these test results at least qualitatively suggest that some type of soil failure does occur beneath the toe of a pile. There is room for debate whether this observed failure is better analyzed using a local-failure bearing-capacity mechanism or an SCE mechanism or perhaps something else but the fact remains that some mechanism other than soil stiffness alone, as Fellenius suggests, is at work here.

That having been said, the implications, at least, of Fellenius' opinion do have merit for consideration for implementation into practice. Specifically, Fellenius has argued that, in general, piles (and presumably all types of DFEs) should be designed primarily on the basis of piles settlement combined with a consideration of structural resistance within the pile shaft so that there is an adequate structural margin of safety. Whether this structural safety is assessed using ASD or LRFD concepts is just a detail. The radical departure and paradigm shift of Fellenius' argument from the status quo is that geotechnical ULS resistance is removed as a primary design parameter. The merit in this design approach is that it removes the need to come up with a single value for ULS toe resistance.

It is worth noting that while the specifics of Fellenius' position (i.e. that there is no soil-failure mechanism of any kind beneath the toe of a DFE) are open to discussion and debate, the implications of his position (i.e. that the geotechnical aspects of deep-foundation design should be based primarily on settlement considerations) are actually in consonance with the state of practice for shallow-foundation design since the 1950s. Thus, the implications of Fellenius' position have actually been embraced by geotechnical and foundation engineers for almost 70 years now so are really nothing new.

In summary and conclusion, given the observed ambiguities in the load-settlement behavior exhibited by many piles, it is clear that defining an unambiguous, single-valued ULS load using any analytical methodology, whether indirect or direct, is fundamentally and inherently impossible in many cases. This is because while shaft resistance (at least due to sliding friction) generally has a well-defined ULS value, toe resistance does not. Note that this implies that interpreting a unique, single-valued ULS load from load-test data is equally ambiguous, a subject that is explored in detail later in this chapter.

As a consequence of this reality and independent of whether or not one agrees with Fellenius that bearing failure (including SCE) in the classical sense never develops beneath the toe of a DFE bearing in soil, Fellenius' position that pile assessment should be based on settlement and structural resistance considerations alone has merit in the writer's opinion. The writer proposed an approximate technique for estimating pile settlement in Horvath (2002) and illustrated its application in Horvath and Trochalides (2004) and Horvath et al.

(2004a, 2004b). Mayne (2007) proposed a more-formal treatment of settlement applicable to all types of DFEs that is based on a theory-of-linear-elasticity boundary-value problem and concomitant solution that the writer has called the *Randolph-Wroth Problem*. This problem and its several modifications and extensions by Niazi are a significant element of this monograph, and are presented and discussed in considerable detail in Chapter 5.

Unfortunately, this trend toward recognizing, at long last, that the concept of a well-defined ULS load for DFEs bearing entirely in soil and loaded in axial compression is often ambiguous and ill-defined and thus needs to be replaced by a better analytical approach is hampered by the global trend toward using only LRFD design concepts for deep foundations in general. By its nature, LRFD is ULS-based in its concept and formulation and tends to treat displacements and deformations as a secondary issue, if at all.

While LRFD may work well for structural materials where material failure is accompanied by an obvious, unambiguous, and permanent material rupture with a complete loss of material strength, it does not, in the writer's opinion, translate well to soil where failure is marked by relatively large displacements and, most importantly, not a complete loss of strength but simply a 'maxing out' of strength to an infinitely sustainable residual level (the writer admits to not being a 'fan' of LRFD for geotechnical applications in general).

Furthermore, although a failure surface or zone may be clearly evident within a failed soil mass, the soil particles themselves are, for the most part, intact so that if the soil particles within the failure zone were to be removed and then put back into place all evidence of prior failure would be obliterated. There is no permanent material rupture as with solid materials. More importantly, the soil strength from these 'failed' soil particles can be made the same or even better than that which existed at the time of failure by following appropriate placement and compaction protocols.

The net result of all this is that the future of an approach to pile design that is based first and foremost on considerations of pile settlement and structural resistance in a global design world dominated by and, it appears, singularly obsessed with LRFD is unclear.

4.2.5 Temporal Effects on Resistance

A general discussion of the time-dependent variation in axial-compressive pile resistance, typically an increase but a decrease can occur, is beyond the intent and scope of this monograph. The aspects of temporal effects discussed here are limited to those that directly impact the topics and content discussed subsequently, primarily in Chapter 5 that contains a detailed assessment of existing resistance-forecasting methodologies.

The length of time after installation has long been recognized in principle as influencing axial-compressive pile resistance although historically it was judged to be strength related and thus felt to be of practical significance only for fine-grained soils. This is due to the obvious consolidation effects subsequent to the undrained soil displacements, excess pore-pressure generation, and localized soil remolding that occurs as the result of pile driving, and the concomitant increase in pile resistance over time as the excess porewater pressures that were generated during driving dissipate. However, the seminal paper by York et al. (1994) established that at least in some cases (significantly, for the theme of this monograph, when tapered piles are involved), time-dependent increases in resistance (colloquially called *setup* or *soil freeze*) can be significant for coarse-grained soils as well. This is important as the benefit of using tapered piles has been most studied, and is thus most widely established and recognized, for coarse-grained soil conditions, primarily sand. The benefit, if any, of tapered piles with fine-grained soils has received minimal attention to date and is thus much less established and historically has often been considered to be nil.

The most significant practical issue and challenge concerning temporal changes in axial-compressive resistance for tapered piles in coarse-grained soils is quantifying the increase in resistance-forecasting methodologies. To date, this has not been done, at least to any significant, widespread extent, for methods based on the static approach, whether indirect or direct. This is because static methodologies are, in general, based on correlations with traditional static load tests where loads are placed mechanically on the head of the pile. To develop a mathematical relationship for the increase in resistance over time would require repeated load tests on the same pile. Not only would this be costly but there are significant issues with repeated loading and unloading of the pile-ground system that would complicate interpretation of the load-test results (Fellenius 2019).

As an aside, the fact that piles in coarse-grained soil can exhibit an increase in resistance over time calls into question most, if not all, of the static resistance-forecasting methodologies, both indirect and direct, that have been developed since the 1950s and up to the present. This issue is discussed in greater detail later in this chapter for reasons given there but the concern raised here is that rarely is the time after driving stated for load-test results used to develop these methodologies. Thus, some of the scatter that is typically found in the databases for these methodologies is likely due simply to variations in time after driving when the load tests were performed.

With regard to the dynamic approach to resistance forecasting, note that, by definition, temporal effects cannot be included in such methodologies as they equate pile resistance at the time of driving with, by implication, immediate post-driving resistance. That having been said, dynamic measurements made in the field at the time of driving have been the primary tool used to measure temporal increases in resistance by comparing interpreted resistance at the *end of initial driving* (EOID) to resistance interpreted at one of more *restrike* events performed after EOID. However, note that this involves two or more separate sets of dynamic measurements performed however many hours or days apart as desired, not one set of measurements extrapolated in time.

The final issue raised here is the physical mechanism behind the increase in pile resistance in coarse-grained soils. With specific reference to the measured increases at JFKIA documented by York at al. (1994), the writer has first-hand recollections of discussions and speculation at the time (circa 1990) that perhaps there was some temporary, localized liquefaction of portions of the bearing stratum during driving. As a result, it was hypothesized that the increased resistance over time after driving was simply due to post-driving dissipation of excess pore pressures and a concomitant increase in soil <u>strength</u>. The fact that significant portions of the bearing stratum at JFKIA consist of saturated medium-to-fine and fine 'clean' sands with relatively low *relative densities*, D_r, and values of *state parameter*, ψ, that suggest contractive volumetric behavior is consistent with this hypothesis.

Countering this is the argument put forth by Fellenius et al. (2000) who found that for the same basic set of problem parameters (tapered piles driven at JFKIA) that the temporal increase in pile resistance that they observed was due to increased soil <u>stiffness</u>. This suggests that understanding the physical mechanism behind the temporal increase in resistance for piles bearing primarily in coarse-grained soil is neither fully understood nor a settled issue at this point in time. This also suggests that if pile-ground system stiffness is the correct mechanism that governs temporal gains in resistance, taking a stiffness-based approach to assessing pile performance as Fellenius and others (including the writer) have argued for some time and as promoted in this monograph via a comprehensive presentation of methodologies in Chapter 5 would be fruitful. Note that taking a stiffness-based approach to DFE design in general is also consistent with the preceding discussion concerning the difficulty in many cases of identifying a single-valued ULS load in static load tests.

4.2.6 Resistance-Verification Methodologies

4.2.6.1 Background and Overview

The most important objective of this monograph is to review, critique, and curate existing resistance-forecasting methodologies for tapered piles and from this distillation of the state of knowledge provide a framework for an improved resistance-forecasting methodology. With this in mind, it is critical to note that an essential component of efforts to both critique and develop analytical methodologies for forecasting the resistance of all types of DFEs is creating a reliable database of load-settlement ground-truth. Such data are used to both assess calculated results from an indirect-method procedure and calibrate direct-method empirical models.

Typically, this ground-truth takes the form of load tests performed on DFEs either during or after installation. There is now a wide variety of static, quasi-static, and dynamic load-test methods that are used for this purpose. As a result, there is more variation than ever in technologies and concomitant outcomes confronting stakeholders involved with deep foundations.

Regardless of the specific load-test approach that is used, there is a tendency for both practitioners and researchers alike to view load-test results as being a unique (i.e. single-valued), absolute, 'correct' answer, without consideration or question of the various elements and issues that can influence the particular test and test protocol used and, therefore, the accuracy of the measured results.

The reality is that there is no such thing as a load test that will provide <u>the</u> unique, absolute, 'correct' answer for a given DFE. The best that can ever be hoped for is that a load test will provide a result that lies within some range of 'correct' answers that exists in reality. There several reasons for this:

- DFE resistance to axial applied loads changes over time to varying degrees and at varying rates due to geomechanics phenomena. This is a natural occurrence and, therefore, beyond human control.

- Every load-test methodology has an inherent, built-in outcome-bias related to its underlying conceptual and theoretical basis. This is especially true of quasi-static and dynamic methods. Because this bias is unique to a methodology, the results from several methodologies would be expected to differ for the same DFE.

- Every load-test methodology has a test protocol that is subject to variation due to how that protocol is implemented on a specific DFE. These variations inherently influence the measured outcomes.

- The outcomes of virtually all load-test methodologies require human (therefore inherently subjective) interpretation to produce the final result. This is true of static, quasi-static, and dynamic methodologies alike. As discussed subsequently, this can result in a surprisingly large range of forecast outcomes from the same load-settlement curve.

The net result of all these factors is that any load-test is simply a snapshot of load-settlement behavior at a particular instant of time after installation of the DFE; for a particular load-application methodology; as implemented and interpreted by a particular person or group of persons. If either the time after installation or loading protocol or interpretation of the measured results is changed, the same DFE will exhibit a different load-settlement outcome. Thus, the actual load-settlement ground-truth of any DFE should be viewed as a range in results although this is rarely, if ever, acknowledged no less done in either practice or research.

This reality that every DFE has a range in load-settlement behaviors and never a unique load-settlement behavior has profound implications for both indirect and direct resistance-forecasting methodologies. Specifically, whenever the calculated outcome of some indirect analytical methodology has differed from a measured result, historically it is almost always assumed that the measured result must be correct therefore the analytical methodology is flawed in some way. However, in reality, there are numerous possible reasons for the deviation in agreement. If the measured results were properly portrayed as a range of results instead of a single value, it may well be that the forecast outcome from some indirect methodology fell within that range and therefore was actually as accurate as could reasonably be expected.

Similarly, all direct resistance-forecasting methodologies are based on statistical correlations of some in-situ test parameter(s) with measured load-test results. In these correlations, the measured results are assumed to be single-valued 'correct' answers. The reality is that every measured result should be more-properly represented as being some range in values. If this reality were taken into account, the resulting direct resistance-forecasting methodology would exhibit a much wider band of uncertainty in forecast outcomes.

The following sections are intended to explore in some detail and highlight key issues that the writer and others have found can impact the results measured in various types of load tests. In some cases, an issue cannot easily or inexpensively be overcome and thus has to be lived with. In other cases, an issue is impossible to assess no less deal with because it involves a load test performed years or decades in the past for which specific important test protocols are simply unknown.

4.2.6.2 Static and Quasi-Static Tests

The *static load test* is widely assumed to be the de-facto 'gold standard' for determining the true geotechnical resistance of any type of DFE to any type of externally applied load and is thus the reference standard against which other load-testing methodologies are compared and sometimes calibrated. Static load tests can be set up to apply an axial force in either direction as well as horizontally. For the purposes of the present discussion, the focus is on the most common variant in which a downward force from dead weights or some type of DFEs or ground anchors acting in tension or even some combination of the two is applied to the head of the DFE being tested.

There are three broad areas of concern with regard to static load tests:

1. Is the tested DFE representative of the intended production DFEs?

2. Do the measured applied loads represent the actual loads delivered to the head of the DFE and do the measured settlements represent the actual settlement of the head of the DFE relative to some fixed point (datum) in space?

3. Assuming that the measured load-settlement behavior is correct, does it represent the true behavior of the DFE or are the results affected by some phenomenon or factor that is not a direct part of the test protocol?

Unfortunately, the axial-compressive static load test as used in foundation engineering is far from standardized in its execution protocol which in and of itself means that the same DFE can exhibit a range of load-settlement behaviors depending on the specific test protocol used. There are several key aspects of the overall test procedure in which variations can affect the final measured outcomes.

The following are ten factors that are known to impact static load tests. The first nine of these factors deal with Questions 1 and 2 stated above, i.e. is the tested DFE a true prototype of the intended production DFEs and do the measured loads and settlements represent the actual loads and settlements. The last factor relates to Question 3 as to whether or not the true load-settlement behavior is being measured.

1. <u>Tested DFE vs. Production DFEs</u>. In the writer's almost 50 years of professional experience, there would have been a time that asking such a question or raising such an issue would have been unnecessary. After all, why would anyone load-test a DFE that did not reflect the DFEs that were to be installed during actual production work? However, recent knowledge of actual projects in the U.S. involving tapered piles has clearly made this issue one that is necessary to bring up. Specifically, the writer is aware of cases where the contractual and building-code requirements for static load testing of tapered steel pipe piles that were to be filled with fluid PCC after driving as is normally done were load tested without the PCC core. Not only did this raise the possibility of overstressing the steel component during load testing but it also changed the axial stiffness of the tested pile. As a result, the measured load-settlement response would not be expected to be representative of the performance of production piles that would be concreted prior to load application. There are many questions raised by this including how and why the licensed professional engineer who would have had to certify such testing allowed this as well as how and why other, higher levels of contractual and code oversight allowed this.

2. <u>Load-generation methodology</u>. This refers to how the downward force applied to the head of a DFE is physically generated. Historically, dead weights supported on a grillage of load-transfer beams were used as the jacking reaction. Over time, the alternative of several tensile reaction elements, each of which is either a DFE or passive (i.e. initially unstressed) ground anchor, was developed. In either case, the load-producing elements create their own system of stress-state changes, primarily horizontal, in the ground adjacent to the DFE being tested. The stress-state changes from the load elements therefore have the potential to impact the resistance of the tested DFE, especially considering the profound influence of horizontal stresses on both the shaft and toe resistances as discussed earlier in this chapter. This issue appears to have gotten very little attention over the years in the published literature, with a theoretical assessment based on the theory of linear elasticity by Poulos and Davis (1980) being a rare exception. The results put forth by Poulos and Davis suggest that this issue should receive a lot more attention and research than it has been given to date.

3. <u>Load-application rate</u>. The traditional method for load application is what is now referred to as the *Maintained Load* (ML) test variant wherein loads are applied in discrete increments. The number of increments and the hold-time per increment can vary widely, especially with respect to the latter which can vary from minutes to days which is three orders of magnitude, i.e. a factor of 1000. The alternative to the ML test is the *Constant Rate of Penetration* (CRP) test in which load is applied as necessary to advance the DFE into the ground at a constant velocity, i.e. settlement rate, as in a cone-penetrometer test.

As discussed by Fellenius (2019), all of these factors, especially variable increment hold-times within the same test, can influence results.

4. <u>Load-application process</u>. Independent of whether an ML or CRP test is performed (but certainly much more common with the former compared to the latter) is whether load application is monotonic up to the point of test termination or includes one or more unload-reload cycles. While unload-reload cycles often seem like a good idea, as discussed by Fellenius (2019), they introduce significant interpretation issues and should always be avoided. The downside of unload-reload cycles centers around the fact that the reloading portion of any cycle, at least up to the point of the largest prior-load magnitude, is testing system response where the ground has been pre-stiffened by the prior loading and thus exhibits less settlement than initial (virgin) loading. This is can be quite significant as noted in the next item.

5. <u>Load-application restart</u>. A subset of the unload-reload issue is the occasional situation where a pile is reloaded within a relatively short period of time (perhaps a day or two) after an initial test which may or may not have included unload-reload cycles. The reason for doing this appears to be entirely disingenuous in the writer's opinion and involves 'gaming the system' with regard to the interpretation of load-test results that is discussed subsequently. Where the writer is aware of this being done (and done just in the last few years, so it is a current practice), and hence the reason for calling it a disingenuous practice, is in situations where a DFE has 'failed' the initial load test because it settled more than some allowable magnitude under some predetermined, target load level (usually twice the desired allowable load or what is referred to colloquially in U.S. practice as 'double design load' meaning twice the intended allowable load (in ASD) or service load (in LRFD)). When the test is redone a relatively short time later (the next day in a recent example known to the writer), the retest is treated not as an unload-reload cycle of the initial test (which it really is) but as an entirely new test starting from zero settlement. Because the foundation-ground system has been stiffened by the prior load test, the same DFE now magically 'passes' the load test by settling less than the allowable magnitude under the very same load level at which it 'failed' in the initial test the day before. While this might satisfy some contract specification or other project requirement, it does so strictly on a letter-of-the-law technicality. The spirit of the contract specification or project requirement is that <u>all</u> DFEs pass a load test on initial loading with virgin system stiffness, not after the system has been preloaded and had its system stiffness increased. In the writer's opinion, this leaves open an interesting ethical question for the design professional(s) involved in approving the results of such testing. As a minimum, it suggests that contract specifications or other project documents be written and enforced to disallow such practice.

6. <u>Test termination</u>. There is wide variation in practice as to when a static load test is terminated. Decision criteria appear to fall into two broad categories. The alternative that is increasingly the more common one is load-based. Specifically, a test is carried out until some predetermined load level is reached. Most often this load is the aforementioned double-design magnitude. The other alternative is settlement-based in that the test is carried out until some relatively large settlement magnitude is reached. As this magnitude is rarely known exactly beforehand, this means that the load test is monitored until the load-settlement curve exhibits yielding where the slope of the curve steepens appreciably over a relatively short range of load magnitude. Historically, this is usually interpreted as indicating that the geotechnical ULS has been reached or is at least being

approached. However, in view of the discussion earlier in this chapter questioning the existence of a well-defined geotechnical ULS, this viewpoint should be re-evaluated. Unfortunately, due to mechanical limits in the load-test measurement hardware, the magnitude of settlement at which a test is terminated is often less than that required to clearly define the post-yield trend.

7. <u>Test instrumentation</u>. In this context, 'instrumentation' means load-displacement measurement devices placed depth-wise within or otherwise along the shaft of the DFE for the purpose of measuring load-and-settlement magnitudes as a function of depth. Fellenius (2019) has opined that:

"Tests on uninstrumented piles are usually a waste of money."

In the writer's opinion, this is an extreme position that is certainly debatable. Arguably more defensible (in the writer's opinion) is the position that any load test that is only carried to a predetermined load level such as double-design is a questionable expenditure of money. Given the significant cost of static load tests, there is a loss of valuable information by not continuing a test until the load-settlement curve exhibits clear signs of yielding and some reasonable magnitude of post-yield settlement of the order of inches (tens of millimetres) is measured. The only exception to this would be for DFEs bearing on or in bedrock where carrying a load test to large settlements is either practically impossible or would be structurally detrimental to the DFE itself or geotechnically detrimental to the rock.

8. <u>Load measurement</u>. Arguably the single most important aspect of any load test is accurate measurement of the force applied to the head of the DFE being tested. Unfortunately, this is an aspect of load testing where engineering practitioners sometimes do not pay enough attention and, as a result, long-deprecated practices continue to be employed on a routine basis simply because construction specifications or other project documents allow them. The specific problem involves the hydraulic jacking system that is used to apply load to the DFE being tested by reacting against the load-frame that was discussed above. For decades, it was considered acceptable to determine the applied force using a calibration with the observed jack pressure. However, at least as early as circa 1970, and possibly earlier, it was known to foundation engineers and other stakeholders that this generally overestimates the applied load by 10 to 20% due to piston-friction issues within the jack (Fellenius 1990). Of note is that the deviation between the theoretical and actual loads increases with increasing force. This behavior was determined by use of an independent electronic load cell as part of the load-application assemblage. Therefore, it has long been considered to be good practice to use a calibrated load cell for every load test, and use the force measurements from the load cell when constructing load-settlement plots. The importance of this cannot be overemphasized as using forces based on jack pressure will always result in an unconservative and potentially unsafe load-settlement plot as the magnitude of load carried by the tested DFE at a given magnitude of settlement will always be overestimated. Note that this has significance not only for current practice but especially for older load-test results that almost certainly did not use a calibrated load cell for force measurement. This is important because it means that basically all of the older load-settlement plots that one sees in the published literature as well as other, non-published results that were undoubtedly used to develop older analytical methodologies,

such as those by Meyerhof and Nordlund discussed earlier, are incorrect. As a result, the resistance-forecasting methodologies, both indirect and direct, derived from them contain inherent errors.

9. <u>Settlement measurement</u>. Accurate measurement of the settlement of the head of the DFE being tested is second in importance only to the accuracy of load measurement. In the writer's experience, settlement measurement is typically treated with a varying level of complacency in routine practice and taken for granted, with minimal, if any, thought given to potential problems. The central issue with the traditional mensuration methodologies that are, and have long been, used for load tests (dial gauges and tensioned wire + scale) is that they require support by what is called a *reference beam*. Problems arise because this beam must be of reasonably short length and thus supported relatively close to both the DFE being tested as well as the support grillage (for dead-weight tests) or tensile reaction elements of the load-delivery system. In either case, there is the potential for settlement of the supports of the reference beam which means that the settlements of the head of the DFE being tested that are inferred from the dial-gauge or scale readings will be less than the actual. The writer has found that this error is typically of the order of 10% (the more-accurate settlement-measurement methodology used by the writer to make this determination is described shortly). The net result is similar to that discussed above for load measurements in that the resulting load-settlement plot will be incorrect on the unconservative and potentially unsafe side as it will show settlements for all load levels that are smaller than actual. In general, this is not as critical as underestimating loads for a given settlement magnitude but it is troubling nonetheless. A secondary issue with the reference beam is that it is subject to thermal-induced displacements that result in phantom inferred vertical displacements of the head of the DFE being tested. As a minimum, sufficient covering of the load-test area must always be provided so that the reference beam is blocked from direct sunlight and thus not subjected to direct heating from solar radiation. Unfortunately, for tests that may extend over hours or days there is little that can be done to protect against thermal movements of the reference beam due to diurnal changes in ambient air temperature. There is a more accurate and foolproof way to measure settlements but it is rarely done in practice due to the cost. This is to use precision optical survey[92] that is referenced to a test-specific benchmark[93] that is set far enough away from the load-test setup so that it will not be affected. On projects in which this advanced mensuration system was installed, conventional dial gauges and tensioned wire + scale were installed as well for redundancy and comparison. In all cases, it was demonstrated that the conventional devices underestimated pile-head settlements, verifying the points raised here.

10. <u>Residual loads from installation</u>. *Residual loads* are defined in this context as the vertical forces acting within, and concomitant axial normal stresses within, the shaft of a DFE after installation even though the external force at the head of the DFE is zero. These loads are broadly similar to the residual stresses that develop in a rolled section of structural steel

---

[92] The writer has been involved in projects where this was achieved by attaching a parallel-plate micrometer to a normal survey level. This increased the precision of the readings from ±0.01 feet (~⅛ in ≅ 3 mm) to ±0.001 feet (~0.01 in ≅ 300 μm).

[93] The writer has been involved in projects where this was achieved by installing a *Borros-Type Anchor* to a sufficient depth in the ground; attaching a survey leveling rod to the steel rod of the anchor that extended above the ground surface; and covering the above-ground assemblage to protect it from direct sunlight and direct solar radiation.

as a result of the manufacturing process (the reason why these are sometimes alternatively called *rolling stresses* or *mill stresses*) although the physical mechanisms involved in the two cases are very different. Fellenius (2015) illustrated that residual loads develop in all types of DFEs regardless of installation methodology and ground conditions. However, of relevance to this monograph is that not only are residual loads most significant for (driven) piles but, all other things being equal, they are more significant for tapered piles as opposed to constant-perimeter piles. This was illustrated in an early, seminal paper by Gregersen et al. (1973) that was summarized with further discussion by Fellenius (2002a). Because of the importance of residual loads to tapered piles, they are discussed in detail later in this chapter and an entire appendix (Appendix H) is devoted to the writer's detailed assessment of the Gregersen et alia case history. The discussion here is limited to the specific relevance of residual loads to resistance-verification methodologies. Historically, residual loads have only been considered when assessing a static load test on an instrumented DFE. An example is the tapered-pile case history presented by Fellenius et al. (2000)[94] that is discussed in detail in Appendix A. However, the precept put forth here by the writer is that residual loads are actually an important consideration for all static load tests, including the non-instrumented ones performed routinely as part of the *Construction Quality Assurance* (CQA) protocols on routine projects. This is because independent of the accuracy of the load and settlement measurements discussed above, a factor that is almost always ignored, not only in practice but research as well, is the fact that residual loads will affect the shape of a load-settlement curve. The reasons for this are explained and illustrated in Fellenius (2015). The specific significance of this is that static load tests in which residual loads are present produce load-settlement curves that are stiffer than the true behavior that an otherwise identical DFE without residual loads would exhibit. This outcome is not trivial or academic as the most common methodology used in the U.S. and elsewhere to interpret the geotechnical ULS from a static load-test curve is the *Davisson Offset Limit Method*. When using this method, and all other things being equal, the stiffer the load-settlement curve, the larger the estimated value of the geotechnical ULS. Thus, a static load test with residual loads will tend to result in a larger estimate of the geotechnical ULS than if the residual loads were not present. This was illustrated for a specific case history in Fellenius (2015). Note that these same comments apply to other settlement-based interpretative methodologies for the geotechnical ULS such as the "30 mm toe movement" criterion noted in Fellenius (2015).

The conclusion drawn from this extended and extensive discussion is that the conventional static load test is far from standard, similar to the lesson learned in the latter decades of the 20th century that the Standard Penetration Test was remarkably not standard. There are many factors that can influence and alter the final, plotted load-settlement result from a static load test that is so routinely taken for granted as being 'correct' and thus reflective of the absolute ground-truth. This has led to the near-universal perception among all stakeholders involved with deep foundations that when 'something' (e.g. the forecast outcome of a resistance-forecasting methodology) differs from a load-test result then that 'something' must be wrong, never the load-test outcome.

In reality, in most cases (certainly those performed on a routine basis for practice as opposed to pure research), load-test results are far from perfect and almost always flawed to some varying, unknown degree. This is especially true of older tests that almost certainly portray loads that are too large because applied loads were interpreted solely based on jack

---

[94] A brief summary discussion is also presented in Fellenius (2002b).

pressure. This reality should be of great concern to all stakeholders (foundation engineers, contractors, owners) as many of the older resistance-forecasting methodologies, both indirect and direct, that are still in routine use in practice were based on databases of results from these flawed tests.

Note that producing an accurate measurement of the load-settlement response of an installed DFE is only part of the process for using that information properly in either practice or research. As discussed earlier in this chapter and in greater detail later in this chapter under the section on design philosophy, the design of DFEs has long focused solely on some well-defined, single-valued ULS geotechnical resistance to which either a safety factor or resistance factors are applied depending on whether ASD or LRFD is used. This means that the load-settlement curve that is produced by a static load test must be interpreted using some analytical methodology.

Unfortunately, there is no shortage of interpretational methodologies for static load tests that have been proposed over the years. This topic has been discussed in detail numerous times by several researchers (Fellenius 1990, 2001; Horvath 2002; Niazi 2014) up to the present with no clear resolution of which method, if any, is 'correct' or even the 'best'. This is not surprising given the extensive prior discussion that showed unequivocally that the assumption that there is always a unique 'failure load' is inherently flawed, especially for piles bearing in coarse-grained soil or on rock.

Therefore, the only logical conclusion is that none of the myriad interpretative methodologies that have appeared in the published literature to date is, strictly speaking, correct although some have a more-rational basis than others. Given the fact that the need in routine practice to define a unique value for the geotechnical ULS whether one really exists or not is not going to disappear anytime soon (although it arguably should), this means that there remains a need for interpretative methodologies.

It is beyond the intent and scope of this monograph to explore what might be 'the best of the lot' of these interpretative methodologies. As discussed later in this chapter, the writer feels that the profession needs to move on to better approaches to the problem, even in routine practice. That having been said, some comments are deemed to be necessary with regard to one of these methods, the aforementioned Davisson Offset Limit Method or simply 'Davisson Method' as it is often referred to colloquially in practice This is by far the most commonly used interpretative method in practice, at least in the U.S., and more so currently than ever.

The Davisson Method was first proposed in the early 1970s by the late Prof. M. Thomas 'Tom' Davisson[95]. Because the method has been around for so long, few recall the specifics of its intended use which are outlined here for background-informational purposes.

To begin with, the method was intended to be used only with (driven) piles that derive all resistance from toe resistance (NeSmith and Siegel 2009). That is a very limited subset of DFEs and arguably one that rarely actually exists although it might be approached in some cases, e.g. a pile driven through very soft clay and end bearing on bedrock. Of relevance to this monograph, the Davisson Method would never apply to tapered piles that are always friction/floating piles where toe resistance might comprise 20% of the ULS geotechnical resistance.

Furthermore, and more importantly, the Davisson Method is arguably only applicable to dynamic analyses and thus not static load tests. This is because Davisson was an early and passionate devotee of the 1-D wave equation as a resistance-forecasting tool for piles. In fact, and with reference to comments made earlier in this chapter, he was one of those who

---

[95] Davisson's first name was Melvin although the writer only ever saw Davisson use the initial of his first name in any published work.

gravitated toward using the 1-D wave equation to the exclusion of resistance-forecasting methodologies based on the static approach. Davisson routinely rationalized and explained away the difference between forecast results using the 1-D wave equation and measured results in a static load test as being due to "soil freeze" or, less commonly, "soil relaxation", i.e. some geomechanics phenomenon that somehow magically happened after a pile was driven. The 1-D wave-equation analysis itself was never deemed to be incorrect.

This focus by Davisson on 1-D wave-equation analyses is relevant as the magnitude of toe displacement (settlement) that Davisson assumed constituted the geotechnical ULS under the dynamic conditions simulated by a 1-D wave-equation analysis was only 0.15 inches (3800 μm) plus 0.83% of the toe diameter. For a nominal toe diameter of 8 inches (200 mm) that is common with tapered steel pipe piles such as the *Monotube* and *TAPERTUBE*, the sum of these two contributions amounts to approximately ¼ inch (6 mm) of toe displacement/settlement or 3% of the toe diameter. This is substantially less than the 10-to-15% of toe diameter criterion used by other settlement-based methodologies for defining the geotechnical ULS.

As is well known, in the Davisson Method this magnitude of settlement is added to calculated values of the assumed elastic shortening of the pile as a structural member under various load levels (again, assuming that all applied load goes to the toe) to create the 'failure line' that is superimposed on a load-settlement plot from a static load test. Where this failure-line crosses the measured load-settlement curve, the corresponding applied load is deemed to be the 'failure load' (geotechnical ULS) of the DFE.

It appears that this relatively small magnitude of toe displacement/settlement to produce soil failure beneath the toe (3% in the above example) was what Davisson assumed was the *quake* (vertical displacement) associated with the geotechnical ULS beneath a pile toe in a 1-D wave-equation analysis. Note that this was based on the original, simplistic Smith rheological model of an axial spring + dashpot/damper that was used in the original versions of the 1-D wave equation to simulate toe resistance. The scientific basis for this quake assumption is both thin and arbitrary.

In the writer's recent observations, there is an irrational (in the writer's opinion) fealty to the Davisson Method by design professionals even though it is usually used incorrectly and always produces overly conservative results. The combined simplicity and extreme conservatism of the method no doubt contribute to its enduring and increasing popularity. Furthermore, because the method associates the geotechnical ULS with a relatively small magnitude of settlement, there is also the perceived 'benefit' that it requires static load tests to be conducted to relatively small settlement levels.

In the writer's opinion, there is a potentially dangerous consequence to the fact that the Davisson Method uses settlement magnitude alone as the criterion for the geotechnical ULS. Surprisingly (given the potential consequences), this issue never discussed. This is the fact that the Davisson Method is readily and easily susceptible to being manipulated and 'gamed' as discussed above under load-test comment No. 4 dealing with restart testing. It is possible for a pile to 'fail' a static load test then be retested the next day and pass because the retest is treated as a separate load test, not as an unload-reload cycle of the initial test which it is in reality. As noted previously but worthy of repetition because it is a deeply troubling trend (in the writer's opinion), the writer has seen this done in practice, including in the recent past.

The final comment about interpreting results from static load tests is for situations where instrumented tests (as defined previously in this chapter) are performed and involves residual loads (comment No. 9 above). Fellenius (2016) presents a detailed discussion about this subject and the need to correct for estimated residual loads whenever an instrumented load test is performed. Failure to correct for residual loads results in implied resistance mobilization and concomitant load distribution within a pile that is grossly incorrect. This was demonstrated specifically for tapered piles by Fellenius et al. (2000).

The issue of residual loads and their impact on the interpretation of results from instrumented static load tests is as much a cautionary note for past tests as it is for present and future tests. There are undoubtedly countless instrumented load tests that were performed on piles going back at least to the 1950s that were used to develop and/or calibrate resistance-forecasting methodologies, both indirect and direct. Along the lines of comments made above under item No. 8 dealing with load measurement concerning incorrect pile-load estimates affecting resistance-forecasting methodologies, it is likely that at least some older instrumented load tests were misinterpreted because the phenomenon of residual loads was either unknown (at least to those performing the load tests as residual loads were noted in the published literature as early as the late 1950s) or not considered at the time. Unfortunately, these misinterpreted results have undoubtedly found their way into resistance-forecasting methodologies, some of which are likely in current use. The implications are obvious in terms of incorrect calculated outcomes from these flawed analytical methodologies.

To complete the discussion of static load testing, mention is made of the *Osterberg Cell*® (*O-Cell*)[96] and *Statnamic* test. The former is a device for static load testing that requires embedment within the DFE to be tested so is largely limited to drilled elements and is not discussed further.

The latter is a quasi-static load-application device that is suitable to use with (driven) piles in general where a downward force is applied at the head of DFE but only for a very limited duration of time that is much shorter than a conventional static load test but longer than the millisecond impact from pile driving. Fellenius (2020) uses the generic term *long-duration impulse testing* for the *Statnamic* and similar test methodologies developed by others such as the *Fundex* and *StatRapid* tests that use mechanical impact as the energy source as opposed to the explosive-like chemical reaction of the *Statnamic* test. The key assumption is that the response to any and all of these impulse-loading tests of the DFE being tested is closer to that of static loading (where load-induced stresses are created along the entire shaft at the same time) as opposed to either pile driving or post-installation dynamic testing (where load-induced stresses are felt only along some limited portion of the shaft at any given instant of time as a 1-D stress wave travels down and back up the shaft). Note, however, that there is always a dynamic component to the applied load in these impulse-type tests that must be analyzed-out of the raw data to produce an equivalent static load-settlement curve. As a result, the *Fundex*, *Statnamic*, and *StatRapid* tests are all assumed to be quasi-static in nature.

A discussion of the accuracy of the interpreted quasi-static load-settlement curve from these impulse-type tests compared to that from a true static load test performed on the same DFE is beyond the scope of this monograph. Some discussion along these lines can be found in Fellenius (2020).

4.2.6.3 Dynamic Tests

As noted in Chapter 1, resistance-forecasting methodologies based on some hypothesized dynamic model of pile driving predated static-model methods based on geomechanics principles by a century, 1850s vs. 1950s. However, field-based resistance-

---

[96] It appears that the U.S. patent(s) covering this device have expired in recent years and that competitive devices have entered the U.S. and global marketplace. Discussion of these competitive devices is beyond the scope of this monograph.

verification methods based on dynamic measurements took far longer to develop (circa 1970) as it was necessary for technology in the form of electronics and digital computers to evolve to the point where they could function in the demanding environment of pile driving.

Smith's 1960 publication of the 1-D wave-equation model and solution for pile driving paved the way for the development of 1-D wave-based methodologies for on-pile measurements combined with both real-time and post-driving processing by the end of that decade. The result is current technologies such as the *Pile Driving Analyzer®* (*PDA*) and *CAPWAP®*.

The writer has first-hand experience with these technologies dating back to the earliest days of their use (1972) and the perception developed by the writer over the last several decades is that there are consistent, persistent issues with regard to their accuracy when used with tapered piles. This is despite the fact that these methods have undergone countless generations of refinement since they were originally developed approximately 50 years ago. The specific issue with respect to tapered piles is that the ULS total resistance as forecast by these dynamics-based methodologies does not always correlate well with that from traditional static load tests in those cases where results from both are available for the same pile. While temporal effects may account for some of this discrepancy, it is the writer's opinion that there are other, potentially more-significant issues as well.

At the top of the list is the fact that, based on what is now known about the correct physical mechanism by which the tapered portion of a pile develops resistance (wedging, as modeled using CCX during pile driving followed by CCE during axial-compressive load application), the accuracy of the basic 1-D wave-equation model can be called into question in this application. The 1-D wave-equation model, from its inception in the 1930s and development in the 1950s to the present, assumes that resistance along a pile shaft can be modeled by a simple combination of an axial spring and dashpot (damper) that act solely in response to displacements and velocities parallel to the pile axis. Stated another way, these simple mechanical elements that are used to model all soil and soil-pile interaction effects can only model the traditional shaft resistance mechanism of sliding friction. They cannot account for the radial displacements along the tapered portion of a pile shaft that occur during driving that define the shaft resistance due to wedging mechanism.

Furthermore, because the 1-D wave model only accounts for what happens during pile driving, by definition it cannot account for the additional cavity expansion and additional resistance that develops during axial-compressive external load application after pile installation. This means that even if future research modifies the elastic-wave model to account for the 3-D effects of wedging that occur during pile driving, this 3-D wave-equation model could fundamentally and inherently never account for the additional resistance that develops as the result of post-installation external-load application.

Note that research has also revealed other issues when applying the 1-D wave model to driving tapered piles. For example, Goble and Hery (1984) addressed the fact that residual loads (an issue already noted and discussed earlier in this chapter with respect to the interpretation of static load tests on instrumented piles) can have a profound effect on the calculated *bearing graphs* of ULS total resistance vs. pile-driving blowcount that have long been the primary forecast outcome of any dynamic analysis, whether using the 1-D wave equation or a traditional dynamic formula. Thus, a 1-D wave-equation analysis for tapered piles should always include steps to take residual loads into account although by Goble and Hery's own admission this only partially accounts for the above-mentioned discrepancy between forecast and actual ULS total resistance with tapered piles.

In summary and conclusion, it is the writer's opinion that there is ample theoretical basis for being cautious about resistance-verification results from any 1-D wave-based technologies such as the *PDA* and *CAPWAP* when they are applied to tapered piles. The simple reason is that the physical and mathematical model on which these methodologies are based is not in consonance with the current, modern understanding of how a tapered pile behaves when subjected to an axial-compressive force, whether it be under static or dynamic conditions. Furthermore, as explained in Fellenius (2002a), the depth-wise load distributions produced by a *CAPWAP* analysis can be significantly in error as they do not account for the effects of residual loads.

That having been said, there is no doubt that 1-D wave-based technologies, the *PDA* in particular, have been instrumental in advancing the state of knowledge related to tapered piles. For example, all that has been learned about the temporal increase in ULS total resistance for piles in coarse-grained soil that was discussed earlier in this chapter is based solely on dynamic measurements made at different restrike times after and compared to the end of initial driving (EOID).

Furthermore, given the cost effectiveness of 'load testing' a pile using devices such as the *PDA* for real-time resistance forecasting and *CAPWAP* analyses for more-detailed after-the-fact assessments as opposed to performing any type of static or quasi-static load test, it is obvious that these 1-D-wave-based technologies will continue to be used into the foreseeable future. The point being made here is that they should be used intelligently. This includes recognizing the fact that there are multiple issues to consider when these devices and technologies are used with tapered piles.

4.3 STRUCTURAL AND SOIL-STRUCTURE INTERACTION ISSUES

4.3.1 Background and Overview

All foundations, shallow and deep, are inherently a soil-structure interaction[97] (SSI) problem. However, the geomechanics aspects of foundations tend to be much more numerous and dominant compared to structural aspects so have historically received the bulk of the attention. Furthermore, foundation design codes have traditionally been intentionally biased toward overdesigning the structural component relative to the geotechnical component. This is based on what can be termed the *rationale of intended consequence* meaning that there is an intentional, logical reason for doing things a certain way so that a certain sequence of events will (hopefully) occur.

Specifically in this case, if a foundation-system ULS were to occur (civil engineers must always design to account for the fact that the ULS is theoretically always possible in any system), design and building codes have long been formulated so that the ULS would intentionally involve the geotechnical component first. The geotechnical ULS in most cases tends to be preceded by relatively large displacements that presumably would produce a visible warning to all concerned that something was amiss and required attention. On the other hand, a structural ULS might not be as visually apparent, especially if it were brittle in nature due to the structural materials and rate of loading involved.

That having been said, with DFEs in general and tapered piles in particular there are two broad areas where structural and SSI issues can be significant considerations and thus require some discussion of substance:

---

[97] As discussed in Horvath (2018b), the writer does not restrict this term to applications involving seismic loading only as is done by some. The usage in this monograph, as in Horvath (2018b), includes any application where a structural element of some kind is in contact with, and thus interacts with, the ground.

- <u>Residual loads</u>, a post-installation phenomenon that affects all DFEs to some extent (Fellenius 2015) and whose existence was noted earlier in this chapter in the context of their effects on load-settlement curves of static load tests.

- The *Poisson Effect* that exists with all DFEs but has three distinct aspects (more than any other type of DFE) when applied to tapered piles. There is also a potential for the Poisson Effect to be influenced by residual loads, a topic that does not appear to have received attention to date. The Poisson Effect was noted earlier in this chapter in a limited context but is discussed here in greater detail.

Each of these topics is discussed in detail in the following sections.

4.3.2 Residual Loads from Installation

4.3.2.1 Introduction and Overview

The issue of residual loads was identified and discussed to a limited extent earlier in this chapter, primarily in the context of their influence on interpreting results obtained in the traditional static load test as well as the effect they have on 1-D wave-equation analyses for tapered steel pipe piles in particular. However, a broader discussion of residual loads was deferred until now because they are fundamentally an SSI problem that involves:

- the structural (axial) stiffness of a DFE;

- the structural geometry of a DFE, i.e. constant-perimeter or tapered;

- the stiffness of the ground into which a DFE is installed; and

- the method of DFE installation or in-situ creation.

In summary, residual loads involve both the geotechnical and structural aspects of a DFE to a significant degree so they need to be addressed holistically.

4.3.2.2 Background

It is beyond the scope of this monograph to provide a comprehensive documentation, review, and assessment of the published body of work to date on residual loads in DFEs that spans more than 60 years at this point in time. Only selected publications are cited and discussed here. Additional uncited references on this subject can be found in the reference section at the end of this monograph.

It appears that the existence of residual loads was recognized relatively early in the evolution of modern foundation engineering that began after World War Two. Mansur and Kaufman (1958) mentioned them which means that there was knowledge of them at the time that the first resistance-forecasting methodologies based on the static approach were being developed by Meyerhof and others in the 1950s. Nordlund alluded to their existence in his seminal 1963 paper on tapered piles although he made no attempt to measure or quantify them in any way. Goble and Hery (1984) referenced publications by Kerisel (1964), Leonards (1970), and Vesic (1970) that reportedly noted the existence of residual loads. Hunter and

Davisson (1969) represents one of the earliest, if not <u>the</u> earliest, attempts to quantify residual loads, albeit indirectly, based on static load tests of full-scale piles.

Even more noteworthy relative to the contents of this monograph (and thus discussed in detail in Appendix H) is the early work of Gregersen et al. (1973) who not only directly measured residual loads on full-scale piles but included tapered piles, something that appears to have been ignored by all other researchers investigating residual loads. The Gregersen et alia paper appears to have been missed by virtually all subsequent researchers of residual loads (including the writer, until very recently) as their paper was not cited in any paper reviewed by the writer in preparation of this monograph except for Fellenius (2002a).

Targeted research into residual loads increased substantially in the later 1970s. This was likely due to the increasing availability of digital computers in general (at least in academia) and the development of software for the 1-D wave equation in particular. This research continued into the 1980s and included at least one Master thesis (Hery 1983) and one doctoral dissertation (Darrag 1987). Residual-load research related specifically to piles tailed off considerably after the 1980s but continues to appear on occasion in the 21st century, e.g. Zhang and Wang (2007).

Initially, research into residual loads was limited to (driven) piles. Not only were they the predominant type of DFE in the U.S. at the time (very few drilled-DFE alternatives existed in the U.S. circa 1970 and those that did were very regional or otherwise limited in their usage) but the prevailing assumption was that traditional impact driving was an essential causative factor in the development of residual loads. It has since come to be appreciated that jacked and drilled DFEs can also have residual loads (Siegel and McGillivray 2009, Flynn et al. 2012, Fellenius 2015), an important fact given that:

- the use of drilled shafts as an alternative to piles increased substantially beginning in the 1980s in the U.S. and, subsequently, around the world;

- many other drilled-DFE technologies have been developed and increased their market share in recent decades; and

- jacked piles have become popular in some regions of the world where subsurface conditions are conducive to their use.

Consequently, the fact that residual loads exist with DFEs in general has been a seminal outcome from this collective cited body of work.

4.3.2.3 Analytical Methodologies for Forecasting and Interpretation

Residual-load research to date has, for the most part, been concentrated on developing analytical methodologies in two broad areas:

- forecasting the theoretical residual-load distribution in piles and

- estimating the actual residual-load distribution in any type of DFE subjected to an instrumented (as defined earlier in this chapter) static load test.

The former appears to have been done earlier so is discussed first.

Early research in the 1980s focused primarily on the development of analytical methodologies of for forecasting residual loads. Some of the key publications of this genre are:

- <u>Hery (1983)</u> developed and used a modified version of the then-current (in the U.S.) commercial software package for 1-D wave-equation analysis named *WEAP* (*W*ave *E*quation *A*nalysis *P*rogram). This modified program, named *CUWEAP* (*C*olorado *U*niversity *W*ave *E*quation *A*nalysis *P*rogram), was used to produce the improved (but still imperfect) forecast outcomes for tapered steel pipe piles that were reported in Goble and Hery (1984) that was noted earlier in this chapter. Of relevance is the fact that *CUWEAP* eventually evolved into the commercial software package *GRLWEAP* (*G*oble *R*ausche *L*ikins *W*ave *E*quation *A*nalysis *P*rogram) that remains in widespread use (in the U.S. at least) to the present.

- <u>Briaud and Tucker (1984)</u> developed a highly simplified method based on SPT N-values.

- <u>Poulos (1987)</u> developed a comprehensive methodology based on theory-of-linear-elasticity solutions that required proprietary computer software for solution.

- <u>Darrag and Lovell (1989)</u> developed what was intended to be a simplified (in the sense of not requiring some proprietary computer software), chart-based analytical methodology that was based largely on the observation that residual loads (in piles driven in coarse-grained soil at least) tend to have the same generic depth-wise distribution. They then used the aforementioned *CUWEAP* program to develop a series of charts so that a design professional could quantify the geometry and magnitude of the generic residual-load distribution curve for a particular application. The concept of making use of the fact that the depth-wise distribution of residual loads in piles in coarse-grained soil tends to be qualitatively the same was resurrected decades later by <u>Wang and Zhang (2008)</u> who also resurrected elements of the aforementioned Tucker and Briaud method.

It is unclear (to the writer at least) what the intended usage of these residual-load forecasts was in most cases, i.e. what was the average design professional in practice to do with this new-found information? The only clear quantitative usage was in the 1-D wave-equation software that was obviously intended to produce more-accurate bearing graphs as noted earlier in this chapter. Qualitatively, the various graphs presented in Poulos (1987) showed how residual loads theoretically affected load-settlement curves in both axial compression and uplift but again, there was no clear, obvious indication of the practical value of this.

After this surge of research into residual loads peaked on the cusp of the 1990s, subsequent decades saw the research focus shift to the development of a rational process for interpreting instrumented static load tests for residual loads. It is now understood that for the vast majority of instrumented static load tests on all types of DFEs that the readings taken during post-installation load application do not inherently account for residual loads from installation of the DFE (as discussed in Appendix H, the tests reported by Gregersen et al. (1973) were atypical in that they did directly measure residual loads). Consequently, use of the raw results from instrumented static load tests can lead to highly incorrect interpretations of resistance distributions. This fact and the outline of an analytical methodology for correcting for residual loads can be found in a number of publications by Fellenius over the past 20 years (Fellenius et al. 2000; Fellenius 2002a, 2002b, 2002c, 2015).

4.3.2.4 Summary of Key Findings

A review of the body of published work on residual loads that was cited in the preceding sections indicates that several key behavioral features are supported by both theoretical work (especially that of Poulos (1987) that was unusually detailed and thorough of its kind) and the observation of full-scale DFEs. As such, they form the core of the current body of knowledge with respect to residual loads in DFEs in general and tapered piles in particular:

- Residual loads are more significant for pre-formed DFEs that are either driven or jacked into the ground as opposed to DFEs that are formed in-situ by some drilling process. DFEs that are torqued into the ground such as helical piles and helical tapered piles are much newer technologies, and the writer is unaware of any published work to date concerning residual loads in such DFEs. However, the torqued-in-place process would appear to be much closer to driving and jacking as opposed to drilling in terms of process kinematics. Therefore, a preliminary assumption might be that residual loads in torqued-in-place DFEs may be not insignificant and thus worthy of future research.

- Residual loads for (driven) piles, at least, appear to have qualitatively similar depth-wise variations, although these variations depend primarily on soil type and, secondarily, on pile stiffness. Note that pile stiffness in the context of residual loads does not appear to be absolute but relative to the stiffness of the surrounding ground. Thus, a given pile can display different stiffnesses with respect to residual loads depending on the stiffness of the ground into which it is driven.

- Residual loads for (driven) piles are relatively larger in coarse-grained soils compared to fine-grained soils.

- Residual loads (at least in coarse-grained soil) are significantly greater for tapered piles, even those with a partially tapered geometry, compared to an otherwise identical pile with a constant perimeter.

- Residual loads increase with decreasing pile stiffness. Because overall pile stiffness includes pile length, residual loads tend to increase with increasing pile length. It is important to note that pile stiffness in this context is that of the driven portion of the pile only. This can be important for steel pipe piles that are filled with fluid PCC after installation as the pile during driving is more compressible than the final pile with a PCC core.

- Residual toe loads can be a substantial portion of the theoretical ULS toe resistance. It appears that this is especially true for tapered piles. Fellenius et al. (2000) interpreted that toe resistance for a tapered steel pipe pile was essentially 100% mobilized after initial driving (see Appendix A). Gregersen et al. (1973) also found relatively substantial residual toe resistance for tapered precast-PCC piles (see Appendix H).

- Residual loads can have a significant influence on load-settlement behavior compared to an ideal reference case of no residual loads. Pile-soil system behavior is always stiffer in axial compression and softer in axial tension (uplift) due to residual loads, with the net result that a pile will appear to be substantially less stiff in uplift compared to compression.

- Residual loads are an essential consideration when interpreting data from instrumented static load tests. Failure to correct for residual loads results in incorrect interpretations concerning the mobilization of both shaft and toe resistance (Poulos 1987 and Fellenius et al. 2000 are but two of many publications that have pointed this out). This necessary correction for residual loads was undoubtedly not done many times in the past and may still not be done at times in the present. As has been noted in the literature, failure to correct for residual loads is now recognized as being one of the reasons why the now-deprecated concept of there being a limiting value of shaft resistance due to sliding friction gained traction and short-lived acceptance in the latter part of the 20th century.

4.3.2.5 Closing Observations and Comments

The overall conclusion drawn by the writer with regard to residual loads is that the foundation engineering profession's knowledge about them, in (driven) piles at least, is almost as old as the existence of 'engineered' pile analysis-and-design methodologies (discussed in detail in Chapter 5 following) that began to evolve in the 1950s. However, based on the writer's professional experience that dates back to 1972, it appears that, except for relatively recent recognition that residual loads exist in jacked piles and drilled DFEs as well as (driven) piles, very little mention has been made of the subject in recent decades except for the persistent publications by Fellenius that detail how essential consideration of residual loads is when evaluating static load tests, both instrumented and not, on all types of DFEs.

What is particularly perplexing to the writer is that resistance-forecasting methodologies for DFEs, especially direct methods based on in-situ tests such as the cone penetrometer, continue to be developed at a relatively rapid pace. In addition, both indirect and direct methods have been extended to be able to generate complete load-settlement curves. Yet, it appears that these various analytical methodologies have been developed in the vacuum of a 'perfect world' of no residual loads that simply never exists in reality.

That having been said, in any objective discussion one must always play the proverbial 'devil's advocate' and look at the other side of the coin. It is clear that chronic failure to include residual loads in resistance-forecasting methodologies does not appear to have had any obvious negative impact on the ability of foundation engineers around the world to routinely design deep-foundation systems that appear to function as intended. However, that does not mean that ignoring residual loads when crafting resistance-forecasting methodologies is something that should continue ad infinitum. This is especially true with the new generation of resistance-forecasting methodologies that explicitly consider and include load-settlement behavior.

Furthermore, as illustrated in later chapters, residual loads are not always the benign issue that they have been made out to be at times. While it is true that residual loads do not (in theory) influence the ULS resistance of a DFE, then can significantly affect, for worse and not better, the <u>interpretation</u> of the ULS resistance as well as the structural resistance.

Therefore, the position adopted by the writer in this matter is to recognize that not only do residual loads exist, they appear to be especially significant for tapered piles in coarse-grained soil which is precisely the ground condition in which tapered piles are most commonly used. Consequently, every reasonable effort should be made to consider incorporation of residual loads in future development of resistance-forecasting methodologies for tapered piles. In addition, there are some niche issues that appear to be affected by residual loads as discussed in the following sections dealing with the Poisson Effect.

4.3.3 Poisson Effect

4.3.3.1 Basic Concept

The Poisson Effect in the context of this monograph relates to the fact that most materials used in engineered construction have a value of Poisson's ratio that is non-zero and positive in sign[98]. In DFE applications, this means that even under uniaxial loading the shaft of the DFE will undergo horizontal or radial displacement.

Historically, the Poisson Effect has been ignored in routine analysis and design of DFEs based on the implied assumption that the effects are insignificant. However, as noted earlier in this chapter, some discussion of the subject is appropriate in this monograph for the sake of completeness as well as to identify several issues involving DFEs in general, and tapered piles in particular, where a consideration of the Poisson Effect may be not insignificant.

With this in mind, the Poisson Effect will be considered and discussed in two broad areas:

- its effect on allocation of axial load for DFEs with a composite cross-section consisting of two or more different components and

- its effect on radial interaction at the DFE-ground interface along the DFE shaft.

The theoretical basis and reference used for both of these discussion areas is a theory-of-linear-elasticity boundary-value problem and concomitant solution that was conceived of and developed by the writer in 2009 and published in 2010 (Horvath 2010a). Because this previously published work is a key element of the following discussions, for ease of reference it has been included in this monograph in an updated and expanded form in Appendix E.

4.3.3.2 Axial Load Distribution in Composite Cross-Sections

4.3.3.2.1 Traditional 1-D Analytical Model

Many DFEs have a composite (in the definition used in this monograph) cross-section that consists of two and, more recently, three distinct material components. With tapered piles, this most commonly occurs with tapered steel pipe piles such as the *Monotube* and *TAPERTUBE* that consist of a relatively substantial steel pipe with a PCC core.

Structural analysis or design of such a composite section requires allocating axial forces within the DFE to the components and then calculating the axial normal stress in each component. Historically, this is done using what the writer calls a '1-D analytical model' in Appendix E. This simple model treats the composite section as a basic problem in elementary physics in which each component is modeled as a linear-elastic axial spring. The stiffness of

---

[98] The latter is not a trivial statement. There are materials such as polymeric foams that exhibit a negative Poisson ratio, i.e. they actually neck (decrease in width or diameter in the horizontal direction) when loaded in uniaxial compression.

each spring per unit length of the DFE shaft is just *AE*, the product of the cross-sectional area and Young's modulus of each component. Solution is typically obtained by assuming:

1. All springs act in parallel.

2. All springs displace the same magnitude at a given magnitude of load.

3. **Each spring is initially unstressed** (highlighted for reasons noted subsequently).

These assumptions are noteworthy for reasons discussed subsequently (in particular, Assumption #3 is incorrect to varying extents due to residual loads) but for now the only comment made is that it is clear that the Poisson Effect is ignored as the explicit assumption is made that only axial behavior is important for allocating axial forces.

4.3.3.2.2 New 3-D Analytical Model

In the real world, simply ignoring something does not mean that it will conveniently not exist, occur, or be insignificant. Thus, even though the traditional 1-D model used to allocate loads and calculate axial normal stresses in a composite DFE cross-section ignores the Poisson Effect, it will still occur in any DFE subjected to axial loading. So, it is of interest to explore the impact of neglecting the Poisson Effect with DFEs.

DFEs with a composite cross-section are typically constructed so that all components are in intimate contact, at least initially and prior to external load application. This means that when axial loads are applied in either compression or uplift and horizontal or radial displacements of each component of the cross-section develop, there has to be both horizontal/radial stress and displacement compatibility at the interface between each component of the DFE. Furthermore, there has to be horizontal/radial stress and displacement compatibility at the interface between the exterior of the DFE shaft and adjacent ground.

While such a problem has been amenable for decades to analyze with the finite-element method, the writer is not aware of any simpler-but-rigorous analytical methodology that could be used for this purpose in routine practice. With this in mind, the writer crafted a closed-form solution based on the theory of linear elasticity using a basic solution that was published in Poulos and Davis (1974). The writer's solution methodology can be solved as a standalone problem using generic spreadsheet software or it can be incorporated into an analytical methodology that has a broader focus, e.g. resistance forecasting.

As can be seen from the exemplar analyses that the writer performed in 2009; published in Horvath (2010a); and that are republished in Appendix E, the conclusion reached over a decade ago was that in most cases involving traditional DFE materials of steel and PCC, the results of a rigorous 3-D model and analysis that include the Poisson Effect do not differ significantly from results obtained using the traditional 1-D model and analysis. The exemplar analyses presented in Appendix E also show that the effect of horizontal/radial interaction with the ground adjacent to a DFE shaft does not impact calculated results for apportioning axial loads between and among the components of a composite DFE.

That having been said, there may well be situations where the Poisson Effect is not insignificant. Several cases relevant to this monograph where this be true are discussed in the following sections. In addition, it would appear to be prudent in the future to include the Poisson Effect whenever some novel DFE or novel application of a traditional DFE is considered rather than to assume a-priori that 3-D effects are insignificant. As noted above, the analytical steps necessary to consider Poisson Effects routinely in practice are not onerous to implement so are not considered an analytical burden.

4.3.3.2.3 Additional New Issues

An important caveat to the preceding discussion and conclusions concerning the Poisson Effect is that they are based on the traditional assumption that after installation but prior to external load application that all DFE components have zero axial force/stress. This is the universal assumed starting point for both the 1-D and 3-D models discussed above.

However, as has been discussed a number of times throughout this chapter, it is now well-established that there will always be residual axial loads in all DFEs after installation. This situation is actually the most critical for tapered steel pipe piles such as the *Monotube* and *TAPERTUBE* because:

- It is now well-known that residual loads are relatively most significant for (driven) piles compared to other types of DFEs.

- Residual loads are more significant for tapered piles compared to otherwise identical piles of constant diameter.

- Tapered pipe piles such as the *Monotube* and *TAPERTUBE* are concreted after driving. Consequently, after pile completion but prior to external load application the PCC core will be unstressed while the steel pipe is already stressed, significantly in some cases. Specific numerical examples of this are presented later in this monograph.

Thus, when a completed tapered steel pipe pile is externally loaded after installation is complete, the applied loads will be <u>additive</u> to the compressive stresses that already exist in the steel component. Note that this is potentially problematic for the much more common case of compressive loading on a pile as the axial normal stresses in the steel component will be larger in magnitude than presumed based on an analysis (1-D or 3-D does not matter in this case) that assumed zero initial stresses in both the steel and PCC components.

The writer is not aware of any published work that has mentioned no less investigated this issue. The fact that there is no known structural failure of a deep-foundation system that can be specifically linked to this issue does not mean that this issue should not at least be investigated. It is quite possible that there are piles where the steel component has yielded which would simply transfer load to the under-stressed PCC core that was still exhibiting elastic behavior.

4.3.3.3 Radial Interaction at Shaft-Soil Interface

4.3.3.3.1 Generic Issues

As noted in the preceding sections, the 3-D model for axial-load allocation that was developed by the writer in 2009 assumes that the radial stress acting along the shaft-ground interface is a problem parameter that is both known beforehand (i.e. it is an input parameter) and remains constant in magnitude no matter how much the axial load in the DFE may change due to external load application in either compression or uplift. Based on this, the radial displacement of the shaft-ground interface is a calculated problem outcome.

Because this radial interface stress generally has negligible practical impact on the calculated allocations of axial load within a composite cross-section (this can be seen in the exemplar problems presented in Horvath (2010a) and Appendix E), the potential implications of these assumptions were not considered further in the writer's work back in 2009 when the model and its solution were developed. However, for the goals of this monograph, the Poisson Effect at the shaft-ground interface requires some discussion. In particular, the assumption that the writer made back in 2009 that the radial stress at the shaft-ground interface does not change even when the radial displacement at this interface changes needs to be revisited. Separate consideration is given to shafts with a constant-perimeter and those with a taper as the theoretical effects are subtly different between the two.

4.3.3.3.2 Constant-Perimeter Shaft

It has long been assumed that the post-installation horizontal or radial stress acting along the shaft-ground interface remains constant in magnitude under axial loading, whether it is compressive or tensile in nature. However, the 3-D model for axial-load allocation developed by the writer clearly indicates that there will always be some radial displacement of the shaft-ground interface, specifically, outward under compressive loading and inward under uplift loading. This is to be expected from the Poisson Effect.

This fact in and of itself would appear to invalidate any assumption that radial (and horizontal as well for DFEs with a non-circular cross-sectional geometry although the writer's 3-D model and concomitant solution only apply to DFEs with a circular cross-section) stresses at the shaft-ground interface remain constant under axial-load application. The question then becomes are these radial/horizontal interface displacements of sufficient magnitude to cause a <u>significant</u> change in the radial/horizontal stress.

It appears that for DFEs with a circular cross-sectional geometry at least this question could be addressed by realizing that this is really a problem in cylindrical-cavity expansion. While the basic mechanism for this has already been discussed in a qualitative manner earlier in this chapter, a much more detailed theoretical discussion including how to quantify this behavioral mechanism in practice is addressed later in this monograph. Therefore, further discussion of the Poisson Effect on the shaft resistance of DFEs with a constant-perimeter is deferred until that point. However, one final comment will be made before moving on.

A separate, but related, issue that is not directly germane to the present discussion but may be impacted by it is whether or not the unit shaft resistance due to sliding friction, r_{ss}, is the same in compressive or uplift loading. There have been conflicting opinions as well as data to support these opinions that have been published over the years (the current position as expressed by Fellenius is that there is no difference).

It should be obvious that the Poisson Effect may play some role in this issue. If it is found that the radial/horizontal displacement of the shaft-ground interface can, at least in some cases, be sufficiently large in magnitude so that it causes a change in the radial/horizontal stress (and, therefore, r_{ss}) then that should provide an answer.

Note, however, that it may well be that there is no one-size-fits-all answer to this issue as it should be obvious that the magnitude of this interface displacement is highly dependent on both the axial stiffness of the DFE shaft and the magnitude of the axial load. Therefore, it may well be that in some cases the interface displacement is sufficiently large in magnitude so as to affect the value of r_{ss} but in other cases it is not. This could well explain why different researchers have come up with different data and concomitant opinions over the years. Each side may be 'correct' for the particular set of problem parameters that they studied.

4.3.3.3.3 Tapered Shaft

The issue of how the Poisson Effect potentially impacts the unit shaft resistance, r_{st}, along a tapered section of a DFE is conceptually more straightforward than it is for a constant-perimeter section. This is because axial loading of a tapered section already presumes that there will be an effective increase in DFE diameter due to taper-related geometric effects as depicted in Figure 4.1. Consequently, the Poisson Effect is simply additive to that which will occur anyway.

The same is true for tapered piles subjected to uplift loads. As proposed earlier in this chapter, the application of uplift loads to tapered piles should be modeled as a problem in cylindrical-cavity contraction in which case the Poisson Effect is simply additive to that.

4.4 DESIGN PHILOSOPHY AND GOALS

The geotechnical ULS for all types of DFEs loaded in axial compression is typically visualized or idealized as a plunging load-settlement behavior, i.e. essentially unrestrained, unlimited settlement under constant load, after passing through a zone of yielding that is characterized by a relatively rapid increase in the slope of the load-settlement curve.

As has been noted in several places of this chapter in different contexts, DFEs generally do not exhibit such a well-defined, unambiguous geotechnical ULS, especially for friction/floating piles in coarse-grained soil. Rather, they continue to exhibit some increasing resistance with increasing settlement. This means that in those cases an attempt to define a unique, single-valued load that is the unambiguous geotechnical ULS using some graphical or other technique is arbitrary at best, and an exercise in futility and waste of time at worst.

A separate but related issue is that the physical concept of the ULS of soil is completely different from that of the ULS of a solid material in multiple key aspects. With soil, there is no permanent rupture of anything and there is no complete loss of strength as with solid-material failure. Rather, the ULS in soil is characterized by large-scale displacements caused by soil particles sliding relative to each other (yet still providing frictional resistance as they do so) that ceases once a new state of equilibrium has been reached for the affected soil mass. The soil particles themselves are, in general, none the worse for the wear as a consequence of 'failure'. In general, if the soil particles were excavated and then placed back into the ground there would be no evidence of the prior failure surface on which the soil-particle displacement had occurred.

As a first approximation, the soil mass in the new state of equilibrium after failure has occurred has the same strength it did as it was 'failing'. An example of this is when a traditional static load test with one or more unload-reload cycles is performed on a DFE. Even if the DFE was loaded to a relatively large settlement on initial loading, i.e. a condition that can broadly be interpreted as the geotechnical ULS, upon unloading then reloading it will return to the same load level at which unloading began via a relatively flat load-settlement path before resuming the steep load-settlement path it was on prior to unloading.

The point being reiterated here (as it was already made earlier in this chapter) is that the behavior of all types of foundations is, with few exceptions, geotechnically governed by settlement and the SLS and not some ULS resistance that is usually referred to as 'bearing capacity'. Fellenius (1999a) illustrated this for both spread footings and DFEs. The need to explicitly consider settlement in design has long been recognized for all types of shallow foundations (footings, mats/rafts, and slabs-on-grade) in addition to some types of DFEs such as drilled shafts. In the latter case, it has been recognized for decades that the relatively large diameters of many drilled shafts mean that settlement limits that are acceptable for performance of the structure being supported by the shaft(s) are reached well before any appreciable toe resistance is mobilized.

Unfortunately, this reality of the critical role that settlement and the SLS plays in the performance of a wide variety of foundations has been slow to be recognized for (driven) piles where the focus since the advent of modern foundation engineering in the 1950s has always been on defining a single-valued ULS load. The ongoing migration from ASD to LRFD design for DFEs has only reinforced this historically narrow focus and, in the process, set back efforts to refocus the attention on the geotechnical SLS and not the geotechnical ULS.

Nevertheless, this monograph adopts the position that routine DFE design in general, and tapered pile design in particular, should always be based on an explicit consideration of settlements and the geotechnical SLS together with the structural ULS. Note that this does not ignore the fact that piles bearing entirely within soil can exhibit characteristics of a geotechnical ULS. It simply recognizes the fact that the geotechnical Limit State, i.e. 'failure', for a DFE is fundamentally a settlement issue given the uniquely different way in which masses of soil fail compared to solid materials. Specifically, a DFE that reaches the geotechnical Limit State in axial compression is still capable of carrying load. The geotechnical Limit State simply means that the downside or penalty for carrying that load is a relatively large magnitude of settlement that is intolerable for most structures.

This settlement-focused perspective is actually in consonance with the modern perspective of dealing with all types of DFEs subjected to downdrag and the drag load that results from this. As illustrated in Figure 2.1 and as Fellenius has argued repeatedly since at least the early 1980s (Fellenius 1984, 1988, 1997, 1998, 2000, 2006, 2019; Tan and Fellenius 2016), the sign of shaft resistance due to the traditional mechanism of sliding friction is reversible as was noted and discussed earlier in this chapter. Whether the ground resists or adds to axial-compressive load within a DFE is solely a question of relative axial displacement between the DFE and adjacent ground.

It is also of interest to note that this modern perspective of the geotechnical Limit State being fundamentally a displacement issue can be found in the current way in which other problems in geotechnical engineering that were previously viewed solely from a ULS perspective are currently addressed. For example, free-standing rigid retaining walls subjected to seismic loading are now often allowed to 'fail' geotechnically and undergo limited rigid-body translation (sliding) on their base during a seismic event as long as adequate structural integrity is maintained. Traditionally, such structures were designed to maintain some margin of safety against all geotechnical modes of failure, even under loading from extreme events.

In summary and conclusion, this monograph adopts the position that DFE design in general and tapered-pile design in particular should follow the *performance-based* design philosophy that is increasingly being adopted in other aspects of geotechnical and foundation engineering, especially when seismic events and concomitant loading are concerned. This simply means that the overall performance of the structure is the focus. In almost all cases, performance is delimited by displacements and deformations. In keeping with this, a necessary goal of modern DFE design practice is to be able to calculate a complete load-settlement curve on a routine basis. How this can be done at present for tapered piles is thus one focus of Chapter 5.

This page intentionally left blank.

Chapter 5

Analytical Methodologies for Resistance Forecasting

5.1 INTRODUCTION

As noted in the opening chapter of this monograph, the overall technology of tapered piles is more generic at present than at any time in its past. As a result, it is more important than ever that all stakeholders in a project (design engineer, piling supplier, pile-driving contractor) be conversant in the analytical methodologies for forecasting the resistance of tapered piles for axial loading[99], especially in compression but increasingly in uplift as well. This broad, widespread need for analytical knowledge is especially true as larger projects in the U.S. increasingly employ a design-build contracting approach. Consequently, the focus of this chapter is a detailed presentation and discussion of analytical methods investigated and, in some cases, developed or extended by the writer for forecasting the resistance of tapered piles.

Note that this chapter does not explicitly address resistance verification (Construction Quality Control, CQA) during the construction phase. The writer's research over the years into tapered piles has not explicitly focused on this topical area so there is no significant body of original work that the writer has generated with regard to resistance verification. However, the writer has made numerous first-hand observations and developed opinions concerning resistance verification, both with respect to verification methodology and interpretation of verification results. These were already presented for the most part in Chapter 4. However, there is some additional indirect commentary concerning field-verification methodologies that is presented incidentally throughout this chapter.

The overall goal of this chapter is to put forth a comprehensive suite of analytical methodologies for resistance forecasting that range from the state of practice to the still-evolving state of knowledge (art). The latter draw on the wide range of geotechnical issues that were introduced and discussed previously in a generic fashion in Chapter 4. The intention is for this suite of presented analytical methods to serve as a reference and resource for all stakeholders (foundation engineers, piling manufacturers, pile-driving contractors, academic researchers) involved with tapered piles.

The inevitable questions for which all tapered-pile stakeholders are likely to seek answers from such a presentation are which analytical methods are 'correct' and, of these, which is the 'best'. Unfortunately, there is an underlying complication that impacts and thwarts answering these questions with any level of confidence and certainty. This troublesome complication is the issue of how any analytical methodology for resistance forecasting is either verified (in the case of indirect methods based on theory) or developed

---

[99] In the writer's opinion, there is nothing inherently unique about tapered piles subjected to lateral loading at the head of the pile other than the fact that the pile perimeter is not depth-wise constant in this case. However, commercially available analytical software for laterally loaded DFEs is nowadays capable of dealing with non-uniform flexural stiffness so this is a situation that is easily handled in practice. Consequently, this monograph does not explicitly address lateral loading of tapered piles as the writer does not perceive a need for this. That having been said, there is a modest body of published work that can be found on the specific subject of laterally loaded tapered piles. Some of these publications are listed in the reference and bibliography at the end of this monograph.

in the first place (in the case of direct methods that rely entirely on empirical correlations with in-situ test measurements). This essential verification or development process requires ground-truth in the form of measurement of the actual resistance of a pile or group of piles comprising a database as the case may be.

Historically and to the present, this acquisition of ground-truth has generally been considered to be a routine, almost trivial, activity to be performed at the lowest possible cost and thus taken for granted in the sense that measurement of actual resistance is presumed to be relatively simple and straightforward. However, as discussed at length in Chapter 4, nowadays there are several resistance-measurement methodologies based on different concepts (static, quasi-static, and dynamic) and there is no inherent reason to assume agreement between or among them when applied to the same pile. Even with the traditional static pile load test, as discussed in Chapter 4 there are many test-setup, measurement, and load-application variables and protocol variations that can and will affect the measured outcomes.

The result is that the ground-truth measured using any given field-test procedure should never be considered <u>the</u> correct, single-valued answer but simply one value in a range of potential answers that depend on and vary with the overall testing methodology used. This reality was emphasized in Chapter 4. Unfortunately, this reality complicates the desired assessment of the accuracy of resistance-forecasting methodologies as there never is just one single-valued 'correct' answer that can be used as the comparative metric for assessing accuracy of an indirect analytical methodology or creating an empirical relationship for a direct analytical methodology.

Note that this reality concerning the inherent non-unique nature of ground-truth for deep-foundation resistance is particularly troubling for direct (empirical) methodologies that, by definition, make use of a database of ground-truths to develop the methodology in the first place. Empirical methodologies require the up-front assumptions that the actual resistance is unique and single-valued for any given DFE and that the measured resistances in the database are these unique values. Thus, when some mathematical relationship is fitted to the database this relationship is assumed to be accurate within some error defined by statistical metrics produced by the curve-fitting process.

However, in reality the error inherent in the derived mathematical relationship is always much greater than it appears. This is because the ground-truth resistances that compromise the database should be viewed not as single values but as but one value in some unknown range in values. Unfortunately, the actual, larger error of the empirical relationship is unknown as the error in the underlying ground-truth database is not quantified. Again, this is just reiterating points made already in Chapter 4. However, these are critical observations relative to the contents of the present chapter so warrant repetition for emphasis.

5.2 PROLOGUE

To transition from the relatively abstract, theoretical discussions of geomechanics and structural issues potentially affecting tapered-pile behavior that were discussed in Chapter 4 and lay the groundwork for a critical assessment of real-world resistance-forecasting methodologies presented in this chapter, it is useful to summarize what the writer feels are essential considerations and elements of an 'ideal' resistance-forecasting methodology for tapered piles.

This 'wish list' is recognized as being inherently subjective in nature and, therefore, it is likely that others will have different opinions. Nevertheless, it does at least provide a list of initial talking points that can be modified as desired by others pursuing the subject.

As will be seen at the conclusion of this chapter, no existing resistance-forecasting methodology satisfies all of the writer's criteria. Therefore, the criteria presented here will serve as the schema or framework for an improved resistance-forecasting methodology presented in Chapter 7.

To begin with, an ideal resistance-forecasting methodology for tapered piles needs to have as its central tenet the fact that the stress-state in the ground at all stages of the overall problem is the single most important consideration. Therefore, three distinct aspects related to stress-state need to be considered:

1. The site-specific K_o profile prior to pile installation.

2. K_h after pile installation, most easily done using an appropriate K_h/K_o ratio. Ideally, there should be consideration of how this K_h/K_o ratio might change over time between the end of driving and prior to the first external load application.

3. Changes to K_h as the result of externally applied axial loading in both compression and uplift. Again, the K_h/K_o ratio concept is useful for expressing these changes.

Items #2 and #3 above reflect what can broadly be termed 'taper benefit' which, as can be seen, occurs in two distinct stages. As will be seen, no current resistance-forecasting methodology explicitly considers these two distinct behavioral stages of tapered piles.

Problem parameters that likely to play a role in quantifying this two-stage taper benefit include:

- soil type,

- soil consistency,

- soil strength,

- overburden stress,

- pile cross-sectional geometry (circular vs. square),

- taper angle,

- toe diameter, and

- residual loads.

With regard to soil type, tapered piles have historically been viewed as being most effective and, at times, only effective in coarse-grained soils. As a result, some resistance-forecasting methodologies only allow for use in coarse-grained soil conditions. However, an ideal resistance-forecasting methodology should be more inclusive and allow for fine-grained soil conditions.

Another key tenet of an ideal resistance-forecasting methodology is its ability to explicitly consider settlement and link it to the forecast total resistance. Of necessity, this will require an explicit consideration of the depth-wise distribution and magnitudes of residual loads which, as will be seen, are quite significant for tapered piles.

Note that an explicit consideration of settlement means that whatever model and analytical solution are used to forecast toe resistance need to allow for variation of toe resistance with toe settlement. This is a significant departure from the historical approach to toe resistance that only focuses on a single ULS value with no reference to toe settlement. Another consideration related to toe resistance is whether or not it is affected by taper angle as has been suggested by Manandhar et alia.

There are a number of secondary issues that should at least be considered for inclusion in an improved resistance-forecasting methodology for tapered piles. These include a consideration of Poisson Effect on resistance.

Finally, implementation of this ambitious wish-list should keep in mind the computer software needed for its implementation. There is no doubt that much can be accomplished nowadays with purpose-written commercial software with a graphical user interface. However, consideration should be given to developing an analytical methodology that is amenable to solution using simpler, generic software such as a spreadsheet.

5.3 OVERVIEW

5.3.1 Background

As noted in Chapter 1, civil engineers have been formally engaged in resistance forecasting for (driven) piles since at least the middle of the 19th century. This is almost as long as the profession of civil engineering has been formally established, in the U.S. at least. In the 170 years since then, resistance forecasting for piles has evolved along multiple paths with the end result that the current states of knowledge and practice are both complex and fragmented. Consequently, before proceeding further, it is necessary to define an organizational structure for this chapter in order to create a coherent framework for material presentation and discussion.

To begin with, as noted earlier in this monograph there are two broad, conceptual approaches used for resistance forecasting for piles: dynamic and static. Although the former was first defined approximately 100 years before the latter, since the 1950s these two approaches have co-existed and evolved more or less independently of each other along parallel developmental paths. The current state of practice is that foundation engineers and other stakeholders typically use some combination of the two approaches on a given project although project-specific conditions as well as either organizational or personal preferences/biases can skew the usage toward one approach or the other.

The position taken in this monograph is not to champion one approach over the other but to present the pros and cons of each as objectively as possible so that stakeholders can perform their own assessment and from this make their own informed decision as to how to use some blended, balanced combination of these two approaches on a given project. That having been said, the quantity of material presented in this chapter is skewed heavily toward the static approach for the simple reason is that this has been the area where the writer has always concentrated his interest and concomitant research.

5.3.2 Chapter Organization

For historical reasons and to understand how the state of knowledge related to pile resistance evolved in the minds of foundation engineers over the last 170 years, the dynamic approach to resistance forecasting is discussed first. This is followed by a discussion of the static approach to resistance forecasting.

Because the discussions of both the dynamic and static approaches include a substantial amount of material, summary overviews of each are first presented in the sections immediately following so that a 'big picture' of each approach can be seen before delving into the granular-level detail that comprises the bulk of this chapter. Following these summary overviews is a focused discussion of site characterization as it applies specifically to the analytical methodologies that are presented subsequently.

5.3.3 Dynamic Approach

This approach in all its forms is based on the overall concept that the amount of physical work, i.e. kinetic energy, imparted to the pile-ground system by the pile-driving hammer during the pile-driving process translates into potential energy in the form of ULS total resistance to an axial-compressive applied load (force) at the head of the pile after driving. This overall concept is intuitively attractive as it seems logical that the pile-ground system 'owes something' in the form of resistance to an externally applied axial-compressive force (and, by logical extension, an uplift force as well) in return for the amount of energy expended getting the pile into the ground.

As an aside, it is of interest to note that the dynamic method is inherently structured to only provide forecasts of axial-compressive ULS resistance. This is because during pile driving there is resistance both along the shaft and at the toe so the two sources of resistance are lumped together in dynamic forecasting methodologies. However, some versions of the dynamic method can be deconstructed to separate out the shaft and toe components of ULS resistance so that a forecast of ULS resistance to uplift load can be obtained.

The dynamic approach was reportedly first formalized mathematically, in the English language at least, in the middle of the 19th century. What has changed in the 170 years since then is the physical model that is hypothesized to represent the pile-ground system and the pile-driving process. For about the first 80 years (i.e. until circa 1930), it was assumed that a pile could be modeled as a single, rigid object, i.e. a classical 'rigid body', with the implication that the entire pile moves as one when impacted by a pile hammer.

A tectonic paradigm shift concerning the hypothesized physics of pile driving evolved beginning in the 1930s. However, it does not appear that the concept gained much technology transfer and concomitant traction until the 1950s when there was a synergistic combination of the right person in the right organization at the right point in time to effect tectonic changes in deep-foundation practice that persist to the present. This new concept in the physics of pile driving posited that piles do _not_ behave as a single rigid body during conventional impact-driving. Rather, piles always exhibit substantial axial compressibility during driving.

The essential outcome of this axial compressibility is that a pile does _not_ move into the ground as a single rigid object when impacted by a pile hammer but rather discrete portions of the pile are stressed, compressed, and move incrementally and sequentially in a top/down/back-to-the-top fashion. This has been likened to the way in which a caterpillar moves across the ground by articulating portions of its body incrementally and sequentially. However, the pile-driving process occurs in a millisecond timeframe that cannot be seen by human vision which is why a pile appears to move rigidly to the human eye.

Understanding the correct physical process that occurs during pile driving led to the development and eventual formal publication of E. A. L. Smith's 1-D wave-equation model in the civil engineering literature in 1960[100]. The hypothesized behavior that forms the basis of

---

[100] Smith was a mechanical engineer by profession and employed in the construction industry. Consequently, elements of his work were actually published several years earlier in mechanical engineering and construction engineering venues.

the 1-D wave-equation model was confirmed, at least for constant-perimeter piles, by the development of field-measurement techniques in the late 1960s by the late Prof. George G. Goble. The use of these field-measurement techniques has grown substantially since then and continue in global use to the present.

The technically significant outcome of recognizing pile compressibility during driving is that this compressibility should not be ignored if the pile-driving process is to be modeled with any accuracy. Nevertheless, despite the overwhelming body of evidence that has existed for almost 100 years now as to the correct physical mechanism associated with pile driving, foundation engineers have been slow to completely abandon use of the older rigid model that is simply incorrect. Not only do some industry sectors, especially state departments of transportation in the U.S., continue to use long-deprecated dynamic formulas based on the 19th century rigid model in the 21st century, some seek to extend the life of such formulas by aligning them with modern LRFD design methods (Allen 2005). The motivation for this is apparently to have a relatively simple tool for CQA that can be used in the field by technician-level inspectors and not require any advanced measurement hardware and data-reduction software that would likely require the involvement (and cost) of a design professional.

It is relevant to note that Goble's late-1960s field-measurement techniques, often referred to colloquially and generically nowadays as *dynamic measurements*, that were crucial to confirming 1-D wave mechanics as the operative physical model during impact-driving of piles have evolved into what is arguably the most common method, dynamic or static, for resistance-verification in the world today. Because the physical model embedded in the dynamic-measurements methodology is essentially Smith's 1-D wave-equation model, it is important to note that many of the comments made in this chapter concerning the 1-D wave equation as a resistance-forecasting methodology are also applicable to dynamic measurements as a resistance-verification methodology.

In summary, except for outliers such as the continued use of outdated dynamic formulas in some sectors of practice, the current state of the dynamic approach to assessing pile resistance, whether for forecasting or field verification, involves analytical methodologies and algorithms that have coalesced around the 1-D wave equation. There are assumptions in the 1-D wave-equation model that have critical implications for piles in general and tapered piles in particular. These are addressed later in this chapter.

5.3.4 Static Approach

The static approach to resistance forecasting for DFEs in general is based on an assumed installed geometry of a DFE combined with assumed axial-force load-resistance mechanisms between the ground and DFE. Geomechanics principles are then applied to these resistance mechanisms with varying degrees of theoretical rigor to evaluate them.

The resistance visualization using the two traditional resistance mechanisms of shaft resistance due to sliding friction and toe resistance due to end bearing are shown in Figure 2.1. The case was made in Chapter 4 for a third resistance mechanism of shaft resistance due to taper based on the physical mechanism of wedging as modeled using cylindrical-cavity mechanics (cavity creation followed by cavity expansion or cavity contraction depending on whether the applied loading is compressive or tensile in nature) for use with tapered piles.

The static approach began to evolve in the 1950s, a full century after the emergence of the dynamic approach in U.S. published literature. From the start and continuing to the present, the static approach has been pursued by researchers along two parallel conceptual paths:

- The more theoretically rigorous is called the *indirect* (sometimes alternatively called *rational*) approach wherein site-specific soil properties, primarily strength-related, are first determined by site characterization. These properties are then used in traditional geomechanics solutions to calculate ULS values of unit stresses for shaft sliding friction, toe bearing capacity (classical or some modern alternative), and shaft wedging (when tapered piles are involved), i.e. the parameters $r_{ss(ult)}$, $r_{t(ult)}$, and $r_{st(ult)}$ as defined in Chapter 4. These solutions are predominantly stress-based although any type of cavity-mechanics theory for either toe end bearing or shaft wedging also requires an estimate of soil stiffness, preferably shear modulus, G.

- Less theoretically rigorous but increasingly popular as global use of in-situ testing has increased over the years is the *direct* approach wherein the measured parameters from a site-characterization tool such as the cone penetrometer are used directly in empirically derived equations to provide estimates of $r_{ss(ult)}$ and $r_{t(ult)}$. No such relationship has been developed to date for $r_{st(ult)}$ but there is one known direct method (the Togliani Method that is discussed later in this chapter) that was developed specifically for tapered piles. However, it uses the Fellenius donut-on-a-stick model (Figure 4.2) for tapered-pile resistance so only uses $r_{ss(ult)}$ and $r_{t(ult)}$.

For approximately the first 50 years that the static approach was used (i.e. up until the early years of the 21st century), the common element shared by both the indirect and direct approaches was that only ULS resistances were calculated. This is because both the ASD or LRFD design methods have always been based conceptually on applying 'safety' in some fashion to the ULS total resistance. This resistance is always taken to be the simple arithmetic sum of the ULS shaft resistance due to sliding and the ULS toe resistance due to end bearing (Fellenius (2016) provides a recent discussion of this conceptual approach).

The point being emphasized here is that there is generally no indication of the magnitude of DFE settlement (head or toe) that is associated with this ULS total resistance. In fact, one can arguably interpret this ULS total resistance as always being accompanied by a plunging-type failure mechanism wherein settlement is, by implication, infinite.

Changes with regard to this traditional fixation on only ULS total resistance that have been evolving in recent years are discussed subsequently but it is relevant to note at this point that the vast majority of routine design still considers only ULS total resistance. In fact, if anything, the transition to using LRFD in recent years has reinforced and literally enforced (through mandatory design codes in many parts of the world) this traditional way of thinking.

Unfortunately, this simplistic approach of summing shaft and toe ULS resistances makes no consideration of several significant behavioral facts concerning these resistance components:

- Overall, the two resistance vs. displacement curves for the shaft and toe develop very differently, at least when all shaft resistance is due to sliding friction. Consequently, the appropriate summation of the combined resistance is not straightforward and, in general, is ill-defined because of issues related to toe resistance as explained subsequently. Fellenius (2018) discusses this point in some detail.

- Shaft resistance due to sliding friction can have a range of behaviors (Fellenius 2016, 2018) but usually tends to be strain-softening in nature, reaching a peak at relatively small displacement magnitudes (fraction of one inch/several millimetres) that are more or less independent of soil type as well as deep-foundation type and diameter.

- Toe resistance tends to be work-hardening in nature and in many (all in some opinions, e.g. Fellenius as discussed in Chapter 4) cases does not appear to reach a peak, at least at the displacement levels to which static load tests are typically carried out. This lends support to Fellenius' hypothesis that a classical bearing-capacity mechanism does not develop beneath any type of concentrically loaded foundation, deep or shallow, as a well-defined peak resistance followed by strain softening would be expected as the soil's shear strength transitioned from peak to constant-volume (critical-state) conditions as the writer discussed in Horvath (2002). In addition, the development of toe resistance is dependent not on absolute displacement as with shaft resistance but the ratio of settlement to toe diameter as clearly shown in Fellenius (2016) as well as several earlier publications by Fellenius so this is not new knowledge. The development of toe resistance also depends on soil type as well as DFE type.

- The development of both shaft and toe resistances as the results of post-installation axial loads in both compression and uplift is complicated by the existence of residual loads. As discussed in Chapter 4, residual loads are generally of greatest importance for (driven) piles in general and tapered piles in particular. In some cases, the presence of residual loads can have a profound effect on the mobilization of shaft and toe resistances. This is illustrated for a case history involving instrumented piles, both tapered and non-tapered, that is discussed at length in Appendix H.

Tapered piles add another layer of complexity to the overall problem as there is now a third resistance mechanism, wedging, to consider. There has been no known published research of tapered piles that illustrates the actual resistance vs. displacement (both axial and radial in this case) behavioral relationship, either theoretical or actual, as there is for the two traditional resistance mechanisms as noted above. In principle, it should be possible to generate theoretical resistance-displacement curves at least using solutions based on cylindrical-cavity mechanics that is now the preferred physical model for the wedging mechanism. However, this is not known to have been done to date.

The writer (Horvath 2002) opined that since the physical model that is used for the wedging mechanism (cylindrical-cavity creation and subsequent further expansion) involves shear-strength mobilization that one would expect to see a distinct peak resistance followed by strain-softening behavior as the shear-strength parameters transitioned from peak to constant-volume/critical-state values. However, subsequent research (Salgado and Prezzi 2007, Cook 2010) has demonstrated that the stress-state around a cylindrical cavity is very complex and thus relatively simple patterns of shear-strength mobilization (i.e. peak to constant volume/critical state) that are applicable to many other geotechnical and foundation applications do not apply to cavity mechanics. In any event, it remains unknown at this point time how the resistance vs. axial displacement behavior of the tapered portion of a pile compares to that for the constant-perimeter portion of a pile and pile toe.

Since the beginning of the 21st century, there have been two significant evolutionary changes in the overall implementation of the static approach in practice that have moved it beyond the traditional ULS resistance-only framework in use since the 1950s. The first change is one promoted by Mayne beginning in circa 2007 in which solutions to a theory-of-linear-elasticity problem, referred to in recent publications by the writer (Horvath 2016a, 2016b) and defined earlier in this monograph as the Randolph-Wroth Problem, are used to construct a complete load-settlement curve.

It is relevant to note that Mayne's application of the Randolph-Wroth Problem is applicable to all types of DFEs, not just piles. It is also important to note that Mayne's analytical methodology is what the writer defines as being a two-step procedure to generating a load-settlement curve in that it requires a separate, a-priori estimate of ULS total resistance using whatever resistance-forecasting methodology the analyst chooses, indirect or direct. The solutions to the Randolph-Wroth Problem used by Maybe simply 'fill in the blanks' between the origin of the load-settlement curve and the ULS total resistance that is approached, but never reached, in an asymptotic fashion that preserves the previously mentioned fact that calculated values of the ULS total resistance historically imply a plunging-type failure that is accompanied by infinite settlement.

Even though Mayne's work related to the Randolph-Wroth Problem is essentially an embellishment of the traditional approach to static resistance-forecasting based on soil strength, it still represents a significant advancement in the state of knowledge as it provides at least a first-order approximation of a rational, theoretical basis for a settlement-based/SLS approach to deep-foundation design that Fellenius and others (including the writer) have championed for some time now. Fellenius (2016) outlines the compelling argument for basing the geotechnical aspects of all deep-foundation design on settlement and the SLS as opposed to solely ULS total resistance.

The second evolutionary change concerning the static approach to occur in the 21st century is much more of a tectonic paradigm shift in its concept and still very much in its infancy. This is the fully integrated, one-step load-settlement concept promoted by Mayne's protégé, Prof. Fawad S. Niazi, since circa 2011. The significant paradigm shift with Niazi's concept is that a complete load-settlement curve is generated but with no need to forecast the ULS total resistance beforehand using a separate analytical methodology as with Mayne's earlier, two-step concept. This is because in Niazi's analytical methodology, load and settlement are linked empirically based solely on site characterization using an sCPTu sounding. Niazi's empirical load-settlement relationship was developed from a zero-based database of sites where both an sCPTu sounding and static load test (not necessarily to the geotechnical ULS) were performed, with the field data analyzed using the aforementioned Randolph-Wroth Problem solution that is essentially applied in reverse.

While Niazi's integrated load-settlement methodology may appear to be a 'dream-come-true' analytical tool for resistance forecasting for all types of DFEs, the writer's assessment of this method to date has raised some very serious questions about its theoretical basis. This is because of the fact that Niazi's methodology is based on the Randolph-Wroth Problem and thus has some inherent, theoretical issues that are built into that problem. These concerns are discussed in detail later in this chapter. Nevertheless, there are some potentially useful elements to Niazi's concept that make it of potential use as long as one is clearly aware of its shortcomings. This is also discussed in detail later in this chapter.

A lesser point overall but certainly one that is central and thus significant to the theme of this monograph is that it does not appear that any type of tapered pile was included in Niazi's database. To a certain extent this was the consequence of Niazi's developing an analytical methodology that requires an sCPTu sounding. While this technology has been around for some time, it is still not yet in the mainstream of use in routine practice on deep-foundation projects, at least in the U.S. where tapered-pile usage is and has long been significant compared to elsewhere in the world. As a result, Niazi's concept is not usable for any type of tapered pile at the present time although the writer will illustrate how this can be overcome in a relatively straightforward, easy manner in the future.

5.4 SITE CHARACTERIZATION

Regardless of which resistance-forecasting approach or method is to be used, the first step on any project is to perform a site-characterization assessment. In the writer's opinion, the current 'gold standard' for projects where tapered piles are planned to be used or at least potentially used is to employ a combination of conventional borings with SPT sampling for soil-identification purposes and cone-penetrometer soundings.

With regard to the latter, some combination of CPTu and sCPTu soundings should always be used regardless of the size of the project, with the latter favored during the early stages of the investigation. This is because only site-specific shear-wave velocity, V_s, measurements can define the 'true' small-strain shear modulus, G_{max}, profile. Experience indicates that even for sites that appear to have essentially the same soil profile and geological age throughout the site, the degree of 'structure' (behavioral age, cementation), which will influence both V_s and G_{max}, can vary significantly (Horvath 2016b). Consequently, it is useful to define any site variations in structure as early as possible in a project's evolution.

As an aside and separate from issues related to pile design, it is now known that soil structure can influence a site's potential for seismic liquefaction (Horvath 2016b). Therefore, on projects where seismic liquefaction is a design consideration, defining the structure of site soils, which can only be done accurately using site-specific V_s data from sCPTu soundings, is another compelling reason to not only use the sCPTu for at least some of the cone-penetrometer soundings but to employ them as early as possible during project design.

Returning now to the issue of pile design, there are myriad empirical relationships for soil properties based on cone-penetrometer data (q_c, f_s, u_2, V_s) that have been published over the years. These relationships can be used to estimate dozens of soil properties and related parameters as illustrated in Horvath (2015) although plots of the depth-wise variation in K_o and G_{max} are the most useful for tapered-pile resistance forecasting as will be seen later in this chapter as well as elsewhere throughout this monograph.

The writer has found that the most useful and up-to-date cone-penetrometer relationships tend to be published by either Prof. Paul W. Mayne or Dr. Peter K. Robertson although significant niche correlations and contributions to the state of knowledge have been made by many others (note that in many cases, the work of others tends to be curated into the work of Prof. Mayne and Dr. Robertson). Unfortunately, there is no one reference that the writer considers to be comprehensive on this subject due to the fact that developing empirical correlations between cone-penetrometer data and soil properties and parameters is a very active area of research. As a result, publications with 'new-and-improved' empirical correlations appear on a regular basis. However, Kulhawy and Mayne (1990) is noteworthy as being perhaps the first attempt to publish a comprehensive treatment of the subject that included results from a wide variety of laboratory and in-situ test methods in addition to the cone penetrometer. Later efforts by Mayne (2007) and Robertson and Cabal (2015) are focused solely on the cone penetrometer.

There are two important facts to keep in mind:

- As noted above, the field of developing empirical relationships for soil properties based on cone-penetrometer data is an active, ongoing area of global research, and revised or new empirical correlations frequently appear in the published literature. As a result, published information tends to go out of date with regularity.

- The empirical relationships for a given soil property or parameter tend not be unique. In addition, different researchers take different approaches for developing an empirical correlation for the same soil property, and even the same researcher sometimes proposes (in the same publication no less) multiple correlations for a given soil property. As a result, a design professional always needs to make their own evaluation of which correlation to use or, in some cases, use several and then arrive at some conclusion from an evaluation of the different results. The writer has frequently done this as can be seen in Horvath (2014b, 2015, 2016a, 2016b) which cite the specific equations and references used in all cases.

The writer is well aware of the fact that for various reasons it may not be possible to achieve the desired site characterization that includes sCPTu soundings and concomitant site-specific V_s profiles for a given project. Furthermore, for sites that have seen prior subsurface exploration, there may be results from earlier borings and/or less-sophisticated cone-penetrometer soundings that a design professional may want to use to create as detailed a picture of soil properties as possible.

The writer's research has shown that reasonable results in terms of estimating relevant soil properties can still be achieved using CPTu and even CPT data as well as SPT N-values, at least for sites comprised predominantly of coarse-grained soil. A hierarchy of three different, lesser levels of data quality have been identified in this regard, presented here in order of decreasing overall accuracy:

- If only CPTu data are available, it is still possible to use the same state-of-knowledge empirical relationships for virtually all soil properties as when sCPTu data are available as most of these relationships do not require field-measured values of V_s. Furthermore, empirical correlations exist to estimate V_s and G_{max} using CPTu data but with the important caveat that these relationships were developed from sites underlain by 'well-behaved' soils that were geologically 'young' (= Holocene Epoch) and thus by definition behaviorally 'young' as well, i.e. with no structure due to aging and/or cementation. Thus, sites where the soils exhibit some degree of structure due to aging and/or cementation may not be properly assessed in terms of V_s and G_{max}. Note that it is impossible to tell whether or not effects of structure are present using CPTu data alone unless there is supplemental information, e.g. published geological knowledge, as to the geological age. However, even this is not foolproof as a soil deposit may be geologically 'aged' but behaviorally 'young' because its age was 'reset' one or more times in the past due to any number of natural and human-induced phenomena. This issue was explored and discussed in detail in Horvath (2016b) using the Pleistocene soils underlying JFKIA as an example.

- If only CPT data are available and pseudo-u_2 values can be reasonably estimated ($u_2 = u_o$ which means this is only valid for coarse-grained soils with hydrostatic groundwater conditions), virtually all of what was said regarding sites with only CPTu data applies here as well. The writer illustrated this in Horvath (2014, 2015, 2016a, 2016b).

- If only SPT N-values are available and only coarse-grained soils are involved then an empirical correlation between N-values and q_c can be used. From this, a limited suite of soil properties can be estimated. The writer first developed an analytical algorithm for this circa 1990 and made numerous improvements to it over the years as new empirical correlations became available in the published literature and otherwise. The most recent version of this algorithm was published in Horvath (2011). As shown in Horvath (2002), this approach can provide surprisingly good results, at least in some cases, although this statement most definitely should <u>not</u> be read or interpreted as justification for maintaining the decades-old status quo of only drilling borings and performing the SPT as site characterization on projects involving tapered piles. Part of the reason why the writer has suspended any further development and publication using SPT N-values as pseudo-q_c values is to <u>not</u> promote and even deprecate continued use of this approach as it should only be used when there is no other alternative.

5.5 DYNAMIC APPROACH

5.5.1 Rigid-Pile Model

5.5.1.1 Overview

A single, basic concept from classical physics involving the interaction between two rigid bodies or masses (the pile and the pile-hammer ram in the case of foundation piling) was used as the theoretical basis for the earliest known efforts to model the pile-driving process using the dynamic approach to resistance forecasting[101]. Chellis (1961) contains a detailed derivation of the algebraic equation that is the rigorous mathematical expression of this concept. However, it does not appear that this rigorous solution itself was ever identified and proposed as a specific, named analytical model (referred to in foundation engineering literature and in this monograph as *dynamic formulas*[102]) for use in practice.

In terms of the historical evolution and timeline during which numerous researchers used this physical concept of rigid-body interaction as a starting point for a physical and mathematical model for pile driving and each developed a dynamic formula as the outcome, it appears that two distinct approaches to problem-solving were used. Each approach assumed that the two rigid bodies interacted in a different manner.

The first approach involved the simple, basic concept of equating the assumed quantity of work done by each of the two rigid bodies and the second involved the assumption of actual physical impact between the two rigid bodies. Stated another way, the first approach only <u>implied</u> a physical interaction between the two rigid bodies in that the amount of work done by each of the two components was the same for each blow of a pile hammer whereas the second approach attempted to actually model the ram-pile impact for each blow of a pile hammer, with different approximations assumed for different resulting formulas.

As it turns out, the former work-based approach is, mathematically at least, equivalent to a much-simplified approximation of the latter impact-based approach so there is a common element between the two approaches. Nevertheless, despite the common thread between these two approaches (arriving at the same conclusion despite taking different paths to get there), the presentation and discussion that follows is broken into two separate sections for clarity. The latter section also presents the unifying synthesis of these divergent approaches to problem solution.

As will be seen, originally a much-simplified version of the exact algebraic equation defining this physical concept of rigid-body interaction was used for the first known dynamic formula circa 1850 and then over the span of several decades succeeding dynamic formulas proposed by others added terms to the original, ultra-simple formula so that the resulting new formulas became increasingly more complex. In other words, problem solutions started simple and then progressively got more complex. Later still, the reverse course of action was pursued, i.e. the problem-solving approach taken was to start complex and then make simplifications to develop a 'new and improved' dynamic formula. Again, the overarching theme that encompasses all of this work is that of divergent approaches to solving the same problem and, thus, essentially winding up in the same place.

Not surprisingly, almost all of the dozens of different dynamic formulas that resulted from these efforts over the course of more than a century are identified by the surname of the

---

[101] As will be seen, the problem solutions that were developed can also be used for resistance verification as part of a CQA program although the focus in this chapter is on forecasting.

[102] As noted in Chapter 1, the term *pile-driving formulas* also appears in the published literature.

researcher who is credited with developing the solution. In a few cases, the same solution has more than one name associated with it or has the name of some organization. These names are useful for keeping track of what otherwise is a relatively complex and confusing process of R&D that was spread out over a period of more than 100 years.

In summary, the dynamic formulas that are based on the older, simpler work-based approach and that are discussed in the first section that follows can be shown to be highly simplified cases of the later, more complete (and complex) impact-based version of this physics concept that are discussed in the second section following. In other words, the dozens of different dynamic formulas identified to date can each be shown to be a simplified version of the same rigorous problem and solution. However, it appears that this unifying perspective was not known until several decades after the older, simpler dynamic formulas based on work concepts had been proposed.

The common, and, as it turns out, most significant, element of the physical concept employed for all dynamic formulas is that the pile is assumed to be a rigid body. The implication of this is that the entire pile from head to toe and everything in between displaces the same amount in a direction that is parallel to the longitudinal axis of the pile and at the same time when the pile is impacted during driving. This was unequivocally proven to be incorrect by the late 1960s for all piles in all cases but this has not stopped use of dynamic formulas based on the rigid-body model to the present, a fact that was noted earlier in this chapter as well as earlier chapters.

5.5.1.2 Work Equation and Derivatives

The first attempts to use the rigid-body dynamic model for resistance forecasting date back to the mid-19th century (Chellis 1961, Likins et al. 2012). The overall physical model used is based on pure rigid-body physics with no explicit geotechnical influence or consideration. It invokes the simplistic, intuitive physical concept that the work done by a pile-driving hammer driving a pile into the ground is perfectly transmitted into the pile and thus equal to the work done by a pile in the form of pile resistance. As such, it will simply be referred to in this monograph as the *Work Equation*.

It is relevant to note that the algebraic expression of the Work Equation, which is presented subsequently, can be formulated for use in either design mode for resistance forecasting prior to pile installation or in analysis mode for resistance verification (CQA) during or after pile installation. The focus here is on the former although use for the latter is noted.

In this application, 'work' is defined in the classic kinetic-energy sense as a force applied to a rigid mass times the rectilinear displacement of the rigid mass to which that force is applied. So, the work done by the hammer (a simple drop-hammer in that era) is simply the weight (= mass times gravity) of the ram times the distance the ram free-falls before impacting the head of the pile (also assumed to be a rigid mass).

For mechanized hammers, the work is the actual energy delivered to the pile head. Optimistically, this is the so-called *rated energy* of the hammer but realistically this should be the rated energy reduced by some dimensionless hammer-efficiency factor that is less than one. In both cases, the hammer energy has dimensions of force times length, e.g. kilonewton-metres, foot-pounds, etc.

The complementary work done by the pile is simply its geotechnical resistance (a vertical force in this case, R_u in Figure 2.1) times the *set* of the pile under a single hammer blow. 'Set' in a pile-driving context means the distance the pile head <u>permanently</u> displaces into the ground under a single hammer blow[103] or the arithmetic mean of some contiguous number of hammer blows. Note that because the pile is assumed to be a rigid body, the set is assumed to represent the displacement of the entire pile. Note also that the set is the reciprocal of the number of blows required to drive a pile some specified distance (foot, inch, metre, etc.). This is important as it is far easier to measure the number blows per some unit length, e.g. blows per foot, as opposed to actually trying to measure the set (which was, and still is, actually done at times, e.g. Damen and Denes (2017)).

Expressed algebraically, the Work Equation can be expressed for a drop hammer as:

$$R_u \cdot s = W \cdot h \tag{5.1a}$$

and for a mechanized hammer as:

$$R_u \cdot s = E_{actual} = \eta \cdot E_{rated} \tag{5.1b}$$

where:
- R_u = ULS resistance,
- s = set of pile head,
- W = weight of the ram of a drop hammer,
- h = free-fall distance of the ram of a drop-hammer,
- E_{actual} = the actual energy delivered to the pile head when using a mechanized hammer,
- η = a dimensionless efficiency factor ($0 \leq \eta \leq 1$) for a mechanized hammer, and
- E_{rated} = the rated energy of a mechanized hammer.

Note that the dimensions of all dimensioned parameters must be consistent.

Equations 5.1a and 5.1b reflect the basic equality of work done by the two rigid bodies involved and as such are simply the mathematical expression of the assumed problem physics. However, for actual use of these equations, the variables are better rearranged depending on whether the equations are to be used for resistance forecasting or resistance verification (CQA).

In the former (forecasting) mode that is of greater interest in the present discussion, typically there is some 'target' ULS resistance that is the minimum acceptable value desired for a pile in some project-specific application. Consequently, the unknown variable in this case is pile set as presumably the characteristics of the hammer to be used are known. Thus, the equation for the largest acceptable value of set to produce the desired ULS resistance becomes:

$$s = \frac{W \cdot h}{R_u} = \frac{E_{actual}}{R_u} = \frac{\eta \cdot E_{rated}}{R_u} \tag{5.2}$$

with appropriate problem parameters used depending on whether a drop or mechanized hammer is used.

In the latter mode of resistance verification (CQA), pile sets become a known variable as the reciprocal parameter (blows per unit length) is observed and recorded during driving. Therefore, pile resistance as a function of measured set becomes the unknown variable:

---

[103] Under a given hammer blow, a pile will displace downward into the ground and then rebound (displace upward) to some degree. As a result, the set is less than the maximum distance that the pile displaced downward under the hammer blow. Note that this rebound is the underlying causal mechanism of the drag load that results in residual loads within the pile after driving.

$$R_u = \frac{W \cdot h}{s} = \frac{E_{actual}}{s} = \frac{\eta \cdot E_{rated}}{s} . \tag{5.3}$$

Note that in this case, E_{actual} is, nowadays, often known with some precision as the result of making dynamic measurements during driving.

It is not known to what extent, if any, the Work Equation in its fundamental form as expressed in Equations 5.2 and 5.3 was ever used in practice. What is known is that circa 1851 the resistance-verification formulation (Equation 5.3) was more commonly used but with an arbitrary safety factor of eight[104] in the denominator (i.e. "s" was replaced by "$8 \cdot s$") so that the calculated resistance was an 'allowable', R_{all}, not the ULS value, R_u. This was in consonance of the widespread use of ASD at the time. In this modified form (R_u replaced by R_{all} and s replaced by $8s$), Equation 5.3 it is known as the *Sanders Formula*.

As time went on, other researchers suggested 'improved' (relative to the Sanders Formula) versions of the Work Equation that ranged from the trivial (e.g. the *Merriman Formula* used a safety factor of six instead of eight) to the somewhat more complex beginning with the *Trautwine Formula*. In addition, the variables in the resulting algebraic equation were often specified to have specific, but inconsistent, units so that the resulting equation was no longer dimensionally consistent. For example, in Imperial units (as were used exclusively in U.S. practice in the 19th century), the pile-set was generally expressed in inches whereas the hammer energy was expressed in foot-pounds. The resulting unit-correction factors explain some of the variations in constant coefficients found in the various formulas.

It appears that all of the various modifications to the original Work Equation were subjective and empirical in nature although there were, apparently, overall goals to eliminate some of the clear shortcomings of the basic Work Equation, not the least of which is that a zero set ('refusal') implied infinite pile resistance which was clearly known even at the time to never to be the case. The desired correction of having zero set produce a finite pile resistance was achieved by adding an arbitrary constant to the set in Equation 5.3. This explains why various formulas have terms such as $(s + 0.1)$[105] in them, so that even if s were zero R_u would still have a finite value. When such modifications were done, the safety factor (actually its reciprocal) was moved to the numerator in Equation 5.3.

Of all the several derivative versions of the Work Equation that were proposed throughout the latter half of the 19th century, the one that achieved lasting prominence that can reasonably be called geotechnical immortality that is right up there with the (in)famous *coefficient of subgrade reaction* in soil-structure interaction (SSI) analyses is the *Wellington Formula*. This formula is far better known by its alternative name, the *Engineering News (EN) Formula* (and, all too often, incorrectly as the *Engineering News-Record Formula*), as the result of Wellington first publishing his formula in the *Engineering News* construction-trade publication in 1888. For this reason, the Engineering News Formula can realistically be considered the ultimate evolutionary step in derivatives of the Work Equation.

5.5.1.3 Rigid-Body Impact Equation and Derivatives

The later and more rigorous interpretation of the rigid-body model for pile driving is that involving actual impact of the two assumed rigid bodies as opposed to equating the work done at implied impact. Chellis (1961) showed that the actual impact of two rigid bodies is

---

[104] Why this was chosen to be the desirable safety factor is not known to the writer.

[105] In the most common case where the quantity $(s + 0.1)$ is used the set has dimensions of inches. This means that 0.1 also has dimensions of inches ($\cong$ 3 mm). Why this particular value was chosen as the set offset is unknown to the writer.

much more complex than what is implied by the simple Work Equation (Equations 5.1a and 5.1b) as it involves parameters such as the *coefficient of restitution* after impact. He provided a detailed derivation of the algebraic equation that incorporates all of the various parameters that go into the complete model of this process. The resulting equation is substantially more complex with many more variables compared to the Work Equation.

Subsequent to deriving the rigorous algebraic equation for rigid-body impact, Chellis went on to identify the algebraic equations for 22 progressively more-simplified versions of the rigorous equation, including identifying the exact simplifying assumptions involved in each. Each of these 22 equations represents a dynamic formula that is identified with the name of a specific researcher or organization that presumably developed or otherwise used the equation. Perhaps the most interesting aspect of Chellis' exercise is that at the end of the simplification process, one winds up with an equation that is identical to the original Work Equation (Equations 5.1a and 5.1b). This is a useful academic exercise as it clearly shows that all dynamic formulas, including the ubiquitous Wellington/Engineering News Formula, are a simplification and approximation of the same solution to a problem in rigid-body physics even though the temporal evolution of dynamic formulas did not occur this way.

As an aside, throughout the latter half of the 20th century and continuing to the present there has been a constant stream of publications (primarily journal and conference papers, with some occasional research reports) that address the accuracy of dynamic formulas from different perspectives. Some publications address the accuracy of one or more formulas relative to statically determined ground-truth such as traditional static load tests (e.g. Tavenas and Audy 1972, Fragaszy et al. 1989) and some publications address the accuracy relative to dynamically determined ground-truth such as 1-D wave-equation analyses (e.g. Mosley and Raamot 1970) and, more recently, dynamic field measurements (e.g. Long et al. 2009). However, despite this overall substantial body of work it does not appear that there has been any attempt to consider performance of dynamic formulas specifically with regard to tapered piles, either to assess the accuracy of dynamic formulas for tapered piles in an absolute sense or relative to piles with a constant perimeter.

In any event, the conclusion drawn by the writer from this observation involving Chellis' work is that if the pile-driving process is modeled as a problem involving two rigid bodies (the pile-hammer ram and pile), there is only one rigorous algebraic equation defining the interaction between these two bodies. However, this one equation has been approximated in numerous ways over the course of approximately 100 years between the middle of the 19th century and the middle of the 20th century with the result that more than two dozen different dynamic formulas, each of which is a unique, simplified approximation of that one rigorous algebraic equation, have been identified in the published literature to date.

Note that the actual number of dynamic formulas that one can find in the published literature is even greater than Chellis' count which was correct as of circa 1960. His count does not include the fact that in recent decades metric-unit equivalents of some formulas have been developed and that some formulas continue to be 'tweaked' to suit a specific need or purpose. Examples of the latter include altering a formula to produce a factored pile resistance to be compliant with LRFD concepts (Allen 2005) or changing the coefficients in, say, the Wellington/Engineering News Formula to provide better statistical correlation with static load tests for site- and project-specific conditions. The latter was done, for example, in the early 1970s with regard to timber piles at JFKIA where a site- and application-specific version of the Wellington/Engineering News Formula was developed. But the important fact to remember is that at the end of the day, all dynamic formulas are just variations on the same theme.

5.5.2 Compressible-Pile Model (One-Dimensional Wave Equation)

The fact that so many versions of dynamic formulas were proposed over the years and continue to be modified to the present (a timeframe spanning across three contiguous centuries) indicates that none was found to be clearly technically superior on either a relative or absolute basis. It should be kept in mind that the fact that certain dynamic formulas became predominant in practice, e.g. the Wellington/Engineering News Formula in the U.S., is not a hallmark of their technical merit but simply their ease of use or some other subjective, non-technical criterion. In any event, it appears that by the early decades of the 20th century, i.e. decades before the emergence of modern foundation engineering in the 1950s, the proliferation of dynamic formulas and the fact that they often yielded widely varying resistance forecasts for the same pile was causing increased concern about the overall reliability of dynamic formulas as a group.

The published record indicates that circa 1930 (approximately 80 years after dynamic formulas first appeared) was the watershed point in time with regard to the evolution of methodologies for the dynamic approach to resistance forecasting and verification. By circa 1930, added to the growing concern about the overall reliability of dynamic formulas was the fact that the use of piles (not just tapered) constructed of materials other than wood was rapidly becoming much more common as discussed in Chapter 3. Of specific relevance to the present discussion was the increasing use of precast-PCC piles, mostly with a constant perimeter, especially in the United Kingdom and U.S. This was coincident with the very active U.S. patent history that was also discussed in Chapter 3.

Cracking of precast-PCC piles was sometimes observed during driving. In retrospect based on what we know now, this is not the least bit surprising as the piles at that time were not prestressed but normally reinforced, at most, as was noted in Chapter 3. A specific, significant shortcoming of dynamic formulas as a group was that they could not provide insight as to the root cause of this cracking. As noted previously, all dynamic formulas assume a rigid pile that implies that all sections of a pile experience the same compressive-only driving stresses at the same time so selective cracking along a pile shaft (as was being observed at the time in practice) while none occurs at the pile head is not something that should occur unless, of course, there are localized physical defects in the pile cross-section.

This disconnect between observed reality and the forecasting ability of dynamic formulas with regard to driving stresses led to research in the U.K. that included an entirely new look at the physics involved in pile driving. Out of this research came the first known understanding that the original physical model of rigid-body impact is not correct for impact-driven piles. Rather, any pile subjected to impact driving exhibits at least some degree of axial compressibility during driving that can be modeled using the theory of 1-D stress-wave transmission through a linear-elastic rod. This theory and concomitant solution are commonly referred to nowadays as the 1-D wave equation or, more commonly, simply the 'wave equation'.

Credit for this seminal breakthrough in determining the actual physics and concomitant theoretical model for what happens during pile driving should be given to research performed by David Victor Isaacs (Isaacs 1931), with significant additional work from Glanville et al. (1938) and possibly others. Note that this circa-1930s work also included the earliest known attempts at making dynamic measurements during pile driving.

Before proceeding further with a focused attention on the 1-D wave equation as applied to tapered piles, there are two broad comments to make. First, as noted in the preceding section, although dynamic formulas evolved along two different developmental paths, each one can be shown to be some simplified version of the same rigorous solution to the same problem in rigid-body physics (Chellis 1961). This implies that a fundamental flaw or flaws in the assumed underlying physical model affects all dynamic formulas. Indeed, as is now well-known, the fatal flaw is that piles simply never behave as a rigid body when installed using conventional impact driving. Thus, it is reasonable to state that dynamic formulas as a group were clearly shown to be 'bad science' by the published research of the 1930s and should have been abandoned decades ago simply because they do not capture the true physics of installing piles using traditional impact driving. However, the use of dynamic formulas has not only endured but, at least to some extent, continues to be promoted for use in practice to the present (Allen (2005) is but one example). This appears to be primarily due to the elegant simplicity of dynamic formulas that foundation engineering practitioners and others who use them seem to find irresistibly appealing to the extent that they knowingly choose ease of use at the expense of scientific fact and accuracy.

The second comment is that it is curious (to the writer at least) that the seminal work related to the 1-D wave equation by Isaacs, Glanville et alia, and possibly others in the U.K. throughout the 1930s has received virtually no mention and concomitant recognition, at least in literature emanating from the U.S., whenever the subject of the origins of the 1-D wave equation for pile-driving applications is discussed. Such recognition in the U.S.-sourced literature typically begin with a citation of Smith (1960) which actually even overlooks Smith's earlier publications (Smith 1954, 1957) as well as the discussions in Smith (1962).

Note that this second comment is not meant to deprecate or trivialize Smith's contributions to advancing the use of the 1-D wave equation for pile driving in both research and practice but it does appear that the actual origins of the wave equation were at least some 30 years earlier than is often presented in the literature, i.e. 1930, not 1960. It is possible that the relatively slow international dissemination of technical knowledge that was typical of much of the 20th century coupled with the interruption and disruption of the Second World War between 1939 and 1945 played significant roles in hobbling global technology transfer.

However, and arguably more importantly, as is now well-known that solution of the 1-D wave equation requires a digital computer. Such devices were simply not available in the 1930s. Smith was in the proverbial right place at the right time a few decades later in this regard. To begin with, he was the Chief Mechanical Engineer of Raymond so would have had unlimited access to a database of pile load test results to use as the basis of comparison for any 1-D wave-equation analyses he performed. In addition, and even more importantly, Smith apparently had access to very early IBM computers that became commercially available in 1952 (both Raymond and IBM were headquartered in New York City at the time). This would have been crucial to his being able to perform analyses to compare with his database of measured pile resistances. Finally, by the time Smith (1960) appeared in publication, it was the cusp of mainframe digital computers such as the IBM Model 360 emerging as a tool that was at least available to academic researchers and large businesses and other organizations with the financial resources to afford one. Thus, as the 1960s progressed it became increasingly possible for others to replicate and ultimately advance Smith's work.

The best-known example of this early synergy between the 1-D wave equation and digital computers is the research conducted during the 1960s at the Texas Transportation Institute (TTI) of Texas A&M University to develop software that practitioners could use to perform 1-D wave-equation analyses for pile-driving simulation. This research culminated in a comprehensive report by Lowery et al. (1969) that contained the FORTRAN IV source code for what was later referred to as the *TTI* (named after the aforementioned Texas Transportation Institute) version of wave-equation software to distinguish it from subsequent programs such as *WEAP* (developed by the late Prof. George G. Goble and associates under contract to the Federal Highway Administration, FHWA) and its proprietary, commercial successors, *CUWEAP* and *GRLWEAP*, that were mentioned earlier in this monograph and are the mainstays of practice to the present.

The writer had early, first-hand experience using the original *TTI* program with tapered piles, specifically, creosoted timber piles. The *TTI* program had been acquired by the writer's employer at the time (PANYNJ) by at least mid-1972. The PANYNJ was a large enough organization to have its own first-generation mainframe computer in-house, complete with remote terminals in selected work areas within the Engineering Department. The *TTI* program was used extensively during the 1972-1973 timeframe to evaluate test piles (referred to as *indicator piles* at the time within the PANYNJ) driven as part of the preliminary design phase of the (never-built) IAB-STRAP project at JFKIA[106]. The results of these wave-equation analyses were presented and discussed in detail in an internal report[107] that was generated for this test-pile program.

As an aside, this same early 1970s PANYNJ internal report for the IAB-STRAP project indicator-pile program also contained an extensive discussion of tweaking the coefficients in the Wellington/Engineering News Formula to better match results from both static load tests as well as 1-D wave-equation analyses from this indicator-pile program (the formula in its original form showed poor correlation). At that time, the Wellington/Engineering News Formula was generally the sole resistance-verification (CQA) tool used for routine timber-pile installation in New York City as the New York City Building Code allowed this formula to be the sole resistance-verification metric for any pile that had an allowable capacity that did not exceed 80 kips (356 kN). Consequently, there was great financial incentive to allow continued use of the Wellington/Engineering News Formula but in an improved form for greater confidence in its outcomes.

Unfortunately, the *TTI* program was not used in this early 1970s PANYNJ project to make pre-driving resistance forecasts. Rather, the program was only used after the fact when not only were the results of high-quality static load tests to the nominal geotechnical ULS available but also the first-ever (at least for the PANYNJ and likely for the New York City metropolitan area as well) use of on-pile dynamic measurements as well. The latter were critical for providing values of actual energies delivered to the piles. This allowed accurate assumptions of hammer efficiencies (which were, much to the astonishment of all involved at the time, as low as approximately 35% for a compressed-air double-acting hammer). In summary, on this particular project the 1-D wave equation was not used as a pre-driving forecasting tool but was simply calibrated post-driving to match the measured results.

In any event, the writer has no recollection if this was the first project on which PANYNJ engineers used the *TTI* program but has recollection (confirmed by the aforementioned PANYNJ internal report) of a substantial number of trial-and-error wave-equation analyses that were performed to essentially get the 1-D wave-equation results to match the field observations and measurements on timber piles. As a result, the values used for various input parameters for the *TTI* program (the aforementioned hammer efficiency as well as the pile-soil springs and dashpots in particular) deviated to varying extents from what were considered typical or recommended values at the time.

The relevant point being made here is that the results from the 1-D wave-equation analyses would not have shown the correlations with measured results that were depicted in the aforementioned internal report had the 1-D wave-equation analyses been performed a-priori using typical parameter values recommended at the time. The logic for using the 1-D

---

[106] The various PANYNJ test-pile programs at JFKIA of which the writer is aware between 1972 and circa 2000 are discussed in detail in Horvath and Trochalides (2004) and Horvath (2014).

[107] *Report on Indicator Pile Program; JFK-IAB-STRAP.* The Port Authority of New York and New Jersey, Engineering Department, Soils and Foundations Division, undated (but circa 1973).

wave equation in this manner, i.e. with input parameters modified by trial and error using ground-truth, for this indicator-pile program at JFKIA appears to have been solely to develop site- and project-specific values for input parameters so that the *TTI* program could be used with confidence as a resistance-forecasting tool during final design as well as production pile driving (which, as it turned out, never occurred as the project was cancelled prior to the start of construction). Had the 1-D wave equation been used as a resistance-forecasting tool beforehand for the indicator-pile program there would have been poor correlation between forecast and measured results, the latter from high-quality static load tests.

The writer had numerous opportunities in later years and to the present to see the 1-D wave equation used in its more traditional manner as a pre-construction resistance-forecasting tool. All of the writer's later experiences have involved *Monotube* and *TAPERTUBE* brand tapered steel pipe piles and the analyses were all performed using the commercially available *GRLWEAP* program. In the U.S. at least, this program has for many years been considered to be the de-facto standard of practice with regard to 1-D wave-equation software.

The writer's overall observation over the course of almost 50 years is that 1-D wave-equation analyses using more or less standard values of input parameters tend to underestimate the actual ULS total resistance of tapered piles although this generalized assessment is admittedly fraught with the multi-faceted problem of defining a single-valued ULS total resistance for comparison to the 1-D wave-equation results as was discussed at length in Chapter 4. Temporal issues are also a sticking point as, in fairness to the 1-D wave equation, it is only intended to simulate resistance at the time of driving so cannot be expected to be able to forecast the time-dependent increase in ULS total resistance that develops with tapered piles driven into predominantly coarse-grained soil.

That having been said, it is the writer's opinion that it is a fair statement that there appear to be particular issues related to tapered piles with regard to 1-D wave-equation analyses. These issues have been noted and subsequently acknowledged by others going back to a relatively early timeframe in the use of the 1-D wave equation in practice when the microcomputer ('personal computer') was still in its infancy as a civil engineering tool, and the computational power of a mainframe computer was still required to execute software such as the 1-D wave equation (e.g. Goble and Hery 1984).

While Goble and Hery addressed one important problem with regard to the 1-D wave equation and tapered piles, the writer feels that a far-larger, more-fundamental problem exists that does not appear to have ever generated any discussion. This has to do with the basic 1-D nature of wave-equation theory itself.

Specifically, the physical spring + dashpot model that is used to simulate interaction between the pile shaft and adjacent ground only offers shaft resistance through the traditional mechanism of sliding friction. This resistance mechanism is mobilized by longitudinal displacements of the pile shaft as the assumed 1-D stress wave travels along the pile shaft. Thus, this model inherently fails to include the fact that as a tapered pile displaces longitudinally it also displaces radially to produce the wedging mechanism (during compression) or radial-contraction mechanism (during tension) that were discussed at length in Chapter 4. In order to properly model the wedging and contraction that occurs with tapered piles there should be additional spring + dashpot components in the physical model that are oriented radially in order to react to and accommodate this radial displacement component. Note that simply modeling a tapered pile as a series of constant-diameter cylinders of varying diameters like a pile with a generic stepped-taper geometry to approximate the tapered geometry (as is typically done) does not even begin to approximate the wedging and contraction mechanisms and thus cannot replace the need to explicitly model the radial component of pile-shaft displacement.

5.5.3 Summary

The foregoing discussions have highlighted the fact that both conceptual models (rigid and compressible pile) employed in the dynamic approach for resistance forecasting are flawed when used with tapered piles. The traditional rigid-pile assumption is simply incorrect for all types of piles by virtue of the fact that all piles exhibit some axial compressibility during driving. As a result, the use of any of the more than two dozen dynamic formulas that have been derived from the basic rigid-pile model should have been discontinued decades ago. However, the simplicity of these formulas, especially when employed for resistance verification (CQA) as opposed to resistance forecasting, has proven to be stubbornly attractive in routine practice, especially to public agencies such as state departments of transportation in the U.S. who are always trying to do more with less when it comes to human resources. These formulas are remarkably simple to use in the field and with minimal technical education required by the user.

On the other hand, the compressible-pile assumption as embodied by the 1-D wave equation is basically a correct concept for pile driving in general because it inherently accounts for pile compressibility. However, the 1-D wave equation is seriously deficient in its historical and current form of implementation in practice with respect to tapered piles. This is due to the 1-D wave equation not having components of the Smith-type physical model (spring + dashpot) that is used to simulate soil resistance along the pile shaft that are oriented in a manner that can approximate the wedging and radial-contraction mechanisms that develop with tapered piles. As such, current formulations of the 1-D wave equation, whether explicitly in analytical software or indirectly in the software used for dynamic field measurements, would not be expected to correctly capture tapered-pile behavior during driving.

Modifying the physical model used along the pile shaft in 1-D wave-equation software to simulate the wedging and radial-contraction mechanisms would appear to be both a reasonable and productive avenue of research. However, to the best of the writer's knowledge it is one that, curiously, has not been pursued to date. The reason is likely simple pragmatic: the lack of an entity willing to fund the necessary R&D. Much of the deep-foundation research performed in the U.S in recent decades has been funded by the FHWA, sometimes in collaboration with states. As noted earlier in this chapter, the FHWA funded the original public version of the *WEAP* 1-D wave-equation program that remains, in a much- and frequently updated form, the state of art to the present more than 40 years later. Understandably, the FHWA has historically been loath to fund research that involves proprietary technologies, believing that the stakeholders of a proprietary technology should fund R&D as they are the ones will benefit financially from its use. As explained in earlier chapters, tapered piles have, to a significant extent, always been viewed as a proprietary geotechnology, at least in the U.S.

With regard to tapered piles, only Raymond has ever shown an interest in supporting advances in tapered-pile technology on a consistent, long-term basis. Unfortunately, although Raymond internally supported seminal early research into piles in general and tapered piles in particular through the published work of their employees:

- E. A. L. Smith, a Chief Mechanical Engineer in the 1950s;

- R. L. Nordlund, a Chief Civil Engineer in the 1950s;

- Tonis 'Tony' Raamot, a Chief Civil Engineer in the 1960s;

- Ernest T. Mosley, an employee in the 1960s;

by the time that significant advances in both computers and electronics that would eventually revolutionize all aspects of deep foundations were beginning to occur in the 1970s, Raymond had all but exited the tapered-pile market. No business entity that deals with tapered piles in the U.S. has picked up the slack left by Raymond's substantial void. One might wonder why Union Metal or its corporate successor, Monotube, did not do more in this regard but with the apparent commercial demise of Monotube (and a business entity willing to pick up and continue the *Monotube* pile line not apparent at the time this monograph was finalized in early 2020) this is a moot point.

There is one final issue to note with respect to dynamic methods that consider pile compressibility, i.e. the 1-D wave equation and the dynamic field measurements that derive from the 1-D wave equation. It affects all piles, not just tapered piles, but is important to bring up as it affects dynamic methodologies that might be used for both resistance forecasting and resistance verification (CQA).

There is a natural inclination to assume that the single-valued resistance that is related to a certain pile blow-count (i.e. the reciprocal of pile set) is the (geotechnical) ULS total resistance. While this is true in many, if not most, cases, there is a significant exception to this. This occurs when the energy transmitted to the pile by the hammer impact is insufficient to move all parts of the pile into the ground a sufficient distance so that all soil resistance is fully mobilized at every point along the pile shaft as well as at the pile toe. Thus, the resistance mobilized is less than the geotechnical ULS and is, in reality, the <u>mechanical</u> ULS resistance of the pile-driving system. This condition, which is generally referred to as *refusal*, is a reflection of the energy limitations of the driving system relative to the pile compressibility and not the (geotechnical) resistance limitations of the ground.

The fact that any pile-driving system has mechanical limitations as to how much energy it can deliver to the head of a pile and whether this energy can exceed the soil resistance has been well-known since the 19th century. As noted earlier in this chapter, it was the reason that more than 100 years ago an arbitrary constant was added to the pile-set in the Work Equation, to remove the illogical outcome of Equation 5.3 whereby $s = 0$ (i.e. refusal) meant $R_u = \infty$. However, this issue is raised here as, in the writer's experience, it tends to be forgotten from time to time by users of the 1-D wave equation and the dynamic field measurements that derive from it.

5.6 STATIC APPROACH

5.6.1 Resistance-Only Methods (Step I of Two-Step Procedure)

5.6.1.1 Background

Historically and to the present, application of the static approach for resistance forecasting to DFEs in general has focused almost exclusively on analytical methods that produce an estimate of ULS total resistance only. This is because regardless of whether ASD or LRFD is used, the usual foundation design process, especially for terrestrial piles, consists solely of incorporating 'safety' against the geotechnical ULS without any explicit consideration the geotechnical SLS, i.e. settlement. Implicit in this approach is the expectation that the settlement that will occur, at least under service-load conditions, will always be acceptably small in magnitude (a fraction of one inch/several millimetres is usually assumed) so need not be calculated explicitly.

Even with the push in recent years by Fellenius, Mayne, Niazi, and others (including the writer) to routinely calculate and use settlement as part of the DFE-design process, the reality is that the calculation of ULS resistance still plays an essential role as it delimits the upper end of the load-settlement curve for most of the load-settlement methodologies that have been developed to date. Thus, even when DFE settlements are considered the calculation of the ULS resistance is still the first step of a two-step process.

Another pragmatic consideration is that the traditional ULS-only design approach has resulted in DFE-supported structures that have, for the most part, performed acceptably. This simple fact has been and remains a powerful disincentive to change routine practice to a more SLS-centric approach. In the writer's experience, design methodologies in routine foundation engineering practice that produce acceptably performing structures are generally slow to even just be improved, no less replaced entirely, despite there being evidence that doing so would result in a more logical, more accurate, and cost-effective end result.

The bottom line is that the need for calculating ULS total resistance for DFEs in general and tapered piles in particular is not going to disappear anytime soon (if ever), even if the profession eventually moves more toward explicitly incorporating DFE settlements into the design process. Thus, the following sections that constitute a significant portion of this chapter contain a comprehensive documentation and discussion of the many analytical methodologies that have been or could be used for forecasting the ULS total resistance only of tapered piles.

Unfortunately, this need to calculate a single-valued ULS total resistance for design purposes conflicts with the reality of actual DFE behavior. As has been discussed several times in this monograph, the reality is that many DFEs do not exhibit the plunging load-settlement behavior that would be consistent with the existence of a 'true', single-valued geotechnical ULS. This is primarily due to the fact that toe resistance often does not display a well-defined ULS resistance as shaft resistance, or at least shaft resistance due to sliding friction, generally does. This means that toe resistance is, for most DFEs, settlement-dependent, even to magnitudes of settlement that are well beyond what would typically be tolerable for geotechnical SLS conditions. Consequently, assessing the accuracy of the analytical methods for forecasting ULS total resistance of tapered piles that are presented in the following sections, which would typically consist of comparing calculated to measured results, is not straightforward.

5.6.1.2 Indirect/Rational Methods

5.6.1.2.1 Overview

Throughout the latter decades of the 20th century, the only indirect resistance-forecasting method for tapered piles that existed was the Nordlund Method. Even to the present, to the extent that indirect methods for tapered piles are mentioned in a textbook or design guide only the Nordlund Method is mentioned.

Unfortunately, what is very rarely mentioned in publications (NYSDOT (2015) is a rare exception) is that there are two versions of the Nordlund Method and they have co-existed for over 40 years now. Publications typically only reference the original 1963 version (complete with misspelling of Nordlund's surname, an unfortunate artifact of an error by the American Society of Civil Engineers in the original publication) and not the revised 1979 version. The significant issue to note here is that these two versions typically produce significantly different numerical outcomes for the shaft and toe ULS resistances for the same soil conditions and pile. While the differences sometimes balance out, on the whole the revised 1979 version tends to produce a lower, sometimes much lower, ULS total resistance. The difference is usually due to a reduced toe-resistance contribution.

In view of the prominent role that the Nordlund Method plays in practice and the fact that several nuances of the method appear to have been forgotten over the years, a significant portion of the following sections dealing with indirect resistance-only forecasting methodologies is devoted to a detailed presentation and discussion of both versions of the Nordlund Method. This is followed by a presentation and discussion with several alternative indirect analytical methods that have appeared in the 21st century.

5.6.1.2.2 Nordlund Method (Overview)

The first seminal, long-lasting analytical work related to tapered piles was done around the middle of the 20th century by R. L. Nordlund who was a former Chief Civil Engineer of Raymond, the once-famous U.S.-based pile-driving contractor with an international presence. As discussed in earlier chapters, Raymond, in all of its various corporate names, was the company that literally invented and commercially developed the first non-timber tapered piles in the U.S. beginning in the late 19th century. For most of the 20th century, they had proprietary tapered (*Raymond Standard*) and partially tapered (*Raymond Step-Taper*) piles that only they designed, provided, and installed on a project-specific turnkey basis. Therefore, in Nordlund's professional-engineering capacity at Raymond he clearly had opportunity, motive, and incentive to further the understanding of how tapered piles develop resistance. It was clearly in Raymond's corporate interest to support in-house research that would result in a better understanding of how to reliably forecast ULS total resistance for their piles before pricing and undertaking a project.

This assumed employer-driven focus of Nordlund's R&D work is evidenced by the fact that what is now referred to as the Nordlund Method is primarily a methodology for what are generally referred to colloquially as friction or floating piles that bear either entirely or predominantly in coarse-grained soil. It has long been known from experience that tapered piles best demonstrate their cost-effectiveness in such applications. Tapered piles offer no benefit and, in fact, are arguably handicapped (due to relatively small toe areas) when end bearing on rock, and their benefit in fine-grained soils has not been conclusively demonstrated one way or the other, at least in the published literature.

That having been said, the Nordlund Method can be used in subsurface stratigraphies where fine-grained strata may be present and play a minor role. In such cases, the contribution of the fine-grained stratum or strata to the calculated ULS total resistance is dealt with very simplistically based on the assumed undrained shear strength of the fine-grained soils.

In any event, it is likely that Nordlund's thoughts concerning an analytical methodology specifically applicable to tapered piles developed and evolved over a period of many years of observing the performance of all types of piles in static load tests. However, it is clear from both the content and cited references in his original 1963 paper that it was concepts based on modern soil mechanics that were published in the post-World War Two era that gave a scientific structure to, and ultimately crystallized, Nordlund's thinking on the subject that resulted in the publication of his analytical methodology (Nordlund 1963). The early decades of modern soil mechanics, from the 1920s to the 1950s and, especially, the years immediately after World War Two, saw relatively rapid advances in many aspects of geotechnology. This technological growth was instrumental in turning foundation engineering from what had largely been an art into something that at least had a basis in science although elements of art based on experience are still important to the present.

The assumptions and analytical components of what is referred to universally as the Nordlund Method are presented and discussed here in detail. There are several reasons why this is necessary:

- Despite the fact that it was first published almost 60 years ago, the Nordlund Method is still widely used, at least in the U.S. where it is often used for road-related projects. This is because it has long been sanctioned and promoted by the FHWA for use in routine practice for (driven) piles in general. This has been done over the years via numerous publications with a design-manual format (e.g. Hannigan et al. 1998a, 1998b), the most recent of which is the three-volume (Hannigan et al. 2016a, 2016b, 2016c) *Geotechnical Engineering Circular* (GEC) *No. 12* that was released in November 2016.

- The method has been used in FHWA-sponsored research up to the relatively recent present (e.g. Pando et al. 2006).

- Independent of FHWA-sanctioned usage, some states include the Nordlund Method in their own design manuals, including for general use with non-tapered piles. In the extreme case of NYSDOT (2015), the Nordlund Method is intended for use with <u>only</u> non-tapered piles.

- The Nordlund Method was incorporated into a *Windows*-based computer program named *DRIVEN* that was developed under contract for, and subsequently technically supported by, the FHWA. It appears that Version 1.0 of this software was first released circa 1998 (Mathias and Cribbs 1998) and that this computer code was subsequently revised at least through Version 1.2, with *Windows XP* being the latest operating system (OS) for which *DRIVEN* was explicitly supported. Although this software is no longer available from or supported by the FHWA and it will not operate on a current 64-bit *Windows*-OS system unless one uses a workaround process such as the *DOSBox* shell, *DRIVEN* is known by the writer to still be used in practice, especially for tapered piles, and even outside of the U.S. The reason is that although this program was developed primarily for use by organizations explicitly supported by the FHWA's mission (e.g. state departments of transportation, Federal lands, etc.), it had an unlimited, no-cost public distribution so likely propagated widely and far beyond its target audience. This widespread distribution was, and likely still is, aided by the fact that this program and its written documentation are still available at no cost from alternative industry-related sources that can be found on the Web. Furthermore, it appears that an updated successor program named *DrivenPile* was developed by, and is available commercially from, the private sector that again is readily found by searching the Web[108] .

- Apparently unknown to many geotechnical engineers[109] is the fact that Nordlund presented an updated and revised version of his analytical method in 1979, 16 years after publication of the original version. The venue for presentation of this updated version was a continuing-education short course, at which he was apparently one of the speakers, that was hosted by a university in the Midwestern U.S. that was very active hosting such

---

[108] The physical address listed on the website for this software suggests that the same person(s) who developed the original *DRIVEN* program under an FHWA contract are now marketing the current commercial version, *DrivenPile*, independently and directly.

[109] Despite a career-long interest in tapered piles that dates back to 1972, the writer was included in this group until early 2017.

courses during that timeframe. Normally, work presented at such venues in the pre-Internet and pre-Web era got minimal circulation beyond the printed notes distributed to course attendees (which would typically number in the low double-digits based on the writer's first-hand experience attending an event on a different topic that was hosted by the same institution during that timeframe). However, it appears that Nordlund's 1979 presentation eventually became known to at least a limited extended audience as it was incorporated into the original 1998 release of the *DRIVEN* program. In addition, the complete paper he authored in 1979 was republished in its original form as part of other work, with NYSDOT (2015) being one known example. As discussed in the following section, which is devoted to a detailed presentation and discussion of what the writer terms the Revised Nordlund Method, the revisions presented by Nordlund in 1979 were, overall, not insignificant as was noted earlier in this chapter. The difference between the ULS total resistance calculated using the original and revised versions of the Nordlund Method can be substantial, especially for longer piles. This is because the Revised Nordlund Method tends to increase shallow shaft resistance and decrease both deeper shaft and toe resistances compared to the Original Nordlund Method.

The point made here is that what is collectively referred to as the Nordlund Method is, in reality, not a single analytical methodology. The overall method actually has different analytical options to choose from when applying the method to a specific pile, tapered or otherwise[110]. By Nordlund's own admission in his 1979 paper, these different options can result in significantly different forecasts of ULS total resistance for the same pile in some cases. This, combined with the fact that the Nordlund Method continues to enjoy continued promotion and usage via its incorporation into computer software, means that users clearly need to understand all aspects of this analytical methodology. This requires a detailed presentation and commentary on the governing equations presented in Nordlund (1963) followed by a detailed discussion of the changes made in the 1979 revision. The one deviation herein from Nordlund's original papers is that the following derivations use notation that differs somewhat from Nordlund's original notation in order to be consistent with notation used throughout this monograph.

5.6.1.2.3 Nordlund Method (Original 1963 Version)

To begin with, Nordlund assumed that tapered-pile resistance comes from only the two traditional mechanisms of shaft resistance due to sliding friction and toe resistance due to end bearing as defined by classical bearing-capacity theory. Thus, Nordlund did not propose treating shaft resistance along a tapered portion of a pile as a distinct third resistance mechanism of shaft resistance due to taper as has long been championed by the writer. However, Nordlund did make a significant contribution in that direction by at least proposing a shaft-friction model that was explicitly related to pile taper. This was an improvement over earlier work by others such as Meyerhof in the 1950s who simply proposed an empirical increase in the ULS unit shaft resistance due to sliding friction without suggesting how or why this occurred with tapered piles.

Using the basic concepts and notation introduced in Chapter 4, Nordlund (1963) rigorously expressed the ultimate total resistance, R_u, as the sum of shaft, $R_{u(shaft)}$, and toe, $R_{u(toe)}$, components:

---

[110] The fact that the NYSDOT (2015) design manual uses the Nordlund Method only for non-tapered piles indicates the popularity of the method beyond its originally intended application.

$$R_u = R_{u(shaft)} + R_{u(toe)} =$$

$$\sum_{z=0}^{z=L} [K_\delta \cdot \sigma'_{vo} \cdot \sin(\alpha + \delta) \cdot \sec\alpha \cdot C_z \cdot \Delta z] + [N_q \cdot A_{toe} \cdot \sigma'_{vo(toe)}] \tag{5.4}$$

where:

- z = depth along the pile shaft relative to the ground surface ($z = 0$);
- L = total embedded length of the pile;
- K_δ = a dimensionless, <u>non</u>-horizontal earth pressure coefficient as uniquely defined by Nordlund (see further discussion following);
- σ'_{vo} = vertical effective overburden stress along the pile shaft;
- α = taper angle as defined in Figure 2.2 (note that Nordlund used the notation ω);
- δ = Mohr-Coulomb friction angle along the assumed planar interface between the material comprising the exterior of the pile shaft and adjacent ground <u>but as uniquely defined by Nordlund</u> (see discussion following);
- C_z = the <u>minimum</u> pile circumference of an arbitrary pile segment of length Δz;
- N_q = a dimensionless bearing-capacity factor;
- A_{toe} = cross-sectional area of the pile toe; and
- $\sigma'_{vo(toe)}$ = vertical effective overburden stress at the depth of the pile toe.

There are four things to note in particular with regard to Equation 5.4. First and foremost is that this equation was formulated assuming coarse-grained soils where shear strength and other parameters would be governed by effective stress and the Mohr-Coulomb strength parameter ϕ only, i.e. the Mohr-Coulomb strength parameter $c = 0$. This makes sense when one understands that tapered piles have historically been assumed to offer no taper-benefit in applications where fine-grained soils are the dominant bearing soils (whether or not this is indeed true is not addressed here).

However, in situations where a fine-grained soil stratum of limited thickness occurs along the pile shaft or if the toe were bearing in fine-grained soil[111], a simplistic $\phi = 0°$ analysis based on assumed undrained 'adhesion' along the shaft (with no taper influence) or undrained shear strength beneath the toe would be used instead. Note that in the latter case, the $(N_q \cdot \sigma'_{vo(toe)})$ 'embedment' term in Equation 5.4 is dropped (a commonly made approximation) and the 'soil-strength' term, $(N_c \cdot c)$, appears in its place, where $N_c = 9$ is assumed and $c = s_{u(toe)}$, the undrained shear strength immediately beneath the pile toe.

The second issue of note is that the parameter K_δ is <u>not</u> equivalent to the lateral earth pressure coefficient, K_h. This was noted and discussed previously in Chapter 4. As defined by Nordlund, the line of action of K_δ is skewed downward at an angle of $(\alpha + \delta)$ below the horizontal. This is consistent with his use of the passive state of lateral earth pressure theory as the basis for determining the baseline magnitude of K_δ. However, as shown in Chapter 4, K_δ and K_h are geometrically related:

$$K_h = K_\delta \cdot \cos(\alpha + \delta) \tag{5.5a}$$

$$K_\delta = K_h \cdot \sec(\alpha + \delta) . \tag{5.5b}$$

---

[111] Typically, quite stiff to the point that geologists often map such soils not as sediments but as 'poorly lithified sedimentary rock'. Case-history examples of such subsurface conditions are presented in Nordlund (1963).

The third item of note is that Equation 5.4 includes two approximations that are commonly and reasonably made for piles in terrestrial applications which was of primary interest to Nordlund[112] as they tend to numerically cancel each other out. One is neglecting the so-called 'soil-weight' term involving the N_γ bearing-capacity factor in the toe-resistance component that should appear on the right-hand side of Equation 5.4 as an <u>addition</u>. For coarse-grained soil conditions and toe-perimeter dimensions normally associated with tapered piles, the magnitude of this term is insignificant compared to that resulting from the N_q term. The second approximation involves pile weight that should appear in the right-hand side of Equation 5.4 as a <u>subtraction</u>. Nordlund simply neglected this term. However, for tapered piles the magnitude of pile weight is relatively small and insignificant compared to the shaft and toe resistances. In addition, neglecting pile weight is, in most cases, almost exactly compensated by the aforementioned approximation involving the toe-resistance term, i.e. neglecting the N_γ term.

As an aside, for the special case in the Nordlund Method of a pile toe bearing in fine-grained soil, dropping the N_q term when the N_c term is added has the same practical effect of cancelling out the ignored pile weight.

The fourth and final item of note is that Nordlund did not actually use Equation 5.4. He observed that the taper angle, α, was always relatively small in magnitude. Specifically, it was generally less than 1° for any commercially available (in the U.S. at least) tapered pile that was in use at the time (primarily the late 1950s) and was thus considered by Nordlund. This observation is still generally true in the present (again, in the U.S. at least) although *TAPERTUBE* piles with α as large as 1.6° have appeared in product literature in recent years. In addition, some locally produced precast-PCC piles outside the U.S. appear to have taper angles considerably greater than 1° (see Figure 3.10).

In any event, Nordlund assumed that Equation 5.4 could be simplified somewhat by assuming $\alpha = 0°$ but <u>only</u> relative to the angle δ and with respect to the spatial orientation of K_δ as reflected in Equations 5.5a and 5.5b. This issue was also noted and discussed previously in Chapter 4. Thus, Equation 5.4 simplifies to:

$$R_u = \sum_{z=0}^{z=L} [K_\delta \cdot \sigma'_{vo} \cdot \sin \delta \cdot C_z \cdot \Delta z] + \left[N_q \cdot A_{toe} \cdot \sigma'_{vo(toe)} \right] . \tag{5.6}$$

Nordlund also noted that if Equation 5.5b (with α taken to be = 0°) is substituted into Equation 5.6 then the following equation, which is the one based on K_h that is normally used with non-tapered piles, is obtained:

$$R_u = \sum_{z=0}^{z=L} [K_h \cdot \sigma'_{vo} \cdot \tan \delta \cdot C_z \cdot \Delta z] + \left[N_q \cdot A_{toe} \cdot \sigma'_{vo(toe)} \right] . \tag{5.7}$$

Note, however, Nordlund did not pursue further use of Equation 5.7. Rather, he used only Equation 5.6 based on K_δ in both the original (1963) and revised (1979) versions of his resistance-forecasting methodology. This is important to note as not only does Equation 5.6 explicitly include the $\alpha \approx 0°$ assumption in its derivation but several of the parameters that

---

[112] In the latter decades of the 20th century, Raymond had developed and was actively promoting use of the *Raymond Concrete Cylinder Pile* in relatively shallow-water marine applications. This was essentially a precast-prestressed/post-tensioned-PCC pipe pile with an open toe and relatively large diameter.

appear in this equation have, to varying extents, an empirical influence, if not explicit basis, that derives from a surprisingly small database[113] of piles with $0° \leq \alpha < 1°$. Therefore, the cautions raised here with regard to any tapered pile with $\alpha > 1°$ are:

- The Nordlund Method in both its original and revised versions should be used with care as the method parameters used were not calibrated against a database that includes such taper angles.

- Strictly speaking, the more-rigorous and accurate relationship (Equation 5.4) developed, but not used, by Nordlund should not be used unless new parameter assessments are made.

The genesis of these comments is that, as with any analytical methodology with empirical elements that are based in whole or in part on some database, caution is required whenever the methodology is used for an application that falls outside of either the database or related assumptions that were made when developing the methodology.

Before exploring separately and in detail the shaft- and toe-resistance parameters in Equation 5.6, the one overarching issue that applies to both shaft- and toe-resistance calculations is the methodology used by Nordlund for determining the friction angle of the soil, ϕ, that is used to evaluate, either directly or indirectly, most of the equation parameters for coarse-grained soils. This is because Nordlund was very explicit on how to do this and, as it turns out, this is one of the areas in which the original and revised versions of his method differ. Thus, following Nordlund's methodology for assessing ϕ should be considered an integral part of correctly using either version of his method in any application.

Limiting the discussion for now to what will hereinafter be referred to as the *Original Nordlund Method*, Nordlund used the empirical correlation between SPT N-values and ϕ that was published in Peck et al. (1953)[114]. Thus, this is referred to hereinafter as the *Peck-Hanson-Thornburn* (PHT) relationship. Nordlund did not reproduce the graphic illustrating this relationship in Nordlund (1963) but did so in Nordlund (1979). A tabular approximation of this empirical curved-line relationship that was presumably incorporated into the *DRIVEN* program appears in Mathias and Cribbs (1998).

Note that the PHT relationship uses what is assumed to be the 'field' N-value, N_{field}, as would have been consistent with the state of practice for performing the SPT in the 1950s. This assumption was made by the writer as Peck et al. (1953) was intended to be a practice-oriented foundation engineering textbook, one of the first of its kind in the English language. Corrections to N_{field} for issues such as the efficiency of the SPT drive system, overburden stress, etc. as would be standard practice nowadays simply did not exist in the 1950s. However, Nordlund was prescient as he did speculate in 1963 that, based on published work by Gibbs and Holtz (1956), overburden stress possibly influenced N_{field} although the state of

---

[113] The entire database of ground-truth on which the 1963 Original Nordlund Method is based is extremely limited. It includes only 41 piles total, of which 15 have a depth-wise constant perimeter: closed-toe steel pipe (11) and steel H (4). Of the 26 tapered piles, only one is timber and the remaining 25 are steel shell or pipe filled with PCC after driving. These 25 piles are distributed among continuously tapered (3); pseudo-continuously (i.e. step) tapered (15); and partially tapered *Monotube* (7) sub-types. This database was apparently expanded somewhat for the 1979 Revised Nordlund Method to include ACIP piles but by how much is not known to the writer.

[114] Nordlund also stated that he used N-values to estimate shaft "adhesion" and toe "cohesion" (i.e. undrained shear strength) in fine-grained soils but did not given any indication of what the relationship was or where it was published, if at all.

knowledge circa 1960 did not, in his opinion, allow for a rational way to apply a depth correction so he ignored doing so.

However, by the time of Nordlund (1979), the need for N_{field} correction, at least for overburden stress, was becoming more widely known and accepted in geotechnical engineering practice, and suggestions on how to do this were becoming available in the published literature. As will be seen under the discussion of the Revised Nordlund Method in the following section, the correction of N_{field} for overburden stress is one of the primary ways the original and revised versions of the Nordlund Method differ.

Note also that by 1979 the issue of the effect of different SPT drive systems on the energy actually delivered during performance of the SPT (typically using the metric of *drive-system efficiency* expressed as a percent) was also becoming more widely appreciated and known. However, normalizing N_{field} to some standardized reference level such as N_{60} (i.e. 60% drive-system efficiency) is not part of the Nordlund Method in either of its versions.

As an aside with regard to the value of ϕ in the PHT relationship, the nature of ϕ was not precisely defined in the context of present-day knowledge. That is, is it a tangent peak value at some operative stress level; a secant peak value to some operative stress level; or the constant-volume (critical-state) value. In all fairness, such nuanced aspects of the shear strength of coarse-grained soils were not recognized in the early 1950s timeframe when this book was published but it is at least relevant to note. Thus, all that can be said is that the PHT relationship simply relates ϕ to N, with ϕ increasing from approximately 28° to 43° as N increase from 5 to 60. As a result, Nordlund assumed that ϕ was a constant for a given value of N.

An area of concern with respect to using either version of the Nordlund Method in practice is that it is clear from both Nordlund (1963) and Nordlund (1979) that he applied a very healthy dose of subjectivity in the form of engineering judgment when choosing the N-value to use with the PHT relationship. Nordlund went so far as to say that in some cases one might use the sum of the <u>first two</u> values recorded for each 6-inch (150 mm) drive of the Standard Split Spoon as opposed to the long-standardized <u>last two</u>[115] depending on how the individual numbers "looked" (exact quote). What exactly constituted 'looked' in this context was not defined or explained by Nordlund. In addition, Nordlund indicated that any "elevated" blow-count due to the presence of gravel should either be ignored or arbitrarily reduced before using the PHT relationship. Again, guidance as to what constituted an 'elevated' N-value and exactly how to deal with it were not explained by Nordlund.

Considering all these issues, the writer suggests that to be consistent with both versions of the Nordlund Method as well as the state of knowledge that existed when the various documents used by Nordlund were published, the N-value used for estimating ϕ be what nowadays would be called N_{45}, i.e. an N_{field} value obtained with a SPT drive-system efficiency of 45%. This is consistent with the cathead-plus-rope drive system that was widely used in the U.S. well into the 1970s at least. Thus, to use either version of the Nordlund Method nowadays the recorded N_{field} values should be converted to N_{45} values using the following equation:

$$N_{45} = N_{field} \cdot \frac{ER_{field}}{45} \tag{5.8}$$

where ER_{field} is the measured or estimated SPT drive-system efficiency under which N_{field} was recorded, expressed as a percent between 0 and 100. When using the Original Nordlund

---

[115] In those days, the Standard Split Spoon was only driven 18 inches (450 mm), not the 24 inches (600 mm) that has become more common in recent decades.

Method that is the focus of the present discussion, the values of N_{45} are used without further adjustment to estimate ϕ using the PHT relationship.

Attention is turned now to discussing the specific parameters used in Equation 5.6 for the shaft- and toe-resistance components. Considering first the shaft resistance, the only parameters requiring elaboration are K_δ and δ.

To provide a baseline and starting point for the discussion that follows (which will get quite complex), first considered is the case of a non-tapered, i.e. constant-perimeter, pile. In that case, the usual generic equation (5.7) applies so the relevant parameters are K_h and δ.

To begin with, these parameters are typically assumed to be independent of each other. As discussed in detail in Chapter 4, K_h (the broad conceptual equivalent of K_δ in non-tapered applications as can be seen from Equations 5.5) is now understood to be a function of the pre-existing K_o and the effect of driving the pile into the ground (which involves the relative volumetric displacement of soil among other things) as expressed by the ratio K_h/K_o. The evolution and use of this ratio were discussed at length in Chapter 4.

On the other hand, δ is solely a function of the pile material at the pile-soil interface and ϕ. For a given pile-soil interface, the ratio δ/ϕ is typically assumed to be constant although, in reality, it varies within some relatively narrow range as ϕ transitions from the peak to constant-volume value. Again, this was discussed at length in Chapter 4.

However, Nordlund viewed and consequently defined K_δ and δ in such a way that they are interrelated. In the writer's opinion, this is unfortunate because it represents a significant paradigm shift in the way of thinking that foundation engineers are used to and thus familiar with. Specifically, Nordlund postulated that the absolute volume of the pile per unit length, V, explicitly influenced each of these parameters. The reasoning presented by Nordlund is discussed separately for each parameter beginning with K_δ.

To begin with, Nordlund did not, in general, differentiate between pile installation and post-installation axial-compressive loading to the implied geotechnical ULS with respect to their respective effects on the stress state within the ground. Thus, the values of K_δ presented in his two publications inherently lump the effects of these two distinct stages together in a manner that is not possible to separate, at least for tapered piles.

The one exception to this is for constant-perimeter piles. As discussed in Chapter 4, it has historically been assumed that while driving a constant-perimeter pile into the ground will cause an increase in the radial/horizontal stress along the pile shaft, that stress will remain unchanged under axial loading in either compression or uplift. Because Nordlund included values of K_δ for constant-perimeter piles, i.e. taper angle, α, = 0, in the figures in his publications, it is reasonable to assume that these values reflect the effects of installation.

With respect to the more general case when taper angle, α, is greater than zero, Nordlund postulated that as the tapered portion of a pile displaces downward into the soil under the combined effects of pile installation followed by external loading in axial compression to the geotechnical ULS, the resulting radial or lateral (depending on the cross-sectional geometry of the pile) displacement of the pile-soil interface at an arbitrary depth such as that defined by x-x' in Figure 4.1 could be modeled as an equivalent infinitely long (i.e. 2-D) rigid wall, i.e. earth-retaining structure, being rotated into the soil. The pseudo angle of rotation of this 'wall' is the same as the taper angle, α, of the pile. The earth pressure acting on the pile shaft at an arbitrary depth, which Nordlund assumed started from the at-rest state ($= K_o \cdot \sigma'_{vo}$), is defined by $K_\delta \cdot \sigma'_{vo}$.

Thus, the analytical heart of the Nordlund Method is a series of empirical curves of K_δ vs. α (the same curves are used in the original and revised versions of his method). How Nordlund developed these curves is complex and subjective but necessary to present and discuss here in order to understand his method.

To begin with, consistent with this 2-D wall model, there will always be some finite value of wall rotation (defined here as α_{max}) at which K_δ reaches a maximum value (defined here as $K_{\delta(max)}$) that Nordlund assumed was defined by K_p, the coefficient of passive earth pressure calculated from some earth-pressure theory. Note that this implies that there is <u>always</u> some value of pile taper angle beyond which taper offers no additional benefit in terms of increased radial/horizontal stress (and concomitant ULS unit shaft resistance) acting along the pile shaft. This implication is discussed further subsequently.

In any event, Nordlund chose to use one of the then-(relatively) new 'exact' lateral earth pressure theories, specifically, the one developed by Caquot and Kerisel in 1948, to define K_p. It is important to note that the basic solution for K_p as a function of ϕ that was developed by Caquot and Kerisel was for $\delta = \phi$. Thus, a correction is required for the more-common case of $\delta < \phi$.

To generate intermediate values of K_δ for α values starting at zero and up to α_{max}, Nordlund made several assumptions:

- The baseline, reference value of K_o is independent of ϕ and is always equal to 0.5. Nordlund based this on cited published work by Terzaghi and Tschebotarioff. Of course, knowing what we know now about geomechanics and how the stress state of a soil mass is affected by its stress history, we know that this 'one-size-fits-all' value of K_o is a gross simplification of reality and simply incorrect. Nevertheless, using Equation 5.5b, this means that the minimum theoretical value of $K_\delta = K_{\delta(min)} = K_o \cdot \sec \delta = 0.5 \cdot \sec \delta$. However, as discussed subsequently, this theoretical minimum value was rarely used as an actual minimum value due to an empirical modification made by Nordlund.

- The measured curves of lateral earth pressure vs. wall rotation that Terzaghi generated based on full-scale retaining wall tests performed in the early decades of the 20th century were used by Nordlund to define α_{max}, the limiting value of α at which $K_{\delta(max)}$ occurs. Terzaghi's results, as reproduced in Nordlund (1963), indicate that α_{max} is strongly dependent on ϕ, with α_{max} decreasing from about 2° to 0.5° as ϕ increases from 25° to 40°. Note that the same comments made earlier in this section concerning the ill-defined nature of the reported ϕ values in Peck, Hanson, and Thornburn's work applies here to Terzaghi's work. As an aside, a practical implication of this assumption by Nordlund is that a taper angle greater than approximately 1° is of no benefit. Coincidentally or not, until recently tapered steel shell/pipe piles in the U.S. did not have a taper angle greater than 0.95°. Although speculative by the writer, it is at least plausible that this arbitrary limit of taper angles of commercially available piles was influenced by Nordlund's work, at least to some extent. It certainly seems likely in the case of Raymond piles.

- The same curves from the Terzaghi retaining-wall tests were used by Nordlund to develop the shapes of the curves for $K_{\delta(min)} < K_\delta < K_{\delta(max)}$.

- The curves for K_δ, especially at the low end of the range and beginning with $K_{\delta(min)}$, were further and empirically modified by Nordlund as a function of V, the aforementioned pile volume per unit length of pile. Nordlund reasoned that the toe diameter of a pile displaced soil independent of and in addition to taper effects, in a sense acting as an equivalent taper and contributing to the value of K_δ. Thus, all things being equal, Nordlund reasoned that the larger the toe diameter, the larger the initial (i.e. $\alpha = 0°$) value of K_δ. As a result, in the final version of the K_δ vs. α curves some of the values of $K_{\delta(min)}$ were increased above the theoretical minimum value of $K_{\delta(min)} = (K_o \cdot \sec \delta) = (0.5 \cdot \sec \delta)$ noted above.

Unfortunately, Nordlund was not entirely clear as to how he quantified this volumetric correction and he did not provide a worked example to show how the correction was arrived at for some sample value of ϕ (he did, however, show a sample calculation for the base curve that did not include volumetric effects). Rather, Nordlund simply showed the final curves that resulted from this subjective volumetric correction. As it turns out, for typical volumetric values of actual tapered piles formerly and currently used in U.S. practice ($V \cong 1 \pm 0.7$ ft³/ft = 0.093 ± 0.065 m³/m), the effect of pile volume is overall relatively modest; tends to be most significant for $\alpha = 0°$, i.e. non-tapered piles; decreases rapidly with increasing α; and disappears completely at values of α that are generally well below α_{max}. Note that because this volumetric correction is greatest when $\alpha = 0°$, this has the net effect of at least partially negating Nordlund's basic assumption that K_h is not affected by the pile-installation process and remains equal to the pre-driving value of K_o. Consequently, after all is said and done, Nordlund is, to some degree, in consonance with current thinking that is based on using a $K_h/K_o > 1$ for the post-installation stress state around a (driven) pile although Nordlund reached that conclusion in a convoluted way.

- The correction (reduction) factors that Caquot and Kerisel developed for K_p for cases where $0 \leq \delta/\phi < 1$ were used to develop values of $K_{\delta(max)}$ for use in a given application.

- The same Caquot-Kerisel reduction factors for $0 \leq \delta/\phi < 1$ were applied to all cases, i.e. $K_{\delta(min)} < K_\delta < K_{\delta(max)}$, not just for $K_{\delta(max)} = K_p$ for which these factors were developed. There was no justification given by Nordlund for this assumption other than that there was apparently nothing better for him to assume.

- For reasons that will become clear shortly, Nordlund developed curves for $\delta/\phi > 1$ although his methodology for doing so was not completely explained by him in his publications.

The conclusion drawn from all this is that on the one hand Nordlund deserves credit for weaving together a combination of theory and actual observations that was very limited in breadth and depth relative to today's knowledge-base of geotechnical engineering in order to develop his desired final result that was revolutionary for its time. The problem is that there were some subjective, empirical modifications made along the way by Nordlund that he did not fully explain or illustrate so that they could be critiqued objectively by others. Thus, there is a significant, troublesome (to the writer at least) 'act of faith' (in Nordlund's experience) component to the parameters used in either version of the Nordlund Method.

Considering next the parameter δ (actually δ/ϕ was used by Nordlund), Nordlund opined that it was a function not only of the material along the pile-soil interface but the pile volume per unit length, V, as well. His reason for including the latter parameter was that he assumed that the process of driving a pile would be expected to densify the soil and increase ϕ. Because the value of ϕ used in his calculation process was that <u>prior</u> to driving (because it was based on <u>pre-driving</u> N-values), the only way to include the hypothesized soil-densification effects was to use a value of δ/ϕ that was fictitiously larger than the actual value and, in some cases, greater than one. This was the underlying reason that Nordlund had for developing K_δ correction factors for $\delta/\phi > 1$.

Unfortunately, once again Nordlund did not provide complete details as to how he arrived at the δ/ϕ values shown in Nordlund (1963) that were presented as δ/ϕ vs. V curves for several different types of tapered and non-tapered piles. In some cases, the relationships shown in that paper were clearly highly speculative, e.g. the curve relating δ/ϕ and V for timber piles was based on exactly <u>one</u> pile and the curve for precast-PCC piles was based on <u>no</u> piles of that type. That a person can purport to reliably show (as Nordlund did) a non-linear relationship using exactly <u>one</u> datapoint or, worse, **<u>no</u>** datapoint is highly questionable in the writer's opinion.

The conclusion drawn by the writer with respect to Nordlund's treatment of δ is that the overall results are highly questionable for two reason:

- Nordlund does not provide adequate supporting data for how he arrived at his results, the extreme case being how could he logically show a curved relationship based on a single piece of data as in the case of timber piles.

- All δ/ϕ curves show substantial variation, in the range of approximately 50-100% increase over the range in pile volumes shown. Since δ/ϕ would be expected to remain constant for a given type of pile and soil, this means that increases in ϕ due to driving in the range of 50-100% are implied to occur. In the writer's opinion, this seems extreme and unlikely.

That having been said, the simple fact is that these δ/ϕ relationships are integral parts of the overall 'black box' that is the Nordlund Method. They must be used as is.

Moving on to toe resistance, the only parameter requiring discussion is the N_q term that is the so-called 'embedment' term[116] from the traditional three-term bearing-capacity theory. Nordlund cited several bearing-capacity solutions that had been published between 1934 and 1961 (but not Terzaghi's which was the first) and selected one developed by Berezantzev, Khrisoforov, and Golubkov, apparently largely on the basis of the fact that the solution explicitly considered the D/B ratio (in this application, the depth of the pile toe relative to the ground surface divided by the width of an equivalent-square pile toe) as part of the solution. That the embedment of deep foundations affected bearing capacity was apparently becoming appreciated at the time Nordlund was finalizing the details of his analytical methodology in the early 1960s, although likely not for the reasons involving soil compressibility that were appreciated in later years (Kulhawy 1984).

In summary up to this point, the single most important issue to note about the Original Nordlund Method is that it requires use of what would now be considered a primitive methodology for estimating ill-defined, pre-driving ϕ values of the soil using uncorrected SPT N_{field} values. These ϕ values are then used to estimate model parameters for both shaft and toe ULS resistances.

Furthermore, while the K_δ and δ/ϕ parameters for shaft resistance have a basis in theory, they contain substantial empirical modifications based on a combination of assumptions and observed behavior from a relatively small number of pile load tests. Unfortunately, most of these empirical modifications were incompletely documented in Nordlund (1963) so that it is not possible to independently and objectively assess the logic behind most of the empiricisms.

In all fairness, there is, no doubt, decades of experience working with tapered piles that are reflected in Nordlund's original published work, at least in a broad, underlying way that defies explicit citation or simple explanation. While the positive influence that the depth and breadth of Nordlund's professional experience may have had on his analytical

---

[116] So named because it reflects the contribution to gross ultimate bearing capacity of the vertical overburden stress (effective or total depending on the analysis performed) due to embedment of foundation level (in the case of deep foundations this is the depth of the toe) below the ground or water (if higher) surface as appropriate.

methodology is not being deprecated here, the fact remains that the method as published in 1963 does not lend itself to piecemeal modification, e.g. using more-modern concepts and methods for estimating ϕ, no matter how rational or logical such modifications might be in and of themselves. Rather, the Original Nordlund Method must be used as is, including back-dating N_{field} values to N_{45} as necessary, as all of the components were presumably calibrated to these N-values.

As a final comment, most of the parameters that are required to use the Original Nordlund Method can be estimated from figures presented in Nordlund (1963). However, these figures are hand-drawn and -lettered and were reproduced in a fairly small size in the original paper. This makes use of the Original Nordlund Method by manual calculation a fairly time-consuming process, especially to scale-off numbers with reasonable accuracy.

In addition, because the relationship between plotted variables in each figure is non-linear and the curves do not have an easily defined mathematical shape (e.g. hyperbola, etc.), this means that there was undoubtedly a certain amount of approximation involved in digitizing these curves when the *DRIVEN* program was created. It appears from the documentation from the original version of this program (Mathias and Cribbs 1998) that the developers scaled off values at several points along a curve and then used some unspecified algorithm (possibly just an assumption of linearity) to interpolate between the scaled values. Therefore, while the *DRIVEN* program makes the Original Nordlund Method easier to use it may not necessarily be inherently more accurate than manual calculation using Nordlund's original 1963 paper.

5.6.1.2.4 Nordlund Method (Revised 1979 Version)

Because the Revised Nordlund Method that was presented in Nordlund (1979) is essentially the original method with relatively few changes, it is easiest and most efficient to only discuss the differences between the two versions.

First and foremost because it affects both shaft- and toe-resistance calculations (Nordlund himself called this the "nub of the problem"), Nordlund updated his original 1963 position on the issue of correcting N_{field} for overburden stress. As noted in the preceding section, Nordlund (1963) acknowledged then-new evidence that N_{field} varied with depth, all else being equal. However, Nordlund felt that the state of knowledge circa 1960 was insufficient for a rational correction to be made to N_{field}.

However, the situation had changed significantly by 1979 as the body of knowledge concerning all the variables that impact the SPT had grown considerably by then. For whatever reason, however, Nordlund chose only to include a correction for overburden stress and not driving efficiency or any other SPT variable in the revised version of his methodology.

A figure showing the correction factor to be applied to N_{field} was published in Nordlund (1979) although its source was not indicated. However, the non-linear relationship shown for the correction factor has a familiar shape:

- an initial value of 2.0;

- a value of 1.0 at a vertical effective overburden stress of 1 ton/ft² (note that this is the Imperial unit 'short ton' that is equal to 2,000 pounds so the stress is approximately equal to 100 kPa); and

- values that decrease below 1.0 with increasing depth and stress.

Unfortunately, in addition to neglecting a correction for SPT drive-system efficiency, Nordlund retained his original suggestion that the N_{field} value used could, at the discretion of the design professional, be the sum of the first two drive increments instead of the long-standard second and third increments. In addition, he indicated that recorded SPT drive-increment values could arbitrarily be reduced or ignored entirely if "elevated" results due to gravel were suspected. Thus, all of the vague subjectivity relative to the SPT N-value to actually use that was inherent in the Original Nordlund Method remained in the revised version.

In summary, when using the Revised Nordlund Method it is still necessary (in the writer's opinion) to use $N_{field} = N_{45}$ as in the original method. However, it is then necessary to correct this value for vertical effective overburden stress before using the N-value to estimate ϕ for resistance calculations. Note that this depth-related correction would tend to reduce calculated resistances for both the toe and shaft for longer piles. This is because N-values below a depth of approximately 15 to 30 feet (5 to 10 m) in coarse-grained soil (depending on the depth of the groundwater table) would be reduced in the revised version compared to the original version. Note that this would most affect not only the pile toe but the tapered portion of piles such as the *Monotube* and *TAPERTUBE* that have a constant-diameter upper section (extension) and tapered lower section.

With specific regard to shaft resistance in the Revised Nordlund Method, Nordlund made a slight change relative to the original (1963) version to two of the curves in the plot of δ/ϕ versus pile volume, V. In addition, he added a curve and concomitant usage suggestions for what he called "augercast" piles but what would now be referred to generically, in the U.S. at least, as *augered cast-in-place* (ACIP) piles. However, one improvement in his 1979 paper was that the figures for the shaft-resistance parameters were larger and more-professionally drawn compared to the 1963 originals.

On the negative side, Nordlund provided no additional discussion or numerical illustration of how he developed the volume-related empirical modifications to theory that are incorporated in the K_δ and δ/ϕ parameters in both versions of his method. In fact, he simply referred back to the original 1963 paper for background information in this regard, stating that all was explained in the original 1963 paper (which it most certainly was not, in the writer's opinion, as discussed in the preceding section).

With specific regard to toe resistance, Nordlund continued to use the 1961 solution to the traditional bearing-capacity problem developed by Berezantzev at alia. However, Nordlund was aware of circa-1960s research by Vesic and circa-1970s research by Meyerhof that suggested there was a limit to the gross ultimate bearing pressure that could develop beneath the toe of a DFE. A copy of a figure relating values of such a limiting pressure to ϕ that was attributed to Meyerhof was included in Nordlund (1979).

Nordlund incorporated the concept of a limiting toe bearing pressure into his revised methodology by placing an arbitrary limit of 3,000 lb/ft^2 (145 kPa) on the vertical effective overburden stress, $\sigma'_{vo(toe)}$, to be used for the toe-resistance component in Equation 5.6. To put this 3,000 psf value in perspective, it corresponds to a depth of approximately 25 to 50 feet (8 to 15 m) in coarse-grained soil depending on the depth of the groundwater table. The net result is that using this revised methodology, longer piles tend to have less, sometimes substantially less, forecast toe resistance compared to the Original Nordlund Method.

Note that this direct reduction in toe resistance is in addition to the indirect reduction due to the aforementioned stress/depth correction for N-values that would tend to reduce the N_q parameter in Equation 5.6. Note also that the N-value used for forecasting toe resistance is always corrected using the <u>actual</u> value of the vertical effective overburden stress. This can result in an inconsistency in logic for toe-resistance calculations wherein the <u>actual</u> vertical effective overburden stress is used to correct the N-value but then the

<u>artificially delimited</u> vertical effective overburden stress is used in the final resistance calculation.

In summary, the Revised Nordlund Method contains two substantive changes, one affecting the SPT N-values (and, as a result, ϕ and thus both shaft and toe resistance indirectly) and the other toe resistance directly, that have the potential to significantly reduce the forecast ULS total resistance compared to the Original Nordlund Method, especially for longer piles. This was confirmed by a very limited comparison between the two versions of the method that was presented at the very end of Nordlund (1979). This comparison also included an equally limited comparison to measured resistances that had been reported in a paper presented by others the year before (1978) at a regional geotechnical conference. Overall, the revised version of Nordlund's method appeared to provide better agreement with these 1978 measured results compared to the original version.

One rather interesting and surprising aspect of this 1978 case-history comparison in Nordlund (1979) was that two tapered precast-PCC piles with a square cross-section (no further details given) were included in the group. Apparently, the only way that Nordlund could get a reasonable correlation between measured and calculated (using either version of his method) resistances was by ignoring taper effects completely. Presumably, even the Revised Nordlund Method substantially overestimated the resistance when taper was considered.

Nordlund suggested in very general terms that some geomechanics effects related to the square pile geometry might be the cause for this disconnect between observations and outcomes using his methodology. The implication was that tapered piles with a square cross-section exhibited behavior that was substantially different from tapered piles with the more-common circular cross-section, an issue that was noted earlier in Chapter 4. This raises the legitimate question of whether the wedging mechanism illustrated in Figure 4.1 that is now recognized for tapered piles plays out differently depending on the cross-sectional geometry of the 'wedge', i.e. tapered pile. In particular, the cylindrical-cavity model for wedging that is clearly appropriate for circular tapered piles is just as clearly questionable for square circular piles.

5.6.1.2.5 Modified Nordlund Method

The writer became aware of Kodikara and Moore's landmark work of modeling the unique wedging mechanism associated with shaft resistance due to taper using cylindrical-cavity mechanics at the time Kodikara and Moore (1993) was published although, in retrospect, it is now apparent that Kodikara and Moore did not go far enough and include the cavity creation that occurs during pile driving in their resistance-forecasting methodology. Nevertheless, it was immediately obvious that their research was the first significant advancement in understanding the unique behavioral mechanism of tapered piles in the 30 years since the Original Nordlund Method had been published and, overall, eclipsed and replaced Nordlund's work. This was because a 3-D cavity-mechanics model is conceptually superior to Nordlund's 2-D retaining-wall model based on classical lateral earth pressure theory for this application.

Of equal importance and significance in the writer's opinion is the fact that the cylindrical-cavity model made it clear that the physical mechanism by which tapered piles develop resistance along their shaft is distinctly different from the traditional shaft-resistance mechanism of simple sliding friction. As such, Kodikara and Moore's work clearly supported what the writer has termed the *third resistance mechanism* for DFEs in general.

Unfortunately, at the time (1993) it was equally clear to the writer that:

- the Kodikara-Moore Method is analytically very complex compared to resistance-forecasting methodologies for DFEs in general;

- had some assumptions that require modification (discussed subsequently);

- required in its original form as presented by Kodikara and Moore a purpose-written computer code for application; and

- would thus take some considerable effort to develop into an analytical tool for use in routine practice that would not scare-off the average design professional due to its analytical complexity.

The writer thus concluded at the time (1993) that while implementation of the Kodikara-Moore Method or some version of it into routine practice was certainly a long-term goal, it was not a goal that was going to occur anytime soon (this turned out to be an understatement). Thus, there was clearly a near-term need for an interim improved methodology for forecasting the resistance of tapered piles.

To achieve this goal, the writer incorporated several key components from Nordlund (1963) into an integrated analytical methodology first presented in Horvath (2002) that also drew heavily on geomechanics concepts promulgated by Kulhawy (1984):

- There is a maximum taper angle beyond which no additional taper benefit accrues and this angle is approximately a linear function of ϕ.

- The maximum taper benefit (in terms of the increase in radial stress acting along the pile shaft, as reflected in the K_h/K_o ratio) relative to the no-taper baseline case is a factor of about six and this factor is, for all practical purposes, independent of ϕ for the values of ϕ typically encountered in practice.

- The variation in taper benefit, as reflected in the K_h/K_o ratio, between the no-taper baseline case and aforementioned maximum taper angle can reasonably be assumed to be linear.

All other assumptions included in either version of the Nordlund Method were discarded. This allowed the use of numerous, modern site-characterization and geomechanics concepts including:

- Developing soil properties, especially for strength (ϕ) and pre-driving stress-state (K_o), from state-of-art site characterization based on the cone penetrometer.

- Recognizing that the post-installation lateral earth pressure coefficient, K_h, is best expressed as the product $[(K_h/K_o) \cdot K_o]$ as was discussed in Chapter 4.

- Allowing for both shaft and toe resistances to be calculated using either a stress-dependent secant value of ϕ_{peak} or ϕ_{cv} as desired.

- Using Hansen's traditional bearing-capacity solution together with Vesic's compressibility/rigidity factors for estimating the toe-resistance component.

The resulting analytical method, which was outlined in detail in Horvath (2002) together with a worked example for actual piles, is hereinafter referred to as the *Modified Nordlund Method* as it was broadly based on key concepts put forth by Nordlund. An obvious question is the accuracy of the writer's Modified Nordlund Method in forecasting ULS total resistance, both in an absolute sense as well as relative to other methods, especially both versions of the Nordlund Method. This is addressed in detail later in this chapter.

Additional research in recent years by the writer during the development of a tetralogy of white papers (Horvath 2014b, 2015, 2016a, 2016b) that had tapered piles as one of the primary topics included the consideration of published material by others that was related to or of possible use with tapered piles. As a result, the writer became aware of a series of papers published by Professors Suman Manandhar and Noriyuki Yasufuku that involved research into tapered piles. Their work is discussed in detail later in this chapter in a separate section but is mentioned here as one significant component of their work involved the use of a spherical-cavity expansion (SCE) model for forecasting the ULS toe resistance. In fact, as will be seen, they developed their toe-resistance model specifically with tapered piles in mind as they assumed that the shaft taper angle influenced the geometry of the failure surface that develops beneath the toe.

The Manandhar-Yasufuku toe-resistance model based on SCE theory was used by the writer in Horvath (2015) as an alternative for forecasting toe resistance in the writer's Modified Nordlund Method. This was in lieu of the writer's original toe-resistance model based on a composite Hansen-Vesic traditional bearing-capacity theory as modified for soil rigidity/compressibility effects.

Note that the writer's Modified Nordlund Model is not limited with respect to the analytical methodology used for calculating the toe-resistance component. In addition to a traditional bearing-capacity theory and the Manandhar-Yasufuku toe-resistance model that have been used to date, any other toe-resistance methodology with a theoretical basis (this is in keeping with the fact that the Modified Nordlund Method is an indirect method) can be used. Other examples explored in this monograph are Vesic's original SCE solution without the modifications proposed by Manandhar and Yasufuku and the *NTH Limit Plasticity Method*, both of which are discussed in Appendix G.

A final comment with respect to the writer's Modified Nordlund Method is that the central element of the method related to taper benefit, i.e. how the K_h/K_o ratio increases with increasing taper angle, draws heavily on material that Nordlund presented in his 1963 paper and repeated in his 1979 lecture notes. This material collectively suggests that there is no apparent increase in taper benefit for taper angles greater than approximately 1°.

Historically, this was a moot point as the taper angles of both timber and steel piles (the two Raymond steel-shell products as well as two steel-pipe products, *Monotube* and *TAPERTUBE*) that dominated U.S. practice throughout the 20th century did not exceed 0.95°. However, with the *TAPERTUBE* product line now listing options with taper angles up to 1.6° and no limit on the taper angle that can be used with precast-PCC piles that tend to be locally designed and cast, this clearly defined limit on taper benefit that is inherent in all three versions of Nordlund's Method is clearly something that requires fundamental research to either confirm or modify as appropriate.

5.6.1.2.6 Bjerrum-Burland-Fellenius (Original) β Method

As illustrated in Chapter 4, the K_h and tan δ parameters in the generic equation (Equation 4.6) that defines the unit stress for shaft resistance due to the traditional mechanism of sliding friction can be lumped together into a single dimensionless parameter that is traditionally given the notation β. The earliest instances of doing this in practice are credited in the published literature to the late Dr. Laurits Bjerrum in the 1960s and Prof. John Burland in the 1970s.

Arguably the most significant aspect of the using the β parameter that is important to note at the outset is that it treats all soils equally on a fundamental effective-stress basis. At the time it was first propose in the 1960s and to the present, it remains a distinct conceptual approach to DFE resistance forecasting that traditionally treats coarse-grained soils using effective stresses but then switches to a total-stress approach using undrained shear strengths for fine-grained soils.

While this method was thus been termed the *Bjerrum-Burland β Method* in earlier published literature, in recent decades it subsequently became more associated with Fellenius (to the point where the writer has seen it referred to colloquially in recent years as the *Fellenius β Method*) who championed its use in numerous papers and textbooks in the final decades of the 20th century. Although Fellenius still mentions this method in recent publications such as Fellenius (2019), since the cusp of the 21st century Fellenius has focused more on using a direct method called the *UniCone Method* that is based on CPTu data and is discussed later in this chapter with other direct methodologies.

It is of interest to note that one of the benefits of the Bjerrum-Burland-Fellenius β Method is that a choice concerning the nature of the shearing mechanism along the shaft of a deep foundation, i.e. the issue of whether the operative friction angle along the shaft is δ for the traditional planar failure surface defined by the shaft-soil interface or φ for a shear zone around the shaft as was noted in Chapter 4 and is discussed further in Appendix A, is sidestepped as this component is incorporated into β.

It is also of interest to note that β can be related to Nordlund's parameters K_δ and sin δ in Equation 5.6 as follows:

$$\beta = K_\delta \cdot \sin \delta_{NORDLUND} = K_\delta \cdot \sin \left[\left(\frac{\delta}{\phi} \right)_{NORDLUND} \cdot \phi_{NORDLUND} \right] \qquad (5.9)$$

where both the δ/φ ratio and φ <u>must</u> be evaluated using the explicit guidelines of either the original or revised version of the Nordlund Method as desired. This is emphasized with subscripts in Equation 5.9 as δ in the original definition of β (Equation 4.6) is conceptually not the same as δ in Equation 5.6 that is used with both the Original and Revised Nordlund Method. This is because Nordlund has specific protocols for evaluating both the δ/φ ratio and φ as was discussed in detail earlier in this chapter.

It should be noted that the *Original β Method* (as it will be referred to hereinafter in this monograph for reasons that will become clear and certainly not to trivialize the seminal contributions of Bjerrum, Burland, or Fellenius) is not, strictly speaking, an indirect method for reasons given in Chapter 4, i.e. the values of β are not determined by any rigorous application of geomechanics concept or theory. Rather, ranges of values that appear in the published literature have been determined by back-calculation from instrumented static load tests[117] and the values used in a particular analysis are typically chosen by the design professional from these published ranges.

---

[117] It is essential that the results interpreted from instrumented static load tests be corrected for residual loads for the reasons discussed in Appendix A. If this is not done, the interpreted β values can be significantly in error as illustrated in Fellenius (2015) for several case histories.

Nevertheless, the writer chose to include the Original β Method as an indirect method for several reasons specific to the method, some of which were previously noted generically in Chapter 4:

- It is clearly not a direct method as the measured results from an in-situ test are not used to directly evaluate β. Therefore, if for no other reason it is placed in the indirect-method category by default.

- It does have a theoretical component (albeit very modest) that aligns it with indirect methods. As shown in Equation 4.6, the design professional has to calculate the vertical effective overburden stress in order to use the Original β Method.

- It is a method for forecasting shaft resistance only so must be coupled with some other method for toe resistance. Direct methods always produce forecasts for both shaft and toe resistance. Thus, in principle, a theoretical methodology based on classical bearing capacity theory, spherical-cavity-expansion (SCE) theory as originally proposed by Vesic or modified by Manandhar and Yasufuku, or some other solution could be used. However, in his publications Fellenius always uses an empirical, dimensionless parameter with the notation N_t for forecasting the toe-resistance component. Values of this parameter are also derived from instrumented static load tests. As a result, this empirical approach for forecasting toe resistance is nowadays generally considered to be an explicit component of the Original β Method.

Of relevance to this monograph is that, to the best of the writer's knowledge, the only values of β that have been reported in the literature to date are for constant-perimeter piles. The Original β Method can, in principle, be used with tapered piles by using Fellenius' donut-on-a-stick model (Figure 4.2) although, as noted in Chapter 4, the writer is not aware of any published work that illustrates the accuracy of the results from using this model for tapered piles. For this reason, the use of this model with tapered piles can, and should, be seriously questioned.

In the writer's opinion, there is no need to resort to using gimmicks such as Fellenius' donut-on-a-stick model as there is no reason why the β concept cannot be extended to include tapered piles in a more-rigorous manner. This is outlined in the following section.

5.6.1.2.7 Extended β Method

In concept, there is no reason why explicit β values for tapered piles could not be developed using the same back-calculation strategy used for constant-perimeter piles. This was first proposed by the writer some years ago (Horvath and Trochalides 2004) and is a simple, straightforward process as illustrated in Appendix A using a case history involving *Monotube* piles. In this case, the β values are a function of not only soil type but also taper angle and pile type as well as the pile surface in contact with the soil can be PCC, steel, or wood.

Developing β values specifically for use with tapered piles is defined here as the *Extended β Method* to distinguish it from merely using the Original β Method and β values for non-tapered piles with the donut-on-a-stick model.

5.6.1.2.8 Modified β Method

A significant concern with using the Original β Method, even for constant-perimeter piles as originally intended, is that the actual range and scatter of back-calculated β values tends to be much greater (an entire order of magnitude in some cases) at relatively shallow depths (defined as approximately the uppermost 5-10 metres (15-30 ft) in this case) compared to the relatively narrow ranges of published β values that are found in many references. This over-simplification of reality has been noted by Fellenius and is illustrated in Fellenius (2019).

Based on initial, preliminary work by the writer (Horvath and Trochalides 2004), it appears that the above-defined Extended β Method when applied to tapered piles is similarly affected by this data-scatter issue, even when the tapered portion of the pile extends below a depth of 10 metres (30 ft) below the ground surface. In the case of tapered piles (*Monotube*, *TAPERTUBE*, and timber) with a relatively wide range of taper angles driven at JFKIA that were investigated by Horvath and Trochalides (2004), the data scatter resulted in difficulty with defining a trend in β as a function of taper angle and pile material.

As a means of at least reducing the obfuscation of any trends in these data, the writer suggested a modification to both the Original and Extended β Methods for the explicit purpose of minimizing the observed parameter variation (Horvath and Trochalides 2004). Specifically, the suggestion made was to modify the analytical methodology by replacing β with the expression $[(\beta/K_o) \cdot K_o]$ that was previously defined and suggested for general use in Chapter 4 (Equation 4.16). This is based on the premise that at least part of the variation in β values, especially at relatively shallow depths, is due to the natural variation in pre-installation K_o values. It is well-known that if there is going to be a variation of K_o at a site, that variation is typically most pronounced at relatively shallow depths.

As defined in Chapter 4, in this application K_o is the site-specific value (in actual applications, this will generally be a depth-wise-variable range in values) that exists prior to pile driving. Nowadays, these site-specific values of K_o can routinely be estimated as part of the usual site-characterization process (Horvath 2011, 2014b, 2015). This leaves the β/K_o ratio dependent largely on the pile type, including taper angle where relevant. Of course, this concept could be extended to tapered and non-tapered DFEs in general in which case the β/K_o ratio would also depend on the installation methodology of the DFE.

Note that this suggestion to use the β/K_o ratio is conceptually similar to Kulhawy's correlation of deep foundation type with the ratio K_h/K_o that is part of the writer's Simplified Nordlund Method as, by definition:

$$\frac{\beta}{K_o} = \frac{K_h \cdot \tan \delta}{K_o} = \left(\frac{K_h}{K_o}\right) \cdot \tan \delta =$$

$$\left(\frac{K_\delta}{K_o}\right) \cdot \sin \delta_{NORDLUND} = \left(\frac{K_\delta}{K_o}\right) \cdot \sin\left[\left(\frac{\delta}{\phi}\right)_{NORDLUND} \cdot \phi_{NORDLUND}\right].$$

(5.10)

Note that the relationship of the β/K_o ratio to Nordlund's analytical parameters is also shown in this equation for the sake of completeness, building on what was shown previously in Equation 5.9.

To expand on what was noted above, the underlying reason why this suggested modification to the Original and Extended β Methods, which is hereinafter referred to as the *Modified β Method*, is helpful with reducing the observed wide variation of β at relatively shallow depths (and greater depths as well with tapered piles) is that K_o often varies substantially at relatively shallow depths before becoming less variable at greater depths. An order-of-magnitude variation in K_o between shallow and deep depths at a site is not uncommon as was noted by Kulhawy and Mayne (1990). This behavior was found by the writer at JFKIA (Horvath 2002, 2014, 2015) and is illustrated with a specific case-history example in Appendix A of this monograph where K_o varied from almost 1.4 at a shallow depth to approximately 0.4 at greater depths.

With the way in which the Original and Extended β Methods are structured, depth-wise variations in K_o get built in to the back-calculated values of β in addition to deep foundation typology (including taper angle where relevant) and soil type. Thus, by normalizing β to K_o as opposed to explicitly incorporating K_o within β the natural variation that can occur in K_o is removed from β and leaves the $β/K_o$ ratio to be dependent solely on the type of DFE (including taper angle where relevant) and soil.

This suggested modification was explored on a very limited basis in Horvath and Trochalides (2004) and showed promise. In particular, while the expected dependence of β on taper angle was poorly defined for the Extended β Method, the expected dependence of the $β/K_o$ ratio on taper angle was much more clearly defined for the Modified β Method. An additional case-history example using a *Monotube* pile is presented in Appendix A.

While it appears that using the Modified β Method with $β/K_o$ ratios offers the potential for improved, i.e. more accurate, ULS-shaft-resistance forecasting compared to using the either the Original or Extended β Methods with β alone, the one restriction is that using the Modified β Method requires a site-specific assessment of the depth-wise variation of K_o within the expected depth of pile installation. While this can be done routinely using either an algorithm presented by the writer (Horvath (2011) contains the most recent version) or more-recent empirical correlations presented and discussed in Horvath (2014b, 2015), it is most easily and accurately done using cone-penetrometer data. Using only SPT N-values provides considerably fewer data points to work with, each of which is inherently less accurate than actual q_c data from a cone penetrometer because of the series of assumptions required to convert N_{field} values first to N_{60} values and then to quasi-q_c values. In addition, the writer's algorithm (Horvath 2011) for converting N-values to quasi-q_c values has only been developed for coarse-grained soils.

While performing cone-penetrometer soundings as part of a site-characterization assessment is no longer exceptional or unusual in most parts of the world, it is still not routine in some areas (the New York City metropolitan area being one with which the writer is intimately familiar) for a variety of reasons, with historical biases and prejudices based on outdated perceptions being high on the list. There is also a tendency among practitioners to do only the minimum of what building or other applicable design codes require and no more. For example, the New York City Building Code requires traditional borings with SPT sampling. The net result is that potential use of the Modified β Method with $β/K_o$ ratios is likely to encounter resistance due to human inertia against change as well as doing anything beyond the bare minimum that is legally required.

5.6.1.2.9 Manandhar-Yasufuku Solution and Methods for Toe Resistance

In recent years, Prof. Suman Manandhar conducted doctoral (Manandhar 2010) and post-doctoral research into tapered piles while at Kyushu University in Japan. This work, which consisted of both an analytical/theoretical component as well as the physical testing of small-scale model piles under 1-g conditions in a laboratory environment, was apparently done under the direction and supervision of Prof. Noriyuki Yasufuku. Consequently, and in view of the fact that almost all publications related to this work have Manandhar and Yasufuku as co-authors, in this monograph specific outcomes from this research are credited to both.

However, it is relevant to note that Manandhar's and Yasufuku's work was performed with the assistance of several others both at Kyushu University as well as Saga University in Japan where Manandhar was later affiliated. Consequently, it appears reasonable to credit the broader outcomes of this collaborative effort to a team, with Prof. Manandhar being the primary principal of this team.

The work-scope of Manandhar et alia included a relatively small, incremental improvement to one parameter used in the Kodikara-Moore Method although publications (Manandhar and Yasufuku 2011b, 2013) are misleading (in the writer's opinion) because they essentially included the entire theoretical derivation from Kodikara and Moore (1993) without making it sufficiently clear (again, in the writer's opinion) that this was not work that was original to the authors, i.e. Manandhar and Yasufuku. In any event, both the Kodikara-Moore Method and the relatively trivial Manandhar-Yasufuku modification of it are discussed in detail later in this chapter in the section dealing with integrated (one-step) load-settlement methods.

The discussion here is limited to another aspect of Manandhar et alia's work that involved the development of a new physical and mathematical model and concomitant solution for toe resistance (in coarse-grained soils only) that was presented in Manandhar and Yasufuku (2012) as well as in a later, more comprehensive paper (Manandhar and Yasufuku 2013) as well as a *PowerPoint* presentation that was given in the U.S. in 2013 to the North Carolina Department of Transportation (Manandhar et al. 2013). This toe-bearing model is an indirect resistance-forecasting methodology that is completely independent of Manandhar et alia's modification to the Kodikara-Moore Method and can, in principle, be used with any resistance-forecasting methodology for shaft resistance (as noted previously, the writer used it in recent years with the Modified Nordlund Method) which is why it is being discussed in this section of the present chapter.

The Manandhar-Yasufuku toe-resistance model and solution is believed to be the first developed specifically with tapered piles in mind (although it can be used with constant-diameter piles as well) as the geometry of the hypothesized failure surface that develops beneath the toe is assumed to be influenced by the taper angle of the pile shaft immediately above the toe. This model and solution also deviates from traditional deep-foundation practice as the mathematical model used is based on Vesic's spherical-cavity expansion (SCE) theory as opposed to the solutions to traditional bearing-capacity theory (such as Berezantzev et alia that was used by Nordlund or the composite Hansen-Vesic solution used by the writer initially for the Modified Nordlund Method) that assume a composite failure surface built around the development of a triangular 'failure wedge' directly beneath the pile toe. More significantly, as discussed in Chapter 4, SCE theory as applied to deep-foundation toe bearing implies that radial stresses control the process whereas traditional bearing capacity implies that vertical effective overburden stress governs.

Note, however, that Manandhar et alia's use of SCE as a physical and mathematical model for deep-foundation toe bearing was neither innovative nor unique as this concept appears to have originated several decades earlier (circa 1970) with Vesic and was subsequently explored by several others such as Baligh (1976) and Yu and Houlsby (1991). The only unique aspects to Manandhar et alia's work in this regard was modifying the geometry of the assumed spherical-cavity geometry to account for the hypothesized influence of pile taper as well as the several empirical relationships that the assumed for quantifying the SCE model parameters.

While the use of cavity-expansion theories in general, i.e. both spherical and cylindrical, in foundation engineering is appealing in many ways, as noted previously in Chapter 4 the difficulty has always been and remains the fact that:

- soil stiffness in the form of a modulus (Young's or shear) in addition to soil shear strength must always be considered in order to achieve a solution;

- solutions typically require a single value for both the stiffness and shear-strength parameters;

- in reality, both stiffness and, to a lesser extent, shear strength vary as a function of deep-foundation loading between some initial state and the ULS; and

- the calculated results are highly sensitive to the value of stiffness in particular which makes choosing some average (secant) value of stiffness to use in some solution both highly subjective and extremely problematic.

This final point in particular should be kept in mind when using the Manandhar-Yasufuku toe-resistance solution as very specific assumptions and empirical relationships regarding both soil stiffness and shear strength that were made by these researchers are incorporated into the various equations that must be used as part of their analytical methodologies that are based on this solution. This is similar to both the Original and Revised Nordlund Methods that require specific correlations and empirical relationships defined by Nordlund to be used.

Another feature of the Manandhar-Yasufuku toe-resistance solution is that it includes both a basic analytical methodology (referred to in this monograph as the *Basic Manandhar-Yasufuku Toe-Resistance Method*) for a single-valued ULS unit toe resistance, $r_{t(ult)}$ (which was given the notation q_{pcal} by them), as well as allows for an enhanced analytical alternative (referred to in this monograph as the *Enhanced Manandhar-Yasufuku Toe-Resistance Method*) for forecasting intermediate values of the unit toe resistance, r_t (that they called q_{cal}), where $0 < q_{cal} < q_{pcal}$. The empirical equation for generating these intermediate values is a function of a normalized, dimensionless ratio of toe settlement divided by toe diameter. This is an attempt to recognize that toe resistance is not an all-or-nothing proposition but is mobilized incrementally with increasing toe settlement as was discussed in Chapter 4.

Note, however, that because the empirical equation used by Manandhar et alia for this purpose has the form of a Kondner hyperbola, it means that the ULS unit toe resistance, q_{pcal}, is approached asymptotically with increasing toe settlement but never reached. This is broadly in consonance with Fellenius' oft-stated opinion that a singular, ULS value of bearing capacity does not, in general, exist but that toe resistance continues to increase with increasing toe settlement and without limit, albeit at a much-reduced rate of resistance increase.

Overall, there are a number of significant issues to be raised at this point concerning the Manandhar-Yasufuku Toe-Resistance Method in both its basic and enhanced versions, beyond the fact that its use is limited to piles bearing in coarse-grained soil (which is the predominant case with tapered piles so certainly not a critical constraint):

- First and foremost is the fact that the assumed influence of pile taper on the geometry of the SCE model used by Manandhar and Yasufuku and, therefore, the ULS toe resistance was simply a geometric hypothesis. It was not, in the writer's opinion, adequately supported by any evidence other than inconclusive measurements made on very small model piles with geometries that bear no resemblance to real piles that were tested under 1-*g* conditions in a laboratory sandbox. The bowling-ball analogy noted in Chapter 4 is certainly relevant in this case.

- The suite of empirical equations that must be used to evaluate the SCE-model parameters for shear modulus, etc. are based on technology that is either antiquated (SPT field *N*-values, N_{field}) or difficult to measure under the best of circumstances and certainly problematic in routine practice (minimum and maximum void ratios, e_{min} and e_{max}).

- The curve-fitting parameters used to define the shape of the Kondner hyperbola used in the Enhanced Manandhar-Yasufuku Toe-Resistance Model are arbitrary and create an unrealistically soft toe response. This issue is explored in detail in Appendix B.

5.6.1.2.10 El Naggar-Sakr Method: Prologue

The first use of cylindrical-cavity mechanics as a physical and theoretical model for tapered piles is credited to Kodikara and Moore. Circa 1990, they used this theory as the basis for an integrated (one-step) load-settlement forecasting methodology, the Kodikara-Moore Method, that was published in Kodikara and Moore (1993) and is discussed at length later in this chapter along with other one-step resistance-forecasting methodologies.

Beginning in the late 1990s, Prof. M. Hesham El Naggar at then-UWO in London, ON, Canada and a succession of graduate students under his direction conducted research into various types of tapered DFEs. They first studied tapered (driven) piles, then tapered drilled shafts, and, most recently, helical tapered piles. Thus, collectively to date they have considered tapered DFEs that are driven, drilled, and torqued into place.

It is the writer's speculation that Prof. El Naggar's initial interest in tapered DFEs may well have been inspired or at least influenced by the fact that he was pursuing his Ph.D. at UWO in the early 1990s. This coincided with the timeframe when Prof. Moore (of Kodikara and Moore) had moved from Australia to Canada and was on the faculty of UWO and co-authoring his paper on the Kodikara-Moore Method with Prof. Kodikara. Regardless of what the source of the initial inspiration or interest was to Prof. El Naggar, the almost two decades of work by El Naggar et alia at UWO/Western is one of the very few sustained research efforts into tapered DFEs of which the writer is aware.

In several published works beginning with El Naggar and Sakr (2000), Prof. El Naggar and his graduate students simplified elements of the Kodikara-Moore Method and used them to develop a resistance-only forecasting methodology for tapered DFEs. Although a number of UWO students have apparently made contributions to this analytical method since the late 1990s, in deference to the fact that the method was first published in El Naggar and Sakr (2000) this resistance-forecasting methodology for tapered piles that is a much-simplified derivative of the Kodikara-Moore Method is referred to in this monograph as the *El Naggar-Sakr Method*.

The El Naggar-Sakr Method should rightfully be discussed in this section of the monograph that deals with resistance-only forecasting methodologies. However, given the fact that this method derives largely from the Kodikara-Moore Method, a detailed discussion of the El Naggar-Sakr Method is deferred until later in this chapter after the Kodikara-Moore Method has been presented. This is because understanding the theoretical elements of the El Naggar-Sakr Method first requires a detailed understanding of the Kodikara-Moore Method.

5.6.1.2.11 Comments re Methodology Assessment

A number of modifications and extensions to existing indirect resistance-only methodologies that were developed or proposed by the writer were presented in the preceding sections. An essential element for rationally and methodically advancing each of these ideas into use in practice is a database of actual field data to provide the necessary ground-truth. However, an important practical consideration is that the level of sophistication of field data required for this purpose varies significantly depending on the particular analytical methodology involved.

For example, to assess the forecasting accuracy of the writer's Modified Nordlund Method, the minimum ground-truth required is the ULS total resistance. Putting aside for a moment the fact that defining a single-value geotechnical ULS in a traditional static load test has its own complications and ambiguities as discussed in Chapter 4, this means that a basic static load test in which only the load vs. settlement relationship at the head of the pile is recorded is adequate for this purpose. The results of dynamic measurements are also of potential use although there are issues (also discussed in Chapter 4) concerning the accuracy of these methods as a group when applied to tapered piles. There are also significant temporal issues associated with dynamic measurements so that restrike data obtained some period of time after the end of initial driving (EOID) are generally more representative of the long-term ULS total resistance compared to EOID data.

Because only the most basic field data are required, the writer's Modified Nordlund Method has already been assessed to a limited extent using a modest database consisting of both tapered and non-tapered piles at JFKIA. The results of this assessment have appeared in several publications over the last two decades (Horvath 2002, 2003c, 2015; Horvath and Trochalides 2004; Horvath et al. 2004a, 2004b). However, the writer is of the opinion that there is always room for improvement of any analytical methodology, even ones developed decades ago such as the Original and Revised Nordlund methods, and thus there is an open-ended need for performing additional assessments of any analytical methodology, if only to enhance the level of confidence in the method.

On the other hand, significantly more-sophisticated ground-truth in the form of results from high-quality instrumented static load tests that have been corrected properly for residual loads are required to develop the necessary database of taper-specific β values for the Extended β Method and β/K_o ratios for the Modified β Method, both of which were proposed by the writer as a logical extension and improvement, respectively, of the Original (Bjerrum-Burland-Fellenius) β Method. This is because depth-wise values of ULS resistance are essential for back-calculating both the β and β/K_o parameters. It is possible that the results from certain types of on-pile dynamic testing and concomitant post-processing such as the *CAPWAP* method that produce a depth-wise estimate of ULS resistance could also be useful, within the caveats expressed above that dynamic methods based on the 1-D wave equation as a group are theoretically problematic with tapered piles.

Note that developing a database of β/K_o ratios for the Modified β Method also requires a depth-wise assessment of pre-driving K_o values at each load-test location. Ideally, this would be done using cone-penetrometer data although SPT N-values could be used if necessary but only for coarse-grained soils and the results would be inherently less reliable in any event.

As an aside, it is relevant to note that assessment of the writer's Modified Nordlund Method also benefits from depth-wise estimates of ultimate resistance, something that has been done to only a very limited extent to date due to the extreme paucity of the necessary data (Horvath 2015 and Appendix A of this monograph). This is because a depth-wise assessment of ULS resistance allows for a detailed assessment of the accuracy of the assumed K_h/K_o ratios along the pile shaft as well as the apportionment of ULS shaft and toe resistances, neither of which is possible when only a measurement of the ULS total resistance is available as with a basic static load test.

Unfortunately, in the writer's experience to date, generating the necessary depth-wise resistance data that are essential for developing the Extended and Modified β Methods as well as highly desirable for a better understanding of the Modified Nordlund Method has proved to be easier said than done. The primary detriment is simply acquiring the necessary ground-truth. The writer has found that the various stakeholders (manufacturers and suppliers of tapered piles, contractors, foundation engineers, owners), in the U.S. at least, who might have or have access to such data have been largely uncooperative with knowledge-sharing.

To the extent that instrumented pile load tests were performed in the past (almost certainly <u>not</u> on timber piles due to the technical difficulties of doing so and more likely on *Monotube* piles as opposed to *TAPERTUBE* piles as the former have been commercially available for much longer than the latter), the writer can state with some certainty that access to quality load-test data has proven to be difficult to the point of impossible. There appear to be any number of reasons for this but most appear to involve various aspects of business-related confidentiality. This is especially true with regard to *Monotube* and *TAPERTUBE* piles that are largely direct competitors in the same niche market. Thus, there is apparently a feeling that anything of technical value that is gleaned from one brand of pile could be used to benefit the other brand of pile in the future. The end result is that there is a long-standing culture of knowledge-hoarding by whomever acquired that knowledge.

With regard to the future, there appears to be little incentive on the part of any stakeholder involved with tapered piles, in the U.S. at least, for doing anything more than the bare minimum of uninstrumented static load testing to meet some statutory or project requirement. This is a departure from the past where certain enlightened owners such as the PANYNJ that were more forward-looking than most in a technical sense recognized the long-term benefit to doing more than the statutory minimum in order to advance the state of knowledge for their own future benefit if nothing else.

As an aside, it is relevant to comment further (some comments were made in earlier chapters) on the role, or, more accurately, lack thereof, that the FHWA and state departments of transportation have played in supporting meaningful research into tapered piles. In the writer's experience, the road-building community in the U.S. has, in general, shown a lack of interest in tapered piles although there have always been localized exceptions to this. The reasons appear to be twofold.

First, timber piles, which undoubtedly supported countless older road-related structures such as bridges and viaducts, have long been felt to be outdated technology due to their relatively low per-pile ULS total resistance compared to any number of DFE alternatives as well as other factors.

Second, the various PCC-filled tapered steel shell and pipe piles have, with justification, been seen as proprietary and thus problematic from a contractual perspective. This latter aspect readily explains the extent to which H-piles, which have long been viewed as the epitome of a generic pile in U.S. practice, have been and continue to be used in the U.S. for road-related projects, even in applications where their technical-and cost-effectiveness could legitimately be called into question.

The net result is that although the FHWA and other government agencies have supported extensive research, including instrumented static load tests, of many types of DFEs over the last several decades, none of this has benefited tapered piles directly or explicitly.

This is because government-supported R&D of DFEs had typically been only for technologies that are deemed to be generic.

Moving on from the issue of data availability, it is important to note that the quality of interpreted data from any instrumented static load tests on tapered piles is extremely important. In fact, quality is arguably much more important than the quantity of data per se. In the writer's opinion, having poor data is always much worse than none at all. This is because poor data provides a false sense of knowledge that can result in misleading or incorrect conclusions that might compromise public safety.

In the context of tapered piles, the issue of data-quality means that the raw measured results in any instrumented static load test need to be carefully evaluated and empirically corrected as necessary for residual loads so that accurate estimates of depth-wise resistance can be determined. This is something that Nordlund pointed out as far back as his original 1963 paper and was confirmed more recently by Fellenius et al. (2000) in their assessment of two *Monotube* piles installed at JFKIA in the mid-1990s. Fellenius et alia also made it clear that this correction is quite complicated and detailed and is much more subjective for tapered piles compared to constant-perimeter piles due to the geometry of tapered piles.

The consensus derived by the writer from the contents of Nordlund (1963) and Fellenius et al. (2000) is that this correction for residual loads is very important for all types of PCC-filled tapered steel shell- or pipe piles driven entirely or predominantly into coarse-grained soil. The full extent to which this also occurs with other types of tapered piles (timber, precast-PCC) or with fine-grained soils is unknown at this time although the writer's appraisal of the Gregersen et al. (1973) case history involving precast-PCC piles that is presented in Appendix H of this monograph does provide some insight.

It appears that the failure to correct for residual loads most affects the inferred relative apportionment of ULS resistance between shaft and toe to the extent that raw, uncorrected results often tend to imply little-to-no toe resistance, with Fellenius et al. (2000) presenting a very typical example of this. In reality, given the relatively modest toe diameter of most tapered piles (a nominal 8 inches (200 mm) is typical for timber in terrestrial applications as well as for both the *Monotube* and *TAPERTUBE* piles), the percentage of ULS resistance carried by the toe relative to the entire shaft is surprisingly high as discussed subsequently.

The only enhanced set of tapered-pile field data that the writer has found to date that meets the above-described criteria (and only minimally at that) for making depth-wise parameter assessments along the pile shaft was presented by Fellenius et al. (2000) for two partially tapered steel pipe piles. This case history involved two essentially identical Type 5NJ8x14 *Monotube* piles that have a taper angle, $\alpha = 0.57°$. Each pile was driven to an embedded length of approximately 20 metres (66 ft) below the ground surface at JFKIA in 1994[118]. The piles were driven approximately 8 metres (25 feet) apart so can reasonably be assumed to have identical subsurface profiles yet each act independently of the other under externally applied axial-compressive load.

As discussed in Horvath (2015), there were, unfortunately, installation and testing issues involving both piles that compromised and limited the utility of the reported results, especially for one of the two piles. In addition, the writer had to scale the CPTu and static load test load-settlement data from the published figures. This made the writer's interpretation of these data less than ideal compared to having access to the original data in tabular or digital form.

---

[118] Fellenius et al. (2000) used SI units primary in their paper. This is maintained for the discussion of these piles in this monograph except with regard to the dimensions of the *Monotube* piles that are exact in Imperial units.

Nevertheless, useful results were still obtained in the writer's opinion. Thus, it is of interest to present the writer's assessment of these results in detail in this monograph as the majority of the writer's information has not been published previously. In addition, the assessment process that is illustrated is generic so can be used by anyone to evaluate other piles as well as to evaluate resistance-forecasting methodologies other than those considered here. This assessment can be found in Appendix A.

5.6.1.3 Direct Methods

5.6.1.3.1 Introduction and Background

From the very beginning of the development of direct resistance-forecasting methodologies for deep foundations in the 1950s, there has been an attraction to linking site-specific measurements obtained using a penetrometer-type site characterization tool to the resistance of DFEs. Not only were such tools the dominant in-situ testing devices at that time, they were then, and remain to the present, intuitive for applications with DFEs. This is not surprising given the obvious geometric similarities between penetrometers in general and DFEs subjected to axial-compressive loading.

What is interesting is that forecasting methodologies based on both the SPT and cone penetrometer (a basic CPT initially) emerged contemporaneously in the 1950s. This is surprising as at that time cone penetrometers were neither widely known nor widely available outside of the Netherlands where they were pioneered in the 1930s and developed and refined in the years following. In fact, well into the latter decades of the 20th century cone penetrometers were most often referred to colloquially in the U.S. as the 'Dutch cone' and considered to be a novelty tool used mainly in the Low Countries[119]. However, it was not until the development and global spread of the CPTu in more-recent decades that resistance-forecasting methodologies based on the cone penetrometer have advanced to the point where they now dominate direct methodologies for resistance forecasting for all types of DFEs.

Despite the extensive research into CPTu- and, more recently, sCPTu-based resistance-forecasting methodologies in recent decades, a distinguishing feature of direct methods for resistance forecasting for DFEs is that they have either ignored tapered piles completely or dealt with them essentially as an afterthought using the donut-on-a-stick model that was discussed in Chapter 4. As already discussed, this model has neither a physical basis in reality nor a body of supportive published research to justify its use. Rather, it is merely an arbitrary, unproven way (albeit one that is visually and intuitively satisfying) to get direct methodologies that were developed for DFEs with a depth-wise constant perimeter to produce results for tapered piles.

In view of the ease-of-use attraction that direct methods have relative to indirect methods, and in consideration of the shortcomings in the current state of knowledge for direct methods, there is both incentive and opportunity for improvement with regard to direct methods as applied to tapered piles. Some initial thoughts and suggestions in this regard by the writer are made in the following sections.

---

[119] There was significant regionality to in-situ testing devices for much of the 20th century based on the country or countries of origin and subsequent R&D, e.g. the pressuremeter (PMT) in France and later the U.K.; the dilatometer (DMT), for many years called the 'Marchetti dilatometer', in Italy; etc. To some extent, this reflected the fact that certain in-situ testing devices were well-suited for the subsurface conditions of a given country or region, so there was a tendency initially to invest in R&D to support their development and use in that country or region.

5.6.1.3.2 Meyerhof Method (Original 1956 Version)

In 1956, Meyerhof published a paper in the American Society of Civil Engineers (ASCE) *Journal of the Soil Mechanics and Foundations Division* that addressed the broad subject of using the measured results from various types of penetrometer soundings for estimating both the USD and ASD resistance of both shallow foundations (spread footings) and pile foundations bearing on or in coarse-grained soil[120]. He considered the following three in-situ testing devices, and measured result(s) from each in empirical algebraic relationships for the shaft and toe resistances of piles:

- The SPT N-value. Although Meyerhof was not specific, given the timeframe of his paper and using the same logic presented in the earlier discussion of Nordlund's original circa-1960 work, it can be inferred that these would be N_{field} values and best correlate to what would now be called N_{45}.

- What Meyerhof called a *static cone penetration test* (SCPT) that was what we would now call the CPT. Given the timeframe of Meyerhof's work, this was more likely the original mechanical version although early versions of electric cones were in use at that time. Meyerhof used both the uncorrected tip resistance, q_c, as well as the sleeve friction, f_s, for his resistance correlations.

- What Meyerhof called a *dynamic cone penetration test* (DCPT) that was apparently a device with a solid conical tip that had the same geometry and dimensions as the CPT of that era (i.e. 60° apex angle and 10 cm² projected tip area) that was impact-driven, not pushed as with the CPT, into the ground in the same manner as the SPT and with the same drive system and theoretical delivered energy (350 ft-lb/476 J). Blowcounts on the DCPT were recorded in the same 6-inch/150-mm interval manner as with the SPT. Meyerhof reported that in the ground conditions he studied, the DCPT yielded blowcounts (expressed as the integer number of blows per foot (300 mm) of penetration) that were approximately twice those recorded in the SPT, i.e. 2 x N_{45}. This was likely due, in part at least, to the fact that the DCPT had a solid tip so no soil could enter the interior of the sampler as with the SPT.

As an aside, Meyerhof also presented suggested correlations between SPT N-values (inferred here to be N_{45} values as noted above) and CPT q_c values. This may be the first attempt to do so, at least in the English-language published literature.

Of relevance to this monograph is that Meyerhof explicitly addressed using the empirical correlations he presented for resistance-forecasting of tapered piles. He recommended that the effects of taper be included by increasing the toe, not shaft, resistance. Specifically, the cross-sectional area of the toe was to be taken to be the cross-sectional area of the pile at a distance one-third of the embedded length of the tapered portion of the pile up from the toe. Note that this is broadly similar to the logic of Fellenius' donut-on-a-stick model discussed in Chapter 4.

Meyerhof did not present any evidence, either theoretical or using a comparison between forecast and measured results, in support of this empirical approach to dealing with tapered piles. However, given Meyerhof's professional reputation, one would like to believe

---

[120] This paper was reprinted in Meyerhof (1982) which is the reference used by the writer when researching this monograph.

that he had some basis for making this recommendation, even though that basis was not expressed in published work in any explicit manner.

5.6.1.3.3 Meyerhof Method (Revised 1975 Version)

Meyerhof revisited the subject of piles in coarse-grained soil approximately 20 years later when he presented ASCE's 1975 Terzaghi Lecture. A paper based on this lecture was published by ASCE in 1976 in the *Journal of the Geotechnical Engineering Division* and again a decade later in a compendium volume of early Terzaghi Lectures. Once again, the writer used Meyerhof (1982) as the specific reference for this lecture-based paper.

Of relevance to this monograph is that as he had done earlier in 1956, Meyerhof again explicitly addressed tapered piles in the revised (1975) empirical relationships that he presented for resistance-forecasting for DFEs in general. However, his 1975 recommendation for tapered piles was conceptually very different than that of 1956. Specifically, the taper benefit was to be included by increasing the shaft, not toe, resistance as previously. He suggested unilaterally using ULS unit shaft resistances that were 1.5 times that of a constant-perimeter pile. Thus, the taper benefit, as defined in this monograph, was assumed to be a constant value of 1.5.

Note that this was a one-size-fits-all approach that does not take taper angle into consideration. For comparison, the *Monotube* Type J (taper angle, α, = 0.57°) piles studied by Fellenius et al. (2000) and evaluated extensively by the writer in Appendix A had taper benefits evaluated several different ways that were of the order of three, i.e. approximately twice Meyerhof's recommendation. Again, Meyerhof did not present any supporting information or evidence for his revised (1975) recommendation for taper benefit or explain why he took a completely different approach to dealing with taper benefit, i.e. shaft vs. toe.

5.6.1.3.4 Original (Eslami-Fellenius) UniCone Method

While still employed full-time in academia at the University of Ottawa, then-Professor Bengt H. Fellenius apparently supervised and guided a doctoral student, Abolfazl Eslami, who developed a CPTu-based direct method for forecasting the ULS resistance of piles that was named the *UniCone Method* (Eslami and Fellenius 1997). Fellenius continues to include this method in his current publications (e.g. Fellenius 2019) although the UniCone name has been dropped and Fellenius now refers to it as the *Eslami and Fellenius Method*. However, for reasons that will become clear shortly, this method is referred to hereinafter in this monograph as the *Original UniCone Method*.

In the Original UniCone Method, the ULS shaft and toe unit resistances, $r_{s(ult)}$ and $r_{t(ult)}$ respectively, are both correlated to a parameter defined as q_E (in some literature this parameter is written q_e). This parameter is the corrected (for localized differential porewater pressure) CPTu tip resistance, q_t, that has been further corrected for global porewater pressure ($q_E = q_t - u_2$) so is essentially the net effective stress acting on the cone tip (hence the subscript 'E' or 'e').

For toe resistance, the value of q_E used for resistance forecasting is averaged over a prescribed depth above and below the toe, a common analytical tactic used for evaluating toe resistance of DFEs as well. In this case, Eslami and Fellenius specified "$+8d_t$ to $-4d_t$", i.e. from eight toe diameters (d_t)[121] above (+) the toe down to four toe diameters below (-) the toe. Non-dimensional coefficients, C_s and C_t respectively, are used to correlate the shaft and toe

---

[121] Depending on the specific publication, alternative notations such as d or d_b are sometimes used.

unit stresses to q_E and are thus the empirical 'heart' of the methodology. The C_s (shaft) parameter is particularly complex as it depends on both soil and DFE types.

Relevant to the topic of this monograph is that tapered piles are not explicitly considered in the Original UniCone Method. Presumably, the generic donut-on-a stick model is intended to be used with tapered piles.

It is of interest to note that to the present, which is more than 20 years after initial publication of the Original UniCone Method in Eslami's 1996 doctoral dissertation, Fellenius suggests that the $r_{s(ult)}$ and $r_{t(ult)}$ values calculated using the Original UniCone Method be used to back-calculate β and N_t (shaft and toe parameters respectively) for the Original β Method. The stated purpose of this is to add to, and presumably enhance and improve, the database of β and N_t values. This implies that Fellenius feels that indirect resistance-forecasting methodologies based on effective stress such as the Original β Method are still a viable approach to use for both resistance forecasting and as an analytical complement or supplement to direct methodologies.

5.6.1.3.5 Modified (Niazi) UniCone Method

Prof. Fawad S. Niazi devoted an entire chapter in his doctoral dissertation (Niazi 2014) to developing what he variously and synonymously referred to as a "modified" or "enhanced" version of the Original UniCone Method[122]. Given that the outcome of Niazi's efforts was, in reality, a wholesale reworking of the UniCone concept, in the writer's opinion 'modified' is clearly the more accurate term to use here. Consequently, in this monograph Niazi's version of the UniCone concept is referred to as the *Modified UniCone Method*.

The primary intent of the Modified UniCone Method appears to be to directly link the evaluation of the dimensionless shaft- and toe-resistance correlation coefficients, which were renamed C_{se} and C_{te} respectively by Niazi, to parameters that can be evaluated using the measured data from CPTu soundings. Niazi used the *Soil Behavior Type (SBT) Index*, I_c, as the primary parameter for this[123]. In concept, this is a significant improvement over the Original UniCone Method as it removes the subjectivity of assuming discrete values of the UniCone correlation coefficients based on soil type as is necessary with the Original UniCone Method. However, the DFE type is still a variable in the Modified UniCone Method and thus there are correction factors (simple constants) for DFE typology that need to be used as well.

The downside of the Modified UniCone Method is that in order to develop empirical relationships for seamlessly and continuously linking C_{se} and C_{te} to I_c, Niazi could not make use of more than two decades' worth of data and experience that are presumably reflected in the current version of the Original UniCone Method, e.g. as it appears in Fellenius (2019). Rather, Niazi had to rely on a totally new database that was assembled by him for this purpose and presented in Niazi (2014). Given the fact that this new database had to cover the full range of soil types as well as DFE typology necessary to develop the Modified UniCone Method, there was a limit as to what was included. Thus, there is clearly room for future

---

[122] A summary of the outcome of Niazi's work on this topic is presented in Niazi and Mayne (2015a).

[123] Robertson, who pioneered the concept of estimating 'soil behavior type' as opposed to 'soil type' as had been done historically with cone penetrometer data, subsequently suggested that the *Normalized Soil Behavior Type (SBTn) Index*, I_{cn}, be used instead of I_c. However, because Niazi used I_c in developing his correlations, this parameter must be used with the Modified UniCone Method to be consistent. In any case, limited research by the writer (Horvath 2015) showed that there is not much difference numerically between I_c and I_{cn} for a given set of CPTu data, at least when coarse-grained soils are involved. It is also of interest to note that it is somewhat more difficult to calculate I_{cn} compared to I_c as an iterative procedure is required for the former but not the latter.

improvement to the Modified UniCone Method by expanding on the relatively limited database developed by Niazi.

A small detail but one worth noting to avoid errors in usage is that Niazi sometimes expressed the C_{se} parameter for shaft resistance as a percent (which is technically a dimension) and sometimes in its basic, dimensionless decimal form. The reason for using the percent alternative appears to be a subjective issue based on visual appearance in a table or graph. The value of C_{se} is always less than one and expressing it as a percent makes it greater than one which tends to result in graphical axes that are easier to read. In any event, the point being raised here is that care is required when using the Modified UniCone Method to use a numerical form of the C_{se} parameter that is consistent within a given set of calculations.

As an aside, in the process of creating the Modified UniCone Method, the evaluation of the toe parameter, C_{te}, was made substantially more complicated than the evaluation of C_t in the Original UniCone Method where $C_t = 1$ is used in most cases (Fellenius 2019) although Niazi did retain the $+8d_t$ to $-4d_t$ averaging of the original version. However, because the entire C_{te} evaluation process lends itself to solution using a spreadsheet (or application-specific software if desired), the additional computational effort has no practical significance.

As noted above and illustrated in detail in Niazi (2014), an entirely new database of DFEs was used to create the empirical relationships for C_{se} and C_{te} in the Modified UniCone Method. Unfortunately, it does not appear that tapered piles were included in this database. Moreover, because the statistical assessments that resulted in the empirical equations for C_{se} and C_{te} were done for specific types of DFEs, it is not clear that using these empirical relationships together with the donut-on-a-stick model would be even an approximate way to apply the Modified UniCone Method to tapered piles. This means an alternative approach needs to be created and this is discussed in the following section.

5.6.1.3.6 Extended UniCone Method

Although neither the Original nor Modified UniCone Method explicitly considers tapered piles, there is no reason why the generic UniCone concept cannot be extended to tapered piles explicitly rather than resorting to using an unproven analytical gimmick such as the donut-on-a-stick model (which, as noted in the preceding section, cannot be used with Niazi's Modified UniCone Method in any event). In this section, suggestions are made for explicitly and rigorously extending the generic UniCone concept to tapered piles.

The writer refers to this proposed extension of the UniCone concept to explicitly include tapered piles as the *Extended UniCone Method*, with the shaft- and toe-resistance correlation coefficients referred to as C_{sx} and C_{tx} respectively to differentiate them from the coefficients defined previously for the Original and Modified UniCone Methods.

Note that the primary way in which C_{sx} is expected to differ from C_s and C_{se} defined previously is that taper angle needs to be considered in some fashion along with soil type and pile typology (in this case really the pile material). Whether taper is considered by using taper angle alone in some way, or taper angle together with some influence of pile diameter in general or toe diameter in particular as well, remains to be seen.

As an aside, the genesis of the latter comment concerning the potential influence of pile or toe diameter on taper benefit is the Fellenius et al. (2000) case history involving two *Monotube* piles installed at JFKIA that is evaluated in detail in Appendix A. The interpreted shaft-resistance distributions for these piles were evaluated by the writer using several different (but theoretically related) analytical parameters (r_s, r_s/σ'_{vo}, K_h, K_h/K_o, β, β/K_o) that are associated with various indirect resistance-forecasting methodologies such as the writer's Modified Nordlund Method and several versions of the β Method. As can be seen in several figures in Appendix A that relate to these assessments, it appears that the actual taper benefit, i.e. the relative increase in radial stress or lateral earth pressure acting along the pile shaft, for the same taper angle is not depth-wise constant as is often assumed but increases in a nominal linear fashion with decreasing pile diameter.

As discussed at length in Appendix A, this behavior is theoretically plausible when viewed from the perspective of the hypothesized cylindrical-cavity expansion (CCE) as being the operative physical model for the shaft-resistance mechanism of wedging that occurs when a tapered pile is externally loaded in axial compression after installation. The reasoning for this begins by recognizing that when a tapered pile is loaded in axial compression, for a given magnitude of pile settlement the radial displacement of the shaft-soil interface, and concomitant apparent increase in shaft diameter, is, as a first-order approximation, uniform along the tapered section of a pile, assuming that the taper angle is depth-wise constant (in reality, it will vary slightly due to both elastic compression and the Poisson Effect of the pile shaft). This is depicted qualitatively in Figure 4.1 that applies to both the cylindrical-cavity creation (CCX) that occurs during pile installation and the CCE that occurs during post-installation external loading in axial compression.

Although the <u>absolute increase</u> in apparent shaft diameter within the tapered section of the pile is approximately uniform in magnitude, the concomitant <u>incremental increase</u> in apparent shaft diameter relative to the original shaft diameter is never uniform. Rather, it varies depth-wise along the tapered section of the pile. So, for example, at the top of the tapered section the <u>relative increase</u> in apparent shaft diameter is less than at the bottom of the tapered section even though the <u>absolute increase</u> is (approximately) the same.

A simple numerical example is useful to illustrate the point being made here. Using the *Monotube* piles discussed in Appendix A as an exemplar, a 1-inch (25 mm) settlement of the tapered section would produce a uniform apparent increase in pile diameter of 0.20 inches (510 μm) along the entire tapered section. However, the <u>relative</u> increase in apparent pile diameter would be 0.25% at the bottom of the tapered section (= pile toe) but only 0.14% at the top of the tapered section (= transition to the constant-diameter section/extension).

Thus, because CCE theory relates the increase in radial stress acting on the cavity wall (shaft-soil interface in this case) to the <u>relative</u> increase in cavity radius, it stands to reason that the increase in radial stress will be greater where the relative radius increase is greater, i.e. at the toe. While this hypothesis is consistent with results presented in Appendix A for the Fellenius et alia case history, additional research is clearly needed to demonstrate that this behavior occurs with all tapered piles.

Returning now to the writer's proposed Extended UniCone Method, in order to create and maintain a parameter-identity separate from that of both the Original and Modified UniCone Methods, a new toe-resistance correlation coefficient, C_{tx}, is defined to provide for a difference from C_t and C_{te}. This is to allow for the possibility that taper influences toe resistance in addition to shaft resistance as has been postulated, but not conclusively proven, by Manandhar and Yasufuku.

As noted earlier in this chapter, further research is needed to determine conclusively whether or not this is the case. In the writer's opinion, the small-scale, 1-*g* model-testing results presented by Manandhar and Yasufuku (Manandhar and Yasufuku 2012, 2013; Manandhar et al. 2013) are not suasive in this regard given:

- the overall scaling issues inherent with small-scale geotechnical models under 1-*g* loading conditions, i.e. the bowling-ball effect;

- the unrealistic geometry of some of the model piles used; and

- the way that the model piles were installed within the soil which was a manner that bears no resemblance to actual deep-foundation installation methods.

Unfortunately, it will take a substantial amount of instrumented pile load test data on a variety of tapered piles to create a sufficient database that will allow a meaningful statistical assessment (similar to what Niazi did for the Modified UniCone Method) to produce empirical relationships for C_{sx} and C_{tx}. However, to at least explore the Extended UniCone Method in a preliminary way, it is of use to analyze the data for the two piles at JFKIA that were discussed by Fellenius et al. (2000). This assessment is presented in Appendix A.

5.6.1.3.7 Togliani Method

Gianni Togliani developed a direct resistance-forecasting methodology earlier in the 21st century that is noteworthy because it was formulated specifically with tapered piles in mind. To the best of the writer's knowledge, what is referred to hereinafter as the *Togliani Method* is unique in this regard although Togliani implied that elements in his analytical methodology that are related to taper may have had some influence from a circa-1990 published work titled "*Manuale dei Piloti*" by Ferruccio Gambini (Togliani 2010). The writer was unable to independently verify this claim but, taken at face value, it would imply that some type of tapered pile (specifically, a continuously tapered, precast-PCC 'pipe' pile with an enclosed toe) has been in commercial use in Italy since the late 20th century. This type of pile was noted in Chapter 3.

An accessible reference, and the one used by the writer, that contains the numerous empirical relationships that taken together comprise the Togliani Method is Togliani (2010) although there are indications that this method may have been published as early as 2008. Unfortunately, Togliani (2010) does not provide any information related to how the method was developed; its accuracy as a forecasting tool by using comparisons to ground-truth; and other technical details that would allow an independent assessment and understanding of the methodology and instill confidence in its use by anyone other than Togliani.

The key aspects of the Togliani Method that can be deduced from Togliani (2010) are that it uses only the most basic, uncorrected results from a cone-penetrometer sounding, q_c and f_s, plus one derived parameter, R_f, for both the shaft and toe resistance forecasts. Therefore, the method can be used with older CPT data if desired.

In addition, the method is built around the donut-on-a-stick model so that the forecast shaft resistance of the tapered portion of a pile is expressed as the sum of two components:

- *Shaft Resistance, R_s,* which in this context is limited to the resistance provided by the constant-diameter 'stick' part of the model and

- *Taper Resistance, R_c,* which in this context is the resistance due to the 'donut' part of the model.

Any constant-perimeter portion of a pile would, of course, only have the Shaft Resistance component

Some additional elements of note for the Togliani Method are:

- All calculated resistances are ULS values.

- The empirical equations for the Taper Resistance component of tapered shaft resistance include toe diameter as a variable in a manner that is separate from, and in addition to, calculating the area of the 'donut'. Interestingly, this is in consonance with the observation made earlier in this chapter (as well as in Chapter 4 and Appendix A) that taper benefit appears to be related to both taper angle and toe diameter, not just taper angle alone.

- The cone-penetrometer data used for calculating toe resistance are averaged over the same $+8d_t$ to $-4d_t$ distance (with d_t = toe diameter) as used in both the Original (Eslami-Fellenius) and Modified (Niazi) UniCone Methods.

- Togliani assumed that the Shaft Resistance (as he narrowly defined it), Taper Resistance, and toe resistance were each fully mobilized, i.e. reached their ULS values, at different magnitudes of settlement. That the overall shaft resistance (= 'Shaft' + 'Taper' components) is mobilized more rapidly than toe resistance is well-known and widely accepted. However, with regard to the overall shaft resistance, it is not clear how one can separate out the settlements required to mobilize the stick ('Shaft') and donut ('Taper') components of the donut-on-a-stick model that Togliani used, and he gave no supporting information for his statement that the stick and donut components mobilized at different rates. Rather, he simply presented it as fact. Some inference as to Togliani's perspective as to the relative magnitudes of settlement required to mobilize each of these three artificial resistance components can be drawn from his stated opinion that an 'allowable', i.e. ASD, pile capacity based solely on (geotechnical)-resistance considerations (i.e. no consideration of structural resistance) would be the sum of 50% of the 'Shaft' (stick), 40% of the 'Taper' (donut), and 33% of the toe ULS resistances. This implies that the taper benefit (technically the shaft resistance due to wedging) requires somewhat more settlement to mobilize compared to the basic shaft resistance due to sliding friction but less settlement than toe resistance. In any event, he provided no supporting evidence as to how he arrived at this apportionment of resistance components.

There are two relevant issues to address with regard to the efficacy of the Togliani Method as a potentially viable analytical methodology for use with tapered piles:

1. The <u>absolute accuracy</u> of the ULS total resistance that is forecast using the method.

2. The <u>relative accuracy</u> of the depth-wise distribution of ULS resistance that is forecast using the method.

The former is essential if the Togliani Method is to be considered for use in routine practice. The latter is largely of academic interest but there are practical implications as well. This is because the ability to correctly forecast at least the relative depth-wise distribution of resistance reflects on the broader issue of the efficacy of the donut-on-a-stick model for use with resistance-forecasting methodologies, both indirect and direct, when applied to tapered piles. As noted earlier in this chapter, the generic use of the donut-on-a-stick model to replicate the wedging mechanism that occurs with tapered piles has been actively promoted by Fellenius and others for some time now.

With regard to the primary issue of <u>absolute</u> overall forecasting accuracy, prior to writing this monograph the writer performed a limited assessment of the Togliani Method for several timber and *Monotube* piles at JFKIA, with a summary of the results published in Horvath (2015). In each case analyzed, the Togliani Method produced a forecast of ULS total resistance that was greater than any other resistance-forecasting method studied and greater, usually by a substantial margin, than the measured (in a static load test) maximum total resistance. Note that all of the piles used for this assessment were subjected to conventional, i.e. uninstrumented, static load tests and loaded to magnitudes of pile-head settlement that were judged to reasonably represent the geotechnical ULS.

With regard to the secondary issue of the <u>relative</u> accuracy of resistance distributions and its reciprocal, the axial-compressive force distribution within the pile shaft, prior to publication of this monograph the writer studied this issue for one *Monotube* pile at JFKIA that was designated Pile P-5A, with the results presented in Horvath (2015). This pile was chosen as it had the same overall dimensions and approximately the same embedded length of the two *Monotube* piles designated Pile #2 and Pile #3, that were the focus of the study presented by Fellenius et al. (2000) and for which ground-truth in the form of depth-wise variations in the axial-compressive forces with the piles and concomitant distributions of resistance along the pile shafts and at the pile toes was available[124].

The results of this earlier assessment of Pile P-5A are shown in Figure 5.1 that is an updated version of a figure that appeared originally in Horvath (2015). Depicted in this figure are the relative (normalized in each case to Q_{max}, the maximum applied or forecast load for each pile[125]) distributions of load, Q, for the following four cases:

- The <u>measured results</u> for the two piles (designated Pile #2 and Pile #3) discussed in Fellenius et al. (2000). How these plotted results were developed by the writer and the approximations involved are discussed in detail in Appendix A.

- The <u>calculated results</u> for Pile P-5A obtained using the writer's Modified Nordlund Method that was discussed earlier in this chapter.

- The <u>calculated results</u> for Pile P-5A obtained by the writer using the Togliani Method.

In all cases, the transition from the constant-diameter upper section (extension) to the tapered lower section occurred at a depth of approximately 40 feet (12 m).

Note that the two calculated results for Pile P-5A were based on a site-characterization assessment made using a CPT sounding (designated CTP-5) that had been performed in the vicinity of this pile, not the CPTu sounding contained in Fellenius et al. (2000) that was presumably performed in the vicinity of their Pile #2 and Pile #3. However, based on extensive research by the writer that has included a review of dozens of CPT, CPTu, and sCPTu soundings performed at JFKIA between the late 1980s and up to the present, the subsurface conditions within the Central Terminal Area (CTA) of JFKIA where all of the piles in question (P-5A, #2, #3) were driven are exceptionally uniform. Consequently, for the purposes of the assessment presented in Figure 5.1 any differences between these two cone-penetrometer soundings is not believed to be significant.

---

[124] For the purposes of the writer's 2015 white paper, Pile P-5A was chosen as an analytical stand-in for Piles #2 and #3 in Fellenius et al. (2000) as the quality of both the cone-penetrometer and static load test data available to the writer were significantly better for Pile P-5A compared to Piles #2 and #3. The issue of cone-penetrometer data quality was especially important for making as accurate and fair an assessment of the Togliani Method as was possible.

[125] For the two sets of calculated results for Pile P-5A, $Q = Q_u$, i.e. the ultimate (ULS) load or capacity of the pile, as per Figure 2.1. However, as discussed in Appendix A, for the measured results for Piles #2 and #3, Q is simply the maximum reached in each static load test although these maximums are likely close to Q_u, especially for Pile #2.

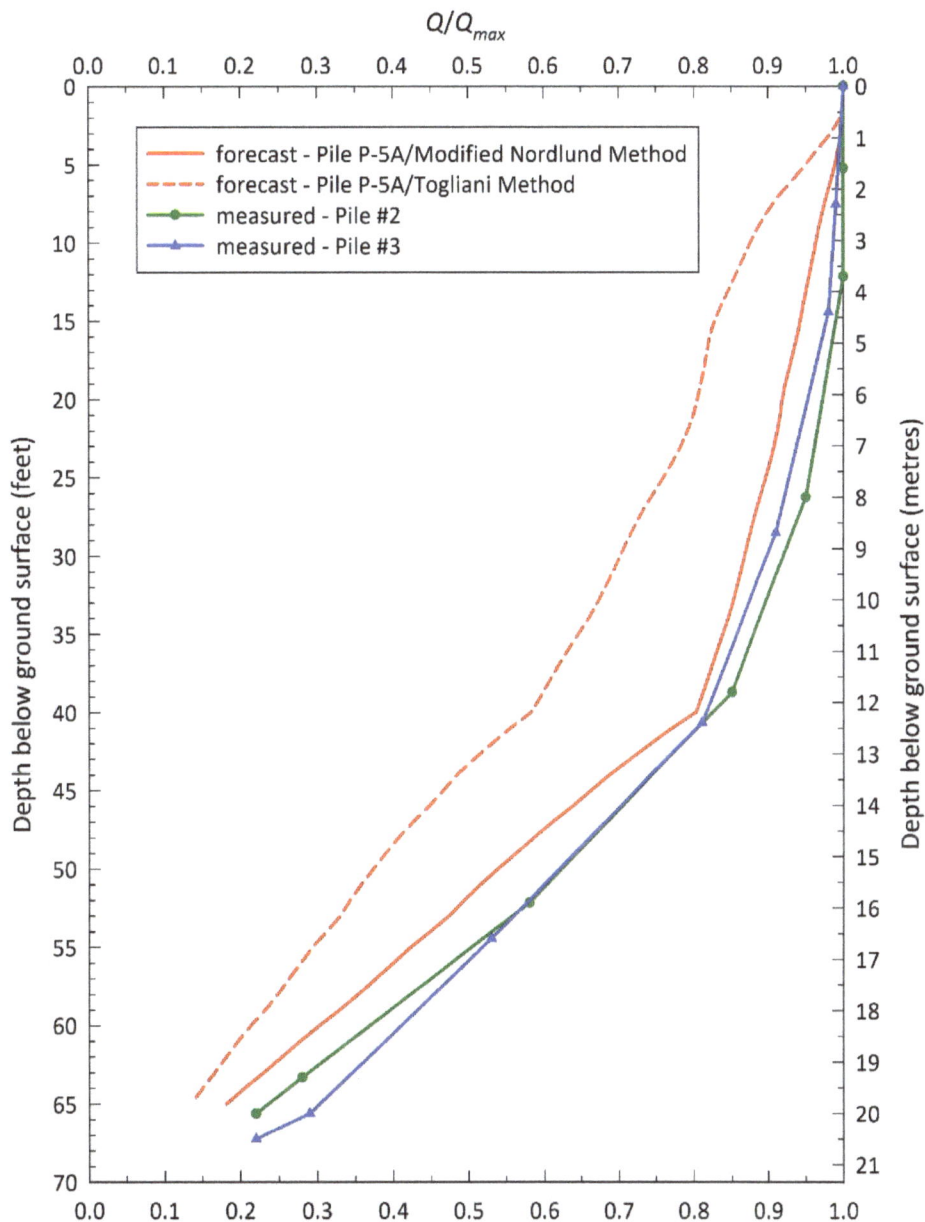

Figure 5.1. Depth-Wise Distribution of Forecast vs. Measured Axial-Compressive Forces in *Monotube* Piles at JFKIA [modified from Horvath (2015)].

With regard to the results shown in Figure 5.1, the following comments are made:

- Because the depth-wise variation in the axial-compressive force distribution within a DFE shaft is always the reciprocal or mirror-image of the depth-wise variation in the resistance that the ground provides to a DFE (see Figure A.1 in Appendix A for an illustration of this), corrected (for residual load) axial-force measurements made in instrumented DFEs (as in this case) are always an acceptable means for drawing conclusions with respect to resistance forecasting.

- The measured results for Pile #2 and Pile #3 are essentially identical despite problems that were encountered during the installation of these piles as is discussed in detail in Appendix A. Therefore, there is a high degree of confidence that the measure results are as 'correct' as can reasonably be obtained and are thus an acceptable reference standard or ground-truth for the forecast results.

- The forecast for the writer's Modified Nordlund Method are in good agreement with the measured results despite the fact that different piles were used. Subsequent to the publication of Horvath (2015) and in preparation of this monograph, the writer performed specific calculations using the Modified Nordlund Method for the two piles (Pile #2 and Pile #3) presented in Fellenius et al. (2000) and using the measured results from the CPTu sounding presented in that paper. These results are shown in Appendix A in Figures A.6 and A.7 and display the same good overall agreement with the measured results that are shown in Figure 5.1. This provides additional supporting proof that the writer's Modified Nordlund Method can capture the unique behavior of tapered piles that include both a tapered lower section and constant-diameter upper section (referred to as an 'extension' in Monotube product literature as discussed in Chapter 3). Specifically, there is a distinctive 'kink' in the axial-force distribution curve at the transition between the two sections of the piles (a depth of approximately 40 feet/12 m in Figure 5.1) that is the result of the significant relative increase in the ULS unit shaft resistance for the tapered section. In this case for the *Monotube* Type J piles that have the intermediate (taper angle, α, = 0.57°) taper, this taper benefit is of the order of three (average) as discussed in Appendix A.

- The forecast results for Pile P-5A using the Togliani Method did not produce this signature kink in axial-force distribution. In fact, the forecast load distribution varied almost linearly with depth from head to toe indicating no meaningful difference in soil resistance between the tapered and constant-diameter sections. This suggests that the donut-on-a-stick model, however intuitively appealing and easy to use it may be, does not adequately capture the distinctive wedging behavior of tapered piles.

- The forecast results for Pile P-5A using the writer's Modified Nordlund Method produced a forecast of relative toe resistance (approximately 20% of Q_{max}) that was closer to that measured for Pile #2 and Pile #3 than the forecast using the Togliani Method.

- Given the relative lengths of the three piles (approximately 65 feet/20 m) and the relatively modest toe diameter of each (8 inches/200 mm), the relative magnitude of total resistance provided by the toe (approximately 20% of R_u) is perhaps surprisingly high. This indicates that resistance-forecasting methodologies for tapered piles need to pay attention to the toe-resistance component in addition to the shaft-resistance component.

For the sake of completeness, subsequent to the publication of Horvath (2015) and in preparation for this monograph the writer performed an analysis using the Togliani Method for the Fellenius et alia Pile #3[126]. The outcomes from a site-characterization assessment made using the CPTu sounding contained in the Fellenius et alia paper were used for this analysis. This work was done to verify that the writer's earlier (2015) qualitative outcomes

---

[126] The Togliani Method was only applied to Fellenius et alia's Pile #3 as the behavior of Pile #2 is essentially identical (as can be seen in Figure 5.1) and, for reasons discussed in Appendix A, there were problems encountered during the driving of Pile #2 in 1994.

for the Togliani Method shown for Pile P-5A in Figure 5.1 were reproducible and thus indicative of generic results produced by the Togliani Method, at least for the piles and conditions considered at JFKIA. The outcomes from this more-recent assessment are contained in Appendix A and shown in Figure A.15.

The conclusion drawn from Figure A.15 is the same as that drawn from Figure 5.1, i.e. the qualitative match between the forecast made using the Togliani Method and the measured results for the depth-wise variation in axial-compressive forces in the pile and, by implication, the reciprocal results of the depth-wise variation of the resistance provided by the ground, is overall poor.

In summary and conclusion with regard to the Togliani Method, available information suggests that it be used with considerable caution until such time as more English-language information becomes available concerning its theoretical basis; the process by which it was developed; and a statistical assessment of forecast vs. measured ULS total resistance.

These comments also highlight the broader need to objectively assess the donut-on-a-stick model for use with other resistance-forecasting methodologies. Togliani is only the latest in a line of researchers who have proposed using this model for tapered piles for both indirect and direct resistance-forecasting methodologies. To the best of the writer's knowledge, no one has ever presented a theoretically sound argument to support using this model or, more importantly, provided an objective assessment comparing measured and forecast (using some specific resistance-forecasting methodology that incorporates the donut-on-a-stick model) results. Until this is done, all available evidence suggests that the donut-on-a-stick model is a poor substitute for explicitly and correctly modeling the wedging mechanism by which tapered DFEs in general develop resistance.

5.6.2 Settlement-Only Methods (Step II of Two-Step Procedure)

5.6.2.1 Background

The settlement of piles and, especially, pile groups (alternatively called *clusters* in some publications) was recognized as a design consideration from the earliest days of developing analytical methodologies for DFE resistance in the 1950s. This is not surprising as from the beginning of modern foundation engineering in the 20th century settlement of shallow foundations was recognized as a design consideration as well.

However, unlike with shallow foundations where settlement was soon found to almost always govern design (a fact that resulted in decades of research that continues to the present and has produced countless settlement-forecasting methodologies), routine practice for (driven) piles in terrestrial applications followed a different path[127]. The observation in routine practice that simply applying an ASD safety factor to an assumed ULS total resistance for a pile produced designs that performed 'acceptably' (to the point that almost to the end of the 20th century some design professionals simplistically believed that pile-supported foundations do not settle), at least in terrestrial applications, combined with the difficulty of forecasting pile settlement accurately resulted in a strength-based state of practice for piles that solely considers ULS total resistance.

---

[127] The need to explicitly consider settlements in design has long been recognized for piles in deep-water marine applications (because of the considerable pile lengths typically involved as well as the fact that the piles are typically an integral part of the superstructure) as well as drilled shafts (because of their relatively large diameters and the fact that toe resistance is only partially mobilized at service-load levels).

This strength-based focus of pile design remains to the present and, if anything, has only been reinforced and emphasized by the evolution to using LRFD as opposed to ASD in pile design. This means that the assumption in practice that there <u>must</u> be a unique, single-valued ULS total resistance for every pile is unquestioned as codes effectively require that such a resistance exists whether or not it actually does.

The following quotation from a recent publication by Dr. Bengt H. Fellenius succinctly puts this way of thinking, i.e. doing things simply because they always have been done this way, in perspective:

"In every science-oriented set of know-how, such as geotechnical engineering, there is a set of concepts held as true, never questioned, only amended and developed within the original framework. In a sense they are what Richard Dawkins ("The Selfish Gene", Oxford University Press, 1976) named 'memes', that is, self-replicating concepts, ideas, or styles that spread from person to person within a culture."

- from "Fallacies in Piled Foundation Design", Geotec Hanoi 2016

The genesis of Fellenius' comments is that for some years now there has been increasing appreciation and concomitant promotion of the fact that the consideration of settlement and the SLS should at least be a part of, if not the primary basis for, the routine state of practice for <u>all</u> DFE design. The primary reason for this is the ambiguity and uncertainty of defining a single-valued ULS toe (and, therefore, total) resistance in many cases, especially those involving friction/floating DFEs in coarse-grained soil, a fact that has been noted and discussed at length earlier in this monograph.

There are other compelling reasons for considering DFE settlement in routine design. This includes appreciation of the fact that performance-based, application-specific design often makes technical and economic sense as opposed to blind adherence to a one-size-fits-all design approach.

This new settlement-based focus for DFE design and its genesis are succinctly stated in the following quotation also attributed to Fellenius who, more than any other individual to date, has proselytized the routine consideration of DFE settlement and the SLS as opposed to ULS total resistance alone as the basis of the preferred approach to pile design:

"Don't get stuck in the capacity singularity, rise above the pack and let settlement issues govern your foundation design."

- From the Deep Foundation Institute 2nd Annual Osterberg Memorial Lecture in 2012, per Niazi (2014, Page 13)

Unfortunately, because designing piles based on strength and ULS considerations alone is one of the more-entrenched 'memes' of foundation engineering, it is unlikely that pile design will undergo a paradigm shift to a settlement-based focus anytime soon. Nevertheless, because settlement analysis of piles does represent a forward-looking concept, it is important to devote a portion of this monograph to discussing settlement analyses for tapered piles.

5.6.2.2 Overview

The settlement-related material presented and discussed in this monograph is limited to individual DFE settlement. The discussion of group/cluster settlements is a separate issue that is beyond the scope of this monograph. This is due to the fact that group settlement is:

- largely independent of DFE type;

- primarily a geomechanics issue in that it is largely governed by the compressibility of strata that underlie the DFE group; and

- already reasonably well understood and handled analytically, e.g. see Fellenius (2019), although there are intra-group effects and issues with respect to load distribution among DFEs in a group that are still in the process of being researched (Fellenius 2018).

Since the beginning of the 21st century, the topic of individual DFE settlement has seen a marked increase in research attention that addressed and successfully overcame the practical difficulties that had for decades restricted the use of theoretical solutions based on the theory of elasticity. As will be seen later in this chapter, this same, relatively recent research subsequently morphed and broadened in scope to not only include all types of DFEs but to provide for a much broader range of forecast outcomes.

The focus at this point of the monograph is on the initial outcomes of this relatively recent research that are referred to herein as *settlement-only methods* that are intended to complement and supplement any conventional strength-based analytical methodology, either indirect or direct, that produces a forecast of ULS total resistance. Essentially, the function of settlement-only methodologies is to 'fill in the blanks', i.e. produce a complete load-settlement curve between zero applied load and a ULS total resistance that is forecast separately in advance. Consequently, it is important to recognize that settlement-only methods are not stand-alone analytical methodologies and must always be used in conjunction with a traditional resistance-forecasting methodology. Furthermore, the accuracy of the final outcome of using a settlement-only method is largely dictated by the accuracy of the resistance-forecasting methodology with which it is coupled.

The writer previously (Horvath 2016b) defined this the *two-step procedure* as the ULS-resistance forecasting and settlement-curve generation are two distinct and theoretically independent steps in the overall process of generating a complete load-settlement curve. Furthermore, using this two-step nomenclature is useful as it inherently allows for generically defining a *one-step procedure* wherein resistance and settlement are forecast as part of a single, integrated, seamless analytical methodology. One-step procedures are presented later in this chapter and, as will be seen, represent the current state of art and reflect the future but still require considerable R&D to develop fully. On the other hand, the generic two-step procedure represents the current state of practice that can be used with ease and a degree of confidence in the present.

The settlement-only methods addressed in this monograph that can be used as Step II of a two-step procedure with any of the Step I resistance-forecasting methodologies presented earlier in this chapter fall into two broad categories of models for creating the desired load-settlement relationship:

- empirical and

- theoretical.

Individual models within these categories are discussed in the following sections.

5.6.2.3 Empirical Models

5.6.2.3.1 Overview

Empirical models for forecasting the settlement of tapered piles can be developed in a variety of ways. They can be based on some arbitrary assumptions that are supported by some underlying theoretical logic or they can be developed solely from a database of field observations. The latter is the approach taken decades ago in the U.S. for the FHWA-supported development of design methods for drilled shafts as it was learned early on that the relatively large diameter of drilled shafts often made settlement under service loads the controlling design factor, not ULS total resistance.

Discussed in the following sections are two empirical settlement-only models developed by the writer specifically with tapered piles in mind although both can be used with constant-perimeter piles as illustrated in Horvath (2015). Note that this presentation is not exhaustive as there may well be other empirical models already developed by others for tapered piles, or other empirical models that were developed for constant-perimeter piles that could be adapted to and adopted for tapered piles. Rather, the presentation here is more to show the process of how one can develop an empirical settlement model for tapered piles than it is to present a comprehensive summary of actual models that have already been developed for this purpose.

5.6.2.3.2 Tri-Linear Settlement Model

One of the tasks pursued during the writer's circa-2000 period of research into tapered piles, which was focused on the then-new *TAPERTUBE* pile, was development of a relatively simple settlement model for simulating the results of a traditional static load test to the geotechnical ULS that was more accurate and representative of actual load-settlement behavior than the basic bi-linear elastoplastic model that generations of civil engineers have applied to a wide variety of materials and systems. The development goal of this settlement model, hereinafter referred to as the *Tri-Linear Settlement Model*, was to be able to construct a first-order approximation of the load-settlement curve for a tapered pile that allowed for modeling the final unloading portion of a test curve if desired. There was an intended emphasis on making the model relatively easy to use with minimal effort so that it would be attractive to use in routine practice.

It is important to note that while this model was initially developed to complement and enhance the writer's Modified Nordlund Method for forecasting ULS total resistance (which the writer also developed during the same circa-2000 timeframe), the Tri-Linear Settlement Model can be used with any analytical methodology for resistance forecasting, either indirect or direct. Thus, this settlement model is completely general in this regard.

The assumptions made by the writer when developing the Tri-Linear Settlement Model were detailed in several papers (Horvath 2003c; Horvath and Trochalides 2004; Horvath et al. 2004a, 2004b) in which the model was presented and the forecast results compared to measured results from a conventional static load test that had been performed on a tapered pile during the same circa-2000 timeframe. These assumptions are restated here for the sake of completeness and ease of reference.

The Tri-Linear Settlement Model assumes that the basic load-settlement curve consists of three straight-line segments that are defined by the following three points. Note that in all cases the settlement is that of the head of the pile:

1. The origin, i.e. zero load and zero settlement initially.

2. A load that is equal in magnitude to the sum of:
    o   the peak-strength ULS shaft resistance within the constant-diameter portion of the pile,
    o   the peak-strength ULS shaft resistance within the tapered portion of the pile, and
    o   10% of the peak-strength ULS toe resistance
    is assumed to correspond to a settlement of 0.12 inches (3 mm) plus the calculated elastic compression of the pile shaft under that load. This is based on the observation that constant-perimeter shaft resistance is mobilized at approximately ⅛ inch (3 mm) relative pile-soil displacement and that approximately 10% of the ULS toe resistance is mobilized at that point. For lack of a better assumption, it was assumed that the peak resistance within the tapered portion of the pile would be mobilized at the same time that the peak resistance was mobilized along the constant-perimeter portion of the pile.

3. A load that is equal in magnitude to the sum of:
    o   the constant-volume (critical-state) strength ULS shaft resistance within the constant-diameter portion of the pile,
    o   the peak-strength ULS shaft resistance within the tapered portion of the pile, and
    o   the peak-strength ULS toe resistance
    is assumed to correspond to a settlement of 15% of the toe diameter plus the elastic compression of the pile shaft under that load. This is based on the rule of thumb suggested by several researchers in the past that a toe settlement equal to 15% of the toe diameter is necessary to mobilize the ULS toe resistance. Note that it was assumed that the shear strength did **not** reduce from the peak to constant-volume/critical-state value along the tapered portion of the shaft as it did along the constant-perimeter portion of the shaft.

Classical plunging load-settlement behavior, i.e. unlimited settlement under constant load, is implied beyond the last point. Alternatively, the load-settlement curve can be brought back to zero load at which point the assumed residual settlement is equal to 15% of the toe diameter of the pile. Note that this implies that all elastic compression of the pile shaft is recovered when the pile is unloaded.

Figure 5.2 is a reproduction of a figure that originally appeared in the circa 2003-2004 references by the writer and others that were cited above. It illustrates an exemplar application of the Tri-Linear Settlement Model (including the third-line-segment alternative of simulating the unloading portion of the load-settlement curve as opposed to simulating plunging behavior) together with the Simplified Nordlund Method to an early, prototype version of the *TAPERTUBE* pile at JFKIA[128]. Also shown in this figure for comparison are the actual load-test results for the pile.

---

[128] As noted above, the focus of the writer's circa-2000 research that was published in the 2003-2004 timeframe was the then-new *TAPERTUBE* pile. The application of the combination of the writer's Tri-Linear Settlement Model and the writer's Simplified Nordlund Method to several timber, *Monotube*, and *TAPERTUBE* piles at JFKIA was illustrated in Horvath (2015). In each case, a comparison to measured static load test results was also shown.

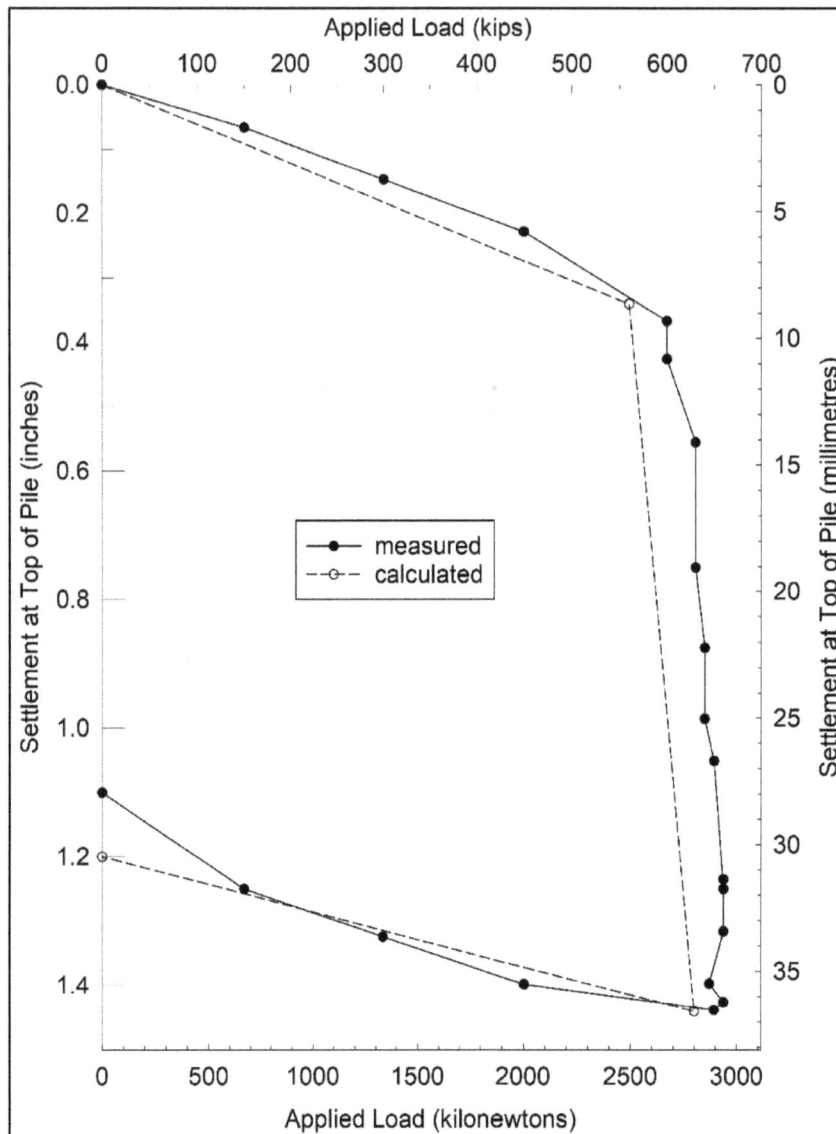

**Figure 5.2. Exemplar Application of Tri-Linear Settlement Model with
Modified Nordlund Method [previously published in Horvath (2003c) and elsewhere].**

While the overall agreement shown in Figure 5.2 is quite good, it should be kept in mind that any settlement-only model that is used as Step II of the two-step procedure is only as good at forecasting settlement as the match between forecast and measured ULS total resistances provided by the first (resistance-forecasting) step as was noted earlier. No settlement-only model can make up for deficiencies in the forecast ULS resistance. Obviously, the agreement in the case of the pile shown in Figure 5.2 was very good using the writer's Modified Nordlund Method and this was one reason why this particular pile was used as an illustration of the Tri-Linear Settlement Model settlement-forecasting methodology. However, subsequent analyses performed approximately a decade later on a wider range of tapered piles at JFKIA showed a wider range of accuracy (Horvath 2015).

As discussed in Chapter 4, subsequent to the original development of the Tri-Linear Settlement Model circa 2000 a lot has been learned via the published literature about the significance of residual loads on piles in general, more so than with any other type of DFE. Of particular relevance to this monograph is the fact that residual loads tend to be especially significant for:

- coarse-grained soils and

- tapered piles.

Both of these conditions are reflected in the exemplar application shown in Figure 5.2 as well as the several applications published subsequently in Horvath (2015).

Based on this new-found appreciation of residual loads in tapered piles and, more importantly, the profound effect that residual loads can have of the shape of the load-settlement curves measured in static load tests (Fellenius 2015), it is clear that the behavioral assumptions made to define the inflection points of the writer's Tri-Linear Settlement Model are subject to question. For example, a key assumption in defining the second point (the first point at the origin is trivial and obvious) is that 10% of the ultimate toe resistance has been mobilized. However, as discussed in Appendix A, for the two instrumented *Monotube* piles at JFKIA studied by Fellenius et al. (2000) it appears that toe resistance was 100% mobilized at the end of initial driving for both piles as the raw measured results indicated essentially no toe resistance during the subsequent static load test. Thus, how residual loads should be accounted for in a settlement-forecasting methodology is certainly a subject for future R&D.

There is actually a broader philosophical issue to consider and that is whether a settlement-forecasting methodology should aim to replicate the theoretical load-settlement behavior in the absence of residual loads (which a real pile would likely never exhibit) or the real-world load-settlement behavior that reflects residual loads, noting that for the same pile in the same soil conditions the residual loads may vary depending on any number of installation variables. As illustrated by Fellenius (2015), the difference between the hypothetical, no-residual-load load-settlement behavior and the actual load-settlement behavior when significant residual loads are present can be both qualitatively and quantitatively very different in two significant ways:

- The actual static load test may suggest a relatively well-defined plunging type geotechnical ULS when none exists in reality (note that the *TAPERTUBE* pile shown in Figure 5.2 may well be an example of this), simply because the toe resistance was already fully mobilized and the load test is largely reflecting shaft resistance only.

- The actual static load test tends to reflect much stiffer initial pile response than would exist in the absence of residual loads. This is not insignificant or trivial as most criteria used in practice for defining the geotechnical ULS in a static load test are settlement-based, a fact noted earlier in this monograph. Therefore, the artificially stiffer response due to residual loads will produce larger values of this implied ULS load than would be interpreted under theoretical load-settlement conditions in the absence of residual loads.

Resolving these issues is well beyond the scope of this monograph as the issues raised are philosophical in addition to technical, and thus require significant consensus rethinking of what is expected from the traditional static load test that has always been the gold standard of ground-truth for all types of DFEs. Consequently, this is a discussion that needs to be taken up by all stakeholders in deep foundations on a global basis.

5.6.2.3.3 Composite JSH-M&Y Settlement Model

A recent development by the writer that was first noted in Horvath (2015) is the *Composite JSH-M&Y Settlement Model* that combines elements of the writer's original Tri-Linear Settlement Model and the Enhanced Manandhar-Yasufuku Toe-Resistance Method for ULS toe resistance that was presented and discussed earlier in this chapter. To briefly review, the Enhanced Manandhar-Yasufuku Toe-Resistance Method uses a spherical-cavity expansion (SCE) model to produce an estimate of the ULS toe resistance (which, if used alone, would constitute the Basic Manandhar-Yasufuku Toe-Resistance Method) and complements and supplements this resistance with a load-settlement relationship based on a hyperbolic curve.

The Composite JSH-M&Y Settlement Model retains the writer's original assumption that peak shaft resistance is mobilized at a pile-head settlement of 0.12 inches (3 mm) plus elastic compression of the pile shaft. However, rather than assuming incremental levels of mobilized toe resistance and the concomitant settlement magnitudes at which these levels of toe resistance are achieved as in the original Tri-Linear Settlement Model, the new composite model uses Manandhar and Yasufuku's empirically based Kondner hyperbolic curve to forecast the rate at which toe resistance is mobilized.

It is important to note that the Composite JSH-M&Y Settlement Model has some restrictions that the writer's original Tri-Linear Settlement Model does not. Specifically, not only does the new composite model require that the writer's Simplified Nordlund Method be used to provide the required forecast of ULS total resistance (the original Tri-Linear Settlement Model allows <u>any</u> resistance-forecasting methodology to be used) but the Simplified Nordlund Method must be used with the Manandhar-Yasufuku toe-resistance model based on SCE theory as an alternative to the writer's original toe-resistance model based on traditional Hansen-Vesic bearing-capacity theory, i.e. Hansen's solution with Vesic's compressibility/rigidity factors.

Unfortunately, the exemplar *TAPERTUBE* pile shown in Figure 5.2 to illustrate the application of the writer's original Tri-Linear Settlement Model cannot be used as both an exemplar and comparative example-application for the Composite JSH-M&Y Settlement Model. This is because both the basic and enhanced versions of the Manandhar-Yasufuku Toe-Resistance Method require that cone-penetrometer data be used to provide the input parameters for evaluating the SCE model for ULS toe resistance. Unfortunately, the specific location at JFKIA where the pile shown in Figure 5.2 was driven did not include any known cone-penetrometer soundings as part of the site-characterization assessment. Only traditional borings with SPT sampling were performed. While the writer years ago developed an algorithm for converting SPT N-values to pseudo-q_c values for the purposes of site characterization (Horvath 2002, 2011), this algorithm was judged to be inadequate for application of the Enhanced Manandhar-Yasufuku Toe-Resistance Method.

As an alternative, a *Monotube* pile (designation LT2-172) that was also driven at JFKIA and for which a CPT sounding was performed close to the pile location will be used for illustrative and comparative purposes here. Using Monotube's nomenclature, this pile is a Type 3NJ8x14 with a total embedded length of 55 feet (17 m). Note that this pile was used as one of several exemplar piles in Horvath (2015).

Figure 5.3 shows the measured load-settlement results for pile LT2-172 together with forecast results from three empirical methods that are capable of producing a complete load-settlement curve as part of a two-step procedure:

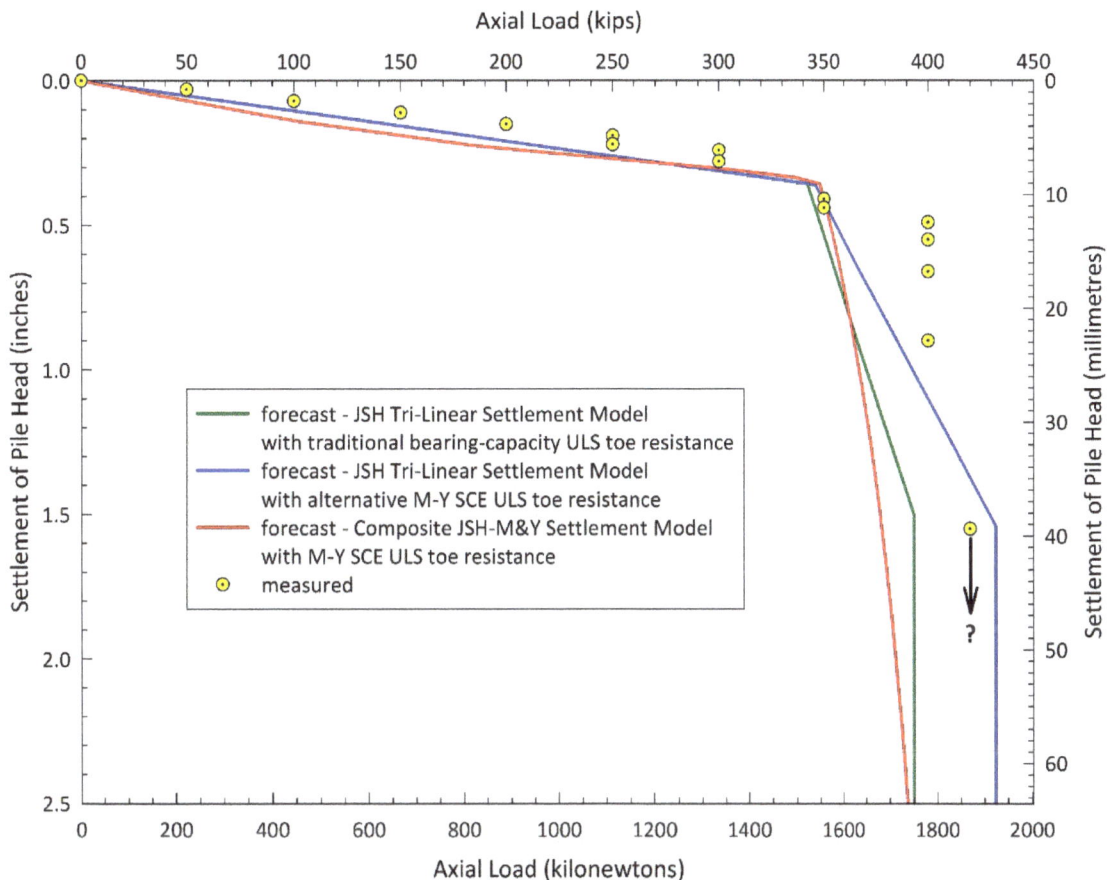

Figure 5.3. Comparison of Forecast vs. Measured Settlements for Pile LT2-172 at JFKIA.

- The writer's original Tri-Linear Settlement Model with a ULS toe resistance based on the composite Hansen-Vesic solutions to traditional bearing-capacity theory.

- The writer's original Tri-Linear Settlement Model with an alternative ULS toe resistance using the Basic Manandhar-Yasufuku Toe-Resistance Method that is based on SCE theory as modified by Manandhar and Yasufuku for use with tapered piles.

- The writer's recently developed Composite JSH-M&Y Settlement Model with ULS toe resistance based on the Enhanced Manandhar-Yasufuku Toe-Resistance Method that is based on SCE theory as modified by Manandhar and Yasufuku for use with tapered piles.

Note that the same calculated outcome obtained using the writer's Modified Nordlund Method was used for forecasting the ULS shaft resistance in all three cases. Consequently, the differences for the three methodologies shown in Figure 5.3 are due solely to the way in which the ULS toe-resistance contribution is handled.

Note also that for the two forecasts made using the writer's original Tri-Linear Settlement Model that the alternative chosen for the third and final straight-line segment was that of plunging failure as opposed to final unloading to zero load as used for the example shown in Figure 5.2. This is because the original load-settlement plot generated by others for the pile shown in Figure 5.3 did not show a final unloading, just an indication of further, unmeasured pile-head settlement beyond the last measured value of approximately 1.55 inches (40 mm), as indicated by the black downward-pointing arrow and question mark.

Before discussing the outcomes in Figure 5.3, some comments concerning the measured results are in order. As was typical during the timeframe (circa 1990) when the pile shown in Figure 5.3 was driven and statically load tested, the PANYNJ, which has held the operational lease of JFKIA[129] since the airport opened for operations in the late 1940s (Horvath 2014b), aperiodically performed what they called 'indicator-pile' programs that were essentially test-pile programs undertaken, in part, to push the technological envelope of deep-foundation selection and design at the facility in which the work was done (JFKIA in this case). The goal was to take into account changing technical needs[130] as well as technical developments in deep foundations in order to optimize foundation design for a major upcoming construction project (Horvath and Trochalides 2004). Thus, the static load tests performed for these indicator-pile programs at JFKIA, which almost always involved <u>uninstrumented</u> static load tests (the static load tests on two <u>instrumented</u> tapered piles discussed by Fellenius et al. (2000) and used in Appendix A were quite unique and anomalous in this regard), were semi-research-oriented in nature. As a result, the load-application protocol specified by the design professionals overseeing these indicator-pile load tests did not follow any standard format and often varied from pile to pile within the same test program as different technical issues were being explored for different piles.

Specifically, for the pile shown in Figure 5.3 a total of five extended 'hold' periods were used, at every 50-kip (223-kN) increment from 250 to 400 kips (1113 to 1780 kN). In addition, two complete, i.e. back to zero load, unload-reload cycles were performed at an applied load of 400 kips (1780 kN). The net result of this loading protocol presents a complex measured load-settlement outcome which is why only the data points for specific loading points are shown in Figure 5.3, not complete load-settlement curves. Note that the variation in load-hold times as well as the unload-reload cycles also complicate interpretation of the results as noted in numerous publications by Fellenius (Fellenius (2014, 2017) are but two recent examples) who has long deprecated both practices, i.e. extended hold times and unload-reload cycles.

In any event, discussing now the overall results shown in Figure 5.3, as has been the writer's experience to date with the Tri-Linear Settlement Model and as reflected in results presented in Horvath (2015)[131], the SCE toe-resistance model developed by Manandhar and Yasufuku as part of the Basic Manandhar-Yasufuku Toe-Resistance Method typically forecasts a larger ULS toe resistance compared to the traditional Hansen-Vesic bearing-capacity solution used originally by the writer as part of the Modified Nordlund Method. As a result, the ULS total resistance forecast using the Manandhar-Yasufuku SCE model is always greater than that forecast using the traditional bearing-capacity model for ULS toe resistance as the ULS shaft resistance used with each toe-resistance model is the same as noted above. Note, however, that the overall forecast results for both cases shown in Figure 5.3 using the Tri-Linear Settlement Model are relatively close and appear to bracket the measured results.

---

[129] The airport property is owned by the City of New York.

[130] Arguably the most significant example of this in the relatively long existence of JFKIA is the emergence of seismic design, both for liquefaction and lateral loading, in the latter years of the 20th century. Of course, the seismic potential had always been there as nature did not change. It was just that the seismic potential for the New York City metropolitan area was not recognized by geoprofessionals in local design practice until the 1980s.

[131] See also the results presented in Figures A.6 and A.7 in Appendix A for the Fellenius et al. (2000) instrumented test piles.

As discussed both earlier in this chapter and in Horvath (2015), the Enhanced Manandhar-Yasufuku Toe-Resistance Method always produces a relatively slow mobilization of toe resistance that implies a relatively soft toe-settlement response. This is clearly reflected in the forecast results shown in Figure 5.3 for the Composite JSH-M&Y Settlement Model that show a much softer load-settlement response than both the measured results (which are somewhat complicated to interpret due to multiple unload-reload cycles and extended load-hold times at higher load levels noted previously) and forecast results from the simpler tri-linear models. This is quite apparent when one notes that the red curve for the Composite JSH-M&Y Settlement Model is asymptotically approaching the plunging blue line.

In conclusion, based on the writer's previous experience that was noted in Horvath (2015)[132] and as typified by Figure 5.3, it does not appear that the version of the hyperbolic model proposed by Manandhar and Yasufuku for use to enhance their SCE-based toe-resistance model for tapered piles merits use. In all cases examined by the writer, their methodology when incorporated into the Composite JSH-M&Y Settlement Model forecast a much softer load-settlement response than that measured in actual load tests. In the writer's opinion, this observation involving actual piles is more significant and persuasive than the results obtained by Manandhar and Yasufuku from 1-g testing of small-scale model piles in a laboratory environment that was presented as validation of their settlement model (Manandhar and Yasufuku 2012).

Reasons that Manandhar and Yasufuku's empirical model for mobilizing toe resistance may compare favorably to results in a laboratory setting, but not actual field conditions, include the fact that the model piles they used in the laboratory did not have a geometry that replicated actual piles and, more importantly, were not installed in a manner and in a stress field that in any way replicated actual field installation and conditions. In particular, it appears from published work by Nordlund (1963), Gregersen et al. (1973), and Fellenius et al. (2000) that is discussed in Chapter 4 and Appendices A and H that residual driving stresses are significant for tapered piles. In the cases presented by Fellenius et alia, the toe resistance was interpreted to be essentially 100% mobilized at the end of driving. The cases noted by Nordlund and Gregersen et alia suggested that toe resistance was at least significantly mobilized at the end of driving. The point being made here is that residual toe loads in actual tapered piles cause a toe load vs. toe settlement response that is very different from that hypothesized by Manandhar and Yasufuku.

The takeaway from this discussion is that the apparent poor performance of the Enhanced Manandhar-Yasufuku Toe-Settlement Method with actual tapered piles may not be the fault of the basic Kondner-type hyperbolic model used by Manandhar and Yasufuku but their assumed model parameters. This suggests that an improved version of their method is possible in principle. The improvement would come from model-parameter correlation with observed behavior of actual piles as opposed to model piles. With this in mind, it is useful to outline the generic process by which the Enhanced Manandhar-Yasufuku Toe-Settlement Method might be improved upon. This presentation can be found in Appendix B.

5.6.2.3.4 Other Models

For the sake of completeness, it is appropriate to mention other empirically derived settlement-only models that have been proposed for use with DFEs that are sufficiently general and generic that they could be applied to tapered piles. Most of the R&D in this area

---

[132] The writer did not plot results from the Composite JSH-M&Y Settlement Model in this referenced document as the results were not materially better than either version of the simpler-to-use tri-linear models.

appears to have been done by Fellenius and his associates who were interested in developing force-displacement relationships with a more-general shape than the basic bi-linear elastoplastic model. This is not surprising as for some time now Fellenius has been the most consistent proponent of the importance of settlement-based assessment of deep foundations for design in routine practice.

While Fellenius has used, mentioned, or at least alluded to settlement-only analytical methodologies in many of his publications, Fellenius (2014) stands out in this regard as this conference paper has an entire appendix devoted to summarizing the five mathematical models (Fellenius refers to them as "functions"), each with its own unique set of empirically derived model parameters, that Fellenius has identified to date for use with deep foundations. These models are as follows, listed in the order presented and with the names used in Fellenius (2014), along with the primary behavioral attribute of each that is most relevant to deep-foundation applications:

- *Ratio Function* that produces a strain-hardening, a.k.a. work-hardening, response that continues ad infinitum and thus does not have a single-valued ULS value.

- Chin-Kondner *Hyperbolic Function* that also produces a strain-hardening response but one that always asymptotically approaches, but of course never reaches, a single-valued ULS value[133].

- van der Veen *Exponential Function* that is essentially a model for plastic material behavior that produces a response that essentially smooths out the kink in the transition from elastic to plastic behavior that is a signature behavioral detail of the basic bi-linear elastoplastic model.

- *Hansen Function* that produces a strain-softening response.

- *Zhang Function* that also produces a strain-softening response.

It is important to note that, as used by Fellenius, these models are always applied independently to the shaft and toe, never to the DFE as a whole, which means that different models can be used for shaft and toe behavior. In fact, the two strain-softening models (Hansen and Zhang) would typically be used only for shaft resistance where peak followed by constant-volume/critical-state strength behavior is often observed. Once the separate force-displacement behaviors of the shaft (what Fellenius calls the *t-z* component and always includes the elastic compression of the shaft as a structural column) and toe (what Fellenius

---

[133] Note that this is the same model used in the Enhanced Manandhar-Yasufuku Toe-Bearing Method although the writer refers to it as the Kondner (not Chin-Kondner) Model in deference to Kondner's seminal paper (Kondner 1963) and contribution to mathematical modeling in geotechnical engineering that remains in widespread use to the present. However, it is of historical interest to note that some authors, including Fellenius, include F. K. Chin's name in the model, apparently in deference to work Chin did in Malaysia circa 1970 that involved use of Kondner's work with the hyperbolic function specifically for modeling the compressive-load vs. settlement relationship of DFEs. Chin (1970, 1972) are the earliest of several references found in the literature of Chin's contributions in this regard. Note that Chin's use of the hyperbolic function to model DFE load-settlement behavior was geared toward developing a methodology for forecasting the ULS total resistance of deep foundations based on load-settlement data obtained in a static load test. This interpretative methodology is referred to as the *Chin Method* in the literature. The Chin Method has subsequently been extended to related applications such as assessing the ULS uplift resistance of ground anchors, e.g. Hanna (1987).

calls the *q-z* component) have been evaluated, the results are simply combined to produce an overall load-settlement behavior at the head of the DFE.

It is relevant to note that if this suite of models is applied to a tapered pile it may be necessary to use different models for the constant-perimeter and tapered portions of the pile shaft. Although there is abundant published research discussing the axial-compressive force-displacement behavior along the shaft of a DFE with a constant perimeter, there is no known similar work for the tapered portion of a pile shaft. However, given the very different physical mechanisms by which shaft resistance due to sliding friction and shaft resistance due to taper (wedging) develop for the constant-perimeter and tapered sections respectively, unless proven otherwise by research it seems logical that the mathematical models used for these two components of tapered-pile shaft resistance will at least have different values for the model parameters even if the same model is used, if not different models completely.

5.6.2.4 Theory of Elasticity

5.6.2.4.1 Background

Research efforts that use theory-of-elasticity solutions to develop settlement-only methodologies for DFEs date back to at least the early 1960s. The solution for *Mindlin Problem No. 1* for the response to a vertical-downward point load located <u>within</u> an elastic half-space, which dates to 1936, appears to have formed the basis for at least some of the published solutions[134]. Details concerning Mindlin's original solution can be found in Poulos and Davis (1974).

Poulos and Davis (1980) present a very complete discussion of the applications-related aspects of elasticity-based work related to DFE settlement forecasting that was done up until that time (1980), including a discussion of estimating Young's modulus in practical applications. Note that this latter aspect was obviously constrained by the state of knowledge at that time and thus does not reflect the significant advances in site characterization, especially using in-situ testing devices such as the cone penetrometer, that have occurred in the four decades since that book was published.

However, the fact that Poulos and Davis addressed the issue of modulus-estimation in practice (always the proverbial 'elephant in the room' whenever any elasticity-based solution is used in geotechnical or foundation engineering) was a significant step in the right direction. The 'highway of geotechnology' is littered with elegant theoretical solutions that have gone nowhere in practice because the essential practical issue of how to adapt a solution

---

[134] There is often confusion in the published literature concerning the contributions of the late Prof. Raymond D. Mindlin to the geomechanics knowledge-base. Specifically, one typically sees reference to "Mindlin's Solution" which implies a single solution to a theory-of-elasticity problem. In reality, Mindlin developed solutions to two distinct boundary-value problems in elasticity that have a place in the geotechnical historical record. The already-noted Mindlin Problem No. 1 is conceptually related to the well-known *Boussinesq Problem* of 1884 that deals with a vertical-downward point load applied to the <u>surface</u> of an elastic half-space. Mindlin applied the same point load to the <u>interior</u> of an elastic half-space in his Problem No. 1. Mindlin Problem No. 1 should not be confused with *Mindlin Problem No. 2*, also from 1936, that deals with a <u>horizontal</u> point load within an elastic half-space. This solution was used in the early 1960s for what may have been the first attempt to apply the theory of elasticity to the laterally loaded deep foundation problem (Spillers and Stoll 1964). However, this effort by Spillers and Stoll was quickly eclipsed by the well-known *p-y curve* concept that became increasingly available to use in both research and practice throughout the 1960s and beyond as the digital computer evolved to become the mainstream computational tool in civil engineering that it is today.

that generally requires a single-valued modulus, Young's or shear, to a geotechnical application where modulus is never a single-value was never addressed during the development of that theoretical solution.

Despite a considerable amount of published worked on the subject of using elasticity solutions for forecasting DFE settlement that was done throughout the 1960s and into the 1970s, very little of practical consequence and usage came out of it. This was likely the result of a number of factors including:

- the complexity of the solutions;

- difficulty in accurately estimating Young's modulus throughout the load range of a typical deep foundation; and

- the fact that routine design then, as now, tends to focus on strength and ULS resistance only, especially for piles, so there was little demand from practitioners to develop user-friendly methodologies related to DFE settlements.

Note that this contrasts with shallow foundations, especially footings, where design has long been settlement driven. As a result, considerable resources have been invested for decades beginning in the 1960s to develop practice-oriented settlement methodologies for footings.

However, since the beginning of the 21st century there has been renewed interest in using the theory of elasticity for forecasting DFE settlements. The significant difference this time is that this renewed activity has been accompanied by an effort to make the resulting analytical methodologies practice-oriented and user-friendly by basing them on modern site-characterization tools that are available worldwide, as well as solution methodologies that can easily be implemented using universally available tools such as spreadsheet software. Because this new research reflects the current state of knowledge, it is discussed in detail in the following section.

5.6.2.4.2 The Randolph-Wroth Problem: Formulation and Generic Solution

Beginning in the late 1970s, Prof. Mark F. Randolph and the late Prof. C. Peter Wroth published several related papers that presented a theory-of-elasticity solution for the boundary-value problem of the settlement of a single DFE subjected to axial-compressive loading. The writer termed this collective body of publications (citations of specific references can be found in Mayne (2007) which also provides an excellent summary and primer on the subject) the *Randolph-Wroth Problem* (Horvath 2016a).

The physical model used by Randolph and Wroth is that the ground surrounding a DFE consists of a series of imaginary, contiguous, concentric, vertical cylinders of linear-elastic material that transfer the assumed vertical shear stress, τ_x, acting along the DFE shaft-ground interface radially by a process of vertical shearing within and between cylinders. **It is important to note that perfect adhesion, i.e. no relative slippage, is assumed at all times along the interface between the DFE shaft and adjacent ground** as well as between all of the imaginary cylinders that constitute the ground around the shaft of the DFE.

Settlement of these concentric cylinders that is caused by these vertical shear stresses attenuates progressing radially away from the DFE shaft-ground interface. At some radial distance from the DFE shaft-ground interface, the settlement magnitude becomes effectively zero. In very broad conceptual terms, this is the same as the concept often used with shallow foundations of a 'significant stressed depth' beneath an applied vertical load on the ground whereby at some depth the settlements caused by that load become negligibly small and are assumed to be zero in magnitude for analytical convenience.

The basic assumptions and parameters of the Randolph-Wroth Problem are shown in Figure 5.4 for the original problem formulation based on using Young's modulus, E, to characterize the ground stiffness. The notation used by Mayne in various publications and presentations beginning in 2007 and as cited in Horvath (2016a) is used in this figure.

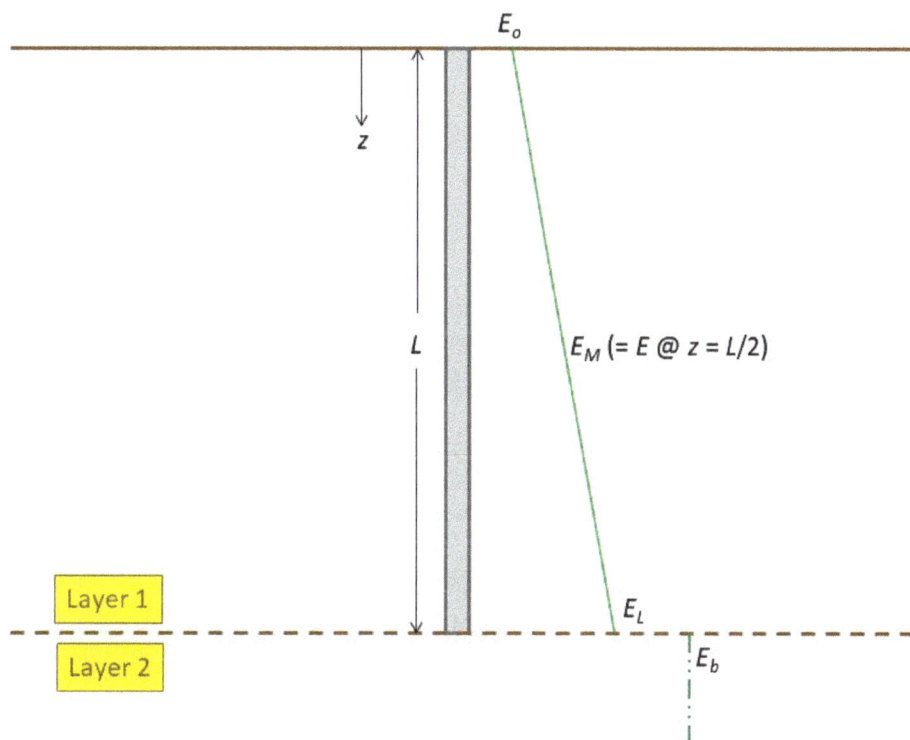

Figure 5.4. Randolph-Wroth Problem - Basic Geometry and Parameter Definition.

Before proceeding further, it is important to make the perhaps obvious (but significant and thus necessary in view of uses of the solutions to the Randolph-Wroth Problem that are discussed in short order) observation that all of the modulus values shown in Figure 5.4 are assumed to be constant in magnitude and independent of applied loads. This is simply an inherent characteristic of theory-of-linear-elasticity solutions that has often inhibited their use in geotechnical and foundation engineering applications as noted in the preceding section.

It is also important to note that, in this case, the parameter E_o shown in Figure 5.4 is not the alternative notation for the small-strain Young's modulus, E_{max}. Rather, it is simply the value of Young's modulus at the ground surface (and can be zero, if desired).

In any event, the Randolph-Wroth Problem is sometimes referred to in the literature as a "two-layer problem" as it allows for the material underlying the toe (called the "base" in the original problem formulation, hence the subscript "b" for the Young's modulus of Layer 2) to have a stiffness that is distinctly different from that anywhere along the shaft. Note, however, that E_b is always assumed to be depth-wise constant in magnitude.

The basic form of the solution to the Randolph-Wroth Problem as given by Mayne (2007) but using the parameter notation shown in Figure 5.4 is:

$$w_t = \frac{Q_t \cdot I_\rho}{d \cdot E_L} \tag{5.11}$$

where:

- w_t = the settlement of the head ("top") of the DFE caused by Q_t;
- Q_t = the downward force applied at the head (called the "top" in the original problem formulation, hence the subscript "t") of the DFE;
- I_ρ = a dimensionless influence factor composed of various problem variables (modulus variations, etc.) that differs significantly for the rigid- vs. compressible-pile solutions, with the compressible-pile solution being significantly more complex algebraically;
- d = the (constant) diameter of the DFE.

For reasons that will become clear subsequently, when the solution to the Randolph-Wroth Problem is applied to the entire DFE as shown in Figure 5.4 this is referred to in this monograph as the *whole-pile solution*.

Note that the overall form of the solution shown in Equation 5.11 is essentially identical to that for shallow-foundation settlement using classical theory-of-elasticity solutions so should be broadly familiar to foundation engineers. Also, although not shown there is also an equation that is a corollary to Equation 5.11 that allows for the toe ("base") settlement, w_b (in the notation used by Mayne), of a DFE to be calculated directly when the compressible-pile solution is used (the head, w_t, and toe, w_b, settlements are obviously the same when the rigid-pile solution is used). This allows the elastic compression of the DFE to be easily determined (= w_t - w_b).

As with all theory-of-elasticity solutions that involve displacements, the problematic issue that always presents itself in any practical application is how to make a theoretical solution that requires a single-valued modulus or single relative distribution of moduli (as shown in Figure 5.4 for the Randolph-Wroth Problem) work within the reality of actual, non-linear soil behavior where moduli are both <u>stress dependent</u> and <u>stress-level dependent</u>. As noted in the preceding section, this is the reality check/deal breaker that has historically limited the application of the theory of elasticity to geotechnical and foundation engineering applications unless an incremental, tangent-modulus numerical-solution approach such as the finite-element method (FEM) is used.

In this context, 'stress dependent' means that, all other things being equal, modulus increases with depth solely due to increased confining stress. However, the rate at which modulus increases with depth can be highly variable depending on a number of factors. In addition, changes in soil stratigraphy with depth also introduce complications.

On the other hand, 'stress-level dependent' means at a given depth, how far along the stress-strain curve to shear failure (defined in terms of stresses) is some load-level of interest. Alternatively, especially in applications involving dynamics but increasingly with nominally static applications such as DFEs as discussed subsequently, the concept of stress-level dependency is replaced by 'strain-level dependent' as the shear-strain magnitude is used instead of shear stress and strength. Thus, in simple terms, stress- or strain-level dependency can be viewed as starting from some initial value of modulus defined at either zero shear stress or a 'small' level of shear strain (at which linear-elastic behavior can be assumed to exist) that is followed by a reduction in modulus (the term *modulus degradation* is often used in this context) with increasing shear-stress-to-shear-strength level or shear-strain magnitude.

Fortunately, the formulation of the Randolph-Wroth Problem inherently addresses the first issue of stress dependency, at least in an approximate way, by allowing for a linear increase in modulus within Layer 1 as well as allowing for an even stiffer (albeit constant-value) modulus within Layer 2. However, this still leaves the second issue of dealing with stress-level dependency. How this can be handled is discussed in the following sections.

5.6.2.4.3 The Randolph-Wroth Problem: Mayne's Whole-Pile Solution

As is well-known, there are two broad ways to deal with the stress-level dependency of modulus analytically: either a tangent- or secant-modulus approach. The basic, generic concepts for each are summarized in Chapter 2 as well as explained in detail in the literature, e.g. Duncan and Change (1970), and are not addressed here. To date, only the secant-modulus approach appears to have been used to solve the Randolph-Wroth Problem.

Prof. Paul W. Mayne deserves credit for crafting a methodology for solving the Randolph-Wroth Problem using repetitive application of a secant Young's modulus that is calculated using an empirical equation first proposed for use with the pressuremeter by Fahey and Carter (1993). The *Fahey-Carter Modulus-Degradation Model* has the form of a hyperbolic relationship with two curve-fitting parameters, f and g, as shown in the following equation:

$$\frac{E_{L(operative)}}{E_{L(max)}} = 1 - f\left(\frac{Q_t}{Q_{t(ult)}}\right)^g \tag{5.12}$$

where:
- $E_{L(operative)}$ = the load-specific operative (secant) value of E_L (= E at the bottom of Layer 1 as shown in Figure 5.4) that occurs under an arbitrary applied load, Q_t, at the head of the DFE that is less than the ultimate load, $Q_{t(ult)}$, defined subsequently;
- $E_{L(max)}$ = the maximum value of E_L which Mayne defined as that existing under assumed linear-elastic small-strain conditions, i.e. E_{max}. Mayne (2007) indicated that $E_{L(max)}$ should be calculated from shear-wave velocities measured in an sCPTu sounding. Alternatively, if only CPTu data are available, the writer has found that a first-order approximation of E_{max} can be made using empirical correlations published in Robertson (2012). However, in all cases, E_{max} is that existing prior to installation of the DFE;
- f, g = the hyperbolic-model parameter that are used to fit Equation 5.12 to the data for a particular application. The roles played by these two parameters in terms of shaping the hyperbolic curve generated by Equation 5.12 are discussed in detail in Appendix D. However, for now, suffice it to say that f controls the end-point of the curve in the limit as the ratio $Q_t/Q_{t(ult)}$, in this case, approaches a value of one while g controls the overall shape (degree of curvature) of the curve; and
- $Q_{t(ult)}$ = the maximum achievable value of Q_t, i.e. the ULS total resistance of the DFE (R_u in Figure 2.1). This value must always be calculated independently beforehand using some indirect or direct resistance-forecasting methodology.

As an aside, the use of a <u>secant</u>-modulus approach to model what is essentially a continuous, non-linear relationship (load vs. settlement in this case) is relatively unusual. As noted in Chapter 2, a <u>tangent</u>-modulus approach has historically been the preferred approach to producing a continuous, non-linear relationship so that each succeeding step in an analysis can build on and add to the cumulative result of the preceding steps. On the other hand, a secant-modulus approach is typically used to arrive at some single-valued final result in an iterative, trial-and-error basis such as the shallow-foundation bearing-capacity example cited in Chapter 2.

Obviously, it is possible to use a secant-modulus approach to model a continuous, non-linear relationship as Mayne illustrated here with the Randolph-Wroth Problem. However, it should be clearly understood that the implication of doing so is that each point on the resulting continuous curve that is created is presumed to be independent of every other point. Therefore, the final result is not a progressive accumulation of intermediate results but an aggregation of independent results. Note that this is not meant to deprecate Mayne's clever use of a secant-modulus approach in this application. Rather, the intent here is simply to make clear the fact that a curve generated using an iterative application of the secant-modulus approach may not yield the same results as those obtained using a cumulative application of the tangent-modulus approach for the same problem.

In any event, Mayne suggested that $f = 1$ and $g = 0.3 \pm 0.1$ (with 0.3 being the default value) were the appropriate values to use for the hyperbolic-model parameters in Equation 5.12 for the DFE load-settlement application. To date, there does not appear to have been any systematic investigation into determining whether it is possible to determine site-specific and/or DFE-specific values of these parameters. Rather, the default values suggested by Mayne are simply used for the most part.

A limited investigation by the writer (Horvath 2016a) suggested that the g-parameter in particular likely varies in some as-yet-undetermined manner that is likely related to both ground conditions and DFE typology. This is consistent with Mayne's published findings and suggests that research into developing some rational methodology for determining site- and application-specific values of at least g, if not f as well, would be a worthwhile effort.

To implement Mayne's solution methodology into practice, the following steps are followed. Note that the computational process is easily incorporated into a spreadsheet for solution:

- The depth-wise magnitudes of the small-strain Young's modulus, E_{max}, are determined from a site-specific site-characterization program that preferably incorporates one or more sCPTu soundings into the program. A best-fit straight line is fit through the data over the length of the DFE to be analyzed (the functions available in spreadsheet software work well for this) and the value of $E_{L(max)}$ (see Figure 5.4) determined.

- $Q_{t(ult)}$ (= R_u in Figure 2.1) is calculated for the DFE to be analyzed using a resistance-forecasting methodology of the design professional's choosing.

- An arbitrary value of the dimensionless $Q_t/Q_{t(ult)}$ ratio between 0 and 1 is assumed.

- Equation 5.12 is used (the hyperbolic-model parameters f and g are assumed to be known beforehand, typically by simply using the aforementioned default values of 1 and 0.3 respectively) to calculate the corresponding dimensionless $E_{L(operative)}/E_{L(max)}$ ratio.

- The value of $E_{L(operative)}$ is calculated using the value of $E_{L(max)}$ that was determined beforehand.

- The value of Q_t is calculated based on the assumed $Q_t/Q_{t(ult)}$ ratio and the value of $Q_{t(ult)}$ that was determined beforehand.

- The value of w_t is calculated by using these values of Q_t and $E_{L(operative)}$ in Equation 5.11 along with the desired version (rigid or compressible pile) of the I_ρ variable. Equations for I_ρ can be found in Mayne (2007) and other publications (be careful of notational variations from one publication to the next!).

- The process is then repeated using other assumed values for the $Q_t/Q_{t(ult)}$ ratio until a sufficient number of Q_t vs. w_t data pairs have been calculated in order to create the desired load-settlement curve for the head of the DFE being analyzed.

Note that if the compressible-pile solution is chosen, the toe settlement, w_b, can also be calculated for each assumed value of the $Q_t/Q_{t(ult)}$ ratio. This will generate Q_t vs. w_b data pairs so that a curve of applied load vs. toe settlement can also be generated. In addition, a plot of elastic compression of the DFE as a function of applied load, Q_t, can also be generated.

5.6.2.4.4 The Randolph-Wroth Problem: Niazi's Whole-Pile Solution

A deceptively modest, but extremely useful. improvement to Mayne's original whole-pile solution using Randolph-Wroth Problem solutions was presented in Niazi et al. (2010), Niazi (2013, 2014), and Niazi and Mayne (2015b). Prof. Fawad S. Niazi, who was working as a doctoral student under Mayne's guidance and oversight at the time, reformulated the Randolph-Wroth Problem using the shear modulus, G, to define the ground stiffness instead of the Young's modulus, E, but otherwise maintaining the basic geometric elements and problem-parameter definitions shown in Figure 5.4 as well as the solution form of Equation 5.11. There were corresponding changes in the equations for both the rigid and compressible-pile versions of the I_ρ variable. In addition, in Equation 5.12 the dimensionless ratio $G_{L(operative)}/G_{L(max)}$ is used.

The changes made by Niazi in the formulation of the Randolph-Wroth Problem solutions are logical. Recall that the basic physical model of the ground surrounding and underlying the DFE on which the Randolph-Wroth Problem is based is that of a series of concentric cylinders deforming in shear. Because the overall DFE-ground and intra-ground load-transfer mechanism that is assumed to be occurring within and between these cylinders is one of vertical shearing, it is obvious that shear modulus, G, is inherently the preferred metric for the cylinder stiffness as opposed to Young's modulus, E. Shear modulus simply better reflects the stiffness of the load-transfer mechanism that is assumed to be occurring within the concentric-cylinder model. Fortuitously, the small-strain shear modulus, G_{max}, is nowadays relatively easy to determine using an sCPTu sounding or other in-situ testing methodologies so this makes using shear modulus straightforward.

This seemingly modest alteration in the formulation of the Randolph-Wroth Problem solutions has actually turned out to be a very useful, practical modification in the writer's opinion. This is because the profile of the small-strain shear modulus, G_{max}, from shear-wave velocities obtained using sCPTu soundings can be used directly as opposed to having to convert the G_{max} data to E_{max} data as in Mayne's original solution formulation[135]. In addition, if only CPTu or even CPT data are available it is still relatively easy to get at least a first-order estimate of the G_{max} profile using empirical correlations (Horvath 2015), a benefit that is illustrated subsequently. This is because far more research has been devoted in recent years to developing cone penetrometer-based empirical correlations for G_{max} as opposed to E_{max}.

---

[135] The need to assume a Poisson ratio for the soil, v_s, does not disappear as v_s appears in the equations for I_ρ in the shear-modulus formulation of the Randolph-Wroth Problem solutions.

Before proceeding further with the discussion of the Randolph-Wroth Problem and its solutions using Niazi's shear-modulus formulation, it is of interest to illustrate its original, basic application using the Fahey-Carter Modulus-Degradation Model as put forth by Mayne and outlined above but using Niazi's alternative solution formulation based on shear modulus. In addition to illustrating the basic outcomes that can be expected, this example sets the stage for a more-in-depth critique that is presented subsequently.

The following example was presented originally in Horvath (2016a) and involves a constant-diameter steel pipe pile that was driven with a closed toe and filled with PCC after installation but prior to static load testing[136]. A relatively basic, conventional pile with constant diameter was intentionally chosen for this exemplar application in order to more clearly illustrate behavioral implications of using the Randolph-Wroth Problem in its conventional form without the complications introduced by tapered piles and their depth-wise variable diameter that is not explicitly accounted for, and thus not readily accommodated, in the Randolph-Wroth Problem.

The example pile, which is designated Pipe Pile 2A, was driven and load tested within the Central Terminal Area (CTA) of JFKIA in late 1988 as part of the test-pile program associated with the PANYNJ's *JFK 2000* building campaign (Horvath and Trochalides 2004)[137]. The pile is 12.75 inches (324 mm) outside diameter with a 0.5-inch (13 mm) thick wall. The pile toe at the end of driving was 85 feet (25.9 m) below the ground surface which is relatively and unusually deep even by current seismic-influenced design requirements at JFKIA.

The subsurface conditions at the location of Pipe Pile 2A are typical of those found within the JFKIA CTA as shown in Figure A.2 in Appendix A as well as discussed in much greater detail in Horvath (2014), with updates in Horvath (2015, 2016b). CPT sounding CTP-2 was the closest to this pile so was used for all site-characterization assessments as well as resistance- and settlement-related analyses performed by the writer that involved this pile.

Comparing the depth of the pile toe to the log for CTP-2 indicates that the toe was embedded near the top of the dense gravelly sand zone with the UGA bearing stratum. Based on current geological interpretations as presented in Moss (2015) and Moss and Canale (2017), this dense zone toward the bottom of the UGA is now interpreted to be the Merrick Formation that is of Pleistocene geological age[138].

---

[136] This is not a trivial statement. As discussed in Chapter 4, concreting a closed-toe steel pipe pile, tapered or non-tapered, after installation but prior to external load application is something that can no longer be taken for granted.

[137] Although tapered piles have historically been the DFE of choice at JFKIA and remain so to the present, at different times in the history of JFKIA, which dates back to the early 1940s, other types of piles as well as drilled shafts have, from time to time, been installed on a trial basis to assess both their technical performance as well as cost-effectiveness for potential production use at the airport (Horvath and Trochalides 2004, Horvath 2014). This was the case for Pipe Pile 2A that was installed at a time when the need to install much deeper DFEs than had been installed up to that point in time to account for seismic liquefaction within the Upper Glacial Aquifer (UGA) bearing stratum was being addressed for the first time. The need to resist seismic-induced lateral loads at foundation level was another technical issue that also evolved at the same time.

[138] The distinction between *geological age* and *behavioral age* is pointed out and highlighted here as the two are not necessarily the same. This is because although the geological age never changes, the behavioral age can be 'reset' without limit by any number of natural and human activities (Horvath 2016b). Of relevance to the present discussion is that the shear-wave velocity, V_s, measured in an sCPTu sounding, and hence the value of G_{max} calculated from it, is very sensitive to behavioral age. This is but one of many reasons why site-specific sCPTu soundings are always recommended for any project as a complement and enhancement to CPTu soundings.

As an aside, the cone-penetrometer soundings performed at JFKIA circa 1988, of which CTP-2 was but one of several, are believed to have been the first ever performed at JFKIA as well as among the earliest performed in the New York City metropolitan area as this region was, and still is, slow to recognize and embrace this site-characterization technology. Note that CTP-2 was only a CPT meaning that only q_c-f_s data pairs were obtained at each electronic data-sampling interval (1 foot (305 mm) in this case).

However, this is of fortuitous benefit to the present discussion as it illustrates how at least a first-order estimate of G_{max} can be obtained without direct measurements of shear-wave velocity, V_s, in an sCPTu sounding and using substantially older cone-penetrometer technology (if that is all that is available for use), albeit with some assumptions and approximations concerning the u_2 pore pressures as well as the behavioral age of the coarse-grained soils encountered. Specifically, $u_2 = u_o$ and a 'young' (Holocene-equivalent) behavioral age were assumed within all coarse-grained soils that comprise almost all of the site subsurface soils in this case as can be seen in Figure A.2.

In any event, Figure 5.5a shows the writer's interpretation of the G_{max} profile for CTP-2 using the "Best Estimate" empirical correlation given in Robinson (2012)[139]. This correlation assumes or at least implies a 'well-behaved' (i.e. non-structured) soil with a young (Holocene-equivalent) behavioral age. Note that although the coarse-grained soils that comprise the bearing stratum for DFEs at JFKIA are of Pleistocene geological age, for the reasons discussed at length in Horvath (2016b) and summarized below they often display behavior consistent with a Holocene behavioral age.

The two red lines in Figure 5.5b shows the interpretation of the best-fit straight lines (this was done using a built-in mathematical function in *Excel 2010*) for both along the shaft (Layer 1 in Figure 5.4) and below the toe (Layer 2 in Figure 5.4) of Pipe Pile 2A. Note that the recommendation of Niazi and Mayne (2015c) of using the arithmetic mean of G_{max} for four toe diameters below the toe (a distance of 51 inches (1295 mm) in this case that sampled four CPT datapoints as indicated in Figure 5.5) for determining G_b (E_b for Layer #2 in Figure 5.4) was used by the writer.

Because use of the Randolph-Wroth Problem solutions as developed by Mayne (based on Young's modulus) and modified by Niazi (to use shear modulus) hinges on the accuracy of the interpreted G_{max} data, a logical question to ask at this point is how reliable is the Robinson (2012) empirical correlation that was used in this case in the absence of site-specific shear-wave velocity measurements. To begin with, as already noted, this specific correlation assumes that the soil is both behaviorally young, i.e. from the Holocene Epoch, as well as well-behaved which, in this context, means devoid of structure due to aging and cementation. While the latter applies to the UGA stratum at JFKIA, the former does not as the three geological formations (Bellmore/Wantagh/Merrick) that comprise the Upper Glacial Aquifer (UGA) that has always been the DFE bearing stratum at JFKIA are interpreted as being from the Pleistocene Epoch.

Fortunately, recent years have seen several sCPTu soundings performed within the CTA at JFKIA so an assessment of the validity of empirical correlations for G_{max} that assume no aging effects can be made rigorously. This was done in Horvath (2016b) and a summary of the findings is included here for ease of reference.

---

[139] Note that Robinson uses the more globally common notation "G_o" for G_{max} in this cited reference. The reasons for using the latter notation in this monograph were discussed in Chapter 2.

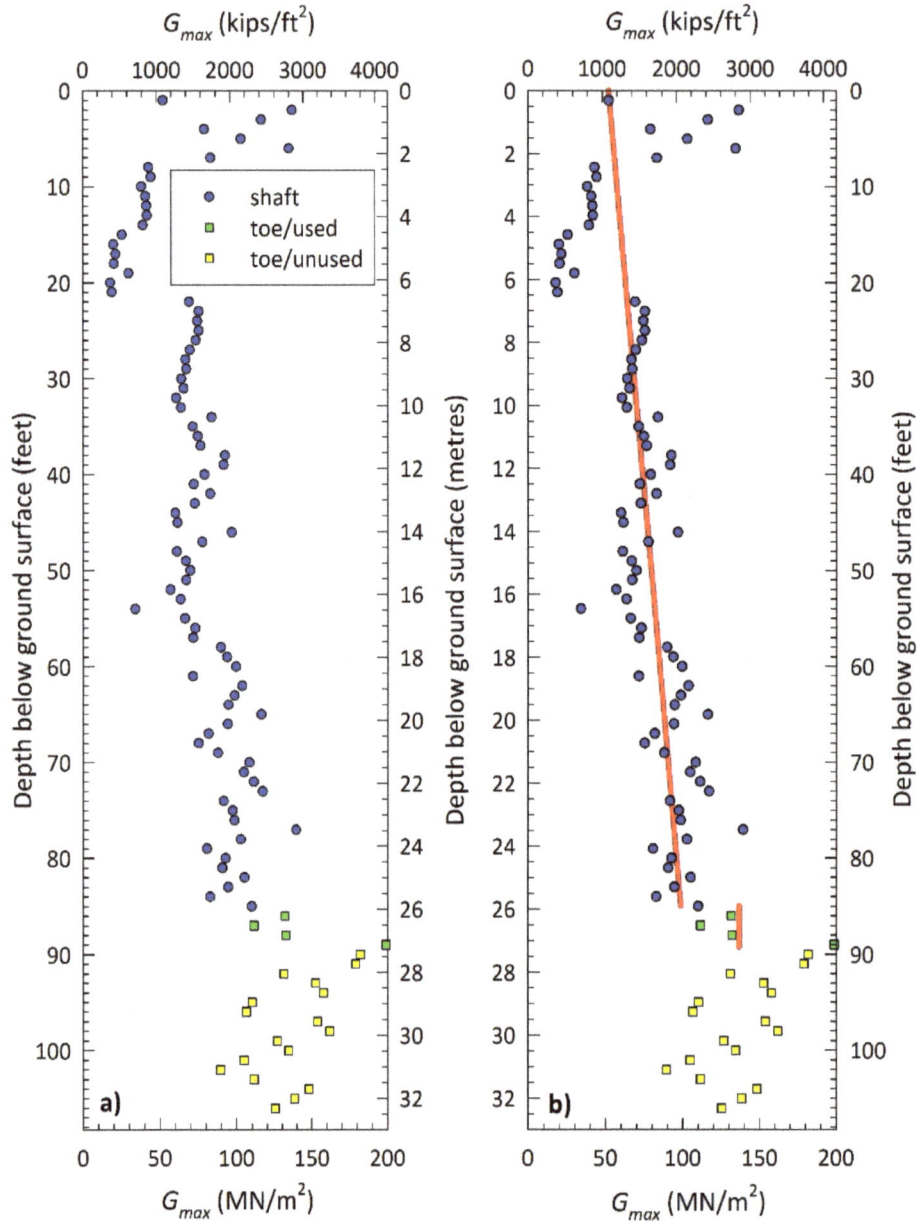

Figure 5.5. Small-Strain Shear Moduli for CPT CTP-2 at JFKIA as Interpreted for Pipe Pile 2A.

Figures 5.6 and 5.7 show the comparisons between:

- measured and empirically estimated values of the shear-wave velocity, V_s;

- calculated (based on measured) and empirically estimated normalized shear-wave velocity, V_{s1}; and

- calculated (based on measured) and empirically estimated small-strain shear modulus, G_{max}

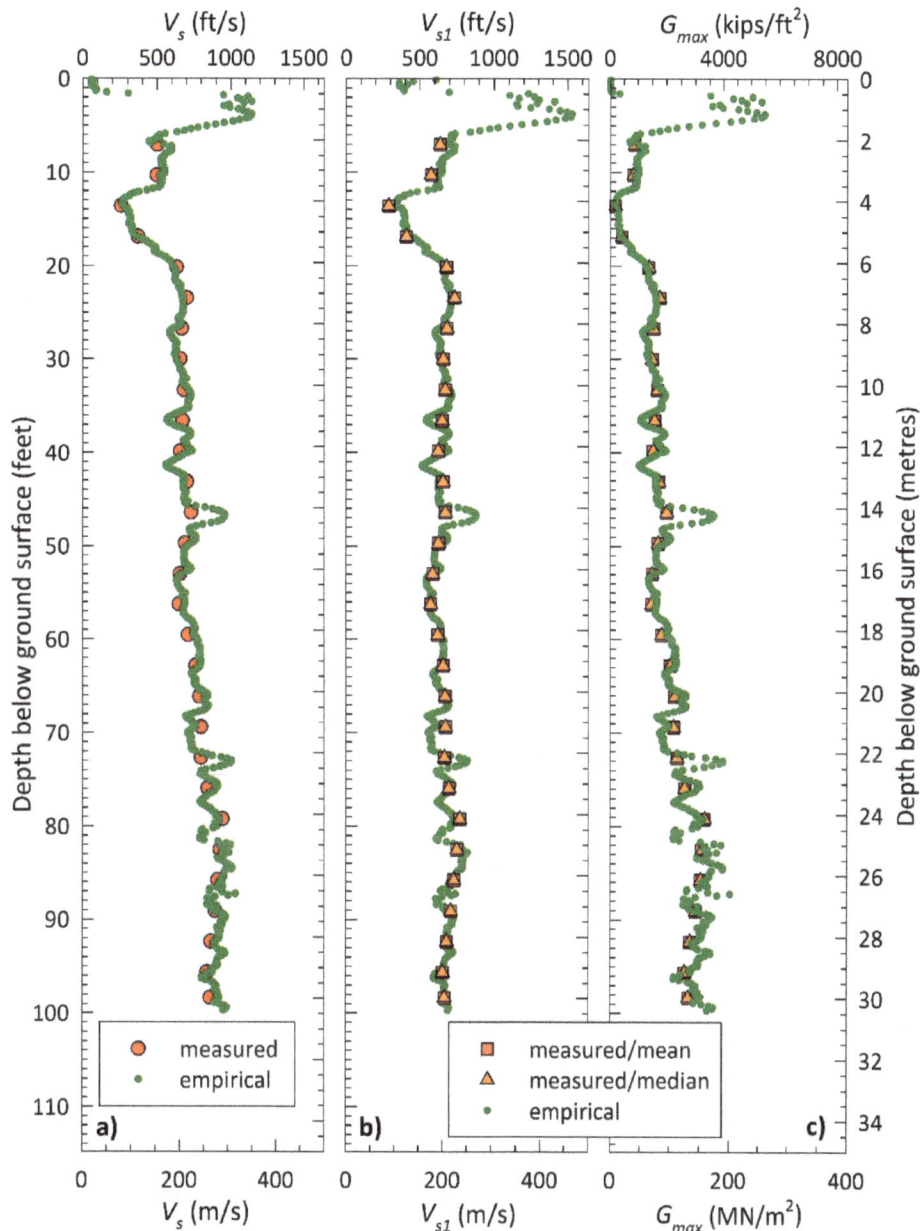

Figure 5.6. Measured vs. Empirical V_s, V_{s1}, and G_{max} for JFKIA-CTA Site for sCPTu Sounding A.

for two sCPTu soundings (designated simply 'A' and 'B' for the purposes of the writer's studies) that were performed during the same timeframe (21st century) within the same terminal area (the exact location is client-confidential) within the JFKIA CTA.

Note that there were several sCPTu soundings performed at this terminal site during the same timeframe. The outcomes shown in these figures were intentionally chosen by the writer to bracket the conditions found, i.e. all the other soundings produced results that were either similar to or in between the results shown in these figures.

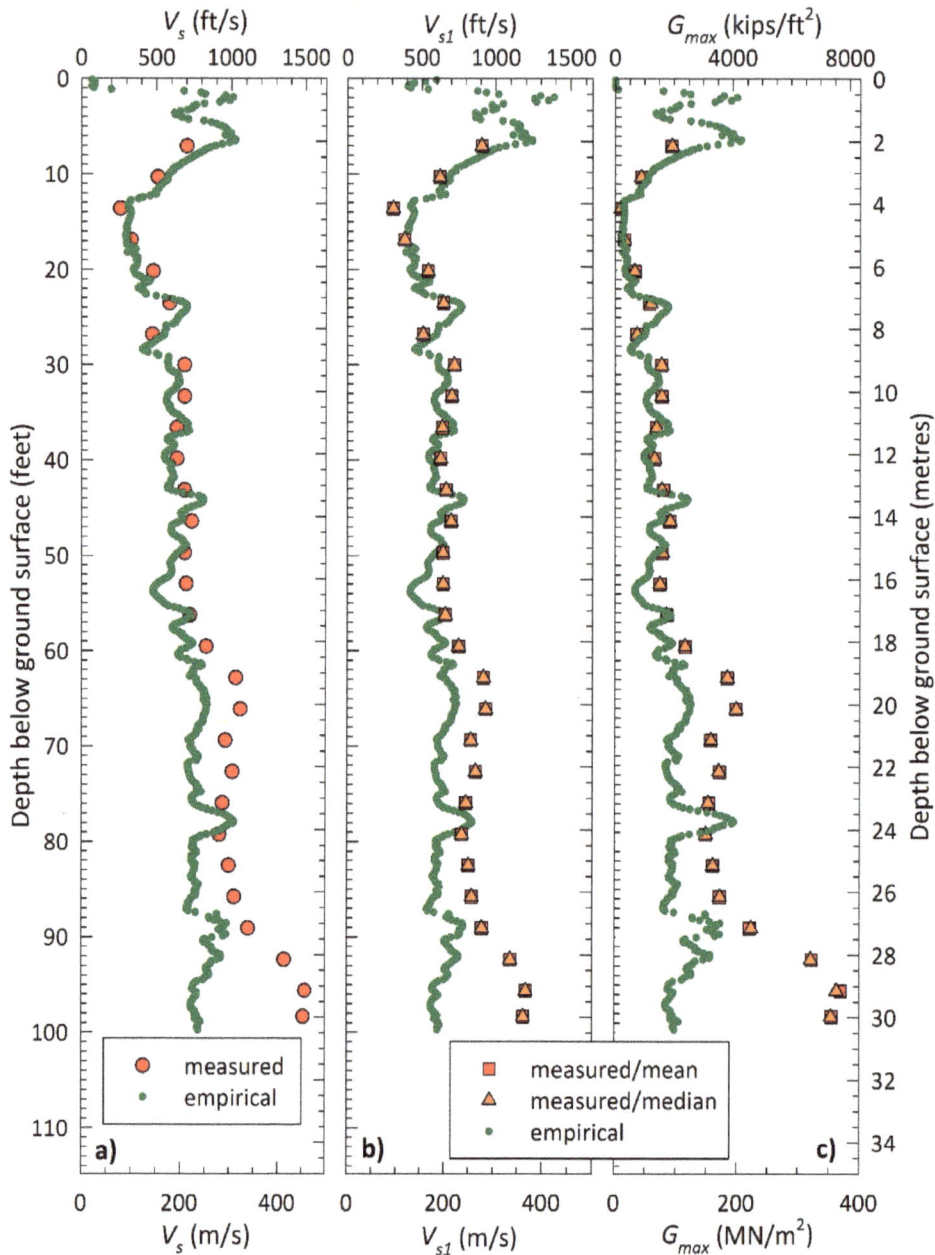

Figure 5.7. Measured vs. Empirical V_s, V_{s1}, and G_{max} for JFKIA-CTA Site for sCPTu Sounding B.

Before discussing the results shown in these figures, two comments are in order concerning the datapoints shown. First, with regard to the empirical results, these are based on the q_c-f_s-u_2 data. The cone used in this case (A_{tip} = 15 cm^2) recorded data every 50 millimetres (2 in) so the calculated V_s, V_{s1}, and G_{max}, datapoints are relatively closely spaced.

With regard to the measured results, an sCPTu sounding only directly measures V_s which is shown in Part a) of each figure. In this case, the vertical spacing between V_s measurements was 1 metre (3.3 ft). Other parameters such as V_{S1} and G_{max} must be calculated from the measured V_s, with the former requiring an estimate of the vertical effective overburden, σ'_{vo}, and the latter requiring an estimate of the soil density, ρ.

In this case, the writer calculated both σ'_{vo} and ρ using the q_c-f_s-u_2 data. This means that for every measured value of V_s there were 20 pieces of calculated σ'_{vo} and ρ results to use in order to calculate the corresponding values of V_{S1} and G_{max}. For many years, the writer used an arithmetic mean for such calculations. However, it was recently suggested to the writer (Dr. Mark Styler, personal communication, 2016) that using a median, as opposed to mean, value of some group of values that are either measured or derived from measurements in a cone-penetrometer sounding provides a more accurate result as it better smooths out any erratic peaks in individual values that can occur. Consequently, in Figures 5.6 and 5.7 the writer chose to calculate and plot both mean- and median-based estimates of V_{S1} and G_{max} for each piece of measured V_s data, hence the dual symbols. As can be seen, in this case there is essentially no difference between the mean and median results which is why the data points essentially plot one on top of the other.

Turning now to a discussion of the results shown in Figures 5.6 and 5.7, Sounding A in Figure 5.6 shows excellent correlation between the measured and empirically forecast results for the entire sounded depth which is likely close to the entire thickness of the UGA DFE bearing stratum at this location within JFKIA. This implies that the soils within the UGA are behaviorally young (= Holocene equivalent) despite being of Pleistocene age. This further implies that the behavioral 'clock' of these soils was 'reset' at some time in the past, at least once but possibly more than once. There are several natural and human-induced mechanisms that alone or in combination could have caused this/these reset/resets (Horvath 2016b), with one or more episodes of seismic liquefaction being high on the list.

On the other hand, Sounding B in Figure 4.7 shows increasing deviation between the measured and forecast results starting at a depth of approximately 60 feet (18 m). This is a classic example of a soil deposit that is showing signs of structure due to either aging or cementation or some combination of the two, with simple aging being the much more likely cause in this case. This is because an increasing depth within the UGA represents traveling farther back in geological time within the Pleistocene Epoch.

In any event, the conclusion drawn from this extended discussion is that in the absence of site-specific shear-wave-velocity data as in the case of the exemplar pile (Pipe Pile 2A) used here to illustrate the application of the basic Randolph-Wroth Problem solutions, use of an empirical correlation for forecasting G_{max} as in Figure 5.5 is, overall, reasonable for the JFKIA Central Terminal Area. Although the empirical correlation used (Robinson 2012) is for behaviorally young (Holocene-equivalent) soils, for whatever reason(s) the behavioral clock of the Pleistocene UGA soils at JFKIA, at least for the upper 50± feet (15± m) within the Bellmore Formation if not for the entire stratum thickness, has been reset one or more times in the geological past. Therefore, the interpreted design G_{max} profile depicted by the two red lines in Figure 5.5b is judged to be reasonably correct for the purposes of the present discussion.

Continuing on now with the exemplar application of the Randolph-Wroth Problem to Pipe Pile 2A, the reason that this is called the whole-pile solution is that the modulus-degradation calculated using Equation 5.12 (or its equivalent using shear modulus as in this case) is applied to all moduli along the entire pile shaft as well as beneath the pile toe, i.e. both Layer 1 and Layer 2 in Figure 5.4, simultaneously. Stated another way, the entire distribution of moduli shown in Figure 5.4 is assumed to degrade (reduce) in the same relative proportion at the same time under the same applied load.

Unfortunately, this uniform modulus-degradation behavior is completely at odds with what theoretically occurs for all types of DFEs in all types of ground conditions. As has been noted several times already in this monograph, it is both long- and well-proven that shaft resistance is fully mobilized at magnitudes of settlement well below that required to mobilize significant portions of toe resistance (the issue of residual loads and their effect on toe resistance is neglected for the time being). Within the context of the current discussion, this means that, in reality, the moduli along the shaft (within Layer 1 in Figure 5.4) degrade much faster than the modulus beneath the toe (within Layer 2 in Figure 5.4). Furthermore, even along the shaft, the degradation of modulus is not uniform and will generally proceed from head to toe as the applied load is increased in a classical example of progressive failure.

The consideration of residual loads, which is now understood to be essential for piles, especially tapered piles, bearing primarily in coarse-grained soil as in this case complicates the subject of modulus degradation significantly. This is because all of the current understanding of the relative mobilization of shaft and toe resistances is based on an assumption of an idealized DFE that has no a-priori mobilization of resistance, i.e. both shaft and toe resistances are starting from zero after pile installation. This is certainly not the case when there are significant residual loads. Clearly, fundamental research is required into what happens with respect to modulus degradation and DFEs when significant residual loads exist.

Continuing on with the present discussion, the writer has performed a relatively extensive assessment of the Randolph-Wroth Problem whole-pile solutions using Niazi's formulation based on shear modulus (Horvath 2016a, 2016b). The conclusion reached is that the overall load-settlement curve for the head of a pile generated by this methodology can be acceptably accurate with the important caveat that the results are only as good as the accuracy of the forecast value of the ULS total resistance, $Q_{t(ult)}$ (R_u in Figure 2.1). It should always be remembered that the Randolph-Wroth Problem is a <u>settlement-only</u> methodology that can only 'fill in the blanks' (Step II) between the origin of the load-settlement plot and the ULS total resistance that is calculated separately beforehand (Step I) as part of a two-step procedure. Consequently, the Randolph-Wroth Problem solutions can never make up for an inaccurate forecast of the ULS total resistance that must be done beforehand using some indirect or direct resistance-forecasting methodology of the analyst's choosing.

Getting back to the exemplar pile being discussed here, Figure 5.8 shows the pile-head settlement for Pipe Pile 2A that was measured in a conventional maintained-load static load test[140] as well forecast using both the rigid- and compressible-pile solutions of the Randolph-Wroth Problem. The forecast of the ULS total resistance, $Q_{t(ult)}$, that was used in conjunction with the Randolph-Wroth Problem whole-pile solutions was made using the writer's Modified Nordlund Method, with the ULS toe resistance calculated using the traditional Hansen bearing capacity solution with Vesic's rigidity/compressibility factors.

---

[140] With regard to the measured data shown in Figures 5.8 and 5.9, as noted elsewhere in this monograph for other static load tests performed at JFKIA in the latter decades of the 20th century that are used in this monograph for reference purposes, a non-standard testing protocol was used that varied from one pile to another. This reflects the fact that the cited load testing was performed as part of quasi-research, pre-production testing programs (what the PANYNJ often referred to as indicator-pile programs) as opposed to Construction Quality Assurance (CQA) load tests performed during production pile driving. As a result, the design professionals in charge of these indicator-pile programs varied the testing protocols to meet program goals. The test protocols generally included: a) extended hold times of the order of several hours or more at higher load levels that were being considered as 'double-design' load levels for production piles as well as b) one or more unload-reload cycles before final loading to what was presumed to be the geotechnical ULS which itself was not consistently defined from one test to another. For example, for the load test for Pipe Pile 2A that is shown in Figures 5.8 and 5.9, there was one intermediate unload-reload cycle and five extended load-holds before the geotechnical ULS was declared at almost 2 inches (50 mm) of pile-head settlement during the final load-hold. This non-standard loading protocol explains why two or more data points are often shown for the same load level and thus should not be taken as plotting errors. Although all data are shown for any extended-hold loads, for clarity of presentation only the beginning and ending data are shown for any unload-reload cycles in this and other, similar figures in this monograph.

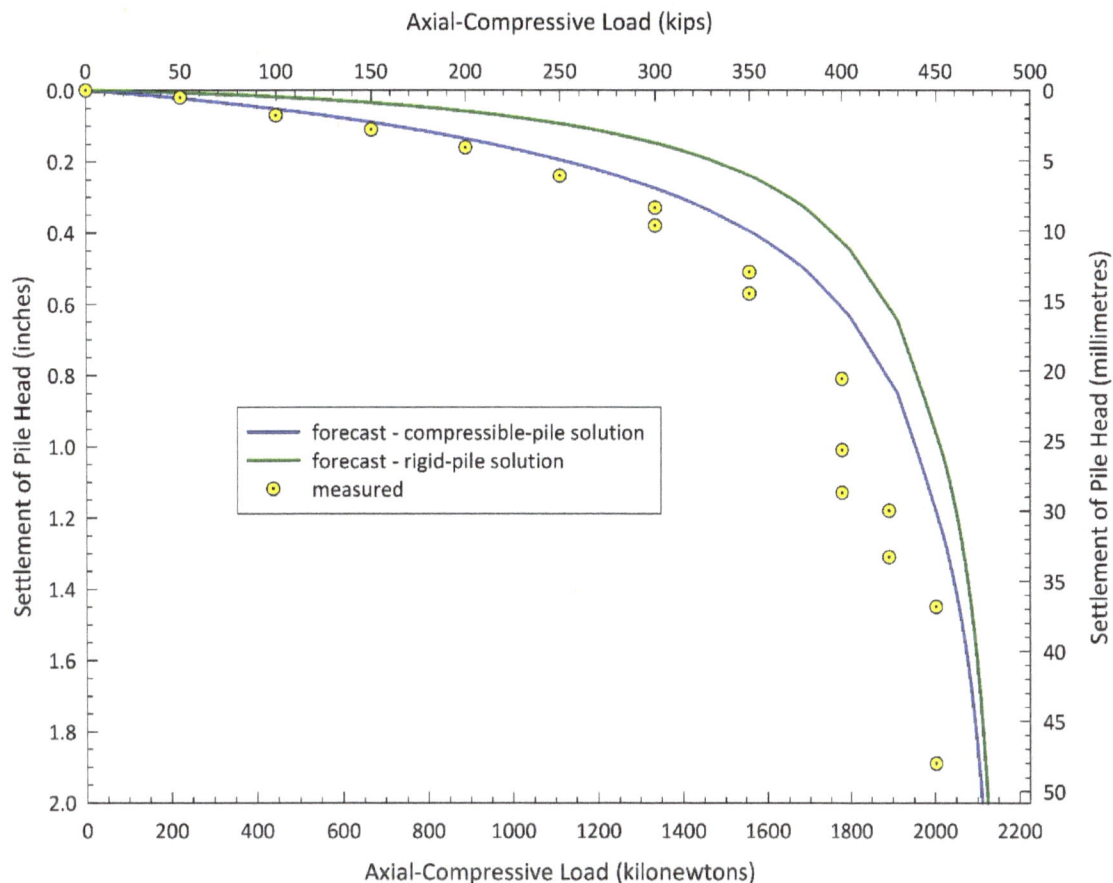

Figure 5.8. Measured vs. Forecast Pile-Head Settlements for Pipe Pile 2A at JFKIA.

There is a noticeable difference between the compressible- and rigid-pile outcomes in this example, with the compressible-pile forecast more closely matching the measured results. Nowadays, with solution of the Randolph-Wroth Problem something that would always be done using a computer, the additional computational effort required to produce the compressible-pile solution is no longer an issue as it was in the past. Thus, there is no reason to perform a rigid-pile solution other than for comparative interest as is the case here.

Figure 5.9 shows the same results as in Figure 5.8 but with the addition of the forecast toe settlement using the compressible-pile solution. Although not shown for clarity, the theoretical elastic compression of the pile shaft is simply the vertical distance between the forecast head and toe settlements obtained using the compressible-pile solutions at any given load level. At the maximum load shown (approximately 470 kips/2100 kN), the theoretical elastic compression of the pile shaft is approximately 0.3 inches (8 mm).

Figure 5.10 shows the influence of varying the g parameter by ±0.1 about the default value of 0.3 as has been suggested by Mayne. Results are shown only for pile-head settlements for the compressible-pile solution. As can be seen, the result of varying the g parameter are relatively significant in this case, and the value of $g = 0.2$ provides a superior fit of the resulting curve to the measured data. This reinforces the writer's suggestion made earlier that research into correlating the two hyperbolic-model parameters associated with the Fahey-Carter Modulus-Degradation Model as implemented into the Randolph-Wroth Problem whole-pile solutions with site-specific soil properties is warranted.

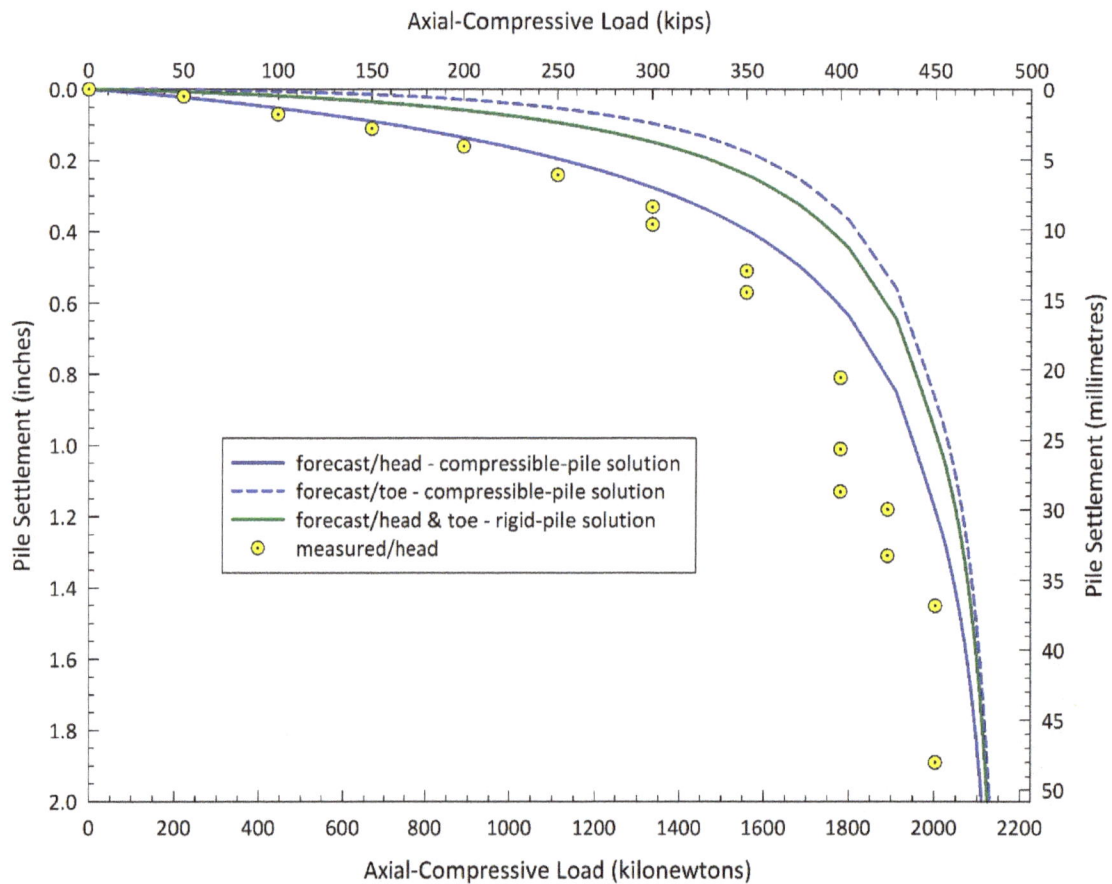

Figure 5.9. Measured vs. Forecast Head and Toe Settlements for Pipe Pile 2A at JFKIA.

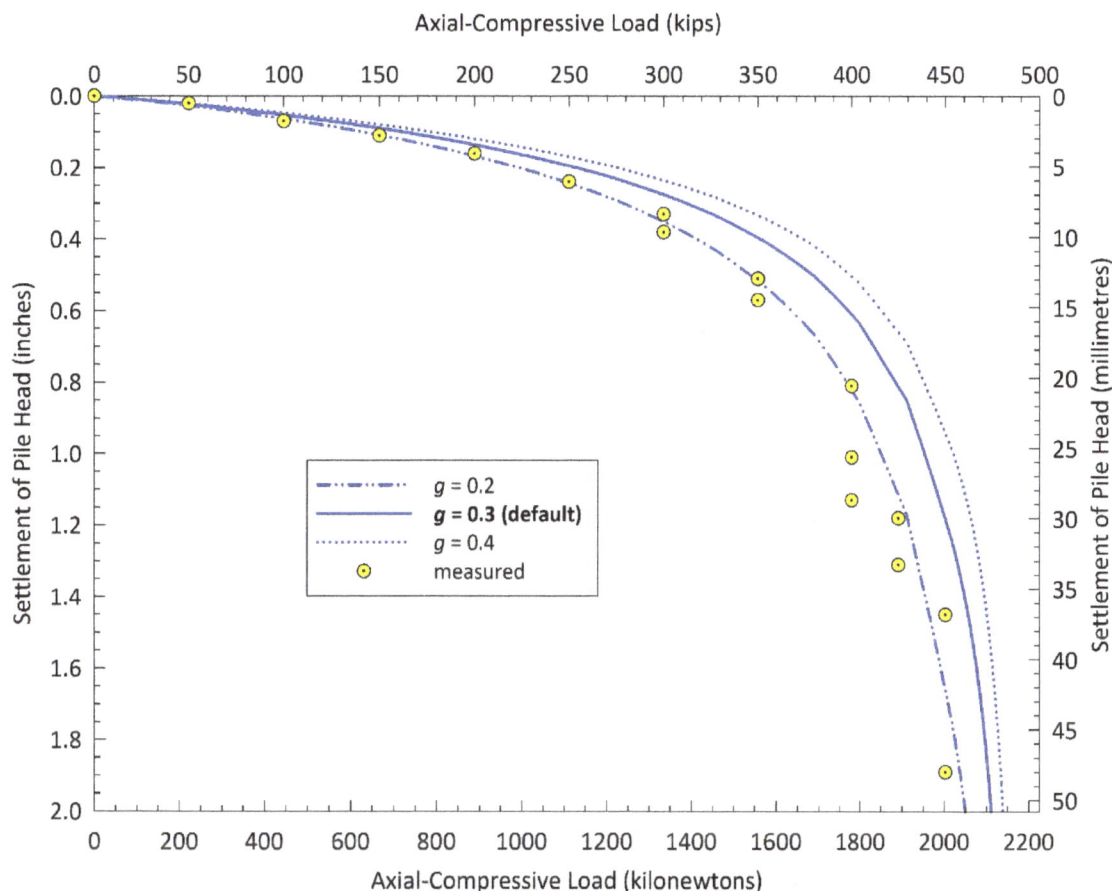

Figure 5.10. Sensitivity of Forecast Pile-Head Settlements to Hyperbolic-Model Parameters for Pipe Pile 2A at JFKIA.

Despite the overall reasonable agreement between the forecast and measured results for the example shown in the preceding figures, the writer has found that the Randolph-Wroth Problem whole-pile solution will not, in general, provide accurate separate assessments of the shaft and toe resistances. This analytical methodology consistently produces shaft resistances that are much too large and exceed the ULS shaft resistance that is calculated independently. This seemingly impossible outcome is, of course, possible because the Randolph-Wroth model is based on the assumption of linear-elastic material behavior where material failure simply never occurs. Conversely, the Randolph-Wroth Problem solution forecasts toe resistances that are much less than the ULS toe resistance, even at relatively large toe settlements where the geotechnical ULS might reasonably be assumed to occur.

These discrepancies tend to be more pronounced for DFEs with a toe that bears in coarse-grained soil where toe resistance tends to be a relatively significant portion of the total resistance despite the relatively small tip area relative to the circumferential area of the shaft. On the other hand, these discrepancies tend to be less when the toe is bearing in fine-grained soil.

Because the writer's pile-related research since circa 2000 has focused on coarse-grained soils, these discrepancies were readily apparent to the writer during the course of research into the Randolph-Wroth Problem over the last several years. On the other hand, case-history examples for the Randolph-Wroth Problem presented by others, e.g. Niazi et al. (2010), have tended to use a DFE bearing entirely in fine-grained soil so that that the discrepancy perhaps did not exhibit itself to those researchers and authors.

There are a number of reasons for this consistent discrepancy. The aforementioned modulus-degradation linkage between shaft and toe is certainly is one of them as is the fact that the Randolph-Wroth Problem formulation assumes that the DFE shaft and adjacent ground will <u>always</u> be perfectly bonded along the entire shaft-ground contact and thus settle the same amount because perfectly elastic conditions exist without limit. Thus, the relative slippage that, in reality, always develops at some point along the shaft-ground interface due to shear failure is simply not allowed to occur.

The specific implications of this behavior for Pipe Pile 2A that is being used here as an exemplar of the Randolph-Wroth Problem whole-pile solutions is illustrated in Figure 5.11. Shown in this figure using solid lines are the ULS shaft and toe resistances, $R_{s(ult)}$ and $R_{t(ult)}$, forecast using the writer's Modified Nordlund Method as noted previously. These lines are vertical as they are the results of a resistance-only methodology and, therefore, inherently independent of settlement. In this case, the ULS resistances provided by the shaft and toe are apportioned 48%-52% respectively. Also shown in this figure are the separate shaft and toe load-settlement curves obtained using the Randolph-Wroth Problem whole-pile/compressible-pile solution. In the limit, the shaft and toe resistances are apportioned 96%-4%, respectively, which is clearly drastically different.

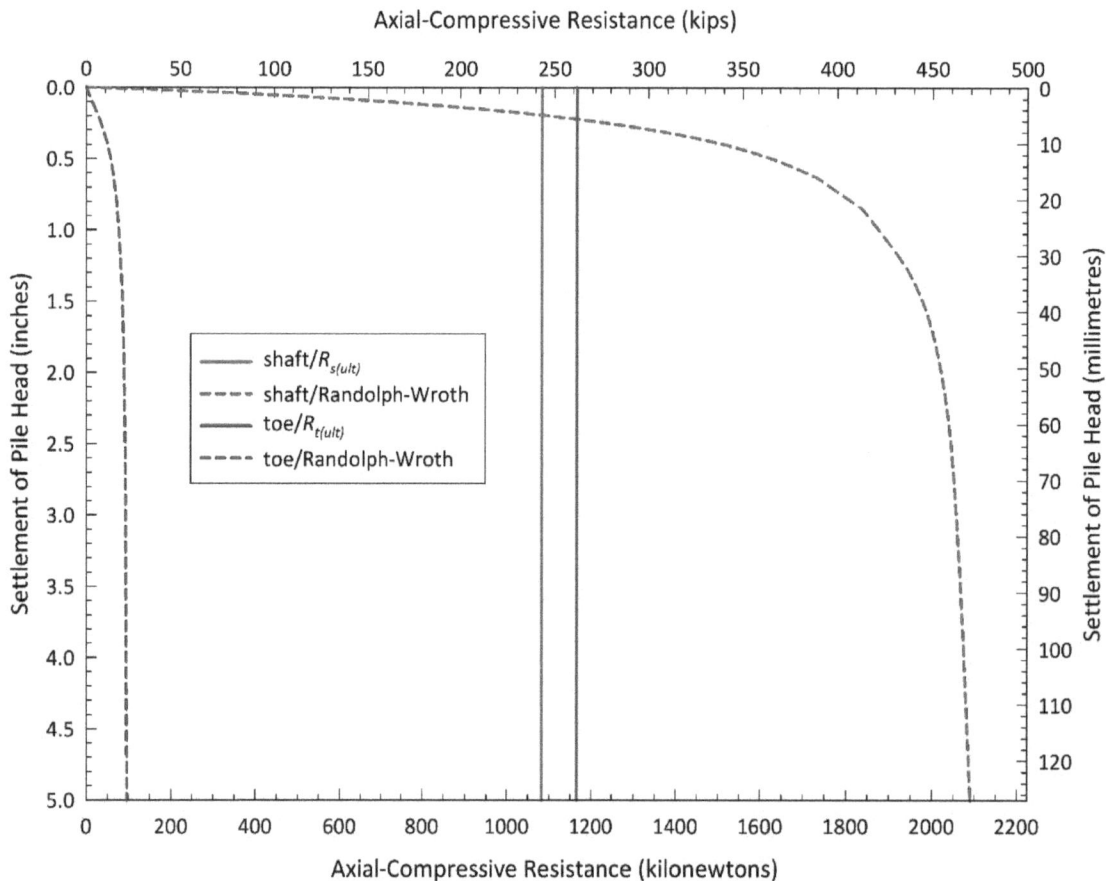

Figure 5.11. Forecast Allocations of Shaft and Toe Resistances for Pipe Pile 2A at JFKIA.

Another perspective of this mismatch in the separate shaft-toe forecast outcomes of the Randolph Wroth Problem whole-pile compressible-pile solution is the elastic compression of the pile shaft. Figure 5.12 shows the forecast axial compression as a function of load for Pipe Pile 2A. Results are shown for the Randolph-Wroth Problem compressible-pile solution as well as the writer's simplistic Tri-Linear Settlement Model that was discussed earlier in this chapter. Also shown are the theoretical results for two limiting cases:

- all load carried by the toe (the PL/AE case) and

- all load carried by the shaft with an assumed uniform ULS unit shaft resistance (the (PL/2)/AE case).

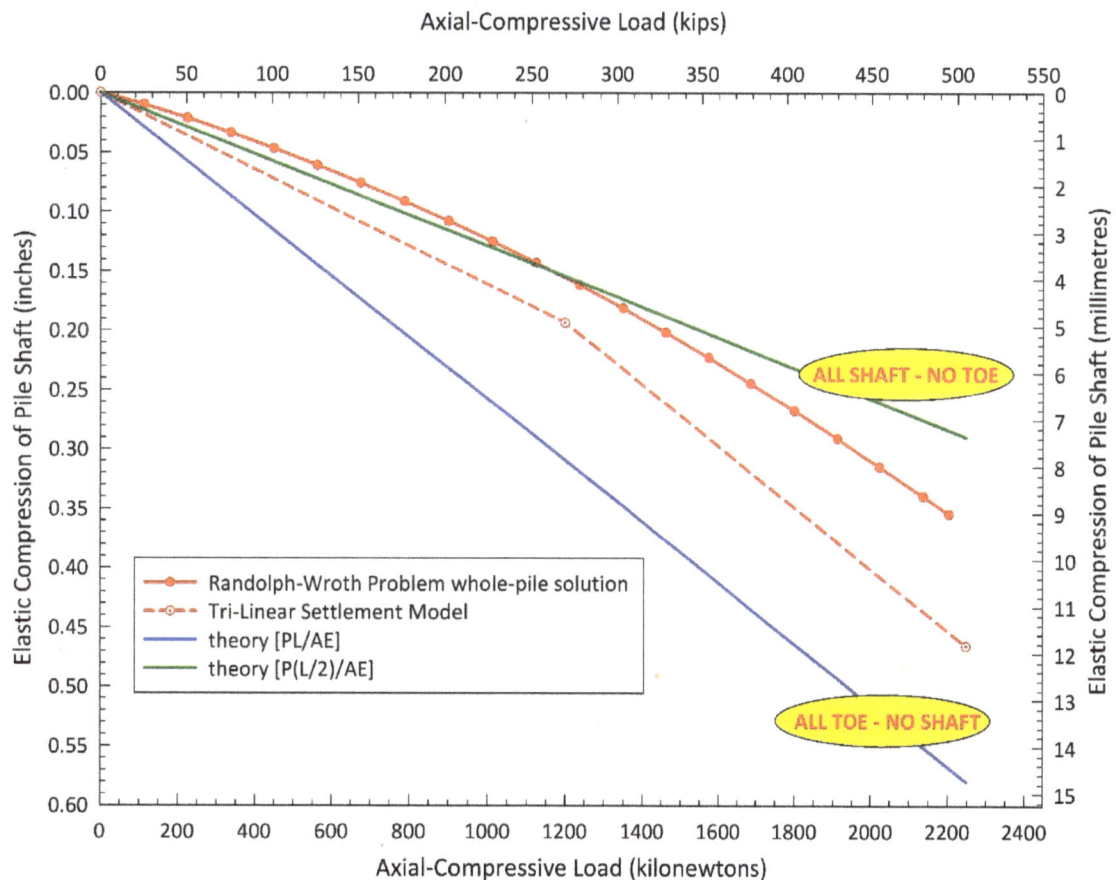

Figure 5.12. Forecast Elastic Compression of Pile Shaft for Different Assumptions for Pipe Pile 2A at JFKIA.

It is clear that the forecast shaft compression from the Randolph-Wroth Problem solution tracks the theoretical all-shaft/no-toe case for approximately one-half of the total load range. This is not surprising given the results shown previously in Figure 5.11.

The conclusion drawn by the writer is that the Randolph-Wroth Problem whole-pile solutions should only be used for forecasting the overall load-settlement response. This is because the separate forecasts of shaft and toe response are unreliable, at least when the toe is bearing in or on a relatively stiff material such as coarse-grained soil. As will be seen in the discussion of integrated load-settlement methods later in this chapter, this has some very significant implications that do not appear to have been appreciated to date.

5.6.2.4.5 The Randolph-Wroth Problem: Niazi's Stacked-Pile Solution

The fact that arguably the most significant flaw in the Randolph-Wroth Problem whole-pile solutions as initially proposed by Mayne in 2007 is its linking shaft and toe modulus degradation was apparently not lost on Mayne. One element of Niazi's circa-2010 doctoral research under Mayne's direction was the development of what Niazi called the *stacked-pile solution*. This involves dividing a DFE into multiple artificial segments of arbitrary and not-necessarily-equal length, with each artificial segment modeled using a Randolph-Wroth Problem whole-pile solution. Thus, the stacked-pile solution is a physically and mathematically linked assemblage of contiguous whole-pile solutions. The intent of doing this is to simulate the development of the progressive failure that develops, at least in principle, in actual DFEs. With the stacked-pile solution, the simulation of progressive failure is a. top-down mobilization of shaft resistance within one whole-pile segment and then another that eventually leads to the mobilization of the actual toe resistance which is only part of the bottom-most shaft segment.

Figure 5.13b[141] shows the writer's re-interpretation (admittedly but one of an infinite number of theoretical interpretations) of the small-strain shear moduli, G_{max}, for Pipe Pile 2A (shown previously in Figure 5.5b interpreted for the whole-pile solution) for the stacked-pile solution. Note that the pile shaft is now divided into five analytical zones (red solid lines 1 through 5 inclusive), each with an assumed constant[142] value of G_{max}, whereas previously a single line (orange dashed line labeled "WP" in Figure 5.13b which is the same as the red solid line in Figure 5.5b) was best-fit to the data along the pile shaft. Note also that the best-fit line of G_{max} = constant below the pile toe is the same red line in both Figures 5.5b and 5.13b (it is labeled "6" in the latter figure).

It appears that Niazi presented the stacked-pile-solution concept in only one publication[143] (Niazi et al. 2010). Curiously, he did not even mention it in passing, no less include it in any detail, in his doctoral dissertation (Niazi 2014).

---

[141] This figure was originally published in Horvath (2016a) which just contained a preliminary overview of the stacked-pile concept. It was published again in Horvath (2016b) that was devoted almost exclusively to exploring the stacked-pile concept in detail.

[142] Niazi stipulated that each shaft segment chosen in the stacked-pile solution should have a depth-wise constant value of G_{max}. However, to the extent that the writer understands the stacked-pile solution, this stipulation appears to be arbitrary. Note that the modulus beneath the toe must have a constant magnitude in both the whole- and stacked-pile solutions as this is stipulated by the original Randolph-Wroth Problem formulation as shown in Figure 5.4.

[143] This comment applies to Niazi's first use of this concept as part of a two-step procedure forecasting methodology where the Randolph-Wroth Problem was used together with the Fahey-Carter Modulus-Degradation Model for the settlement-only portion of the overall forecasting methodology. In Niazi et al. (2010), the writers used results from five different shaft-resistance methodologies and three different toe-resistance methodologies in order to come up with the necessary forecast of ULS total resistance that is required in the two-step-approach process. This point is emphasized as the 2010 publication by Niazi et alia should not be confused with the fact that Niazi later used elements of the stacked-pile concept as part of a composite, one-step-approach load-settlement forecasting methodology. This is a distinctly different application of the stacked-pile concept that is presented and discussed later in this chapter as part of the overall treatment of integrated (one-step) load-settlement methods.

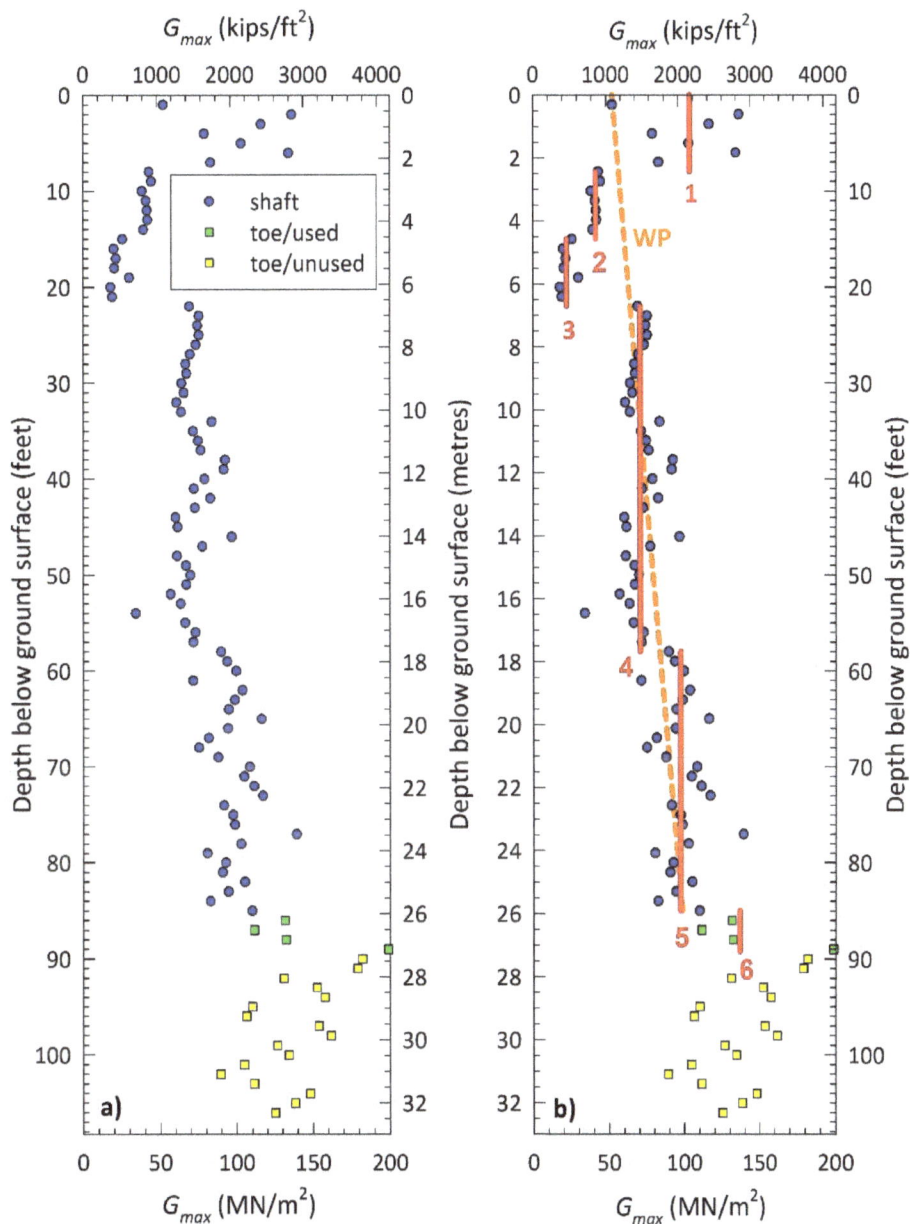

Figure 5.13. Stacked-Pile vs. Whole-Pile Small-Strain Shear Moduli for CPT CTP-2 at JFKIA as Interpreted for Pipe Pile 2A.

In the writer's opinion, there were many gaps in the presentation given in Niazi et al. (2010) of how one actually goes about using the stacked-pile concept in practice although it was clear that the level of effort required was significantly greater than that for the baseline whole-pile solutions. Furthermore, Niazi only illustrated the application of the stacked-pile concept to only one DFE, what he termed a "bored pile" ('drilled shaft' in U.S terminology) bearing in predominantly fine-grained soil. Curiously, he did not compare results from the whole- and stacked-pile solutions so that the increment of forecasting improvement presumably inherent in the stacked-pile concept could be seen.

The writer attempted to implement the stacked-pile concept as part of a broad assessment of the Randolph-Wroth Problem (Horvath 2016b) but without success as the omissions in technical content in Niazi et al. (2010) could not be resolved despite extensive effort by the writer. Multiple communication attempts by the writer over an extended period of time to communicate with Prof. Niazi for the purpose of clarifying and ultimately resolving several ambiguous technical issues were unsuccessful as Prof. Niazi repeatedly indicated that he was 'too busy'.

However, the writer's efforts to implement the stacked-pile solution, although unsuccessful, nevertheless made it clear that the overarching technical issue when implementing the stacked-pile solution is what continuity assumption is invoked between the artificial segments into which the DFE shaft is divided (segments 1 through 5, inclusive, in Figure 5.13b). Specifically, while there should clearly be compatibility of both vertical forces and settlements between artificial segments, e.g. the 'toe' of Segment 2 in Figure 5.13b is also the 'head' of Segment 3, it appears that it is not possible to do so simultaneously within the inherent constraints of the Randolph-Wroth Problem. The writer found that if continuity of forces is used the forecast outcome will differ from that obtained by assuming continuity of settlements. This appears to be a direct result of the fact that while the Randolph-Wroth Problem whole-pile solution yields reasonable forecasts of overall head and toe settlements, the shaft component of resistance is over-forecast while the toe component is under-forecast, sometimes grossly so as in the example illustrated for Pipe Pile 2A in Figure 5.11.

5.6.2.4.6 The Randolph-Wroth Problem: Summary and Conclusion

In summary and conclusion with regard to the Randolph-Wroth Problem, at the present time the baseline whole-pile solutions when applied to an entire DFE appear to be a viable settlement-only methodology that can be used to complement any indirect or direct resistance-forecasting methodology as part of a two-step procedure to creating a load vs. settlement curve for DFEs loaded in axial compression. However, a whole-pile solution should be used to forecast only the mobilization of total resistance, not the separate mobilizations of shaft and toe resistance as the latter are unreliable when viewed separately.

As for the stacked-pile solution of the Randolph-Wroth Problem, further explanation of the implementation of this solution by Niazi or someone else versed in its usage, including a detailed worked example, would appear to be necessary in the writer's opinion. This is because it appears that exactly one (incomplete) example, that of a drilled shaft in fine-grained soil, has ever appeared in the published literature. Comments as to the accuracy of results produced by the stacked-pile solution, both in an absolute sense and relative to the baseline whole-pile solution, cannot be made unless and until there is resolution of several technical issues.

5.6.3 Integrated Load-Settlement Methods (One-Step Procedure)

5.6.3.1 Introduction and Overview

In the context of forecasting the response of DFEs to axial-compressive loads, what are referred to in this monograph as *integrated load-settlement methods* are analytical methodologies that produce forecasts of both total resistance (up to and including a ULS value) and settlement as outcomes of the same, seamless, one-step analytical process. This contrasts to the previously discussed two-step procedure of creating a load-settlement methodology by combining a resistance-only forecasting methodology (Step I) with a settlement-only methodology (Step II) to produce the desired load-settlement outcome. As a group, one-step load-settlement methods can be viewed, at least in principle, as superior to the prior-described two-step procedure of first having to estimate the ULS total resistance and then generating a load-settlement curve using two separate analytical methodologies that have no common or unifying theoretical basis.

The writer has identified two broad approaches that have been used to date for creating one-step load-settlement analytical methods:

1. Some physical model that is readily solvable by computer is either identified from pre-existing models or created for a specific application. The model parameters are then calibrated based on interpreted load tests on actual DFEs, in some cases supplemented by theoretical relationships based on geomechanics. The type of load test (static, quasi-static, or dynamic) and required level of test instrumentation varies with the specific methodology. This approach is conceptually identical to that used for direct methods for resistance forecasting wherein measured results from load tests on actual DFEs are used to calibrate analytical models based on some in-situ test parameter(s).

2. A theoretical model and concomitant solution that does not require a-priori calibration but only site-specific soil properties and project-specific properties of the DFE (physical dimensions and structural properties) to be analyzed. This approach is conceptually identical to that used for indirect methods for resistance forecasting wherein the focus is on developing the appropriate soil properties that are then input into some physical model that is based on theoretical concepts.

Considered in this monograph are two different methods based on Approach #1 and one (but with multiple versions) on Approach #2. Note that the list of methods considered in this monograph is not exhaustive (nor is intended to be so) but is felt to adequately reflect and represent both widely used methods in current practice as well as state-of-art methods, especially those targeted for use with tapered piles.

5.6.3.2 Calibrated Physical Models

5.6.3.2.1 Load-Transfer (t-z/Q-w) Method

The older of the two physical-modeling approaches considered in this monograph is referred to as the *Load-Transfer Method* by Poulos and Davis (1980) although it appears to be more commonly referred to in the literature as the *t-z Method*, with *Q-w Method* being a nomenclature variant found in some U.S. Army Corp of Engineers publications that are available to the public. This method was a natural outgrowth and complement of the well-known *p-y Method* for laterally loaded DFEs as both the *t-z* and *p-y* methods use independent axial springs as the physical model for the force-displacement response of the interface between the DFE and ground, with the springs oriented in the direction of the displacement being modeled (vertical in the case of the *t-z* Method).

Both the *t-z* and *p-y* methods have their developmental roots in the early days (circa 1950s) of the U.S. offshore oil-production industry in the Gulf of Mexico. However, it was the evolution and commercial availability of early mainframe digital computers in academia in the 1960s that resulted in substantial R&D devoted to the *p-y* Method and, to a much lesser extent, the *t-z* Method.

In the writer's professional experience, use of the *t-z* Method never caught on anywhere near the extent of the *p-y* Method, at least in routine terrestrial and shallow-water marine practice. This appears to be a combination of the fact that instrumented pile load tests are required to quantify the necessary model parameters (spring constants) for the *t-z* Method combined with the fact that settlement estimates of individual DFEs has not be done in routine practice historically as has been noted several times in this monograph. This is clearly evident in the recent (2016) version of the FHWA design manual for pile foundations (Hannigan et al. 2016a, 2016b, 2016c) that discusses settlement forecasting for pile groups but essentially makes no mention of settlement forecasting for individual piles.

That having been said, a relatively recent (2006) FHWA-sponsored research report (Pando et al. 2006) contained a very complete discussion of the load-transfer concept[144]. In particular, this report reviews multiple ways for defining and calibrating the pile-ground interaction springs based on various combinations of field measurements and theory.

Relevant to the theme of this monograph is that there does not appear to have been any explicit research devoted to applying the *t-z* Method to tapered piles although there is no fundamental reason why this could not be done. The simpler of two ways to do this would be to develop *t-z* spring relationships for the tapered portions of pile shafts that could be added to the existing database of relationships for the constant-perimeter portions of pile shafts as well as pile toes. Note that in addition to soil type, these relationships would, as a minimum, vary with taper angle and pile material as well.

An alternative approach would be to revise the overall physical model that is used to include both vertical and horizontal/radial springs within the tapered portion of a pile. While this would better model the complex horizontal/radial expansion of the pile-ground interface due to wedging as a pile moves downward and would thus inherently incorporate taper angle, it is not obvious that this would be the preferred approach in practice. The reason is that it would now be necessary to develop ways to calibrate two springs, not one, at every artificial segment into which the pile is divided.

The writer's primary theoretical criticism of the *t-z* Method is broadly similar to that for the *p-y* Method in that each soil-spring used to model the load-displacement (settlement in this case) response of each artificial segment into which a DFE is broken for analytical purposes is assumed to behave independently of adjacent springs. In the parlance of soil-structure interaction (SSI) and concomitant subgrade modeling, the springs are said to be *uncoupled* (Horvath 2018b). This means that each artificial analytical segment is assumed to have no effect on the behavior of the segments above or below it other than that due to compatibility at the interface between adjacent artificial segments. Thus, the load-settlement response of an artificial segment is assumed to be governed primarily by the load on that segment and the soil surrounding that segment.

However, in actual behavior the soil-springs are always *coupled* (Horvath 2018b). In principle, this coupling can be accounted for by proper calibration of uncoupled springs as the coupling effects can be incorporated within the stiffness of the uncoupled spring through a process called *pseudo-coupling* (Horvath 2018b). This approach has been applied successfully to the *p-y* Method. The problem is that this pseudo-coupling is specific to the instrumented DFE used to calibrate the springs and is not, in principle, universally transferable or applicable to other applications. Of course, this becomes less and less of a problem as the database used to calibrate the uncoupled springs becomes larger and larger which is what has happened for the *p-y* Method over the course of more than 50 years of use

---

[144] The following is a direct link (last accessed 15 March 2020) to the report chapter that covers the application of the Load-Transfer Method to a case-history pile in coarse-grained soil:
  www.fhwa.dot.gov/publications/research/infrastructure/structures/04043/07.cfm

and ongoing research and development. However, given the substantially less development of the *t-z* Method over the same period of time, it is likely that there is less confidence in the calibration of the soil springs in the *t-z* Method.

5.6.3.2.2 Niazi Method

Niazi's doctoral research at Georgia Tech under Mayne's tutelage had several distinct components although each was related in some way to the load-displacement behavior of DFEs subjected to axial loads (both compression and uplift were considered by Niazi). Discussed earlier in this chapter were:

- Niazi's direct resistance-only forecasting methodology based on the overall conceptual framework of the UniCone Method that is referred to in this monograph as the Modified UniCone Method.

- Settlement-only analytical methodologies based on the whole- and stacked-pile concepts using the Randolph-Wroth Problem compressible-pile solution and Fahey-Carter Modulus Degradation Model, both expressed using an alternative shear-modulus formulation that better complements and makes use of today's site-characterization capabilities using the sCPTu. These settlement-only methodologies are intended to be used with a resistance-only forecasting methodology as part of a two-step load-settlement forecasting procedure.

Note that, in principle, these two distinctly separate contributions of Niazi's doctoral research could be used together as part of a two-step procedure for forecasting the load-settlement behavior of any type of DFE, with the limitation that the Modified UniCone Method (Step I) in its present form does not accommodate tapered piles as was noted earlier in this chapter. However, it does not appear that Niazi pursued this possibility to any significant extent beyond what was presented in Niazi et al. (2010). One can infer that Niazi perceived the two-step approach for producing a load-settlement curve for a DFE to be a technological dead-end as evidenced by the fact that Niazi did not even mention, no less discuss, this two-step procedure in his final doctoral dissertation (Niazi 2014). Only the Modified UniCone Method was discussed in his dissertation as a standalone topic.

The bulk of Niazi's doctoral dissertation was devoted to the most novel and potentially significant component of Niazi's doctoral research that was the development of an integrated load-settlement methodology (a one-step procedure in the context of the current discussion) based on algebraic rearrangements and clever interpretations of the Randolph-Wroth Problem solutions. In this monograph, this innovative one-step methodology is referred to simply as the *Niazi Method.*

The development of the Niazi Method was presented in detail in Niazi (2014), with a summary presented in Niazi and Mayne (2015c). That summary was recently updated and republished in Niazi and Mayne (2019). However, given the potential use of this method with tapered piles as discussed later in this chapter, an outline of key conceptual elements of the Niazi Method is presented here. This is essential as there are some important assumptions and concomitant caveats that are built into the underlying theoretical framework of the Niazi Method that should be clearly understood by end users of the method. Of note is that the implications of these assumptions and caveats are not highlighted clearly in any of Niazi's publications to date that the writer is aware of.

To understand the conceptual basis of the Niazi Method, it is useful to start by restating the basic, generic equation for both Randolph-Wroth Problem solutions (rigid and compressible pile) that was presented earlier in this chapter (Equation 5.11) but this time using the alternative shear-modulus formulation proposed by Niazi as well as emphasizing the fact that the shear modulus is, in general, an operative value, i.e. varies in magnitude as a function of the load Q_t:

$$w_t = \frac{Q_t \cdot I_{\rho(G)}}{d \cdot G_{L(operative)}}. \tag{5.13}$$

Note that in this case, the dimensionless influence factor, $I_{\rho(G)}$, <u>must</u> be the version based on shear modulus so the variable subscript has been changed from that shown in Equation 5.11 to emphasize this fact.

In the original two-step procedure presented earlier in this chapter in which a Randolph-Wroth Problem solution is coupled with some standalone resistance-forecasting methodology in order to generate a complete load-settlement curve, values of w_t are calculated using Equation 5.13 by assuming arbitrary values of Q_t that are less than the $Q_{t(ult)}$ that must be calculated beforehand using a resistance-forecasting methodology, the choice of which is left up to the design professional. The values of Q_t (actually the dimensionless ratio $Q_t/Q_{t(ult)}$) and $G_{L(operative)}$ (actually the dimensionless ratio $G_{L(operative)}/G_{L(max)}$) are linked together via a <u>stress</u>-based modulus-degradation relationship that requires an a-priori estimate of the ULS total resistance, $Q_{t(ult)}$ (R_u in Figure 2.1). The Fahey-Carter model that was developed originally for pressuremeter data-reduction is the one that has been used exclusively to date for modulus degradation in this application as illustrated in Equation 5.12 (but nowadays using shear modulus instead of Young's modulus as in that equation).

On the other hand, the Niazi Method is fundamentally a <u>strain</u>-based approach so that the form of the Randolph-Wroth Problem solution that is used is as follows after a simple algebraic rearrangement of Equation 5.13:

$$Q_t = \frac{w_t \cdot d \cdot G_{L(operative)}}{I_{\rho(G)}}. \tag{5.14}$$

This equation can be applied in either a whole-pile or stacked-pile analytical process. The former is the original and more-basic application so is discussed first.

In practice, a value of Q_t is calculated using Equation 5.14 by assuming an arbitrary value of w_t. Actually, a value of the dimensionless ratio w_t/d (defined by Niazi as a strain term called γ_p that is expressed as a percent) is used, where d is the DFE diameter (only constant-perimeter DFEs were studied by Niazi). The operative value, i.e. degraded or reduced from the small-strain value $G_{L(max)}$, of $G_{L(operative)}$ (actually, the dimensionless ratio $G_{L(operative)}/G_{L(max)}$ is used) that is used in Equation 5.14 is linked to γ_p via an empirical relationship that is derived from static load tests on actual DFEs. It is through this empirical relationship that ground resistance is built into the problem formulation (the method can be used for DFEs embedded in rock). This process is then repeated until a sufficient number of Q_t vs. w_t data pairs are generated so that a smooth load-settlement curve can be plotted. The writer has found that a simple spreadsheet solution works well for this purpose (Horvath 2016a, 2016b).

In summary up to this point, the empirical relationship linking DFE settlement and operative shear modulus is the key analytical component as well as the most unique, innovative feature of the Niazi Method. This is because it combines both essential elements (resistance forecasting and modulus degradation) that are necessary to solve Equation 5.14 into a single, integrated algebraic relationship rather than relying on separate, discrete analytical methodologies as in a two-step procedure.

Stated another way, what makes the Niazi Method a one-step procedure that does not require a separate assessment of the ULS total resistance, $Q_{t(ult)}$, is that the strain-dependent modulus-degradation relationship used to solve Equation 5.14 is not a generic algebraic relationship such as the Fahey-Carter model (Equation 5.12) whose model parameters must be calculated separately beforehand but a method-specific algebraic relationship that is developed based on a database of actual load-settlement data from static load tests. Thus, all resistances are built into a single analytical methodology.

However, as it turns out, this one-equation-solves-all aspect of the Niazi Method has cons (in the writer's opinion) in addition to the obvious pros. Furthermore, this one-equation-solves-all approach has some implied behaviors that need to be clearly understood by end users. The writer has researched the basic whole-pile solution of the Niazi Method extensively in recent years (Horvath 2016a, 2016b) and the lessons learned from this exercise are summarized in the following presentation and discussion.

To begin with, as with any analytical methodology that is based in whole or in part on an empirical algebraic relationship, the accuracy of the Niazi Method is totally dependent on the database developed to link γ_p (= w_t/d) and $G_{L(operative)}/G_{L(max)}$. Such a database is specific to the Niazi Method because it requires not only measured load-settlement data from static load tests but also an <u>accurate</u> profile of shear-wave velocities, V_s, that were measured prior to installation of the load-tested DFE. The V_s profile is used to calculate the $G_{L(max)}$ profile such as that shown in Figure 5.5a.

It is at this point that the picture is somewhat (and only recently) muddied in the writer's opinion. The writer's reading and understanding of both Niazi (2014) and Niazi and Mayne (2015c) was that <u>only</u> DFE load-test sites at which an sCPTu sounding had been performed were included in Niazi's database. However, the recent paper by Niazi and Mayne (2019) explicitly states that DFE load-test sites were used where the necessary V_s profile was obtained by in-situ testing methodologies other than an sCPTu sounding[145] as well as sites where the V_s profile was estimated using empirical correlations such as were illustrated earlier in this chapter in Figures 5.6 and 5.7.

The use of empirical correlations in particular to estimate V_s is potentially problematic in the writer's opinion. This is because in some cases, e.g. as shown in Figure 5.6, the empirical relationship may accurately reflect the actual values of V_s. However, in other cases and sometimes even at the same site, e.g. as shown in Figure 5.7, the empirical relationship will yield values of V_s that are significantly different from the actual values. Unfortunately, which case applies at a given site or given area within a site can never be known unless <u>actual</u> V_s measurements are made.

Nevertheless, the relevant point being made here is that the overall commercial availability and everyday use of in-situ testing methodologies for generating V_s profiles is relatively recent in the history of foundation engineering. This means that the database that Niazi assembled and used to develop his method is, in the writer's opinion, inherently relatively thin[146], especially when one considers the fact that the Niazi Method includes DFEs consisting of a variety of geometries and structural materials that were installed using a wide variety of driven, jacked, and drilled methodologies in a wide range of ground conditions.

---

[145] Specifically, Niazi and Mayne (2019) mention the *seismic dilatometer* (sDMT), *spectral analysis of surface waves* (SASW), and "standard" (not otherwise explained) downhole tests.
[146] A complete list and description of all the case histories that Niazi used to create his database (or at least his initial database) can be found in Niazi (2014).

However, there are always two sides to every story and in the interest of fairness and objectivity it should be noted that Niazi clearly feels that the database he used was sufficiently comprehensive and robust for the intended purpose, as evidenced by the proactive argument made in Niazi and Mayne (2019). This is clearly one of those cases where a design professional contemplating use of the Niazi Method needs to form their own opinion based on available, published information.

However, what is <u>not</u> open to debate or opinion is fact based on that which has been published by Niazi. Not surprisingly, given the overall novelty of the Niazi Method and the fact that the required database was apparently undergoing development and refinement during the several years that Niazi was conducting research at Georgia Tech, the writer (Horvath 2016a) identified several evolutionary versions of Niazi's DFE database in the published literature:

- An interim-basic version that was apparently released in 2011 (the writer used Niazi (2013) as the reference for this) that used a single empirical equation for all types of DFEs and ground conditions.

- A final-basic version in Niazi (2014) that also used a single empirical equation that included all problem parameters, i.e. DFE typology and ground typology.

- An advanced version in Niazi (2014) that included separate variables (called α_1 and β_1) that allowed for variation in the empirical relationship based on DFE typology.

- An even-more-advanced version in Niazi (2014) that included additional separate variables (called α_2 and β_2) that allowed for additional variation in the empirical relationship based on soil Plasticity Index. This version actually consisted of two separate sub-versions that varied depending on the specific algebraic equation (exponential vs. hyperbolic-tangent) used to fit the data. Based on the most-recent publications to date that discusses this aspect of the Niazi Method (Niazi and Mayne 2015b, 2019), it appears that the hyperbolic-tangent curve fitting reflects Niazi's current preference.

Based on the writer's assessment of the Niazi Method that was presented and discussed in detail in Horvath (2016a), while the advanced version of the empirical γ_p vs. $G_{L(operative)}/G_{L(max)}$ relationship that explicitly considers DFE typology appears to be valid, further refinement of this relationship to consider ground typology based on Plasticity Index alone appears to be highly speculative and unjustified based on the current database (note that the position taken by Niazi in Niazi and Mayne (2019) clearly disagrees with the writer's sentiment in this regard). This is especially true for cases where a DFE penetrates different soil types which tends to be more the rule than the exception in the real world of deep foundations. Picking a single soil type for the purposes of modifying the empirical relationship is highly subjective and problematic in such cases.

Notably (in the context of this monograph) absent from any of these versions of Niazi's database was any kind of steel or PCC tapered pile[147]. On a positive note is the fact that it is relatively easy to both expand Niazi's existing database as well as create a new database as desired. Thus, not only could tapered piles be added to Niazi's original database (this is explored later in this chapter) but on a large project it would be relatively easy to create a database solely for project-specific conditions of DFE types, subsurface conditions, etc.

---

[147] Niazi and Mayne (2019) state that "timber" piles were included in the database but there is no indication of how many. Note that none was listed in Niazi (2014).

Of arguably greater importance than the breadth of the database and parameter structure of the empirical γ_p vs. $G_{L(operative)}/G_{L(max)}$ relationship is the theoretical basis used to create, and thus inherently underlying, this empirical relationship. This is because whatever behavioral characteristics are part of the theoretical basis will carry over into any empirical relationship derived using the theory, a point that is discussed further subsequently.

Not surprisingly, Niazi used the Randolph-Wroth Problem compressible-pile solution as the basis for developing his empirical γ_p vs. $G_{L(operative)}/G_{L(max)}$ relationship. In simple terms, the process involves solving the following rearranged version of Equation 5.13 for operative values of $G_{L(operative)}$ using values of $G_{L(max)}$ calculated from an sCPTu sounding (or alternative methodology as discussed earlier) and Q_t vs. w_t data-pairs from the static load test performed at the site being studied:

$$G_{L(operative)} = \frac{Q_t \cdot I_{\rho(G)}}{d \cdot w_t}. \qquad (5.15)$$

The details for doing this are presented in Niazi (2014), with an abbreviated version in Niazi and Mayne (2015b, 2019). However, for the sake of completeness, a brief outline is presented here using the format of a generic, hypothetical case history and using data from Pipe Pile 2A at JFKIA simply to illustrate typical results.

To begin with, application of the Niazi Method presumes that:

- A site-specific V_s profile is obtained prior to installing the DFE. This can be done using one of several in-situ methodologies enumerated in Niazi and Mayne (2019) although an sCPTu sounding would appear to be the preferred option in the writer's opinion, assuming that subsurface conditions allow it. If all else fails, this profile could be generated using some empirical methodology (with all the uncertainty that entails) although Niazi and Mayne (2019) did not make any specific recommendations or suggestions for this.

- A conventional, i.e. uninstrumented, static load test is performed on the installed DFE. The only data required from the load test are applied loads and corresponding settlements, both measured at the head of the DFE.

To begin the calculation process, the calculated (from V_s data) G_{max} profile is plotted and then interpreted along the length of the load-tested DFE as shown in Figure 5.5 in the same manner as done previously when the Randolph-Wroth Problem solution is used as part of a two-step procedure for generating a load-settlement curve[148]. From this interpretation (Figure 5.5b), $G_{L(max)}$ is determined (approximately 2,100 kips/ft^2 (100 MN/m^2) for the actual pile shown in this figure). The interpretation shown in Figure 5.5b also allows various problem variables within the $I_{\rho(G)}$ parameter that are based on the relative distribution of G_{max} along the shaft and beneath the toe of the DFE to be evaluated.

Q_t vs. w_t data-pairs from the load test are then taken in turn (the value of w_t is used to calculate the corresponding value of $\gamma_p = w_t/d$) and the value of $G_{L(operative)}$ is then calculated using Equation 5.15. Note that because $G_{L(operative)}$ is embedded in the $I_{\rho(G)}$ parameter on the right-hand side of this equation, an iterative solution methodology is required. The specifics for doing this are left up to the design professional (Niazi and Mayne 2019). The calculated

---

[148] Note that the specific empirical correlation used by the writer to generate the G_{max} profile shown in Figure 5.5 skipped over the intermediate step of first calculating V_s and calculated G_{max} directly.

value of $G_{L(operative)}$ is then divided by $G_{L(max)}$ to yield the $G_{L(operative)}/G_{L(max)}$ ratio that corresponds to the previously calculated value of γ_p.

This process is then repeated until all the Q_t vs. w_t data-pairs obtained from the load test have been processed to create a set of γ_p vs. $G_{L(operative)}/G_{L(max)}$ data-pairs. This dataset is then used for further interpretation. It is relevant to note that a load test does not have to be performed to the geotechnical ULS in order to produce usable results. This is because data from any load level provide usable outcomes for the Niazi Method.

For the sake of completeness and academic interest, a variety of related interpretations are presented before illustrating the specific interpretation methodology used to develop the empirical equations for his methodology. To begin with, Niazi found that these calculated outcomes tend to display the signature S-shaped curve found for many geotechnical and foundation behaviors (especially those related to small-amplitude dynamics) when plotted using the axis variables and semi-log scaling shown in Figure 5.14. In fact, Niazi used various concepts from dynamic testing, e.g. the well-known resonant-column test in particular, in formulating and interpreting his empirical relationship between γ_p and $G_{L(operative)}/G_{L(max)}$.

As a result of this influence of small-amplitude dynamics, as shown in Figure 5.14 Niazi normalized γ_p to a parameter he called $\gamma_{p\text{-}ref}$ and assigned $\gamma_{p\text{-}ref}$ an arbitrary value of 0.01 = 1%. Note that this definition of $\gamma_{p\text{-}ref}$ corresponds to a relatively small value of pile-head settlement, w_t. For example, for Pipe Pile 2A that is being used here as an example, the magnitude of w_t corresponding to $\gamma_{p\text{-}ref}$ is only approximately ⅛ inch (3 mm).

Plotted in Figure 5.14 as an exemplar of the outcomes of the above-described process are the:

- continuous curve for the empirical relationship for (driven) piles presented in Niazi (2014) and Niazi and Mayne (2015c) but neglecting correction for soil Plasticity Index and

- writer's interpreted empirical data for Pipe Pile 2A using the load-test data shown in Figure 5.8.

While the load-test data for this pile are far from ideal because of the extended hold times plus unload-reload cycles, the agreement with Niazi's empirical relationship is still remarkably good.

As an aside, the Niazi Method highlights the necessity of making numerous load-test measurements at relatively small load levels that are often overlooked during conventional load testing if one wants to back-calculate the γ_p vs. $G_{L(operative)}/G_{L(max)}$ relationship over the entire five-log-cycle strain range shown in Figure 5.14. This is because the assumed limit of small-strain shear modulus/elastic soil behavior and the transition to degraded shear modulus/inelastic soil behavior occurs at an assumed (by Niazi) value of $\gamma_p/\gamma_{p\text{-}ref} = 0.001$. For Pipe Pile 2A, this is a value of pile-head settlement, w_t, $\cong 0.00013$ inches (3 μm). This is an extraordinarily small number that is well below the precision of mensuration systems used in routine static load testing of DFEs.

Before proceeding further, a relevant question to ask is how does the modulus-degradation relationship proposed by Niazi as shown in Figure 5.14 compare to that obtained using the Fahey-Carter Modulus Reduction Model discussed earlier in this chapter. Although the Fahey-Carter model is stress-based, it is possible to relate it to strain-based behavior for a specific DFE. This was done by the writer for the exemplar Pipe Pile 2A and the outcomes are shown in Figure 5.15.

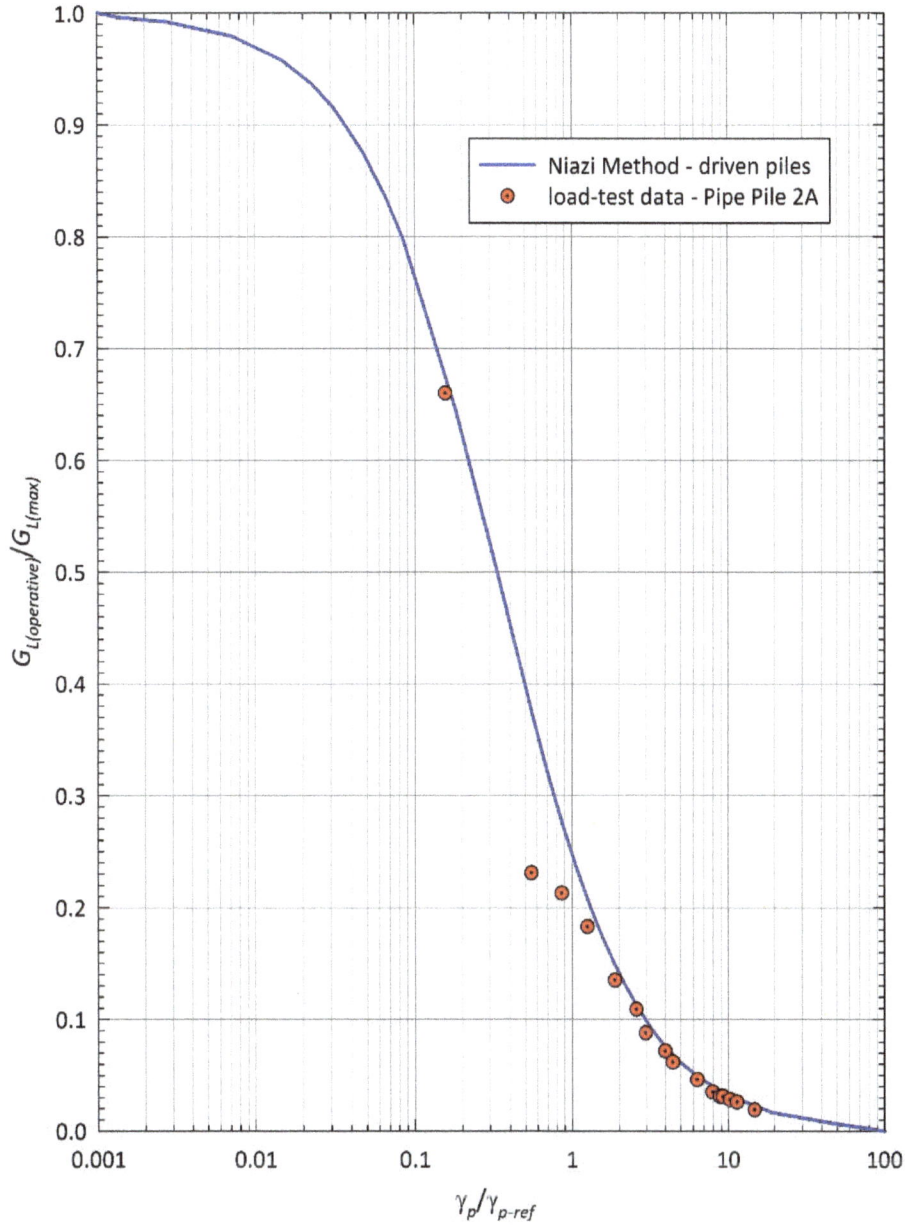

Figure 5.14. Back-Calculated $G_{L(operative)}/G_{L(max)}$ vs. γ_p/γ_{p-ref} Values for Pipe Pile 2A at JFKIA.

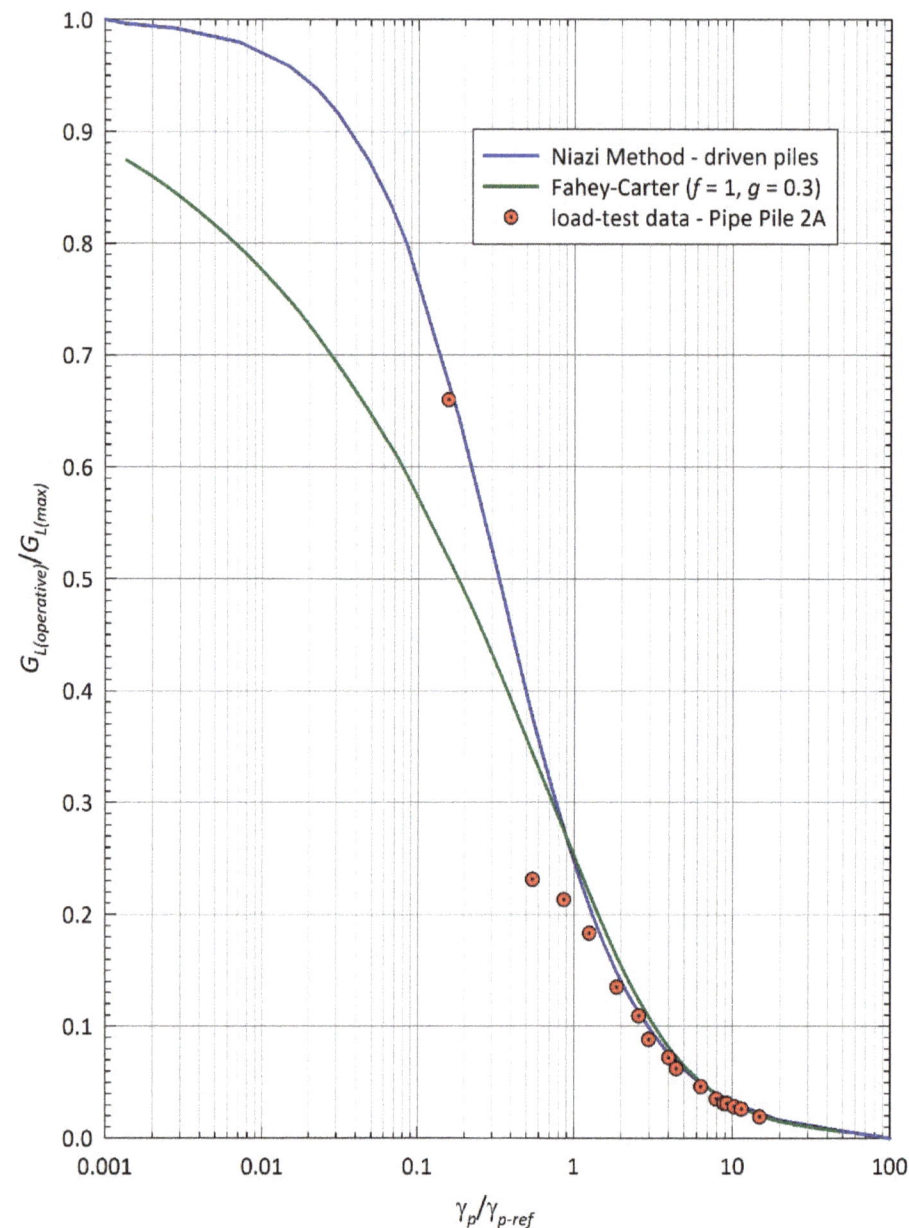

Figure 5.15. Comparison of Theoretical $G_{L(operative)}/G_{L(max)}$ vs. $\gamma_p/\gamma_{p\text{-}ref}$ Relationships for Pipe Pile 2A at JFKIA.

There is clearly a significant difference between the Niazi and Fahey-Carter model results for relatively small displacements, with the latter implying a much 'softer' load-settlement response initially. However, this difference disappears for all intents and purposes as the $\gamma_p/\gamma_{p\text{-}ref}$ ratio approaches a value of 1, i.e. $\gamma_p = 0.01 = 1\%$ which is a pile-head settlement, w_t, of approximately ⅛ inch (3 mm) in this case. As can be seen in Figure 5.15, this range where the two forecasting models produce essentially the same results covers the bulk of the load-test range (Figure 5.8) for Pipe Pile 2A, i.e. from approximately 150 kips (670 kN) on to the maximum load of 450 kips (2000 kN).

Moving on now with the discussion of the development of the Niazi Method, the results shown in Figure 5.14 are only an intermediate stage in the development process for Niazi's empirical γ_p vs. $G_{L(operative)}/G_{L(max)}$ relationship. Niazi found that the $\gamma_p/\gamma_{p\text{-}ref}$ vs. $G_{L(operative)}/G_{L(max)}$ data-pairs formed a linear-trending relationship when plotted using the axis parameters and linear scaling shown in Figure 5.16. The results shown in this figure are the same as those shown in Figure 5.14, i.e. Niazi's overall best-fit trend line for all (driven) piles and the writer's back-calculated results for the load test performed on Pipe Pile 2A at JFKIA. As with Figure 5.14, the load-test results from Pipe Pile 2A are in overall very good agreement with the general trend line the Niazi found for driven piles.

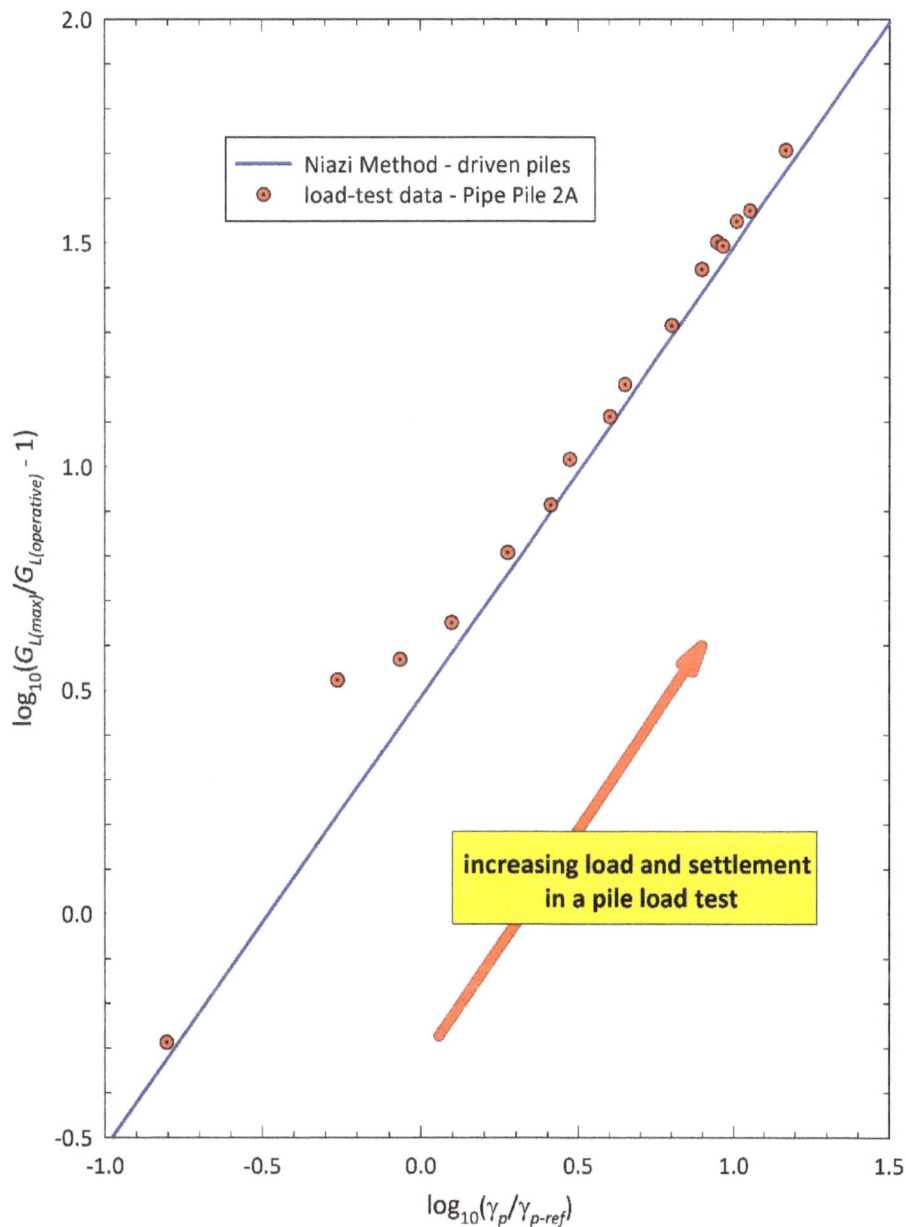

Figure 5.16. Back-Calculated $G_{L(operative)}/G_{L(max)}$ vs. $\gamma_p/\gamma_{p\text{-}ref}$ Values for Pipe Pile 2A at JFKIA.

The several evolutionary versions listed previously of the empirical γ_p vs. $G_{L(operative)}/G_{L(max)}$ relationships that form the basis of the Niazi Method were developed by Niazi by defining best-fit straight lines to plots of data similar to that shown for Pipe Pile 2A in Figure 5.16 and then manipulating the results algebraically into a form shown subsequently. Because the overall process is straightforward to replicate using nothing more than commercially available spreadsheet software, as noted earlier in this presentation it is possible to either extend the Niazi Method to DFE types not considered by Niazi, e.g. tapered piles, helical piles, and micropiles[149], or create a site- and project-specific database for larger projects involving any type of DFE, even those embedded in rock.

The writer has done this extension of the Niazi Method in a preliminary manner for tapered piles, primarily as a proof-of-concept exercise as opposed to developing a final result that is intended for routine use in practice. However, before presenting the results of this R&D work there are some additional topics related to the basic Niazi Method that require discussion.

To begin with, as with the original application of the Randolph-Wroth Problem solutions using the Fahey-Carter Modulus-Degradation Model as part of a two-step procedure for generating a complete load-settlement curve, the Niazi Method can be applied using either the basic whole-pile approach or the stacked-pile approach that Niazi developed in an attempt to overcome the problem of concurrent shaft- and toe-modulus degradation that is inherent in a whole-pile analysis. However, the writer's attempts to use the Niazi Method for stacked-pile analysis (Horvath 2016b) encountered the same unresolved problems that were discussed earlier in this chapter for stacked-pile analysis using the Fahey-Carter model. In the writer's opinion, Niazi's publications (Niazi 2014, Niazi and Mayne 2015c) that discuss use of the Niazi Method in a stacked-pile analysis do not contain the worked-example details sufficient to illustrate and educate one in use of the concept. It is of interest to note that in his most recent publication on the subject (Niazi and Mayne 2019), Niazi has dropped all mention of the stacked-pile analytical alternative.

A broader issue that is independent of whether a whole-pile or stacked-pile approach is used is that the most obvious limitation of the Niazi Method in general is that it should only be used for sites where a site-specific V_s profile has been defined prior to installation of the DFE, preferably using an appropriate in-situ testing device such as the sCPTu. This is necessary in order to define the site-specific G_{max} profile. In theory, for sites lacking actual V_s data it is possible to estimate the G_{max} profile using empirical correlations applied to CPTu data as was illustrated earlier in this chapter. Note, however, there is a very real risk of creating an empirically based G_{max} profile that deviates from the actual, especially in those cases where the site soils exhibit some structure as the result of aging and/or cementation. This was discussed in Horvath (2016b) for JFKIA and is illustrated in Figure 5.7c (aging, not cementation, is believed to be the root cause of the soil structure in the JFKIA example).

However, the writer's earlier research into the Niazi Method indicated that a far more serious flaw is its being based on a Randolph-Wroth Problem solution in the first place. As discussed earlier in this chapter and illustrated in Figure 5.11 as well as in detail in Horvath (2016a), the Randolph-Wroth Problem does a very poor job of properly allocating shaft and toe resistances in certain applications such as those involving coarse-grained soils where toe resistance is a relatively significant fraction of the total resistance even though the toe area is small relative to the shaft circumferential area. Unfortunately, Niazi did not explore a wide range of actual applications of his methodology in any of his publications that might have

---

[149] Also known as *minipiles*, *pin piles*, *root piles*, and the original Italian-language term used for them: *pali radice*. Development of this deep-foundation concept is generally attributed to the late Dr. Fernando Lizzi of Fondedile in the years immediately following World War Two.

revealed this. In fact, in all of the publications known to and reviewed by the writer, Niazi used the same case history (a drilled shaft/bored pile in fine-grained soil) to illustrate both the stacked-pile concept using the Fahey-Carter stress/force-based modulus-degradation model as part of a two-step procedure (Niazi et al. 2010) and his own strain-based modulus-degradation model and one-step procedure (Niazi and Mayne 2015c).

In summary and conclusion, the Niazi Method for integrated, one-step load-settlement forecasting is sufficiently intriguing so as to warrant further investigation and evaluation by comparison to measured load-settlement behavior but with the overall caution to not rely on this method as the sole methodology for forecasting the ULS total resistance unless and until a much greater level of confidence has been gained through experience. Note, however, that Niazi clearly has a very different opinion in this regard (Niazi and Mayne 2019). In the writer's opinion, it is not reassuring that in the known (to the writer) published literature to date Niazi used exactly <u>one</u> case history to illustrate the claimed forecasting accuracy of his method.

In addition to this general caution, there are the following specific cautions at the present time and for the foreseeable future:

- It appears reasonable to use the Niazi Method only at sites where the G_{max} profile is known explicitly from an sCPTu sounding or other in-situ device that measures V_s. Site-specific shear moduli were essential elements of developing the Niazi Method so this must be respected.

- Shear-modulus degradation should be evaluated using only the equation that accounts for DFE typology. A concise reference for this is Niazi and Mayne (2015b). Based on the writer's limited investigation as reported in Horvath (2016a), it appears that the database Niazi used to develop his method was insufficient to allow for reliable extension to account for soil typology (in the form of Plasticity Index) in addition to DFE typology although Niazi clearly feels otherwise (Niazi and Mayne 2019).

- Only a basic whole-pile analysis should be performed. In the writer's opinion, Niazi has not satisfactorily illustrated application of the stacked-pile concept to a wide variety of DFE types and, more importantly, soil conditions. Furthermore, in the writer's first-hand experience it is unclear how to properly implement a stacked-pile analysis. Niazi's claims that this can be done with nothing more than generic spreadsheet software but the details are elusive, at least to the writer.

- Only forecasts of total resistance obtained using the Niazi Method should be considered potentially reliable. The underlying Randolph-Wroth Problem solution appears to do a poor job of allocating shaft and toe resistances correctly in some applications although these errors appear to be compensating so that the total resistance can be reasonable.

5.6.3.2.3 Extended Niazi Method: Background and Overview

As noted in the preceding section, it appears that Niazi did not include any tapered piles in his database that was used to create the several versions of empirical equations used for forecasting shear-modulus degradation as a function of head-settlement of a DFE that is the key analytical component of the Niazi Method. In principle, there is no reason why tapered piles could not be added to the database or placed in their own database to extend the Niazi Method. Niazi actually clearly explains the process for plotting and interpreting load-settlement data from load tests in detail in Niazi (2014) and with a summary in Niazi and Mayne (2015b) so the procedural blueprint for extending his methodology readily exists and can readily be followed. This process was also outlined in the preceding section using the exemplar of Pipe Pile 2A at JFKIA.

That having been said, there are several unique factors associated with tapered piles that need to be understand and properly dealt with in any extension of Niazi's one-step procedure in its basic whole-pile analytical form. First and foremost is the fact that the boundary conditions for the Randolph-Wroth Problem that is used as the theoretical basis for the Niazi Method assume a DFE of depth-wise constant perimeter (Figure 5.4). For piles that are tapered over their entire length (timber, most precast-PCC, and certain steel shell piles such as the *Raymond Standard* and *Raymond Step-Taper*), it would be a theoretically defensible first-order approximation to use an average perimeter. However, for steel pipe piles such as the *Monotube* and *TAPERTUBE* as well as some types of precast-PCC piles (see Figure 3.11) that always consist of an upper constant-perimeter section and a lower tapered section the correct approach to use is not obvious. It might be tempting as a first-order approximation to simply use an overall average perimeter for such piles as well and, for lack of a better alternative, the writer did this for results shown subsequently. The problem is that this mixes the shaft-resistance contributions for the constant-perimeter and tapered sections and actually gives more 'weight' in terms of resistance contribution to the constant-perimeter section which is the opposite of the relative contributions in reality. This fact is illustrated in the detailed assessment of an instrumented *Monotube* pile that is presented in Appendix A.

How to deal with this issue in a rational manner will require further study. Niazi's stacked-pile concept might be particularly useful in such cases. However, as discussed in the preceding section, there are still some significant unanswered questions concerning implementation of this analytical concept that must be addressed before this can be explored.

Additional factors that need to be considered and included explicitly in any expanded database for the Niazi Method clearly include taper angle and perhaps toe diameter as well. As noted earlier in this chapter, there is some evidence that for a given taper angle the taper benefit is related to the toe diameter.

Material type may be a factor as well although this is likely less important than taper angle. Unknown at this time is whether or not cross-sectional geometry is a significant factor. The vast majority of tapered piles used historically have had a circular cross-section either naturally or by design but precast-PCC tapered piles with a square cross-section are known to exist as illustrated in Chapter 3. As discussed earlier in this chapter, based on limited, fragmentary information provided by Nordlund (1979), it appears that tapered piles with a square cross-section may behave resistance-wise in a manner that is different from tapered pile with a circular cross-section, all other factors being equal.

When all of these factors are considered, it is clear that simply lumping all tapered piles together into one database to be interpreted to produce the necessary empirical equation for shear-modulus degradation as a function of pile-head settlement is a gross oversimplification of the problem. It appears that, as a minimum, separate databases should be developed based on taper angle and pile type although this would not resolve the above-described issues of dealing with piles such as the *Monotube* and *TAPERTUBE* that have both constant-perimeter and tapered sections.

There is one additional consideration and concomitant caution that the writer has discovered during the course of initial, preliminary research performed over the past few years for developing what will be referred to in this monograph as the *Extended Niazi Method* for tapered piles, the specifics of which are discussed subsequently. Note that this consideration is not specifically related to tapered piles and is thus worthy of note for any attempt to extend the Niazi Method beyond Niazi's original database for any type of DFE, e.g. helical piles and micropiles as noted earlier.

As discussed in Niazi (2014) and Niazi and Mayne (2015b), a plot of the type illustrated by Figure 5.16 requires a best-fit linear interpretation of the datapoints in order to develop the coefficients of the empirical equation used when applying the method in practice. Depending on the quality of the load-test data used, there can be considerable scatter in the plot for data associated with small load and settlement levels, e.g. the lower-left-hand portion of Figure 5.16. As will be seen, this was true for the data used by the writer that were all from older load tests for which only hand-drawn load-settlement plots, which required manual scaling of data values by the writer, were available. Note, however, that such scatter did not appear in the plots in Niazi (2014) and Niazi and Mayne (2015b), presumably because Niazi had access to original load-test records in which the load-settlement data were tabulated, digitized, or otherwise available with greater precision.

In any event, when interpreting the plotted data for a best-fit line, the inclusion of all data in the process can produce unreasonable results because of the aforementioned scatter at small load and settlement levels. This is especially true if the built-in mathematical function in software such as a spreadsheet is used to produce the linear fit as such software typically treats each piece of data with equal weight for curve-fitting purposes. The alternatives are to either manually fit a line to the data using engineering judgment or use only data from larger loads and settlements, where the scatter tends to be less, in any type of automated process. This is illustrated and discussed further subsequently.

5.6.3.2.4 Extended Niazi Method: Proof of Concept

Over the last few years, the writer performed initial, preliminary research into developing the Extended Niazi Method to include tapered piles using the following modest database of tapered piles driven in predominantly coarse-grained soil conditions at JFKIA between 1972 and approximately 2000:

- five timber piles with overall average taper angles[150] between 0.2° and 0.3°;

- nine *Monotube* piles with the intermediate Type J tapered lower portion that has a taper angle of 0.57°; and

- one *Monotube* pile with the maximum Type Y tapered lower portion that has a taper angle of 0.95° plus four *TAPERTUBE* piles with the same taper angle.

Given the very modest size of this database combined with the fact that sCPTu soundings were unavailable at the locations of these so G_{max} profiles had to be estimated based on empirical correlations using CPT data, the goal of the writer's research was primarily proof of concept in nature. Consequently, the outcomes were, and are, not intended to be used for future design purposes but rather to qualitatively explore broad trends.

Initially, a rigorous extension of the Niazi Method was explored by the writer. How this was done is best understood by starting with the most general advanced form of the

---

[150] As discussed in Chapter 3, the creosoted Southern Yellow Pine timber piles driven at JFKIA in the circa-1970s timeframe that were used for the writer's research reported in this monograph typically exhibited taper angles that varied along the pile shaft approximately ±⅓ relative to the overall average angle.

empirical modulus-degradation relationship that is the heart of the Niazi Method (Niazi 2014, Niazi and Mayne (2015b)):

$$\frac{G_{L(operative)}}{G_{L(max)}} = \frac{1}{\left\{1 + 3.634\,\alpha_1\,\alpha_2\left[\left(\frac{\gamma_p}{\gamma_{p-ref}}\right)^{0.942\beta_1\beta_2}\right]\right\}} \qquad (5.16)$$

where α_1 and β_1 are empirical variables that reflect DFE typology, and α_2 and β_2 are empirical variables that reflect soil typology (specifically, Plasticity Index).

To relate this equation to what was shown earlier in this chapter with respect to the Niazi Method, all of the numerical parameters and variables in Equation 5.16 reflect the usual $ax + b = 0$ variables that define the best-fit straight line through the data on a plot such as shown in Figure 5.16. Specifically:

- The quantity $(0.942\,\beta_1\,\beta_2)$ is the slope a of the best-fit straight line in Figure 5.16.

- The $\log_{10}$ of the quantity $(3.634\,\alpha_1\,\alpha_2)$ is the intercept b of the best-fit straight line in Figure 5.16 with the $\log_{10}(\gamma_p/\gamma_{p-ref}) = 0$ axis.

One approach to developing the Extended Niazi Method for tapered piles would be to develop new values of the variables for DFE typology, α_1 and β_1, and assume that the variables for soil type, α_2 and β_2, are unaffected[151]. However, the writer chose an alternative approach of using Niazi's basic values for (driven) piles, $\alpha_1 = 0.837$ and $\beta_1 = 1.068$, and adding new variables, α_3 and β_3, that are intended to include all taper-related effects (taper angle, toe diameter, pile material, etc.). Using this alternative approach, the rigorous form of the modulus-degradation equation for the Extended Niazi Method becomes:

$$\frac{G_{L(operative)}}{G_{L(max)}} = \frac{1}{\left\{1 + 3.634\,\alpha_1\,\alpha_2\,\alpha_3\left[\left(\frac{\gamma_p}{\gamma_{p-ref}}\right)^{0.942\beta_1\beta_2\beta_3}\right]\right\}}. \qquad (5.17)$$

Note that for constant-perimeter piles, $\alpha_3 = \beta_3 = 1$ and Equation 5.16 is recovered as expected.

Using Niazi's α_1 and β_1 values for (driven) piles and neglecting the soil-typology correction (in which case $\alpha_2 = \beta_2 = 1$), Equation 5.17 simplifies to:

$$\frac{G_{L(operative)}}{G_{L(max)}} = \frac{1}{\left\{1 + 3.042\,\alpha_3\left[\left(\frac{\gamma_p}{\gamma_{p-ref}}\right)^{1.006\beta_3}\right]\right\}}. \qquad (5.18)$$

Quantifying α_3 and β_3 is then a straightforward application of the same process that Niazi used to create his original database that was outlined earlier in this chapter. For each of the 19 piles at JFKIA that was assessed by the writer, the following sequence of events was performed:

---

[151] As discussed earlier in this chapter, it is the writer's opinion and position that in the present stage of development of the Niazi Method a correction for soil typology appears to be unsupported by the database so this correction can be ignored in which case $\alpha_2 = \beta_2 = 1$ always.

- A G_{max} profile was estimated using CPT data and empirical correlations, then interpreted as shown in Figure 5.5 for $G_{L(max)}$.

- For each piece of load-settlement data obtained in a static load test, $G_{L(operative)}$ was evaluated using Equation 5.15.

- For each piece of load-settlement data, $\gamma_p/\gamma_{p\text{-}ref}$ vs. $G_{L(operative)}/G_{L(max)}$ data-pairs were evaluated and plotted in the format shown in Figure 5.16. The mathematical functions built into *Excel* were then used to determine best-fit straight lines through data. From the slope and intercept of these lines, the α_3 and β_3 variables were quantified.

Figure 5.17 shows the results of this process applied to creating a database of the aforementioned 19 tapered piles driven at JFKIA between 1972 and approximately 2000. The three different colors used for the symbols in this figure reflect the three taper angles represented by these piles. Also shown in this figure is the line defining the best fit for all types of constant-perimeter (driven) piles for the original Niazi Method and neglecting soil typology, i.e. Equation 5.18 with the writer's proposed taper-correction factors, α_3 and β_3, equal to one. This is the same line shown previously in Figure 5.16.

As it turned out, the aforementioned scatter in the data for relatively small loads and settlements (values of $\log_{10}(\gamma_p/\gamma_{p\text{-}ref})$ less than approximately +0.5) resulted in best-fit lines (not shown) through the tapered-pile data that were judged by the writer to be unreasonable. Specifically, the trends of the lines were not parallel to Niazi's line for constant-perimeter piles as one would expect as the line-fitting process was dominated by the widely scattered data in the lower-left-hand portion of Figure 5.17. Nevertheless, two broad trends can still be inferred from the results shown in this figure:

- As a group, tapered piles exhibit behavior that differs from Niazi's best-fit trend line for constant-perimeter piles.

- There is definite segregation in the tapered-pile results based on taper angle, with the deviation in the data from Niazi's trend line for constant-perimeter piles increasing with increasing taper angle (from yellow to cyan (light blue) to dark blue in the symbols used in Figure 5.17). This is more apparent in Figure 5.18 that is an enlarged view of the right half of Figure 5.17 showing data at larger loads and settlements (approximately ¼ inch (6 mm) and greater in this case) in the load tests. The trend with increasing taper angle is clearly to shift Niazi's trendline for constant-perimeter (driven) piles in the direction noted by the red arrow in Figure 5.18.

Although not obvious from Figures 5.17 and 5.18, the interpreted data for tapered piles indicates greater overall pile-soil system stiffness, i.e. less degradation in shear modulus for a given magnitude of pile-head settlement, than for constant-perimeter piles. This is more clearly seen by replotting the complete results shown in Figure 5.17 using the format of Figure 5.14 which is the most common format used historically in geotechnical and foundation engineering for depicting modulus degradation. This is done in Figure 5.19.

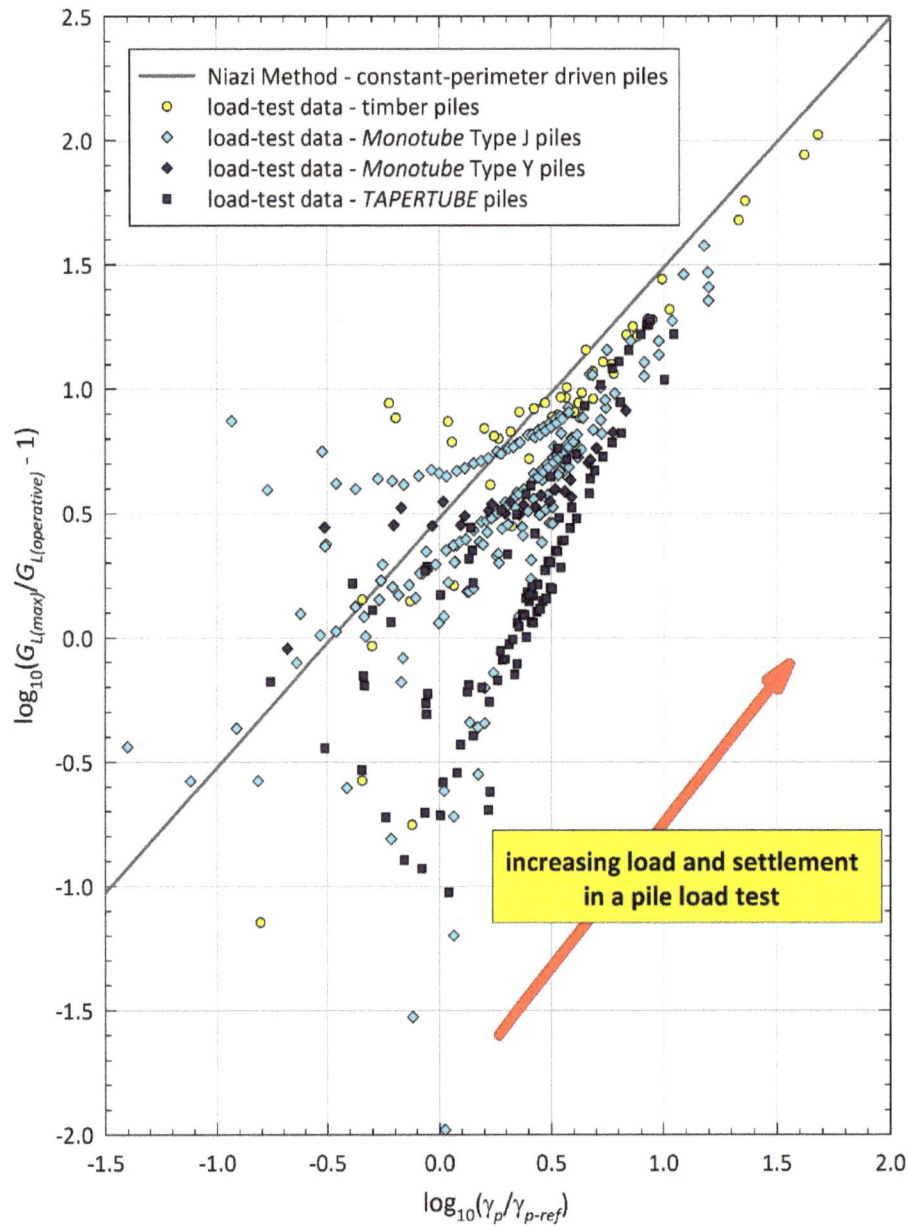

Figure 5.17. Niazi-Type (Strain-Based) Shear-Modulus Degradation for JFKIA Tapered-Pile Load-Test Data (Log-Log Plotting).

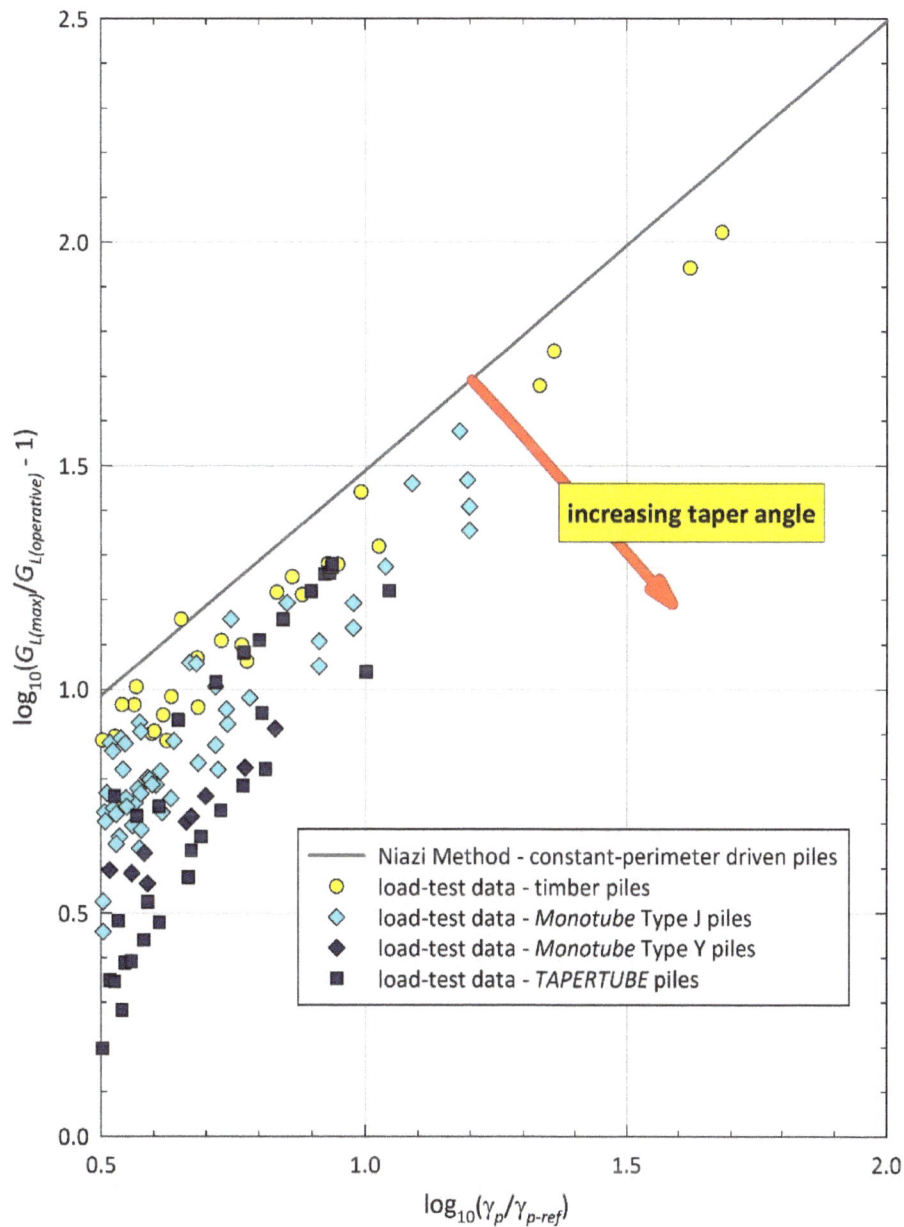

Figure 5.18. Niazi-Type (Strain-Based) Shear-Modulus Degradation for JFKIA Tapered-Pile Load-Test Data (Log-Log Plotting/Enlarged View).

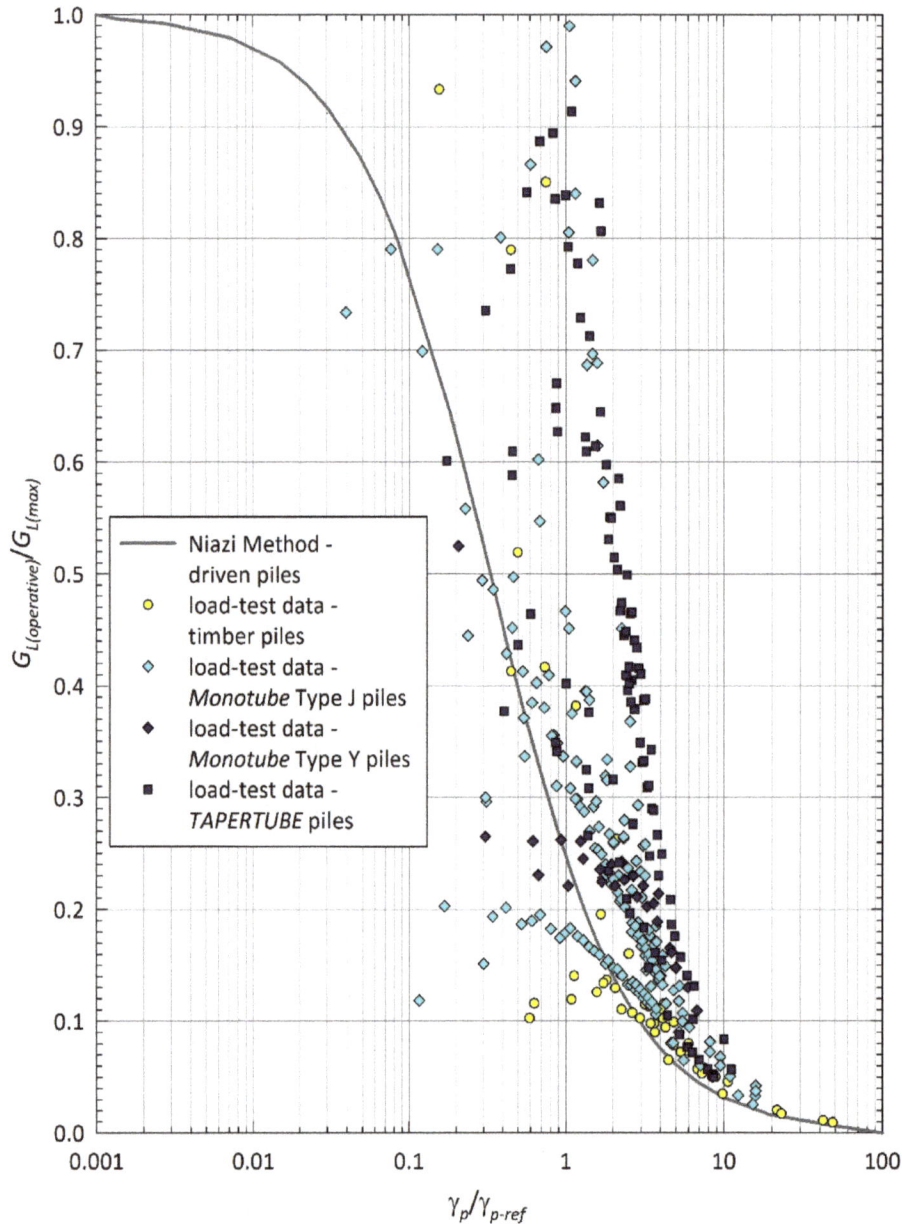

Figure 5.19. Niazi-Type (Strain-Based) Shear-Modulus Degradation for JFKIA Tapered-Pile Load-Test Data (Semi-Log Plotting).

The influence of taper on the outcomes compared to the baseline results for constant-perimeter (driven) piles is much more apparent in the plotting format shown in Figure 5.19. For a given level of normalized strain (which is actually pile-head settlement, w_t, in this case), tapered piles as a group exhibit less modulus degradation, i.e. increased pile-soil system stiffness, relative to Niazi's baseline relationship for constant-perimeter (driven) piles although the effect is not uniform. Specifically, the pile-soil system stiffness is much more pronounced for the tapered steel pipe piles, especially those with the larger taper angles (the two different dark-blue symbols), compared to the timber piles (yellow symbols). This suggests that pile-soil system stiffness increases with increasing taper angle (this is also apparent in Figures 5.17 and 5.18) although the differences between and among the tapered-pile types decreases substantially with increasing strain/pile-head settlement. Furthermore, the pile-soil system stiffness of all tapered piles collapses toward the trend line for constant-perimeter piles at the relatively large normalized strain levels/pile-head settlements that reflect the geotechnical ULS. For reference purposes, a pile-head settlement, w_t, of one inch (25 mm) is represented by $(\gamma_p/\gamma_{p\text{-}ref}) = 12.5$ in Figure 5.19 and $\log_{10}(\gamma_p/\gamma_{p\text{-}ref}) = 1.1$ in Figures 5.17 and 5.18.

The overall conclusion drawn from this proof-of-concept exercise is that tapered piles exhibit a stiffer response to axial-compressive loads compared to constant-perimeter piles. This translates into increased resistance for tapered piles relative to constant-perimeter piles for a given level of strain/pile-head settlement. Unfortunately, due to the limited number (19) of tapered piles in the writer's database and the significant scatter in the plotted results, it was not possible to quantify reliable values for the writer's proposed taper variables α_3 and β_3 in Equations 5.17 and 5.18.

It is of interest to speculate as to the reason(s) why tapered piles exhibit greater pile-soil system stiffness compared to constant-perimeter piles. One possibility is that the physical mechanism of wedging by which tapered piles develop shaft resistance is simply a stiffer resistance mechanism compared to shaft resistance due to sliding friction. It is also possible that residual load may play a role, especially since the published literature clearly indicates that tapered piles develop relatively greater magnitudes of residual loads compared to any other type of DFE. The steel pipe piles (*Monotube* and *TAPERTUBE*) tend to be much longer and certainly stiffer structurally than timber piles. Therefore, it would be expected that residual loads are proportionately greater for steel piles compared to timber piles although it would be difficult to instrument a timber pile to prove this one way or the other.

5.6.3.3 Cylindrical-Cavity Expansion Theory

5.6.3.3.1 Kodikara-Moore Method: Background

Despite the game-changing conceptual breakthrough of linking cylindrical-cavity expansion (CCE) and the wedging mechanism of tapered piles that is reflected in the Kodikara-Moore Method, the entire gamut of published technical literature since 1993 (research reports, technical papers in scholarly journals and at conferences, design manuals, informative articles in trade magazines) has, with relatively few exceptions, not even mentioned their published work no less made use of it in either practice or follow-on research. One likely reason for this is that although both Professors Kodikara and Moore are still active in academia (in Australia and Canada respectively), it does not appear that either conducted any substantive follow-on research or publication on the topic of tapered DFEs that would have assisted in promoting their ideas. As the writer has learned after almost five decades of professional practice, a new and novel idea generally requires repeated publication in every venue possible over a period of years if it has any hope of becoming noticed no less established in practice[152].

---

[152] One of the rare exceptions to this general observation and certainly one of significant relevance to this monograph is the Nordlund Method which, for all practical purposes, only appeared in publication one time, in 1963. Nordlund (1963) has been widely cited in countless publications, including to the present, although most curiously perpetuate the unfortunate misspelling of his surname in the original paper.

The writer became aware of Kodikara and Moore's seminal paper at the time of its publication in 1993 and immediately recognized the significance of their work. Because of the writer's interest in tapered piles that dated back to 1972, the writer immediately contacted both authors via postal mail[153] to inquire about the availability of the computer code that Kodikara used to perform the analyses reflected in the paper as well as to inform the authors about commercially available tapered piles, primarily timber and *Monotube* at that time, that they had failed to even mention, no less consider, in their research.

The writer eventually received polite, but effectively dead-end, replies from each author. In retrospect, this is not surprising given the fact that time has shown that neither author apparently performed any significant follow-on research related to tapered DFEs although Prof. Moore, who by that time had moved from Australia to Canada, may have had an indirect role in encouraging work related to tapered DFEs by others in Canada (primarily El Naggar and his generations of graduate students) as discussed subsequently.

Throughout the 1990s, the writer was actively and fully involved in research related to cellular geosynthetics (geofoams and geocombs). Consequently, it was not circa 2000 that the writer, motivated by the then-recent development of the *TAPERTUBE* pile and its initial use at JFKIA, again became actively involved in researching tapered piles. This research produced a series of research reports, conference papers, and white papers beginning with Horvath (2002) and continuing to and culminating in the present with the publication of this monograph.

The central theme of all of these publications is that the writer realized that Kodikara and Moore's work defined what for many years the writer called the *third capacity mechanism of side resistance from cylindrical cavity expansion*, a concept and phrase that first appeared in print in 2002 (Horvath 2002). The basic concept is still valid but, in the terminology adopted in this monograph, it is now defined as the *third resistance mechanism of shaft resistance due to wedging*. The physical mechanism of wedging is now modeled using the theoretical concepts of cylindrical-cavity mechanics that have been broadened to include cavity creation, CCX (during pile driving); cavity expansion, CCE (during post-installation axial-compressive loading); and cavity contraction, CCC (during post-installation uplift loading).

With specific regard to the Kodikara-Moore Method, the writer's research that began on the cusp of the 21st century included another, closer look at their work. This was done with the initial intent of developing a computer code that would allow the writer to use the Kodikara-Moore Method in both research and practice. However, it quickly became apparent that the theoretical simplifications that were made by Kodikara and Moore in developing their method (discussed in detail subsequently) were problematic for several reasons. In addition, it also became apparent that the mathematical relationships that they used are, overall, quite complex, involving operations such as the evaluation and differentiation of infinite series. While certainly not insurmountable in the larger scheme of things, this fact alone makes use of the Kodikara-Moore Method daunting to the average foundation engineering practitioner.

Thus, the writer's attention was first directed toward developing the Simplified Nordlund Method as an interim improvement for resistance forecasting of tapered piles. It was only during the several years of concentrated research that resulted in publication of this monograph that the writer focused on the Kodikara-Moore Method more deeply than ever and realized that it had broader problematic issues than the writer first thought in the past.

---

[153] It should be kept in mind that in 1993 computer-based communication tools such as the Internet and email that are taken for granted nowadays simply were not widely available to the general public although that changed by the late 1990s.

5.6.3.3.2 Kodikara-Moore Method: Overview

The concept of CCE on which the entire Kodikara-Moore Method is based was introduced and discussed in broad terms in Chapter 4. This was done to both acknowledge Kodikara and Moore's seminal contribution to the current state of knowledge related to tapered DFEs (the Kodikara-Moore Method is not limited to piles as is the Nordlund Method) as well as to define the concept's rightful place as the superior physical model proposed and mathematized to date for forecasting resistance to axial-compressive loading by tapered DFEs in general and tapered piles in particular.

Although CCE decades ago replaced (at least in principle) Nordlund's early-1960s concept of passive lateral earth pressure development as being the way to visualize the wedging behavior that is unique to tapered piles subjected to axial-compressive loading, there has been, and continues to be, a curious disconnect with foundation engineering practice. Unfortunately, most stakeholders involved with tapered piles, especially design professionals, appear to be ignorant of even the existence of Kodikara and Moore's work. For example, current design-oriented documents for piles that have been generated by government agencies such as NYSDOT (2015) and the FHWA/NHI[154] three-volume design-manual set issued in 2016 do not even mention Kodikara and Moore's work no less attempt to make use of it as a practical analysis and design tool.

Although the CCE model only produces forecasts of shaft resistance, Kodikara and Moore incorporated this model into an integrated, one-step load-settlement forecasting methodology. This is because the resistance to any type of expanding cavity, cylindrical or spherical, is never fixed in magnitude but a function of normal displacement parallel to the direction of cavity expansion. In the case of a cylindrical cavity that is used to model the wedging that occurs along the shaft of a tapered DFE, the relevant displacement is in the radial direction.

However, the shaft geometry of a tapered DFE is such that radial displacement of the DFE-ground interface is a function of DFE settlement as illustrated conceptually in Figure 4.1. As can be seen in this graphic, any settlement of a tapered DFE results in radial expansion of the DFE-ground interface. Consequently, using the CCE model as the cornerstone of an integrated, one-step load-settlement forecasting methodology is a logical extension of the basic concept as DFE settlement must be considered explicitly anyway when evaluating the effects of cavity expansion.

Before undertaking a detailed assessment of the Kodikara-Moore Method, it is of interest to highlight a significant conceptual difference between the Kodikara-Moore and Nordlund methods in addition to the obvious difference in the physical model that each chose to replicate the wedging mechanism of tapered-pile behavior under axial-compressive loading. As will be seen, this conceptual difference impacts perceptions related to the important practical design-related issue of taper benefit.

---

[154] The National Highway Institute (NHI) is a training function of the FHWA Office of Technical Services (OTS). The NHI essentially functions as a non-degree-granting, post-graduate, continuing-education educational institution for professional involved in highway engineering. The NHI functions on the business model of bringing the campus to the student (in part because some states simply will not pay for their employees to go out of state for any kind of training) and currently offers courses that use both traditional on-site, in-person teaching by a contract instructor as well as Web-based instruction. In the interest of full disclosure, in the 1990s the writer was a paid peer-reviewer of an FHWA R&D contract involving ground modification and improvement that was intended to be developed into an NHI course or courses after an initial period of beta-testing. However, the writer has never been involved directly with the NHI in any manner.

To begin the discussion, it is useful to reiterate the point made previously in Chapter 4 that cavity <u>expansion</u> (as is being discussed here) is not the same physically or analytically as cavity <u>creation</u>. In the Kodikara-Moore Method, Kodikara and Moore did not attempt to model the process of installing a tapered DFE into the ground and thus initially <u>creating</u> a cavity where none existed. As discussed in the following section, it appears that they essentially imagined a tapered DFE as already being in the ground in the not-uncommon practice referred to in geomechanics R&D as *wished-in-place* modeling. Thus, the process of cavity <u>expansion</u> as incorporated into the Kodikara-Moore Method only models the post-installation radial displacement of the pile-soil interface as the direct result of axial-compressive loading applied to the head of the DFE.

This point is emphasized here as Kodikara and Moore did not even mention, no less highlight, the conceptual limitations of their resistance-forecasting methodology in their 1993 paper. In fact, they glossed over the installation aspects of tapered DFEs and said nothing about the fact that installing a tapered DFE into the ground may contribute to the overall taper benefit and may possibly produce the bulk of the taper benefit depending on the specific type of tapered DFE and its installation process. Thus, by inference, they attribute all taper benefit to post-installation externally applied loading (even then, only a portion of the loading as discussed subsequently) which, in the writer's opinion, has led not only Kodikara and Moore but others (discussed subsequently) to draw what appears to be significantly incorrect conclusions concerning taper benefit.

It follows, then, that all taper benefit that is produced analytically by the Kodikara-Moore Method is due solely to the settlement of a tapered DFE under applied axial-compressive loading. This means that, in this case, taper angle alone does not define taper benefit as the same magnitude of cavity expansion could, in theory, be achieved regardless of taper angle as long as there was sufficient settlement. Stated another way, there are theoretically an infinite number of combinations of taper angle and pile settlement that can produce a given magnitude of radial displacement of the pile-soil interface. So, in this case taper benefit is a complex function of both taper angle and settlement (and toe diameter as well as cavity pressure is relative to the initial cavity radius). This also raises the possibility that settlement can actually govern in actual applications.

On the other hand and in sharp contrast to these implications of the Kodikara-Moore Method, the Nordlund Method in both its original and revised versions implies that taper benefit derives from the combination of the pile-installation process and subsequent external loading in axial compression to the geotechnical ULS[155] although no attempt was made to separate out the relative contribution of each component. Specifically, the driving of a tapered pile into the ground followed by loading to the geotechnical ULS in axial compression was viewed as being mathematically equivalent to rotating a rigid wall about its base (which corresponds to the toe of the pile in this case) and, in the process, generating lateral earth pressures acting along the pile shaft that are greater than those that would theoretically occur with a non-tapered pile. Thus, in Nordlund's model taper benefit is related primarily to taper angle as this angle uniquely defines how much 'wall rotation' occurs (there is some influence of toe diameter due to the fact that several model parameters are dependent on the absolute volume of the pile per unit length). Settlement of the installed pile under applied load is implied but is not considered explicitly in the Nordlund Method as it is in the Kodikara-Moore

---

[155] That the Nordlund Method implies post-installation axial-compressive loading to the geotechnical ULS derives from the basic equation for resistance that Nordlund used to develop his analytical methodology (Equation 5.4). It is abundantly clear from this equation that it was assumed that post-installation axial-compressive loading would be sufficient to mobilize ULS values of both shaft and toe resistance.

Method. This is consistent with the Nordlund Method being a resistance-only forecasting methodology.

The practical outcome of these two different perspectives as to the genesis of taper benefit, i.e. exclusively and incrementally (depending on settlement under applied load) post-installation (Kodikara and Moore) vs. the synergistic result of both installation plus loading to the geotechnical ULS (Nordlund), is the inference as to whether or not there is a limit to taper benefit. Stated another way, is there a maximum taper angle beyond which there is no additional benefit in terms of increased resistance.

The Nordlund Method is unequivocal on this and the answer is 'yes'. This is because Nordlund assumed that once the passive earth pressure state was fully mobilized at some finite value of pseudo-wall rotation that further rotation provided no increase in lateral earth pressure which is consistent with that theory. This is why the charts in both the original and revised versions of the Nordlund Method show a definite value of what can be called the 'limiting value of taper angle' that is a function of the ϕ angle of the soil. Specifically, the limiting value of taper angle goes from approximately 2° to 1° to 0.5° as ϕ varies from 30° to 35° to 40°.

The Kodikara-Moore Method presents a more-nuanced perspective and concomitant conclusion on this subject. On the one hand, as noted in Chapter 4 there is always a limiting radial stress, called the *limiting pressure*, to cavity expansion in general. This limiting value is a function of problem-specific soil properties as well as the cavity radius. However, in any given application this limiting pressure cannot be uniquely related to a taper angle for two interrelated factors:

- Limiting pressure is related to relative cavity expansion, i.e. the magnitude of radial expansion of the cylindrical cavity relative to its initial radius.

- The magnitude of radial expansion is related to a combination of taper angle and settlement as noted previously.

The conclusion is that toe diameter, taper angle, and magnitude of settlement each and collectively influence the limiting cavity pressure in CCE. Thus, there is no limiting taper angle per se in the Kodikara-Moore Method as there is unequivocally in the Nordlund Method as, in theory, the limiting cavity pressure in CCE can be reached for any taper angle greater than zero depending on the initial cavity radius and how much settlement and concomitant radial expansion occur.

As an aside, it is of interest to observe that in the Kodikara-Moore Method toe diameter plays an explicit role as together with taper angle these two problem parameters define the initial radius of the cylindrical cavity that is the shaft-ground interface at any given point along the tapered portion of a DFE. As has been noted now several times throughout this monograph, toe diameter is emerging as a key variable that influences the resistance of tapered piles. Perhaps this is not surprising as Nordlund concluded in his original 1963 paper and reaffirmed it in his 1979 presentation notes that pile volume per unit length of pile shaft, which is clearly related to toe diameter, was a variable that influenced the overall taper benefit in his analytical methodology.

For the sake of completeness, it is relevant to note an additional consideration on the subject of a limiting taper benefit with regard to the Kodikara-Moore Method. At some point of increasing taper angle, the physical model of CCE will cease to be a reasonable approximation of reality, something that Kodikara and Moore alluded to in their 1993 paper. The writer is not aware of any published research to date that indicates at what taper angle this transition from CCE to some other physical model would occur. However, it is noted that a taper angle of 5° was the largest considered in any of the plotted results of parametric studies presented in Kodikara and Moore (1993). It is unclear as to whether they thought that this angle was the limit of viability for their CCE model or was just some arbitrary choice.

There is one final point to be made concerning cavity-expansion theories in general that has potential relevance to the current discussion. In problems involving cavity creation, it has sometimes proven to be adequate to simply use the theoretical limiting pressure as representing the upper-bound of cavity pressure and ignoring the displacements and concomitant intermediate values of cavity pressure required to get to that point. One example of this that dates back to Vesic's pioneering research in the 1970s is spherical-cavity expansion (SCE) used as a bearing-capacity model for both shallow and deep foundations (toe resistance in the latter case as discussed previously in Chapter 4). Thus, in principle it is possible to develop a simplified, resistance-only forecasting methodology based on CCE theory in which settlements are ignored and the limiting cavity pressure is used. However, this concept is not pursued further in this monograph.

5.6.3.3.3 Kodikara-Moore Method: Assessment and Critique

The first thing that should be noted when assessing Kodikara and Moore's work in detail is that it appears to reflect the knowledge-base of Kodikara's graduate education and post-graduate professional employment in Australia. Consequently, the term "pile" was used in Kodikara and Moore's 1993 paper in its broadest definition as is common outside of the U.S. to include all types of DFEs as opposed to the narrower definition of preformed, impact-driven elements as adopted for this monograph. In addition, both Kodikara and Moore appear to have been totally ignorant of the numerous types of tapered piles that either had been or were still in use, at least in some parts of the world, up to that point in time (circa 1990), including timber piles[156] (this was confirmed when the writer corresponded with each author individually). They did reference Nordlund's 1963 paper but were ignorant of the correct spelling of Nordlund's name and did not make use of any of the case histories cited in Nordlund's paper.

As a consequence of Kodikara and Moore's collective knowledge-deficit with regard to tapered-pile types, the only references to specific tapered DFEs made in their 1993 paper were two examples from the former Soviet Union that they apparently found in the literature. Neither of the DFEs from the former Soviet Union was a typical tapered (driven) pile by U.S. standards. One was a drilled shaft/bored pile and the other a precast-PCC (driven) pile but with a square cross-section, both of which were noted previously in Chapter 3. As a result, only the former was used as a case-history comparison for the Kodikara-Moore Method in their 1993 paper, the latter being discarded due to the fact that the square cross-sectional geometry was physically inconsistent with their CCE model.

Following up on an important fact noted in the preceding section concerning how taper benefit was defined by Nordlund vs. Kodikara and Moore, the latter state with regard to basic assumptions made in their 1993 paper that:

"It is also assumed that the pile is preformed [sic] *in an ideal elastic-plastic ground..."*

---

[156] Although Kodikara's co-author in the 1993 paper, Prof. Ian D. Moore, was working in Canada at the time of the publication of this paper, all of Moore's professional education was acquired in Australia so his knowledge, or lack thereof, with regard to commercially available types of tapered piles was presumably similar to that of Kodikara.

which is somewhat cryptic as to its precise meaning. It is unclear if "preformed" is intended to explicitly mean 'cast in place' as in the case of a bored pile/drilled shaft or simply 'wished in place', with the latter implying that installation effects were ignored in their model. In either case, it appears that the outcome, if not explicit intent, of the stated assumption was that neither the pre-installation stress-state in the ground nor how that stress-state might be modified by the DFE-installation process were modeled explicitly as part of their analytical methodology.

As will be seen, Kodikara and Moore began their analytical process with the assumption that the post-installation stress-state in the ground is somehow known a-priori by the design professional. Thus, for all practical intents and purposes this is effectively a wished-in-place analytical strategy as was noted in the preceding section.

Although the focus of this monograph is tapered (driven) piles, in keeping with the fact that the Kodikara-Moore Method is inherently applicable to all types of tapered DFEs and the fact that the exemplar application used by Kodikara and Moore in their 1993 was a tapered bored pile/drilled shaft, the following presentation is intentionally generalized to include all types of tapered DFEs. However, in keeping with the standard terminology adopted for this monograph, the term 'pile' will only mean driven pile and is thus not used in its more-general meaning as in the Kodikara and Moore paper.

Furthermore, the Kodikara-Moore Method is not restricted as to soil type as the Nordlund Method essentially is[157]. In fact, they did not even limit their method to soil but allowed for any type of geomaterial that had both Mohr-Coulomb strength parameters, ϕ and c, that were non-zero simultaneously. Thus, the Kodikara-Moore Method is arguably applicable even in rock. Therefore, the following presentation allows for a more-general ground condition.

The primary theoretical component of the Kodikara-Moore Method is a model for the settlement-dependent development of shaft resistance in what Kodikara and Moore defined as three distinct phases. Although the model was crafted specifically with tapered DFEs in mind, when the taper angle is zero the governing equations for the model default to a solution that is appropriate for a constant-diameter DFE. Consequently, the Kodikara-Moore Method can be used for piles such as the *Monotube* and *TAPERTUBE* that have both tapered and constant-diameter portions although it is not applicable to any type of DFE with a square cross-section

The three assumed phases of the development of shaft resistance are as follows:

1. **Phase I:** This covers the initial assumed elastic response in which the DFE and adjacent ground are assumed to settle the same magnitude, i.e. there is no slippage or differential settlement between shaft and ground along the assumed planar shaft-ground interface. This behavior is assumed to continue until the calculated shear stress at the shaft-ground interface equals the interface yield stress as defined using the usual Mohr-Coulomb strength parameters, ϕ and c, of the ground. Note the fact that in any frictional interface there must be some relative displacement (slippage) in order to mobilize shear stress is ignored. In theoretical terms, rigid-plastic interface behavior is assumed whereas in reality it is elastoplastic. Thus, the equations governing Phase I are the same as those proposed by Randolph and Wroth that form the basis of the Randolph-Wroth Problem

---

[157] As noted earlier in this chapter, Nordlund assumed that taper benefit only accrued from coarse-grained soil along the piles shaft. However, he did allow for the fact that a tapered pile might pass through a relatively thin stratum of fine-grained soil (in which case shaft resistance would just be due to some simple adhesion that was not influenced by taper) or bear on a relatively stiff stratum of fine-grained soil.

discussed extensively earlier in this chapter. Of interest is that Kodikara and Moore assumed that settlements during Phase I are always sufficiently small so that no significant radial expansion of the shaft-soil interface with concomitant increase in radial stress occurs. This means that <u>no</u> taper benefit occurs during Phase I.

2. **Phase II:** Once the shear strength along the shaft-ground interface is fully mobilized, relative shaft-ground displacement (slippage) with concomitant DFE settlement greater than the settlement of the adjacent ground is assumed to occur along the interface. Note that the inability to allow or account for this slippage is a major conceptual and theoretical shortcoming of the Randolph-Wroth Problem as was noted earlier in this chapter. The Randolph-Wroth Problem formulation essentially assumes that DFE behavior remains locked-in to Kodikara and Moore's Phase I mode of behavior indefinitely, albeit with an ever-degrading (reducing) ground stiffness (modulus) when a modulus-degradation scheme is used as part of the analytical process. In any event, initial cavity expansion under elastic conditions with concomitant radial-stress increase is assumed to occur during Phase II so this marks the beginning of taper benefit. This phase continues until the radial stress reaches the yield stress of the ground (in this case, the ground adjacent to the shaft-ground interface, not the interface itself as in Phase I) that is again based on the classical Mohr-Coulomb strength parameters. Of note is that Kodikara and Moore showed that when the taper angle is zero, the equations governing Phase II become those normally associated with a constant-diameter DFE undergoing settlement due to continuous DFE-ground interface slippage under constant shear and normal stresses along the shaft-ground interface.

3. **Phase III:** This phase represents the post-yield elastoplastic expansion of the cylindrical cavity as the result of additional DFE settlement. Phase III differs from Phase II in that there is a plastic, i.e. yielded, zone of the ground surrounding the DFE shaft. The thickness of this plastic zone increases with increasing radial expansion of the shaft-ground interface which, of course, is a function of increasing DFE settlement. Out of the three theoretical solutions to the CCE problem that Kodikara and Moore deemed potentially appropriate to use for modeling Phase III, they selected a solution developed by Yu and Houlsby as the one that was the most complete and accurate for the intended purpose, albeit at the expense of being by far the most complex mathematically of the three alternatives.

As noted in Chapter 4, CCE only models the shaft resistance of a tapered DFE during externally applied axial-compressive loading. Toe resistance in the Kodikara-Moore Method was modeled as a purely linear-elastic process using the same solution used by Randolph and Wroth that was incorporated into the Randolph-Wroth Problem. Kodikara and Moore opined, without any supportive evidence, that purely linear-elastic behavior for toe resistance was adequate for the intended overall purpose.

Kodikara and Moore structured the overall problem solution by first developing a total-differential equation that was based on vertical-force equilibrium and vertical normal strains for the overall DFE. This equation is solved by dividing the DFE shaft into multiple artificial segments and applying an iterative calculation and solution process that was outlined in an appendix to Kodikara and Moore (1993).

In broad terms, each cycle of this iterative process begins by applying a monotonically increasing value of assumed settlement to the DFE toe which directly produces the corresponding vertical force (load) at the toe. The governing equilibrium equation is then solved for each artificial segment of the DFE shaft, working upward from the DFE toe to the

DFE head. In this process, the calculated settlement at the top of each artificial segment becomes the assumed known settlement for the bottom of the artificial segment directly above it. The vertical force (load) in each artificial segment comes out of the solution process and in this way the settlement and load at the head of the DFE are calculated for each assumed value of toe settlement. The cycle is repeated as often as desired using an increasing magnitude of assumed toe settlement with each cycle.

The primary outcomes of this process are the data necessary to create complete load-settlement curves for both the head and toe of the analyzed DFE. Although Kodikara and Moore did not note this explicitly no less illustrate it in their 1993 paper, because the radial stress from the ground that is acting on the exterior each artificial segment of the DFE shaft is calculated for each calculation cycle, these data are available for use as well. Consequently, additional useful outcomes would appear to be plots of depth-wise variations in these radial stresses and/or lateral earth pressure coefficients, K_h or K_h/K_o, at different stages of settlement and loading.

In the writer's opinion, the Kodikara-Moore Method has a number of rather serious theoretical deficiencies:

- Installation effects are not considered explicitly as part of the analytical model which implies that the DFE is simply wished into place. This means that the all-important post-installation stress state in the ground adjacent to the DFE (see the following item) must be determined by the design professional in some unspecified manner as an input parameter. This also means that all taper benefit is assumed to accrue only from the wedging due to post-installation external loading of the DFE, not any wedging that occurs during DFE installation unless the design professional somehow estimates those effects and includes them in the input initial stress state. This would seem to contradict both measured (Gregersen et al. 1973) and inferred (Fellenius et al. 2000) residual loads on tapered piles that clearly indicate that there is not-insignificant taper benefit solely from (driven) pile installation and before any external loads are applied.

- The initial radial stress in the ground, which Kodikara labeled σ_o and can be either a total or effective stress depending on how a problem is formulated with regard to shear strength, is assumed to be depth-wise constant along the pile shaft. As noted in the preceding item, this post-installation stress has to incorporate both the geostatic (pre-installation) stress state as well as installation effects. More importantly, because this post-installation stress will almost always be depth-wise variable (simply because overburden stresses are inherently depth-wise variable), Kodikara and Moore provided no direction as to how a depth-wise constant value was to be rationally determined.

- The shaft of the DFE is assumed to be unstressed initially in the axial direction which means that residual loads are ignored.

- A homogeneous support medium for the DFE is assumed. This means that all material properties for both the stiffness (modulus) and strength of the ground are assumed to be the same along the pile shaft and beneath the pile toe.

- The stiffness of the ground, which is formulated in terms of the shear modulus, G, and Poisson's ratio, ν, remains constant during all stages of loading.

- The Mohr-Coulomb strength parameters, ϕ and c, along with the dilation angle, ψ, defining plastic deformations all remain constant during all stages of loading.

- The load-settlement behavior of the toe is assumed to remain linear-elastic during all stages of loading.

While any one of these assumptions is problematic on their own in the writer's opinion, based on the extensive discussion earlier in this chapter of settlement-only analytical methodologies, the assumption that the shear modulus is both depth- and load-wise constant stands out as being the most egregious given how important this parameter is to any problem in cavity mechanic (Salgado and Prezzi 2007, Cook 2010). It appears that Kodikara and Moore were aware of the implications of this assumption, at least to some extent. In the one case-history comparison presented in their 1993 paper, the sensitivity to the assume value of G, which Kodikara and Moore stated was the assumed "secant" value at "failure" of the ground, was explored to a limited extent (see Figure 4 in their paper). Note that this same case history was used to show the overall good agreement between the Kodikara-Moore Method and measured results.

Unfortunately, in the writer's opinion, this case history used by Kodikara and Moore leaves quite a bit to be desired as being the one and only example to 'prove' the viability and forecasting accuracy of their analytical methodology. To begin with, the case-history DFE was a drilled shaft/bored pile, not a (driven) pile. This immediately raises important questions of how a tapered drilled shaft was successfully constructed, its structural integrity and dimensions verified, etc. Furthermore, this drilled shaft was only 4500 millimetres (14.8 ft) long with reported head and toe diameters of 600 and 400 millimetres (24 and 16 in) respectively for a theoretical taper angle of 1.3°. While this might be a typical tapered DFE used in some parts of the world, it is much shorter and somewhat more tapered than tapered piles used historically in the U.S. Thus, this case history was, in the writer's opinion, a rather poor exemplar of how the various assumptions in the Kodikara-Moore Method that were itemized above would perform with a more typical tapered (driven) pile that is perhaps three or four times longer.

In summary and conclusion, despite some significant idealizations in its formulation, the Kodikara-Moore Method in general, and the concept of CCE that it incorporates as its key element, represents what is arguably the most significant, noteworthy contribution to date relative to the understanding of how tapered DFEs behave in axial compression. Furthermore, the theoretical deficiencies that these idealizations produce appear, in principle at least, possible to overcome. This possibility is pursued further in the following two sections.

5.6.3.3.4 Kodikara-Moore Method: Modifications by Manandhar and Yasufuku

Professors Suman Manandhar and Noriyuki Yasufuku made two distinct contributions to the knowledge-base related to tapered DFEs:

1. A toe-resistance model based on spherical-cavity expansion (SCE) theory (Manandhar and Yasufuku 2011a, 2012. 2013). While not the first researchers to use SCE as a bearing-capacity model (Vesic had done this 40 years before them), the unique aspect of Manandhar and Yasufuku's model is that it assumes that shaft taper influences the geometry of the toe-failure mechanism and thus the calculated outcome. This model and the two different versions (basic and enhanced) of the analytical method derived from it

were discussed earlier in this chapter as it is a standalone analytical methodology that can be used with any resistance-forecasting methodology for shaft resistance.

2. Relatively modest modifications to calculating shaft resistance in the original Kodikara-Moore Method (Manandhar and Yasufuku 2011b, 2013). These modifications are the subject of the current discussion.

The initial shaft-resistance modification to the Kodikara-Moore Method made by Manandhar and Yasufuku is to allow the stress-dependent component of ground strength in the Mohr-Coulomb failure criterion (the well-known ϕ angle) to vary as a function of mean effective stress. The mean effective stress acting at any point along the shaft varies as a function of both depth (overburden stress) and applied loading and concomitant settlement as the de-facto cylindrical 'cavity' of a tapered DFE expands radially.

As noted in the preceding section, in the original Kodikara-Moore Method ϕ is assumed to be a single-value problem constant that varies neither with depth nor applied load. It is implied by Manandhar and Yasufuku that in the original Kodikara-Moore Method one should use the constant-volume (critical-state) value of ϕ for a given soil as being representative of behavior under large-strain conditions although Kodikara and Moore made no such statement in their 1993 paper. In any event, Manandhar and Yasufuku's position can be interpreted and stated as meaning that the dilatancy component of ϕ, ϕ_d, has a not-insignificant effect on the calculated outcomes.

Manandhar and Yasufuku accounted for dilatancy effects in their modified analytical methodology by simply incorporating Bolton's well-known empirical relationship based on mean effective stress, relative density, and ϕ_{cv} that was noted in Chapter 4 into the original calculation algorithm that was developed by Kodikara and Moore. Essentially, an iterative step is added to the solution algorithm to match the mean effective stress and ϕ at a given operative stress level.

Unfortunately, Manandhar and Yasufuku did not provide any direct comparative results in their papers to illustrate the relative significance of making this modification compared to a benchmark outcome using a constant value of ϕ for some exemplar problem involving a tapered DFE, whether actual or assumed. Therefore, the practical significance of their modification cannot be assessed directly and objectively in this manner. The authors did, however, include the results of a very limited parametric study in their 2011 paper. Specifically, for one specific case they illustrated the effect of varying the dilatancy angle, ψ, relative to zero dilatancy. There was an effect but how this relates to behavior on a broad scale involving an actual DFE is unclear.

It is relevant to note that the modifications to the Kodikara-Moore Method that were presented in Manandhar and Yasufuku (2011b) did not include any changes to the algorithm for calculating toe resistance. They retained the original assumption of purely elastic behavior that was made by Kodikara and Moore. This is significant as the problem-solution algorithm used by Kodikara and Moore and essentially copied by Manandhar and Yasufuku with minor modification for calculating the operative value of ϕ begins each analytical cycle by assuming a magnitude of toe settlement and calculating the concomitant toe-resistance force and then progressively propagating these force-displacement results upward along the shaft to the head.

As noted previously, Manandhar and Yasufuku presented the results of their own original work related to toe resistance in 2011. As discussed earlier in this chapter, their SCE model for toe-resistance allowed for both basic and enhanced forecasting of:

- ULS toe resistance only (what the writer termed the Basic Manandhar-Yasufuku Toe-Resistance Method) and

- full elastoplastic behavior by incorporating an empirical hyperbolic relationship for load-settlement behavior (what the writer termed the Enhanced Manandhar-Yasufuku Toe-Resistance Method).

Not surprisingly, Manandhar and Yasufuku subsequently suggested (Manandhar and Yasufuku 2013) coupling their enhanced toe-bearing methodology with Kodikara and Moore's methodology for shaft resistance (but allowing for stress-dependent variation in ϕ) into what is defined herein as the *Modified Kodikara-Moore Method*. The capabilities of this method are summarized as follows:

- Shaft resistance based on a cylindrical-cavity-expansion (CCE) model in the three-stage, elastic-to-plastic behavioral process outlined earlier in this chapter. This is essentially the solution developed by Kodikara and Moore but allowing for stress-dependent variation in the operative value of ϕ at each analytical segment along the shaft with each stage of applied load.

- Toe resistance based on an SCE model developed by Manandhar and Yasufuku that accounts for pile taper. Toe resistance is not assumed to be a fixed ULS value but develops from zero as a function of toe settlement based on an empirical hyperbolic model developed by Manandhar and Yasufuku that is critiqued at length in Appendix B of this monograph. Note that toe resistance develops very slowly using this model and will generally be a relatively small fraction of the ULS value under any reasonable magnitude of toe settlement. Note also that this toe-bearing model must be used with empirically defined model parameters defined by Manandhar and Yasufuku.

As a final, collective comment concerning the published work of Manandhar and Yasufuku, it is important to note that much of the ground-truth that the authors used to support, i.e. 'prove', their theoretical work, especially that related to their toe-bearing model, is a series of laboratory tests on small-scale model piles under 1-*g* conditions. Of particular relevance is the fact that the two tapered piles they studied had geometries that did not resemble any known commercially available tapered pile. Specifically, the model piles had a constant-diameter lower section (to accommodate a load cell to measure toe load) and a tapered upper section. Furthermore, these model piles were essentially wished-in-place by placing sand around them. Consequently, no installation effects, as would exist with actual tapered piles, were modeled.

In light of these facts, it is the writer's opinion that the results and conclusions drawn from these laboratory tests, which are repeated in virtually all of the publications by Manandhar and Yasufuku that the writer has reviewed during development of this monograph, should be viewed with caution. This is because of well-known 'bowling ball' scaling issues when relatively small physical models (the test piles in this case were only 500 millimetres (18 in) long and of the order of 25 millimetres (1 in) in diameter) are used with full-scale soil particles under 1-*g* stress conditions. Therefore, unless and until the forecast outcomes of the Modified Kodikara-Moore Method are compare to actual tapered piles, the calculated outcomes of this method, as with those of the original Kodikara-Moore Method, should be viewed with the appropriate caution.

5.6.3.3.5 Kodikara-Moore Method: Potential Additional Modifications

Despite the relatively complex mathematics of the Kodikara-Moore Method that have apparently made it unattractive for use by practitioners for almost three decades now, the basic methodology still has merit as the basis for both research and potential future use by a vendor of commercial software who could embed the method behind a user-friendly graphical user-interface. However, in the writer's opinion there are several significant technical issues that were enumerated and discussed earlier in this chapter in the critique of the Kodikara-Moore Method that should be addressed with the goal of revising the overall methodology.

Manandhar and Yasufuku made some modest advances in this regard with their modifications related to the operative value of ϕ along the shaft and toe resistance. However, there are significant issues with their enhanced toe-resistance model as it develops resistance at much too slow a rate in the writer's opinion, as discussed in Appendix B. Nevertheless, linking toe resistance to toe settlement is certainly a step in the right direction away from the traditional assumption of a fixed-value toe resistance that is typically assumed to be some ULS value.

In addition to making further improvements to the load-settlement model for the toe, the most pressing improvement required for the Kodikara-Moore Method in the writer's opinion is to allow for both depth- and load-wise variation in the shear modulus, G. Both theoretical work (Cook 2010) as well as the applications-oriented research of Mayne and Niazi using the Randolph-Wroth Problem solutions have shown conclusively that cavity mechanics in particular and DFE load-settlement in general require consideration of variable moduli. How this implementation of modulus degradation is implemented is the subject of future study.

However, any such changes to the Kodikara-Moore Method would still not address the fact that the method only includes Stage II taper benefit due to post-installation, externally applied, axial-compressive loading and, thus, leaves unresolved the inherent failure of the method to explicitly model the DFE-installation process that is necessary to include the Stage I taper benefit that occurs with tapered (driven) piles at least. Add to this the fact that residual loads are ignored. This suggests that rather than tinker ad infinitum with the Kodikara-Moore Method that a better, more-efficient approach would be to develop a new one-step resistance-forecasting methodology from scratch. This is addressed in Chapter 7.

5.6.3.3.6 El Naggar-Sakr Method: Background

As noted earlier in this chapter, the Kodikara-Moore Method does not appear to be widely known or cited in published work no less used in research and practice. However, the writer found two notable exceptions to this.

One is the previously discussed doctoral work of Prof. Manandhar that was conducted under the direction and supervision of Prof. Yasufuku in Japan. However, the published work of Manandhar, Yasufuku, and several of their associates and research collaborators appears to have been of relatively limited scope and duration to date. As discussed earlier, their two modifications to the Kodikara-Moore Method were very modest, with the more-substantive one being the use of an elastoplastic toe-resistance model, referred to as the Manandhar-Yasufuku Toe-Resistance Method in this monograph, in lieu of the simplistic linear-elastic model used by Kodikara and Moore. While the writer has significant issues with the Manandhar-Yasufuku toe-resistance methodology as discussed at length in Appendix B, it is at least a step in the right direction of recognizing that the toe resistance for DFEs is settlement-dependent as Fellenius has been advocating for some years now, not a single, fixed value as has been assumed historically.

The second and much more substantive exception is the research conducted at The University of Western Ontario/Western University that began in the late 1990s and apparently continues to the present. This work has been done by a succession of graduate students working under the overall supervision and direction of Prof. M. Hesham El Naggar. Their collective work and the resistance-only forecasting methodology that it produced, which is referred to as the El Naggar-Sakr Method in this monograph, was noted earlier in this chapter but a discussion was deferred to this point because it required a detailed presentation of the Kodikara-Moore Method as background for understanding the genesis and basis of the El Naggar-Sakr Method.

As an aside, the writer speculates that El Naggar's interest in tapered DFEs may have been at least inspired, if not instigated, by Prof. Ian D. Moore who was on the faculty of UWO in the early 1990s when:

- El Naggar was doing his doctoral work there and

- Kodikara and Moore (1993) was published.

In any event, the published record of the work of El Naggar et alia related to tapered DFEs appears to have had its origins in the master's thesis of Wei (1998). Over the years, the work at UWO has covered:

- both coarse- and fine-grained soils;

- (driven) piles, drilled shafts, and (most recently) torqued-in-place tapered helical piles;

- different DFE materials, including traditional ones such as metal and PCC and non-traditional ones such as fiber(glass)-reinforced polymer (FRP);

- novel pile-driving techniques such as *toe driving* where the toe of the pile is impacted directly by use of a special mandrel and essentially drags a shell-like pile shaft into the ground as opposed to applying the installation force to the head of the pile via impact or jacking (a necessity when the shaft consists of non-traditional materials such as FRP that are not sufficiently robust for direct impact driving); and

- both axial and transverse (lateral) post-installation load application at the head of the DFE.

Of specific relevance here is that in several published works beginning with El Naggar and Sakr (2000) that dealt specifically with tapered (driven) piles, El Naggar and his students used substantial, key portions of the Kodikara-Moore Method to develop a resistance-only forecasting methodology for tapered piles that has since been broadened to include a wide range of tapered DFEs. Although a number of UWO students apparently made contributions to this analytical method since the late 1990s, in deference to the fact that the method was first published in El Naggar and Sakr (2000) this method is referred to in this monograph as the El Naggar-Sakr Method.

In discussing the El Naggar-Sakr Method in the following sections and Appendix C of this monograph, the term 'pile' is used in keeping with the primary focus of this monograph as well as the initial development of this analytical methodology for use with (driven) piles. However, as evidenced by the fact that El Naggar and subsequent generations of graduate students broadened their interest to other types of tapered DFEs, it is noted that the El Naggar-Sakr Method can be used with any typed of tapered DFE.

5.6.3.3.7 El Naggar-Sakr Method: Overview

The El Naggar-Sakr Method in its original form is a resistance-only forecasting methodology which is why it was mentioned earlier in this chapter in a prologue fashion. Furthermore, in its original form it applies to coarse-grained soil conditions only, similar to the assumption made by Nordlund. However, because it derives from the Kodikara-Moore Method, which is inherently a one-step procedure for load-settlement forecasting, and could, with some modifications, be generalized back into a one-step procedure if desired, detailed discussion of the El Naggar-Sakr Method was deferred to this point in the chapter.

To begin with, toe resistance in the El Naggar-Sakr Method is treated in a very simplistic, traditional manner assuming rigid-plastic behavior of the soil underlying the toe and using a simplified, semi-empirical version of classical bearing-capacity theory as was apparently recommended in a then-current (1992) version of the *Canadian Foundation Engineering Manual*. This document was developed by the Canadian Geotechnical Society and will not be discussed further other than to note that El Naggar and Sakr's assumption of rigid-plastic toe behavior was a complete paradigm shift from the Kodikara-Moore Method that assumed linear-elastic behavior of the toe[158]. In that sense, El Naggar and Sakr's assumption was both a technological step forward (in that it allowed for the geotechnical ULS at the toe, something that Kodikara and Moore did not) but also a technological step backward (in that toe resistance was assumed to be constant in magnitude and not a function of toe settlement). Obviously, if the El Naggar-Sakr Method were ever generalized into a one-step procedure for load-settlement forecasting the toe-resistance model would have to be changed.

In the writer's opinion, the innovative aspect of the El Naggar-Sakr Method is the methodology for forecasting shaft resistance and only that is discussed in detail in this monograph. The basic theoretical assumption is that the ULS unit shaft resistance, $r_{s(ult)}$, which El Naggar and Sakr called *skin friction* with the notation τ_s, is given by the following equation:

$$\tau_s = K_t K_s \tan \delta \, \sigma_v' . \tag{5.19}$$

Note that the notation in this equation is that used by El Naggar and Sakr and not the standardized notation adopted for this monograph as it is important for the present discussion to work with the problem variables as defined by them. Specifically:

- K_t = dimensionless 'taper-benefit' coefficient that is > 1 for tapered piles ($K_t = 1$ for the baseline-reference case of a constant-diameter pile). As will be seen, K_t is, in theory, not a problem-constant but a function of pile settlement. However, El Naggar and Sakr chose to make this parameter a problem-constant as they evaluated it at only one arbitrary

---

[158] For the sake of completeness, it is noted that a subsequent paper (Sakr et al. 2004a) that used the El Naggar-Sakr Method stated that spherical-cavity expansions (SCE) theory, specifically, the solution of Yu and Houlsby (1991), could be used be used as an alternative for forecasting the ULS toe resistance.

settlement magnitude that they assume constitutes the geotechnical ULS. It is primarily for this reason that the El Naggar-Sakr Method is a resistance-only forecasting methodology.

- K_s = dimensionless <u>post-installation</u> coefficient of lateral earth pressure acting along the pile shaft that would exist for a baseline-reference of a constant-diameter pile in the same soil conditions. This parameter is assumed to include the effects of the site-specific, pre-installation K_o value (although El Naggar and Sakr did not state this explicitly) plus the pile-installation effects on the stress-state in the ground. Thus, while it is, in theory, a problem-constant, it would be expected to vary from application to application depending on the site stress history, soil conditions, and specific type of pile that is installed.

- δ = shaft-soil interface friction angle along the pile shaft. This is assumed to be a problem constant.

- σ'_v = vertical effective stress along the pile shaft that is implied to be the overburden stress. It is allowed to vary depth-wise along the pile shaft.

The most fundamental issue to address is the manner in which the El Naggar-Sakr Method accounts for taper effects and concomitant benefits compared to that used in the Nordlund and Kodikara-Moore methods. Because each of these three methods uses different parameters in their mathematical formulation of ULS unit shaft resistance, the comparison is not obvious and thus requires some elaboration.

To begin with, although the El Naggar-Sakr Method uses two coefficients of lateral earth pressure in Equation 5.19, only one of them, K_t, relates to taper effects. As will be see, this parameter depends only on post-installation applied loading to the assumed geotechnical ULS. Although the other parameter, K_s, is assumed to reflect both the pre-installation stress-state, i.e. K_o effects, as well as pile-installation effects, it is explicitly assumed to reflect these combined effects for the baseline-reference case of a constant-perimeter pile of the same basic type in the same conditions, i.e. if a tapered steel pipe pile is being analyzed then K_s must be that of a constant-diameter steel pipe pile of similar dimensions. Thus, in simple terms the El Naggar-Sakr Method clearly attributes all taper effects and benefits to what occurs during post-installation loading to the assumed geotechnical ULS and reflects those effects and benefits in the K_t parameter.

On the other hand, as discussed in detail earlier in this chapter:

- Both versions of the Nordlund Method and its derivatives such as the writer's Simplified Nordlund Method attribute all taper effects and benefits to what occurs during driving <u>plus</u> loading to the geotechnical ULS, with no attempt to separate the relative contributions of driving effects and loading effects or to explicitly consider the pre-installation stress-state, i.e. K_o effects.

- Strictly speaking, the Kodikara-Moore Method attributes all taper effects and benefits to post-installation external load application up to and including the geotechnical ULS. However, it does allow for an assumption of the post-installation/pre-load-application radial stress state, σ_o, that theoretically allows for consideration of both the pre-installation stress-state, i.e. K_o effects, as well as installation effects although it places the entire burden for estimating σ_o on the design professional.

Thus, the way in which the El Naggar-Sakr Method accounts for taper benefit falls somewhere in between the Nordlund and Kodikara-Moore methods in terms of theoretical rigor. Furthermore, all three methods fall well short of the theoretical ideal of explicitly accounting for taper benefit, both as the result of pile installation (Stage I taper benefit) as well as subsequent external load application (Stage II taper benefit), as an explicit, inherent part of the overall analytical methodology. There appears to be a reason that El Naggar and Sakr formulated their analytical methodology this way but before exploring this some comments concerning the baseline K_s parameter are in order.

In the publication that presented the El Naggar-Sakr Method for the first time in 2000 as well as in subsequent publications such as Sakr et al. (2004a) that used the method, the K_s parameter was back-calculated using measurements obtained on an instrumented constant-diameter DFE that had been installed along with tapered DFEs in the particular test program that was the focus of the publication. Note that all of these tests were performed on small-scale model piles in a laboratory environment under 1-g conditions that were broadly similar to those used by Manandhar et alia in their work. As for what a practitioner should do in routine practice when they would have to come up with a value for K_s on their own, there was no clear direction or suggestion given by El Naggar and Sakr other than to use "available correlations" for both K_s (and δ as well).

Unfortunately, this is rather vague, unhelpful direction and, in the writer's opinion, certainly one shortcoming of the El Naggar-Sakr Method with respect to its utility in routine practice. In this regard, El Naggar and Sakr were no more helpful to potential users of their method than Kodikara and Moore who assumed that the value of σ_o in their method (which is essentially the same as the product of El Naggar and Sakr's K_s and σ'_v) would somehow be magically known to users of their method.

Considering now the K_t parameter that is the most novel element of the El Naggar-Sakr Method, the conceptual function of this parameter is to incorporate all of the unique load-bearing behaviors due to taper as they affect shaft resistance. Although not stated explicitly by El Naggar and Sakr, by virtue of the theoretical assumptions they made for calculating values of the K_t parameter (discussed subsequently), they assumed, by implication, that all taper benefit occurs only during post-installation applied axial-compressive loading as was noted previously. No taper benefit is assumed to occur as the result of pile installation as the lateral earth pressure parameter that defines installation effects, K_s, is assumed to be the same for both tapered and constant-perimeter piles of the same basic type at a given site.

For evaluating K_t, El Naggar and Sakr recognized that cylindrical-cavity expansion (CCE) represents the correct physical model for the wedging behavior of a tapered pile in response to post-installation axial-compressive loading. For this purpose, they elected to use some, but not all, of the theoretical and analytical concepts developed by Kodikara and Moore in their methodology. Specifically, El Naggar and Sakr reasoned that for the relatively small taper angles (approximately 1° or less) of interest to them, soil yielding due to the expanding cylinder (shaft-soil interface in reality) would not occur under any reasonable level of pile settlement.

Unfortunately, El Naggar and Sakr did not provide any data or other evidence to support this blanket assumption. The limited information presented in Kodikara and Moore (1993) related to the development of their three assumed behavioral phases as a function of DFE settlement suggests that this assumption might be reasonable under working loads but becomes progressively approximate as settlements increase. In any event, El Naggar and Sakr thus assumed that only Phases I and II of the three-phase behavioral model postulated by Kodikara and Moore that was discussed earlier in this chapter would develop.

The description of Phases I and II for the development of shaft resistance in the Kodikara-Moore Method that were given earlier in this chapter are repeated here for ease of reference:

1. **Phase I:** This covers the initial assumed elastic response in which the DFE and adjacent ground are assumed to settle the same magnitude, i.e. there is no slippage or differential settlement between shaft and ground along the assumed planar shaft-ground interface. This behavior is assumed to continue until the calculated shear stress at the shaft-ground interface equals the interface yield stress as defined using the usual Mohr-Coulomb strength parameters, ϕ and c, of the ground. Note the fact that in any frictional interface there must be some relative displacement (slippage) in order to mobilize shear stress is ignored. In theoretical terms, rigid-plastic interface behavior is assumed whereas in reality it is elastoplastic. Thus, the equations governing Phase I are the same as those proposed by Randolph and Wroth that form the basis of the Randolph-Wroth Problem discussed extensively earlier in this chapter. Of interest is that Kodikara and Moore assumed that settlements during Phase I are always sufficiently small so that no significant radial expansion of the shaft-soil interface with concomitant increase in radial stress occurs. This means that <u>no</u> taper benefit occurs during Phase I.

2. **Phase II:** Once the shear strength along the shaft-ground interface is fully mobilized, relative shaft-ground displacement (slippage) with concomitant DFE settlement greater than the settlement of the adjacent ground is assumed to occur along the interface. Note that the inability to allow or account for this slippage is a major conceptual and theoretical shortcoming of the Randolph-Wroth Problem as was noted earlier in this chapter. The Randolph-Wroth Problem formulation essentially assumes that DFE behavior remains locked-in to Kodikara and Moore's Phase I mode of behavior indefinitely, albeit with an ever-degrading (reducing) ground stiffness (modulus) when a modulus-degradation scheme is used as part of the analytical process. In any event, initial cavity expansion under elastic conditions with concomitant radial-stress increase is assumed to occur during Phase II so this marks the beginning of taper benefit. This phase continues until the radial stress reaches the yield stress of the ground (in this case, the ground adjacent to the shaft-ground interface, not the interface itself as in Phase I) that is again based on the classical Mohr-Coulomb strength parameters. Of note is that Kodikara and Moore showed that when the taper angle is zero, the equations governing Phase II become those normally associated with a constant-diameter DFE undergoing settlement due to continuous DFE-ground interface slippage under constant shear and normal stresses along the shaft-ground interface.

El Naggar and Sakr (2000) illustrated how the theoretical equations for Phases I and II of the Kodikara-Moore Method can be used to develop an algebraic relationship for the K_t parameter. This development is somewhat lengthy but is important for purposes of this monograph that are presented in later chapters so is presented in detail in Appendix C.

5.6.3.3.8 El Naggar-Sakr Method: Assessment and Critique

Some critique of the El Naggar-Sakr Method was made during the course of presenting an overview in the preceding section as well as in Appendix C. A more-formal assessment and critique of this analytical methodology is presented in this section.

As can be seen in either Appendix C or any of the publications where El Naggar and his colleagues presented development details of the El Naggar-Sakr Method (El Naggar and Sakr 2000, Sakr et al. 2004a), in any specific application the taper-benefit parameter, K_t, is never constant in magnitude but a function of pile settlement. This is to be expected as the original Kodikara-Moore Method is an integrated, one-step load-settlement forecasting methodology whereby the benefit of taper is assumed to increase as DFE settlement increases under post-installation externally applied axial-compressive load. As discussed earlier in this chapter, solution of the Kodikara-Moore Method is a cumulative process based on progressive assumptions of toe settlement that for each calculation cycle produce an estimate of settlement and concomitant applied load at the head of the DFE being analyzed.

However, El Naggar and Sakr chose to use a single value of K_t in their methodology, specifically, the value at what they defined as the geotechnical ULS. Curiously, they used a relatively archaic, simplistic metric for defining the geotechnical ULS, specifically, settlement at 10% of the pile diameter (the average diameter in the case of the continuously tapered piles studied in El Naggar and Sakr (2000)) as proposed by Terzaghi (1942) almost 60 years earlier (almost 80 years ago at the time this monograph was published). Not to deprecate Terzaghi's genius, but one might have thought that a more-modern criterion would have been selected.

Nevertheless, the basic formulation of the El Naggar-Sakr Method makes it relatively easy to generalize it by making K_t a settlement-dependent problem parameter and thus allow for it to be used as part of an integrated, one-step load-settlement methodology in the manner of the Kodikara-Moore Method from which it was derived. However, this was not explored by the writer as there are too many other negatives with the El Naggar-Sakr Method that render such an effort of dubious value in the overall scheme of things. As was concluded earlier in this chapter with respect to the Kodikara-Moore Method, rather than engage in applying multiple 'tweaks' to the El Naggar-Sakr Method in an attempt to improve on it, it appears much more logical to simply craft an improved one-step resistance-forecasting methodology for tapered piles beginning from scratch with a clean slate as is done in Chapter 7.

The most substantial criticism of the El Naggar-Sakr Method is that a constant shear modulus, G, is assumed when evaluating K_t. This is not surprising as Kodikara and Moore as well as Manandhar and Yasufuku did the same thing as noted earlier in this chapter. Clearly, this is incorrect based on the current state of knowledge. Even if yield of the soil around a tapered pile does not occur and thus remains in the elastic range as El Naggar and Sakr assumed, the soil behavior would be expected to be in the non-linear elastic range.

Cook (2010) clearly illustrated that around an expanding cylindrical cavity there are, in general, three distinct zones of soil behavior as one progresses radially outward away from the expanding interface between cylinder and soil:

1. A <u>plastic zone</u> where behavior is governed not by soil stiffness but by the operative value of soil shear strength as defined by ϕ that varies from ϕ_{cv} adjacent to the DFE shaft to ϕ_{peak} at the edge of the plastic zone.

2. A <u>non-linear elastic zone</u> where the shear modulus, G, increases from some operative value to the small-strain value, G_{max} (Cook used the alternative notation of G_o).

3. A <u>linear-elastic zone</u> where the shear modulus remains at its small-strain value, G_{max}.

Note that even if the plastic zone does not develop as assumed by El Naggar and Sakr, the non-linear elastic zone would still develop.

Unfortunately, as with the K_s parameter, El Naggar and Sakr provided no meaningful guidance as to how to choose problem-specific values for the several soil properties that are required to use their analytical methodology. This includes, primarily, any two of the three elastic material properties of the soil:

- E (Young's modulus),

- G (shear modulus), and

- ν (Poisson's ratio),

that are required to evaluate K_t. El Naggar and Sakr (2000) noted that they chose 'cookbook' values from a then-current foundation engineering textbook that, based on the writer's first-hand experience as a teacher of undergraduate geotechnical engineering for over three decades, was in widespread use at the time (circa 2000) in the U.S., at least as an entry-level textbook.

Chapter 6

Case-History Comparison of Resistance-Forecasting Methodologies

6.1 INTRODUCTION

6.1.1 Objective

At this point, it is of interest to present an exemplar application and comparison of the various resistance-forecasting methodologies for tapered piles that were presented and discussed in Chapter 5. Note that this is solely intended to be an illustration of how different analytical methods would be applied to a typical tapered pile in practice. It is not intended to be a ranking of the absolute and relative accuracy of these methods from which definitive conclusions could be drawn to guide future practice. Such an undertaking would require comparisons to a large database of piles with the appropriate statistical assessments necessary to produce the desired objective outcomes. This is well beyond the writer's resources and the scope of this monograph.

For maximum benefit from this exercise, the writer decided that the calculations needed to be performed for an actual tapered pile with the necessary load vs. settlement ground-truth for comparison. This pile should ideally have the following characteristics, ranked in the order of importance (in the writer's opinion) for the intended goals of this chapter:

1. Driven at a site where coarse-grained soils are at least predominant as these have historically been the subsurface conditions where tapered piles perform best.

2. A conventional boring with SPT data performed close to the pile location to provide visual identification of the soil as well as N-values to use with older analytical methodologies such as the Nordlund Method.

3. An sCPTu sounding performed close to the pile location to provide data for state-of-knowledge site characterization as well as site-specific V_s and G_{max} data.

4. Internally instrumented and subjected to a conventional static load test so that the true distribution of axial resistance and residual loads can be estimated. Furthermore, this load test must have been carried to a point that can reasonably be assumed to be the geotechnical ULS.

6.1.2 Case Histories Considered

Arguably the most obvious potential sources of the necessary ground-truth for the goals of this chapter are the stakeholders involved in the manufacturing, sale, and installation of tapered piles. One would naturally expect that their business self-interest in the success of their product(s) would lead these entities to acquire such information as a normal course of doing business.

Unfortunately, for the reasons discussed earlier in this monograph, the relevant stakeholders in the U.S. were variously unable or unwilling to provide the writer with the necessary information. This meant that selection had to be based on either the writer's available information related to JFKIA or the published literature.

The writer identified two potential case histories in the published literature, Gregersen et al. (1973) and Fellenius et al. (2000). These case histories actually have considerable use for the broader aspects of this monograph so are discussed at length in Appendix H and Appendix A, respectively, of this monograph.

6.1.3 Selection

Unfortunately, neither the JFKIA site nor either of the two above-referenced case histories satisfied all four criteria listed at the beginning of this chapter. On the positive side, all of these choices were sites with predominantly coarse-grained soil so any one of them satisfied the most-important Criterion #1. Therefore, to meet the goals of this chapter the writer had to make compromises with the remaining three criteria that were revised and consolidated as follows:

1. A conventional boring with SPT N-values had to be performed in the vicinity of the pile so that the Nordlund Method could be used.

2. At least a CPT sounding had to be performed in the vicinity of the pile. Based on the writer's experience as presented in Horvath (2002) and numerous subsequent publications up to and including Horvath (2016b), the basic q_c and f_s data provided by a CPT can be extended by assuming $u_2 \cong u_o$ as is not unreasonable for coarse-grained soils. This assumption allows CPTu-based empirical correlations for soil properties to be used. If it is further assumed that the soil is 'well-behaved', i.e. is behaviorally young and exhibits no structure from aging or cementation, then the small-strain shear modulus, G_{max}, can reliably be estimated from CPT (faux-CPTu) data as well that eliminates the need for a site-specific sCPTu sounding.

3. At least a conventional static load test with load and settlement readings at the pile head was performed on the pile. Furthermore, the pile must have been loaded to a point that can reasonably be assumed to be the geotechnical ULS.

Based on these criteria, the writer selected a *Monotube* pile designated Pile P-5A that was driven at JFKIA in the late 1980s. This pile is representative of what can be considered the second generation of widely used foundation piling at JFKIA that essentially replaced the first-generation timber piles that dated back to the original airport construction in the 1940s[159]. This pile has the added benefit of being essentially identical to the two instrumented *Monotube* piles studied by Fellenius et al. (2000) that are discussed at length in Appendix A. Consequently, some of the measured and interpreted results from the Fellenius et alia piles, especially with respect to residual loads, can be used for the purposes of the present chapter as was done previously in Chapter 5 where Pile P-5A was used in conjunction with the Fellenius et alia piles to critique the Togliani Method.

---

[159] This second generation of JFKIA foundation piling was, in turn, superseded beginning in circa 2000 by a third generation: *TAPERTUBE* piles with greater taper angle ($\alpha = 0.95°$) as well as *Monotube* piles in some cases, also with $\alpha = 0.95°$.

6.2 PILE AND SUBSURFACE DATA

Pile P-5A was used extensively by the writer over the past 20 years in pile-related research that was published in Horvath (2002, 2014b, 2015, 2016a, 2016b) and was also used in Chapter 5 as part of the critique of the Togliani Method. However, relevant details for this pile are repeated here for ease of reference:

- *Monotube* pile nomenclature = 3NJ8x14 that means:
  - all components are 3-gauge steel (thickness = 0.2391 inches = 6073 μm)[160]
  - Type N constant-diameter (= 14 in/356 mm) extension
  - Type J tapered section → 0.57° taper angle
  - toe diameter = 8 inches (203 mm)
  - diameter at top of tapered section = 14 inches (356 mm)
  - length of tapered section = 25 feet (7.6 m)
- total length of pile installed = 64 feet (19.5 m).

For comparison, the Fellenius et alia piles were 5NJ8x14 and 66 feet (20 m) embedded length.

A conventional boring with SPT sampling (No. 3-256) and a CPT sounding (No. CTP-5) were performed in the vicinity of Pile P-5A during the same late-1980s timeframe in which Pile P-5A was driven and load tested. The measured data in each, which are shown side by side in Figure 6.1 for visual comparison, are quite typical of the relatively uniform subsurface conditions within the Central Terminal Area (CTA) at JFKIA in which Pile P-5A was driven (Horvath 2014b). As such, these data have been used in several publications by the writer as far back as Horvath (2002) to illustrate exemplar subsurface conditions within the JFKIA CTA.

There are several items to note with respect to the field data shown in Figure 6.1:

- The soil stratum designated "MTM" is the acronym for *marine tidal marsh*. This is the colloquial term used for the Holocene organic soils deposited in a salt- or brackish-water environment within the various bays and tidal estuaries within the New York City metropolitan area, in this case within Jamaica Bay that connects to the Atlantic Ocean. These soils primarily consist of organic clays, with occasional peat layers in areas that were sufficiently above tidal levels to support sustained growth of phragmites and other natural vegetation.

- The Holocene sand-fill stratum, which was deposited by hydraulic filling during the early-1940s construction of the airport (Horvath 2014b), is an unconfined, groundwater-table aquifer that is sufficiently far from the current Jamaica Bay shoreline within the entire JFKIA CTA that there are no tidal effects (the mean tidal range within Jamaica Bay is approximately 5 feet (1500 mm).

- Although the Pleistocene sand stratum (technically the Upper Glacial Aquifer) is a confined aquifer, the MTM stratum is an imperfect confining layer (Horvath 2014b). Consequently, it is sufficient for routine foundation design to assume that hydrostatic groundwater conditions, as defined by the indicated groundwater level, exist throughout the entire geologic profile shown in Figure 6.1.

---

[160] The stated precision of the pile wall thickness is that given in Monotube product literature accessed during the timeframe in which this monograph was written. Whether such precisions are actually achieved in routine production piles is a question that should be directed to Monotube.

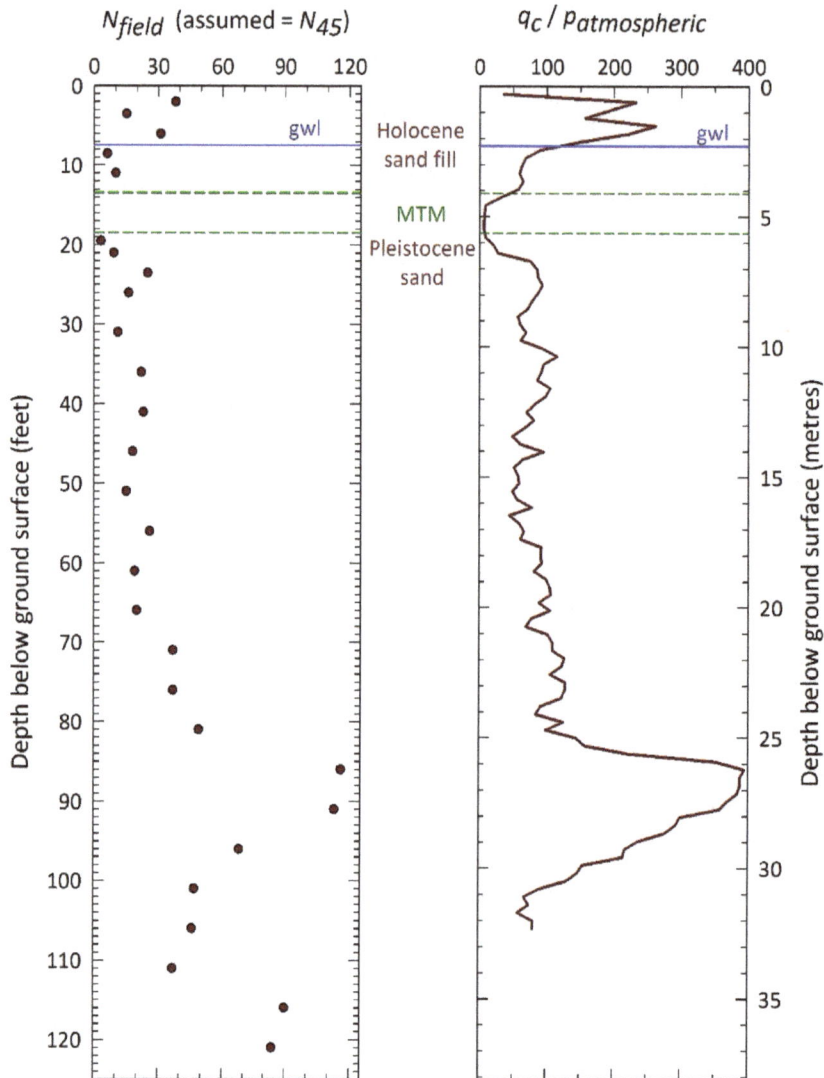

Figure 6.1. Measured Field Data - Boring 3-256 and CPT CTP-5 within JFKIA CTA.

- The field or raw SPT N-values, N_{field}, are assumed to be N_{45} values based on the drive-hammer methodology in widespread use in the New York City metropolitan area at that time (no measurements of actual hammer energy and concomitant drive-system efficiency are known by the writer to have been made for JFKIA borings in the late 1980s). In the writer's opinion, this actually proved to be a benefit for the goals of this chapter as the SPT-sampling technology in use when this boring was made was a throwback to the circa-1960 timeframe in which Nordlund developed the original version of his resistance-forecasting methodology.

- Site-characterization assessments made by the writer, especially in recent years, require at least u_2 pore-pressure measurements and, in some cases, shear-wave velocities (V_s). Obviously, neither of these basic parameters is measured in a CPT sounding which was the technology used at JFKIA in the late 1980s (and considered the state of practice for its time, at least in the New York City area that has been historically very slow to embrace cone penetrometers in routine practice). Consequently, the writer assumed pseudo-u_2

values for both the Holocene sand fill and Pleistocene sand. These values were equal to the hydrostatic water pressure, u_o, as defined by the groundwater level in the Holocene sand-fill stratum. The reasonableness of this assumption has been confirmed by the writer in recent years as the result of access to dozens of CPTu and sCPTu soundings performed within the JFKIA CTA. This can be independently verified, if desired, by referencing the CPTu sounding shown in Fellenius et al. (2000).

6.3 ANALYSES

6.3.1 Resistance-Forecasting Methods Used

The following resistance-forecasting methods were used to analyze Pile P-5A:

- Nordlund Method:
  - Original (1963) version
  - Revised (1979) version

- Simplified Nordlund Method using the 'best estimate' (per Horvath 2002) combination of ϕ_{cv} for the constant-diameter shaft resistance; ϕ_{peak} for the tapered shaft resistance; and ϕ_{peak} for the toe resistance, with toe resistance calculated using three different analytical methodologies:
  - Hansen-Vesic traditional bearing-capacity solution (per Horvath 2002)
  - Basic Manandhar-Yasufuku Toe-Resistance Method (per Horvath 2015)
  - Vesic spherical-cavity expansion (SCE) solution (as introduced in this monograph and discussed in detail in Appendix G)

- Togliani Method.

As can be seen, all the methods chosen are resistance-only methodologies, although both indirect and direct methods are represented.

Some comments are in order to explain the rationale behind these choices:

- First and foremost was the self-imposed (by the writer) requirement that a selected methodology must be accessible and usable to the average practitioner in its current form and at the present time.

- No settlement-only methodologies, which could be used as part of a two-step procedure for generating a complete load-settlement curve, are included. The writer has explored two-step procedures in earlier publications going back to the simplistic, empirical Tri-Linear Settlement Model developed almost 20 years ago and explored in considerable detail along with various forms of the Randolph-Wroth Problem solutions in Horvath (2015, 2016a, 2016b). However, as discussed at length in these publications as well as Chapter 5, the Randolph-Wroth Problem formulation has some serious errors in terms of allocating shaft vs. toe resistances. As a result, while settlement-only methods can add an informative, useful element to the final forecast outcomes, they are, in the writer's opinion, of questionable value in routine practice, especially in light of the fact that residual loads are, in general, significant for all piles but especially tapered piles. The effects of residual loads are simply not even remotely replicated by the Randolph-Wroth Problem solutions.

- None of the integrated (one-step procedure) load-settlement methods are included. The simple reason is that although such methods are the desired state of art to allow pile design based on a consideration of settlements, not just the geotechnical ULS, there is no one-step procedure at the current time that is usable in practice in the writer's opinion. At one time, the Kodikara-Moore Method and its derivatives developed by Manandhar et alia and El Naggar et alia appeared, to the writer at least, to be the long-sought-after ideal analytical methodology or at least the basic framework for such. However, detailed research performed during the preparation of this monograph clearly indicated that the key assumption made by Kodikara and Moore and accepted by Manandhar et alia and El Naggar et alia that the wedging mechanism for tapered piles develops only from relative displacement between the pile and adjacent ground during axial-compressive load application after installation is flawed. Furthermore, none of these researchers provides any usable guidance as to how an end user of their analytical methodology can rationally select a single value of the operative shear modulus for a highly complex problem that involves zones of plasticity, non-linear elasticity, and small-strain linear elasticity. The bottom line is that the Kodikara-Moore Method; its modified version developed by Manandhar and Yasufuku; and its simplified rendition as contained in the El Naggar-Sakr Method, are simply not practical, usable analytical methodologies in their current forms in the writer's opinion.

6.3.2 Results

The results of the analyses performed using the resistance-forecasting methodologies listed in the preceding section are shown in Figure 6.2 as vertical lines as the forecast ULS resistances are independent of settlement. Also shown in this figure are the measured load-settlement data from the static load test. Note that the measured results are shown as a series of individual datapoints as opposed to a continuous load-settlement curve, with a plunging-type behavior (at which point the load test was terminated due to hardware limitations) for the final datapoint as indicated by the vertical red arrow.

As discussed in detail in Horvath (2015) and noted for other, similar load-test results presented elsewhere in this monograph, the static load tests that were performed at JFKIA in the latter decades of the 20th century under the auspices of the PANYNJ's Engineering Department, of which the test on Pile P-5A was but one of many, were ML-type tests of an unconventional nature in terms of the load-application protocols used. There were typically multiple unload-reload cycles at intermediate load levels as well as extended intermediate hold-times during the course of a test. This is because the foundation designers were experimentally trying to establish an 'allowable design capacity' for various types of piles by loading to various target 'double-design' load levels.

As discussed in Horvath and Trochalides (2004) and Horvath (2014b), in the late 1980s *Monotube* piles were a then-new alternative to the over four decades of timber-pile usage at JFKIA and as such represented a significant paradigm shift in terms of deep-foundation design at JFKIA. That what was being sought from *Monotube* piles both in terms of embedment depth and resistance was a tectonic shift from over 40 years of using timber piles can be seen with reference to Figure 6.2 and noting that timber piles at JFKIA were historically never designed to have more than 120 kips (534 kN) ULS resistance and 60 kips (267 kN) allowable capacity in axial compression. As can be seen in Figure 6.2, the actual ULS resistance of Pile P-5A was approximately four times that sought for the timber piles in use at JFKIA up to that point in time.

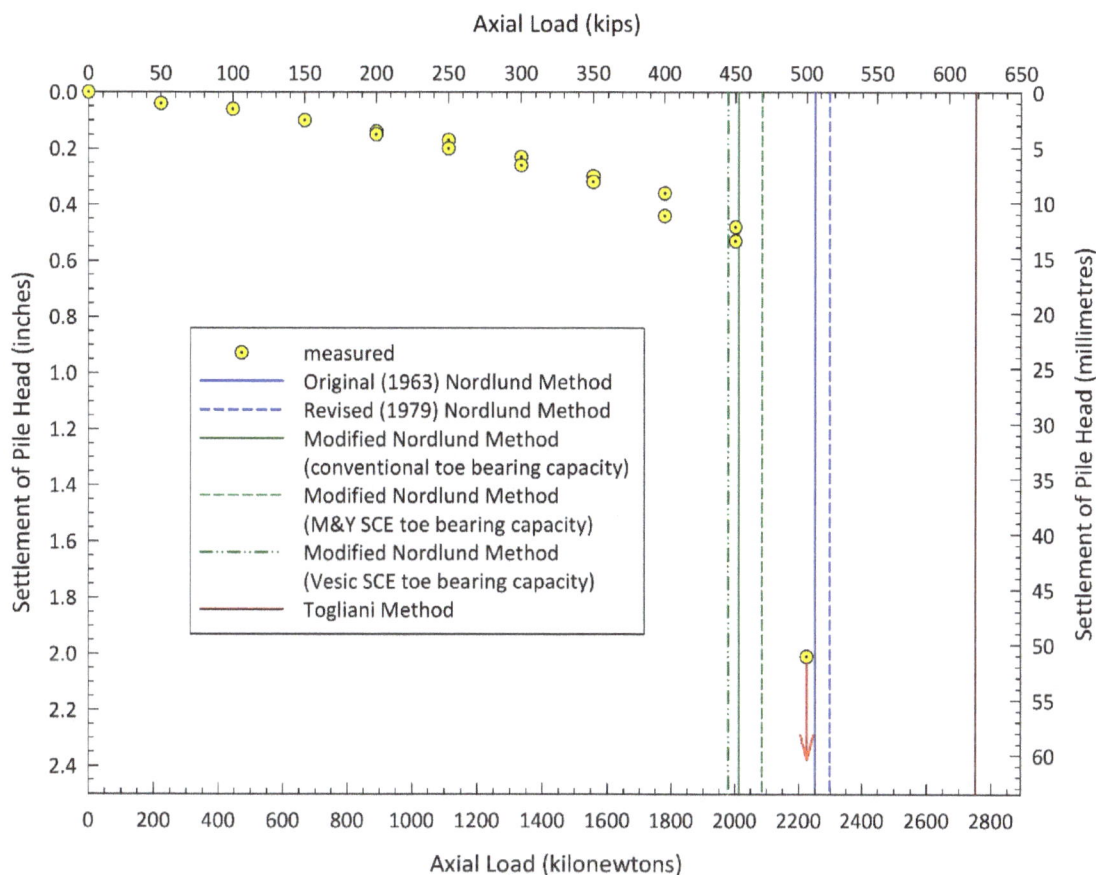

Figure 6.2. Forecast vs. Measured Results for Pile P-5A at JFKIA.

6.3.3 Discussion

While the results in Figure 6.2 are self-explanatory, they do not provide a complete picture of, and concomitant insight into, each of the analytical methodologies that was used. There are further data presentations that provide substantially more insight into the inner workings of each method.

To begin with, Table 6.1 shows the breakdown of the calculated ULS values of the resistance components for each analytical methodology. Note that the Togliani Method is the only method included in this assessment that uses the donut-on-a-stick model for the tapered portion of the pile. Therefore, the tapered component of resistance for this method (278 kips/126 kN) can actually be broken down even further from that shown in table. Specifically, the calculated 'stick' contribution is 130 kips (577 kN) while the calculated 'donut' contribution is 148 kips (659 kN). This is close to being a 50-50 split in the resistance allocation.

The values in Table 6.1 reveal a substantial variation in the relative contribution of each of the three resistance components between the different analytical methods although there is internal consistency in resistance allocations within different versions of a given method. This pattern is more easily seen in the graphic displays of Figure 6.3 that show the relative contribution of each resistance component using a bar-graph format. Also shown using red dashed lines are the actual contribution percentages interpreted by Fellenius et al. (2000) for two essentially identical (including installed length) *Monotube* piles at JFKIA.

Table 6.1. Breakdown of Calculated ULS Values of Resistance Components for Pile P-5A at JFKIA (Tabular Format).

Resistance-Forecasting Methodology		ULS Resistance, kips (kN)			
		shaft		toe	total
		constant diameter	tapered		
Nordlund	Original (1963)	58 (260)	410 (1826)	37 (165)	506 (2251)
	Revised (1979)	80 (356)	414 (1844)	22 (98)	516 (2297)
Modified Nordlund	toe: traditional	107 (477)	275 (1224)	70 (312)	452 (2013)
	toe: M&Y SCE	107 (477)	275 (1224)	87 (385)	469 (2086)
	toe: Vesic SCE	107 (477)	275 (1224)	62 (277)	445 (1978)
Togliani		262 (1165)	278 (1236)	79 (351)	618 (2751)

<u>Note</u>: Total resistance may differ slightly from the sum of the resistance components. This is due to roundoff of figures from actual calculations that were performed with greater precision to minimize accumulated roundoff errors during intermediate stages of calculations.

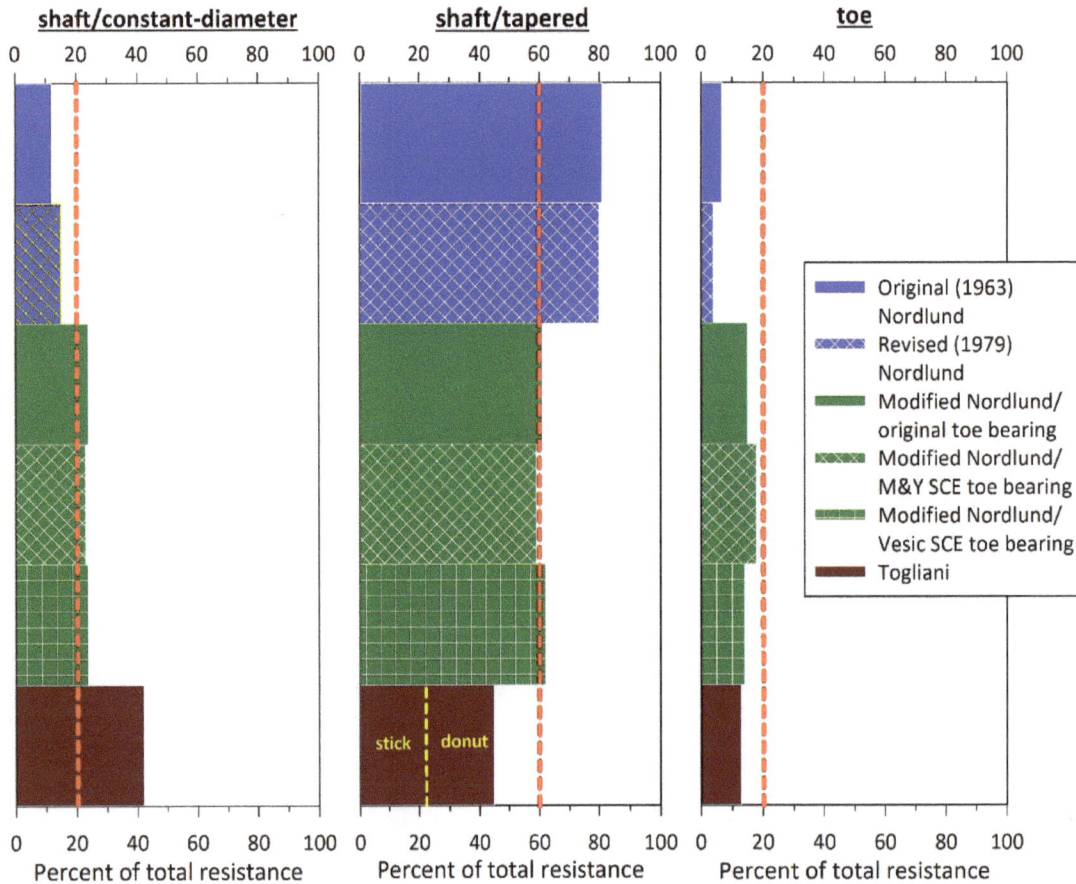

Figure 6.3. Breakdown of Calculated ULS Values of Resistance Components for Pile P-5A at JFKIA (Graphical Format).

The broad trends exhibited by the different analytical methodologies considered are:

- Overall, the Nordlund Method produced results that allocate much more resistance to the tapered portion of the pile than the other two methods, primarily at the expense of toe resistance. The net result of Nordlund's 1979 revision, which uses N_1 (= N_{field} values normalized to a vertical effective overburden stress of 1 ton/ft^2 = 2 kips/ft^2 = 95.8 kPa) instead of N_{field} values and limits the vertical effective stress at the toe to 3 ksf (144 kPa), is to shift more resistance to shallower depths.

- The writer's Modified Nordlund Method tracks the observed distribution of resistance observed in the Fellenius et alia piles well although toe resistances are slightly under-forecast.

- The Togliani Method allocates substantially more resistance to the constant-diameter portion of the shaft than the other methods, at the expense of both tapered-shaft and toe resistances.

Another insightful graphical technique for portraying the data in Table 6.1 is shown in Figure 6.4 that depicts the depth-wise attenuation of axial-compressive forces within the pile, normalized to the forecast capacity (= ULS total resistance). This is the same plotting format used elsewhere in this monograph. As shown in Figure A.1 in Appendix A, this plotting format is a mirror image of the depth-wise distribution of resistance.

Note that because the curve shown for each analytical method in Figure 6.4 is relative to the maximum forecast capacity for that methodology, the absolute variation in forecast capacities (ULS total resistances) between methods is removed. This allows for an apples-to-apples relative comparison between analytical methodologies.

Because the two instrumented test piles evaluated by Fellenius et alia were essentially identical to Pile P-5A that is under current discussion, the approximate inferred '20-60-20' allocation of capacity (ULS resistance) to constant-diameter shaft, tapered shaft, and toe that was found to exist by Fellenius et alia is also shown in Figure 6.4 and serves as a presumed ground-truth to which the various analytical methodologies shown in Figure 6.4 can be compared.

The results presented in Figure 6.4 are consistent with results and discussions elsewhere in this monograph, and can be summarized as follows:

- Partially tapered piles with the traditional placement of the tapered section of the pile in the lower portion of the overall pile and embedded in reasonably homogeneous coarse-grained soil conditions exhibit a distinctive kink in the depth-wise distribution of resistance and its complement, the axial-compressive force in the pile, at least under applied loads approaching or at what can be considered the geotechnical ULS.

- Both the Nordlund and the writer's Modified Nordlund methods replicate this kink although the limited comparisons made to date indicate that the Modified Nordlund Method does a noticeably better job of matching reality.

- The Togliani Method produces results that do not exhibit the characteristic kink but are for all practical purposes depth-wise linear with a constant slope.

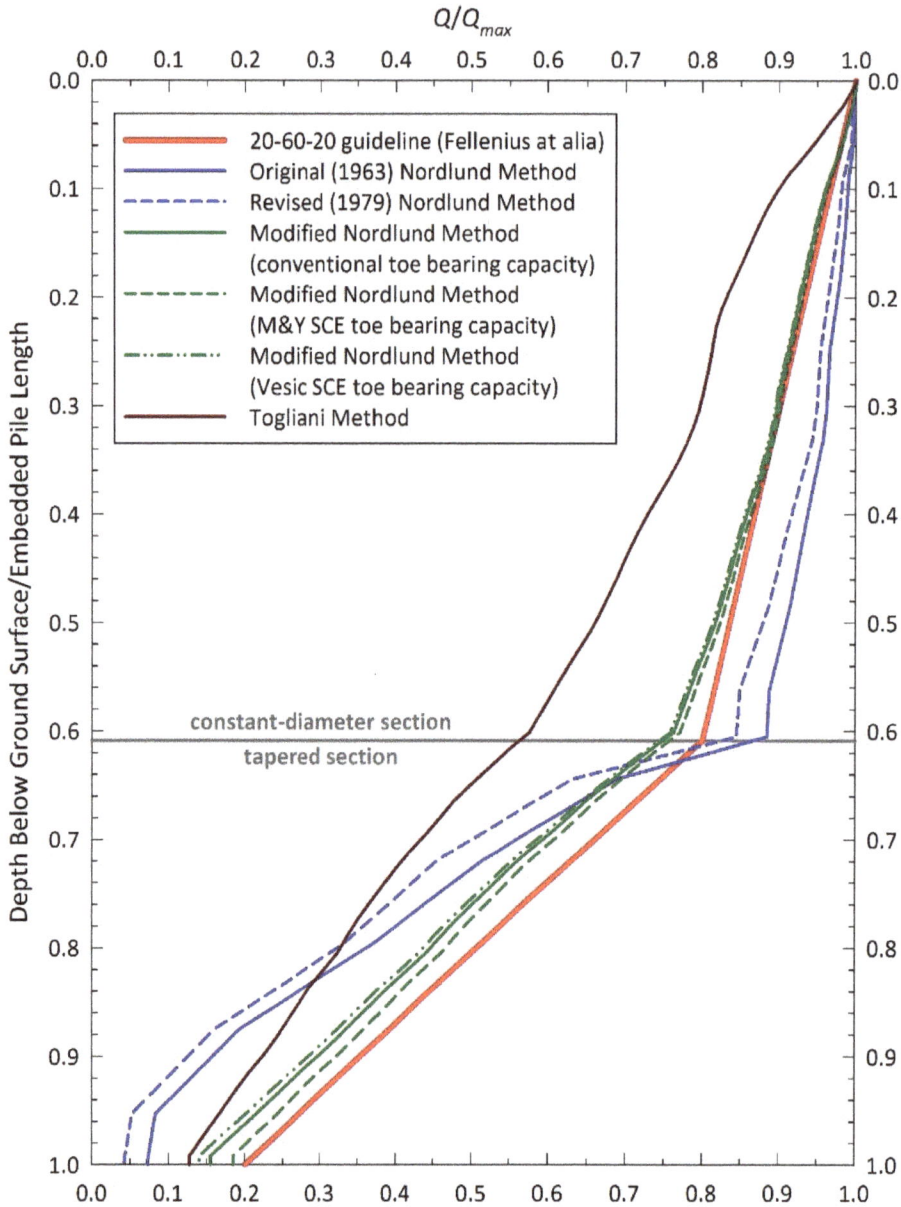

Figure 6.4. Depth-Wise Distributions of Forecast Normalized Axial-Compressive Forces in Pile P-5A at JFKIA.

6.4 CLOSING COMMENTS

As stated at the outset of this chapter, the material presented in this chapter is not intended to be a definitive critique and ranking of resistance-forecasting methodologies for tapered piles. To the best of the writer's knowledge, such a study has never been done and to do so adequately would require many instrumented piles load tested to the geotechnical ULS in order to provide the body of ground-truth necessary to support a statistical assessment.

Nevertheless, there are some useful generalities and observations that can be made about the analytical methodologies used in this chapter based on the writer's prior experience using them to perform comparative resistance forecasting for a modest database of timber, *Monotube*, and *TAPERTUBE* piles at JFKIA that was reported in earlier publications (Horvath 2002, 2014b, 2015,2016a, 2016b). Some of these comments and observations have been made in this chapter and others are presented in Chapter 9 that concludes this monograph.

This page intentionally left blank.

Chapter 7

Schema for a One-Step Procedure for Resistance Forecasting

7.1 INTRODUCTION

Having reviewed the current state of knowledge related to resistance-forecasting methodologies for the response of tapered piles to axial-compressive loading in Chapter 5 and considered which of these are actually of practical, usable value at the present time and in their current form in Chapter 6, the conclusion reached by the writer is that none of the currently available methodologies is ideal in its present form. Further, none of them addresses the response of tapered piles to uplift loading that has become increasingly more common in practice as the result of increased design for extreme events (earthquakes, floods) combined with more-efficient structural design (making for lighter, more slender structures) due to the ubiquitous use of digital computers and concomitant software. The basis for these conclusions is discussed subsequently.

Although none of the existing analytical methodologies for tapered piles is ideal in its current form, it is reasonable to ask if any is 'good enough' for continued use in routine practice, at least with respect to axial-compressive loading. Certainly, there are those who feel that a state of good-enough currently exists as there are government agencies in the U.S. (FHWA, NYSDOT) that continue to support and promote use of the Nordlund Method more than 50 years after it was developed and first published. However, it is unclear to the writer if these agencies fully understand some of the subtler nuances implied by the Nordlund Method such as the required use of N_{field} values that reflect circa-1960 SPT practices and the use of correlation between N-values and ϕ that reflect simplistic 1950s 'cookbook' values.

It is obvious that the answer to this question of good-enough is largely subjective, even when there are factual comparisons of measured vs. forecast results such presented as in Chapter 6 and earlier publications by the writer (Horvath 2002, 2015, 2016a, 2016b) that show existing forecasting methodologies to be deficient in various ways and to varying extents. This is because the answer to the perennial question of how close a match between theory and reality is considered 'close enough' and thus acceptable to use in routine practice is always subjective in civil engineering and will vary from one design professional or academic researcher to another. As a minimum, a researcher who published an analytical methodology must have thought that it was not only 'good enough' but an improvement over the status quo that existed at the time their work was published. In the past, the writer certainly felt that way about the writer's Simplified Nordlund Method. And presumably Kodikara and Moore, El Naggar et alia, Manandhar et alia, and Togliani felt (and may still feel) this way about their contributions to the subject of tapered piles.

However, technology always moves on with time and given the changes in subsurface-exploration technology, site-characterization sophistication, and the overall state of knowledge concerning the load-settlement behavior of DFEs in general in the almost 20 years since the Simplified Nordlund Method was developed by the writer, the conclusion reached by the writer during preparation of this monograph is unequivocally that we (the collective stakeholders in foundation design and research) can do better when it comes to forecasting the resistance of tapered piles to externally applied axial loads.

7.2 OBJECTIVE

The objective of this chapter is to present a schema or framework for an improved resistance-forecasting methodology for the behavior of tapered piles under axial loading. The intention is that this schema will be used to develop a practice-oriented computer-based solution methodology in the future.

While this schema obviously reflects the writer's preferences and opinions, every reasonable effort is made to first state, as objectively as possible, why a specific technical issue was chosen or not for inclusion in the proposed schema and then to follow this up with technically based reasoning as to why a given theoretical solution or analytical approach is suggested as the way to solve the issue. The intention of presenting this comprehensive rationale for the presented schema is to create a clear understanding of the what, why, and how of the writer's decision making. This is to provide a well-documented starting point for future actualization and implementation of the schema as well as a rational basis for other researchers to disagree with the writer and develop their own schema as desired.

7.3 BACKGROUND

The writer's opinion that current resistance-forecasting methodologies for tapered piles are no longer 'good enough' for use in routine practice is based on numerous factors. However, the single most important one is the observation that there is compelling evidence that taper benefit derives from two distinct, successive, cumulative stages of occurrence of the wedging mechanism that is unique to tapered (driven) piles but not tapered DFEs in general. No existing analytical methodology adequately and accurately reflects this reality. There are several additional criticisms that can be made with regard to existing forecasting methodologies, such as failure to incorporate shear-modulus degradation, but the failure to account for the wedging that occurs during both installation and post-installation externally applied axial-compressive loading in a holistic manner is overarching in the writer's opinion.

Because this two-stage occurrence of wedging is such as crucial, central, distinguishing feature of tapered (driven) piles, and formal recognition of this behavior represents a significant paradigm shift from all resistance-forecasting methodologies developed to date, it is useful to explain this behavior in detail before getting involved in the many specifics of the schema presented in this chapter.

This two-stage wedging is readily understood in terms of the lateral earth pressure coefficient, K_h, acting along the pile shaft and, in particular, the K_h/K_o ratio as the latter is a useful metric for relating relative change in radial earth pressure:

- The pre-installation stress-state in the ground is defined by $K_h = K_o$. K_o is typically depth-wise variable and the result of a site's unique stress history that can be very complex. Furthermore, K_o will reflect any prior episodes of human activity, some of which may not be obvious, e.g. past groundwater lowering for water supply or construction that has since returned to a normal level, and not be documented or otherwise known. Therefore, the magnitude and distribution of K_o should never be assumed or taken for granted no matter how 'greenfield' or virgin a site may appear on the surface. For example, a never-developed site may have been affected by groundwater lowering on an adjacent site at some time in the past. Thus, K_o should always be determined on a site-specific basis prior to construction.

- The pre-installation stress-state in the ground (as defined by $K_h/K_o = 1$) increases due to pile driving (as reflected in a post-installation $K_h/K_o > 1$), at least when conventional impact driving for the entire embedded length of a pile is used. Note that this increase does <u>not</u> occur with drilled DFEs, tapered or otherwise. The writer's hypothesis is that the increased K_h/K_o ratio that results from installation is <u>always</u> greater for tapered piles than an otherwise identical constant-perimeter pile in the same conditions. This difference is hypothesized as being due to the differences in what occurs during the physical process of <u>cylindrical-cavity creation</u> (CCX) when a pile is driven into the ground. A pile toe opens an initial cavity in the ground for all types of piles. However, the difference occurs when the toe has passed a given depth in the ground. With a tapered pile of conventional geometry where the shaft taper begins just above the pile toe, it is intuitively obvious that further cavity creation occurs as the pile is driven further into the ground. This is Stage I of the wedging mechanism. Neither additional cavity creation nor wedging occurs with a constant-perimeter pile as the initial cavity created by the toe simply remains fixed in diameter during additional driving. Because the additional cavity creation and wedging along the pile shaft clearly are unique to tapered piles (but <u>not</u> tapered DFEs in general), this means that after installation but prior to the application of external axial loading the shaft of a tapered pile is subjected to a larger radial stress than a comparable constant-perimeter pile.

- The additional, Stage II wedging and concomitant increase in radial stress (as reflected in a further increase in the K_h/K_o ratio) acting along the tapered portion of the pile shaft occurs as an axial-compressive load is externally applied to the pile at some point after installation. Note that this is the <u>only</u> wedging and concomitant taper benefit that occurs with tapered DFEs created by a drilling process. This wedging stage is well modeled using <u>cylindrical-cavity expansion</u> (CCE) theory as the additional radial expansion of the shaft-ground interface due to pile settlement under axial-compressive external loading is occurring to a pre-existing cavity in the ground (the result of the DFE installation). By comparison, a constant-perimeter pile experiences no further increase in the K_h/K_o ratio along its shaft under post-installation application of an axial-compressive external load as the diameter of the cavity created by pile installation does not change. Furthermore, it is the writer's hypothesis that this Stage II is the simple geometric result of the <u>total</u> pile settlement at a point, i.e. it is related to the tangent of the taper angle, α, of the pile. Note that this hypothesis is distinctly different from that of Kodikara and Moore (and El Naggar et alia and Manandhar et alia) who assumed that the radial expansion was related to the <u>differential</u> settlement between the DFE shaft and adjacent ground that is always smaller in magnitude. This assumption, combined with the fact that Stage I wedging that occurs as a result of driving is not explicitly considered, is likely why the taper benefit that is implied in various figures found in Kodikara and Moore (1993) is well below what would be expected based on the performance of real tapered piles (keep in mind that Kodikara and Moore looked at exactly one real tapered DFE in their paper and it was a relatively short drilled shaft in a soil with a substantial fines content, hardly a good example to use as an exemplar of the benefit of tapered (driven) piles in coarse-grained soil).

Note that clearly defining the conceptually and mathematically distinct steps of cavity <u>creation</u> during pile driving and cavity <u>expansion</u> during post-installation axial-compressive external load application (something not done heretofore in any resistance-forecasting methodology for tapered piles) allows for rational extension of cavity-mechanics concepts to include post-installation externally applied uplift loading (also something not done heretofore in any resistance-forecasting methodology for tapered piles). It is intuitively obvious that the upward-vertical displacement of a tapered-pile shaft relative to the surrounding ground causes a <u>decrease</u> in the diameter of the cavity created during pile installation and that was expanded on if there were compressive loads applied externally prior to the application of uplift loads.

This unwedging of the shaft-ground interface along the tapered-pile shaft as the result of uplift loading can be modeled using <u>cylindrical-cavity contraction</u> (CCC). By doing so, this offers the potential to rationally quantify the radial stresses acting along the pile shaft and, from this, rationally forecast the resistance to uplift loading, something that historically has eluded the resistance-forecasting of tapered piles. Perhaps most importantly, it will once and for all dispel the long-held perception by some (many? most?) that tapered piles will somehow magically pop out of the ground when subjected to uplift loads and thus have minimal uplift resistance. The Gregersen et al. (1973) case history involving tapered piles (both continuously and partially tapered) that is evaluated at length in Appendix H clearly shows that this common perception is grossly incorrect.

In summary, the root-cause shortcoming of all tapered-pile R&D to date, and thus the root-cause shortcoming of all resistance-forecasting methodologies that have come out of this R&D, is that the 3-D physical mechanism of wedging has either not been modeled correctly (this applies to all the versions and modifications of the Nordlund Method) or modeled incompletely (the Kodikara-Moore Method and the derivative work of El Naggar et alia and Manandhar et alia). Even more problematic are methodologies such as the Togliani Method that are crafted around a gimmicky physical model (donut on a stick) that does not remotely replicate the actual wedging mechanisms of a tapered pile at all. The theoretically sound theories of cylindrical-cavity mechanics offer the potential to address all aspects of tapered-pile installation and post-installation external load application in either compression or uplift in a unified manner.

7.4 OVERVIEW

In developing the schema presented in this chapter, the writer first and foremost considered the overarching need to explicitly consider the separate taper-based effects from both the initial pile installation and subsequent external loading, both in compression and uplift, as explained in the preceding section. Furthermore, these taper effects need to be framed within the physical mechanism of wedging (and its partial reversal in the case of uplift loading) as is now accepted as the physical mechanism that best portrays what happens both when a tapered pile is driven and when it is subsequently loaded axially. This means that a gimmick or workaround such as the donut-on-a-stick model is not considered adequate for use, no matter how appealing it may be on some level or who promotes its use. Also considered inadequate are older physical models for wedging such as 2-D passive earth pressure used by Nordlund. Cylindrical-cavity mechanics, which correctly models the 3-D nature of wedging (at least for piles with a circular cross-section), is far superior for this purpose and is thus deemed to be an essential component of the presented schema.

Next considered was the list of desirable behavioral features that was given at the beginning of Chapter 5. This list was compiled based on the writer's critical assessment of all the geomechanical and structural issues affecting or impacting the behavior of tapered piles in any manner that were presented and discussed in Chapter 4. Obviously, the geo-structural contents of Chapter 4 reflect the writer's personal assessment of the current state of knowledge that is relevant to tapered piles, even tangentially or incidentally. However, the point being made here is that anyone looking to develop their own schema for an improved resistance-forecasting methodology for tapered piles should logically start by enumerating what they feel are the significant geo-structural issues controlling tapered-pile behavior and then perform an assessment of these issues to come up with their own list of desirable behavioral features.

In consideration of the fact that the schema outlined in this chapter includes behavioral issues not included in any existing methodology as well as the fact that tapered piles can have a variable geometry (typically partially tapered but also continuously tapered but with non-uniform taper angle as in the case of timber piles), the writer feels that at the present time and state of knowledge that developing an indirect method is preferable to a direct method. This method-philosophy continues and builds on the precedent established in 1963 by the first true analytical methodology for tapered piles (the Original Nordlund Method) and continued through the Kodikara-Moore Method and all of its derivatives by El Naggar et alia and Manandhar et alia.

It is recognized that direct methods such as the Togliani Method that use some measured parameter(s) from an in-situ test such as a cone penetrometer to directly forecast both ULS unit shaft and ULS unit toe resistances are the essence of simplicity and thus very attractive to use in practice. However, indirect methods inherently provide superior insight into what is actually occurring from geomechanics and structural mechanics perspectives during all stages of pile installation and subsequent applied loading. This ability to 'see' what is going on theoretically and mathematically as opposed to simply getting a final result from some 'black-box' direct methodology is considered by the writer to be essential at the present time.

Similar reasoning was used to eliminate from consideration any extension of the Niazi Method to tapered piles. While the one-step load-settlement end result of the Niazi Method is very attractive and preliminary research by the writer has shown that it can readily be extended to include tapered piles, its theoretical basis in the Randolph-Wroth Problem that assumes perfect contact with <u>no</u> slippage between the shaft of a DFE and surrounding ground during <u>all</u> stages of loading is systemically problematic in the writer's opinion. Also problematic is the fact that applied loads are consistently over-allocated to shaft resistance (at the expense of toe resistance) well beyond the ULS shaft resistance that one might calculate using a strength-based resistance-forecasting methodology. The fact that the shaft vs. toe allocation errors often compensate and result in an acceptable total resistance does not ameliorate the writer's negative, deal-breaking opinion in this regard.

It is important to note that implementation of the schema presented herein should not be assumed to be an all-or-nothing proposition. In the writer's opinion, the overarching goal at this time should be to advance the state of practice for resistance-forecasting of tapered piles as soon as possible and in stages as the necessary R&D progresses rather than waiting until some unknown, ill-defined time in the future when 'everything' is known. To quote from the 1997 film "Wag the Dog" (which itself is an adaptation of a quote attributed to General George S. Patton):

"A good plan today is better than a perfect plan tomorrow."

This logic reflects the reality that advances in the state of art tend to be incremental and piecemeal in nature. Consequently, suggestions for specific incremental improvements are noted where appropriate throughout the subsequent presentations in this chapter.

Another important consideration is that any resistance-forecasting methodology eventually developed based on the schema presented in this chapter needs to strike a balance between theoretical rigor and ease of use so that practitioners will actually use the methodology. This need is abundantly clear from the history of resistance-forecasting methodologies for tapered piles. The Nordlund Method, which is almost 60 years old at this point in time and based on a simplistic behavioral model for wedging (2-D passive lateral earth pressure on an imaginary rigid wall) and antiquated site characterization (SPT N-values), both of which date from the early decades of the 20th century, is still the mainstay of practice, in the U.S. at least. This is despite the fact that the mechanism of 3-D wedging as modeled using cylindrical-cavity mechanics was identified as a superior model for tapered piles in the early 1990s and the use of spherical-cavity mechanics for toe resistance was identified even earlier, in the early 1970s. However, the fact that neither meaningful guidance on soil-property selection was ever provided nor commercially available computer software was ever developed to make the relatively complex mathematics and three-phase analytical method developed by Kodikara and Moore palatable for everyday foundation design is clear evidence that their methodology is simply too daunting to practitioners. Note that as discussed in Horvath (2002), even the writer was put off by the practical challenges and intricacies of the Kodikara-Moore Method so developed the Simplified Nordlund Method as an interim methodology.

Further to this point is the fact that the El Naggar-Sakr Method, which is a much-simplified version of the Kodikara-Moore Method that only produces a forecast of the ULS total resistance, does not appear to have developed any critical-mass of acceptance and usage, at least among practitioners, despite being in existence for approximately 20 years at this point in time. In the writer's opinion, the lack of clear direction from El Naggar et alia as to how critical soil properties such as a single-value shear modulus should be determined in practice when using the El Naggar-Sakr Method is but one impediment to its use.

To conclude this overview of what is to come in the remainder of this chapter, there is no doubt that further improvements to resistance-forecasting methodologies for tapered piles can be made beyond those presented and discussed herein. Ultimately, it may also be possible to develop a robust and reliable direct method based on cone-penetrometer data that adequately incorporates all the behavioral nuances of tapered piles. Thus, the material presented in this chapter should be viewed only as a blueprint for the near-term future and not the final word on the subject.

7.5 SCHEMA DETAILS

7.5.1 Introduction and Overview

The details of a schema for an improved analytical methodology for tapered piles that constitute the remainder of this chapter are broadly structured along the lines of what the writer feels are desired behavioral features of such a methodology. Also included are some thoughts concerning the structure and execution of a solution algorithm.

The list of desirable behavioral features that was presented at the beginning of Chapter 5 was used as the basis for this purpose. As noted in Chapter 5, the writer developed this list based on an overall assessment of the myriad topics that were presented and discussed in Chapter 4. Some of these topics are quite general in nature and thus applicable to a broader range of geotechnical and/or foundation problems and applications than just tapered piles. There is also interaction and interdependent behavior between and among these topics. However, when curated and synergized, the body of knowledge presented in Chapter 4 yields a state-of-art interpretation of the numerous factors that, in the writer's opinion, influence the installation and axial loading of tapered piles.

7.5.2 Basic Problem Geometry

7.5.2.1 Pile Cross-Sectional Geometry

As illustrated in Chapter 3, tapered piles can have either a circular or square cross-sectional geometry although the former have long predominated in practice, especially in the U.S. It is presumably for this pragmatic reason that resistance-forecasting methodologies for tapered piles have not, to date, included cross-sectional geometry as an explicit problem variable or, as in the case of the Kodikara-Moore Method and the derivative works of El Naggar et alia and Manandhar et alia, have indirectly limited the analytical methodology to piles with a circular cross-section by virtue of the physical model used as the basis of the methodology, i.e. cylindrical-cavity expansion.

However, currently available information and knowledge as noted earlier in this monograph clearly indicate that cross-sectional geometry likely influences taper benefit. This was first noted anecdotally by Nordlund, especially in his later (1979) publication, with the implication that the Nordlund Method in either of its versions is only reliable for piles with a circular cross-section.

More recently and significantly, the identification of wedging as the operative physical mechanism by which tapered piles develop shaft resistance has provided a sound theoretical basis for expecting cross-sectional geometry to play a clear role. This is because the current, and proposed continued, use of cylindrical-cavity mechanics as the preferred physical and concomitant mathematical model of the wedging mechanism is clearly applicable only to piles with a circular cross-section.

It is not immediately obvious what mathematical model is appropriate for the wedging that occurs with piles with a square cross-section. Wedges with a non-circular cross-sectional geometry that are used in real-world applications for felling trees or splitting firewood have a geometry such that the taper exists for only two opposing faces of the wedge and thus produces the wedging action in only one horizontal direction. However, non-circular tapered piles typically have a cross-sectional geometry such that wedging occurs in two orthogonal directions, presumably more or less equally. The currently unanswered question, then, is what sort of physical behavior occurs within the circular quadrant between each face of a square wedge and how the overall displacement pattern that develops is best modeled mathematically (expansion of a right-circular cylinder clearly does not appear to be appropriate).

In the writer's opinion, fundamental research is required to first and foremost define what is occurring physically as a tapered pile with a non-circular cross-section is driven and then loaded in axial compression, i.e. what does wedging due to four planar surfaces being expanded simultaneously and horizontally into a soil mass look like. Only then can a mathematical model for this physical mechanism be proposed and developed and eventually extended to include the unwedging associated with uplift loading.

Therefore, for the time being any improved resistance-forecasting methodology for tapered piles should be restricted to piles with a circular cross-section as the physical and mathematical understanding of such piles is much farther advanced. However, future extension to piles with a non-circular cross-sectional geometry should be a long-term goal. Precast-PCC piles are economical alternatives to steel and wood in many parts of the world, including the U.S. Precast-PCC piles, whether tapered or not, often have a square cross-section. Consequently, there is economic incentive to undertake research to understand the behavior of tapered piles with a square cross-section as they may prove to be economical deep-foundation alternatives in many parts of the world.

7.5.2.2 Direction of External Load Application

Historically, resistance-forecasting methods for DFEs in general have focused on axial-compressive loads with good reason as these have always predominated in real-world applications. More specifically, and as noted earlier in this monograph, the relatively few analytical methodologies developed to date specifically for tapered piles have only considered axial-compressive loads.

That having been said, the writer feels that an improved resistance-forecasting methodology for tapered piles should be capable of handling uplift loads as well as compressive. To elaborate on comments made earlier in this regard, in the writer's opinion, which derives from first-hand experience in this case, there are several reasons for this based on changes in both needs and capabilities that have evolved over time:

- The need and concomitant frequency to design deep foundations for uplift loads has increased in recent decades. This is a combination of the increasing need to design for extreme events such as earthquakes and hurricanes; the ability to efficiently design high-rise structures with minimal dead loads; and developers wanting ever taller and more-slender structures ('sliver buildings') on building lots that historically would not or could not be developed for high-rise buildings.

- In the past, tapered piles were sometimes eliminated from consideration as a deep-foundation alternative based on the simplistic assumption that they had no significant uplift resistance as they would somehow instantaneously lift out of a tapered hole in the ground if subjected to an uplift load. While this was clearly overly simplistic, there was no rational analytical methodology available to disprove it.

- Cavity mechanics provides the long-needed analytical tool to be able to be able to rationally evaluate the response of a tapered pile subjected to uplift. As noted in Chapter 4, there does not appear to be a reason why cylindrical-cavity <u>expansion</u> (CCE) theory could not be reversed and used as cylindrical-cavity <u>contraction</u> CCC theory. In fact, Yu and Houlsby noted this explicitly in their 1991 paper in which they referred to CCC as the "unloading" complement to CCE, which they referred to as "loading". The intended goal of developing CCC for tapered piles is to demonstrate that as uplift loads are applied to the head of a tapered pile and the shaft displaces upward, the shaft does not completely and instantaneously separate from the surrounding soil and lose all resistance to uplift. Rather, there is an orderly contraction (unwedging) of the shaft-soil interface with concomitant orderly decrease in the radial stress as defined by a reduction in the K_h/K_o ratio. Thus, although the unit shaft resistance decreases as uplift load is applied because of the decrease in K_h, this resistance does not immediately go to zero. In fact, it can be argued that it could never go completely to zero. The reason is that cylindrical-cavity mechanics is conceptually identical to lateral earth pressure theory, where lateral displacement of a rigid wall <u>away</u> from a mass of retained soil (the reverse of Nordlund's assumption of a fictitious wall displacing <u>into</u> a soil mass) causes the lateral earth pressure to decrease but never to zero. Rather, lateral earth pressures reduce in an orderly manner from the initial value until the active earth pressure state is reached. At that point, lateral earth pressures remain at the active-state level no matter how much additional lateral displacement may occur. Admittedly, the issue of a 3-D cavity is more complicated than that of a 2-D lateral earth pressure problem because of the ability of soil to arch horizontally around a circular cavity. But the point is made nonetheless.

Based on these considerations, any improved resistance-forecasting method for tapered piles should be capable of analyzing both axial-compressive and axial-uplift loads. However, as with the issue of cross-sectional geometry, it may be desirable from a pragmatic perspective to limit initial developmental efforts to compressive loading and defer uplift loading to a later date. This is because less is known about CCC at present compared to CCE so fundamental research into cavity contraction is required.

Finally, and for the sake of completeness in the present discussion, mention should be made with respect to lateral and/or moment loads applied to the head of a tapered pile even though this issue was addressed earlier in this monograph. Unlike with externally applied axial loading, there is nothing behaviorally unique about tapered piles subjected to lateral/moment loading. The response of any DFE to lateral/moment loading involves soil-structure interaction (SSI) effects that are unrelated to the resistance mechanisms that develop in response to axially applied loading. The only issue that needs to be addressed in modeling a tapered pile's response to lateral/moment loading is the depth-wise variation in flexural stiffness that is due to the depth-wise variation in pile cross-sectional geometry.

7.5.3 Overall Behavior

7.5.3.1 Soil Conditions

7.5.3.1.1 Soil Type and Behavior

Historically, the niche for which tapered piles were found to be most cost effective was as so-called 'friction' or 'floating' piles in coarse-grained soil although they often have to penetrate through one or more minor layers of fill of various composition and/or fine-grained soil in the process. The one exception was for the thousands of years when timber piles were used in all types of soil and even for end bearing on rock but this was because they were the only type of pile available.

One of the questions that remains unanswered conclusively to the present is whether or not there is taper benefit for a pile embedded entirely, or at least primarily, in fine-grained soil. The writer is aware of anecdotal and unpublished second-hand comments claiming both 'no' and 'yes' to this question but is not aware of any published work that supports either answer conclusively. We do know that Nordlund crafted his resistance-forecasting method solely for coarse-grained soil conditions. At the time Nordlund was finalizing his analytical methodology in the early 1960s, his employer, Raymond, already had several decades of widespread commercial experience driving tapered piles. One can argue that if Raymond had found tapered piles to be commercially cost effective in fine-grained soils then the Original Nordlund Method would have reflected this fact.

Moving on in time, although Kodikara and Moore developed their resistance-forecasting methodology based on ground that was allowed to have non-zero values of both Mohr-Coulomb strength parameters, ϕ and c, it is the writer's opinion that this was done solely for academic generality and not for any relation to reality. Both authors were involved in academic research at the time and in subsequent written communications initiated by the writer both acknowledged the fact that they were largely unfamiliar with actual tapered-pile types and practices at the time they conducted their research. This is reflected in the fact that in their 1993 paper they used a tapered drilled shaft (bored pile) in fine-grained soil (despite the fact that tapered driven piles in coarse-grained soil have always been predominant) as the sole exemplar of a tapered DFE against which calculated outcomes from their analytical methodology were compared. Thus, it does not appear that Kodikara and Moore's assumption of c-ϕ shear-strength parameters was motivated by any practical need or purpose. This is also supported by the fact that El Naggar et alia assumed $c = 0$ in their subsequent publications that developed a simplified version of the Kodikara-Moore Method.

Nevertheless, it is suggested that any improved resistance-forecasting methodology for tapered piles allow for layered soil systems (something that Kodikara and Moore did not do) where the shear strength of any given layer is defined by either c or ϕ but not both. This implies that any layer of fine-grained soil is assumed to behave in an undrained manner so that the chosen value of cohesion reflects its undrained shear strength. This overall approach is consistent with most resistance-forecasting methodologies for DFEs in general since the advent of engineered foundation design in the 1950s. The only exceptions to this are methodologies such as the β Method that use an effective-stress basis for all soil types. Furthermore, explicitly allowing for fine-grained soil strata will provide a basis for future research into the taper benefit, or lack thereof, in such soils.

7.5.3.1.2 Temporal Issues

As discussed in Chapter 4, one of the most significant findings relative to tapered piles that occurred in the latter decades of the 20th century is the fact that piles, especially tapered piles, in coarse-grained soil often exhibit significant time-dependent increases in axial-compressive resistance in the days and weeks after driving. Therefore, this time dependency needs to at least be acknowledged and discussed.

To begin with, the writer is not aware of any resistance-forecasting methodology for any type of DFE that explicitly considers time as an explicit variable. If for no other reason, this is likely due to the fact that there is no consensus as to the causal mechanism for this increase in resistance. Initially, the thinking was that the increase was <u>strength</u> related, perhaps due to reconsolidation of loose, fine sands that were temporarily liquified during pile driving[161]. However, more-recent thinking, such as expressed in Fellenius et al. (2000), is that the increase is <u>stiffness</u> related. In the writer's opinion, residual loads and temporal changes thereto may play a role in this phenomenon of temporal increase in resistance.

Until this issue is resolved, an appropriate time-dependent model cannot be postulated and verified. Therefore, for the indefinite future the issue of time dependency of axial resistance for tapered piles is something that must at least be recognized for its existence. The best approach for now appears to be to focus on longer-term, as opposed to end-of-initial-driving, resistance so that any comparisons between forecast and measured results should always be done for piles that have had at least several weeks between the time that they are driven and tested in some manner.

7.5.3.2 Load-Settlement

A recurring theme throughout this monograph is that there needs to be a fundamental paradigm shift in thinking concerning the axial-compressive behavior of all types of DFEs as it applies to routine analysis and design. Specifically, there is a need to routinely consider

---

[161] The writer was professionally acquainted with (and, for a while in the early 1970s, worked under the supervision of) Donald L. 'Don' York, P.E. who was the principal author of York et al. (1994). This paper was one of the first, if not <u>the</u> first, to note and discuss the issue of temporal increase in resistance for tapered piles in coarse-grained soils. The writer had discussions with Mr. York in the 1990s on this topic and the comments made herein reflect the writer's recollection of these discussions.

complete load-settlement behavior as opposed to focusing only on a single-valued ULS total resistance that is inherently ambiguous to define, even when using the long-accepted gold-standard of a static load test.

Therefore, whatever improved resistance-forecasting methodology is developed for tapered piles should be capable of generating a complete load-settlement curve as an intrinsic part of the methodology. However, to what extent the complete curve that results in used in any given application is a separate issue that is not addressed in this monograph.

In the writer's opinion, it is highly preferable that the load-settlement curve be generated as part of an integrated, one-step analytical procedure along the lines of that incorporated into the Kodikara-Moore Method as opposed to a two-step procedure that requires separate, unlinked calculations of ULS total resistance and settlement. Although there is a certain attraction to two-step procedures as a group because they allow unlimited combinations of resistance-only and settlement-only methodologies, there are a number of reasons why a one-step procedure is preferred.

To begin with, there are significant issues with all of the settlement-only methodologies that are of potential use as part of a two-step procedure and were presented and discussed in Chapter 5. In particular, the Randolph-Wroth Problem has an inherent inability to properly allocate shaft and toe resistances across the entire spectrum of ground conditions and applied loads, even when modulus degradation is incorporated into the solution methodology. Furthermore, the formulation of the Randolph-Wroth Problem inherently assumes that the pile and soil remain attached along the full length of the shaft during all stages of loading. Stated another way, relative slip between the pile shaft and adjacent ground can never occur. This is clearly unrealistic, no matter how much modulus reduction occurs, which is why Kodikara and Moore used the Randolph-Wroth Problem solution only for Phase I of their assumed three-phase pile response as discussed in Appendix C. Phases II and III of the Kodikara-Moore Method assume that relative pile-soil slip along the shaft develops.

7.5.3.3 Residual Loads

7.5.3.3.1 Introduction and Background

Unfortunately, a significant complication in implementing an accurate one-step load-settlement forecasting capability for tapered piles, and DFEs in general, on a routine basis is another important paradigm shift that needs to be made in both DFE research and everyday practice. This is recognizing (actually, re-recognizing is more accurate) that the residual loads that result after DFE installation in general, and driven- or jacked-pile installation in particular, are the rule, not the exception. Therefore, the existence of these loads needs to be recognized on a routine basis as ignoring them does not make them disappear.

Recognition of residual loads in the published literature dates back to at least Mansur and Kaufman (1958). Of relevance to the subject of this monograph is that residual loads were noted by Nordlund (1963) who cited Mansur and Kaufman's 1958 paper. The topic of forecasting residual loads in constant-perimeter (driven) piles was a relatively active area of research and concomitant publication for a period of time in the latter part of the 20th century. However, it appears that residual loads have, for the most part, fallen below the radar of both foundation engineering practitioners and researchers alike. One reason might be the fact that residual loads do not, in theory, affect the magnitude of the ULS total resistance that, historically and to the present, is generally the primary, if not only, geotechnical outcome of interest in routine practice. As has been noted several times throughout this monograph, the global migration from ASD to LRFD in routine practice has made focus on ULS total resistance more important and central to DFE design than ever. Consequently, it might be argued that designs based solely on the ULS total resistance can safely ignore residual loads.

However, it turns out that the situation is much more complicated than this simplistic rationalizing-away of residual loads from routine consideration. While it is true that residual loads do not affect the ULS total resistance in principle, the presence of residual loads can impact many aspects of deep foundations, both geotechnical and structural, including those that are related to accurate interpretations of ULS resistance. So, the reality is that residual loads can and do affect ULS total resistance when it is based on any type of physical load testing, whether static or dynamic.

Thus, it is essential to discuss residual loads and how to deal with them in the current context of developing an improved resistance-forecasting methodology for tapered piles. This is because such loads tend to be proportionately greater for tapered piles compared to constant-perimeter piles, all other problem variables being equal. Gregersen et al. (1973) present a clear and unambiguous example of this using full-scale piles as explored in depth in Appendix H. Furthermore, curation of available information during preparation of this monograph suggests that residual loads are more significant for tapered piles than any other type of DFE, especially with regard to the toe resistance. There are indications that the toe resistance is significantly, if not fully, mobilized at the end of driving tapered piles, at least in coarse-grained soil.

7.5.3.3.2 Overview

In the simplest terms, the presence of residual loads means that a pile is not in a geotechnical resistance-free state externally, and thus not in a structural stress-free state internally, when axial loads are first applied to the head of the pile after driving. Historically, resistance-forecasting methodologies and associated structural-design procedures for all types of DFEs assume a zero-resistance/zero-stress initial condition for the DFE. The writer is not aware of any resistance-forecasting or structural-design methodology in routine use for any type of DFE that explicitly assumes and considers a non-zero stress-state after installation and prior to external load application.

It appears that the only exceptions to this (and indirectly at that) are methods such as the Niazi Method that are calibrated using static load-test data. To the extent that the DFEs in the load-test database had residual loads (and most likely did, at least to some degree), this behavior is reflected in the method parameters that were quantified using this database. However, because the effects of residual load were masked and buried in the overall behavior, it is impossible to separate out that portion of the overall load-settlement behavior that is due to residual loads or to identify how the presence of residual loads influenced load-settlement behavior compared to that of an ideal, resistance-and-stress-free DFE.

The problem with neglecting residual loads and assuming a zero-resistance/zero-stress initial condition is that the overall stiffness of the DFE-soil system is not properly modeled. This can be likened to the difference between a normally reinforced PCC member and a prestressed version of the same member. In the absence of externally applied loads, the two members may look identical and be in overall static equilibrium. However, the way in which each member reacts to externally applied load and the concomitant internal stresses within each member are quite different. Thus, while residual loads in a DFE do not affect the ULS total resistance (at least in theory), they do affect the intermediate load-displacement stiffness and concomitant behavior. Furthermore, the stresses within the DFE are not properly modeled either so there are structural issues in addition to geotechnical issues.

Geotechnical and structural issues related to residual loads are discussed in the following two sections. These discussions are followed by a section in which the issue of how to deal with residual loads in the schema presented in this chapter is addressed.

7.5.3.3.3 Geotechnical Issues

For DFEs with a depth-wise constant perimeter, the traditional assumption of a zero-resistance initial condition always results in the conclusion that shaft resistance is fully mobilized well before any significant toe resistance is mobilized as axial-compressive loads are applied to the head of the DFE. This is primarily due to the significant disparity in the DFE-ground system stiffness of the two traditional resistance mechanisms for DFEs, shaft resistance due to sliding friction and toe resistance. The structural stiffness of the DFE shaft also plays a role but this is typically quite minor compared to the inherent difference in stiffness between these two resistance mechanisms.

The situation is currently much less settled for tapered piles as there has not been sufficient research to date to define how the third resistance mechanism for DFEs, shaft resistance due to wedging, is mobilized compared to the two traditional resistance mechanisms. Initial indications are that shaft resistance due to wedging continues to develop well after shaft resistance due to sliding friction has reached its maximum.

The presence of residual loads, especially for tapered piles, completely changes this traditional, simplistic picture of DFE response to axial-compressive loads. This is because residual loads typically result in significant toe resistance already mobilized, at least for driven and jacked piles. For example, for the two *Monotube* piles at JFKIA discussed in Appendix A, the toe resistance was essentially 100% mobilized at the end of driving. At the same time, a substantial portion of the shaft (a portion of the tapered section plus virtually all of the constant-diameter section) was subjected to drag load as shown in Figure 2.1. The precast-PCC Holmen Island test piles, both tapered and constant-perimeter, discussed in Appendix H exhibited similar behavior.

As a result of residual loads, when an external load is applied at the head of a pile after installation, this applied load is taken up primarily or even exclusively along the shaft by first reversing the zones of drag load to resistance and then mobilizing additional resistance farther down the shaft. Depending on how much of the toe resistance was already mobilized by residual loads, in extreme cases such as *Monotube* piles at JFKIA that are discussed in Appendix A there was no further additional resistance provided by the toe.

Because both resistance and drag load along the shaft requires substantially less relative displacement for mobilization compared to toe resistance, the net result is that the load-settlement response of a DFE with relatively significant residual loads tends to be much stiffer than the simplistic no-residual-load assumption results because most (and, in some extreme cases, all) of the DFE settlement and resulting resistance mobilization is occurring along the shaft. The toe is minimally involved in the process.

In addition, the fact that some of the shaft loading is occurring along the unload-reload portion of the load-displacement curve adds to overall system stiffness as reload stiffness is typically greater than initial or virgin-loading stiffness. Thus, the load-settlement response of the traditional resistance mechanism of shaft resistance due to sliding friction totally dominates the axial-compressive stiffness of a pile with significant residual loads.

It is useful at this point to digress to explain the point made earlier that residual loads can, in fact, influence the interpretation of ULS total resistance although residual loads do not actually alter the ULS total resistance, at least in theory. The reason is that the most common methods used in practice for interpreting ULS total resistance from load tests are based on pile-head settlement. This includes the Davisson Offset Method that predominates U.S. practice as well as methods used in other countries and regions that are based on some fixed percentage of the DFE diameter. Because residual loads cause a much stiffer pile-head response compared to a DFE with no residual loads, it is obvious that a given magnitude of pile-head settlement that is assumed to define the geotechnical ULS will correlate to a larger magnitude of applied load compared to the theoretical case of no residual loads. Note that this is not just a theoretical or hypothetical exercise. Fellenius (2015) demonstrated this conclusively using actual static load test data both without and with correction for residual loads and reload vs. virgin-loading stiffness.

The important point being made here is that the conventional wisdom that residual loads can be safely ignored when determining ULS total resistance for all types of DFEs is incorrect, at least when the determination of ULS total resistance is based on interpretation of a physical load test as is generally the case in both routine practice as well as academic research. Furthermore, this error of interpreted geotechnical ULS is always on the unconservative side (i.e. the DFE is presumed to have a larger geotechnical-ULS value than it actually has) so should be of concern to all stakeholders involved with deep foundations as it implies that DFEs have less 'safety' than is routinely imagined or assumed.

7.5.3.3.4 Structural Issues

Another consideration with respect to incorporating residual loads into a resistance-forecasting methodology concerns the axial normal stresses within the pile shaft. This is because the presence of residual loads causes these stresses to not start from zero as is traditionally assumed when residual loads are neglected. Note that this is independent of whether a 1-D or 3-D structural model (as discussed in Appendix E) is used to apportion axial forces and concomitant stresses in a composite-pile cross-section consisting of more than one material.

The effect of residual loads on pile stresses is particularly important when piles consist of a closed-toe steel pipe that is filled with PCC after driving as is often the case for tapered piles such as the *Monotube* and *TAPERTUBE* that are widely used in the U.S. For such piles, only the steel pipe will have initial stresses from the residual loads. The fluid PCC that is used to fill the pipe after driving is always initially stress-free once it hardens and prior to external load application.

As a result, conventional methods for apportioning stresses to the pile materials under applied-load conditions as discussed in Appendix E will always yield incorrect results as the two materials are starting from very different initial-stress conditions (the methods discussed in Appendix E always assume that all materials are initially stress free). Specifically, the compressive stresses in the steel pipe will be underestimated and the compressive stresses in the PCC core will be overestimated. The relative magnitudes of overstress and understress will vary on a case-by-case basis and as a function of depth along a pile as well so no generalizations cannot be made, at least at this time.

That having been said, given that this issue does not appear to have been addressed at all in the published literature to date, it is at least of interest to look at a specific example using actual ground-truth. The interpreted results from Pile #3, a *Monotube* pile, in Fellenius et al. (2000) that is discussed in Appendix A are used for this purpose.

To begin with, Figure 7.1 shows the depth-wise distribution of axial forces in the pile as interpreted by Fellenius et alia. Three distributions are shown in this figure and all reflect conditions at the maximum applied load (2500 kN = 560 kips) in the conventional static load test that was performed on this pile:

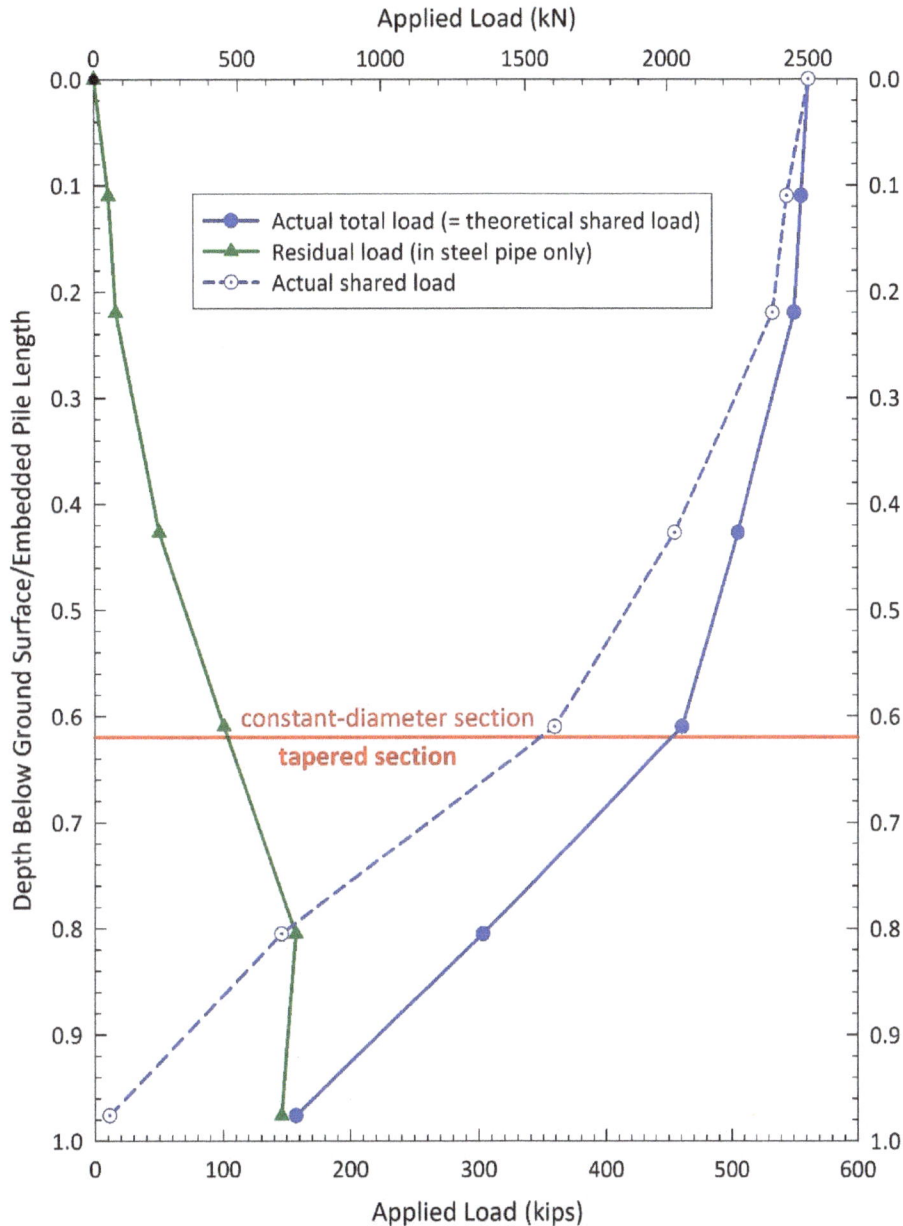

Figure 7.1. Axial Forces within Pile Shaft - Fellenius et al. (2000) Pile #3 at JFKIA.

1. The <u>actual total load (= theoretical apportioned load)</u>. If all shaft components were stress-free after driving and prior to load testing as is typically assumed, this is the force that would be apportioned between the steel pipe and PCC core in the usual manner using either a 1-D or 3-D method along the lines of those discussed in Appendix E.

2. The <u>residual load</u>. This is the force locked-in to the pile after driving and prior to concreting and thus creating normal stresses in the steel pipe only.

3. The <u>actual apportioned load</u>. This is the difference between the actual total load (#1) and that which is already being carried by the steel pipe due to residual loads (#2). In reality,

it is only this difference in forces (#3) that is apportioned between the steel pipe and PCC core when loads are applied in the static load test.

The results shown in this figure are rather striking. The residual loads are a significant portion of the actual total load within the tapered portion of the pile, with the two curves essentially joining at the toe (this implies that toe resistance was essentially 100% mobilized at the end of driving). The relative significance of residual loads can be seen in Figure 7.2 where residual loads are plotted as a percentage of the maximum loads ultimately applied to the pile during the static load test.

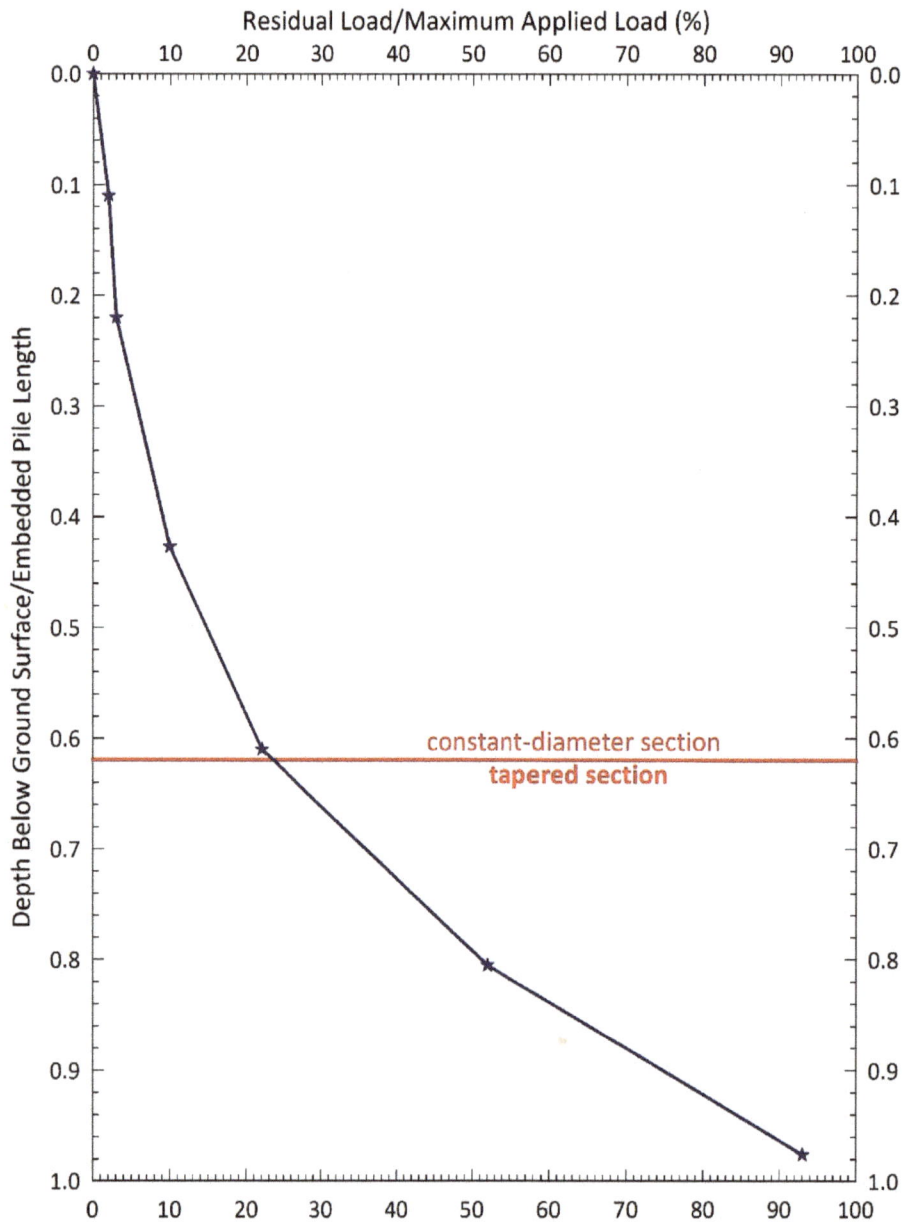

Figure 7.2. Residual Load as a Percentage of Maximum Applied Load Fellenius et al. (2000) Pile #3 at JFKIA.

Another insight from Figure 7.1 is that the actual shared load is what the instrumentation data in an instrumented pile such as the *Monotube* piles studied by Fellenius et al. (2000) would indicate if these data were not corrected for residual load. It is obvious from this figure that when residual loads are significant, uncorrected results from an instrumented pile can present a very misleading picture of the distribution of resistance provided by the ground. Fellenius and others have raised this issue in numerous publications over the years (Fellenius et al. (2000) is but one of many examples) so this is certainly not new information. In this case, one would conclude, quite erroneously, that the toe provides essentially none of the ULS resistance when it is actually providing approximately 20% of the ULS total resistance.

Even more enlightening and of greater practical importance are the axal-compressive normal stresses within the pile materials. Figure 7.3a shows the stresses within the steel pipe for the same three load components shown in Figure 7.1:

1. The <u>theoretical</u> stresses based on the traditional assumption that there are no residual loads so that <u>all</u> externally applied loads are shared with the PCC core.

2. The <u>stresses</u> caused by the estimated residual loads.

3. The <u>actual</u> stresses under externally applied loads from the load test. These stresses are the sum of the stresses due to residual loads plus stresses from the actual shared loads that were shown in Figure 7.1.

Figure 7.3b shows the stresses within the PCC core. Because residual loads no not impact the core, there are only two stress components:

1. The <u>theoretical</u> stresses based on the traditional assumption that there are no residual loads so <u>all</u> externally applied loads are shared with the steel pipe.

2. The <u>actual</u> stresses under externally applied loads from the load test. These stresses are due solely to the actual shared loads that were shown in Figure 7.1.

Note that in both parts of Figure 7.3 the horizontal red line indicates the transition between the constant-diameter and tapered sections as in Figure 7.1. Labeling of this line was intentionally omitted in Figure 7.3 to avoid visual clutter as these plots are visually 'busy' enough as it is.

Discussing first the stresses in the steel pipe (Figure 7.3a), it is clear that the entire length of the pile is stressed in compression by the residual loads. As a result, the stresses within the entire steel pipe are greater under the externally applied load-test load than would be expected based on the normal assumption of an initially stress-free pipe.

The magnitudes of these overstresses are particularly significant within the tapered portion of the pile. This is the combined result of a 'perfect storm' of conditions that occurred within the tapered portion of the pile:

• As shown in Figure 7.2, a significant percentage (20% to almost 100%) of the eventual total shaft resistance was already mobilized by residual loads prior to external load application during the load test.

• The steel cross-sectional area decreases linearly with depth.

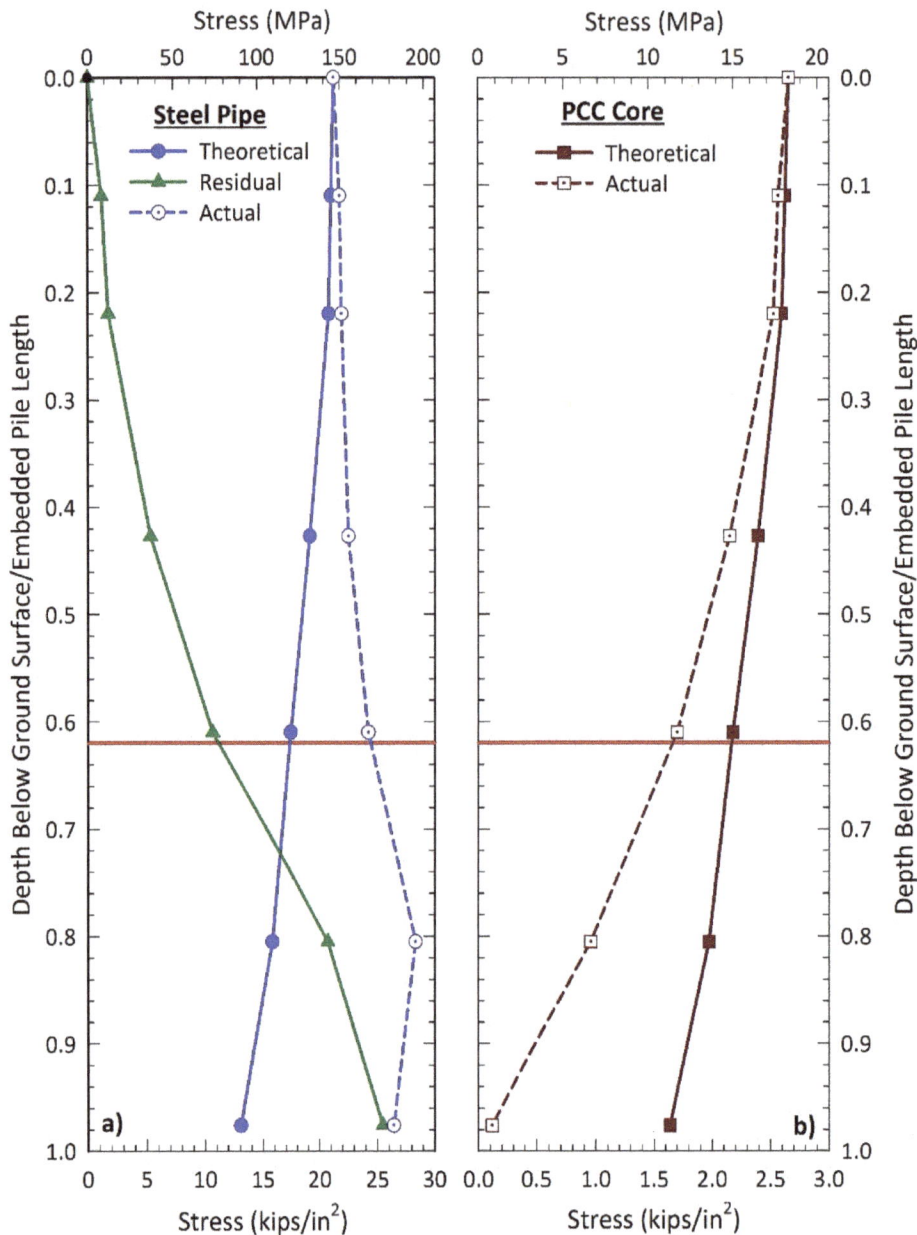

Figure 7.3. Normal Compressive Stresses in Pile - Fellenius et al. (2000) Pile #3 at JFKIA.

The net result of these two facts is that when the load test was performed portions of the steel within the tapered section were overstressed by a factor of approximately two in some areas. This represents a serious erosion of the margin of safety that the traditional stress-free assumption would imply existed.

As expected, the stress state in the PCC core (Figure 7.3b) reflects the fact that the steel pipe is carrying much more load than was intended. For the entire length of the pile, the PCC compressive stresses are less than those expected based on the traditional zero-initial-stress assumption. In particular, within the tapered portion of the pile the compressive stresses in the PCC core decrease to almost zero near the toe.

In conclusion, any future efforts to include the effect of residual loads on the geotechnical aspects of DFEs as discussed in the preceding section should logically include the effect of residual loads on the structural components as well. However, and more importantly, it should be recognized that the issue of residual loads has broad implications for DFEs in general and, therefore, deserves renewed attention in both research and practice. Specifically, the traditional belief that residual loads do not impact ULS total resistance and can, therefore, be safely ignored in traditional resistance-forecasting methodologies for DFEs in general is only partially correct. While the ULS geotechnical aspects may not be affected by residual loads, the ULS structural aspects can be significantly impacted. This is especially true for driven and jacked piles that consist of a steel pipe that is filled with fluid PCC after the pipe is installed in the ground. Perceived safety margins for the structural components of such piles can be significantly in error unless the effects of residual loads are taken into account.

7.5.3.3.5 Relevance for Proposed Schema

Returning now to the primary issue at hand, it is clear that residual loads are more significant for tapered piles than for any other type of DFE. Therefore, any improved resistance-forecasting methodology for tapered piles should account for the presence of such loads for two important reasons:

- They impact the geotechnical load-settlement behavior even though they do not, in principle, affect the actual geotechnical ULS total resistance.

- They profoundly affect the pile as a structural member in terms of axial-compressive normal stresses within the pile materials, especially when composite piles such as steel pipe piles that are concreted after driving are used. Material stresses in a pile are a clear-cut safety and reliability issue.

The problem is that at the present time there is no relatively simple, reliable way to estimate residual loads for tapered piles prior to driving that would be amenable for use in routine practice. There have been attempts going back to the late 20th century to develop residual-load forecasting methodologies but they all involved constant-perimeter piles (Briaud and Tucker 1984, Poulos 1987, Darrag and Lovell 1989). It is clear from the work of Gregersen at al. (1973) that the shaft-resistance mechanism of wedging that is unique to tapered piles produces a pattern of residual loads that is quite different from that which results from the traditional shaft-resistance mechanism of sliding friction. These differences are illustrated in Appendix H.

Therefore, it is clear that the outcomes of this earlier research into residual loads cannot simply be resurrected and repurposed in their original forms for tapered piles. New research specifically with tapered piles in mind needs to be conducted. Based on this research, it may be possible to modify earlier forecasting methodologies for residual loads, or it may be necessary to develop an entirely new methodology.

Looking ahead as to what this required future research might entail, the primary issue with regard to residual loads and tapered piles would appear to be how to define and quantify the initial stress state of cylindrical-cavity mechanics around the tapered portion of a pile under conditions of both shaft resistance and drag load as defined in Figure 2.1. Then, it needs to be determined how this initial stress state responds to externally applied loads that will cause an interface-shear-stress reversal in the case of drag load.

As noted earlier in this manuscript, while the overall model of a cylindrical cavity is the best physical and mathematical model identified to date for the shaft-resistance mechanism of wedging, the cavity-mechanics solutions developed and used to date have been ones for an assumed cylinder that remains spatially fixed with respect to the longitudinal axis of the cylinder as the cylinder is expanded, e.g. as in the case of a pressuremeter test (which is a primary reason why geotechnical applications of cylindrical-cavity expansion was studied in the first place). This means that all displacements are only occurring in the radial direction of the cylinder.

On the other hand, with tapered piles there is always relative pile-soil displacement and concomitant shearing in the nominally vertical direction that is occurring coincident with cylinder expansion in the radial direction. This nominally vertical shearing occurs during driving (cavity creation, CCX), axial-compressive external loading (cavity expansion, CCE), and uplift loading (cavity contraction, CCC). The writer is not aware of any research to date that has explored how the complex stress state of cylindrical-cavity mechanics is affected by shear stress along the cavity wall. Logic suggests that there is rotation of principal stresses that is occurring in the vicinity of the pile-soil interface and that this may have an effect.

Note that when crafting future research, defining a residual-load distribution along a pile is only part of the problem. Because both shaft and toe resistances are mobilized to some degree along the entire pile, it will be necessary to define the shear moduli that correspond to this condition. In some cases, these moduli will need to be appropriate for further development of resistance (along the lower portion of the shaft as well as at the toe) and in other cases these moduli need to be appropriate for unloading followed by reloading (along the upper portion of the shaft that is subjected to drag load). This can be seen conceptually in Fellenius (2015) as these same issues apply to the traditional resistance mechanisms of shaft resistance due to sliding friction and toe resistance.

7.5.3.4 Site Characterization

7.5.3.4.1 Introduction and Background

The Kodikara-Moore Method, as well its derivatives developed by El Naggar et alia and Manandhar et alia, pay little or no explicit attention to the issue of site characterization. The necessary soil properties used in their methodologies are simply assumed to magically exist, with little or no direction or assistance given to users of the methodology.

The position taken in this monograph is that explicitly incorporating site characterization for necessary site- and application-specific soil properties is an essential component of any improved resistance-forecasting methodology for tapered piles. For over 30 years, the writer has promoted the concept of more closely coupling or integrating site characterization and foundation analysis into a single, seamless process to improve the forecasting accuracy of forecasting foundation behavior. In fact, the first published work by the writer along these lines (Horvath 1989) dealt with deep foundations. Thus, the proposed program presented here reflects the writer's thinking on the subject that are the result of more than three decades of ongoing development and refinement.

7.5.3.4.2 Overview

In the writer's almost five decades of first-hand professional experience related to tapered piles, there is no shortage of stakeholders who, even to the present day, insist that nothing more sophisticated than SPT N-values (and uncorrected field values, N_{field}, at that) should be required or needed for the routine design of tapered piles. However, it is the writer's unequivocal, unambiguous opinion that the forecasting capabilities with respect to tapered piles will never improve beyond what Nordlund proposed almost 60 years ago if this position is maintained.

Thus, the position taken in this monograph is that any improved resistance-forecasting methodology for tapered piles should take full advantage of the more-advanced site-characterization tools that are available nowadays in routine practice, with the cone penetrometer being at the top of the list. Furthermore, this usage includes both the hardware for data collection (primarily CPTu and sCPTu soundings) and the dozens of theoretical and empirical relationships that have been, and continue to be, developed to use cone-penetrometer data to estimate dozens of soil properties and related analytical parameters.

Note that this is not a new position adopted by the writer. The writer rejected the SPT-only approach for DFE analysis and design in general as far back as the late 1980s (Horvath 1989), a point in time at which CPT soundings had become widely available in U.S. practice. The writer's position that the cone penetrometer is the site-characterization tool of choice for most soil sites was made specifically for tapered piles with the publication of Horvath (2002). By that time, the state of practice with regard to cone-penetrometer soundings had advanced to routine use of CPTu soundings which substantially broadened the number of soil properties and problem parameters that could be estimated compared to the CPT. Although the writer included an algorithm for converting SPT N-values to pseudo-CPT(u) data for coarse-grained soils only (Horvath 2002, 2011), this was only ever intended to be a bridge from past to present to allow assessment of older case histories where only SPT data were available. The writer never intended to support or condone the use of only SPT data for new projects where CPTu or sCPTu soundings could be made.

7.5.3.4.3 Proposed Program

A site-characterization program that is based, at least in part, on CPTu and, preferably, sCPTu soundings is required to be performed. The data from these soundings is assumed to be processed using published empirical correlations combined with basic theoretical relationships to produce depth-wise profiles of the following soil properties and parameters:

- σ'_{vo}

- K_o

- D_r (relative density, for coarse-grained soils)

- ϕ_{peak} and ϕ_{cv} (for coarse-grained soils)

- s_u and ϕ (for fine-grained soils, if any)

- G_{max}

- Irr (for coarse-grained soils)

- Ir (for fine-grained soils, if any).

Developing empirical relationships between cone-penetrometer data and soil properties and parameters has been for decades, and continues to be, a very active area of ongoing research worldwide. As a result, new publications related to this appear frequently and consistently. Consequently, it is not a simple and straightforward task to cite universal references that are temporally static for the soil properties and parameters enumerated above as this is an ever-changing landscape.

However, it is relevant to note that for some time now the writer has had a personal preference for using the published works of Prof. Paul Mayne (and colleagues who are typically graduate students at Georgia Tech) and Dr. Peter K. Robertson. Each of these researcher-authors has published not only their original work but has done, and continues to do, an excellent job of curating the published work of others and incorporating these curations into their own work and publications. The overall end results of these curations go a long way toward creating and presenting a coherent picture of what, at times, can be a very complex picture, especially when fine-grained soils are involved.

The writer acknowledges that there will always be some stakeholder-segment of foundation engineering practice that claims that they or their client or customer can only 'afford' standard borings and SPT sampling on some given project, and that performing cone-penetrometer soundings is out of the question. The writer's simple reply is that one cannot be expected to produce 21st-century resistance-forecasting results using early-20th-century technology. The Nordlund Method remains available for those who choose, for whatever reason(s), to live in the past and remain constrained by technology that had its origins more than a century ago.

7.5.3.5 Shear Modulus and Modulus Degradation

7.5.3.5.1 The Role of Soil Modulus

The concepts of *cavity creation* and *cavity expansion* have emerged as the physical basis for important mathematical developments for defining and quantifying all aspects of both DFE installation and post-installation axial-compressive loading. These specific applications of cavity mechanics, along with the less-well-known *cavity contraction* for uplift loading, figure prominently in any state-of-art schema for resistance forecasting for tapered piles.

One thing that distinguishes cavity mechanics from most other concepts used in geotechnical and foundation engineering is that <u>both</u> stiffness <u>and</u> strength play essential roles in the resulting mathematical models and their solutions. This is a significant paradigm shift from traditional mathematical models used in geotechnical and foundation engineering that are typically based on <u>either</u> strength <u>or</u> stiffness alone but <u>not</u> both at the same time.

Arguably the single most important soil property used in cavity mechanics is soil stiffness in the form of a material modulus. Sometimes modulus appears in the solution explicitly and sometimes it is incorporated into either the rigidity index, I_r, or reduced rigidity index, I_{rr}, as a matter of computational convenience. Rigidity index (in the broadest sense) is a useful concept because it contains both fundamental stiffness and strength parameters.

The earliest applications of cavity mechanics to geotechnical and foundation engineering, such as Vesic's work in the early 1970s that was related to DFE toe resistance, used Young's modulus as the soil-stiffness parameter. However, subsequent research to the present has coalesced around the use of shear modulus. Not only is shear modulus behaviorally more relevant to cavity mechanics compared to Young's modulus, recent advances in site-characterization technology have made shear modulus much easier to quantify on a site-specific basis. Therefore, the schema presented herein presumes that only shear modulus is used to define soil stiffness.

7.5.3.5.2 Modulus Degradation

There is now abundant theoretical evidence that shear-modulus degradation (reduction) starting from an initial small-strain, linear-elastic value (defined as G_{max} in this monograph but, more commonly, G_o in most other publications) is an essential consideration whenever cavity mechanics is applied to a geotechnical problem. This simply relates to the fact that soil stiffness is both stress and stress-level dependent as discussed in Chapter 4.

In general, modulus degradation can take on two, broad forms. This is illustrated here using the parameter of toe resistance:

1. If a single-valued outcome is sought from cavity mechanics, e.g. using Vesic's spherical-cavity expansion (SCE) solution to forecast only the ULS toe resistance of a pile, then it is possible in both concept and practice to define a single-valued secant shear modulus or (reduced) rigidity index that is acceptably accurate for the intended purpose.

2. If a complete forecast of toe settlement as a function of toe load is required, then shear-modulus degradation must be applied in an incremental way using either secant or tangent values of modulus depending on how the solution algorithm is formulated. Another decision that must be made on a case-by-case basis is whether a strain- or stress-based degradation model is used. As discussed in Appendix D, this choice largely depends on the particular application and whether it is easier to frame modulus degradation in terms of strains or stresses.

7.5.3.5.3 Relevance for Proposed Schema

In the writer's opinion, the use of a constant-value, load-independent shear modulus is the single biggest shortcoming of existing resistance-forecasting methodologies based on cylindrical-cavity mechanics that are otherwise conceptually a step in the right direction in terms of developing an improved resistance-forecasting methodology for tapered piles. Furthermore, this shortcoming was made worse in the exemplar problems (and implied guidance) contained in the various publications by Kodikara and Moore, El Naggar et alia, and Manandhar et alia for two reasons:

- The shear modulus was assumed to be depth-wise constant in magnitude.

- No constructive, meaningful guidance was given as to how a user should rationally select a value of shear modulus in a project-specific application. Note that the writer does not consider 'cookbook' values presented in an elementary, undergraduate textbook, as was done by El Naggar et alia, to be either meaningful or rational.

Consequently, prior published works related to using cylindrical-cavity expansions (CCE) for the shaft resistance for tapered piles were judged to be of no value in terms of providing either precedence or guidance for either shear-modulus selection or shear-modulus degradation. This means that a rational process for evaluation of shear modulus in an improved resistance-forecasting methodology for tapered piles needs to be developed from scratch.

Because shear modulus can be used for difference resistance components (constant-diameter and tapered portions of the shaft, toe) and each component will have a different physical and mathematical model, this means that there will likely be different modulus-

312

degradation models used. However, the one element these degradation schemes have in common is that the site-specific, small-strain, linear-elastic value, G_{max}, that existed prior to pile installation serves as the baseline reference value. This is because G_{max} is nowadays universally recognized as a standard frame of reference for many problems that involve shearing in soil. Furthermore, this soil property is straightforward to obtain not only in research but in routine practice as well at any site using either a sCPTu sounding or other geophysical in-situ testing tool, depending on the nature of the subgrade material.

The desire to include shear-modulus degradation in an improved resistance-forecasting methodology for tapered piles is complicated by the fact that even for a given physical and mathematical model (e.g. Vesic's SCE model for toe resistance) there is no single mathematical relationship for this. Appendix D contains a detailed assessment of the various modulus-degradation relationships that were identified during the preparation of this monograph. The findings presented in Appendix D and its subappendices D1 and D2 can be summarized as follows:

- A hyperbola is the most common mathematical curve-form by far that is used to model modulus degradation in general and shear-modulus degradation in particular. However, there is no one form of a hyperbolic relationship that is used to model modulus degradation. Rather, there are different forms that vary in the number and nature of model parameters that are used to control the shape (curvature) and endpoint of the hyperbola. The latter issue is relevant as a basic hyperbolic relationship will, by definition, never reach where it is going but only approach its final value asymptotically. Consequently, it has proven useful in some geotechnical applications to introduce a model parameter to force the hyperbola to reach its ultimate value in some finite space.

- A hyperbolic modulus-degradation relationship can be interpreted to produce either a secant or tangent value of the plotted modulus in the context of the generic discussion in Chapter 2 of the secant and tangent interpretation of geomechanics parameters. In general, published works related to geotechnical and foundation engineering almost always use the secant-modulus interpretation but more times than not they do not make this distinction clear. Clarity on this issue is important as the modulus interpretation that is chosen must always be consistent with the way that the interpreted modulus is used in some analytical methodology. In general, when the end-use analytical method is either a one-time-only or a repetitive-trial type of analysis using a single equation, a secant modulus is appropriate. A relevant example of this that was presented in Chapter 5 is the two-step-procedure use of the Fahey-Carter version of the hyperbolic model together with the Randolph-Wroth Problem solutions. Together they are used to create a load-settlement curve for a ULS total resistance determined beforehand using some standalone resistance-forecasting methodology. On the other hand, when the end-use analytical methodology is an incremental, step-by-step process where calculated results are sequentially additive, or in situations where a series of analytical steps must be considered, a tangent modulus is appropriate. A well-known example for the past five decades is finite-element analyses (FEAs) of force-displacement problems. An example relevant to this monograph would be the Kodikara-Moore Method in all of its forms although none of the researchers (Kodikara and Moore, El Naggar et alia, Manandhar et alia) who have used this method to date have done so as they have all used single values of shear modulus for an entire analysis..

- While modulus is always the dependent variable in any modulus-degradation equation, the parameter to select as the independent variable to correlate with modulus is subjective. While shear strain was the original independent variable and is still the de-facto default choice, other parameters can be chosen to better suit a given application. For example, Fahey and Carter used stress/strength (the reciprocal of ASD safety factor could be used as well) and Niazi used strain but in his case strain was not shear strain but essentially normal strain as it was related to the settlement of the head of a DFE.

• When strain of some kind is used as the independent correlation variable for modulus degradation, there is no consistency between using a dimensionless decimal value or a percent (which is technically a dimension). The distinction becomes important because correct use of whatever mathematical relationship is derived for modulus degradation requires that strain be input in the same format, decimal or percent, used to derive the mathematical relationship, unless the developer has specified otherwise.

• Separate from, but related to, the problem parameters chosen as the dependent and independent variables in the modulus-degradation equation is the normalization and concomitant non-dimensionalization of these variables. It has long been a given to normalize the operative value of a modulus to its corresponding small-strain value (G_{max} in the case of shear modulus). More recently, it has become common to do the same to the other correlation variable (shear strain, stress, etc.) as well. However, unlike with modulus, the parameter chosen to normalize and non-dimensionalize the independent correlation variable is quite subjective. For example, when shear strain is the independent correlation variable, it is typically normalized to some 'reference value' of shear strain. However, what this 'reference value' corresponds to is completely arbitrary and subjective. It could be a specific magnitude of shear strain (1% is common for obvious reasons of simplifying the end result) or it could be the magnitude of shear strain that correlates with a certain fraction or percentage of modulus degradation (50% or 0.5 is common) or even something else. While this offers useful flexibility for the analyst to choose a reference value that best suits some particular application, it also means that the end results are not inherently transferable to some other application.

Before closing out this discussion of shear modulus and shear-modulus degradation, some additional comments are offered concerning the Fahey-Carter Modulus-Degradation Model. While this is certainly not the only degradation model of potential use with tapered piles, the secant-modulus solution of this stress-based model has proven to be very popular in geotechnical and foundation engineering, including DFEs in recent years as can be seen in Mayne's use of this model that was discussed in Chapter 5.

To begin with, it has become common to routinely use what have become de-facto default values of the f and g model parameters, 1.0 and 0.3±0.1 respectively. It should be recognized that these are not 'one size fits all' values and, in fact, may be inappropriate for some applications. This is especially true of the f parameter that dictates the endpoint of the hyperbola (the g parameter governs the degree of curvature of the hyperbolic curve).

It is strongly recommended that researchers who are contemplating use of Fahey and Carter's work read their original 1993 paper that introduced their model. This will provide a much better sense of the relative influence of values of f and g as the authors present and discuss several graphics that show the result of parametric variations to illustrate the broad implications of chosen values of the f and g model parameters.

Fahey and Carter's 1993 paper also illustrates how the values of f and g can be varied to match a set of experimental data. Mayne did this to a very limited extent in some of his later publications illustrating the use of the Fahey-Carter Model for DFE load-settlement forecasting but only to explore the ±0.1 variation of the g (curve-shape) parameter. The broader variation of both f and *g* is something to keep in mind if the Fahey-Carter Model is used in an analytical methodology for tapered piles. Rather than assume values of *f* and *g* beforehand, it would appear to be more prudent to conduct research into varying the values of these parameters to match the form of the behavior that is being modeled mathematically. Some thoughts along these lines are presented for modulus-degradation models in general in Appendix D.

Finally, Fahey and Carter's 1993 paper also presents the tangent-modulus formulation of their model in addition to the secant-modulus formulation. Publications that use the Fahey-Carter Model tend to only give the much-more-common secant-modulus solution, generally without explicitly mentioning that it is a secant modulus or mentioning that a tangent-modulus form is also available.

7.5.3.6 Shear-Strength Dependence on Operative Stress Level

It has been well-known and well-established in the geomechanics knowledge-base for decades that:

- Shear strength can always be fundamentally interpreted using effective stress and an effective-stress friction angle, ϕ (sometimes the notation ϕ' is used to emphasize this), although how ϕ is used as part of a particular failure criterion (Mohr-Coulomb, etc.) varies.

- Most soils exhibit a peak strength, ϕ_{peak}, followed by strain-softening to a constant-volume or critical-state strength, ϕ_{cv}, with the difference reflected in the dilatancy angle, ϕ_d, as expressed mathematically in Equation 4.7. Soils composed of clay minerals can 'soften' even further under large strains and concomitant particle reorientation to a residual-strength state defined by $\phi_{residual}$.

- The constant-volume friction angle can reasonably be assumed to be constant in magnitude for a given soil.

- The peak friction angle is never constant in magnitude, even for a given soil, and varies with void ratio and stress level.

The clear and unambiguous conclusion drawn from this assessment of the state of knowledge is that the stress-dependency of shear strength is something that should be incorporated into any modern resistance-forecasting methodology for DFEs in general. However, this has historically been resisted or simply ignored, even with analytical methodologies developed in the relatively recent past for tapered piles such as the Kodikara-Moore Method and the El Naggar-Sakr Method.

The underlying cause for this disconnect between well-established geomechanics knowledge and foundation engineering practice is that mathematical models and the solutions that derive from them that are used in both practice and research typically require a single-valued ϕ. While this is easy enough to do for constant-volume strength conditions, peak-strength conditions require some thought depending on the analytical approach used.

As discussed in Chapter 2, the central element in this thought process involves the concept of an *operative stress level* that is defined in this monograph as being either a single-valued level of effective stress or a range of effective stresses at or over which some geomechanics problem is being analyzed. Which of these two cases, i.e. single-valued or range, is being analyzed determines whether it is more appropriate to use a tangent value of the peak friction angle, $\phi_{peak(tangent)}$, that is essentially the slope of the failure envelope at the assumed single-valued operative stress level or a secant value of the peak friction angle, $\phi_{peak(secant)}$, that is the average over some stress range.

Unlike with shear-modulus degradation that was discussed in the preceding sections, there is a body of prior research and knowledge to draw on concerning shear strength and piles. This experience indicates that for pile analysis in general and tapered piles in particular, for both shaft and toe resistance it is logical to calculate and use a value of $\phi_{peak(tangent)}$ at some single-valued operative stress level. The writer followed this approach during the original development of the Simplified Nordlund Method (Horvath 2002) and Manandhar and Yasufuku incorporated this into their modified approach for calculating shaft resistance using the Kodikara-Moore Method. In both of these cases, Equation 4.7 was used together with Bolton's empirical relationship for calculating ϕ_d using the mean effective stress at failure at the relevant pile-soil interface, i.e. along the shaft or at the toe, combined with relative density of the soil at that location.

Note that using Equation 4.7 together with the well-known and widely used relationship in Bolton (1986) is not the only approach for estimating $\phi_{peak(tangent)}$. Estimating relative density, as is necessary with Bolton's empirical relationship, has historically been problematic in both practice and research although empirical correlations developed for cone-penetrometer data have made this more practical to do nowadays on a routine basis. Thus, other approaches for estimating $\phi_{peak(tangent)}$ have been developed, as noted in Cook (2010) for example. This includes an empirical relationship developed by Been and Jeffries (1985) that uses the state parameter, ψ, from Critical-State Soil Mechanics (CSSM) and a theoretical relationship with empirical coefficients developed by Lancelotta (1995) that uses the parameter v_λ that is a specific case of the *specific volume*, v, from CSSM.

The conclusion drawn here is that incorporating the stress-level dependence of shear strength into any improved analytical methodology for tapered piles is both necessary and reasonable to do. How this is actually done, i.e. which of several empirical relationships such as those cited above is used, is a detail to be determined.

7.5.3.7 Structural (Poisson) Effects

7.5.3.7.1 Introduction and Background

The Poisson Effect as it applies to DFEs under axial loading is fundamentally and primarily a structural issue but with secondary structure-soil interaction (SSI) effects. The conclusion that can be drawn from the detailed discussion of the Poisson Effect on a DFE that is presented in Appendix E is that there are three distinct issues of potential interest that have a single theoretical linkage between them:

1. The allocation of axial forces and concomitant axial normal stresses in composite cross-sections consisting of two or more different structural components.

2. For a composite cross-section consisting of a core composed of either Portland-cement concrete or Portland-cement grout, the potential radial-confinement effect on the core by the surrounding shell (used in a generic context here, not in the specific context of shell piles as defined in Chapter 3) that is usually steel. Confinement generally increases the compressive strength of cementitious materials relative to the unconfined compressive strength that is normally used in design.

3. The radial displacement of the interface between a DFE shaft and surrounding ground. This displacement, which will occur even if the DFE consists of only one component/material, has the behavioral form of cylindrical-cavity expansion (CCE) or cylindrical-cavity contraction (CCC) depending on whether the axial loading is compressive or tensile (uplift), respectively. As a result, this radial expansion or contraction will, in principle, cause a change in the radial stress acting along the DFE shaft and thus has the potential to change the unit shaft resistance regardless of whether the shaft is constant-diameter or tapered.

7.5.3.7.2 Assessment and Critique

It is relevant to assess and critique these three outcomes within the context of this monograph in general and developing a schema for an improved resistance-calculation methodology for tapered piles in particular.

The first issue of axial-load and -stress allocation has always been important for all types of DFEs as it is essential for the purposes of forecasting material stresses for design purposes. As discussed in Appendix E, historically this issue has been analyzed using a simple 1-D model that ignores the Poisson Effect. The writer's limited comparison of this traditional 1-D model to a theoretically rigorous 3-D model suggests that the Poisson Effect has minimal influence on axial-force and -stress allocation, at least for the traditional assumption that all materials are stress-free after DFE installation but prior to external load application. Note, however, that it is essential to use the rigorous 3-D model if either or both of the other two above-noted issues are to be explored.

The second issue of radial stresses within a composite-DFE cross-section is of significant interest primarily to a limited segment of the deep-foundation industry involved with micropiles (Fuller et al. 2003). Thus, this issue is of limited relevance to this monograph even though many tapered piles have a PCC core surrounded by a steel 'shell' (= pipe). In any event, based on limited research by the writer that is summarized in Appendix E, it appears that the radial normal stress at the contact between a PCC core and steel confining 'shell' is nil. Again, this conclusion is based on the traditional assumption that all pile materials are initially stress-free prior to post-installation external load application.

The third issue of radial displacement of the exterior of a DFE shaft has historically been mentioned only occasionally in the published literature and even then typically only in the context of being a possible physical mechanism to explain perceived differences in the ULS unit shaft resistance in compression vs. uplift for constant-perimeter DFEs (as noted earlier in this monograph, there is disagreement as to whether there really is such a difference). With tapered piles, this issue of radial displacement due to the Poisson Effect potentially has a broader impact as it is additive to the cylindrical-cavity behavior associated with either pile settlement or uplift-induced displacement.

In conclusion, the rigorous 3-D structural model of a DFE shaft that accounts for the Poisson Effect is relatively easy to incorporate into an analytical methodology (it can be implemented into a spreadsheet if desired) although in the future it should be modified to allow for non-zero initial stresses in the structural components of the DFE to account for residual loads. Furthermore, a 3-D structural model can yield useful behavioral insights into several issues such as core confinement and radial interaction between the DFE and surrounding ground that historically have received relatively little attention. Therefore, in the writer's opinion, there is no reason not to include this 3-D behavior in an improved resistance-forecasting methodology for tapered piles, at least on an R&D basis.

7.5.4 Resistance Calculation

7.5.4.1 Introduction

Having discussed overall behavioral issues that should be considered in the schema for an improved resistance-forecasting methodology for tapered piles, it is next appropriate to address specific analytical issues that are related to calculation of shaft and toe resistance. Note that the comments made in the following sections reflect what the writer feels are the key elements for consideration based on the current state of knowledge. As further research, development, and subsequent implementation is made into these topics, it is likely that these comments will evolve and may need to be modified.

7.5.4.2 Shaft

7.5.4.2.1 Introduction and Overview

As noted at the beginning of this chapter, the single most important element of an improved resistance-forecasting methodology for tapered piles is the recognition that the radial stress state that controls the shaft resistance of the tapered portion of a pile begins with the pre-driving K_o condition and then goes through two distinct stages of wedging-related behavior due to driving then subsequent external compressive-load application. None of the resistance-forecasting methodologies for tapered piles that have been proposed to date acknowledge this two-stage process specifically so this is one area in which the schema presented herein differs significantly from the current state of knowledge.

There are additional significant differences. The best physical model identified to date for the wedging mechanism involves cylindrical-cavity mechanics. While the Kodikara-Moore Method and derivative works by Manandhar et alia and El Naggar et alia are based on cylindrical-cavity mechanics, in the writer's opinion they do not apply it correctly to the tapered-pile problem for two reasons:

1. They assume that cylindrical-cavity expansion (CCE) occurs only after pile-soil slip occurs during post-installation applied loading in axial compression. The basic kinematics of the problem as illustrated in Figure 4.1 suggest that both the cylindrical-cavity creation (CCX) caused by pile driving and the CCE caused by post-installation external load application in compression occur as the result of absolute downward-vertical pile displacement relative to an arbitrary fixed point in space, again, as shown in Figure 4.1.

2. The CCE solution used is one developed for a cylinder such as a pressuremeter that is spatially fixed in the vertical direction so that vertical shear stress between the cylinder and surrounding ground does not enter into the problem free-body diagram and concomitant solution. In reality, both pile driving and post-driving load application involve significant shear stresses at the pile-soil interface. Furthermore, due to residual loads from driving along portions of the shaft, these shear stresses will reverse in direction during the course of post-installation compressive-load application. The significance of these shear stresses is something that requires consideration and investigation.

The schema presented herein considers both of these issues.

7.5.4.2.2 Radial Stresses: Post-Driving/Pre-External Load Application Phase

Discussing first the initial radial stress after pile installation but prior to external load application, this stress is better dealt with not as a single problem parameter as in the Kodikara-Moore Method (called σ_o by Kodikara and Moore but σ'_r here to be consistent with notation used in this monograph) or the subsequent modest modifications made by Manandhar et alia but decomposed into the product of two fundamental components as was done by El Naggar et alia (although these researchers used their own notation that is discussed subsequently):

- a dimensionless lateral earth pressure coefficient, K_h, and

- the vertical effective overburden stress, σ'_{vo}.

Thus:

$$\sigma'_r = K_h \cdot \sigma'_{vo} \tag{7.1}$$

which is a specific version of the more-general Equation 4.4 that has been narrowed slightly in two ways to:

- explicitly reflect overburden stress conditions for the more-general vertical effective stress, σ'_v, and

- reflect the fact that the current discussion is limited to piles with a circular cross-section so radial stresses are used in lieu of the more-general horizontal stress, σ'_h.

It is apparent from Equation 7.1 that:

- σ'_r can never be depth-wise constant as Kodikara and Moore assumed in their exemplar application and parametric studies involving their analytical methodology that they presented in their 1993 paper for the simple reason that σ'_{vo} is never depth-wise constant. This fact was recognized by El Naggar and Sakr in their adaptation of the Kodikara-Moore Method.

- Attention should be given to evaluating K_h in a rational, application- and site-specific manner that takes into account the unique nature of tapered piles, something that was not done by Kodikara and Moore, nor by El Naggar et alia or Manandhar et alia who used the Kodikara-Moore Method to varying extents in their own methodologies.

With regard to the evaluation of K_h, as discussed throughout this monograph, decades of deep-foundations research can be drawn upon to provide guidance as to how to most rationally approach evaluation of this problem parameter in practice. To begin with, the effect of the site-specific, pre-driving coefficient of lateral earth pressure at rest, K_o, should always be included as research has demonstrated that it is always the starting point for changes in lateral earth pressure caused by the installation process of any and all types of DFEs. The effect of K_o on K_h is most easily reflected in, and accounted for, by defining and using the dimensionless ratio K_h/K_o as this clearly defines K_h as starting from and modifying the pre-installation K_o. Seminal research by Kulhawy and others has demonstrated that this ratio can be correlated, as a minimum, with the type of DFE and its installation process.

The actual value of K_h as used in any calculation algorithm is thus most easily expressed as the product of the K_h/K_o ratio and K_o, i.e. $(K_h/K_o) \cdot K_o$. Note that this product is equivalent to the post-installation lateral earth pressure coefficient K_s defined and used in the El Naggar-Sakr Method as illustrated in Appendix C. Although El Naggar et alia did not formally link their K_s parameter to the site-specific, pre-driving K_o, there is no reason why it could not be so linked as El Naggar et alia provided no guidance on the quantification of K_s and thus left this task entirely up to end users of their methodology.

There are three specific comments to be made with respect to the post-driving, pre-load-application value of K_h:

1. In view of the fact that K_h is always dependent on the site-specific, pre-installation value of K_o, an a priori assumption that K_h is depth-wise constant in a problem is clearly not reasonable as a rule. Even if the K_h/K_o <u>ratio</u> is depth-wise constant for a given application (as will be seen subsequently, recent research suggests that it is not), K_o typically varies with depth for most sites. This means that K_h will, in consonance, typically vary with depth for most sites. Nevertheless, El Naggar et alia suggested that their equivalent (to K_h) problem parameter, K_s, is always depth-wise constant although they did allow it to vary with the type of DFE and site-specific subsurface conditions.

2. To the best of the writer's knowledge, no one has developed an analytical methodology specifically for forecasting the K_h/K_o ratio for DFEs of any type. The current state of knowledge in this regard is defined solely and entirely by tabular, 'ballpark' ranges in values in published works by Kulhawy that are now several decades old. The closest analytical effort in this regard appears to have been the development of a forecasting model for K_h only involving constant-diameter (driven) piles with an implied solid cross-section embedded in a homogenous, isotropic sand where K_o is implied to be constant with depth (Sabry and Hanna 2009)[162]. Sabry and Hanna's work is discussed further subsequently but for now it is just noted that the most interesting outcome of this research is that they found that K_h varied significantly and non-linearly with depth. This implies that the K_h/K_o ratio is not constant with depth even in the simplistic, ideal case of K_o constant with depth. It is relevant to note that the interpreted (by the writer) outcomes of the Gregersen et al. (1973) case history involving both tapered and constant-perimeter instrumented piles clearly indicates that, in general, neither K_h nor the K_h/K_o ratio are even close to be constant in magnitude with depth. The plotted results illustrating this are presented in Appendix H.

3. As the writer hypothesized earlier in this monograph, it appears logical to postulate that the inherent geometry of a tapered pile will affect the post-driving, pre-load-application K_h/K_o ratio compared to an otherwise identical constant-diameter pile in the same subsurface conditions. This is because after the toe of a tapered pile creates an initial cavity in the ground, the tapered section of the pile immediately behind the toe will open that cavity further in a nominally cylindrical manner as pile-driving progresses. Thus, unlike with a constant-diameter pile, there is further cavity expansion as a tapered pile is driven. In the writer's opinion, this implies an initial taper benefit in the form of a further

---

[162] These authors called the K_h parameter K_s. Although it does not appear to be intentionally the same as the K_s parameter used by El Naggar et alia, by coincidence it plays essentially the same theoretical role.

increase in K_h. This initial taper benefit is derived before any external load application after installation causes further expansion of a cylindrical cavity and further increase in K_h. Although the above-referenced work of Sabry and Hanna (2009) did not include tapered piles, inferences relative to possible tapered-pile behavior can be drawn from their work. These inferences, combined with the interpreted (by the writer) results of the Gregersen et alia case-history piles discussed in Appendix H, support the writer's working hypothesis that there is an element of taper benefit that occurs simply from driving a tapered pile and before further taper benefit occurs during post-installation external load application.

In summary up to this point, the three main takeaways of practical significance are that:

1. The pre-existing stress-state as defined by K_o influences the post-driving, pre-load-application value of K_h.

2. K_h for all piles, regardless of their geometry, varies with depth as a complex outcome of creating a cylindrical cavity in the ground as the pile is driven. When combined with the well-established fact that K_o typically varies with depth for most sites, this means that the crucial K_h/K_o ratio that defines the stress-state of a pile after installation but prior to external load application should always be expected to be depth-wise variable. This suggests that Kulhawy's observation that the K_h/K_o ratio is depth-wise constant for a given type of DFE, while revolutionary and thus seminal for its time, was simplistic and is now outdated, at least for (driven) piles which, it is acknowledged, were not the focus of Kulhawy's research.

3. The pattern of development of K_h and its effect on the K_h/K_o ratio are likely unique for tapered piles compared to constant-diameter piles but what this pattern may be is not known at the present time.

It is relevant to note that these observations, when taken as a group, markedly contrast with the assumption of El Naggar et alia that K_h (K_s in their notation) is:

•   not explicitly dependent on K_o,

•   depth-wise constant, and

•   the same for tapered and constant-diameter piles in the same subsurface condition.

In view of these three takeaways, in the writer's opinion an improved resistance-forecasting methodology for tapered piles should be based on using the concept of a K_h/K_o ratio for determining K_h for the post-driving, pre-external load application phase. This means that two key problem parameters always need to be determined beforehand for each site- and pile-specific application:

1. a K_o profile needs to be developed and

2. an appropriate depth-wise variation in the K_h/K_o ratio that includes taper benefit as appropriate needs to be defined.

Unfortunately, while the former can increasingly be done relatively easily and reliably using in-situ testing devices, especially the cone penetrometer, the latter is currently not possible except for the simplistic empirical relationships suggested by Kulhawy that, as noted previously, assume that the K_h/K_o ratio is constant with depth and make no allowance for taper benefit. Thus, fundamental research into developing a way to forecast the K_h/K_o ratio is a high priority for the future as the inability at present to forecast the depth-wise variation in K_h/K_o, including due consideration of taper benefit and residual loads, presents a significant limitation on any improved resistance-forecasting methodology for tapered piles at the present time.

However, in an effort to gain at least some preliminary, first-order insight into the depth-wise variation of the K_h/K_o ratio, it is useful to examine the analytical methodology proposed by Sabry and Hanna (2009) in some detail. The details of this are presented in Appendix F.

As can be seen from the discussion at the end of Appendix F, the doctoral research of Sabry (2005), which was given the briefest of summaries in Sabry and Hanna (2009), produced what the writer considers to be unrealistic estimates of K_h. This outcome appears to be the overall result of trying to simulate pile driving using the finite-element method (FEM) with a simple elastoplastic soil model. As discussed by Dijkstra et al. (2011), there are many numerical-analysis challenges when trying to model pile installation (note that Dijkstra et alia simulated a jacked pile that is simpler than a (driven) pile due to the absence of dynamic effects).

In conclusion, the current reality is that it is not possible to reliably estimate the depth-wise variation of either K_h, or the K_h/K_o ratio on a site- and application-specific basis for a constant-perimeter pile no less to do so for a tapered pile. Therefore, the desired goal of being able to define the first of two stages of taper benefit that is due solely to driving a tapered pile into the ground remains an elusive goal at this time. Clearly, this is a high-priority area for future R&D to fulfill this need.

However, before concluding this discussion it is of interest to take a look at the detailed assessment presented in Appendix A for the instrumented pile loads tests that were conducted on two *Monotube* piles at JFKIA (Fellenius et al. 2000). To begin with and with reference to Figure A.4b in Appendix A, it is reasonable to assume that the average value (1.7) of the K_h/K_o ratio back-calculated for the upper constant-diameter section at the conclusion of the static load test applies to the post-driving, pre-load-application condition as well. This is consistent with the historical, consensus assumption that axial-compressive loading of a constant-perimeter pile does not cause an increase in K_h beyond that already caused by pile installation. It is of interest to note that this back-calculated value (1.7) of the K_h/K_o ratio falls nicely within the range of values suggested by Kulhawy for driven piles. Furthermore, it is remarkably (and coincidentally) close to the values of 1.65 and 1.67 that the writer has used for many years with the writer's Modified Nordlund Method.

7.5.4.2.3 Radial Stresses: External Load Application Phase

Considered next is the change in radial stress along the tapered portion of a pile during post-installation external load application. The discussion will focus on the much more common and, therefore, important case of compressive loads although uplift loads are addressed as well.

To be notationally consistent with how the post-driving, pre-load-application radial stresses were handled analytically in the preceding section, this stress change is defined generically using the parameter ΔK_h that reflects the change in lateral earth pressure coefficient over and above K_h as defined previously. Whether ΔK_h reflects an increase or decrease depends on whether the applied load is compressive or tensile (uplift), respectively.

However, it is at this point that a decision has to be made. This is because there are two different conceptual ways in which to use the ΔK_h parameter to define the aggregate[163] radial stress acting along the tapered portion of a pile during axial-compressive loading:

1. As an <u>additive</u> process so that the aggregate lateral earth pressure is reflected in the <u>sum</u> of the post-driving, pre-load application value <u>plus</u> the change due to the wedging (if a compressive load) or contraction (if an uplift load) that occurs during load application. In terms of lateral earth pressure coefficients, this can be expressed as the quantity:

$$[(K_h/K_o) \cdot K_o] + \Delta K_h . \qquad (7.2)$$

Note that in this approach, ΔK_h starts from an initial value of zero and then either increases (due to a compressive load) or decreases (due to an uplift load) in magnitude along the tapered portion of a pile only. Otherwise, ΔK_h remains at value of zero along the constant-diameter portion of a pile, assuming that the Poisson Effect is neglected.

2. As a <u>product</u> that can be expressed notationally as:

$$[(K_h/K_o) \cdot K_o] \cdot \Delta K_h . \qquad (7.3)$$

Note that a subtle difference in this case is that ΔK_h starts from an initial value of one, not zero as in the additive approach, and then either increases (due to a compressive load) or decreases (due to an uplift load) in magnitude along the tapered portion of a pile. Otherwise, it remains at a value of one along the constant-diameter portion of a pile, assuming that the Poisson Effect is neglected.

The former, additive approach was used by Kodikara and Moore although they did not use lateral earth pressure coefficients to define the post-driving, pre-external-load-application radial stress. As noted previously, these authors used the parameter σ_o for radial stress that served the same conceptual function as (but was obviously not equivalent to) the quantity $(K_h/K_o) \cdot K_o$ used here. However, Kodikara and Moore did use parameters that were functionally similar to, but notationally different from, ΔK_h for defining the increase in radial stress during post-driving external compressive-load application (they did not mention, no less consider analytically, uplift loading).

On the other hand, El Naggar and Sakr chose the latter approach involving a product when developing their resistance-forecasting methodology that is based on the Kodikara-Moore Method. In this case, ΔK_h is equivalent to El Naggar and Sakr's K_t parameter that they called the *taper coefficient*. Thus, the product $\Delta K_h \cdot (K_h/K_o) \cdot K_o$ used here is the same as the product $K_t \cdot K_s$ in Equation C.1 that defines the ULS unit shaft resistance in the El Naggar-Sakr Method.

Upon initial consideration, it might appear that the two approaches produce the same results and indeed they can but only under certain circumstances. Thus, the two approaches are not universally identical in their outcomes. This is because the latter (product) approach used by El Naggar and Sakr explicitly assumes that the radial stresses, including any changes

---

[163] 'Aggregate' is used here intentionally as opposed to the linguistically simpler and more common 'total'. This is to avoid any confusion and unintentional conflation with total (radial stress) in the traditional geomechanics sense of being the sum of effective radial stress plus porewater pressure.

due to external loading, are, at all phases of behavior, linearly proportional to the vertical effective overburden stress as shown in Equation C.1 in Appendix C. On the other hand, the former (additive) approach used by Kodikara and Moore is more general and analytically flexible as it does not assume an inherent linkage of load-induced changes to vertical effective overburden stress although this can be done if desired.

Another important point to note is that regardless of the approach used, ΔK_h is not constant as it varies with the applied pile load and concomitant pile settlement or upward displacement, as appropriate to the type of loading (compressive vs. uplift). However, for simplicity in developing their analytical methodology, El Naggar and Sakr assumed that their equivalent (to ΔK_h) parameter K_t was always a problem-specific constant. They defined K_t as the value corresponding to an arbitrary magnitude of settlement equal to one-tenth (= 10%) of the average pile diameter (note that for tapered piles this is not the same as the toe diameter). It appears that the reason they did so was because they intended the El Naggar-Sakr Method to be a resistance-only forecasting methodology, with the outcome intended to represent the ULS total resistance. Because generation of a complete load-settlement curve was not an intended use of the El Naggar-Sakr Method, there was no need to allow for and consider the settlement-dependent variation of K_t.

In conclusion, when developing an improved resistance-forecasting methodology for tapered piles, a not-insignificant decision that will need to be made is whether to frame the definition of ΔK_h in terms of Equation 7.2 or Equation 7.3. The choice should first and foremost be based on theoretical need based on the analytical theories used to model cavity expansion and contraction. As noted above, a solution might be based explicitly on radial stresses so that it becomes easier to frame the outcome in an additive fashion using Equation 7.2. After theoretical issues are considered, the choice then becomes subjective and depends on personal preference.

7.5.4.2.4 Closing Comments

When the writer first read Kodikara and Moore's seminal 1993 paper, the immediate impression was that cylindrical-cavity expansion (CCE) theory provided a much-improved theoretical framework for modeling the response of tapered piles to axial-compressive loading. It appeared that all that remained to be done was to create a computer program to render the relatively complex three-phase solution algorithm presented in Kodikara and Moore (1993) and outlined in detail in Appendix C usable in routine practice.

However, in the more than quarter-century since Kodikara and Moore's paper appeared, much has been published on the subjects of cavity mechanics and the importance of shear-modulus degradation. Furthermore, closer examination of Kodikara and Moore's published work has called into question the key assumption they made that the CCE that provides the unique behavior of tapered piles is limited to relative pile-soil displacement during post-driving external load application only. The taper benefit that can accrue during the cavity creation associated with pile driving is completely ignored analytically.

As a cumulative result of these and other factors, in the writer's opinion, the current state of knowledge with respect to tapered piles has become one of those classic cases where the more that is learned, the less that is actually known. Stated another way, it is the writer's opinion that the Kodikara-Moore Method and the derivative work by El Naggar et alia and Manandhar et alia are no longer considered potentially usable in their original forms for the tapered-pile problem as they each ignore or fail to correctly consider too many critical issues and factors.

The basis of this opinion is that it is now clear that what is broadly referred to in this monograph as 'taper benefit', i.e. the fact that tapered piles have a substantially greater ULS unit shaft resistance compared to a constant-diameter pile in the same conditions, is due to a synergistic combination of:

1. Cylindrical-cavity creation (CCX), which is common to all piles as the pile toe displaces soil and creates a hole in the ground, but in the case of tapered piles includes additional cavity creation as the pile toe advances further into the ground and the tapered shaft immediately behind the toe enlarges the hole that the toe created. It is this latter aspect of further cavity enlargement as the pile is driven that is unique to tapered piles.

2. Additional CCE during external load application after driving.

Note that in both cases there are shear stresses at the interface between the cavity wall and surrounding ground.

While this is two-stage process is now understood to occur in concept, translating this process into the appropriate mathematical model(s) and concomitant solution(s) remains elusive. To begin with, at the present time the complex series of cavity-mechanics events that occurs during driving is not adequately understood as, to the best of the writer's knowledge, it has never been studied to date. Although CCX has been explored for many years for the purpose of modeling cone penetrometers (Salgado and Prezzi 2007), the issue of further widening of the cavity after initial creation, as occurs with tapered piles, has not been studied.

There are additional, potentially significant complications to developing an appropriate mathematical model for this because the pile-driving process is not the same phenomenologically as making a sounding with a cone penetrometer, even though the two processes are often thought of and referred to as being broadly identical in nature. A cone-penetrometer sounding is a monotonic penetration under relatively slow dynamic conditions, i.e. there are no inertia effects that need to be considered. On the other, pile driving is a true dynamic process where inertia effects are important.

Pile driving can be viewed simplistically as a repetitive 'two steps forward, one step backward' process. This is because as the initial compressive stress wave that is due to a pile hammer impacting the head of a pile travels down the pile, the pile shaft displaces downward some gross amount under a combination of pile-soil displacement combined with elastic compression of the pile as a structural member. After that stress wave is reflected back up the pile shaft as either a compressive or tensile wave depending on driving conditions, there will always be some rebound (upward displacement) of the pile shaft. Thus, while there is always some net downward displacement of the pile (this is referred to as the pile *set*) after a given blow of a pile hammer, that set is the net result of some gross downward displacement (the 'two steps forward') minus the rebound (the 'one step backward').

For a tapered pile, this complex behavior translates into cavity <u>expansion</u> during the 'two steps forward' followed by some lesser amount of cavity <u>contraction</u> during the 'one step backward'. So, while the final pile set produces a net cavity expansion compared to what existed prior to the hammer striking the pile, that net expansion did not occur as the result of some monotonic expansion of a cylindrical cavity as is typically assumed when developing solutions in cavity mechanics. Rather, the final cavity expansion is the net result of a larger cavity expansion followed by a partial cavity contraction. Furthermore, all this is occurring with significant shearing occurring at the pile-ground interface (neglected in traditional cavity-mechanics models even though such shearing occurs with cone penetrometers) and with development of significant residual loads in the pile.

Until either confirmed or proven otherwise, it is reasonable to assume that this very complex behavior, which, unlike with cone penetrometers, is occurring under true dynamic conditions where inertia effects are a factor, requires fundamental research into developing an appropriate mathematical model and concomitant solution as opposed to simply applying solutions in cavity mechanics that were developed for far-simpler problems such as pressuremeter testing and cone penetrometers.

Compared to the complex behavior that occurs during pile driving, the behavior under post-driving applied loading appears to be somewhat simpler. This is because it involves monotonic cavity expansion or contraction (depending on the direction of axial loading) under quasi-static conditions. However, there is still shearing at the pile-soil interface to contend with that is neglected by cavity-expansion models and their solutions that have been developed to date. The effect of shear stresses under the drained-strength conditions and with a non-associated flow rule, as would be relevant to coarse-grained soils, is a topic that is need of fundamental research. This is because the limited study to date of the effect of shear stresses on CCE under undrained-strength conditions (appropriate for fine-grained soils) indicates that such stresses are significant (Zhou et al. 2014).

In conclusion, there are now multiple reasons why the Kodikara-Moore Method and, by implication, the derivative works of El Naggar et alia and Manandhar et alia, have, unfortunately, turned out not to be the long-sought improvement to the Nordlund Method for tapered piles. Even though these alternatives to the Nordlund Method are based on a 3-D cylindrical-cavity expansion that is a superior physical model for the shaft-resistance mechanism of wedging compared to Nordlund's 2-D lateral earth pressure model, the way in which CCE was implemented misses the true behavior of tapered piles in several important aspects. This has actually led to researchers drawing conclusions about tapered piles that, in the writer's opinion, are highly questionable. These questionable outcomes include:

- Concluding that taper is beneficial only at relatively low confining stresses. This has led to the recommendation that partially tapered piles should have the tapered section directly beneath the head of the pile with a constant-diameter below that to the toe. This is completely the reverse of the design in widespread use in the U.S. for almost a century, going back to the original development of the *Monotube* pile.

- Concluding that to be effective, taper angles need to be significantly greater than those used historically in the U.S. Specifically, taper angles need to be several degrees, not the fraction of one degree that has proven to be successful.

The root causes of these questionable conclusions are many and have been enumerated previously. However, the single most important cause appears to be the failure to explicitly consider the substantial cavity creation that occurs during pile driving and prior to external load application. Simple arithmetic indicates that the relative magnitude of additional cavity creation that occurs during driving after the initial cavity creation by the pile toe is two to three orders of magnitude greater than the cavity expansion that occurs under subsequent service loads or even a pile load test.

Thus, the assumption made by Kodikara and Moore and propagated by El Naggar et alia and Manandhar et alia that all taper benefit derives from post-driving external load application results in a very distorted picture of reality. In the writer's opinion, it is a classic case of incorrect information being worse than no information at all. It will take substantial research of the very complex two-stage mechanism by which taper benefit actually develops during driving and subsequent external load application. Until then, development of an improved resistance-forecasting methodology for tapered piles will remain an elusive goal.

7.5.4.3 Toe

7.5.4.3.1 Prologue

In the writer's opinion, the position taken and expressed for many years now by Fellenius that a ULS value for toe resistance simply does not exist is an extreme departure from the position taken in the state of practice that has evolved since the 1950s and that remains in widespread use to the present. If anything, the presumption that a ULS value for toe resistance beneath all types of DFEs always exists is stronger now than ever due to the universal emergence of LRFD as the preferred design methodology for DFEs. By definition, LRFD inherently implies that <u>all</u> system materials and components <u>always</u> have some single-value ULS value. Thus, for any number of reasons, completely abandoning the concept of a ULS toe resistance does not appear to be a reasonable position to take in an improved resistance-forecasting methodology for tapered piles, at least for the foreseeable future.

There is, however, an element of validity in Fellenius' position that should not be dismissed and that is that the settlement dependency of toe resistance should not be ignored, even in routine practice. It has long been recognized that the development of toe resistance is much more settlement dependent compared to shaft resistance, at least for constant-perimeter DFEs. Specifically, the magnitude of settlement required to mobilize a condition approaching what most would consider to be a ULS value of toe resistance is at least one order of magnitude greater than that required to fully mobilize shaft resistance due to sliding friction. Note, however, that this assumes initially stress-free virgin-loading conditions and thus ignores residual-load effects that can change things significantly.

In any event, while it might be reasonable in a simple analytical methodology used for resistance forecasting in routine practice to ignore the magnitude of settlement required to mobilize shaft resistance due to sliding friction, it is not reasonable to assume that toe resistance is either all (a ULS value) or nothing as has been done for (driven) piles and many other types of DFEs since the 1950s.

Therefore, the position taken in this monograph is that the analytical methodology used for toe resistance in any improved resistance-forecasting method for tapered piles needs to incorporate explicit consideration of a toe-load vs. toe-settlement relationship that leads to a ULS value of toe resistance, even if that ULS value is only approached asymptotically but never reached (which would be in the spirit of Fellenius' thoughts re toe resistance). However, that still leaves considerable room for subjective judgment as to the specific theory and solution used for this purpose as is discussed in the following sections.

7.5.4.3.2 Background

Because the schema for an improved resistance-forecasting method for tapered piles is based on the concept of using cavity mechanics for evaluating the shaft resistance, it is appropriate to begin the discussion of toe-resistance methodologies by summarizing the detailed discussion in Chapter 5 of how toe resistance was treated by Kodikara and Moore as well as El Naggar et alia and Manandhar et alia in their analytical methodologies that were based on the Kodikara-Moore Method.

The three versions of the Kodikara-Moore Method that have appeared in the published literature to date span a wide spectrum of conceptual ways to deal with toe resistance:

1. Kodikara and Moore were unconventional in their approach and assumed purely linear-elastic behavior with a constant-value shear modulus. Their argument was that this is an acceptable assumption under what might be interpreted to mean service-load conditions. However, because they provided no guidance as to how the required single-valued shear modulus is to be determined on a site- and application-specific basis, the validity of their approach cannot be evaluated.

2. El Naggar et alia used the traditional assumption of rigid-plastic behavior, i.e. a single-valued ULS resistance. They recommended use of an empirical methodology for calculating this ULS resistance that was contained in the then-current (1992) version of the *Canadian Foundation Engineering Manual*. The writer could not perform an independent assessment of the accuracy of this methodology. Although El Naggar et alia did not explicitly consider toe settlements, by virtue of the fact that they assumed that the geotechnical ULS for the shaft was reached at a settlement magnitude equal to 10% of the average pile diameter this indirectly implies the magnitude of settlement at which the ULS toe resistance is achieved.

3. Of all the researchers being discussed in this section, Manandhar et alia dealt with toe resistance in a manner that is closest to that recommended for the writer's schema, at least in broad, conceptual terms. The central element of their approach is a methodology for forecasting a value of ULS toe resistance, what is referred to in this monograph as the Basic Manandhar-Yasufuku Toe-Resistance Method. It is based on the basic elements of Vesic's spherical-cavity expansion (SCE) model and solution for toe resistance. However, the assumed location and geometry of the spherical cavity relative to the pile toe as illustrated in Manandhar and Yasufuku (2012) is substantially different from the location and geometry visualized by Vesic as illustrated in Vesic (1977). Furthermore, Manandhar and Yasufuku assumed that the taper angle, α, influences the overall geometry of the assumed failure mechanism. As is well known and has been commented on by the writer many times over the years in many different contexts, anyone can develop and posit a physical behavioral mechanism in civil engineering. However, proof of some kind must subsequently be developed to demonstrate, in a fashion that can be replicated by others, that such a mechanism actually can occur[164]. In this case, the failure mechanism proposed by Manandhar and Yasufuku is supported, and only indirectly at that, by 1-*g* laboratory tests on small-scale model piles. Some of these model piles have geometries that bear no resemblance to actual tapered piles (they have a tapered upper section and a constant-diameter lower section) and all piles were installed in the sand-filled test box in a manner that did not replicate actual pile driving. In addition, the required soil properties for the Basic Manandhar-Yasufuku Toe-Resistance Method must be quantified using various empirical relationships presented by the authors, some of which are based on SPT *N*-values. Presumably, these are N_{field} values which means that a wide variety of outcomes will be produced depending on the specific hammer-drive system used. Finally, the empirical hyperbolic equation that Manandhar and Yasufuku used to transform the Basic Manandhar-Yasufuku Toe-Resistance Method (that produces only a single-valued ULS resistance outcome) into the Enhanced Manandhar-Yasufuku Toe-Resistance Method

---

[164] A classic example of the failure to do so, and the decades of misguided engineering practice that can result from this, involves cantilever sheet-pile walls. As discussed in Horvath (2014a), generations of geotechnical and structural engineers have designed such walls based on a simplistic, intuitively pleasing ULS failure mechanism that, in reality, is physically questionable and, as it turns out, simply does not occur in reality.

(that produces a toe-load vs. toe-settlement outcome) is questionable for several reasons that are explored in Appendix B.

In conclusion, although these three research efforts covered the range of behaviors (perfectly linear-elastic, perfectly plastic, and something in between) with respect to conceptual approaches to toe resistance, there is still infinite opportunity to vary the specifics of how any of these behaviors is defined and quantified in terms of analytical models, parameter selection, etc. Thus, selection of a toe-resistance methodology for an improved resistance-forecasting methodology for tapered piles very much remains an open, unanswered question.

7.5.4.3.3 Overview

The writer is not aware of any currently available analytical methodology for a unified analysis of a settlement-dependent DFE toe resistance leading up to a well-defined, single-valued ULS resistance. Consequently, the position taken for the schema presented in this chapter is that toe resistance needs to be modeled using some existing ULS-resistance model and solution combined with some methodology for generating an intermediate toe load vs. toe settlement curve. These two parts of an overall toe-resistance methodology are discussed in the following two sections.

7.5.4.3.4 ULS Resistance

To begin with, it is the writer's opinion that there is no conclusive supporting evidence that taper affects the magnitude of ULS toe resistance in any manner when compared to an otherwise identical constant-diameter pile in the same conditions. The 'proof' for such a linkage presented by Manandhar and Yasufuku is far from persuasive. Therefore, the choices for a ULS-resistance methodology to use for tapered piles include all methodologies that have been or could be used for DFEs with a constant perimeter. With this in mind, the primary choices are as follows:

- Some <u>traditional bearing-capacity solution</u> that produces only a single-valued ULS toe resistance. There are at least three sub-categories of such solutions to choose from:
  - A classical solution such as that of Berezantzev et alia as used by Nordlund.
  - A classical solution but with Vesic's rigidity factors as recommended by Kulhawy (1984) in order to deal with soil compressibility in a rational, theoretical manner. This approach, using Hansen's classical solution, was used by the writer for many years beginning with Horvath (2002) and to the present in this monograph.
  - A solution that is out of the current mainstream such as the *NTH Limit Plasticity* solution that is discussed in Appendix G.

- The Vesic <u>spherical-cavity expansion (SCE) solution</u>. This solution was developed originally for use with DFEs (Vesic 1977) but never really gained a critical-mass of acceptance and use in either practice or research, no doubt due to the inherent difficulty found with all cavity-mechanics solutions, i.e. evaluating the soil-stiffness parameters correctly. In recent years, Vesic's SCE solution has found a second life with some researchers such as Prof. Paul Mayne as a methodology for interpreting the tip resistance of cone penetrometers. This has led to significant research into the previously 'missing ingredient' of parameter assessment that is discussed in Appendix G. As a result, it seems appropriate to revisit the used of Vesic's SCE solution with DFEs as it was originally intended to be used.

- A <u>cylindrical-cavity creation (CCX) solution</u>. As discussed in Chapter 4, Prof. Rodrigo Salgado, oftentimes collaborating with others, has, for some years now, been the principal advocate of the hypothesis that advancing a cone penetrometer into the ground is best modeled physically and mathematically as <u>cylindrical</u>-cavity <u>creation</u>. Note that this stands in sharp contrast to the work of Prof. Paul Mayne who uses a <u>spherical</u>-cavity <u>expansion</u> model and concomitant solution for the same problem. As part of their work, Salgado et alia relate the uncorrected tip resistance, q_c, of a cone penetrometer to CCX. It is straightforward to scale this concept up to piles where q_c becomes the ULS unit toe resistance, $r_{t(ult)}$, that is the same as q_{ult} that is often used in bearing-capacity solutions to denote the gross ultimate bearing capacity. Thus, the work of Salgado et alia can be viewed as an alternative theoretical approach to Vesic's use of SCE as the physical and mathematical model for pile toe resistance. For the sake of completeness, it is noted that Cook (2010) presented an alternative cavity solution that can be used for both cavity creation and expansion (Cook focused on cylindrical cavities only). Although Cook modeled his work broadly along the lines of that of Salgado et alia, Cook made several different choices for both the model and its numerical solution. As a result, Cook's outcomes differ somewhat from those of Salgado's work that was published in 2007 (Salgado and Prezzi 2007) and thus offer an alternative to use with DFEs.

The writer has chosen not to identify a clear preference from this selection for the method and solution to use in an improved resistance-forecasting methodology for tapered piles, primarily because the use of CCX solutions remain unresearched in deep-foundation applications. Until further research is performed into the use of CCX solutions for pile toe resistance, a complete assessment and comparison of the above choices cannot be performed.

However, it appears at this time that some method based on cavity mechanics is the most preferable. There are several reasons for this:

- Cavity mechanics offers the best behavioral concept at present for the pattern of resistance that develops beneath the toe of a (driven) pile at least, if not DFEs in general.

- Decades of research into cone penetrometers has clearly established the significant role that radial stresses, <u>not</u> vertical stresses, play in the penetration resistance of relatively slender, deeply embedded cylindrical objects such as cone penetrometers and piles (Salgado and Prezzi 2007). Only cavity mechanics provides a theoretical mechanism for accounting for the essential role of radial stresses. Traditional bearing-capacity solutions are forever locked-in to the assumption that only vertical stresses play a role.

- As illustrated in Appendix G for Vesic's SCE solution, there have been significant advances in site characterization, especially using cone-penetrometer data, that make the historically problematic task of parameter assessment much easier. There is no reason why similar advances could not be developed for the cylindrical-cavity methodologies developed by either Salgado and Prezzi or Cook should either of these prove superior to Vesic's SCE solution when applied to the toe resistance of tapered piles.

Before closing out this section, it is important to discuss an issue that is common to any and all methods for ULS toe resistance that is rarely addressed to the extent that it should. This involves the specifics of evaluating the necessary soil properties for the chosen analytical methodology. As a minimum, this involves ϕ and, for methods based on cavity mechanics, G_{max} and K_o as well. Nowadays, quantification of these parameters is preferably and most often based on some in-situ testing methodology although Salgado and Prezzi (2007) and Cook (2010) chose to use a combination of data from laboratory tests; various theoretical and empirical algebraic relationships; and simple assumptions (in the case of K_o).

Independent of how the relevant soil properties are determined is the depth or range in depths over which the relevant soil properties are used. One could simply use the soil properties at toe depth as Nordlund did for both versions of his method. In the past, when SPT N-values, which typically have a 5-foot (1500-mm) interval spacing in U.S. practice, were used this was arguably sufficiently accurate as one would typically use the N-value at or just below the toe. However, with the greater use of cone penetrometers that nowadays collect data every 20 to 50 millimetres (0.8 to 2 in)[165] and generally exhibit some variation from one datapoint to the next this is no longer defensible and can lead to large variations in calculated ULS resistance over a relatively short vertical distance as the writer has found in his research.

As noted earlier in this monograph for various resistance-forecasting methodologies that were discussed in Chapter 5 as well as in Appendix A, it is common when evaluating the ULS toe resistance that some arbitrary assumption is made for a depth-range of influence for the purposes of averaging (usually arithmetically, not in any weighted manner) the values of soil properties that are input into the ULS-resistance methodology. This depth range is usually expressed as some number of toe diameters, d_t, that goes from above the toe depth to below the toe depth. One of the more common of such depth ranges that is used for both the Original (Eslami-Fellenius) and Modified (Niazi) UniCone methods as well as the Togliani Method goes from eight toe diameters above (+) to four toe diameters below (-), usually seen expressed as $+8d_t$ to $-4d_t$.

In the writer's opinion, the problem is that such ranges are often simply stated as fact without supporting evidence to justify the assumption although Niazi (2014) provides some discussion of this. Thus, it is suggested that whatever averaging metric is ultimately chosen for use in an improved resistance-forecasting methodology for tapered piles be based on some logic as opposed to simply reusing some rule of thumb from the past. This is because it seems logical that the averaging range chosen should be consistent with at least two considerations:

1. the physical model used as the basis for the chosen ULS methodology and

2. the unique characteristics of the in-situ testing tool used for parameter assessment in terms of detecting changes in soil properties.

As an example of how this might be done, the information presented in Vesic (1977) was reviewed for guidance with respect to Item #1. To begin with, for the purposes of this monograph the pile toe is defined as the end of the pile that is driven into the ground but <u>not</u> including any non-flat toe-closure piece. Thus, the signature hemispherical closure of a *Monotube* pile or the rounded-conical closure of a *TAPERTUBE* pile (Figure 3.3) are <u>not</u> considered the toe. Rather, the toe is defined as the point where these cast-steel closure pieces are attached to the bottom of the tapered section.

The reason for adopting this definition is that it has long been established that even for piles with a flat toe closure, a conical wedge of soil will form naturally beneath the flat toe closure and move downward with the pile. As illustrated clearly in Vesic (1977), this conical

---

[165] Even the older CPT soundings performed in the New York City area in the late 1980s recorded data every 12 inches (300 mm).

wedge of soil is considered to be part of the geomechanical failure mechanism <u>below</u> the pile toe. Therefore, any non-flat toe closure that is appended to a pile really acts as part of this conical wedge that would form anyway and is thus effectively part of the failure mechanism, not the pile itself.

Based on this definition and a review of the information presented in Robinsky (1963, as republished in Krinitzsky 1970) and Vesic (1977), the writer argues that a guideline of $+1d_t$ to $-3d_t$, i.e. one toe-diameter (d_t) above the toe to three toe-diameters below the toe, is appropriate to use, at least based on consideration of what is physically occurring. That is because all of the observed failure mechanism associated with the toe is confined within this depth range.

However, this $+1d_t$ to $-3d_t$ depth range needs to be evaluated further in the context of the in-situ testing tool used. Nowadays, this is most likely to be a cone penetrometer. As is well known, the tip resistance of a cone penetrometer does not react instantaneously to a significant change in soil type and/or consistency over a relatively short distance. There is a lag of sorts in sensing such changes. Therefore, it may be prudent to extend this $+1d_t$ to $-3d_t$ depth range somewhat to make some allowance for this resistance-lag of a cone penetrometer. What this allowance should be is a detail and topic for future research. The point being made here is that there should be some scientific thought given to what this range should be as opposed to simply reusing or repurposing some existing rule of thumb whose theoretical basis or justification is unknown.

7.5.4.3.5 Intermediate Load-Settlement Behavior

Having discussed methods for determining the ULS toe resistance, the question remains how to develop a mathematical relationship for the intermediate load-settlement behavior. As discussed earlier in this monograph, because explicit determination of load-settlement behavior has never been part of routine practice for DFEs in general and tapered piles in particular, there is much less precedent to draw on compared to ULS toe resistance.

With specific regard to methods developed for tapered piles, the collective work of Kodikara and Moore, El Naggar et alia, and Manandhar et alia is, in the writer's opinion, unhelpful. Kodikara and Moore assumed linear-elastic behavior which is totally unrealistic and El Naggar et alia essentially ignored the issue altogether when they simply assumed a single-valued (ULS) toe resistance. Manandhar et alia modeled non-linear load-settlement behavior but using a hyperbolic relationship whose curve-defining parameters were arbitrary assumptions with no fundamental linkage to site- and application-specific soil properties.

As explored in Appendix B, Manandhar et alia's assumptions for the parameters that control the shape and endpoint (more technically, how rapidly the hyperbola approaches its limiting ultimate value that is, of course, never reached in finite space) of the hyperbolic relationship appear to be quite unrealistic. Although several alternatives were identified and explored to a limited extent in Appendix B, they all had some deficiency of being purely empirical or requiring a calibration database of actual toe-load vs. toe-settlement behavior that is currently non-existent and not easily developed. Consequently, the writer concluded that the load-settlement model used for the toe-resistance component of an improved resistance-forecasting methodology for tapered piles needs to be thought out using a zero-based decision-making approach.

The writer has adopted the position that any load vs. settlement model for toe resistance should be:

- based on theory and

- utilize site- and application-specific soil properties developed using appropriate in-situ testing devices, especially the cone penetrometer,

to the greatest extent practicable.

With this in mind, the following conceptual outline was developed for an analytical model for toe resistance as a function of toe settlement:

- Explicit recognition and consideration of residual loads is an overarching consideration. There is ample evidence that residual toe loads are very significant for tapered piles. This means that external loading after pile installation will follow a complex load path that needs to be taken into consideration.

- The ULS toe resistance described in the preceding section is used to define the ULS unit toe resistance, $r_{t(ult)}$.

- The theory-of-elasticity solution used by Kodikara and Moore is used to calculate toe resistance but with the important difference that the shear modulus degrades (reduces in magnitude) as a function of toe settlement as opposed to remaining constant as assumed by Kodikara and Moore. Using a theoretical force-displacement model combined with appropriate modulus degradation avoids the issue of having to develop an empirical model based on a hyperbola or some other curve-shape.

- Research is required to determine whether a stress- or strain-based modulus-degradation model is the better approach. A stress-based model such as the Fahey-Carter model could logically use the ULS toe resistance as the reference stress value. Alternatively, a strain-based degradation model could use a strain definition of toe settlement divided by toe diameter, similar to that used by Niazi, with some arbitrary strain level (such as the 1% used by Niazi) as the reference strain value. Note that consideration of the need to accurately model residual-load behavior may play a role in determining which approach, i.e. stress-or strain-based, is better to use in this application.

7.5.5 Solution Algorithm

7.5.5.1 Introduction

An essential consideration that requires careful thought when developing a new resistance-forecasting methodology for tapered piles is the overall structure of the solution algorithm used. The reason is that producing a load-settlement curve in an integrated, one-step process is far more complicated than either simply forecasting the ULS total resistance (as has always dominated routine practice) or applying a settlement-only methodology such as a Randolph-Wroth Problem solution to a previously forecast ULS total resistance as part of a two-step process. This is because a one-step resistance-forecasting methodology implies an incremental analytical process that needs to satisfy both force and displacement compatibility along the entire pile shaft at each step of the calculation. The case of a tapered pile is particularly complex because the forces and displacement compatibility involves both the axial and radial directions.

In consideration of the importance of the solution algorithm to the schema presented in this chapter, it is worthwhile to provide some thoughts in this matter to at least provide a starting point for future consideration.

7.5.5.2 Overview

Although none of the existing resistance-forecasting methodologies for tapered piles meets the writer's criteria for an improved analytical methodology as outlined earlier in this chapter, the underlying algorithm framework of the Kodikara-Moore Method appears to be a sound one and thus of reference value for an improved analytical methodology. This is because the Kodikara-Moore Method is based on the same key behavioral elements as the schema presented in this chapter. Namely, both shaft and toe resistance are assumed to be settlement dependent although the specifics of this dependency differ considerably between the Kodikara-Moore Method and the schema presented here.

Note, however, that this is not to say that a zero-based solution algorithm should not be considered in the future if new research points in that direction, e.g. indicating that some currently unknown physical model is superior to cylindrical-cavity mechanics for modeling the wedging behavior of tapered piles. For example, as discussed earlier in this chapter, available information suggests that cross-sectional geometry plays a role in the behavior of tapered piles and that cylindrical-cavity mechanics is clearly not the best model for all cross-sectional geometries. Future research may dictate that the wedging that occurs with tapered piles with a square cross-section requires its own physical model and concomitant solution that may require a completely different solution algorithm from that proposed by Kodikara and Moore and repurposed here.

7.5.5.3 Structure

The overall structure of the solution algorithm published in Kodikara and Moore (1993) can be characterized as consisting of two incremental, cumulative calculation cycles, one nested inside the other, that are each settlement-based, i.e. settlement is the independent problem variable. The innermost calculation cycle consists of assuming an arbitrary magnitude of toe settlement that directly translates into a toe force by virtue of a linear-elastic model for toe resistance. The calculation cycle then progresses upward along the pile shaft to the pile head by calculating the settlement-based shaft resistances for each artificial segment into which the pile has been divided. Elastic compression of each pile segment is included as well. The outcomes at the end of this calculation cycle are a data-pair consisting of the axial-compressive load and settlement at the head of the pile.

The second calculation cycle involves repeating this toe-to-head calculation cycle as often as desired to create a sufficient number of head load vs. head settlement data pairs in order to generate a load-settlement curve. Note that there are clearly other calculated results such as depth-wise values of mobilized resistance, radial stresses acting on the shaft-ground interface, etc. that are generated by the overall solution process that could be displayed graphically. However, Kodikara and Moore did not pursue possibilities with these other calculated outcomes.

Manandhar and Yasufuku retained this cycle-within-a-cycle solution algorithm but added an additional, iterative cycle within the toe-to-head calculation cycle to allow for determination of the stress-dependent value of ϕ as used to calculate shaft resistance for each artificial segment into which the shaft is divided. They also used a completely different toe-resistance model but retained toe settlement as the independent variable. Thus, the original feature of the Kodikara-Moore Method in which an assumed toe settlement produces a calculated toe force was retained, at least qualitatively. Manandhar and Yasufuku also focused solely on the calculated load-settlement results at the pile head and did not explore other calculated outcomes of potential interest.

It is important to note that whatever physical and mathematical model and its solution are chosen for the improved resistance-forecasting method for tapered piles do not have to follow the work of Kodikara and Moore and Manandhar and Yasufuku exactly, especially with regard to the pile toe. All that matters is that there must be a mathematical relationship between toe force and toe settlement that is defined beforehand. Which parameter is the independent variable, for which an arbitrary value is assumed to initiate a calculation cycle, and which parameter is the dependent variable that is the calculated outcome is simply a detail. Thus, there is no reason why the approach chosen could not be reversed so that toe force becomes the independent variable with toe settlement the dependent variable. Once the magnitude of toe settlement for a given head load vs. head settlement calculation cycle has been calculated, the calculations can proceed upward along the pile shaft using settlement as the independent variable.

It is likely that an improved resistance-forecasting methodology for tapered piles will add additional calculation cycles in the form of iteration cycles within the two cycles used in the Kodikara-Moore Method. In much the same way that Manandhar and Yasufuku iterated the operative ϕ value for each artificial segment of the pile shaft, there will likely be iterations for the complex interaction between the Poisson Effect on the shaft and the radial stress at the pile-soil interface.

It is also suggested that in the future that productive use be made of the numerous problem parameters that are calculated outcomes of the solution process in addition to the load and settlement at the head of the pile. This includes distributions of shaft resistance and its mirror image of axial load within the pile shaft. These distributions are of interest for many reasons, including an objective assessment of whether or not an analytical methodology can reasonably replicate the pattern of resistance and shaft load observed in actual tapered piles. This comparison was used in Chapter 6 and Appendix A as a metric for evaluating the accuracy of the Nordlund Method (both versions), the writer's Simplified Nordlund Method, and the Togliani Method.

7.5.6 Computational Tools

7.5.6.1 Introduction

The writer has seen foundation engineering transition from the sliderule era to the digital-computer era on a first-hand basis. Within the computer era, the transition to computationally powerful microcomputers has allowed for complex calculations to be performed routinely using a wide variety of computational software. Thus, it is of some interest and value to discuss computational tools that can be used to implement the schema discussed in this chapter. This is because in the writer's personal experience working with resistance-forecasting methodologies for tapered piles over the past twenty years, the computational tool that is chosen can actually have a significant influence on the theoretical rigor of the analytical methodology that is eventually chosen. Conversely, if a certain level of theoretical rigor is deemed essential or at least desirable, then this will tend to rank or order the computational tools that can be used.

It is important to note at the outset that the comments made herein with regard to computational tools are focused on developing a proof-of-concept analytical methodology as would be used by an end user who wants to develop software for their personal use or use within a limited network of professional colleagues or students. Obviously, professional software developers who are interested in producing a standalone software package for commercial sale and distribution have completely different goals and concomitant developmental guidelines that are not addressed here.

7.5.6.2 Background

To begin with and for the sake of completeness, it is assumed that any improved resistance-forecasting methodology for tapered piles will:

- require computer software for routine use and

- not make use of any graphical or tabular solutions that cannot be reasonably converted to digital form for incorporation into solution software.

These requirements may seem trivially obvious in today's computational environment but are stated here for clarity and to remove any ambiguity or uncertainty. This is because in the U.S. at least the current de-facto standard resistance-forecasting methodology for tapered piles is still the Nordlund Method. In its original form, which is the only version that most design professionals are aware of, it dates back to 1963. This was an era when the sliderule was the only calculation tool available to the average design professional[166] and graphical (often relatively crudely hand-drawn as in the case of Nordlund (1963)) and tabular solutions were the norm.

There are always two, broad considerations to keep in mind when it comes to developing any computer-based analytical methodology in foundation engineering:

- The computational tool influences choices made for both the geotechnical and structural aspects of the chosen analytical methodology. This is because certain arithmetic operations and processes are more easily and/or efficiently done using certain computational tools compared to others. Nowadays, the issue of linking the graphical portrayal of calculated results is also a consideration, especially when the end-use software is intended for commercial distribution and sale. However, for the proof-of-concept purposes as are being discussed here, it is the execution of arithmetic processes that is the dominant consideration by far. The presumption is that regardless of the computational tool, the end user will always be able to take calculated outcomes and create plots or charts or table using other pieces of software if necessary or desirable.

- The computational tool greatly affects the potential for use of the analytical methodology by both design professionals in practice and academic researchers who may be interested in tweaking an existing program or developing their own program to use. Because the need for computer hardware per se (versus manual calculation using a handheld calculator with figures or tables) is no longer a potential burden as it was until late in the

---

[166] The handheld calculator did not appear commercially in the U.S. until the early 1970s. From personal, first-hand experience, the writer can attest to the fact that the first models could do little more than basic arithmetic functions (addition, subtraction, etc.) and were relatively expensive (approximately the current cost of a mid-range laptop computer when adjusted for inflation).

20th century[167], the issue has shifted entirely to software. The significance of this is, of course, the specific type of software as is discussed in the following section.

The writer's personal experience is, unfortunately, that these two considerations tend to act at cross purposes. As a result, the decision made in a specific case tends to be highly subjective as it requires whomever is developing an analytical methodology to weigh these two competing factors and then make a decision that largely reflects personal experiences and preferences and the personal bias that tends to result from them. Consequently, it is appropriate to present the writer's thought process in the particular case of an improved resistance-forecasting methodology for tapered piles that is being addressed in this chapter.

7.5.6.3 Overview

As noted in the preceding section, the issue of the computational tool to use boils down to a question of computer software. There are two broad paths that can be followed in this regard. The first is purpose-written software using some generic programming language that is suitable for engineering applications such as *FORTRAN*, *C*, *Python*, etc. This category can be further broken down into:

- Software that is self-written by the end user. This implies access to and knowledge of that programming language by the end user. Because personal knowledge is generally limited to the programming language per se, most software of this type does not have graphical interfaces for pre- and post-processing. The biggest issue of relevance to the current discussion is whether or not others will have access to the software and if so, at what financial cost (if any). Nowadays, there are concerns involving legal liability for providing access to such software, even at no cost, as in many legal jurisdictions such as the U.S. a professional who dispenses 'free information' still incurs a professional liability for that information.

- Software written by a commercial software vendor and made available for licensed usage for some fee. Nowadays, such software will typically have graphical interfaces on the front and back ends although the quality may vary considerably.

The second path that can be followed is to base a solution on generic software that is in the public domain (albeit at some cost to the user in most cases) that is adapted for use on an application-specific bases. This category can be further broken down into:

- Generic, all-purpose spreadsheet software such as *Excel* that can be adapted for engineering applications. The downside to this is that spreadsheet software was developed first and foremost for business applications. Although the ability to perform engineering calculations has improved substantially over the years, there are still limitations that can be significant.

---

[167] Early microcomputers in the 1980s typically did not have the inherent capability for the type of intensive mathematical processing associated with programs written for civil engineering applications. The writer can still remember having to pay extra for a 'math chip' with a microcomputer circa 1990 and for some years thereafter.

- Software such as *Maple*, *Mathcad®*, *Mathematica*, *MATLAB®*, and possibly others that the writer is not aware of that is targeted at the engineering/science end user. Note that to the extent that these software packages allow a user to create an application-specific program using some programming language built into the software package there is some conceptual overlap with the self-written software using *FORTRAN*, etc. as discussed above.

There are objective pros and cons to each of these choices and sub-choices beyond those already noted. There are also subjective end-user biases that may influence decisions, at least subliminally if not overtly. For example, the writer's initial introduction to digital computers was in 1967, an era when purpose-written software using *FORTRAN* that was hand-typed onto punch cards and fed into a mainframe computer was the only alternative to a sliderule that was available to a civil engineer. It was not until about three decades after this that microcomputers capable of routinely executing civil engineering software became widely available. Because of this and habits formed over many decades, the writer continued to use self-written *FORTRAN*-based codes for even relatively simple problem solutions that could be done using a spreadsheet until the recent past.

7.5.6.4 Assessment

With more than 50 years of experience using the digital computer as a computational tool for civil engineering applications, the writer has had extensive experience using three of the four computational tools noted in the preceding section:

- self-written software using *FORTRAN*, including using software by others (sometimes with modifications made by the writer);

- commercial software; and

- *Excel*.

Consequently, the writer's thoughts with regard to a computational tool to use with an improved resistance-forecasting methodology for tapered piles are based on these choices. Because of the developmental, proof-of-concept nature of the work presented in this monograph, commercial software by others was obviously not a viable alternative. This, then, left the choices as *FORTRAN* vs. *Excel*.

An important aspect of the goal of this chapter is to develop a schema for an improved analytical methodology that is practice-oriented and that design professionals would find desirable to use. Thus, there was a desire from the start of the multi-year process of researching and writing this monograph to develop an analytical methodology that could be solved using *Excel*. For several years now, the writer has observed that this is a goal of many authors, including academicians and their graduate students, who present analytical methods in their published work. This is because 'solvable by spreadsheet' has become the de-facto standard defining a practice-oriented analytical methodology.

The primary benefit of a solvable-by-spreadsheet methodology is the global availability of such software, often as part of a basic suite of software that is installed on a microcomputer nowadays, and the fact that there is a growing list of alternative (to *Excel*) spreadsheet programs that are available should this be desired by the end user. Thus, in principle, solvable-by-spreadsheet allows the maximum number of end users to implement the proposed analytical methodology in the shortest possible timeframe at the least possible financial cost. Furthermore, within limitations, spreadsheet software allows coordinated plot generation and concomitant mathematical interpretation, e.g. best-fit line, of desired results. Calculated outcomes are also readily available for transfer to other software as simple text if desired.

Another benefit of spreadsheet software that the writer has come to appreciate more and more in recent years as the result of investigating many different resistance-forecasting methodologies of piles in general is the ability to see the consequences of some computational sequence on a real-time basis, with numbers literally changing before one's eyes. This has proven to be a very useful capability and diagnostic tool during the R&D stage of some analytical methodology, even if it means using a spreadsheet to perform some type of calculation such as iteration that is much more efficiently handled by some programming language such as *FORTRAN* in a purpose-written program.

Countering the positives of pursuing a solvable-by-spreadsheet approach, the writer also knew from the start that there were some negatives to this choice. The writer has used *Excel* extensively for many years now for geotechnical and foundation engineering applications, and over the past decade or so has had the opportunity to directly compare it to the more-traditional approach of a purpose-written *FORTRAN* code with which the writer is very familiar. In particular, this comparison was made for several specific geotechnical and foundation engineering applications where the same application was solved using both *Excel* and a *FORTRAN* code.

The writer's experience from this this 'apples-to-apples' type of comparison is that spreadsheet software is overall best suited for use with applications where the solution algorithm is already known and thus well-defined beforehand so that pros and cons of spreadsheet software can be utilized and addressed via upfront planning. A particular example of working efficiently with *Excel* is remembering that it performs calculations from left to right, top to bottom. This becomes critically important when performing some calculation where the final result also appears as an initially unknown variable within the calculation sequence in what Excel calls a *circular reference*. *Excel* can be configured to solve a circular reference iteratively but unless the cells are set up to perform the overall calculation in the proper sequence, an annoying 'forest' of error messages will display in every cell involved in the calculation. On the other hand, in the same scenario it is trivial to create a DO-loop in *FORTRAN* to iterate to some desired precision in the calculated results.

In addition, the writer has found that a spreadsheet is not as flexible to use when there is variation in the number of problem variables such as pieces of cone-penetrometer input data, how many artificial segments are used to model a pile, etc., from one analysis to another. Again, with a purpose-written *FORTRAN* code this is a trivial consideration.

Thus, in the writer's experience, a spreadsheet solution is much less adaptable during a research phase when changes in the solution algorithm are being investigated and made on an ongoing basis. It is often not possible to structure the cell layout of calculations in a manner that is efficient for the spreadsheet software because the organization of the solution steps is simply not known beforehand but evolves as the research progresses.

As this monograph manuscript and its associated theoretical research evolved over time, it became clear to the writer that the solvable-by-spreadsheet goal was unreasonable for an improved resistance-forecasting methodology for tapered piles. This is largely due to the fact that it became clear that the CCE model used by Kodikara and Moore, El Nagger et alia, and Manandhar et alia should not be used as it did not capture or correctly model numerous crucial elements that occur with tapered piles.

Thus, simply making some relatively modest tweaks and incremental improvements to the current state of knowledge as defined by the Kodikara-Moore Method (as the writer imagined for many years was all that was required) would not be adequate. It became clear that a completely new, much more complex cavity-mechanics model such as that used by Salgado and Prezzi (2007) and Cook (2010) would be required to simulate the separate taper benefits of driving and external load application. Such a model requires extensive iterative solution of 1,000 to 2,000 artificial 'shells' surrounding a cylindrical cavity as noted by Cook (2010) and this is for each and every artificial segment into which a tapered pile is divided. Such a complex model is not amenable to solution-by-spreadsheet in the writer's opinion.

In conclusion, the writer reached the conclusion that a purpose-written *FORTRAN* (or similar) code is required for an improved resistance-forecasting method for tapered piles, even at the current proof-of-concept phase of development. This will provide much greater computational flexibility for implementing and investigating different theoretical components of such a methodology in an experimental manner that is consistent with the theoretical complexity of the overall problem that was not initially clear based on Kodikara and Moore (1993).

However, it is important to note that once the scope of an improved analytical methodology for tapered piles is established that there is nothing that would prevent future development of alternative solution tools. First and foremost is the possibility of a spreadsheet solution, especially if simplifications of the very complex cavity-mechanics solutions presented by Salgado and Prezzi (2007) and Cook (2010) can be found. Ultimately, a purpose-written program developed by a commercial software vendor will likely prove desirable as such programs nowadays tend to have graphical user input and graphical displays of calculated outcomes that justify their cost to the end user.

This page intentionally left blank.

Chapter 8

Taper Benefit and Resistance Efficiency

8.1 PROLOGUE

Foundation design has always been, and continues to be, an exercise in finding the optimum combination of technical performance and life-cycle cost. On large projects involving deep foundations where considerable money is at stake and it is desired to refine this optimization process as much as possible, it is common to perform pre-construction test programs at the project site where exemplars of proposed DFE alternatives are installed and load tested.

Specific examples of this that involve tapered piles are the 'indicator-pile' programs that the PANYNJ conducted at JFKIA multiple times in that airport's 75-plus years of existence. This work was summarized and discussed in Horvath and Trochalides (2004) and Horvath (2014b). More recently, since late 2019 the Deep Foundations Institute (DFI) has been hosting a webinar series titled *IT$ Money (Increased Testing $aves Money)* to highlight the fact that pre-construction testing to refine design parameters for production DFEs can be cost-effective, even on relatively small projects.

Within this overall, generic framework of finding the most cost-efficient DFE that satisfies technical requirements on a project-specific basis is the issue of *taper benefit* of tapered piles. This chapter is devoted to a discussion of this specific aspect of the overall process of DFE optimization although, as will be seen, there are some suggestions of broader use for all types of DFEs.

8.2 BACKGROUND

The term 'taper benefit' was defined in Chapter 2 and subsequently used throughout this monograph up to this point to mean or imply the superior load-bearing performance of a tapered pile relative to a comparable constant-perimeter pile when subjected to axial-compressive loads in the same subsurface conditions. Stated another way, the term taper benefit implies that a tapered pile can do more (in terms of load bearing) with less (in terms of the physical amount of pile in the ground) compared to a comparable non-tapered pile.

While the concept of taper benefit is clear, up to this point the usage of this term has mostly been in a descriptive, qualitative context. There were some limited numerical comparisons made in Appendix A using K_h, β, and other parameters but these were more applicable to intra-pile comparisons between the tapered and constant-diameter sections of the partially tapered Monotube piles that were discussed in that appendix. It is desirable to quantify taper benefit in some broader, absolute manner to facilitate making the kind of objective, technical- and cost-based assessments and comparisons by stakeholders in the foundation design process as described in the prologue to this chapter.

Quantifying taper benefit explicitly implies that the load-bearing capability of the non-tapered pile against which a tapered pile is compared always has to be quantified as well in order to make the necessary comparison. Thus, the writer recognized that this presents a broader opportunity to extend the comparative concept of taper benefit to DFEs in general. There are two reasons for this:

- The global availability of non-pile DFEs that are jacked, drilled, or torqued into the ground and against which tapered piles compete in the marketplace has proliferated significantly in recent decades. Thus, it is necessary to allow for consideration of a broad range of deep-foundation alternatives in any comparative assessment that is done in practice, not just other types of (driven) piles.

- It appears that there are certain ground conditions in which non-pile (typically drilled) tapered DFEs can be economically and satisfactorily constructed. Although this monograph stated at the outset that non-pile tapered DFEs were excluded from detailed analytical consideration, such DFEs should be provided for in any quantification of taper benefit to allow for the broadest consideration in practice.

8.3 RESISTANCE EFFICIENCY

8.3.1 Definitions

With this broader focus of taper benefit in mind, the generic concept of *resistance efficiency* is hereby introduced and defined to mean the relative load-bearing efficiency of any DFE under axial loading, either in compression or tension (uplift). Thus, for every DFE there is a *Compressive Resistance Efficiency* (CRE) and a *Tensile Resistance Efficiency* (TRE). This means that the concept of taper benefit of tapered piles is just a special case or subset of CRE and that quantifying taper benefit is a natural outcome of quantifying CRE.

8.3.2 Metrics

8.3.2.1 Overview

The first issue to discuss with respect to resistance efficiency is the generic metric to use to quantify this parameter. Given that foundation cost is the overarching issue in any assessment that would make use of CRE or TRE outcomes, it would appear logical that any metric for quantifying resistance efficiency should include cost as one of its variables. However, a review of the literature indicates that this is not always the case so it is useful to review the different alternative metrics that have been used or proposed for assessing resistance efficiency. It is also useful to discuss additional alternative metrics that might be developed for this purpose.

8.3.2.2 Cost

In the writer's experience, the most meaningful metric for DFE resistance efficiency is cost per unit of axial resistance. In this case, the axial resistance is the ULS value (geotechnical or structural, whichever controls) in compression or uplift as appropriate. In U.S. practice, the typical units used for this metric would be 'U.S. dollars per kip'.

With this metric, the smaller the value, the more efficient the DFE is in providing resistance. Note, however, that the specific value is units-dependent so care needs to be exercised when comparing results obtained using this metric across different currency systems and/or systems of units. Note also that, as discussed later in this chapter, this metric can be misleading in that it does not always lead to selection of the DFE with the lowest true cost.

8.3.2.3 Volume

With tapered piles and the specific intent to quantify taper benefit in mind, Zhan et al. (2012) used the resistance-efficiency metric of applied load per unit volume of installed pile. In U.S. practice, the typical units used would be 'kips per cubic foot' and the larger the value, the more efficient the pile. Note that this is opposite that of the aforementioned cost metric.

In the writer's opinion, two modifications of this definition would be useful. First, it is appropriate to replace the vague reference to 'applied load' with the ULS value of axial resistance (again, geotechnical or structural, whichever controls) in compression or uplift as appropriate.

Second, unpublished research that was performed by the writer in 2016 during preparation of this monograph (and prior to learning about the Zhan et alia paper) went one step further and non-dimensionalized this metric by dividing the resistance per unit volume by the unit weight of water. This simplifies the ability to compare results across different systems of units.

Regardless of whether a volume-based metric for resistance efficiency is non-dimensionalized or not, it is an intuitively pleasing metric. This is because in most cases a tapered pile will have a smaller volume compared to a comparable constant-perimeter pile yet provide at least as much and typically (much) more resistance. Case-history examples of this are provided at the end of this chapter.

However, there are some significant limitations and drawbacks to using a volume-based metric for resistance efficiency:

- The important parameter of cost is not used so, strictly speaking, this metric only produces apples-to-apples comparative results if the different DFE types being compared have the same cost per unit volume installed.

- DFE volume can be misleading as there are an infinite number of combinations of DFE diameter and length that will produce the same volume. This metric does not differentiate between and among such variations. On the other hand, this fact could prove useful in practice for fine-tuning the optimum combination of toe diameter, taper angle, and pile length for a tapered pile. Different combinations with the same total volume could be analyzed to see which one produced the largest value of this metric.

- It is difficult, if not impossible, to apply this metric to DFE types that do not have a readily or easily defined volume, e.g. H-piles and helical piles. This has become an increasingly important issue as DFEs such as helical piles that do not have a traditional solid-cylindrical geometry have become increasingly common and widely used in practice.

8.3.2.4 Length

One obvious alternative to using DFE volume as a parameter in a metric for resistance efficiency is the embedded length of the DFE in which case resistance efficiency is defined as resistance (ULS as previously) as a function of unit length. This removes most of the objections raised in the preceding section concerning the use of DFE volume. The one drawback is that it is not immediately obvious how resistance per unit length, e.g. 'kips per foot' or 'kilonewtons per metre', can be non-dimensionalized for generality and universality. The usual geotechnical engineering metrics for non-dimensionalization (atmospheric pressure, p_{atm}, and unit weight of water, γ_w) do not have the correct dimensions for this application.

8.3.2.5 Other Alternatives

Given the fact that it is now understood that the axial-compressive resistance of DFEs is a complex combination of pile geometry (taper angle where relevant, toe diameter, shaft length) as well as residual loads from installation, it is possible that a more-refined metric for resistance efficiency can be developed that takes into account some or all of these factors. This is a subject that is left for future research.

8.3.3 Additional Considerations

8.3.3.1 Field Measurements vs. Calculations

Regardless of which metric is used for calculating CRE or TRE, the ULS resistance of the DFE is always a required input parameter. The accuracy of any assessment of resistance efficiency is largely dependent on the accuracy of the ULS resistance used in the calculation as the other parameters used in the metric (cost, volume, length, etc.) are essentially known beforehand.

In the vast majority of cases, this ULS resistance used is the geotechnical, not structural, resistance. Consequently, the issue of how the geotechnical ULS resistance is determined, i.e. through field measurement vs. a calculated outcome of a resistance-forecasting methodology, becomes vitally important.

Field measurements made during a pre-construction testing program offer the optimal accuracy in the form of ground-truth with the important caveat that any residual loads from installation need to be corrected for. This is due to the effect that residual loads can have on interpreting the geotechnical ULS using a displacement-based analytical methodology such as the Davisson Offset Method.

On the other hand, using forecast values for the geotechnical ULS results in the obvious near-certainty of having a wide range in accuracy. The recent (2019) and ongoing (2020) DFI *IT$ MONEY* webinar series noted at the beginning of this chapter, which consists of case-history presentations where field testing to determine the geotechnical ULS was done either before or at the beginning of the construction phase of a project to improve upon calculated geotechnical ULS forecasts made during the design phase of the project, has consistently illustrated how far from the ground-truth calculated resistance forecasts can be. This appears to be due to a combination of factors, with overly conservative assumptions concerning soil properties (a depressingly large percentage of foundation engineers still seem to live forever in a world where <u>all</u> coarse-grained soils have $\phi = 30°$) being the primary one.

The point being made here is that when calculated geotechnical ULS values from some resistance-forecasting methodology are used to quantify a resistance-efficiency metric, there is always a very real possibility (probability?) of arriving at an erroneous conclusion concerning which DFE is the most efficient to use on a given project. Thus, on any project where there are relatively significant cost implications related to DFE selection, serious consideration should always be given to performing field testing. The aforementioned DFI *IT$ MONEY* webinar series has repeatedly demonstrated the cost effectiveness of field testing, even on relatively modest-sized projects, as field testing allows for a far more accurate assessment of resistance and, as a result, resistance efficiency.

8.3.3.2 Pile Caps and Related Structural Elements

It is important to note that any CRE and/or TRE assessment metric that focuses on a DFE alone has the potential to be significantly flawed. This is because in most cases there is an intermediate structural member that functions as a load-transfer element between the heads of the DFEs and the superstructure (building, bridge, etc.) being supported by the DFEs. This intermediate structural member is typically called a *pile cap* even when the DFEs are not piles although the member name may be different for certain types of structures.

Piles caps typically have a not-insignificant cost associated with them. Thus, it is possible that a CRE or TRE assessment that focuses solely on the cost of the DFEs involved might reach the conclusion that a larger number of lower-capacity DFEs was more cost-effective than a smaller number of higher-capacity DFEs. Yet, the former alternative would almost certainly require a larger, more-expensive pile cap than the latter (although the latter might require a thicker pile cap due to punching-shear considerations of the heads of the DFEs) so that the overall cost of the foundation system wound up being greater than if the combination of fewer higher-capacity DFEs and a smaller pile cap had been selected.

Related to this is that in some cases the choice of a particular type of DFE can significantly alter the design of, or even the need for, a pile cap or related load-transfer element. For example, one of the selling points of the *Monotube* pile for decades was that in certain applications such as bridges, viaducts, and similar transportation structures, the piles could be extended well above the ground surface to the underside of the bridge, etc. superstructure. The reason was that the inherent vertical fluting of the *Monotube* pile (recall from the discussion in Chapter 3 that the *Monotube* pile is simply a repurposed decorative pole used for street lighting) was marketed as a decorative element that could be painted and incorporated into the overall bridge architecture and design.

The point being made here is that any DFE should always be viewed as but one component of a foundation <u>system</u>. Different DFEs being considered for use on a project may require or allow for different types and sizes of piles caps and related structural elements so that factors in addition to resistance efficiency of the DFE itself should always be considered in order to make a proper assessment of optimizing foundation costs.

The one exception to this general guideline is that pile-cap costs tend to be less relevant if the comparisons being made involve different types of DFEs but each with the same specified resistance. The writer has seen many projects where this was case in practice. Specifically, a foundation design drawing was issued for bidding that showed piles caps and DFE locations within those caps and simply specified that each DFE had to have some minimum specified capacity. The choice of the specific DFE to use to fulfill the specified capacity was then left up to the contractor. In such cases, the cost of the pile cap would be the same for any DFE but with one important caveat, i.e. all of the candidate DFEs need to have the same head diameter. This relates to punching shear which often governs the thickness of a pile cap. If different DFE alternatives have the same capacity but different diameters then the required minimum thickness of the pile cap may not be the same for all alternatives.

8.3.3.3 Code Minimums

Building codes typically require a minimum number of DFEs (three typically) per pile cap for overall stability reasons unless there are grade beams between pile caps to provide that function. The point being made here is that there are many cases involving lightly loaded buildings, such as single-story warehouses, where code minimums often control the number of DFEs per cap, not the column loads from the superstructure that are supported on that pile cap. In such cases, the outcomes of performing a CRE or TRE assessment to find the most cost-effective DFE alternative based on resistance considerations alone may be misleading if they are overridden by code minimums. This is the reverse of the hypothetical example used in the preceding section in that a relatively low-capacity DFE that may appear to be less efficient than a relatively high-capacity DFE may actually turn out to be the more cost-effective alternative overall when code minimums are considered.

8.3.3.4 Environmental Factors

8.3.3.4.1 Introduction

Historically, DFE selection has been based primarily on geo-structural considerations related to load-bearing capability. While there have always been other factors that influence DFE selection, e.g. concern about pile-driving vibrations adjacent to sensitive structures, in recent years new factors related to environmental considerations have become increasingly important. In the writer's opinion, these factors, either alone or together, have the potential to affect most, if not all, DFE selection in the future to the point where they could even dominate DFE selection in some cases and thus render traditional load-based metrics for CRE and TRE moot. Thus, for the sake of completeness it is desirable to at least mention these factors in this chapter.

8.3.3.4.2 Carbon Footprint

Carbon footprint is broadly defined as the total 'greenhouse' gas emissions. primarily carbon dioxide, CO_2, and methane, CH_4, related to some product or process. In the case of DFEs, it would include manufacturing, shipping, and installing the DFE.

The concept of calculating the carbon footprint of construction materials and processes has been underway for some time now. It is likely that this process will only increase in the future and may, at some point, become mandated through building codes or other government regulations. Consequently, it is not unreasonable to foresee a time when carbon footprint becomes the primary metric for DFE selection, with traditional metrics such as CRE and TRE playing only a secondary or supportive role.

8.3.3.4.3 Energy Foundations

Using the ground as a natural heat exchanger is not a new idea by any means. Geothermal energy has been exploited in various ways by humans for thousands of years. A much more recent development is what is referred to in this monograph generically as *energy foundations* but much more commonly in colloquial usage as *energy piles*. Note, however, that because the technology for energy piles developed outside of the U.S., the term "pile" here is used in the broad context noted at the beginning of this monograph to mean and include all DFEs. Thus, to avoid confusion with the narrow definition of 'pile' adopted in this monograph the term energy foundation is used here.

In simple terms, energy foundations are DFEs that incorporate heat-exchange tubing or piping within the DFE shaft. For constructability reasons, such DFEs have been and are, to date, typically drilled shafts (bored piles). Whether the technology evolves in the future to incorporate the necessary heat-exchange hardware into a preformed pile remains to be seen. However, the point being made here is that there are projects now, and will likely be even more in the future, where DFE selection is based primarily on the ability to produce an energy foundation, with load-bearing efficiency being a secondary consideration.

8.4 RESIDUAL LOADS

8.4.1 Background

Although the focus of this chapter is on load-bearing efficiency of DFEs in response to post-installation axial loads applied at the head of the DFE, this monograph has demonstrated that residual loads can be relatively significant in magnitude for tapered piles and, as a consequence of this, have a profound effect on post-driving external load application. Consequently, it appears useful to define the concept of a *Residual Load Mobilization Factor* (RLMF) that is a metric of the magnitude of residual loads relative to some other parameter that is specific to that DFE.

The purpose of the RLMF metric is that it may prove useful for correlating residual loads to the load-bearing performance of a DFE. Since the RLMF is a new concept, research will be required to first determine which definition of this metric is most useful (some potential definitions are presented and discussed in the following section) and then determine how this metric can be used in practical applications. Towards this end, evaluation of the RLMF for some actual case-history piles, both tapered and non-tapered, for which residual loads were either measured or estimated is presented with the exemplar applications at the end of this chapter.

8.4.2 Metric Definitions

It is proposed that any metric for the RLMF be based on the maximum value of the compressive force within the DFE shaft due to residual loads. This quantity is straightforward to either measure or estimate (Fellenius 2015). It then remains to be determined what this maximum residual-load force should be correlated with.

Suggestions for a correlation parameter include:

- The geotechnical ULS in axial loading. This produces a dimensionless value of the RLMF, best expressed as a percent, that would reflect the maximum axial-compressive residual-load force in the DFE as a percent (between 0% and 100%) of the ULS geotechnical resistance that the DFE produces in either compression or uplift as they case may be. This metric definition is useful because it provides explicit insight into the relative amount of the soil's ULS resistance already mobilized simply by installing the DFE in the ground.

- DFE volume. This produces a dimensioned (force per unit length cubed) value of the RLMF so the result should be non-dimensionalized for generality using the unit weight of water. The primary benefit of this metric definition for the RLMF is that it produces a result that is conceptually identical to the CRE and TRE when defined on the basis of volume as discussed previously in this chapter.

- DFE embedded length. This produces a dimensioned (force per unit length) value of the RLMF that is not readily non-dimensionalized. The primary benefit of this metric definition for the RLMF is that it produces a result that is conceptually identical to the CRE and TRE when defined on the basis of length and thus, in principle, avoids some of the potential problems with a volume-based definition that were discussed earlier in this chapter.

As with resistance efficiency, there are undoubtedly other parameters or combinations of parameters against which residual loads can be correlated to produce a metric definition for the RLMF. Exploring these possibilities is left to future research.

8.5 EXEMPLAR APPLICATIONS

8.5.1 Overview

To close out this chapter, examples of calculating and comparing resistance efficiency based on ground-truth for the geotechnical ULS obtained in static load tests are presented in the following sections for two case-history sites. Comments concerning the specific metrics used for determining resistance efficiency, i.e. volume vs. length, as well as various inferences and implications of the calculated outcomes are also presented for each site.

Both sites studied included both tapered and non-tapered piles although the bulk of the piles are tapered. And for the tapered piles, both sites contained both continuously tapered and partially tapered piles. Consequently, these examples are considered to be reasonably comprehensive but prejudiced toward the subject of taper benefit as a subset of the more general topic of resistance efficiency. This is not considered to be a negative as it is in keeping with the theme and goals of this monograph.

8.5.2 Norwegian Geotechnical Institute Test Site - Holmen Island

8.5.2.1 Background

The nominally four instrumented test piles driven and load tested in 1969 at the Norwegian Geotechnical Institute's (now-closed) 'sand' test site on Holmen Island in the Drammensfjord are discussed in detail in Appendix H. The focus of that appendix is the relatively significant residual loads that existed in each of the four piles after installation and prior to load testing, and the profound effect that these loads had on measured performance in the axial-compressive load tests that were performed.

The primary focus in this chapter is on the resistance efficiency calculated for each of these four piles although the newly defined (earlier in this chapter) parameter of Residual Load Mobilization Factor is also explored. This case history is particularly useful for these purposes for several reasons:

- All piles were essentially identical in manufacture (a slightly modified, for the purposes of instrument installation, proprietary type of precast-PCC pile called a *Brynildsen* pile) except for geometry. One of the four precast segments that were used to create these four piles had a modest continuous taper, comparable to that of a timber pile, while the other three segments had a length-wise constant perimeter.

- Because the geo-structural instrumentation was cast into the piles during manufacture, zero readings were taken prior to pile driving. Furthermore, because the piles were extracted after the load-test program was concluded, another set of zero readings was obtained once the piles were removed from the ground. This allowed the residual loads that existed after driving but prior to load testing to be measured directly. This was a relatively rare opportunity because in many cases, such as when a steel pipe pile is driven, the instrumentation is installed and zero readings taken after driving. This means that residual loads can only be estimated, a process that requires some thought and engineering judgment (Fellenius 2002a, 2002b, 2002c). Such was the case with the *Monotube* piles at JFKIA discussed by Fellenius et al. (2000) that are the focus of Appendix A.

- A tapered and non-tapered pile were each driven in relatively close proximity to the exact same depth for each of two different depths. Thus, this case history provides two different comparative cases that differ only in depth of the installed piles where a tapered and non-tapered pile can be directly compared to each other. One could not ask for a more equal, 'textbook' comparison of two piles that were identical except for geometry.

- Identical (in terms of load-application protocols) compressive and uplift load tests were performed on each of the four piles to vertical displacement levels in all cases that can reasonably be assumed to have allowed the geotechnical ULS to occur in each of the eight (total) load tests. Thus, this case history provided a very rare opportunity where both the CRE and TRE (as well as RLMF) can be calculated for the same pile.

- The authors who originally wrote about these test piles (Gregersen et al. 1973) used a consistent analytical methodology to define the geotechnical ULS in both the compression and uplift load tests. This is the parameter Q_{90} that is defined in detail in Appendix H. For consistency in the calculated outcomes presented in the following sections, the writer used these Q_{90} values for defining the geotechnical ULS in both compression and uplift. In general, these Q_{90} values are very close to the actual maximum resistances displayed in the load tests (values of both Q_{90} and the actual maximum resistances are presented in several tables in Appendix H).

- It is reasonable to assume that had this been a commercial project, the cost of manufacturing and installing each pair of comparable (tapered vs. non-tapered) piles would have been more or less the same. Thus, this would be expected to be one of those atypical cases where the volume-based metric for resistance efficiency would be directly convertible to a more useful (in practice) cost-based metric as the cost per unit volume of pile was, for all practical purposes, the same for each comparable pair of piles.

In conclusion, the 1969 Holmen Island test piles are about as ideal as a case history can be for illustrating and quantifying resistance efficiency in general and taper benefit in particular as it allows apples-to-apples comparisons in both compression and uplift. The only downsides to this case history are that the installation protocol used to create four test piles using, essentially, only two different pile locations was unorthodox, and the sand at the site was unusually loose and with a very low small-strain shear stiffness so that residual loads appeared to play a particularly significant role in behavior, under compressive load testing at least. Nevertheless, the positives outweigh the negatives by far and thus make this case history very useful for the purposes of this chapter in particular and monograph as a whole (which is why an entire appendix was devoted to a detailed assessment of these piles).

One further caveat is that the pile volumes calculated and used by the writer for the various parameters presented subsequently are approximate. As discussed in Appendix H, the four pile segments were not perfectly circular in cross-section. This is because diametrically opposed faces of all piles were somewhat flattened to accommodate the extensive geo-structural instrumentation placed inside the pile shafts during manufacturing of the pile segments. The inability to know the exact dimensions of the piles was a modest, but not fatal, limitation for the purposes of Appendix H.

However, for the purposes of this chapter, being able to calculate pile volumes was essential as pile volume is explicitly required to be known for calculation of specific versions of the RLMF, CRE, and TRE metrics. The writer accomplished this by assuming a perfectly circular cross-section so all volumes are somewhat overestimated. Because the same error applied to all piles segments and was likely relatively small to begin with, it is believed that the relative parameter values portrayed are correct.

8.5.2.2 Pile Summary

The four Holmen Island piles are discussed in Appendix H. Only the key elements relative the current discussion of resistance efficiency and residual loads are presented here for ease of reference.

In all of the following discussions, the key elements for comparison are:

- There were two piles driven to an embedment of 8 metres (26 ft)[168]. Pile A had a constant perimeter and Pile C was continuously tapered, with a taper angle, α, = 0.29° which is approximately that of timber piles used for foundation purposes, at least for circa-1970 construction in both Norway and the New York City metropolitan area.

- There were two piles driven to an embedment of 16 metres (52 ft). Pile D/A had a constant perimeter and Pile B/C was partially tapered. Specifically, the upper half of Pile B/C had a constant perimeter while the lower half was continuously tapered with a taper angle, α, = 0.29°.

Thus, the most significant comparison is between the tapered and non-tapered piles at each of the two depths. Of secondary interest is the comparison between results at the two different depths.

8.5.2.3 Residual Load Mobilization Factor (RLMF)

As stated earlier in this chapter, there are several potential ways in which to frame the metric of Residual Load Mobilization Factor, RLMF. All of the suggested definitions presented in this monograph involve use of the maximum axial-compressive force in the pile due to residual loads divided by some other parameter.

The first definition considered is dividing the maximum force due to residual loads by the geotechnical ULS measured in the compressive load test, with the result expressed as a percent. This is arguably of greatest interest as it reflects how much of the DFEs ULS axial-compressive resistance is mobilized already simply by installing the DFE in the ground. The results obtained using this definition are shown in Table 8.1. Recall that the larger the RLMF percentage, the larger the residual loads relative to the geotechnical ULS in axial compression.

---

[168] SI units are primary for this case history as this reflects the fact that metric (but non-SI) units were used in Gregersen et al. (1973).

Table 8.1. Residual Load Mobilization Factor (RLMF) - Force-Ratio Basis
Gregersen et al. (1973) *Brynildsen* Piles at Holmen Island, Norway.

Measured and Calculated Parameters	Piles			
	Intermediate length (8 m/26 ft)		Full length (16 m/52 ft)	
	A	C (tapered)	D/A	B/C (part tapered)
Maximum residual force, kN (kips)	31 (7)	73 (16)	165 (37)	218 (49)
Compressive Q_{90}, kN (kips)	263 (59)	278 (63)	482 (108)	467 (105)
RLMF (force ratio), %	12	26	34	47

The takeaways from the results shown in this table are:

- Tapered piles develop relatively larger residual loads compared to comparable constant-perimeter piles driven to the same depth. This was noted in Fellenius (2015).

- Longer piles develop relatively larger residual loads compared to comparable shorter piles in the same soil conditions. This may, at least in part, be due to the fact that pile compressibility inherently increases as pile length increases.

Next considered are two definitions for RLMF that are based on the geometry of the DFE. One definition divides the maximum force due to residual loads by the DFE volume (and then non-dimensionalizes the result by dividing by the unit weight of water) and the other divides the residual-load force by the embedded length of the pile. The results of both are shown in Table 8.2. As with the force-basis definition of RLMF, the larger the value, the larger the residual load relative to either the pile volume or length as the case may be.

Table 8.2. Residual Load Mobilization Factor (RLMF) - Pile-Geometry Basis
Gregersen et al. (1973) *Brynildsen* Piles at Holmen Island, Norway.

RLMF Definition	Piles			
	Intermediate length (8 m/26 ft)		Full length (16 m/52 ft)	
	A	C (tapered)	D/A	B/C (part tapered)
Volume/γ_w	6	21	17	26
Length, kN/m (kip/ft)	3.9 (0.27)	9.1 (0.62)	10 (0.71)	14 (0.93)

The takeaways from the results shown in this table are:

- The overall trends are the same as those observed for the force-ratio definition of the RLMF.

- The trends are more pronounced for the volume-based definition as both tapered piles have a smaller volume compared to the respective non-tapered pile of the same length.

8.5.2.4 Compressive Resistance Efficiency (CRE) and Taper Benefit

Cost data were not available for the Holmen Island piles so Compressive Resistance Efficiency, CRE, was only calculated based on the pile geometry. Specifically, the geotechnical ULS force (defined here using the Q_{90} values determined by Gregersen et al. (1973)) were divided by the pile volume (with the calculated outcome divided by the unit weight of water for non-dimensionalization) in one case and the pile length in the other case. The results are shown in Table 8.3. Recall that for both the volume and length basis, the larger the CRE value, the more efficient the pile in axial-compressive load bearing.

Table 8.3. Compressive Resistance Efficiency (CRE) - Pile-Geometry Basis
Gregersen et al. (1973) *Brynildsen* Piles at Holmen Island, Norway.

Measured and Calculated Parameters	Piles			
	Intermediate length (8 m/26 ft)		Full length (16 m/52 ft)	
	A	C (tapered)	D/A	B/C (part tapered)
Compressive Q_{90}, kN (kips)	263 (59)	278 (63)	482 (108)	467 (105)
CRE (volume/γ_w basis)	54	78	50	56
CRE (length basis), kN/m (kips/ft)	33 (2.3)	35 (2.4)	30 (2.1)	29 (2.0)

The takeaways from the results shown in this table are:

- Overall, a volume-based definition of CRE appears to clearly distinguish between the performance of tapered and non-tapered piles while a length-based definition of CRE is of much less value in this regard.

- CRE with a volume-based definition does provide a useful metric of taper benefit. In this case, continuously tapered Pile C provided 44% more resistance per unit volume than the otherwise comparable constant-perimeter Pile A. For the partially tapered Pile B/C, the taper benefit relative to constant-perimeter Pile D/A was less dramatic, 12% more resistance per unit volume.

- The two constant-perimeter piles (A and D/A) exhibited similar resistances per unit volume. On the other hand, the continuously tapered Pile C exhibited substantially more resistance per unit volume compared to partially tapered Pile B/C as might be expected.

While these results are interesting given the dearth of similar apples-to-apples comparative case histories in the published literature, the writer cautions about reading too much into them. To begin with, the sand at the Holmen Island site was in a very loose consistency and the geotechnical ULS resistances of the piles were relatively small in absolute magnitude. Furthermore, the pile taper was quite modest, essentially the same as a timber pile and considerably less than the taper that would be used for a *Monotube* or *TAPERTUBE* pile. Nevertheless, given the dearth of case histories involving instrumented taper piles, the Holmen Island case history is of considerable value.

8.5.2.5 Tensile Resistance Efficiency (TRE)

The same volume and length definitions used for calculating the CRE values shown in Table 8.3 were used for calculating the Tensile Resistance Efficiency (TRE) based on the uplift load tests that were performed on the Holmen Island test piles. The results are shown in Table 8.4. Recall that as with CRE, the larger the TRE value, the more efficient the pile in resisting uplift loads.

Table 8.4. Tensile Resistance Efficiency (TRE) - Pile-Geometry Basis
Gregersen et al. (1973) *Brynildsen* Piles at Holmen Island, Norway.

Measured and Calculated Parameters	Piles			
	Intermediate length (8 m/26 ft)		Full length (16 m/52 ft)	
	A	C (tapered)	D/A	B/C (part tapered)
Uplift Q_{90}, kN (kips)	93 (21)	120 (27)	265 (60)	239 (54)
TRE (volume/γ_w basis)	19	34	27	29
TRE (length basis), kN/m (kips/ft)	12 (0.80)	15 (1.0)	17 (1.1)	15 (1.0)

The takeaways from the results shown in this table are:

- First and foremost, and independent of the TRE calculations, these results should now and forever more dispel the notion that tapered piles offer inferior resistance to uplift loads compared to non-tapered piles. The shorter tapered pile (Pile C) actually had almost one-third <u>more</u> uplift resistance than the comparable constant-perimeter pile (Pile A). The longer tapered pile (Pile B/C) had an uplift resistance only 10% less than that of the comparable constant-perimeter pile (Pile D/A).

- As with CRE, a volume-based definition of TRE appears to distinguish clearly between the performance of tapered and non-tapered piles while a length-based definition of TRE is of much less value in this regard.

- As with CRE, TRE with a volume-based definition provides a useful metric for the uplift resistance of tapered piles relative to comparable non-tapered piles. In this case, tapered piles were as good or better than constant-perimeter piles on a volume-for-volume basis.

In conclusion, uplift tests on tapered piles are essentially non-existent in the published literature so it is impossible to tell to what extent the performance of the Holmen Island test piles shown in Table 8.4 reflects the norm of tapered vs. non-tapered pile behavior in uplift loading, especially in consideration of the fact that all four piles were loaded to the geotechnical ULS in compression before the uplift load tests were performed. Nevertheless, the Holmen Island test-pile results provide seminal ground-truth that the current widely held view that tapered piles have significantly diminished uplift-load resistance relative to comparable non-tapered piles is incorrect. The hypothesis that a tapered pile will simply lift out of the tapered hole in the ground that it created when it was driven is simplistic to the point of being grossly unrealistic. The reality is that the wedging mechanism that occurs during the driving of a tapered pile into the ground creates a radial stress state around the pile that is substantially <u>larger</u> in magnitude than that of an otherwise comparable constant-perimeter pile. Thus, when post-installation uplift loading is applied, a tapered pile is starting from a position of <u>greater</u> radial stresses and thus <u>greater</u> axial resistance acting along the shaft-soil interface when compared to a comparable constant-perimeter pile. Consequently, even if these radial stresses reduce somewhat in magnitude due to cylindrical-cavity contraction (CCC) as uplift loads are applied, they are still greater than those acting on an otherwise comparable constant-perimeter pile.

8.5.3 John F. Kennedy International Airport - New York City

8.5.3.1 Background

Tapered piles have been the DFE of choice at JFKIA since initial construction of the airport in the 1940s. As stated in the Preface, it was the writer's early-career involvement with tapered piles at JFKIA almost 50 years ago that can truly be said to be the genesis for this monograph.

Because of the writer's career-long interest in tapered piles, previous publications (Horvath 2002, 2014b; Horvath and Trochalides 2004) adequately cover the temporal evolution of pile selection and design at JFKIA so this history is not repeated here. The paper by Horvath and Trochalides (2004) is particularly relevant to the present discussion as it puts into context how the different groups of piles used for the purposes of this chapter and that are summarized in the following section came to be used. Previous publications (Horvath 2002, 2014b, 2015, 2016a, 2016b) also cover numerous resistance-forecasting issues related to piles at JFKIA that are also not repeated here.

The focus of the presentation in this chapter is the Compressive Resistance Efficiency, CRE. There is also some minor discussion of the Residual Load Mobilization Factor, RLMF, as there were two instrumented load tests on tapered piles at JFKIA that can be assessed for this parameter.

In all cases when computing CRE and RLMF for the current discussion, the maximum applied load achieved in a static load test was used in calculations. While in the vast majority of cases the magnitude of pile-head settlement was sufficiently large so that the geotechnical ULS could be considered to at least be approached if not reached, there are a few cases where this was not the case. Such cases are noted in any tabular or graphical presentation of calculated results.

The JFKIA piles used for the present discussion were driven and load tested over a period of time that spanned approximately 40 years as opposed to being a single test program as with the Holmen Island piles discussed previously. Thus, the nature of calculated outcomes that are presented for the JFKIA piles differ substantially from those of the Holmen Island piles. The JFKIA piles cover a much wider range in pile volumes as well as a range in taper angles, α, that span from zero, i.e. constant-diameter piles, to 0.95° which was, until relatively recently, the maximum taper angle routinely used for manufactured tapered piles in the U.S. Thus, the presented outcomes are not comparisons of pairs of piles as with the Holmen Island piles but reflect trends of CRE as functions of both pile volume as well as taper angle in a very uniform geological setting.

8.5.3.2 Pile Summary

The writer had access to a relatively large database of piles driven at JFKIA between 1972 and 2013 that were designed, overseen, and inspected by a diverse group of professional engineers and technicians from both the PANYNJ that operates JFKIA (the airport property is owned by the City of New York) as well as several different consulting engineering firms. In selecting piles from this group to include in the analyses performed for this chapter, the writer considered the following to be essential criteria:

- A pile had to be driven with conventional impact hammers for essentially its entire embedded length. Within the last few years, there has been a tendency to use a vibratory hammer for approximately two-thirds of the final embedded pile length and then drive the pile the remaining one-third using a conventional impact hammer. To the best of the writer's knowledge, the effect of this mixed installation protocol has never been investigated as to how the resulting pile resistance compares to that of piles that are impact-driven for their entire embedded length.

- Where applicable, a steel pipe pile had to have been concreted prior to load testing. Again, in recent years there has been a trend to sometimes load testing steel pipe piles prior to concreting. In essence, this tests a pile that does not replicate production piles.

- A pile had to have been subjected to conventional static load testing, preferably to sufficient pile-head settlement so that the geotechnical ULS can reasonably be assumed to have been at least approached if not reached. The PANYNJ pioneered the use of dynamic testing of (driven) piles in the New York City metropolitan area in 1972 and such testing has been done routinely at JFKIA ever since. While dynamic testing is a useful Construction Quality Assurance (CQA) tool, it does not match and certainly does not replace the accuracy of static load tests. As discussed earlier in this monograph, Fellenius (2002a) has illustrated that dynamic testing does not completely account for the effect of residual loads in piles which is potentially problematic for tapered piles that constitute the bulk of the database being presented in this chapter.

- Load measurements in static load tests had to have been made using an electronic load cell. Again, as discussed earlier in this monograph, load tests that rely solely on calibrations with jack pressure can be seriously in error with respect to actual loads delivered to the pile head. Unfortunately, there continue to be cases where static load tests that depend solely on jack-pressure calibrations are allowed by owners or their professional engineering representatives and performed by professional engineering testing firms.

With these selection criteria in mind, there are four groups of JFKIA piles driven between 1972 and 1999[169] that were considered for the CRE assessments presented subsequently:

1. A small group (four total) of <u>constant-diameter, closed-toe steel pipe piles</u> that were driven and load tested between 1988 and 1990 as part of the *JFK 2000* test-pile program (Horvath and Trochalides 2004). Although limited in number, they are representative of, and provide baseline data for, $\alpha = 0°$ piles against which the much more numerous tapered piles can be compared.

2. <u>Timber piles</u> driven as part of the 1972-1973 IAB-STRAP test-pile program as well as the aforementioned *JFK 2000* test-pile program (Horvath and Trochalides 2004). There was

---

[169] It is of interest to note that all piles load-tested in the 21st century for which the writer had access to data failed one or more of these four, essential selection criteria.

a total of seven piles that met the writer's aforementioned essential criteria for use. Collectively, these are representative of non-uniformly continuously tapered piles with average taper angles, α_{avg}, in the range of 0.16° to 0.33°. As noted in Chapter 3, such piles typically have a variation in α of $\pm\frac{1}{3}$ α_{avg} along the pile shaft, with minimum taper near the head and maximum taper near the toe. This is efficient as it always put the taper where it provides the greatest benefit and, if the pile is cut off after driving, the part that is removed has the least taper.

3. _Monotube_ piles with the intermediate Type J taper (α = 0.57°). The 10 piles used are representative of partially tapered piles and, for the most part, were driven and load tested between 1988 and 1990 as part of the aforementioned _JFK 2000_ test-pile program (Horvath and Trochalides 2004). In addition, the two piles of this type that were instrumented and load tested in 1994 as reported by Fellenius et al. (2000) and discussed in detail in Appendix A are included in this group.

4. _TAPERTUBE_ piles that were originally developed toward the end of the 20th century, specifically with the subsurface conditions and foundation needs of JFKIA in mind although this type of pile is certainly applicable in a wide range of terrestrial and shallow-water marine applications worldwide. These are another representative example of partially tapered piles. The six piles included here were all driven and load tested between 1998 and 2000 as part of the construction of the _AirTrain_ transit system that was being built at the time, as well as for some passenger-terminal construction within the Central Terminal Area (CTA). Note that there were many more _TAPERTUBE_ piles used at JFKIA after 2000 and up to the present as this has evolved to be the current pile of choice to use at JFKIA. Unfortunately and as noted earlier in this section, piles driven in recent years and for which the writer had access to records did not meet the writer's aforementioned selection criteria for one reason or another.

While there is some variability in pile lengths within this diverse group of piles that were installed over a period of almost 30 years, sufficient for the purposes of the present discussion is that the piles in this database fall into two broad categories depth-wise. Specifically, the timber piles were typically driven to an embedded length of approximately 30 feet (10 m) while most of the tapered and non-tapered steel pipe piles were driven to approximately twice that embedment, i.e. 60 feet (20 m).

As a final comment concerning the database of piles used for presentation in the following sections, the single largest, collective uncertainty is the length of time between driving and load testing. As discussed in Chapter 4, it is now well-known that there are temporal effects on the axial-compressive resistance of piles, especially tapered piles, in coarse-grained soil. In fact, it was work done and knowledge gained as part of the aforementioned _JFK 2000_ load-test program at JFKIA between 1988 and 1990 that ultimately led to one of the seminal technical papers on this subject by York et al. (1994).

Unfortunately, the time between driving and load testing is not known for the vast majority of piles included in the writer's database. Thus, there is a possibility, if not a probability, that some of the Compressive Resistance Efficiencies (CREs) discussed subsequently are too small in magnitude as they underestimate the axial-compressive resistance that the piles could have developed had there been even a modest increase in time, of the order of a few weeks, between driving and load testing. However, this uncertainty does not negate the overall benefit to be derived from the calculated outcomes that are presented.

8.5.3.3 Residual Load Mobilization Factor (RLMF)

Of the 27 JFKIA piles included in the writer's database for use in this chapter, only the two *Monotube* piles reported in Fellenius et al. (2000) were fully instrumented and thus able to be assessed for post-driving residual loads. Of the two, only one (Pile #2) appears to have been on the cusp of the geotechnical ULS when the test was terminated. The other (Pile #3) was likely approaching the beginning of the geotechnical ULS but had not reached the same relative level as Pile #2.

The calculated RLMF results for these two piles are shown in a single table (Table 8.5[170]) that includes results using both the force-ratio basis as well as the two pile-geometry bases (non-dimensionalized pile volume and embedded length of pile) that were used earlier in this chapter for the Holmen Island test piles. Note that the embedded-length definition was included for the sake of completeness even though the results from the Holmen Island test piles suggest that DFE length does not provide insightful results for the RLMF (or CRE or TRE for that matter).

Table 8.5. Residual Load Mobilization Factor (RLMF) - Force-Ratio and Pile-Geometry Bases Fellenius et al. (2000) *Monotube* Piles at JFKIA.

Measured and Calculated Parameters	Piles	
	#2	#3
Maximum residual force, kN (kips)	585 (131)	700 (157)
Maximum applied compressive load, kN (kips)	2000 (449)	2500 (562)
RLMF (force ratio), %	29	28
RLMF (volume/γ_w)	119	145
RLMF (length), kN/m (kip/ft)	29 (2.0)	34 (2.3)

The primary takeaway from the results shown in this table is that although there are significant differences between the Holmen Island and JFKIA sites in terms of both the gradation and consistency of the sandy soils as well as the piles, the RLMF calculated on a force-ratio basis is not significantly different between the two sites (refer to Table 8.1 for the Holmen Island site where the RLMF for the two tapered piles varied between 26-47%). This suggests that the maximum axial-compressive forces generated within the pile shaft as a result of residual loads are a relatively significant portion of the geotechnical ULS, at least for tapered piles. This was noted by Fellenius some years ago (Fellenius 2002a).

8.5.3.4 Compressive Resistance Efficiency (CRE) and Taper Benefit

Because of the number (27) of piles involved, the calculated CREs for the JFKIA piles do not lend themselves to tabular presentation. Rather, two forms of graphical presentation are used instead.

Note that in each of the figures presented subsequently, only the CRE calculated on a non-dimensionalized volume basis is shown. The results presented previously for the Holmen Island test piles suggests that a length-based definition does not produce meaningful results in terms of distinguishing one pile from another. This would likely be the case at JFKIA

---

[170] SI units are primary for this case history as these units were primary in Fellenius et al. (2000).

where there are relatively significant variations in pile volume (as will be seen) but relatively minor variations in pile length (with the exception of timber piles).

To begin with, Figure 8.1 shows the CRE as defined using non-dimensionalized pile volume as a function of pile volume. Note that different symbols are used for the four categories of piles described previously. Also, for a given type of pile and where appropriate, different symbols distinguish between piles that the writer felt displayed behavior consistent with reaching the geotechnical ULS in the static load test versus piles that did not reach that level but were likely close to doing so.

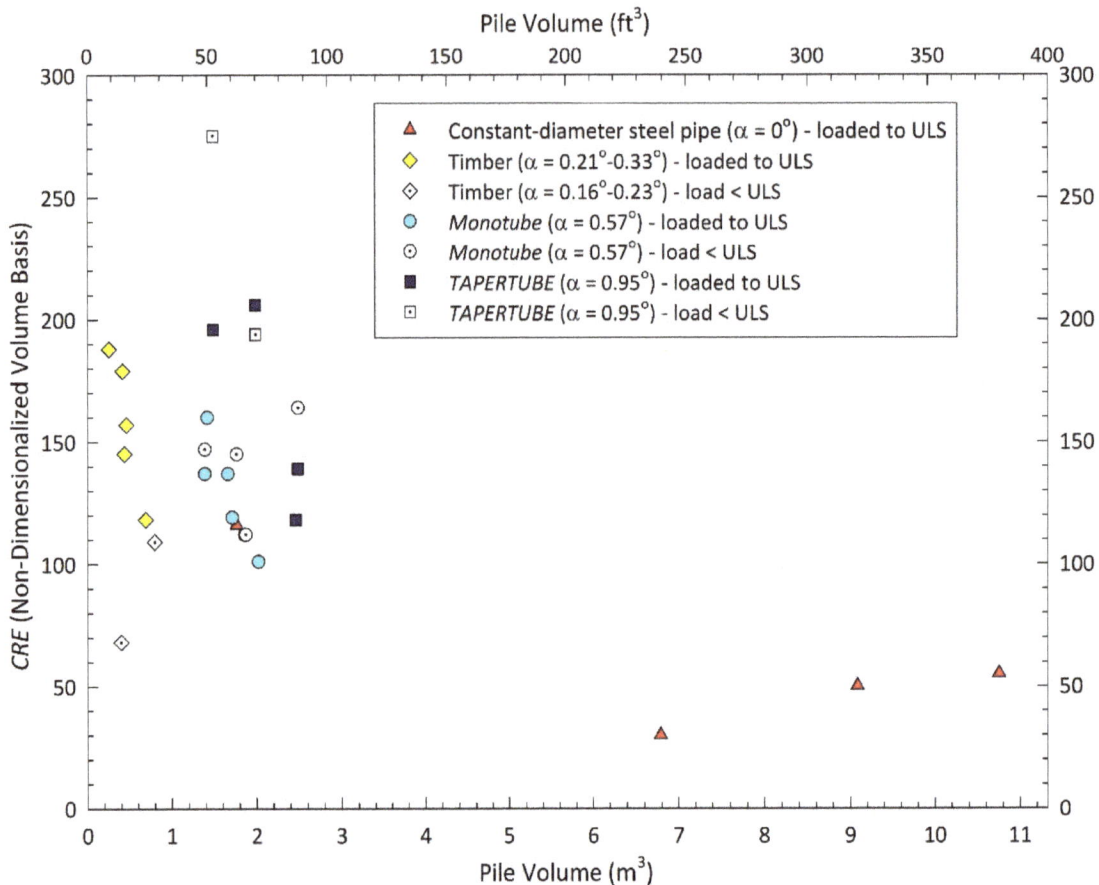

Figure 8.1. Compressive Resistance Efficiency (CRE) vs. Pile Volume - JFKIA Piles.

The primary takeaway from the results shown in this figure is that there is a clear taper benefit for all three types of tapered piles relative to constant-diameter piles although the taper benefit is quite variable and thus exhibits a lot of scatter. This is due, in part, to the relatively wide variation in taper angles for the tapered piles as well as the fact that some tapered piles were continuously tapered while others were partially tapered.

In an effort to remove the scatter due to taper angle, these data were replotted as shown in Figure 8.2, with the same CRE data plotted as a function of taper angle, α. Taper benefit is both more apparent and more clearly defined in this figure. Even though for a given taper angle there is still a wide variation in CRE, there is no mistaking the overall trend of increasing CRE with increasing taper angle.

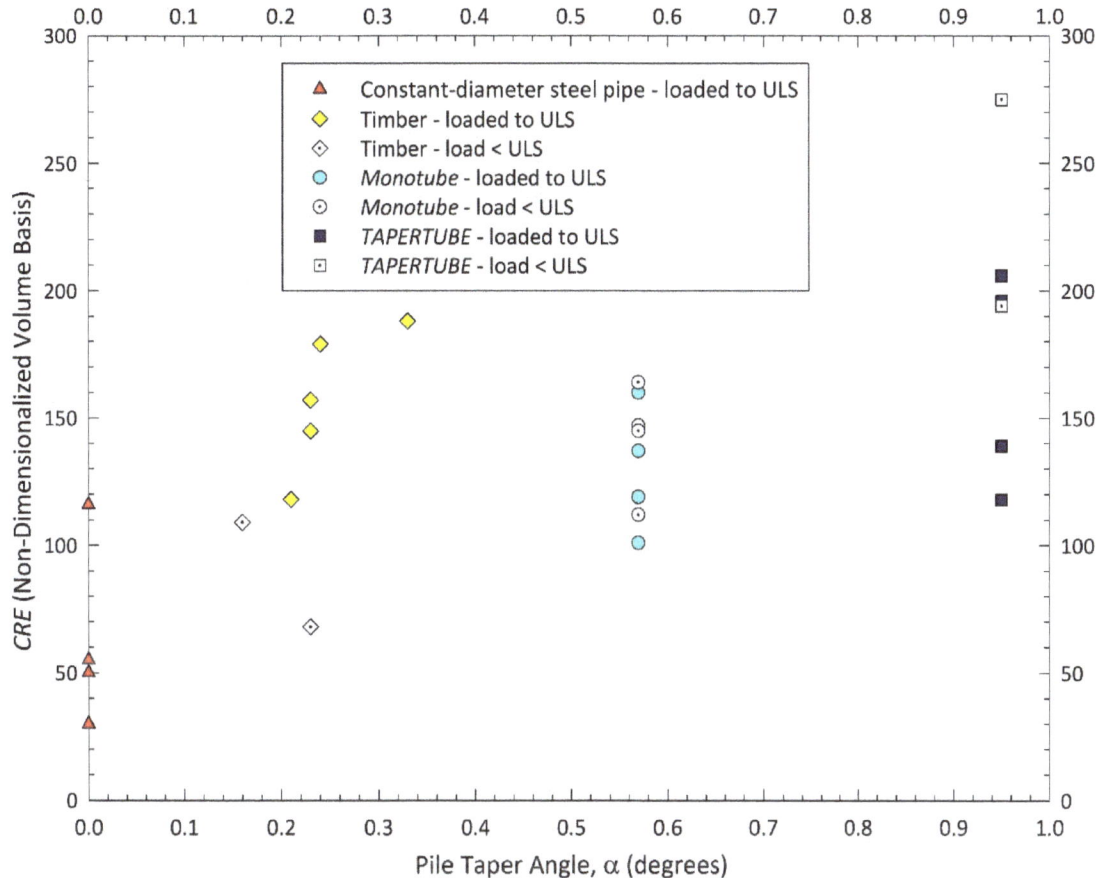

Figure 8.2. Compressive Resistance Efficiency (CRE) vs. Taper Angle - JFKIA Piles.

Although not shown in this figure, the writer used the statistical functions built in to *Excel* to generate best-fit straight lines for the trend of CRE as a function of taper angle, for both the entire data sample as well as only for those data that are for piles that were interpreted by the writer to have reached the geotechnical ULS. The *Coefficient of Determination, R^2*, for the entire data sample was 0.42 and for ULS-only data sample it was 0.31. Neither value is particularly high but they do indicate modest statistical correlation between CRE and taper angle that supports what is, in the writer's opinion, visually apparent in Figure 8.2.

There are two takeaways from Figures 8.1 and 8.2 that the writer found particularly interesting. One is that the lowly, mundane timber pile that humans have used for over 7000 years exhibited CRE values that were comparable to the *Monotube* and *TAPERTUBE* piles. Of course, this is deceiving as the values of ULS resistance of the timber piles were only a fraction (20-25%) of those of the manufactured piles (*Monotube* and *TAPERTUBE*). Consequently, in any practical application many more timber piles (four to five times as many), with the concomitant much larger pile cap, would be required to support a given column load from a building. In addition, there is the fact that these timber piles would bear within a zone of soil that is highly likely to liquefy during a regionally significant earthquake. But it does illustrate the enduring appeal of timber piles in those regions where they are economically available and thus viable.

The other takeaway of particular note is that even a little taper goes a long way toward improving resistance efficiency to axial-compressive loads. This is particularly apparent in Figure 8.2 where there is a noticeable increase in CRE going from the constant-diameter pipe piles to the timber piles. This is consistent with the Holmen Island, Norway case history discussed earlier in this chapter where the precast-PCC piles only had a taper angle that was essentially that of a timber pile yet these piles exhibited resistance efficiency in both compression and uplift that was either superior to or at least comparable to otherwise identical constant-perimeter piles.

It is interesting to note that this observation that a little goes a long way when it comes to piles taper is in complete consonance with quantitative implications of the Nordlund Method which, despite the relatively crude 2-D physical model that Nordlund used for the wedging mechanism of tapered piles, is presumed to reflect decades of real-world experience with tapered piles by Nordlund and his employer, Raymond. The specific aspect of the Nordlund Method of relevance here is the variation in the Nordlund earth pressure coefficient, K_δ, as a function of taper angle (which Nordlund called ω and is used here in lieu of α to be consistent with figures in Nordlund's published works). For a reasonable value of the soil friction angle, ϕ (which will typically be between 35° and 40°), K_δ increases rapidly from its baseline ($\omega = 0°$) value as ω increases in magnitude. Specifically, as can be seen in the appropriate figures in either Nordlund (1963) or Nordlund (1979), even for a typical timber-pile taper angle of the order of 0.25° (average) there is an increase in the value of K_δ of the order of two to four times its baseline ($\omega = 0°$) value. It is of interest to note in Figure 8.2 that the increase in CRE going from the constant-diameter pipe piles to the timber piles is also of the order of two to four times.

Chapter 9

Denouement

9.1 INTRODUCTION AND BACKGROUND

In some ways, research for this monograph began in June 1972 when the writer was first exposed to tapered piles and related, then-novel foundation engineering tools such as the 1-D wave equation and on-pile dynamic testing at the John F. Kennedy International Airport in New York City. Since then, there have been a number of benchmark developments in the writer's professional career that sustained and advanced the writer's interest in tapered piles. None was more prominent and game-changing than the publication of Kodikara and Moore (1993) that, for the first time, connected the theory of cylindrical-cavity mechanics to the reality of the wedging mechanism that occurs when tapered piles are driven into the ground and then loaded in axial compression.

With literally decades of first-hand experience, thought, and reflection relative to tapered piles in hand, when the writer began final development of this monograph in the latter part of 2018 there was a very clear vision and concomitant plan of how this publication effort would evolve in terms of additional research, technical content, and conclusions. As sometimes happens in academic research, the reality of what eventually results when a line of research is pursued to its logical conclusion is quite different from what was imagined when the research began. This monograph is one of those cases.

In summary, what started out to be an effort to make some theoretical improvements to the Kodikara-Moore Method beyond those made earlier by El Naggar et alia and Manandhar et alia and produce a practice-oriented resistance-forecasting methodology for tapered piles to replace the theoretically outdated Nordlund Method did not end up that way. As the writer's research and writing progressed, it became clear that while Kodikara and Moore correctly linked cylindrical-cavity mechanics to tapered piles they did so in a way that was significantly and fatally incomplete and, as a result, produced what the writer believes are incorrect perceptions concerning the behavior of tapered piles. Unfortunately, subsequent research by El Naggar et alia and Manandhar et alia only propagated these errors.

Specifically, Kodikara and Moore (and El Naggar et alia and Manandhar et alia who simply copied the key theoretical elements of Kodikara and Moore's work) assumed up front that all taper benefit derives from the cylindrical-cavity expansion (CCE) that occurs under only a portion of post-installation externally applied compressive loading (Phases II and III of the Kodikara-Moore behavioral model as discussed in Appendix C). While this may be broadly correct for the tapered drilled shaft that Kodikara and Moore used as their sole real-world exemplar of a tapered 'pile' in a 'cohesive' soil, it is far from correct for tapered (driven) piles in coarse-grained soils that are much more widely used as well as the focus of this monograph.

The conclusion reached by the writer is that significant taper benefit for tapered piles that are installed using conventional impact driving derives from the cylindrical-cavity creation (CCX) that occurs as the result of driving the pile into the ground. Because of the geometry of a tapered pile, this cavity creation is hypothesized to be different (in a manner still to be determined by future research) than the cavity creation that occurs when a constant-diameter pile is driven into the ground, at least in part due to the fact that driving a tapered pile continues to expand on the cavity created by the pile toe whereas with a constant-perimeter pile this additional cavity expansion does not occur. Stated another way, there is an <u>enhanced</u> cylindrical-cavity creation that occurs with a tapered pile that does not occur when a constant-diameter pile is driven. This enhanced cylindrical-cavity creation consists of the initial opening of a cavity by the pile toe followed by additional radial expansion of the cavity so can be called *CCXE* to indicate the additional expansion that follows the basic cavity-creation (CCX) associated with all pile toes. This enhanced behavior that is unique to tapered piles results in larger radial stresses acting along the tapered portion of the shaft at least.

While there is additional cylindrical-cavity <u>expansion</u> (CCE) that occurs under post-installation axial-compressive external loading, this occurs with significant interface shear stress between the pile shaft and adjacent ground. Such shear stresses are ignored with the cavity-expansion solution used by Kodikara and Moore as the solution they used was originally developed for applications involving nominally static geomechanics applications such as the pressuremeter test. Furthermore, because of the relatively significant residual loads that develop when driving a tapered pile into coarse-grained soil, there are initial, non-zero values to these interface shear stresses and they can have either an upward (resistance) or downward (drag load) sense depending on the location along the pile shaft.

As a result of the fact that, upon close evaluation, the Kodikara-Moore Method did not turn out to be a usable starting point, this monograph has not, unfortunately, yielded the desired 'new and improved' resistance-forecasting methodology for tapered piles as was envisaged by the writer in 2018. What it has produced in 2020 are:

- Guidelines for what a site- and project-specific site-characterization program needs to produce in terms of soil properties.

- A much clearer understanding of what occurs when tapered piles are driven and then subjected to externally applied axial loads, both compressive and tensile (uplift).

- A schema for future research that will hopefully lead to development of an improved resistance-forecasting methodology for tapered piles.

- Guidance as to which resistance-forecasting methodologies to use in the interim and how to use them properly.

While all of these issues have been addressed individually and incrementally throughout different chapters and appendices of this monograph, the purpose of this closing chapter is to summarize them in a single, unified, coherent manner in the literary form of a denouement. This is so that the current state of knowledge related to the response of tapered piles to axial loading can be seen as a problem that has several distinct theoretical aspects or threads that actually weave together into a single outcome. Because many of these threads are incompletely or poorly understood at this time, this chapter is also intended to suggest an R&D path forward to extend, expand, and improve on the current state of knowledge, with the ultimate goal of one day producing an improved resistance-forecasting methodology for tapered piles that will almost certainly have use with other types of DFEs.

9.2 OVERVIEW

The remainder of this chapter is divided into sections that address the following topics that are presented in a logical order so that one section logically segues into the next:

- Essential soil properties that need to be produced on a site- and project-specific basis for use with resistance-forecasting methodologies, and the testing tools that can produce them.

- The physical mechanism of what occurs during the unique, enhanced cavity creation (CCXE) caused by driving of a tapered pile. This involves Stage I of the shaft-resistance mechanism of wedging and concomitant taper benefit that is unique to tapered <u>driven</u> piles and <u>not</u> tapered DFEs in general.

- Post-driving residual loads and their distribution.

- The physical mechanism of what occurs during the cavity expansion (CCE) or contraction (CCC) when post-driving axial loads (both compressive and uplift loads are addressed) are applied at the head of the pile. This involves Stage II of the shaft-resistance mechanism of wedging and concomitant taper benefit under compressive loading that is applicable to <u>all</u> tapered DFEs (and is the <u>only</u> taper benefit for non-driven tapered DFEs).

- Resistance forecasting using both static and dynamic (wave equation) approaches.

- Resistance verification as a Construction Quality Assurance (CQA) tool both during pile driving using dynamic measurements and after pile installation using various types of load tests.

- Quantification of taper benefit.

Although the material presented focuses on (driven) piles as has been the focus throughout this monograph, there is some discussion about tapered DFEs in general that could be installed or created by means and methods other than traditional impact driving. This is done in order to illustrate how critical the installation process is to affecting taper benefit and to thereby emphasize that observations made using one type of tapered DFE do not necessarily carry over to another type of DFE. For example, Kodikara and Moore (1993) used measured results from a tapered DFE installed by drilling and concreting in place (essentially a drilled shaft/bored pile) to verify the veracity of their resistance-forecasting methodology. In the writer's opinion, this correlation is completely irrelevant to the much more common tapered DFE that is preformed and installed by conventional impact driving.

9.3 SITE CHARACTERIZATION

9.3.1 Overview

The soil properties and related parameters that are required for a resistance-forecasting methodology for tapered DFEs in general will obviously depend on the specifics of the analytical methodology used and the site-specific nature of the soils. The opinions expressed here are based on the assumptions that:

- Any resistance-forecasting methodology will be an indirect or rational method in the context defined in Chapter 5 and based on cavity mechanics.

- Coarse-grained soils are predominant.

- Traditional borings with the Standard Penetration Test will be performed to, as a minimum, provide physical soil samples for traditional soil identification using traditional visual-manual procedures and index-property testing.

Consequently, the opinions expressed in this chapter will need to at least be re-evaluated, if not revised, in the future if any of these assumptions change.

9.3.2 Soil Properties and Parameters

The soil properties and parameters that are required for solutions involving cavity mechanics are depth-wise variations of the following:

- K_o
- G_{max}
- I_{rr}
- ϕ_{peak}
- ϕ_{cv}
- soil consistency.

Historically, the final item would be defined using relative density, D_r. However, there is increasing evidence that the state parameter, ψ, is a more-reliable metric for the consistency of coarse-grained soils so both D_r and ψ should be determined.

9.3.3 Investigative Tools

Given the current state of knowledge, in-situ testing is the preferred approach over laboratory testing for the desired outcomes in this case (some of the soil properties listed above are, for all intents and purposes, impossible to measure in the laboratory on test specimens obtained from undisturbed sampling). The in-situ tool of choice for providing (through empirical correlations in the published literature) all of the requisite soil properties and parameters is the sCPTu although in practice it is satisfactory to use a combination of sCPTu and CPTu soundings for overall efficiency and economy.

In addition, it is assumed that one or more traditional borings with SPT sampling are performed to provide physical samples so that the site-specific soils can be correlated with the Normalized Soil Behavior Type (SBTn) inferred from the cone-penetrometer soundings.

9.3.4 Research Needs

The success of using the cone penetrometer as an in-situ testing tool for soil properties and parameters lies entirely with the empirical correlations between the basic parameters measured in a CPTu or sCPTu sounding (q_c, f_s, u_2) and the desired soil properties and parameters. There has been substantial advancement in this regard over the past several decades but there is always room for continued improvement. This is because the nature of empirical correlations is that they are statistical assessments of some database of measured values. Because there is no limit to how large such databases can or should be, it stands to reason that the potential for improving upon empirical correlations is similarly without limits. Therefore, there is, in principle, always room for improvement for CPTu-/sCPTu-related empirical correlations.

9.4 PILE INSTALLATION

9.4.1 Overview

There are several distinct aspects of pile installation that are addressed individually in this chapter for clarity and focus in the presentation. However, they are, in reality, all inter-related and part of, or directly result from, the overall pile-installation process. Consequently, they are each an essential part of the whole:

- Geomechanics of tapered-pile driving.

- Residual loads.

- Wave mechanics as an analytical tool.

9.4.2 Geomechanics of Tapered-Pile Driving

9.4.2.1 Hypothesized Geo-Mechanism

There has been substantial research into cavity mechanics for the better part of a century now. Salgado and Prezzi (2007) trace cavity-expansion research back to the 1940s when the focus was on indentation of metals.

Within the context of cavity mechanics, it is visually obvious that pile installation using conventional impact driving is a problem in cylindrical-cavity <u>creation</u> (CCX), i.e. a cavity is created in the ground where none existed previously. Research into the geotechnical applications of CCX has focused primarily on modeling what happens when a cone penetrometer is pushed into the ground. However, the extension to pile driving is intuitively obvious and has been noted by researchers given the geometric similarity between the standard cone penetrometer and a pile with a circular cross-section and constant diameter.

The relevance of CCX mechanics to tapered piles, and, indeed, to driven or jacked piles in general, is that CCX solutions offer the potential for forecasting the radial stresses, arguably best expressed using the K_h/K_o ratio concept that has been used throughout this monograph, that exist after pile installation. Being able to forecast these radial stresses is essential for developing a new, more advanced resistance-forecasting methodology for tapered piles. This is because it is now understood that taper benefit is the synergistic result of two distinct stages or components, with Stage I taper benefit being defined by the stress state that exists around a tapered pile after driving.

It is the writer's opinion that this extension of CCX theory to pile driving is not completely correct as usually imagined or stated. CCX theory, such as developed and presented by Salgado and Prezzi (2007), is clearly based on the implied assumption that whatever right-circular cylindrical object is creating the cavity in the ground is doing so in a manner of continuous, monotonic, quasi-static advancement.

Unfortunately, this is not what happens when a pile is driven into the ground by impact. It has been well-known for over a century that:

- Pile driving is far from being a quasi-static process. Ground resistance to an advancing pile has a dynamic component that is velocity-dependent in nature. This dynamic component of resistance must be factored out in order to correctly estimate the static component of resistance. This is conceptually identical to what has been done for some time now when using quasi-static load tests as discussed in Chapter 4.

- Pile driving does not produce monotonic advancement of a pile. Rather, pile driving can be likened to a 'two steps forward, one step backward' process in which there is always some partial rebound or upward retrenchment of the pile after advancing deeper into the ground under each hammer blow.

Independent of and in addition to these issues, all CCX research to date has focused on creation of a cylinder with a constant diameter. This is clearly not the case with tapered piles of conventional construction that are either continuously tapered or partially tapered with the tapered section placed just behind the pile toe. As a consequence of taper, it is hypothesized by the writer that the cavity-creation mechanism associated with tapered piles is unique, what was defined in this monograph as an enhanced version of cylindrical-cavity creation labeled CCXE. With CCXE, the diameter of the cavity continues to expand further once the toe of the pile passes some arbitrary depth, a process that does not occur with constant-diameter piles.

However, this additional expansion that occurs due to pile taper is also affected by the aforementioned 'two steps forward, one step backward' advance-and-retreat behavior exhibited by piles installed by traditional impact driving. This is because the retrenchment that occurs toward the end of each hammer blow likely produces some reduction in the radial-stress state around the pile when tapered piles are involved. This is because the increase in diameter of the cavity created by the initial downward displacement of the pile as a result of the stress wave from the hammer blow traveling from head to toe of the pile is followed by a decrease in cavity diameter as that stress wave returns from the toe to the head of the pile. Thus, there is a complex pattern of additional cavity expansion (beyond the initial cavity creation caused by the toe of the pile) followed by partial cavity contraction within the tapered portion of the pile with each hammer blow.

The bottom line is that the cylindrical-cavity creation process associated with tapered piles appears to be unique to that problem in that it involves an initial, one-time cavity creation followed by repeated cycles of cavity expansion and partial contraction that continue until either pile driving has stopped or the tapered portion of a pile has past some depth in the ground that is under consideration. Furthermore, the additional cycles of expansion followed by partial contraction are occurring under dynamic conditions and with shear stresses acting along the cavity wall, i.e. the interface between the pile shaft and adjacent soil.

As an aside, it is noted for the sake of completeness that the hypothesis presented above clearly does not apply to a tapered DFE constructed by drilling in the manner of a drilled shaft/bored pile. Cavity creation in the case of a drilled shaft does not share any of the behavior of a (driven) pile and thus the stress states at the end of installation of a driven vs. drilled tapered DFE would be expected to be completely different. In fact, the Stage I taper benefit that occurs with tapered (driven) piles is completely absent for a tapered DFE constructed by drilling as the cavity creation would not be materially different from that of a normal drilled shaft with a constant diameter.

9.4.2.2 Research Needs

It is clear from the preceding discussion that the primary research need related to the installation of tapered piles is to develop a cylindrical-cavity creation model that properly considers:

- the geometry of a tapered pile that results in the creation of a cylindrical cavity that does not have a constant diameter but a diameter that has a net increase,

- the interface shear stresses on the cavity wall as the ever-expanding cavity is being created, and

- the dynamic nature with partial rebound of cavity creation by the pile-driving process

in order to more accurately forecast the static radial-stress state around a tapered pile after installation. Such a theoretical model should, by default, also forecast the radial-stress state around a constant-diameter pile so that the difference between the radial-stress states around a tapered and non-tapered pile can be compared to define the Stage I taper benefit.

In addition to this primary, near-term research need, there are several secondary research needs that should be addressed over time:

- The effect of cross-sectional geometry, i.e. circular vs. square, on taper benefit needs to be resolved conclusively. Clearly, a cavity is created when a tapered pile with a square cross-section is driven but it is also clearly not a cylindrical cavity. Therefore, fundamental research is required to define the nature, i.e. geometry, of the cavity created by a pile with a square cross-section and then to model it mathematically.

- The concept of a partially tapered pile with a geometry that is the reverse of what has been used historically, i.e. with the tapered section placed above the constant-diameter section, should be investigated for the sake of completeness given the fact that more than one researcher has concluded that this is the optimum arrangement for a partially tapered pile. The veracity, or lack thereof, of these published conclusions can only be made on the basis of further research and scientific fact beyond that published to date.

- The CCXE model should be extended to fine-grained soil conditions so that the question of whether or not there is a taper benefit in predominantly fine-grained soil conditions can be answered conclusively.

9.4.3 Residual Loads

9.4.3.1 Current State of Knowledge

The presence of significant residual loads in (driven) piles has been recognized in the published literature since at least the 1950s which means that it has been known (albeit, perhaps, not widely) for as long as engineered analysis and design of deep foundations has been practiced. There was a flurry of research activity and concomitant publications into analytical methodologies for forecasting residual-load magnitudes and distributions in the 1980s but in recent years but this does not appear to have had a lasting impact on routine practice or even academic research, at least in the U.S. with which the writer is most familiar.

In recent years, only Fellenius (2002a, 2002b, 2002c, 2015) has published consistently on the subject of residual loads. His work has been noteworthy because he has called attention to the fact that residual loads:

- exist with most types of DFEs, not just (driven) piles, and

- are of significance with tapered piles more than any other type of DFE, driven or otherwise.

The latter observation makes residual loads of particular relevance and importance to this monograph.

It appears that a likely reason why residual loads fell below the radar of routine practice at least is because of the widely held perception, even stated by Fellenius, that residual loads do not affect the magnitude of the geotechnical ULS per se. While this may be correct in principle, it overlooks the fact that residual loads significantly affect both the:

- <u>nature of the load-settlement response</u>. This is because the resistance component of residual loads is always concentrated within the lower portion of a DFE, i.e. a combination of toe resistance and shaft resistance along the lower portion of the shaft. Conversely, the upper portion of the shaft is subjected to drag load. Thus, external compressive-load application at the head of a DFE primarily produces shaft-soil interaction. This is because the drag load acting along the upper portion of the shaft has to be first reduced in magnitude to zero and then ultimately reversed into providing resistance before additional resistance develops along the lower portion of the shaft Additional toe resistance, which may be modest at best depending on what percentage of the ULS toe resistance was mobilized by residual loads, is almost an afterthought.

- <u>shape of the post-installation load-settlement curve</u> due to externally applied loads in axial compression. This is a direct consequence of the above-described concentration of external load-DFE-soil interaction along the DFE shaft. As a result, the shape of the load-settlement curve affects the interpretative methodologies used to determine a single-valued geotechnical ULS using the load-settlement curve obtained in some type of load test. Of great importance is that the interpretative affect tends to be on the unconservative and, therefore, potentially unsafe side (i.e. a larger magnitude of the ULS total resistance is interpreted to exist at the designated magnitude of pile-head settlement than actually exists at the designated magnitude of pile-head settlement) as there is less margin of safety than believed.

Two case histories involving tapered piles and residual loads were explored at length in this monograph:

- the Holmen Island, Norway 1969 *Brynildsen* piles discussed in Appendix H and

- the JFKIA 1994 *Monotube* piles discussed in Appendix A.

These case histories provide significant, relevant ground-truth that residual toe loads can approach 100% of the geotechnical ULS and that significant pile-head settlements that are well beyond the typical 'allowable' range may be required to simply reverse residual-load drag load. Stated another way, in some cases it appears that residual-load effects in the form of drag load along the upper portion of a pile shaft may persist even under maximum service-load conditions. These case histories plus the collective information presented in Fellenius (2002a, 2002b, 2002c, 2015) clearly indicates the residual loads need to be considered in routine practice, at least for (driven) piles if not for other types of DFEs as well.

9.4.3.2 Research Needs

Substantial basic research is required to build on the relatively modest research efforts of the 1980s that were focused on forecasting residual loads in constant-perimeter (driven) piles. The goal should be to develop a reliable residual-load forecasting model for tapered piles that can eventually be incorporated into an overall resistance-forecasting model for tapered piles subjected to external axial load (either compressive or uplift) that includes forecasting displacement as a function of the applied load.

It is clear that this R&D effort will require a substantial number of instrumented piles that are statically load tested. Considerable attention needs to be given to the load-test protocol (this subject is discussed in a subsequent section of this chapter that deals with resistance verification). Most importantly, the load-test results need to be properly interpreted which can be a subjective, time-consuming task, especially for tapered piles, as discussed in by Fellenius et al. (2000).

One issue that does not appear to have ever been addressed but needs to be considered in any future research is the potential variation of residual loads with time after driving. The possibility that any such change may be linked in some way to the increase in resistance to externally applied loads that is now well-proven should also be investigated.

Note that investigation of temporal dependency of residual loads will likely require the use of test piles that have the appropriate structural instrumentation installed and zeroed-out prior to pile driving as occurred with the 1969 Holmen Island test piles discussed in Appendix H. In addition, because Fellenius has opined that the temporal increase in resistance observed for piles driven into coarse-grained soils is related to temporal changes in soil <u>stiffness</u>, not soil <u>strength</u>, it is possible that any temporal dependency of residual loads may also be dependent on temporal changes in the stiffness of the soil around a pile. Thus, it would appear to be useful for a better understanding of how both residual loads and resistance to external loading change over time to perform periodic measurements of soil stiffness after test piles are driven. One way to accomplish this could be to perform sCPTu soundings in relatively close proximity to a test pile at specified times after driving, e.g. one hour, one day, one week, one month, etc.

9.5 POST-INSTALLATION AXIAL LOADING

9.5.1 Overview

It is assumed that the process of pile installation discussed previously creates a nominally cylindrical cavity in the ground so that after installation and just prior to the application of an axial load at the head of the pile:

- The radial stresses acting along the shaft-soil interface (as reflected in the K_h/K_o ratio) are greater than 1, i.e. the radial stresses are greater than the horizontal stresses that existed in the ground under at-rest conditions prior to pile installation.

- There are shear stresses acting along the entire shaft-soil interface due to residual loads so that this interface is not initially stress-free as is traditionally and historically assumed or at least implied in all of the resistance-forecasting methodologies, whether indirect or direct in the context defined in Chapter 4, used in routine practice.

Discussed separately in the following sections are the response to an axial load at the head of a tapered pile that is applied in either a downward (compressive) or upward (tensile/uplift) sense. Separate discussions are required because although they start from the same assumed post-installation state of stress described above, the hypothesized geo-mechanisms for pile resistance are significantly different in each case.

9.5.2 Compressive Loading

9.5.2.1 Hypothesized Geo-Mechanism

Axial-compressive loading at the head of a pile results in settlement throughout the entire length of the pile. Within the tapered portion of the pile, this produces additional wedging and Stage II taper benefit. This behavior is well-modeled using cylindrical-cavity expansion (CCE) as first hypothesized by Kodikara and Moore (1993) but with several very significant differences:

- Basic geometry dictates that CCE begins immediately as the pile settles under load. Kodikara and Moore assumed that there was a "Phase I" behavioral mode where radial expansion and CCE did not occur. This is simply physically impossible.

- Prior to external load application, there are initial, non-zero shear stresses acting along the entire cavity wall, i.e. the interface between the pile shaft and surrounding soil. These stresses can be acting either upward (providing resistance) or downward (the result of drag load) depending on the distribution of residual loads and the specific depth along the pile shaft. The published literature on CCE that has been cited and used by numerous researchers for decades neglects these shear stresses and assumes that the cavity wall at the onset of expansion is stress-free with respect to such shear stresses. The effect of neglecting such shear stresses is unknown at the present time as it appears this subject has gotten minimal research attention to date.

- Depending on which direction these shaft-soil interface shear stresses are acting, the pile settlement due to externally applied compressive load will either first have to reduce and then reverse the direction of these shear stresses (if they are due to residual load-related drag load) or will simply increase these shear stresses in magnitude until the geotechnical ULS is reached (if they are due to residual load-related resistance). Again, the published solutions to CCE assume that there are no such shear-stress changes during cavity expansion so the effect of these stress changes on the calculated outcomes is unknown.

In summary and conclusion, the cylindrical-cavity expansion that is hypothesized to develop during axial-compressive loading of a tapered pile is far more complex than that assumed by Kodikara and Moore and used by El Naggar et alia and Manandhar et alia in terms of the boundary conditions acting along the assumed cavity wall that represents the shaft-soil interface in this case. Specifically, there are shear stresses acting all along the cavity wall both prior to and throughout all stages of compressive-load application. Furthermore, these shear stresses not only are constantly changing in magnitude as a result of load application but along some portions of the shaft they reverse in sign as well.

In addition to and independent of this boundary-condition issue, the assumption made by Kodikara and Moore, El Naggar et alia, and Manandhar et alia that both the shear modulus and Mohr-Coulomb friction angle, ϕ, of the soil surrounding a pile remain constant during cavity expansion is grossly incorrect. As shown in Salgado and Prezzi (2007) and Cook (2010), there is a complex pattern of plastic, non-linear elastic, and linear-elastic soil behavior that encircles an expanding cavity. The variation of the relevant soil properties within each of these behavioral zones (ϕ within the plastic zone and G within the non-linear elastic zone) must be considered.

9.5.2.2 Research Needs

It is clear from the preceding discussion that despite the voluminous treatment already given to cavity expansion in the published literature over the last several decades that cylindrical-cavity expansion needs additional research devoted to developing a solution that includes not only shear stress at the cavity wall (which models the shaft-soil interface in this case) but allows for the direction of this shear stress to reverse during cavity expansion. This solution also needs to consider the three zones of soil behavior (plastic, non-linear elastic, linear elastic) that accompany cylindrical-cavity expansion in general.

9.5.3 Uplift Loading

9.5.3.1 Hypothesized Geo-Mechanism

The 1969 Holmen Island, Norway test-pile program that is reviewed and parsed in Appendix H clearly demonstrates that tapered piles are capable of providing resistance to uplift loads on a par with comparable constant-perimeter piles. This ground-truth should once and for all refute the long-standing, widely held perception that tapered piles are inherently and substantially inferior relative to constant-perimeter piles when subjected to uplift loading. This grossly incorrect perception is based on fanciful, unproven 'imagineering' that the tapered hole in the ground created by installing a tapered pile will somehow and magically remain fixed in that geometry when uplift is applied to the pile head, thus allowing the tapered pile to lift out of the ground with minimal resistance. This hypothesis defies basic geomechanics based on lateral earth pressure theory as well as the experience of anyone who has driven a steel wedge into a block of wood with a sledge hammer and they tried to pull that wedge out by hand.

The hypothesized geo-mechanism of what really happens when a tapered pile is subjected to uplift is the physical mechanism of unwedging that is well-modeled mathematically by cylindrical-cavity contraction (CCC), with shear stresses acting on the cavity boundary that replicates the shaft-soil interface due to the fact that the pile is displacing upward during this process due to the applied uplift load at the pile head. With CCC, radial stresses acting along the pile shaft along the shaft-soil interface gradually reduce from the stresses that existed after pile installation but they do not suddenly decrease to zero so continue to provide resistance to uplift. Using the concept of the K_h/K_o ratio, this ratio gradually reduces from the $K_h/K_o > 1$ that existed after driving the pile into the ground.

Conceptual support for this hypothesized geo-mechanism is that it is physically similar to unloading curves observed in a pressuremeter test. It is also the way in which lateral earth pressures decrease from the at-rest (K_o) state toward the active (K_a) state as an earth-retaining structure moves away from a retained-soil mass. Note that this behavior is a logical complement to Nordlund's assumption that tapered piles subjected to axial-compressive loading can be modeled mathematically as lateral earth pressures increasing from the at-rest (K_o) state toward the passive (K_p) state.

In any event, the CCC model proposed for use in modeling tapered piles subjected to uplift readily explains the actual behavior observed in the Holmen Island test piles. As a tapered pile is subjected to uplift loading, there is a decrease in radial stresses acting along the shaft-soil interface. However, because the decrease is likely to be relatively modest, combined with the fact that the pile is starting from a radial stress (as reflected in the post-installation K_h/K_o ratio) that is larger in magnitude than a comparable non-tapered pile, the radial stress remains significant and, therefore, provides substantial uplift resistance that is comparable to a comparable non-tapered pile.

9.5.3.2 Research Needs

Although there has been theoretical research into CCC, it does not appear to have been as extensive as the research into CCE nor does it appear that CCC solutions developed to date consider shear stresses at the cavity boundary that defines the shaft-soil interface of a tapered pile. Consequently, fundamental research into CCC with a shear-stress boundary condition along the cavity wall is required in order to advance the theoretical understanding of the behavior of tapered piles subjected to uplift loading.

9.6 RESISTANCE FORECASTING

9.6.1 Overview

There are two broad issues to address with respect to resistance-forecasting methodologies for tapered piles:

1. The key elements of improved analytical methods that need to be developed in the future.

2. What existing analytical methods that were discussed in detail in Chapter 5 should or should not be used in the interim.

Within each of these categories, analytical methodologies based on both the static and dynamic approaches as defined in Chapter 5 are addressed. Within the static approach, both indirect and direct methods are addressed.

9.6.2 Improved Analytical Methodologies

9.6.2.1 Overview

The writer's opinion as expressed throughout the earlier chapters of this monograph is that an improved resistance-forecasting methodology for tapered piles should first and foremost be developed as an indirect method using the static approach. The reason is that this is the only way in which all the factors that occur during and after pile installation can be included within an open framework of theoretical geomechanics. Nothing is hidden behind empiricisms or empirically quantified parameters.

Note that is not to say that farther in the future that a direct static approach or dynamic approach could not be developed as a complement to an indirect static approach. It simply means that an indirect static approach is the best way to clearly see how all the complexities involving Stage 1 wedging and taper benefit due to enhanced cylindrical-cavity creation (CCXE) during driving, post-installation residual loads, and Stage 2 wedging and taper benefit due to cylindrical-cavity expansion (CCE) during axial-compressive external loading all combine to produce a final result. As the writer was counseled early in his professional career, you have to learn how to walk before you can learn how to run. The theoretical transparency of an indirect static approach accomplishes that.

9.6.2.2 Static Approach

9.6.2.2.1 Indirect Methods

As noted above, efforts to develop an improved resistance-forecasting methodology for tapered piles should focus initially on an indirect method using the static approach. Chapter 7 was devoted entirely to presenting a schema for such a model so the key elements are only summarized here:

- Model the Stage I wedging and taper benefit that develops during pile installation using enhanced cylindrical-cavity creation (CCXE) that considers the initial cavity created by the pile toe followed by additional cavity expansion by the tapered portion of the shaft as well as the shear stresses acting along the cavity boundary that represents the shaft-soil interface.

- Estimate the depth-wise distribution and magnitude of residual loads that result from pile installation, including any temporal variation.

- Model resistance to axial-compressive external-load application. Consider the initial stress-state acting on the pile from residual loads and model the Stage II wedging and concomitant taper benefit that develops as a result of pile settlement due to the applied loads using cylindrical-cavity expansion (CCE) that considers shear stresses on the cavity boundary that represents the shaft-soil interface. Include any temporal variation.

- Model resistance to axial-tensile (uplift) external-load application. Consider the initial stress-state acting on the pile from residual loads and model the unwedging and concomitant reduction in radial stresses that develops as a result of upward pile displacement due to the applied loads using cylindrical-cavity contraction (CCC) that considers shear stresses on the cavity boundary that represents the shaft-soil interface. Include any temporal variation.

- Be capable of modeling the complete load vs. displacement curve in response to external applied loads in either compression or uplift.

9.6.2.2.2 Direct Methods

As discussed in Chapter 5, direct methods for resistance forecasting using the static approach offer the attraction of simplicity of use, especially when the required input for the analytical methodology comes from some in-situ test such as the cone penetrometer. Given the explosive growth in the use of in-situ testing, especially cone penetrometers, in recent decades, it is no surprise that there has been a similar proliferation in direct methods for resistance forecasting of all types of constant-perimeter DFEs.

As noted above, there is no reason why one or more improved resistance-forecasting methodologies for tapered piles could not be developed in the future using the direct approach. At the present time, it appears that these efforts could be made in two broadly different directions:

1. Development of empirical correlations between the measured data (q_c, f_s, u_2) in a CPTu or sCPTu sounding and the three resistance components of a tapered pile:
   - ➢ *shaft resistance due to sliding friction (constant-perimeter section, if any);*
   - ➢ *shaft resistance due to wedging (tapered section);*
   - ➢ *toe resistance.*
   Note that the tapered section of the shaft should be treated in its actual geometry and not decomposed using some unproven gimmick such as the donut-on-a-stick model as was done in the Togliani Method (which is not recommended for use as discussed in a following section). This means that different correlations will need to be developed as a function of taper angle and perhaps other pile parameters (e.g. toe diameter) as well. Note also that the geotechnical ULS outcomes from such a method could be combined with a settlement-only method such as the Randolph-Wroth Problem solutions in order to provide at least a first-order approximation of complete load-settlement behavior.

2. Development of a stiffness-based analytical methodology by extending the Niazi Method to include tapered piles. As noted in Chapter 5, the writer has done this on a preliminary, proof-of-concept basis called the Extended Niazi Method. The attraction of developing a stiffness-based methodology such as the Extended Niazi Method is that it is a one-step method that inherently produces a load-settlement curve.

9.6.2.3 Dynamic Approach

The 1-D wave equation remains a staple in the toolbox of foundation engineers. However, nowadays it is used more to complement static methods as opposed to being used as an exclusive resistance-forecasting tool as some of the early devotees of the 1-D wave equation proselytized in the 1970s. Consequently, given the fact that the 1-D wave equation has become an integral part of pile design and installation, a near-term research goal should be to modify the 1-D wave equation to better model and thus analyze tapered piles.

The problem with the 1-D wave equation model is that it does not come even close to replicating what occurs when a stress wave travels down the shaft of a tapered pile. The issue here is not whether or not the simplistic Smith spring-and-dashpot model is adequate for modeling soil response to transient-dynamic loading. The problem is that the 1-D wave equation assumes, as its name implies, that all pile displacement occurs along the axial direction of the pile only. Because the spring-and-dashpot combination only produce soil resistance in the direction of pile displacement, this means that the radial component of pile displacement that occurs along the tapered portion of the shaft of a tapered pile is not modeled. Because this radial displacement is not modeled, it does not replicate and capture the wedging phenomenon that is the signature behavior of a tapered pile.

In summary, the 1-D wave equation can only replicate the two traditional resistance mechanisms of a pile, shaft resistance due to sliding friction and toe resistance due to end bearing. It does not and cannot replicate the third resistance mechanism of shaft resistance due to taper-induced wedging.

If the wave-mechanics concept is to be used in the future to more accurately model the installation process of a tapered pile, then it must be modified to more accurately model what occurs when a tapered pile is driven. Ideally, a new, multidimensional wave-equation model should be developed.

A possible approximate alternative would be to try to work with the existing 1-D model. In its present form, soil resistance along the shaft is assumed to be related solely to longitudinal displacement of the shaft. One possibility is to introduce an additional soil-resistance component consisting of a Smith spring-and-dashpot model that acts perpendicular to the longitudinal axis of the shaft. The radial displacement used to produce resistance in this model can be related by simple geometry of the taper angle to the longitudinal displacement of the shaft so that radial displacement does not appear as an explicit problem parameter that would negate the 1-D assumption.

Note that whether a true 3-D model or a modified 1-D model is used, the variable cross-sectional geometry of a tapered pile should always, of course, be modeled.

9.6.3 Existing Analytical Methodologies

9.6.3.1 Overview

It will take a substantial investment in research time and money to develop the improvements to resistance forecasting for tapered piles that were outlined in the preceding sections and discussed in detail in Chapter 7. Therefore, a reasonable question is what analytical methodologies can reasonably be used in the interim.

Due to the lack of available load-test data, published or otherwise, this monograph does not include a large-scale comparison between ground-truth for tapered-pile resistance and calculated outcomes using existing resistance-forecasting methodologies that could be assessed with the usual statistical metrics to form objective conclusions. Nevertheless, some observations can be made based on the writer's experience with various analytical methodologies over the last two decades.

9.6.3.2 Static Approach

9.6.3.2.1 Indirect Methods

The Nordlund Method remains the default resistance-forecasting methodology to use for tapered piles almost 60 years after it was developed. However, it appears that relatively few design professionals and academicians are aware of the fact that there are two versions of this methodology.

The differences between the Original Nordlund Method of 1963 and the Revised Nordlund Method of 1979 are not insignificant conceptually as discussed in general in Chapter 5 and as noted by Nordlund himself in his 1979 presentation notes. Although the two versions produced forecasts for the ULS total resistance that were relatively close in magnitude for the case-history application presented in Chapter 6, a close examination of the shaft and toe results indicates much more significant differences. Specifically, the several changes that Nordlund implemented with his revised version in 1979 have the net, cumulative effect of increasing both shaft and toe resistances at shallower depths and decreasing both shaft and toe resistances at deeper depths. Consequently, the overall trend of the Revised Nordlund Method is to forecast increases in resistance for shorter piles but decreases in resistance for longer piles relative to outcomes produced by the Original Nordlund Method.

Therefore, future, interim use of the Nordlund Method should always involve calculations made using both the Original and Revised versions so that results can be compared, with the design professional using their professional judgment as to whether one result or the other or an arithmetic mean should be used. However, the single most important thing to remember is that this method is heavily dependent on empirical correlations for essential problem parameters that undoubtedly reflect decades of experience of Raymond in general and Nordlund in particular and that, in turn, these problem parameters are heavily dependent on SPT N-values. The point to remember here is that the SPT N-values are those representative of U.S. practice in the 1950s to early 1960s. The writer has interpreted this to be N_{45} values in today's parlance, i.e. N-values obtained with a drive system that averages 45% efficiency. Therefore, in practice today and into the future it will likely be necessary to empirically 'dumb down' (increase) the field N-values (N_{field}) obtained to approximate N_{45} values.

As far as alternatives or complements to the Nordlund Method, the writer's Modified Nordlund Method that was first presented in Horvath (2002) and simplified somewhat since then as discussed earlier in this monograph is a viable alternative. As illustrated in Appendix A, the writer's method produces depth-wise variations in the forecast resistance that agree well with measured results.

The writer has found that the greatest uncertainty in using the Modified Nordlund Method is the calculated toe resistance. The original 2002 version used a classical bearing-capacity solution (Hansen's) with Vesic's correction for soil compressibility/rigidity. This correction is always problematic as it involves selection of an operative value of shear modulus, $G_{operative}$, as well as the reduced rigidity factor, I_{rr}, for the soil although recent advances in CPTu and sCPTu correlations have made direct estimates of I_{rr} possible.

In recent years, the writer has also explored alternative use of:

1. the Basic Manandhar-Yasufuku Toe-Resistance Method (discussed in detail in Appendix B),

2. Vesic's spherical-cavity expansion (SCE) model (discussed in Appendix G), and

3. the NTH Method (also discussed in Appendix G).

Each of these analytical methods or models has its own issues in terms of assumptions to be made concerning model parameters but as discussed throughout this monograph, modern site characterization using cone-penetrometer data and empirical correlations have eased this burden somewhat and reduced the uncertainty in the forecast outcome in the process. At the present time, it appears to be prudent to calculate toe resistance for the Modified Nordlund Method using several methodologies and then compare results. A final decision can then be made using engineering judgment.

As for additional alternative resistance-forecasting methodologies, the writer recommends against using any method intended for constant-perimeter piles together with Fellenius' donut-on-a-stick model that decomposes the tapered portion of a pile into a constant-diameter contribution and an equivalent toe-bearing contribution. As noted earlier in this monograph, as intuitively appealing as this may be, the writer is not aware of any published work that demonstrates that this model produces accurate results.

As discussed in Chapter 5, a better alternative than using some established method for constant-perimeter piles such as the β Method together with the donut-on-a-stick model would be to develop β factors (using a β/K_o ratio would be even better in the writer's opinion) specifically for the tapered portion of a pile. This was discussed in concept in Chapter 5 and an exemplar illustration of how both β and β/K_o can easily be determined from instrumented static load tests was presented in Appendix A.

9.6.3.2.2 Direct Methods

At the present time, the available choices for direct methods for resistance-forecasting of tapered piles using the static approach fall into two broad categories:

1. Any of the numerous methods developed for constant-perimeter piles, with resistance for the tapered portion of the shaft modeled using the generic donut-on-a-stick model.

2. A method developed specifically for tapered piles which at the present time is limited to the Togliani Method (which is fundamentally crafted around the donut-on-a-stick model).

Neither of these approaches can be recommended. The issues concerning the generic donut-on-a-stick model have already been noted numerous times throughout this monograph and the writer has not found the Togliani Method to provide acceptable correlation with measured results in static load tests (Horvath (2015) as well as this monograph).

9.6.3.3 Dynamic Approach

The 1-D wave equation in its present form simply does not even approximate the unique manner in which shaft resistance develops for a tapered pile so cannot be recommend for use although that clearly has not stopped its being used for more than 50 years with tapered piles. Note that the modifications discussed by Goble and Hery (1984) that were done specifically at the request, and with the financial support, of Monotube did not address the larger issue of replicating the Stage I wedging mechanism that occurs when a tapered pile is driven into the ground using conventional impact driving.

9.7 RESISTANCE VERIFICATION

9.7.1 Overview

Addressed in this section are the methodologies used to provide ground-truth for axial-compressive resistance during production driving of tapered piles on a project. Thus, the discussion here does not include instrumented load tests performed strictly for research purposes.

As is typical in current practice, most projects have a few conventional static loads tests with much more extensive use of dynamic, on-pile testing. The latter typically involves a combination of both end-of-initial-driving (EOID) and restrike measurements made some days or weeks later in order to gauge the temporal gain in resistance (what used to be called 'setup' or 'soil freeze' in the past).

9.7.2 Static Approach

The writer has long taken issue with many aspects of conventional static load testing of piles that are related to the basic way in which the applied load and pile-head settlement are measured (Horvath 2002). Research and publication in recent years (Horvath 2015, 2016a, 2016b) as well as discussion in this monograph have added several questionable load-test procedural issues to these basic concerns so that the following are the writer's aggregate thoughts on the subject of static load testing:

- Load tests should only be performed on piles that are in their complete state. This is relevant for many types of tapered piles such as the *Monotube* and *TAPERTUBE* that consist of a closed-toe steel pipe that is filled with fluid Portland-cement concrete after driving.

- Electronic load cells should always be used to measure applied loads. There is ample evidence from many sources over a period of decades that loads based solely on jack-pressure measurements are simply incorrect. More significantly, they tend to be incorrect on the unconservative side meaning that the applied load is overestimated. This is potentially unsafe as it indicates that the resistance of the pile is overestimated and the pile is presumed capable of carrying more load at a given magnitude of pile-head settlement than it actually can.

- The near-universal use of dial gauges and/or tensioned wires to measure pile-head settlements tend to produce incorrect results. This is because the reference beams or supports for these mensuration devices are affected by ground-surface settlement caused by the load-test setup. Again, these errors tend to be on the unconservative and potentially unsafe side because they indicate that the pile settles somewhat less than it actually does for a given magnitude of applied load. The only true and correct settlement-measurement methodology that the writer is aware of is high-precision optical survey that references a deep-seated, site-specific benchmark installed for this purpose. This is costly (primarily for the labor involved in having a licensed land surveyor present during the entire load test) so realistically it is unlikely that the traditional mensuration practice for static load tests will change anytime soon unless some new technology is developed. Perhaps some methodology based on a remotely operated laser that refers to a benchmark placed some distance from the load test can be developed as such devices have proven useful in other areas of geotechnical instrumentation.

- All parts of the load-application and settlement-measurement hardware should be shielded from direct solar radiation in order to minimize heat-induced physical distortion of system components.

- The load test should be conducted to a level of pile-head settlement (several inches or tens of millimetres) that can reasonably be assumed to be the geotechnical ULS. This is to maximize the benefit obtained from a test. Given the cost of performing a traditional static load test, to stop the test short at some arbitrary point such as 'double design load' is a poor economy given the substantial additional information to be obtained by continuing the test to its maximum. The only exception to this is for DFEs supported on or in rock which does not apply to tapered piles.

- Loads should be applied in a monotonic fashion until the conclusion of a test. There should be no unload-reload cycles or extended hold times. How many load increments should be used and how rapidly each increment should be applied are open for discussion.

- No 'gaming the system' in the form of reloading a pile that has 'failed' a load test (typically by settling more at double-design load than the Davisson Offset Method for load-test-data interpretation allows) should be allowed. Such a practice is disingenuous because the reload test is <u>not</u> testing an as-installed pile but one that has been preloaded by the original load test and thus is measuring a stiffer pile-ground system response.

- On large projects or projects where fine-tuning an estimate of pile resistance can produce cost savings, the performance of one or more instrumented static load tests should be considered. As the assessment of the two instrumented *Monotube* piles at JFKIA that is presented in Appendix A shows, a significant amount of useful information concerning residual-load magnitudes and distribution as well as depth-wise pile resistance can be obtained from an instrumented load test. Note, however, that it is essential that any instrumented static load test be properly corrected for residual loads otherwise grossly incorrect conclusions concerning resistance distributions can be drawn.

9.7.2 Dynamic Approach

Resistance verification using on-pile dynamic measurements obtained during hammer impact (which can be during initial driving or at any time thereafter) have become a worldwide staple of pile driving in the approximately 50 years since this technology was first developed. However, it should be kept in mind that this technology (both the real-time results obtained immediately after pile impact as well as the more-detailed assessments subsequently made after the fact) is fundamentally based on 1-D wave mechanics so there is an inherent error built in to this technology as there is with the 1-D wave equation discussed earlier in this chapter.

Separate from this issue is that the depth-wise variation in resistance that is one of the calculated outcomes obtained using the more-detailed assessment made after the fact (e.g. the *CAPWAP®* software from Pile Dynamics, Inc.) does not always correctly account for residual loads. Consequently, this forecast of resistance distribution may be in error compared to the actual distribution (Fellenius 2002a).

9.8 QUANTIFICATION OF RESISTANCE EFFICIENCY AND TAPER BENEFIT

Cost per unit force of axial resistance in either compression or uplift as appropriate remains the overall most useful, all-purpose metric for resistance efficiency of DFEs in general, with the important caveat that the cost of a pile cap or other structural load-transfer element between the DFEs and superstructure that they support must always be factored in to an assessment of foundation costs. Thus, cost per unit force of axial resistance is also an efficient metric to use for evaluating taper benefit whenever tapered piles are being compared to other deep-foundation alternatives. This is because taper benefit is really just a particular application of the concept of resistance efficiency.

However, as demonstrated in Chapter 8, the resistance per unit volume of an embedded DFE can also be a useful metric for resistance efficiency and taper benefit in cases such as piles where the DFE has a well-defined volume and the costs per unit volume of the DFEs being compared are comparable. With respect to tapered piles specifically, resistance per unit volume can be particularly useful when tapered piles of different lengths, tapers, and diameters (both toe and constant-diameter shaft) are being compared to find the most-efficient combination of pile geometries in a given application.

This page intentionally left blank.

Appendix A

Generic Guidelines for Parameter Quantification for Resistance-Forecasting Methodologies

---

A.1 INTRODUCTION

Several of the analytical methods presented in Chapter 5 are the writer's modifications or extensions of established methodologies developed by others. In some cases, the writer's contributions are being published for the first time in this monograph.

As noted in Chapter 5, as part of a rational process for developing these modified/extended analytical methodologies for potential eventual use in practice, it is necessary to create a database of the relevant problem parameters based on appropriate ground-truth. Depending on the specific analytical methodology involved, this ground-truth includes certain soil properties from pre-installation site characterization such as K_o. This is in addition to field data from appropriate post-installation static load tests that might be supplemented with interpreted results from appropriate on-pile dynamic measurements.

The purpose of this appendix is to illustrate the necessary problem-parameter assessment for each of the writer's methodologies presented in Chapter 5 using the results from case histories. The material presented in this chapter is intended to serve as a generic guideline and template for future research by interested stakeholders who might want to replicate the writer's assessment process using their own case-history data. The eventual goal of these collective efforts would be to create a sufficiently populated, robust problem-parameter database that will allow these methodologies to be used in routine practice reliably and with end-user confidence in the forecast outcomes.

Also included in this presentation, where relevant, is generic problem-parameter assessment for established analytical methodologies to see how their forecast results compare to measured results. As the writer stated in Chapter 5, every analytical methodology, no matter how long it has been in use, should be compared to relevant ground-truth whenever possible, if only to reinforce the accuracy and continued viability in practice of an established methodology.

A.2 OVERVIEW

As noted in Chapter 5, all of the problem-parameter evaluations necessary for the several enhancements to analytical methods that have been proposed by the writer require, as a minimum, static load tests on piles that are instrumented for their entire embedded length so that the depth-wise axial-load transfer can be evaluated. Thus, the first topic discussed in this appendix is the specific case history used by the writer for this purpose.

The remainder of this appendix that follows is organized using the same broad divisions as in Chapter 5:

- Resistance-only methods that adhere to the traditional approach first developed in the 1950s of producing a single-valued forecast of the ULS total resistance that is then used in strength-based design approaches, both ASD and LRFD. Most of the resistance-only methods that are discussed in this appendix are indirect in nature but there is also limited discussion of direct methods where relevant.

- <u>Settlement-only methods</u> that are intended to complement and expand on the results produced by resistance-only methods as part of a two-step procedure in order to create a complete load-settlement curve up to the point of a single-valued ULS total resistance. Note that a critical, distinguishing feature of settlement-only methods is that they do not and cannot exist by themselves. Rather, they must always be used in conjunction with a resistance-only method as part of an overall two-step procedure for generating a load-settlement curve. However, in most cases the resistance-only method with which a settlement-only method is linked is not unique. Thus, most settlement-only methods can be used with any of the numerous resistance-only methods and vice versa, with the choice of linked methodologies left to the design professional using them.

- <u>Integrated load-settlement methods</u> that are frequently viewed as 'complete', one-step procedures in that they inherently link the forecasts of settlement and total resistance in a single, integrated analytical methodology. Note that, depending on the specifics of a given analytical methodology, the calculated total resistance may not be a nominally ULS value. Nevertheless, integrated methods can be considered to be the 'gold standard' of analytical methodologies for forecasting the axial load vs. displacement behavior of all types of DFEs using a one-step approach.

A.3 INSTRUMENTED STATIC LOAD TEST CASE HISTORY

A.3.1 Introduction

As noted and discussed in Chapter 5, case histories for instrumented static load tests on tapered piles that are well-documented and, more importantly, properly interpreted and adjusted for residual loads appear to be extremely rare, at least in the published literature. Most of the research-quality instrumented static load tests performed on DFEs in recent decades were funded by U.S.-government agencies such as the FHWA that have not historically supported work involving proprietary products and technologies which, unfortunately, covers most tapered piles other than timber piles. Furthermore, the need to find an appropriate case history for tapered piles that also includes the CPTu or sCPTu data that are necessary, in the writer's opinion, for site characterization and the desired assessments of analytical methodologies further restricts the pool of potential data.

The one documented case history that was found that reasonably met the writer's criteria was presented by Fellenius et al. (2000). The project site they discussed was at JFKIA. This was quite fortuitous as the writer is intimately familiar with the extensive pile-foundation history at this site and has used other JFKIA data for both tapered and constant-perimeter piles as the source material for several research reports (Horvath 2002), conference papers (Horvath 2003b, 2003c; Horvath and Trochalides 2004; Horvath et al. 2004a, 2004b), and, most recently, white papers (Horvath 2014b, 2015, 2016a, 2016b). Data from JFKIA has also been used extensively throughout this monograph, especially for the exemplar applications presented in Chapters 6 and 8. Thus, the writer was able to draw on this broad, personal knowledge-base to both complement and, in some cases, supplement the information presented in Fellenius et al. (2000).

A.3.2 Background

Fellenius et al. (2000) presented a factual summary and their interpretation of results from what was apparently a small, two-pile load-test program performed at JFKIA in 1994. This program appears to have been either a follow-on supplement or complement to the much larger deep-foundation test program that was conducted between 1988 and 1990 during the planning phase of what was called the *JFK 2000* redevelopment program. As discussed by Horvath and Trochalides (2004), this was but one of several similar, large-scale deep-foundation test programs conducted over the years at JFKIA. The purpose of these testing programs was to rationally and systematically advance the state of knowledge related to deep-foundation usage there as both DFE technology and geotechnical (specifically, seismic) design requirements changed there over the years.

Although the specific location of the 1994 test piles was not indicated by Fellenius et alia, it was presumably within the main passenger-terminal area of the airport that is called the Central Terminal Area (CTA). As such, this allowed the writer to draw on almost 50 years of personal geotechnical and foundation engineering knowledge and research related to JFKIA, most of it from within the CTA.

A noteworthy element of the 1994 testing reported on by Fellenius et alia was that the piles involved were instrumented for their entire installed length in order to measure the depth-wise variation in applied axial-compressive load under conventional static load testing. None of the test piles installed and load tested between 1988 and 1990 as part of the primary *JFK 2000* testing program are believed to have been so instrumented.

The purpose of this additional, more-advanced testing in 1994 is believed to have been, at least in part, to investigate in greater detail the then-new observation that at least some types of piles driven at JFKIA (including tapered steel pipe) exhibited relatively significant resistance increases, colloquially called 'setup' or 'soil freeze', within the days and weeks immediately after driving (York et al. 1994). At that time, such a phenomenon was not generally expected for piles that derived virtually all of their resistance from coarse-grained soil so would have been quite the technical novelty and worthy of further detailed investigation.

As to the specifics of the two piles installed and tested in 1994, as the result of outcomes derived from the bulk of the pile driving and load testing (both static and on-pile dynamic) related to the *JFK 2000* test-pile program that had concluded by circa 1990, combined with the then-new need to consider seismic-liquefaction potential for all new foundation designs at JFKIA, by the time of the supplemental testing in 1994 the *Monotube* pile, specifically, the intermediate Type J taper, had emerged as the new pile of choice at JFKIA. This pile type replaced timber piles that had been the deep-foundation mainstay at JFKIA since initial construction began there in the 1940s.

Thus, it is no surprise that the two 1994 test piles, which were referred to by Fellenius et alia simply as "Pile 2" and "Pile 3", had the intermediate (0.57° taper angle, α) *Monotube* Type J taper, with a 25-foot (7.6-m) tapered lower section transitioning from 8- (203-) to 14-inches (356-mm) in diameter. The proprietary *Monotube* Type N constant-diameter extension/upper section (14 inch/356 mm in this case) was used. This is because the use of generic hot-rolled steel pipe as an alternative to the proprietary, cold-formed *Monotube* Type N extension (the reasons for this are discussed in Chapter 3) had not been become fully integrated into local (New York City metropolitan area) practice in 1994 although it would soon be so (Brand 1997). Both the tapered lower section and constant-diameter upper section/extension were 5 Gauge (0.2092 in/5314 μm)[171] wall thickness which is the second thickest of four wall-thickness options for *Monotube* piles.

Both piles were approximately 20 metres (66 feet) installed embedded length. The earlier 1988-1990 testing had determined that was more or less the optimal embedded length for *Monotube* piles in order to meet the design requirements established at that time which were for piles to:

- penetrate beyond the forecast zone(s) of potential seismic liquefaction within the Upper Glacial Aquifer ("Pleistocene sand" in Figure 6.1) bearing stratum at JFKIA and

- have an allowable capacity in the range of 200 to 240 kips (890 to 1070 kN).

With regard to the latter requirement, based on the ASD-based design practice at that time that essentially called for piles to be installed to 'double-design' ULS resistances as a minimum, this would have meant a range of target values of ULS total resistance of twice the desired allowable capacity, i.e. 400 to 480 kips (1780 to 2140 kN).

To put these numbers in a historical perspective, the *Monotube* piles at that time (i.e. circa 1990) were being used as replacements for timber piles that had been used at JFKIA for decades by that time and typically had minimum required values of ULS total resistance of 120 kips (530 kN). So, the *Monotube* piles of circa 1990 were a significant change in terms of ULS resistance. But that improvement, as substantial as it was, was relatively short lived as by a decade later on the cusp of the 21st century the then-new *TAPERTUBE* pile was found to have ULS resistances as large as 1,000 kips (4450 kN) at JFKIA although ULS resistances routinely sought in recent years have been a more modest 800 kips (3560 kN).

A.3.3 Limitations

In the writer's opinion, there are some negatives in the form of technical issues with regard to the two test piles studied by Fellenius et alia that detract somewhat from the utility of the interpreted outcomes for the purposes of this appendix. However, given the lack of suitable alternatives and in considerations of some very definite positives, the writer elected to make the most and best possible use of the data presented by Fellenius et alia despite the perceived shortcomings. However, in the interest of objectivity, it is appropriate to at least summarize the factors that detract from the writer's assessment based on these particular piles.

To begin with, the writer did not have access to tabular or digital versions of the various plotted results presented by Fellenius et alia. This contrasts with data for the primary *JFK 2000* load-test program that was conducted between 1988 and 1990 for which the writer at least had access to tabular data for the numerous CPT soundings that were performed for that test program. This data-access issue was especially problematic for the CPTu data shown in Fellenius et al. (2000) as such data are typically collected on a 20 to 50 millimetre (0.8 to 2 in) spacing depending on the specific cone used. Rather, the writer had to resort to scaling values off the published plots. Thus, the precision of various interpreted results presented throughout this appendix is not as great as could have been achieved had the writer had access to the original data. However, the writer feels that there is still sufficient precision in

---

[171] This is the precision indicated in Monotube product literature that was reviewed by the writer in 2018 during preparation of this monograph. Whether or not the product is or was actually manufactured to that precision can only be answered by Monotube.

the scaled data to make the interpretations and assessments presented later in this appendix of value.

The next issue of note is that Pile #2 encountered PCC fragments at the interface between the bottom of the Holocene sand fill that was placed at the time of the initial airport construction in the early 1940s and top of the Holocene marine tidal marsh (primarily organic clay with some peat) that was the original, pre-construction ground surface throughout most of the airport and virtually all of the CTA (see Figure 6.1). Given the known, complex construction history of the JFKIA property (discussed in detail in Horvath (2014b)) that included localized development for various uses in the decades prior to the airport construction, this is not surprising and was likely material associated with either the recreational development at Idlewild (Long Neck) Point that dates back to the 1800s or the former Idlewild Beach Golf Club that was opened in 1930. Both of these developments were covered over by the early-1940s hydraulic sand fill that was placed to create JFKIA. Without knowing the exact location of the 1994 test piles, it is not possible to correlate their location with the known, specific locations of these or other historical features, the approximate locations of which were shown in Horvath (2014b).

The net result of encountering this obstruction was that the *Monotube* steel pipe of Pile #2 was temporarily removed from the ground; the obstruction removed; and the same pile re-driven at the same location. Why a new pile was not driven at a slightly different location is not known to the writer as what was done seems like a poor choice given the critical, experimental nature of this pile.

The next issue is that only Pile #2 was loaded to a level approaching what might be considered the geotechnical ULS, in this case a head settlement of approximately ¾ inch (18 mm) combined with a distinct slope change in the load-settlement curve. Pile #3 had a maximum head settlement of approximately ½ inch (13 mm) and no indication of slope change in the load-settlement curve before the load-testing apparatus reached its operational limit of 2500 kN (562 kips).

Again, the writer found it curious that these piles would not have been loaded further until substantially greater pile-head settlements were achieved. Loading test piles to relatively large settlements in an attempt to define a relatively unambiguous geotechnical ULS had been standard operating procedure for earlier test-pile programs conducted by the PANYNJ at JFKIA in 1972 and again in 1988-1990 with which the writer is personally familiar. Given the effort and expense expended to install and load test instrumented test piles, one would think that making sure that they were loaded until an unambiguous geotechnical ULS was reached would have been a high priority.

Finally, although Fellenius et alia included plotted results of a CPTu sounding that was presumably performed in the vicinity of the two 1994 test piles, they did not provide a log of a conventional boring with SPT *N*-values. The latter would have been useful for some of the assessments presented subsequently in this appendix. It is not known whether a boring was made close to the Fellenius et alia test piles and they chose to leave it out of their paper, or if no boring was performed which would have been a surprising (to the writer) from typical PANYNJ practice at that time.

Based on the writer's extensive experience and use of data from numerous non-instrumented static load tests performed on other test piles installed and tested between 1988 and 1990 for the *JFK 2000* test-pile program, it was typical to have both CPT[172] and SPT

---

[172] Note that CPT, not CPTu, soundings were performed in 1988 for the main *JFK 2000* test-pile program. The CPTu sounding for the two 1994 test piles discussed by Fellenius et alia was presumably done sometime after 1988 and may well have been the first use of a piezocone at JFKIA, at least by the PANYNJ.

data reasonably close to each test-pile location. This was due, in part, to the fact that 1988 marked the first time that cone-penetrometer soundings of any type had been performed at JFKIA. Therefore, there was the desire to be able to correlate soil stratigraphy defined by traditional borings with measurements made in and interpretations from the CPT soundings. Note that in the late 1980s the ability to infer soil-behavior type based on q_c-f_s data only from a CPT sounding was much more primitive and less reliable than inferences that can be made currently based on q_c-f_s-u_2 data from a CPTu or sCPTu sounding.

In summary, there were several surprising (to the writer at least) errors, omissions, and shortcomings in the small-scale test-pile program studied by Fellenius et alia when compared to practices employed in similar, larger programs in both 1972 and 1988-1990 with which the writer has personal, first-hand experience. Nevertheless, in the writer's opinion the drawbacks reflected in these issues are more than compensated by the significant, state-of-art interpretative efforts made by Fellenius et alia to:

- correctly estimate the operative value of Young's modulus for the PCC used to fill the piles after driving and, even more importantly,

- estimate post-driving residual loads in the piles so that the measured loads, which typically provide a flawed, false representation of the depth-wise variation in resistance, could be corrected to produce a truer representation. Fellenius individually has written at length over the years on this subject, with Fellenius (2015) being a current discussion and summary on the subject.

A.3.4 Primary Outcomes

It is of interest at this point to summarize the key outcomes presented in the Fellenius et alia paper that are of relevance to both the specific goals of this appendix and the overall scope of this monograph:

- The temporal increase in axial-compressive resistance appeared to occur almost entirely within the <u>tapered portion</u> of the pile shaft. Essentially no increase occurred within the constant-diameter portion of the shaft and only a slight increase, at most, occurred at the toe. The increase within the tapered portion was attributed solely to <u>increased soil stiffness</u> over time as opposed to strength gain related to excess pore-pressure dissipation as was one school of thought back circa 1990 when this phenomenon was first observed at JFKIA. Note that this stiffness-based interpretation was the genesis for the writer's observation and suggestion made in Chapter 9 that multiple sCPTu soundings performed in the vicinity of a tapered pile at different times after installation might be a way to measure and quantify this hypothesized increase in soil stiffness as it should be reflected in the measured G_{max} values.

- Interpretation of the inferred changes in the Young's modulus of PCC during post-installation external load application indicated that the resistance had peaked and stabilized within the constant-diameter portion of the shaft whereas resistance continued to increase within the tapered portion of the shaft. This is consistent with the writer's opinion that while the constant-diameter portion of a partially tapered pile exhibits the classical mechanism of shaft resistance due to sliding friction (which requires relatively small relative pile-soil displacements to mobilize fully), the tapered portion exhibits a different shaft-resistance mechanism, that of wedging as modeled using cylindrical-cavity mechanics. While cavity mechanics, whether starting from the initial condition of no cavity or an initial cavity, indicates that the radial stress acting on the cavity always (in theory) has a limiting maximum value, it appears that this limit, at least for the taper angle and range of settlement magnitudes for the two test piles in question, is reached at some as-yet-to-be-defined point well after shaft resistance due to sliding friction peaks.

- Residual loads within the pile after driving were interpreted subsequently to be quite significant. In particular, essentially all toe resistance was already fully mobilized and locked-in at the end of driving. So, it is no surprise that the uncorrected ("false") load-distribution curves shown in Fellenius et alia indicate virtually no resistance at the toe at any load level during the static load test. When interpreted using the concepts and insights presented in Fellenius (2015), this means that the piles would be expected to exhibit a nominally plunging type of load-settlement behavior under load test which was indeed the case for the one test pile (Pile #2) that was loaded to the early stages of the geotechnical ULS as noted previously. Of relevance is that such plunging load-settlement behavior was also observed for several non-instrumented *Monotube* piles that were driven and load-tested at JFKIA between 1988 and 1990 as part of the main test-pile program for the *JFK 2000* redevelopment (these piles have been studied in detail by the writer at various times since the 1990s) as well as for *TAPERTUBE* piles driven and load-tested at JFKIA circa 2000 when this pile was being introduced commercially for the first time (the load test shown in Figure 5.2 is representative of this behavior). This implies that these other tapered piles most likely had similar, if not identical, residual-load distributions.

Figure A.1 shows the distribution of the "true", i.e. after correction by Fellenius et alia for estimated residual loads, axial-compressive loads (forces), Q, in the two circa-1994 *Monotube* instrumented test piles. These loads are normalized to the maximum force, Q_{max}, applied at the head of each pile in the static load test to allow for an apples-to-apples comparison. Also shown in this figure is the complement to the pile loads, the developed total ground resistance, R_{total}, normalized to the maximum resistance developed at the end of each static load test, $R_{total\,(max)}$.

Note that both maximums (applied load and ground resistance) are probably close to the ULS total resistance, R_u (see Figure 2.1), for each pile. Note also that both the pile loads and ground resistance in Figure A.1 are plotted relative to the depth normalized to the embedded pile length so that the two piles can be compared directly. This is because the two piles were driven to slightly different embedded depths (Pile #3 was driven 500 millimetres (20 in) deeper than Pile #2).

Some general observations of interest are:

- The relative load and resistance distributions in each pile are essentially the same despite the fact that:
  o Pile #3 was loaded to 25% greater maximum load than Pile #2 and
  o Pile #2 had installation issues as noted previously that resulted in the uppermost portion of the constant-diameter shaft having minimal soil resistance due to disturbance.

The identical nature of the load distributions lends confidence to the overall correctness of the load distributions as ground-truth for the purposes of this appendix and monograph.

388

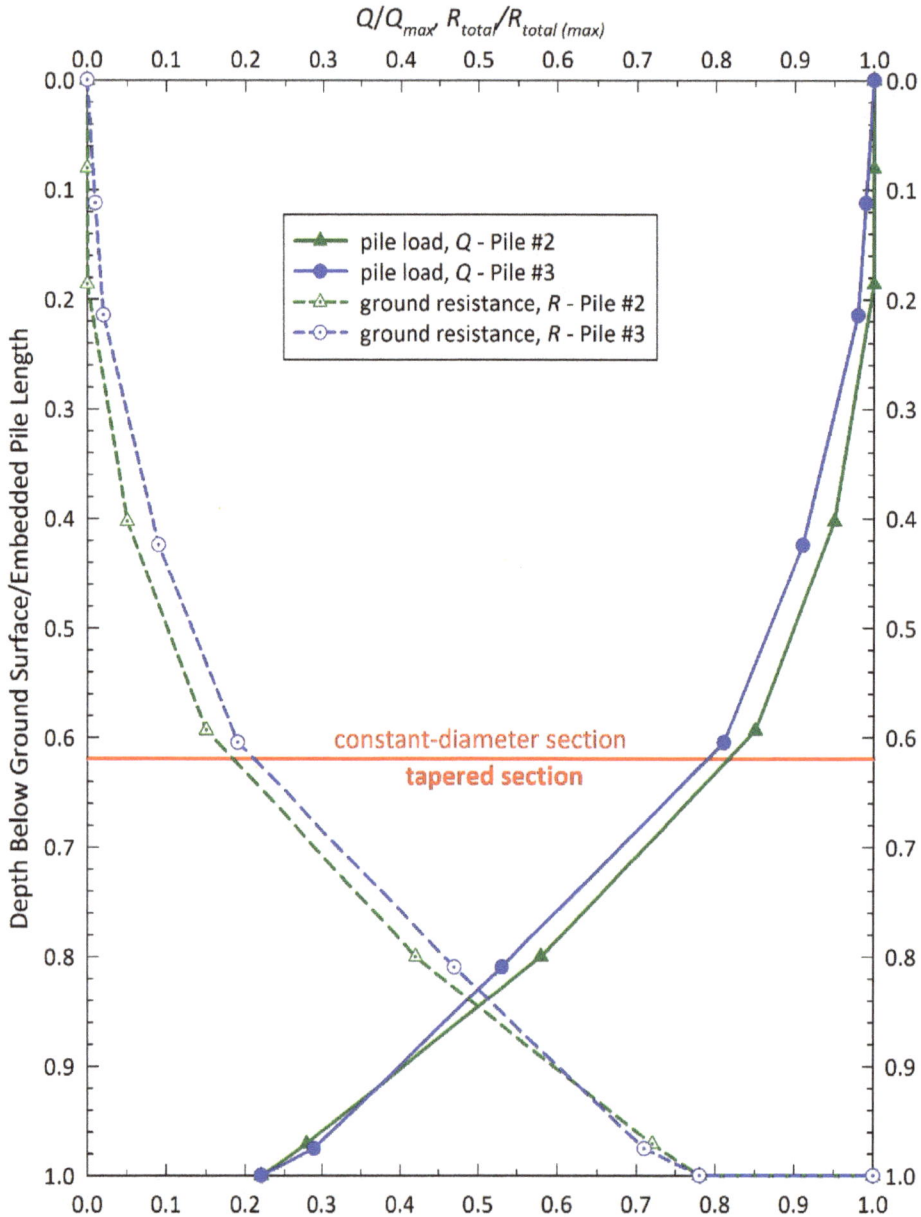

Figure A.1. Depth-Wise Distribution of Relative Corrected ("True") Axial-Compressive Forces and Relative Ground Resistance for Fellenius et al. (2000) *Monotube* **Piles at JFKIA.**

- The tapered section accounts for approximately 60% of the total pile resistance and 80% of the total shaft resistance despite the fact that it comprises only about 40% of the embedded pile length and 30% of the circumferential area of the shaft that is in contact with the surrounding ground. Stated another way, a unit area of the tapered section provides approximately 6.25 times more resistance than a unit area of the constant-diameter section. It is recognized that some of this difference is due to the fact that the average vertical effective overburden stress along the tapered portion of the shaft is approximately 2.5 times that along the constant-diameter portion. But even if this difference were accounted for by scaling one or the other unit-shaft-resistance contributions up or down as desired, the tapered portion still provides about 2.5 times more resistance per unit area than the constant-diameter section under the same vertical overburden stress. Thus, there is clearly a *taper benefit*, a topic that is discussed in more detail in Chapter 8 as there are several ways in which taper benefit can be expressed. That there is a taper benefit is not surprising as the efficiency of tapered piles in coarse-grained soil conditions has long be recognized and established and is the underlying reason that this monograph was written in the first place.

- Toe resistance accounts for approximately 20% of the total resistance which is perhaps a bit surprising considering that:
  o the toe of each pile is only a nominal 8 inches (200 mm) in diameter;
  o the embedded length of each pile is relatively substantial, approximately 65 feet (20 m), which means that there is substantial shaft area in contact with the surrounding ground, almost all of which is coarse-grained soil; and
  o the magnitude of toe settlement at maximum applied load as inferred from the residual pile-head settlement upon unloading at the end of each static load test was less than 10 millimetres (⅜ in) in each case.

However, it should be kept in mind that the residual-load correction process applied by Fellenius et alia inferred that the residual load was concentrated at and near the toe. Consequently, toe resistance was already substantially (close to 100%) mobilized at the conclusion of driving and before any additional load application and concomitant toe displacement occurred during the static load test. Therefore, there should be no quid-pro-quo inference or cause-effect relationship derived or implied from the fact that there was a substantial toe resistance with such relatively small toe settlements during the static load test. The fact is that essentially 100% toe resistance was already mobilized before any external load application and toe settlement resulting from that load application occurred.

The conclusion drawn up to this point is that the instrumented static load tests on two *Monotube* piles presented by Fellenius et al. (2000) are of adequate quality for the assessments presented in the remainder of this appendix. Whatever issues there are with regard to the driving and load testing of these piles is more than overshadowed by the state-of-art interpretation of the measured data that included correction for residual-load effects.

A.4 RESISTANCE-ONLY METHODS

A.4.1 Introduction and Overview

Several of the writer's embellishments to indirect resistance-only forecasting methods depend explicitly on site-specific, depth-wise values of the coefficient of lateral earth pressure at rest, K_o, that exist prior to pile installation. These values of K_o are used to normalize problem parameters such as K_h and β. In addition, even though the direct methods for resistance forecasting that are also addressed in this appendix do not require explicit knowledge of K_o, some basic understanding of the subsurface stratigraphy at the site of the Fellenius et alia case history is useful as background information. Therefore, before discussing specific analytical methods, a general discussion of K_o evaluation for the Fellenius et alia case history site used by the writer is presented as K_o is not a parameter that historically has been part of routine site characterization, especially for coarse-grained soil.

The discussion of specific analytical methods that follows the discussion related to K_o is divided into separate discussions of indirect and direct methods. The discussion of indirect methods is further divided into sections devoted to shaft resistance and toe resistance as it is possible, at the discretion of the foundation designer, to use a given indirect method for shaft resistance with any one of several different indirect methods for toe resistance and vice versa.

Note that this freedom to mix and match analytical methodologies is not true of direct methods. Because they are entirely empirical in nature, the algebraic correlations for shaft and toe resistance for a specific direct method generally must always be used together.

A.4.2 Site Characterization

Figure A.2 shows the writer's interpretation of the variation of pre-driving K_o vs. depth for the one CPTu sounding presented by Fellenius et alia that is applicable to both piles. Results are shown for two empirical relationships proposed by Mayne (2006b, 2007) and down to a depth a few metres below the toes of the piles (the CPTu sounding went to a depth of approximately 30 metres (100 ft) below ground surface). Note that only the methodology presented in Mayne (2006b) is capable of producing results for fine-grained soils.

The results of the two methods are very close (the writer used an arithmetic mean of the two for all calculations related to these two piles that are presented in this appendix) and are similar in both magnitude and distribution to those estimated by the writer for numerous other locations throughout the JFKIA CTA. Note that this type of depth-wise variation in K_o illustrates why normalization of pile-analysis parameters such as β and K_h to K_o that was first suggested in Chapter 4 makes sense as a way to reduce the scatter in parameter values by removing the pre-driving site stress history as a variable.

As noted earlier in this appendix, no conventional boring was included in Fellenius et al. (2000) so the stratigraphy shown in Figure A.2 is based on a combination of the Soil Behavior Types[173] interpreted by the writer for the CPTu sounding plus the writer's general knowledge of the relatively uniform subsurface conditions within the JFKIA CTA that were discussed at length in Horvath (2014b).

Note that for routine foundation and geotechnical analytical purposes, the groundwater conditions at JFKIA within the limits of the depths shown in Figure A.2 are typically assumed, with good approximation, to be hydrostatic relative to the groundwater table that is shown within the Holocene sand fill. This stratum is the groundwater-table or unconfined aquifer throughout most of JFKIA and all of the CTA. This hydrostatic condition is due to the fact that the Holocene organics stratum is an imperfect confining layer (a.k.a. aquitard or aquiclude) for the underlying Pleistocene sand (technically the Upper Glacial Aquifer, UGA[174]). The UGA is a confined artesian aquifer although no longer flowing-artesian as it reportedly was circa 1900. Aquifer communication between the Holocene sand fill and UGA exists because the Holocene organics stratum is relatively thin throughout JFKIA and has been penetrated in countless places by seven decades of deep-foundation installation.

---

[173] Current empirical interpretations of CPTu and sCPTu soundings produce estimates of what is called *Soil Behavior Type* (SBT), more recently upgraded to *Normalized Soil Behavior Type* (SBTn). This is subtly different from earlier CPT and CPTu practice that produced estimates of what was explicitly called *Soil Type*. The reasoning behind this nuanced change in terminology is that because the actual soil is unseen when performing a cone-penetrometer sounding, it is more logical and. arguably more relevant, to infer how a soil will <u>behave</u> under loading, i.e. 'sand-like' vs. 'clay-like', than to focus on the explicit gradation and plasticity of the soil particles as is the focus of traditional soil identification and concomitant soil types.

[174] For the sake of completeness, recent work by Moss (2015) and Moss and Canale (2017) has noted that the UGA actually consists of three distinct formations. However, this detail is not important for the purposes of this appendix.

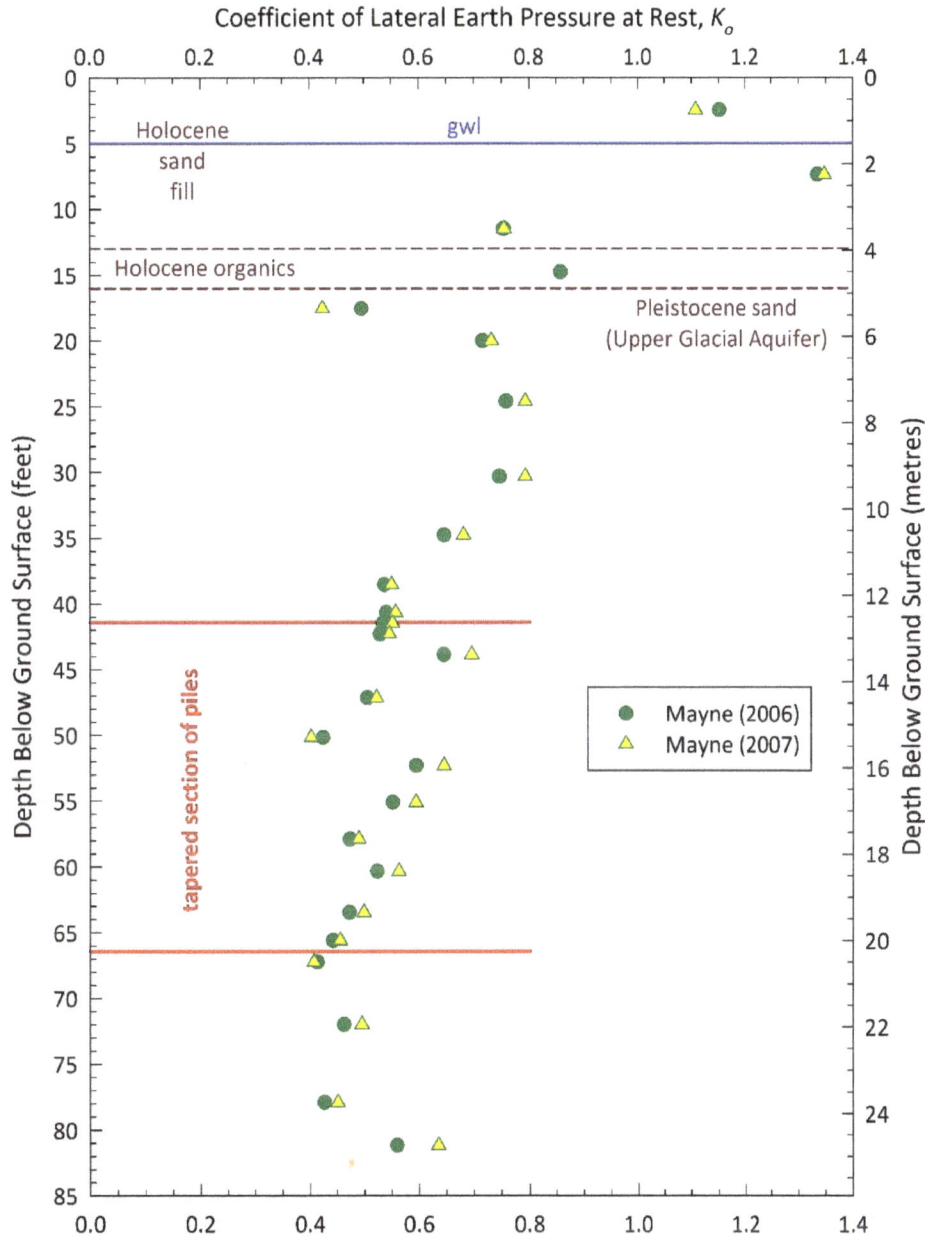

Figure A.2. Estimated Pre-Driving K_o at Site of Fellenius et al. (2000) *Monotube* Piles at JFKIA.

That having been said, in reality the writer has found that careful examination of u_2 data from CPTu and sCPTu soundings performed in recent decades usually shows some slight deviation from perfect hydrostatic conditions, with the Holocene sand fill stratum usually slightly perched relative to the piezometric levels in the UGA within the range of approximately 0 to 1 metres (0 to 3 ft) of head. This head difference between the two aquifers tends to vary with location, likely as a result of numerous factors:

- the varying nature of coverage of the ground surface, i.e. permeable vs. impermeable;

- precipitation;

- tidal cycles for sites close to Jamaica Bay which has approximately a 5-foot (1500-mm) mean tidal range; and

- human-related factors such as leaking water lines and sewerage.

In the particular case of the two *Monotube* piles discussed by Fellenius et al. (2000), the writer interpreted the u_2 data presented in their paper as showing the common perched condition with a variation in piezometric levels between the two aquifers that is toward the upper end of this 0-to-1 metre range. Thus, the groundwater-level line shown in Figure A.2 reflects the groundwater table within the Holocene sand fill and the piezometric level within the UGA is approximately 1 metre (3 ft) below that.

A.4.3 Indirect/Rational Methods

A.4.3.1 Shaft Resistance

A.4.3.1.1 Generic Results

Before considering specific analytical methods, it is of interest to review measured results of a broad, generic nature that are independent of any specific methodology. Specifically, it is of interest to investigate the unit shaft resistance, r_s, that is related to the shaft resistance, R_s, as defined in the following fundamental equation:

$$R_s = \int_0^L r_s(z) \cdot C(z) \cdot dz \tag{A.1}$$

where:
- $C(z)$ = the pile circumference as a function of depth, z, below the ground surface
- L = embedded length of the pile.

Shown first in Figure A.3a are the measured values of unit shaft resistance, r_s, as a function of depth, z, along the pile shaft. These values were back-calculated by the writer using the "true", i.e. corrected for residual load, results that were scaled from Fellenius et al. (2000). Note that both r_s and z have been non-dimensionalized for generality, to atmospheric pressure, p_{atm}, and embedded pile length, L, respectively. Although not indicated in this figure, $r_s = r_{ss}$ (unit shaft resistance due to sliding friction) within the constant-diameter upper section and $r_s = r_{st}$ (unit shaft resistance due to taper) within the tapered lower section.

Note that the r_{ss} and r_{st} values for Pile #2 are essentially ultimate values ($r_{ss(ult)}$ and $r_{st(ult)}$ respectively) whereas for Pile #3 they are likely slightly less than ultimate values. Nevertheless, and despite the aforementioned installation problems affecting Pile #2 plus the inherent inaccuracy that is the result of scaling, the results for the two piles are very similar. This provides a level of confidence in the overall accuracy of these measured results which are used throughout the remainder of this appendix.

It is difficult to detect trends in the results shown in Figure A.3a because the r_s values are strongly influenced by the vertical effective overburden stresses acting along the pile shaft. Thus, a portion of the variability in r_s values is simply due to the increase in overburden stresses with depth. Therefore, Figure A.3b shows the r_s values normalized (and also non-dimensionalized at the same time) to σ'_{vo} instead of p_{atm}. This removes the influence of embedment effects so that the r_s/σ'_{vo} ratio only reflects the effects of:

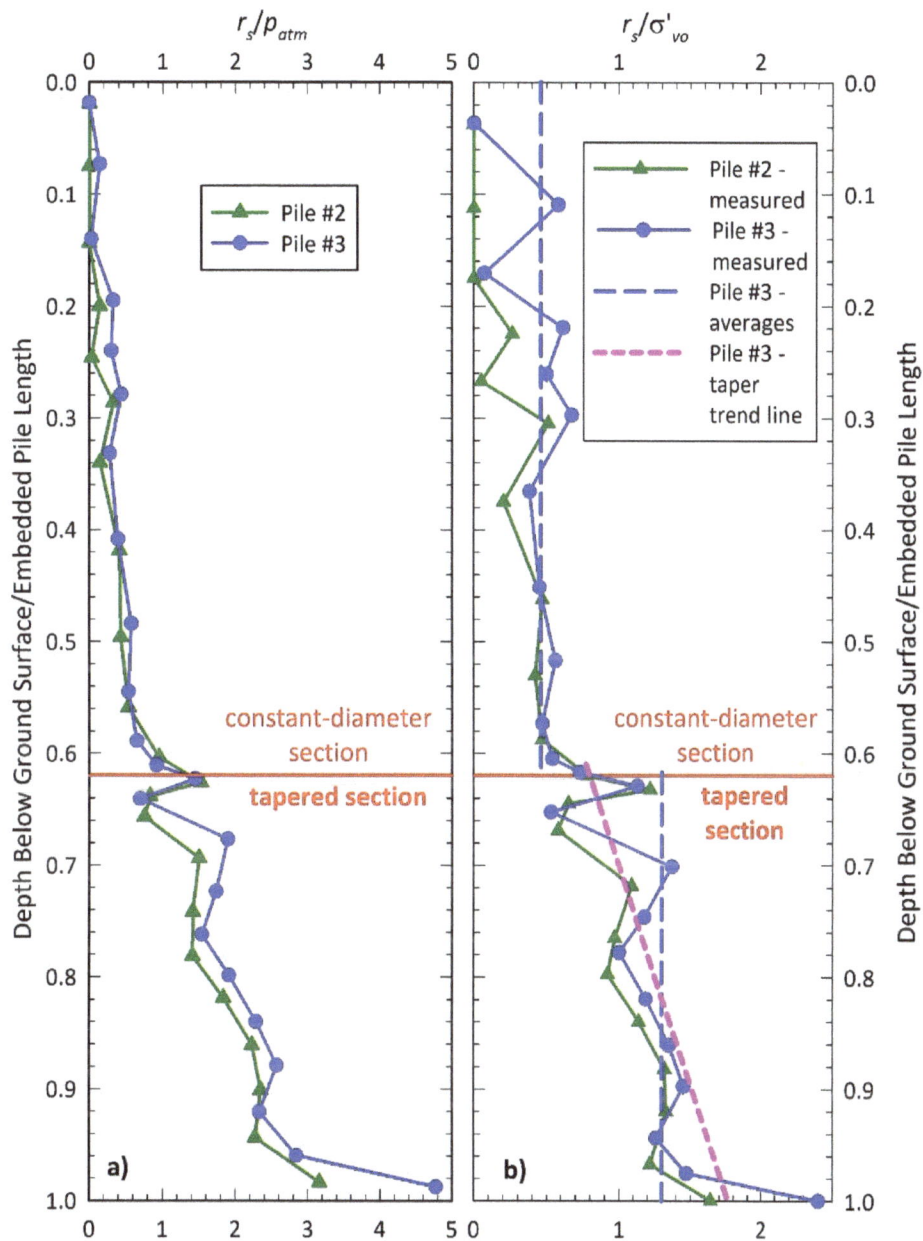

Figure A.3. Depth-Wise Distribution of Unit Shaft Resistance for Fellenius et al. (2000) *Monotube* Piles at JFKIA.

- soil type;

- soil consistency (relative density, D_r, and the state parameter, ψ, in this case) which, as a first-order approximation, can be assumed to be depth-wise uniform in this case; and

- pile typology (tapered vs. constant-diameter sections in this case).

Although there is still some variability in the results shown in Figure A.3b, this appears to be due, in part, to the difficulties in accurately extracting data from the Fellenius et alia paper. Nevertheless, there are clear and reasonably consistent distinctions between trends in r_s/σ'_{vo} values between the tapered and constant-diameter sections of the two piles.

Based on this, the writer calculated arithmetic means (simple averages) for each pile section but only for Pile #3. This is because Pile #2 had effectively zero shaft resistance for some depth below the ground surface due to the obstruction-related issues discussed earlier in this appendix. Consequently, the results within the constant-diameter upper section of Pile #2 are badly compromised by this.

These averages for Pile #3 are shown graphically in Figure A.3b and numerically are:

- r_s/σ'_{vo} = 0.46 within the constant-diameter section of Pile #3 and

- r_s/σ'_{vo} = 1.3 within the tapered section of Pile #3

for a tapered-to-non-tapered ratio (i.e. r_{st}/r_{ss}) of 2.8. Stated another way, all other factors being equal, the tapered section of Pile #3 generated a unit shaft resistance that was, on average, 2.8 times that of the non-tapered section of the same pile in these soil conditions. Note that this ratio of 2.8 is another expression of taper benefit and compares favorably to the 2.5 value of taper benefit noted earlier in this appendix.

Although not shown, the values of the r_s/σ'_{vo} ratio and taper benefit for Pile #2 are slightly less than those for Pile #3 but comparable and thus broadly support the conclusions drawn based on the results for Pile #3.

Before ending this generic discussion and moving on to considering specific analytical methodologies, there is one more observation to make. Historically (and this derives from Nordlund's original, 1963 paper), it has been assumed that taper benefit is only a function of taper angle. This is why in Figure A.3b a single average value of r_s/σ'_{vo} was interpreted within the tapered section of the pile and the calculated taper benefit was presumed to be a single value.

However, a close examination of the results shown in Figure A.3b suggest that taper benefit is, in reality, non-uniform within the tapered section and increases with increasing depth within the tapered section. This is illustrated by the dashed line in Figure A.3b that is labeled "taper trend line". This is simply a linear best-fit to the data of Pile #3. This trend line varies from approximately r_s/σ'_{vo} = 0.75 to r_s/σ'_{vo} = 1.8 (vs. a uniform average of 1.3). This, in turn, produces a taper benefit that varies from approximately 1.6 to 3.9 (vs. a uniform average of 2.8).

The direct conclusion drawn from this trend line is that taper benefit is greatest near the toe and least at the transition from the tapered section to constant-diameter upper section/extension. The indirect conclusion drawn is that toe diameter may affect taper benefit based on the argument presented subsequently.

These conclusions, while a notable departure from long-standing thinking, are actually consistent with the cylindrical-cavity mechanics concept that is used to model the two-stage (driving (Stage I) followed by axial-compressive loading (Stage II)) wedging mechanism of tapered piles. Cavity mechanics indicates that the increase in radial stress acting on a cylinder is a function of the expansion of that cylinder relative to its initial diameter. When a tapered pile is loaded in axial compression, as a first-order approximation the entire tapered section settles the same amount (this ignores elastic compression of the tapered section) and, therefore, if the Poisson Effect is also ignored the entire tapered section will have the same magnitude of radial displacement of the pile-soil interface, i.e. the cylindrical cavity. However, that magnitude of <u>absolute</u> radial displacement will have the <u>greatest relative</u> increase at the bottom of the tapered section, i.e. the toe, and the <u>least relative</u> increase at the top of the tapered section at the transition to the constant-diameter upper section. In the writer's opinion, this is exactly when is being seen with the trend line shown in Figure A.3b.

A.4.3.1.2 Modified Nordlund Method

As discussed in Chapter 5, the writer's Modified Nordlund Method uses the same basic framework as either version of the Nordlund Method in terms of relating taper benefit (as reflected in an increased value of an earth pressure coefficient relative to that of no taper angle) when calculating shaft resistance but with two key differences that are the focus of the present discussion:

- The generic lateral earth pressure coefficient, K_h, is used in lieu of the Nordlund K_δ earth pressure parameter that always acts at some angle below the horizontal.

- K_h is normalized to K_o in order to remove site-specific variability that is due to the pre-construction stress state so that the K_h/K_o ratio becomes a problem parameter that solely reflects the effects of pile geometry, i.e. taper angle or lack thereof, and installation methodology.

To begin with, expanding on Equation A.1 using the fundamental concepts presented in Chapter 4 in general and Equation 4.6 in particular, the generic ULS unit shaft resistance, $r_{s(ult)}$, as a function of depth, z, below the ground surface can be defined in terms of both K_h and K_h/K_o as follows:

$$
\begin{aligned}
r_{s(ult)}(z) &= \int_{z_1}^{z_2} K_h(z) \cdot \tan \delta(z) \cdot \sigma'_{vo}(z) \cdot dz \\
&= \int_{z_1}^{z_2} \left[\left(\frac{K_h}{K_o} \right)(z) \right] \cdot K_o(z) \cdot \tan \delta(z) \cdot \sigma'_{vo}(z) \cdot dz .
\end{aligned}
\tag{A.2}
$$

In this equation, $r_{s(ult)}$ is evaluated over some arbitrary range in depth between z_1 and z_2 where $L \geq z_2 > z_1 \geq 0$ (L = embedded length of the pile and 0 = ground surface). Whether $r_{s(ult)}$ corresponds to $r_{ss(ult)}$, the ULS unit shaft resistance due to sliding friction, or $r_{st(ult)}$, the ULS unit shaft resistance due to taper, depends on whether the pile has a constant perimeter or is tapered over the z_1-z_2 depth range.

Experimental determination of both K_h and K_h/K_o is straightforward using Equation A.2, at least in concept, by simple arithmetic manipulation and solution of Equation A.2 using site-specific data. However, there are complications that can arise that are discussed subsequently,

For the Fellenius et alia case history discussed in this appendix, the values of r_s shown in Figure A.3a were assumed to be ULS values, $r_{s(ult)}$, and used together with estimates of tan δ and σ'_{vo} that were outcomes from the writer's site-characterization assessment using the data for the CPTu sounding contained in Fellenius et al. (2000). The traditional assumption was made that sliding between the pile shaft and soil occurred at the interface between these two dissimilar materials. Consequently, $\delta/\phi = 0.6$ was assumed and the results for the constant-volume (critical-state) strength were used as being consistent with ULS conditions.

Figure A.4a shows the back-calculated depth-wise variation of K_h for both piles. As with the unit shaft resistance shown in Figure A.3, the results for each pile are similar except for the upper portion of Pile #2 that was adversely affected by the aforementioned installation problems.

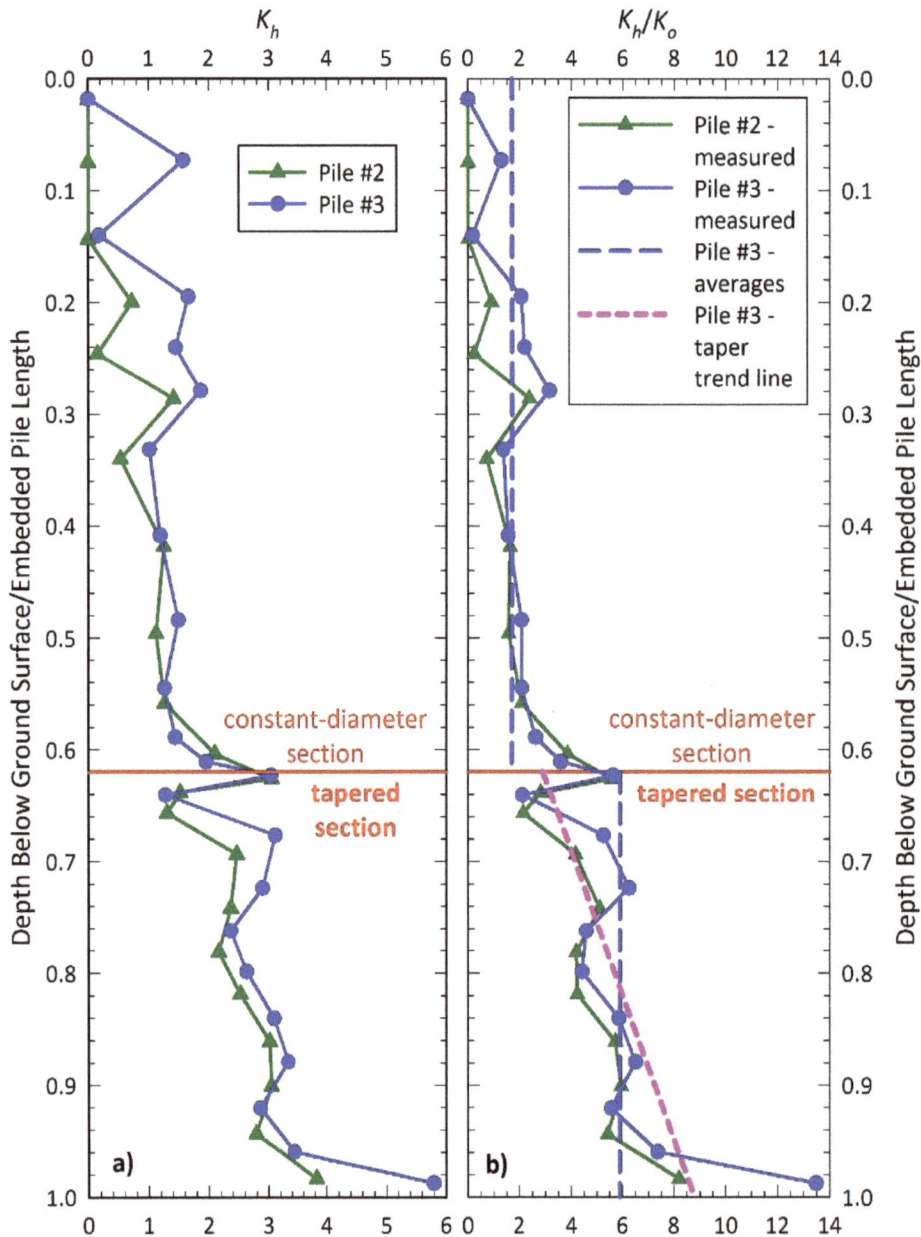

Figure A.4. Depth-Wise Distribution of Back-Calculated K_h and K_h/K_o for Fellenius et al. (2000) *Monotube* Piles at JFKIA.

Figure A.4b shows K_h normalized to K_o using the values for the latter parameter shown in Figure A.2. As intended and expected, this normalization reduces some of the depth-wise variation, especially along the upper portion of the shaft where K_o varies significantly with depth.

Also shown in Figure A.4b are average values of the K_h/K_o ratio, again, only for Pile #3:

- K_h/K_o = 1.7 within the constant-diameter section which agrees well with the range of 1-2 suggested by Kulhawy (1984) for "large-displacement piles".

- K_h/K_o = 5.9 within the tapered section. This results in a ratio of $(K_h/K_o)_{tapered}$ to $(K_h/K_o)_{non-tapered}$ of 3.5 that can be viewed as another metric of taper benefit. This ratio is somewhat higher than the ratio of r_{st}/r_{ss} = 2.8 observed for unit shaft resistance.

- As with unit shaft resistance shown in Figure A.3b, the K_h/K_o ratio within the tapered section is better represented with a trend line that increases linearly with depth as opposed to being a single average value. This trend line is shown in Figure A.4b and varies approximately from K_h/K_o = 2.9 to 8.8.

Because both K_h and K_h/K_o are key parameters that are generated as part of the calculation process when using the Modified Nordlund Method, it is of interest to see how values of these parameters that are generated by this methodology compare to these measured results. Figure A.5a shows this comparison for K_h and Figure A.5b does the same for K_h/K_o. Note that the measured results are shown for both Pile #2 and Pile #3 while the calculated results are shown only for Pile #3 as the Pile #2 results are exactly the same as only one CPTu sounding was made for both piles.

The agreement between measured and calculated results is considered to be good overall. With particular attention to the K_h/K_o parameter shown in Figure A.5b, within the constant-diameter section the writer's Modified Nordlund Method produced a value of 1.65 that compares very favorably to the 1.7 average measured value shown in Figure A.4b. Within the tapered section, the Modified Nordlund Method produced a value of 6.7 which is somewhat greater than the 5.9 average measured for Pile #3 shown in Figure A.4b. The ratio of $(K_h/K_o)_{tapered}$ to $(K_h/K_o)_{non-tapered}$ for the calculated results shown in Figure A.5b is 4.1 which is somewhat greater than the 3.5 for the measured results shown in Figure A.4b.

While it is interesting from a theoretical and research perspective to compare individual problem parameters such as K_h and K_h/K_o, of greater practical interest and relevance is the overall forecasting accuracy of the writer's Modified Nordlund Method. To begin with, Table A.1 summarizes the maximum applied load at the head of each pile, Q_{max}, both measured as reported in Fellenius et al. (2000) and as forecast using the Modified Nordlund Method (note that this method forecasts the ULS resistance from the ground, R_u, but this is the same as the ultimate capacity of the pile in terms of externally applied compressive load, Q_u, as can be seen in Figure 2.1). The two different toe-resistance models used for the forecast results are discussed later in this appendix. Note that for the calculated results and, marginally, Pile #2 Q_{max} is the same as the ultimate value, R_u. For Pile #3, Q_{max} is somewhat less than R_u.

The writer considers the overall agreement between measured and forecast results to be quite good, especially in consideration of the fact that the CPTu data were scaled from published figures. As a result, only a very limited number of data sets and concomitant artificial pile segments were used in the writer's forecasting calculations. Normally, every piece of cone penetrometer data would be used which results in substantially more analytical pile segments with greater precision in the final results. There was also some approximation and potential error introduced by the scaling process itself.

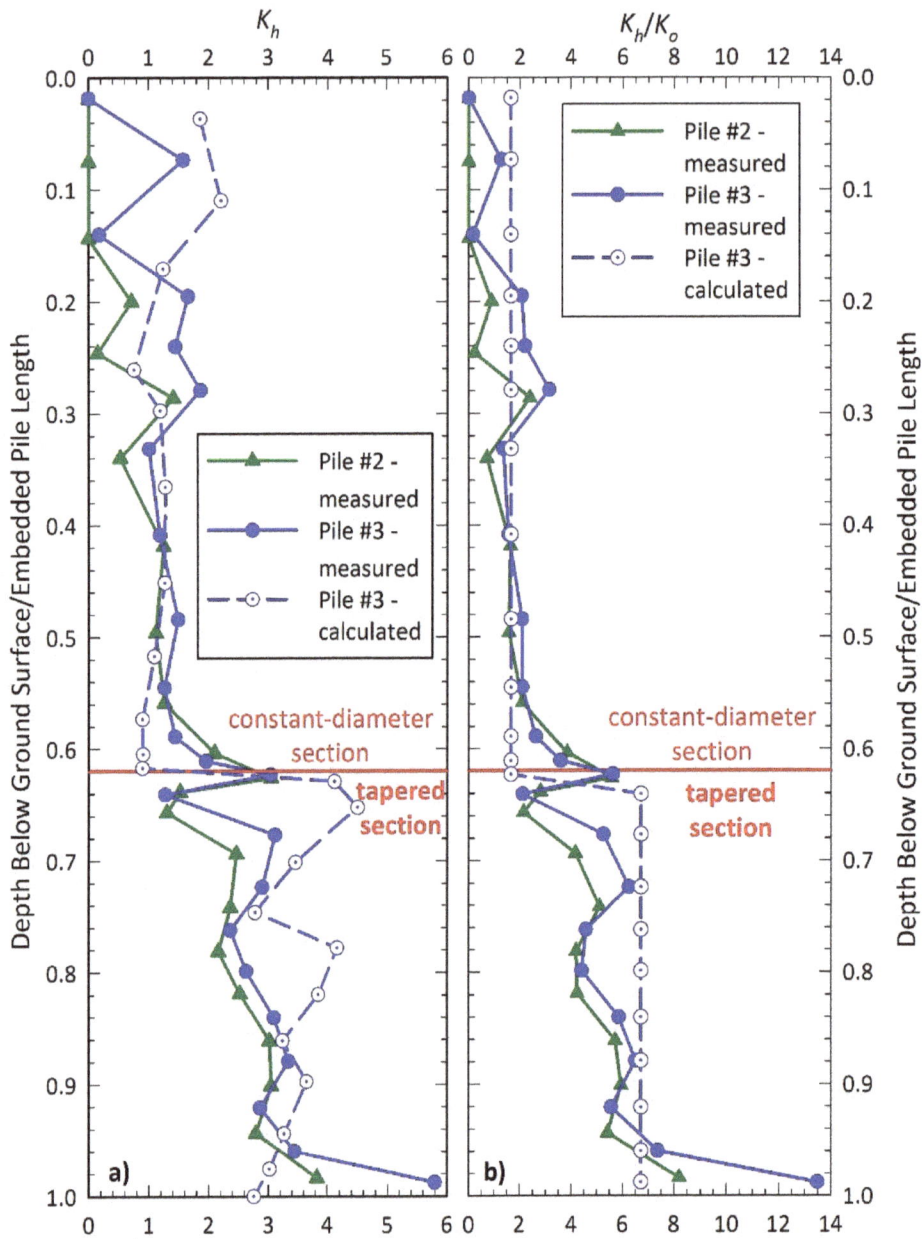

Figure A.5. Calculated vs. Measured Values of K_h and K_h/K_o for Fellenius et al. (2000) *Monotube* Piles at JFKIA.

Table A.1. Measured vs. Forecast Maximum Capacity for Fellenius et al. (2000) *Monotube* Piles at JFKIA.

Pile	Q_{max}, kN (kips)		
	measured	forecast (Modified Nordlund Method)	
		original toe model	M&Y toe model
#2	2000 (449)	2297 (516)	2452 (551)
#3	2500 (562)	2404 (540)	2525 (567)

The implied over-forecasting for Pile #2 is likely due to a combination of the difficulties encountered during pile installation that resulted in a somewhat bent or 'dog-leg' pile with almost no shaft resistance in the upper 4 metres (13 ft), which accounts for 20% of the overall pile length, plus the fact that it was load tested only three days after installation compared to 21 days for Pile #3. As discussed in Chapter 4, due to the now-well-known temporal increase in resistance that occurs with piles driven in coarse-grained soil, some noticeable difference in measured ULS total resistance would be expected between otherwise identical piles tested 3 and 21 days after installation. The writer's forecasting methodology is intended to provide an estimate of long-term resistance.

It is also of interest and significance to see how the depth-wise distribution of resistance compares between forecasts made using the writer's Modified Nordlund Method and measured results. In order to have confidence in an analytical methodology for forecasting deep-foundation resistance, not only should the total resistance be accurate but the individual shaft and toe components should be as well.

These comparisons are shown in Figures A.6 and A.7 for Piles #2 and #3 respectively. Note that the applied axial-compressive loads are normalized to the maximum value that was either measured or forecast as show previously in Table A.1 so that the comparison focuses on the relative load distributions with depth. This is the same conceptual approach that was used previously in Figure A.1 to compare the relative measured load distributions between Piles #2 and #3. Note also that two forecast resistances are shown for each pile, one made using the writer's original toe-resistance methodology based on classical bearing-capacity theory and the other based on SCE theory as modified for tapered piles by Manandhar and Yasufuku (M&Y). The shaft resistances are the same in each case.

In the writer's opinion, the agreement between the forecast and measured results is overall quite good for each pile. This is quite encouraging given the various issues involving the actual piles and the approximations that went into converting the published data into analyzable results. When the results in Figures A.6 and A.7 are taken together with the results in Table A.1, in the writer's opinion this supports the writer's Modified Nordlund Method as a viable analytical tool for tapered piles in soil profiles that consist predominantly of coarse-grained soil. This conclusion is consistent with approximately 20 years of research by the writer and concomitant published works (Horvath 2002, 2003c, 2014b, 2015, 2016a, 2016b; Horvath and Trochalides 2004; Horvath et al. 2004a, 2004b) that were limited to comparisons of ULS total resistance calculated using the Modified Nordlund Method with estimates of geotechnical-ULS loads interpreted from basic, non-instrumented static load tests on a variety of tapered and constant-diameter piles in predominantly coarse-grained soil profiles.

The most significant deviations in the results shown in Figures A.6 and A.7 involve the toe resistance which is somewhat underestimated by the writer's forecasting methodologies relative to the estimated actual resistance indicated by Fellenius et alia. It is relevant to note that both versions of the toe resistances in the writer's analytical method were made using the constant-volume (critical-state) angle of internal friction. This was done intentionally to provide a lower-bound estimate of the ULS toe resistance as would occur under relatively large toe settlements that would normally be associated with the geotechnical ULS. In this case, although the toe settlements of the actual piles are inferred to have been relatively small during the load tests, Fellenius et alia interpreted the residual loads in the piles to indicate that toe resistance was already fully mobilized in each case prior to the start of the load tests. Therefore, assuming that constant-volume/critical-state strength had been fully mobilized was judged by the writer to be a reasonable decision.

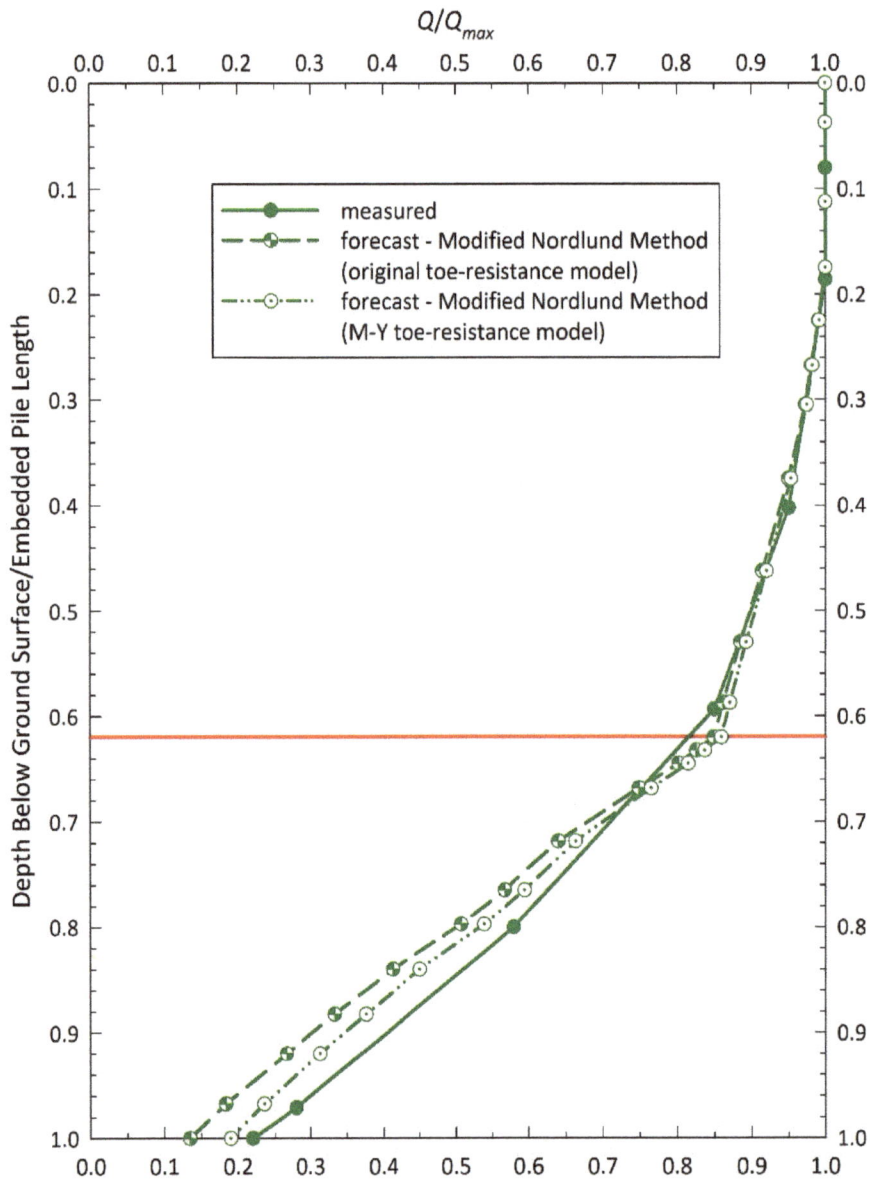

Figure A.6. Depth-Wise Relative Distribution of Forecast vs. Measured Axial-Compressive Forces in Pile for Fellenius et al. (2000) *Monotube* Pile #2 at JFKIA.

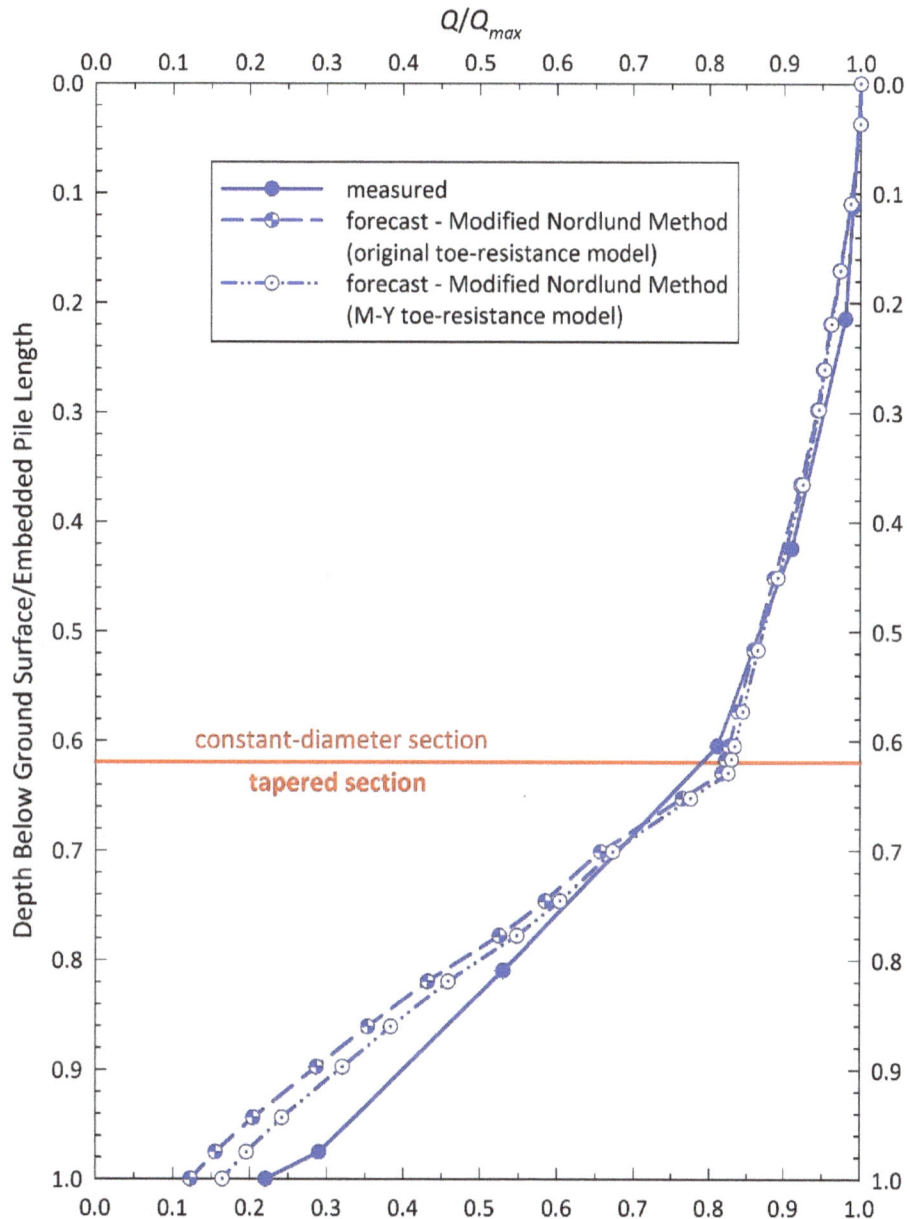

Figure A.7. Depth-Wise Relative Distribution of Forecast vs. Measured Axial-Compressive Forces in Pile for Fellenius et al. (2000) *Monotube* Pile #3 at JFKIA.

In any event, the writer's experience with calculating peak-strength toe resistances, as outlined in detail in Horvath (2002), is that the peak friction angle beneath the toe is only marginally greater than the constant-volume friction angle. This is because the dilatancy angle, which is the difference between the peak and constant-volume values, is relatively small due to the relatively high compressive stresses involved. As a result, in most cases the peak-strength toe resistance is only of the order of 10% greater than the constant-volume-strength toe resistance.

In summary up to this point, for the two *Monotube* piles discussed by Fellenius et alia, the writer's Modified Nordlund Method:

- closely matched the measured shaft resistance within the upper, constant-diameter section of the piles;

- slightly overestimated shaft resistance within the lower, tapered portion of the piles; and

- slightly underestimated toe resistances

with a net overall close match of the total resistance. Furthermore, the overall depth-wise distribution of resistance forecast by the writer's Modified Nordlund Method was a very good match to that measured in the two piles.

There is one final issue to discuss concerning the writer's Modified Nordlund Method. As can be seen in Equation A.2, whether formulated in terms of K_h or K_h/K_o the operational value of the Mohr-Coulomb friction angle acting along the shaft-soil interface must always be considered and evaluated explicitly. Historically, calculation of this parameter was treated as being relatively straightforward to the point of being trivial as it was assumed to be the angle, $\delta(z)$, of the soil sliding on the structural material that comprised the exterior of the DFE shaft in question. This problem parameter was assumed to be measurable in laboratory testing or reasonable to 'guesstimate' in routine practice.

Until recently, the only exception to this generic interpretation of $\delta(z)$ was to be found in both versions of the Nordlund Method. As discussed in Chapter 5, Nordlund chose to include pile-volume into the process of quantifying the magnitude of this angle. As a result, the angle δ as defined by Nordlund is unique to his method and, in general, quite different than any other geotechnical or foundation engineering usage of this parameter.

However, recent seminal research[175] at Purdue University that included detailed photographic and video documentation of soil-particle displacement adjacent to model piles in a 1-g laboratory environment suggests that in most cases vertical shearing occurs not as sliding along a well-defined, 1-D pile-soil interface but as intra-soil shearing within a relatively narrow, but definitely 2-D, zone or 'shear band' adjacent to the pile shaft.

In retrospect, this finding is not surprising. As noted in Chapter 4, the suggestion of a shear-band of some finite thickness as opposed to a distinct shear interface of nominally zero thickness being the operative physical mechanism for the classical shaft-resistance mechanism of sliding friction for constant-perimeter DFEs has been proposed by others for some time now as implied by the following quote from Fellenius (2018, with the writer's comments highlighted in yellow):

> "...the 'movement' [Fellenius is referring to the axial displacement of a DFE shaft relative to the adjacent ground.] is not a slippage, i.e., definite sliding of the pile element against a stable body of soil, but occurs as a shear deformation within some zone or band of soil next to the pile element. Therefore, the movement along the side of the element is the relative movement between the pile element surface and a [sic] outer boundary or shear zone, a somewhat undefined location, actually."

---

[175] A presentation titled "Insights into Pile Design from Experimental and Theoretical Research" that summarized this research was made by Prof. Rodrigo Salgado of Purdue at the 41st annual seminar of the ASCE Geo-Institute Metropolitan Section in New York City on 11 May 2017. Unfortunately, Prof. Salgado did not provide any written documentation of his lecture as part of the seminar proceedings so the comments made in this monograph are based solely on the writer's personal recollection from attending this seminar and talking with Prof. Salgado briefly after his presentation.

However, the research presented by Salgado in 2017 was seminal (in the writer's opinion) in that it provided examples of both physical and numerical-analysis proof in support of the hypothesized behavior. The conclusion drawn from this is that it is the friction angle of the soil, ϕ, and not the interface friction angle, δ, that should somehow be used in any back-calculation involving field measurement or forecasting involving an analytical methodology although the exact manner in which ϕ should be used is unclear (to the writer at least) at this time.

As an aside before pursuing this line of reasoning further, it is emphasized that this concept of a shear band providing shaft resistance is, at this point in time, applicable only to deep foundations with a constant perimeter. It remains to be seen whether or not a shear band develops with tapered piles and the concomitant shaft-resistance mechanism of wedging.

It is noted that the writer did bring this issue to Prof. Salgado's attention in a personal conversation as well as follow-up emails in 2017. However, Prof. Salgado did not proffer an opinion on or response to this specific issue, perhaps due to Prof. Salgado's (presumed) unfamiliarity with tapered piles and/or lack of specific knowledge on the subject. As of the timeframe during which this monograph was finalized for publication in early 2020, the writer has had no further response or follow up from Prof. Salgado.

In any event, in the writer's opinion, this recent research at Purdue represents a significant paradigm shift and divergence from established thinking. Additional research is clearly required, especially with tapered piles, as all piles tested at Purdue to date had a constant diameter. The effect of pile cross-sectional geometry, i.e. circular vs. square, should also be studied for both constant-perimeter and tapered piles.

In addition, the writer's Modified Nordlund Method was crafted based on the assumption that the shearing that develops along a pile shaft is the traditional assumption of 1-D sliding of soil on structural material. Therefore, the comparison between measured and forecast results presented previously in Figures A.4 and A.5 was based on using an assumed soil-pile friction angle, δ. Specifically, $\delta/\phi_{cv} = 0.6$ was assumed for the constant-diameter portion of the piles and $\delta/\phi_{peak} = 0.6$ was assumed within the tapered portion. The rationale for assuming constant-volume (critical-state) versus peak angles along different portions of the piles was discussed in detail in Horvath (2002). In all cases, ϕ values were evaluated using the writer's site-characterization algorithm that was discussed in Horvath (2015).

Given the recent research outcomes from Purdue University as noted above, it is at least of academic interest to reinterpret the measured results from Fellenius et alia assuming that the pile-soil interface friction along the pile shafts is defined by ϕ, not δ. The comparisons of interpreted results using the different friction assumptions and for both K_h and K_h/K_o are shown in Figure A.8 for Pile #3 only (the results for Pile #2 are qualitatively identical and were omitted for clarity).

Note that the purpose of Figure A.8 is not to suggest or infer that one assumption concerning the operative interface friction angle along the pile shaft is correct compared to the other but to highlight the fact that whichever assumption is made results in a significant difference in interpreted results compared to the other assumption. Clearly, additional research needs to be conducted, especially for tapered piles which have not yet been tested at Purdue, before any change in design philosophy is suggested and implemented.

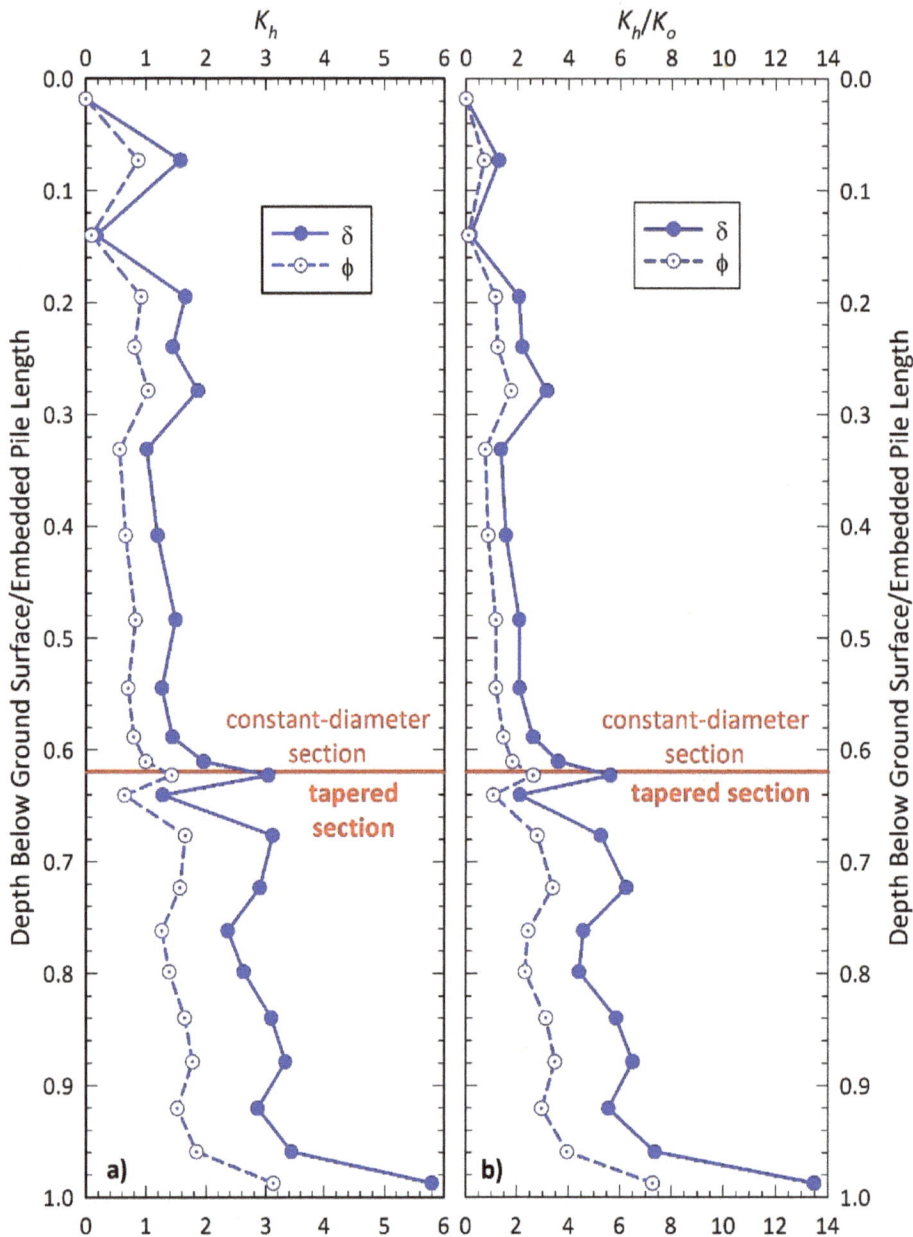

Figures A.8. Effect of Shaft-Friction Assumption for Fellenius et al. (2000) _Monotube_ Pile #3 at JFKIA.

A.4.3.1.3 Extended β Method

What is defined and referred to in Chapter 5 as the Original β Method that is the result of the collective seminal contributions of Bjerrum (1960s), Burland (1970s), and, in more-recent decades, Fellenius. The method combines the dimensionless components of lateral earth pressure, K_h and δ, that are acting along the shaft of a DFE into a single, dimensionless, empirical parameter, β, as follows:

$$r_{s(ult)}(z) = \int_{z_1}^{z_2} K_h(z) \cdot \tan \delta(z) \cdot \sigma'_{vo}(z) \cdot dz = \int_{z_1}^{z_2} \beta(z) \cdot \sigma'_{vo}(z) \cdot dz \qquad \text{(A.3)}$$

where z_1 and z_2 define some arbitrary depth range as was done earlier in this appendix for the writer's Modified Nordlund Method.

As noted in Chapter 5, perhaps the most unique aspect of the β Method is that it treats all soils in the same way, i.e. on an effective-stress basis. It was a radical departure from the state of practice in the 1960s when the concept was first suggested and remains an outlier in the state of practice at the present time as other indirect methods use effective stresses only for coarse-grained soil and resort to a total-stress strength interpretation for fine-grained soils.

Because the value of β reflects the DFE type and installation effects in addition to soil type, tapered piles have historically not been accounted for in the Original β Method as the recommended β values that have appeared over the last 50-plus years in a wide variety of publications have not included explicit mention of tapered piles. Thus, one is forced to resort to using some approximate, indirect, unproven gimmick such as the donut-on-a-stick analytical model proposed by Fellenius for tapered piles that was discussed in Chapter 5. In the writer's opinion, this is wholly unsatisfactory as the efficacy of the donut-on-a-stick model does not appear to have ever been demonstrated in the published literature.

A much more logical and theoretically sounder approach is to expand the database of piles for which β values are available to explicitly include tapered piles. In this way, the methodology is extended directly without resorting to any approximations or workarounds of dubious and questionable accuracy. In addition to the variables of pile material and soil type, it is logical to expect that β values for tapered piles will, as a minimum, be a function of taper angle. Based on the material presented up to this point in this appendix, it is possible that toe diameter may play a role as well.

As noted in Chapter 5, the writer termed such an expanded database of β values the Extended β Method. The purpose of the present discussion is to illustrate how a database for the Extended β Method can be populated using as examples the two *Monotube* piles presented in Fellenius et al. (2000).

Figure A.9a shows the measured values of β that were back-calculated by the writer. The vertical effective overburden stresses necessary to perform the calculations were estimated by the writer based on a site-characterization assessment performed on the CPTu sounding presented in the Fellenius et alia paper. The results between the two piles are reasonably close as might be expected based on results for these two piles shown earlier in this appendix.

Also shown in Figure A.9a are average values of the back-calculated β values for Pile #3 only. These averages are $\beta = 0.44$ within the upper constant-diameter section and $\beta = 1.3$ within the tapered lower section. This means that the average β value within the tapered section is almost three times that within the constant-diameter section. This is another metric for taper benefit and is numerically consistent with tapered-to-non-tapered ratios found for other shaft-resistance parameters earlier in this appendix.

As was found with other parameters that were discussed earlier in this appendix, the measured β values within the tapered section actually display a depth-wise increasing trend as opposed to a single average value. The best-fit trend line for this is also shown in Figure A.9a and the β values vary from approximately 0.8 to 1.8. This means that the taper benefit actually varies from approximately 1.8 to 4.1 within the tapered section as opposed to having a single average value of approximately 3.0.

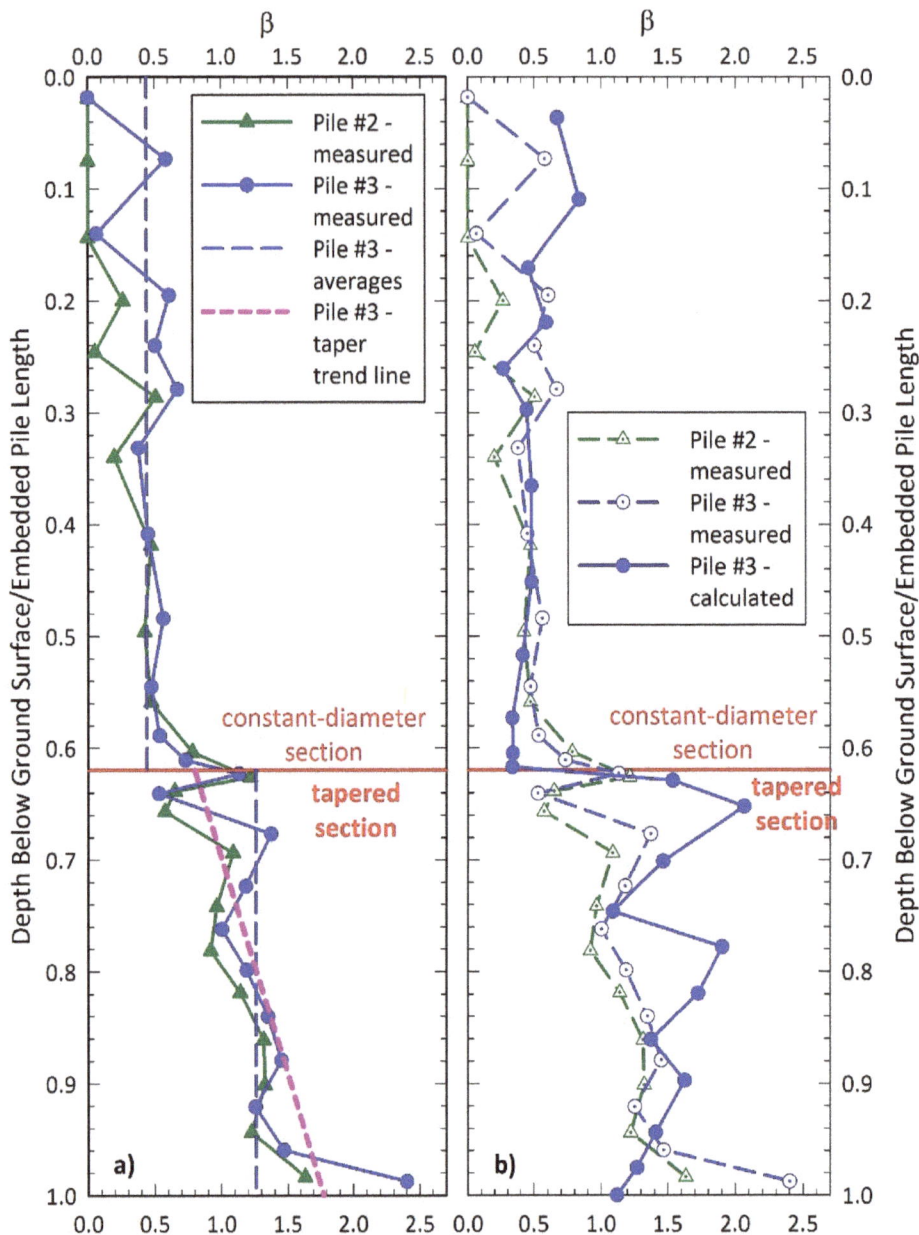

Figure A.9. Depth-Wise Distribution of β Values for Fellenius et al. (2000) *Monotube* Piles at JFKIA.

Overall, the results shown in Figure A.9a support the argument made by the writer that creating an Extended β Method that explicitly considers tapered piles is a necessary endeavor if the benefit of tapered piles is to be assessed correctly in analytical methods.

Historically, β values have only been evaluated using the results of instrumented static load tests. It is important to note that it is essential that the data from such tests be corrected as necessary for residual loads. Failure to do so will produce erroneous values of β as is illustrated in Fellenius (2015) using actual static load test results. In the writer's opinion, it is quite likely (to the point of certainty) that at least some of the β values reported in the literature over the past 50-plus years were based on measured results that were not corrected for residual loads and are thus inherently incorrect. This is an important fact to keep in mind going forward in the use of the β Method in any of its forms.

Because developing a sufficient database of β values for tapered piles in order to make the writer's proposed Extended β Method viable for use in routine practice will take time, it is suggested that an initial database be developed using other analytical methodologies. Such a database could then be revised over time as measured results became available.

It is both relevant and important to note that this suggestion of using other analytical methodologies to populate a database for the Extended β Method has precedent. As discussed in Chapter 5, for some time now Fellenius has been suggesting that the Original UniCone Method (which he now refers to as the Eslami and Fellenius Method), which is a direct resistance-forecasting methodology, be used to generate the desired parameter values, β and N_t, for the Original β Method.

Using a combination of theory and calculated results from another resistance-forecasting methodology to develop, complement, or supplement a database of β values is possible because of the following generic relationship that is the result of rearranging and simplifying Equation A.3:

$$\beta = \frac{r_{s(ult)}}{\sigma'_{vo}}. \tag{A.4}$$

Thus, by taking the ULS unit shaft resistance, $r_{s(ult)}$ (more specifically, $r_{ss(ult)}$ for a constant-perimeter section or $r_{st(ult)}$ for a tapered section) at some arbitrary depth along a pile shaft that is produced by some analytical methodology and dividing by the calculated vertical effective overburden stress at the same depth then an estimate of β at that depth is obtained.

The writer did this for the two piles in the Fellenius et alia case history, with $r_{ss(ult)}$ and $r_{st(ult)}$ calculated using the writer's Modified Nordlund Method. The results are shown in Figure A.9b for Pile #3 only as the results for Pile #2 were essentially the same. The calculated values of β within the constant-diameter section compare favorably with the measured results ($\beta_{average}$ = 0.47 calculated vs. $\beta_{average}$ = 0.44 measured). The agreement within the tapered section is somewhat poorer, $\beta_{average}$ = 1.5 calculated vs. $\beta_{average}$ = 1.3 measured. The modest overestimation of calculated results is consistent with similar overestimations for other analytical parameters within the tapered section that were discussed previously. This is to be expected as all of these parameters are mathematically interrelated. Nevertheless, the writer feels that this theoretical approach for estimating β values has merit and benefit given the difficulty of obtaining measured β values following the traditional route of using instrumented static load tests that have been properly corrected for residual loads.

A.4.3.1.4 Modified β Method

As discussed in Chapters 4 and 5, the writer found that normalizing β values to the pre-driving K_o values helps to eliminate some of the depth-wise variation in plotted values, both measured and forecast, so that trends that reflect either changed soil conditions or changed pile geometry are easier to detect. The benefit of this normalization concept was illustrated earlier in this appendix in Figure A.4 with the use of K_h/K_o as opposed to K_h alone as a key element of the writer's Modified Nordlund Method. Note that evaluating and using specific β/K_o values for tapered piles is an essential part of the Modified β Method that is being discussed in this section as it builds on the concepts presented in the preceding section for the Extended β Method.

Figure A.10a shows the values of β/K_o that were back-calculated by the writer. The K_o values (shown in Figure A.2 for the Fellenius et alia *Monotube* test-pile site at JFKIA) and vertical effective overburden stresses necessary to perform the calculations were estimated by the writer based on a site-characterization assessment performed on the CPTu sounding presented in the Fellenius et alia paper. The results between the two piles are reasonably close as might be expected based on results for these two piles shown earlier in this appendix.

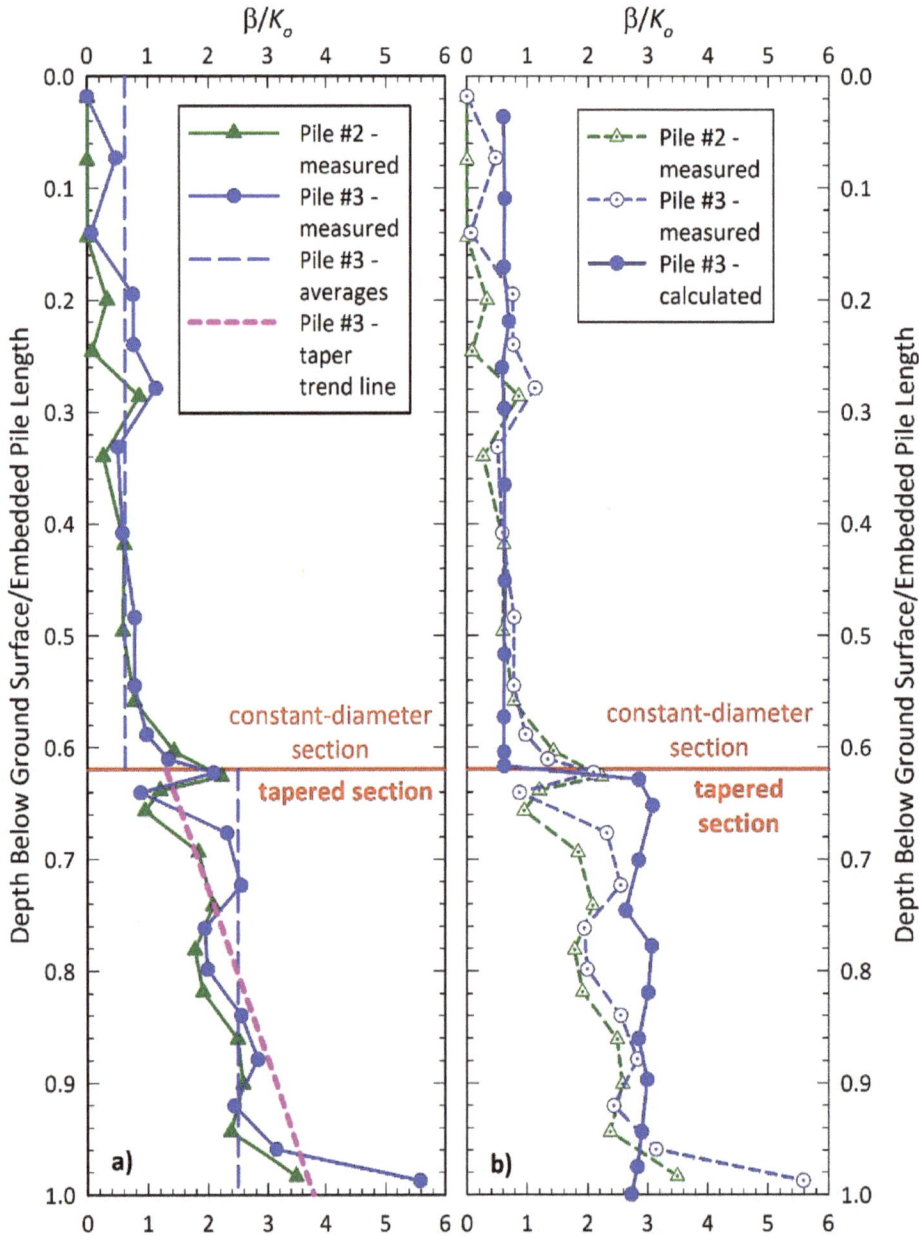

Figure A.10. Depth-Wise Distribution of β/K_o for Fellenius et al. (2000) *Monotube* Piles at JFKIA.

Also shown in Figure A.10a are average values of the back-calculated β/K_o values for Pile #3. These averages are $\beta/K_o = 0.62$ for the constant-diameter section and $\beta/K_o = 2.5$ for the tapered section. This means that the average β/K_o value within the tapered section is approximately four times that within the constant-diameter section. This is another metric for taper benefit and is broadly consistent with tapered-to-non-tapered ratios found for other shaft-resistance parameters earlier in this appendix.

Once again, as was found with other parameters discussed earlier in this appendix the measured β/K_o values within the tapered section actually display a depth-wise increasing trend as opposed to a single average value. The best-fit trend line for this is also shown in Figure A.10a and the β/K_o values vary from approximately 1.3 to 3.8. This means that the taper benefit actually varies from approximately 2.1 to 6.1 within the tapered section as opposed to having a single average value of approximately 4.

For the same reasons and precedent outlined above for the Extended β Method, there is significant benefit to developing confidence in using analytical methodologies to forecast β/K_o values for tapered piles in order to develop a database for the Modified β Method that could be used in the future with confidence in practice. To illustrate this, Figure A.10b compares the calculated results obtained using the writer's Modified Nordlund Method (which can be used to calculate β/K_o values as well) to the measured results. Calculated results are again shown only for Pile #3 as the results for Pile #2 are essentially the same.

The calculated values of β/K_o within the constant-diameter section compare very favorably with the measured results, $(\beta/K_o)_{average} = 0.62$ calculated vs. $(\beta/K_o)_{average} = 0.62$ measured. The agreement within the tapered section is somewhat poorer, $(\beta/K_o)_{average} = 2.9$ calculated vs. $(\beta/K_o)_{average} = 2.5$ measured. The modest overestimation of calculated results is consistent with similar overestimations for other analytical parameters within the tapered section that were discussed previously. This is to be expected as all of these parameters are mathematically interrelated. Nevertheless, the writer feels that this theoretical approach for estimating β/K_o values has merit and benefit given the difficulty of obtaining measured β/K_o values following the traditional route using instrumented static load tests that have been corrected for residual load effects.

A.4.3.1.5 Comments Concerning the β Method

In the writer's opinion, the results presented in the preceding sections discussing variants of the generic β Method support the position first put forth in Horvath and Trochalides (2004) that there are distinct computational benefits to using β/K_o as opposed to β alone. This is true regardless of whether a DFE is tapered or has a constant perimeter. In the specific case-history example explored in this appendix, visual trends that were suggested in Figure A.9 using β alone are better and more clearly defined in Figure A.10 using β/K_o.

This is particularly true with regard to the calculated values. As can be seen in Figure A.10b compared to Figure A.9b, calculated results based on the writer's Modified Nordlund Method show near-uniform distributions of forecast β/K_o values within both the constant-diameter and tapered sections as the effect of the varying initial stress state in the ground have been factored out so that the effects of pile typology (in this case) are more clearly seen.

That having been said, it is also clear from the material presented up to this point that taper benefit, whether expressed as K_h/K_o or β/K_o, clearly is not uniform but has a depth-wise linear trend within the tapered portion of the piles studied. This depth-wise variation was quite a surprise to the writer as it is not replicated by any current analytical methodology such as the Nordlund Method (either version) or the writer's Modified Nordlund Method that assumes that taper angle alone determines taper benefit. There are clearly one or more additional factors that influence taper benefit and the argument based on cylindrical-cavity mechanics that was made earlier in this appendix suggests that toe diameter may be that additional factor or at least one of the additional factors if there turns out to be more than one.

A.4.3.2 Toe Resistance

A.4.3.2.1 Introduction

As discussed at length in Chapter 4, toe resistance is an aspect of DFE resistance that has been undergoing profound reassessment in recent years, even if that reassessment has been largely at the research level and has not yet propagated to the extent of modifying routine practice to any significant extent. Therefore, it is useful to extract as much useful information as possible from the two Fellenius et alia piles with regard to toe resistance to see if the measured results can shed any light on the evolving world of both conceptually and numerically forecasting toe resistance for DFEs.

A.4.3.2.2 Background

To curate and summarize the current state of knowledge up to the present time, toe resistance was historically assumed to behave like shaft resistance in that there is always a well-defined, single-valued ULS value that can be calculated. This was and still is done primarily using traditional bearing-capacity theory that is based on the theory of plasticity and assumed rigid-plastic soil behavior. A failure wedge is assumed to develop beneath the toe of a DFE and this is assumed to be connected to a radial failure zone of varying geometry and complexity that is resisted by <u>vertical</u> overburden stresses. Any discussion and opinion are limited to whose bearing-capacity solution should be used for this purpose. A solution that was developed with deep-foundation applications in mind was and is generally favored. Nordlund (1963) contains a typical discussion along these lines that resulted in his use of the Berezantzev et alia solution that was a then-recent development for the circa-1960 timeframe of Nordlund's original work.

Arguably the most significant contribution to this traditional conceptual approach was by Kulhawy who said that soil compressibility/rigidity effects could not, and thus should not, be ignored for deep foundations in general and that Vesic's rigidity/compressibility factors should be assessed and applied in every deep-foundation application. Unfortunately, this requires bringing soil stiffness into the calculation process, something that always adds a significant layer of complexity and uncertainty due to the inherent non-linear, stress-dependent stress-strain behavior of soil. This fact has always limited the use of Kulhawy's recommendation in practice even though it was made well over 30 years ago.

As far back as circa 1970 there were efforts to develop an alternative physical mechanism for the failure surface that developed beneath the toe of a DFE based on spherical-cavity expansion (SCE) theory. The most significant and noteworthy aspect of SCE theory in the writer's opinion is the fact that <u>radial</u> stresses at the toe govern, <u>not</u> vertical stresses as with traditional bearing-capacity solutions. This was and remains a tectonic paradigm shift in thinking with respect to DFE toe resistance.

Unfortunately, application of an SCE-based solution also requires bringing soil stiffness into the calculation process so the same pragmatic issues that have inhibited routine, widespread use of Vesic's rigidity/compressibility factors with traditional bearing-capacity solutions have inhibited the use of SCE theory in practice for the last 50 years.

While the concept of the existence of a single-valued ULS toe resistance persists in routine practice, it has long been recognized that many friction/floating piles that bear entirely in soil do not exhibit the characteristic plunging load-settlement behavior in axial compression that is consistent with a classical bearing-capacity failure regardless of the failure mechanism, i.e. the traditional wedge or spherical cavity. Rather, toe resistance continues to increase with increasing settlement, albeit at a much-reduced rate.

In recognition of this, Fellenius has taken the extreme position that there is no such thing as a bearing-capacity failure for DFEs. However, the more middle-of-the-road approach adopted by others, including the writer, is that while there is a ULS toe resistance (which is probably better defined using an SCE model), the toe settlement that is required to mobilize it cannot be ignored as was and still is done traditionally. Thus, toe resistance, even in routine practice, is best viewed as being related to toe settlement in some fashion whether one believes that it approaches a single-valued ULS value or not. Consequently, it is not necessary to totally buy in to Fellenius' extreme position in order to agree that DFE toe resistance has a not-insignificant settlement dependency.

In the writer's opinion, this is a very important point as with the continued migration from ASD to LRFD in routine design practice the concept of the unequivocal existence of a single-valued ULS toe resistance for DFEs that is independent of toe settlement is becoming more entrenched in practice than ever. This is because LRFD is, by definition, a strength- and ULS-based based concept and design framework that demands single-valued ULS values for all relevant problem parameters and system components. Period. No questions asked. In this regard, the toe resistance of DFEs does not 'play well with others' and this fact of life simply needs to be recognized in the writer's opinion (as can be inferred, the writer is <u>not</u> a blind-faith fan of or adherent to the LRFD concept in geotechnical and foundation engineering).

The bottom line is that within the confines and constraints of LRFD it simply must be recognized that mobilization of DFE toe resistance is settlement dependent. This has been done for drilled shafts for decades now and there is no reason why this should not be done in some way for piles (and other types of DFEs) as well. One possible approach is to define a single-valued pseudo-ULS toe resistance at some specific, arbitrary magnitude of toe settlement that is based on either a fixed value or some percentage of the toe diameter as discussed in Fellenius (2018). However, addressing the specifics of such an approach is beyond the scope of this monograph and is the subject for future study by others.

A.4.3.2.3 Case-History Results

Addressing now the specific case history that is the focus of this appendix, Figures A.11 and A.12 show measured and forecast toe resistances for Piles #2 and #3 respectively from Fellenius et al. (2000). These figures are relatively 'busy' so require some explanation.

To begin with, the ordinate in both figures is the unit toe resistance, r_t, normalized and non-dimensionalized to atmospheric pressure, p_{atm}, simply for plotting generality. Note that, depending on the specific plotted parameter, this stress-related term may or may not represent a ULS value, $r_{t(ult)}$. The abscissa in both figures is the toe settlement, ρ_{toe}, normalized to the toe diameter, d_{toe}, and expressed non-dimensionally as a decimal. Codified metrics for ULS toe resistance are frequently defined using such a ratio although it is often expressed as a percent which is technically a dimension (Fellenius et al. 2000).

The following explain, and comment on as appropriate, the measured results and analytical methods shown in these figures:

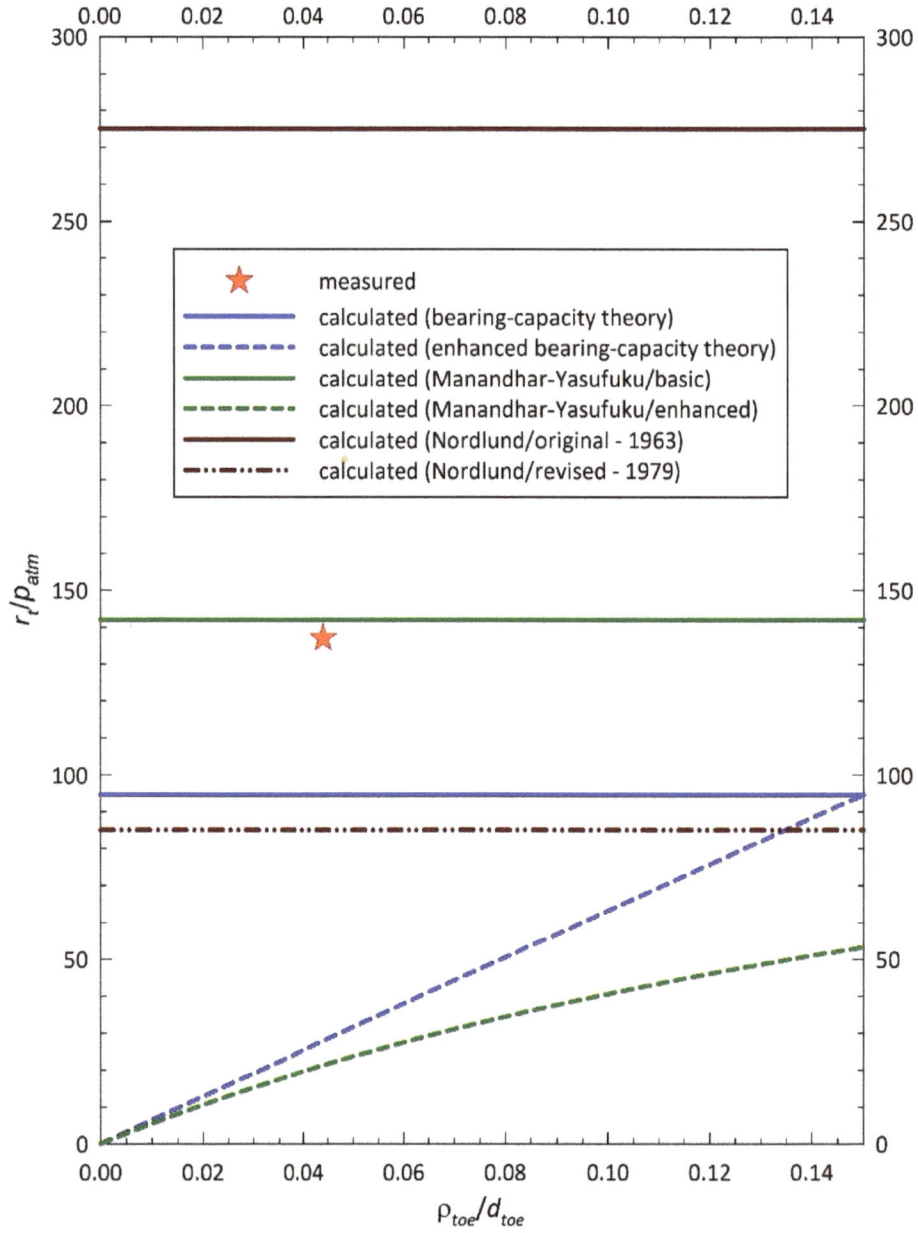

Figure A.11. Toe-Resistance Comparisons for Fellenius et al. (2000) *Monotube* Pile #2 at JFKIA.

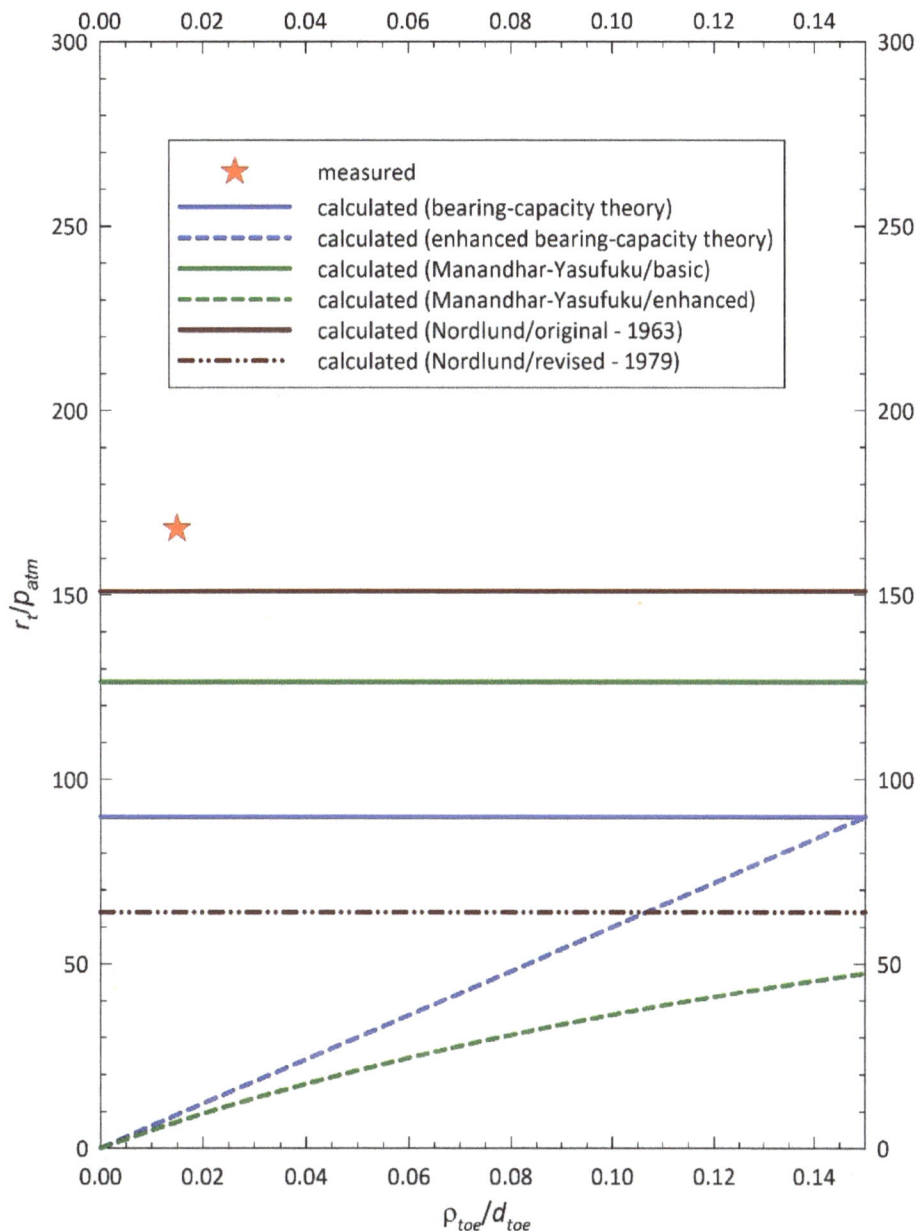

Figure A.12. Toe-Resistance Comparisons for Fellenius et al. (2000) *Monotube* Pile #3 at JFKIA.

- The <u>measured results</u> are based on estimated toe forces scaled from Fellenius et al. (2000) as was done for all of the shaft-related assessments presented earlier in this appendix. As with these shaft-related assessments, the toe resistances reflected in Figures A.11 and A.12 were the "true" values after correcting the measured "false" values for residual loads (the raw measured results indicated essentially no toe forces, even at the maximum applied load-test load that was approaching the ULS total resistance, which, of course, makes no sense). Note also that the assigned (in Figures A.11 and A.12) settlement magnitudes at which these toe resistances occurred are both approximate and misleading. The plotted results are <u>approximate</u> as no specific toe-settlement data were reported in Fellenius et alia so the writer used the residual pile-head settlement after unloading at the end of the static load test as being the same as the maximum toe settlement that occurred during the load test. The plotted results are <u>misleading</u> as all indications are that toe resistance was already 100% mobilized at the end of pile driving and before load testing. Therefore, it can be argued that while the interpreted toe resistance is something close to a ULS value it actually occurred at a much larger magnitude of settlement that happened during pile driving and is thus unknown. Thus, the relatively small toe settlements (approximately 4% and 2% of d_{toe} for Pile #2 and Pile #3 respectively) that are inferred to have occurred during the static load test simply picked up where pile driving left off in terms of pushing the toe farther into the underlying ground. This process is well-illustrated using case histories in Fellenius (2015). So, the writer concludes that while the measured toe-resistance values shown in Figures A.11 and A.12 are likely close to the actual ULS values, the corresponding actual toe settlements at which these resistances occurred are something greater than the corresponding plotted settlement values shown in Figures A.11 and A.12.

- The <u>calculated value using classical bearing-capacity theory</u> (which is inherently independent of settlement magnitude, hence is shown as a horizontal line) is based on Hansen's solution to the classical bearing-capacity problem together with Vesic's compressibility/rigidity factors. Details are given in Horvath (2002). This is the value used for the forecast results with the "original toe-resistance model" shown in Figures A.6 and A.7. The choice of Hansen's solution simply reflects the writer's personal preference based on an assessment of four different solutions (Hansen, Meyerhof, Terzaghi, Vesic) that were studied in Horvath (2000a, 2000b). Note that the results shown in Figures A.11 and A.12 are intentionally lower-bound estimates as they are based on an assumption of constant-volume (critical-state) shear strength, i.e. using ϕ_{cv}. As discussed in Horvath (2002), the alternative is to go through an iterative trial-and-error procedure to find the stress-dependent operative secant value of the peak-strength parameters and from this calculate an upper-bound forecast of this resistance. This is actually a generic process that can be used for many foundation engineering applications and is outlined in detail in Horvath (2000a, 2000b, 2011). In the writer's experience, because of the relatively high stress levels involved beneath the toe of a pile at the geotechnical ULS, the value of $\phi_{peak/secant}$ is typically only slightly greater than ϕ_{cv} so that the forecast ULS toe resistance for peak-strength conditions is not much greater than that for constant-volume/critical-state conditions.

- The <u>calculated value using enhanced bearing-capacity theory</u> reflects an empirical, first-order attempt by the writer to recognize the fact that toe resistance is settlement-dependent so builds up to the ULS value determined in the preceding procedure. The writer developed an empirical model (referred to as the Tri-Linear Settlement Model in this monograph) that was intentionally very simplistic to encourage its use in routine practice (Horvath 2003c; Horvath and Trochalides 2004; Horvath et al. 2004a, 2004b). As discussed in Chapter 5, in developing this model it was assumed that the toe-load vs. toe-settlement response up to the ULS toe resistance of a tapered pile would be a tri-linear relationship in which 10% of the calculated ULS toe resistance would be mobilized at 100% mobilization of shaft resistance (assumed to occur at a pile-head settlement of 0.12 inches (3 mm) plus elastic compression of the pile shaft) and 100% of the calculated ULS toe resistance would be mobilized at a ρ_{toe}/d_{toe} ratio of 0.15 = 15%. Upon reaching 100% of the ULS toe resistance (which is the same as the ULS total resistance for a pile), the assumed post-failure response (i.e. the third line element of the overall tri-linear model) is either plunging in nature, i.e. unlimited settlement occurs under constant load, or represents the unloading curve back to 0% applied load. Note that the calculated ULS toe resistance used is simply that obtained from classical bearing-capacity theory as outlined above. Unfortunately, by happenstance due to the problem parameters involved for the two case history piles considered in this appendix, the tri-linear response up to 100% of the ULS toe resistance appears as a single straight line up to the point of the ULS toe resistance in Figures A.11 and A.12. Consequently, the tri-linear nature of this model is not apparent.

- The <u>Manandhar-Yasufuku (M-Y) Toe-Bearing Method</u> is based on a generic SCE model that was empirically modified by these researchers to account for shaft taper which Manandhar and Yasufuku assumed affects the geometry of the spherical failure surface that develops beneath the toe. They also developed several empirical relationships for evaluating the soil strength and stiffness properties that are required to evaluate an SCE model. The Basic version of their methodology yields a forecast of only the ULS toe resistance without any consideration of toe settlement so the specific results in this case are shown in Figures A.11 and A.12 as horizontal lines. This is the calculated result that was used as an alternative for the toe resistance in the writer's Modified Nordlund Method results shown in Figures A.6 and A.7. The Enhanced version of the Manandhar-Yasufuku Toe-Bearing Method assumes that the load-settlement behavior of the toe follows an empirically based hyperbolic curve which means that their ULS value of toe resistance is never actually reached but is only approached asymptotically. To fit the assumed hyperbolic curve to the data, Manandhar and Yasufuku assumed that that only one-half (50%) of the ULS toe resistance is mobilized at a ρ_{toe}/d_{toe} ratio = 0.25 (25%). The net result of these assumptions is that within the magnitudes of toe settlement likely to be experienced under service-load conditions, the difference in forecast results between the Basic and Enhanced versions of the Manandhar-Yasufuku Toe-Bearing Method is quite large, approximately one order of magnitude, as can be seen in Figures A.11 and A.12.

- The <u>Nordlund Method</u> in both its original (1963) and revised (1979) versions uses the same solution (Berezantzev et alia) for the classical bearing-capacity problem and same overall analytical methodology to provide an estimate of the ULS toe resistance. Toe settlement does not enter the picture in either version so the forecast results are shown in Figures A.11 and A.12 as horizontal lines. Despite the common theoretical basis, as can be seen in these figures the differences in forecast results between the original and revised versions of the Nordlund Method are substantial, with the original version yielding results that are approximately 2½ to 3 times that of the revised version. Given the overall identical analytical process used by the two versions, the obvious question is why is there such a marked difference in calculated outcomes? The reason is due to two independent changes implemented by Nordlund in the revised version that have a surprisingly significant, cumulative effect on the final forecast result. First, in the revised version, Nordlund implemented an overburden-stress correction factor on the SPT N-values[176] (assumed by the writer to be N_{45} values for the reasons explained in Chapter 5)

---

[176] As noted at the beginning of this appendix, Fellenius et al. (2000) did not provide any SPT N-values for the site of Pile #2 and Pile #3. However, the writer's CPTu-based site-characterization methodology produces an estimate of N_{60} for each set of q_c-f_s-u_2 data for coarse-grained soils only.
[Footnote continued at bottom of following page]

used for both shaft- and toe-resistance calculations. The correction factor used is based on normalization of overburden stresses to 1 ton/ft^2 (= 2,000 lb/ft^2 = 95800 Pa), an assumption that became common in practice in the latter decades of the 20th century. At the calculated stress levels at the toes of the two piles under discussion, the correction factor was of the order of 0.75. Because Nordlund used the same empirical relationship between N-values and ϕ in both versions of his method, this correction results in lower design values of ϕ in the revised version. Although the difference is only a few degrees of angle, the effect on the design value of the N_q bearing capacity factor is significant. As is well-known, bearing-capacity factors for traditional bearing-capacity solutions are very sensitive to the value of ϕ, regardless of whose solution is used. In the case of the Berezantzev et alia solution, N_q varies by a factor of approximately 10 as ϕ varies from 30° to 40°. The second factor influencing the forecast ULS toe resistances is that in the revised version of his method, Nordlund implemented an arbitrary not-to-exceed vertical effective overburden stress of 3,000 lb/ft^2 (144 kPa) when calculating toe resistance. The logic for this as stated in Nordlund (1979) was the assumption that vertical effective overburden stresses along piles reached a 'limiting value', a concept in vogue at the time but long since shown to be incorrect as discussed in Chapter 4. In the specific case of the two piles under discussion, this arbitrary cap on vertical stresses resulted in approximately a 30% reduction in the overburden stress used in toe-resistance calculations for the revised version of the Nordlund Method. Because there is a linear relationship between ULS toe resistance and vertical effective overburden stress in both versions of the Nordlund Method, the effect on the calculated outcomes was not insignificant. In summary, the Revised Nordlund Method resulted in a reduced value of the bearing-capacity factor, N_q, due to a reduced value of ϕ that when multiplied by a reduced value of vertical effective overburden stress resulted in a <u>much</u>-reduced forecast of the ULS toe resistance.

It is not possible to draw specific conclusions concerning toe-resistance forecasting methodologies from this discussion as measured results were available for only two piles and one of them (Pile #2) had installation issues that may have affected the measured toe resistance. Also, the measured results were not actual measurements but inferred from a subjective assessment to correct the measured results for assumed residual loads in the piles.

Nevertheless, it is clear that there is a wide divergence in forecast results using analytical methodologies that reflect a representative sampling of the current state of knowledge for forecasting the ULS toe resistance of tapered piles. This suggests that toe resistance is an aspect of pile analysis and design that requires significant new research to better understand the fundamental mechanism(s) involved and then to translate that understanding into workable, usable information for use in routine practice. A key part of future research should be to determine once and for all whether or not toe resistance reaches a ULS value and what analytical model (traditional bearing capacity, SCE, or something else) best replicates the physical mechanism associated with toe resistance. The fact that residual toe loads are significant for piles in general is a further complication that needs to be considered and addressed in future research as well.

[Footnote continued from bottom of preceding page]
From this calculated value of N_{60}, the assumption of a linear relationship between SPT driving efficiency and N-value was used to calculate the N_{45} values used in both versions of the Nordlund Method. Note that because Pile #3 was driven 500 millimetres (20 in) deeper than Pile #2, this resulted in different calculated values of N_{45} at the toes of Piles #2 and #3 which explains why there were different forecast results of toe resistance between these two piles using the Nordlund Method.

However, one conclusion that can be stated with certainty at this time is that the relationship between toe resistance and toe settlement needs to be implemented into routine practice in some fashion. It has been clear for some time now that mobilization of toe resistance is much more dependent on settlement magnitude compared to shaft resistance. While shaft resistance can routinely and simplistically be assumed to be fully mobilized as part of an analysis or design methodology, such is not the case with toe resistance. This means that either toe resistance vs. settlement be explicitly evaluated as part of any methodology or, if a single-valued toe resistance is desired, it should be a value that is consistent with an acceptable settlement magnitude. Unfortunately, the development of significant residual loads that can include full or near-full mobilization of toe resistance as in the case of the Fellenius et alia case-history piles that are the focus of this appendix greatly complicates developing an analytical methodology for toe resistance that explicitly considers settlement.

A.4.4 Direct Methods

A.4.4.1 UniCone Concept and Methods

A.4.4.1.1 Shaft Resistance

To facilitate the discussion that follows concerning the various versions of the UniCone concept that have been developed to date by Eslami and Fellenius (Original UniCone Method) and Niazi (Modified UniCone Method) as well as proposed by the writer (Extended UniCone Method) and how they compare to measured results for the Fellenius et al. (2000) case-history piles that are the focus of this appendix, a generic, dimensionless *shaft-resistance correlation coefficient*, C_{SHAFT}, that is consistent with the UniCone concept is defined as follows:

$$C_{SHAFT} = \frac{r_s}{q_E}. \tag{A.5}$$

Note that this equation yields a dimensionless result that typically has an order-of-magnitude range of 0.001 to 0.01 so it is optional to multiply the result by 100 to yield a value in percent (0.1% to 1% in this case) for plotting convenience.

Note also that no restriction is placed on the unit shaft resistance, r_s, in this definition. It does not have to be a ULS value although in most cases it will be. In addition, it can be determined either by measurement, as shown subsequently using the results from the Fellenius et alia instrumented test piles, or calculated using some resistance-forecasting methodology, also as shown subsequently. Thus, the parameters C_s, C_{se}, and C_{sx} that were defined in Chapter 5 for specific versions of the UniCone concept (Original, Modified, and Extended UniCone Methods respectively) are just special ULS cases of C_{SHAFT}.

First shown in Figure 4.13 is the depth-wise variation of C_{SHAFT} (in decimal form) as measured for the *Monotube* piles discussed by Fellenius et alia. Note that these values are likely on the cusp of being ULS values. The two piles show essentially identical results except for the upper portion of Pile #2 where there was minimal shaft resistance due to the problems that were encountered during initial pile installation.

Also shown in Figure A.13 is the general subsurface stratigraphy depicted in Figure A.2. This is to better understand the genesis of the spike in C_{SHAFT} observed for Pile #3 at a relatively shallow depth that coincides more or less with the relatively thin Holocene organic stratum that consists primarily of organic clay. Note that this spike is absent for Pile #2 because of the aforementioned installation-related disturbance.

418

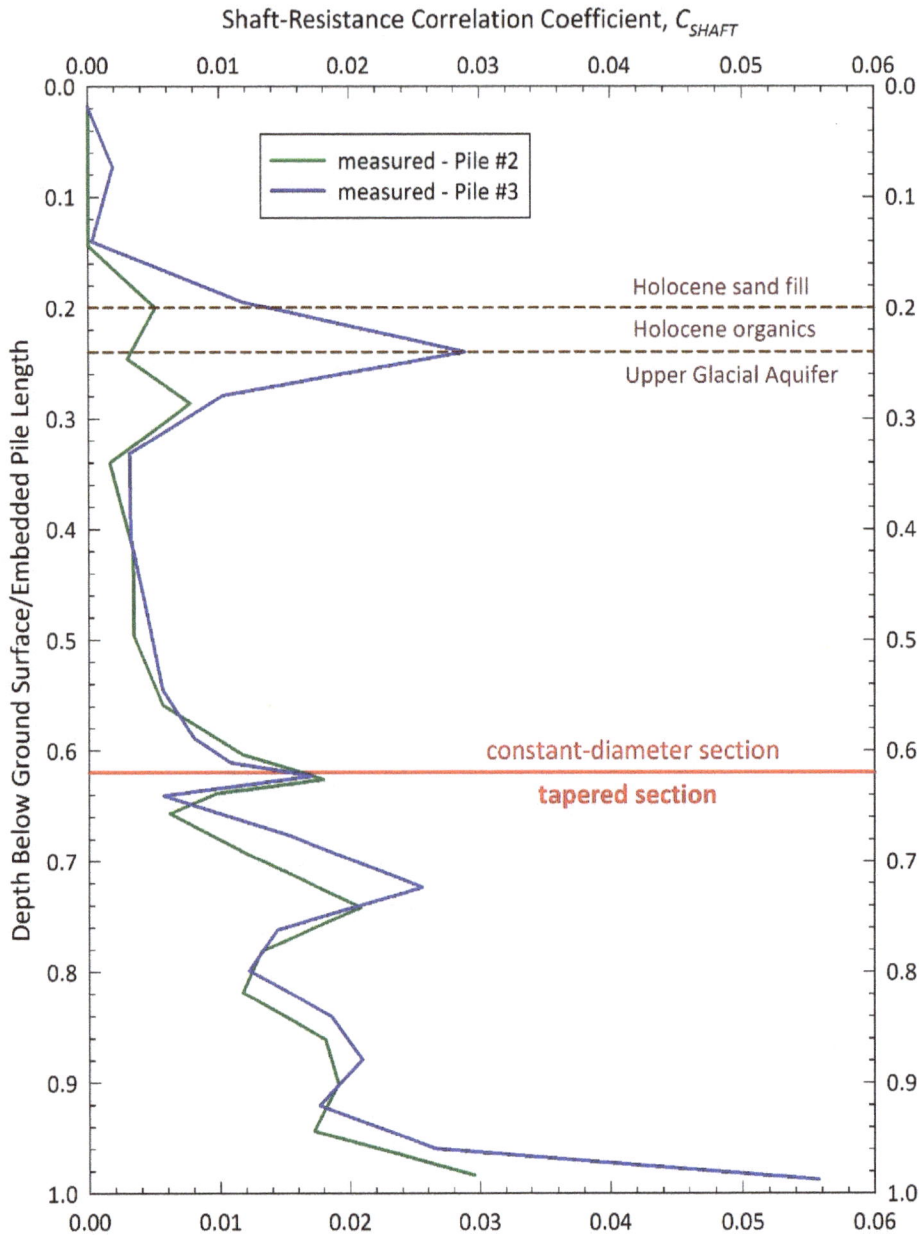

Figure A.13. Depth-Wise Distribution of Measured UniCone-Concept Shaft-Resistance Correlation Coefficients for Fellenius et al. (2000) *Monotube* Piles at JFKIA.

Because of the way in which C_{SHAFT} is defined, there is a marked increase in its value with any significant increase in the measured porewater pressure, u_2, over and above the hydrostatic value, u_o. This is because such increases in u_2 produce a significant decrease in q_E ($= q_t - u_2$), especially if q_t simultaneously decreases as well. This combination of conditions (a drop in q_t accompanied by a rise in u_2) typically occurs when the fines content and the plasticity of the fines increase significantly. This is the case reflected in Figure A.13 when the stratigraphy rapidly changes from a relatively clean sand to organic clay and then back to relatively clean sand all within a distance of approximately one metre (3 ft).

Unfortunately, this signature behavior of the C_{SHAFT} parameter when fine-grained soils are encountered complicates interpretations when tapered piles are involved. As can be seen in Figure A.13, the observed increases in C_{SHAFT} within the tapered portion of both piles are of similar magnitude to that of the constant-diameter portion in fine-grained soil even though the tapered portion of the pile is entirely within clean sand.

Nevertheless, there are some broad trends that can be inferred from the measured results in this figure:

- Discounting the results related to the presence of the Holocene organic stratum, the average value of C_{SHAFT} within the constant-diameter portion of the piles is approximately 0.005 (0.5%) which is consistent with UniCone Method ULS values associated with coarse-grained soils.

- Within the tapered portion of the piles, the average value of C_{SHAFT} (which would be the C_{sx} parameter in the writer's Extended UniCone Method) is approximately 0.02 (2%). This implies a taper benefit of about 4 (= 0.02/0.005) which is of the same order of magnitude as the taper benefit observed using several indirect methods that were discussed earlier in this appendix.

- The measured values of C_{SHAFT} increase in an approximately depth-wise linear manner within the tapered portion of the piles, from approximately 0.01 (1%) to 0.03 (3%). Again, this is consistent with observations made earlier in this chapter where the parameters associated with a variety of indirect methods also exhibited a nominally linear increasing trend with depth within the tapered portions of the piles. This is discussed further subsequently.

Figure A.14 shows a comparison of C_{SHAFT} values between the measured results (only Pile #3 is shown for clarity) and three calculated ULS results from various versions of the UniCone concept that were discussed in Chapter 5:

- The writer's Extended UniCone Method (C_{sx}) intended for use with tapered piles using values calculated from the K_h/K_o assumptions made for the writer's Modified Nordlund Method, site-specific estimates of K_o from Figure A.1, and Equations A.2 and A.5. As can be seen in Figure A.14, there is reasonable agreement with the measured values of C_{SHAFT} within both the constant-diameter and tapered sections of the pile.

- Niazi's Modified UniCone Method (C_{se}) using a direct application of Niazi's empirical equations with the data from the CPTu sounding contained in Fellenius et al. (2000). There is reasonable agreement with the measured results for the constant-diameter portion of the pile. Agreement for the tapered section of the pile was obviously not expected but the calculated results are presented for this section anyway, solely as information. Note that the calculated results track the measured results qualitatively, indicating that it may be possible to modify Niazi's empirical equations for pile taper.

- A range of results for the Eslami-Fellenius Original UniCone Method (C_s). Application of this original version of the UniCone concept relies heavily on the judgment of the design professional using either figures or tables. The writer chose not to go through this subjective process for the purposes of this comparison so shown in Figure A.14 is the range of values (0.004 to 0.01) suggested in Fellenius (2019) for coarse-grained soil.

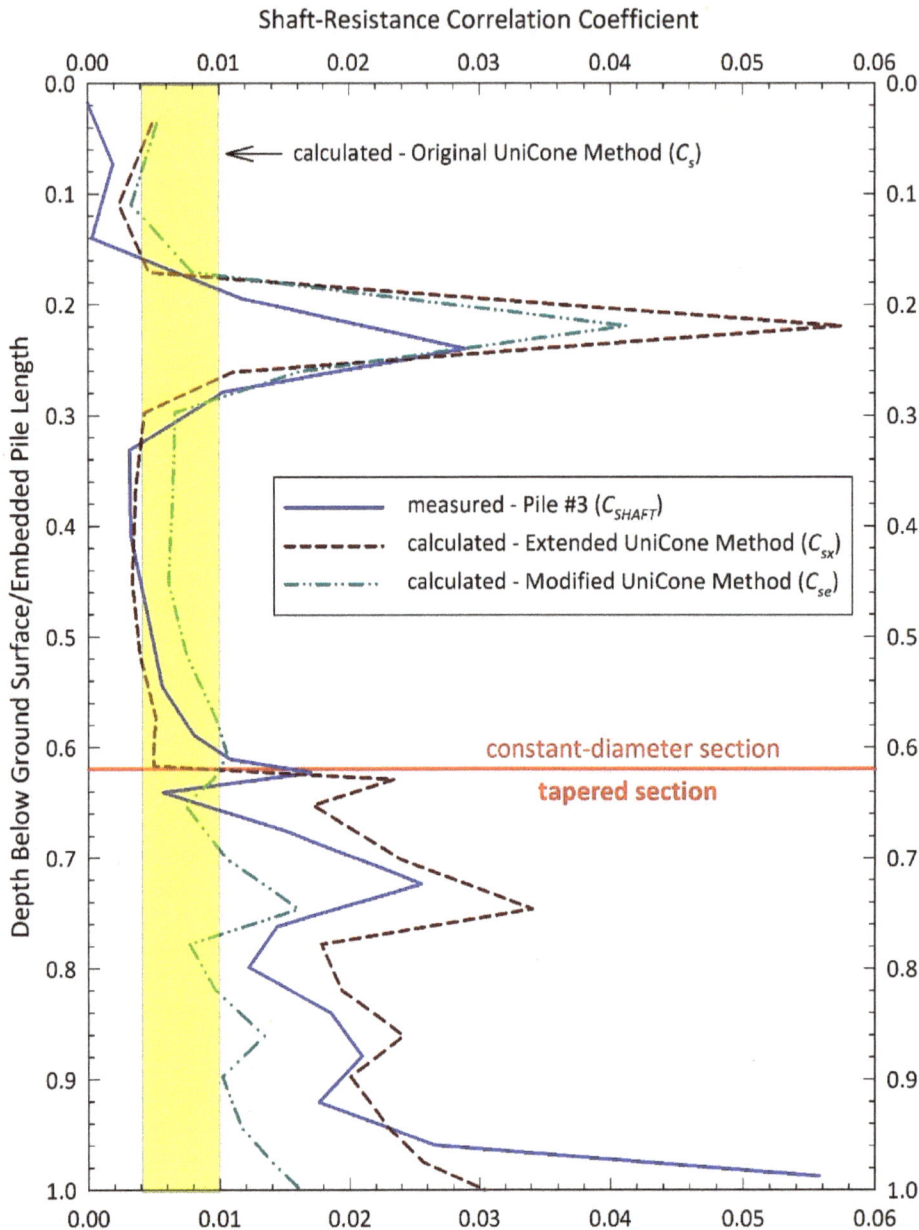

Figure A.14. Depth-Wise Distribution of Measured vs. Calculated UniCone-Concept Shaft-Resistance Correlation Coefficients for Fellenius et al. (2000) *Monotube* Piles at JFKIA.

In summary up to this point, it is clear from the interpretation of the measured results shown in Figures A.13 and A.14 that taper causes an increase in the generic shaft-resistance correlation factor, C_{SHAFT}, for the UniCone concept. Furthermore, C_{SHAFT} increases in a nominally linear fashion within the tapered section. These observed outcomes concerning the effect of taper on C_{SHAFT} are completely consistent with the earlier results for the shaft-resistance parameters such as K_h/K_o and β/K_o that are used in various indirect resistance-forecasting methodologies. More specifically, it appears that taper benefit does not develop and exhibit itself uniformly along the tapered section of a pile as has been believed since the 1950s and until recent years. Rather, taper benefit appears to be influenced by some combination of taper angle and pile diameter, with taper benefit increasing as diameter decreases, all other things being equal.

As noted earlier in this appendix, this observed behavior concerning taper benefit is consistent with cylindrical-cavity mechanics that is the best physical model identified to date for the wedging mechanism by which tapered piles develop shaft resistance. This is because for a given taper angle and magnitude of pile settlement, a relatively larger amount of incremental radial expansion will occur the smaller the initial diameter of the cavity (pile shaft in this case). For example, for the *Monotube* Type J (taper angle, α, = 0.57°) piles that are the ground-truth used throughout this appendix, one inch (25 mm) of settlement applied uniformly to the 25-foot (7.6 m) long tapered section of these piles produces a radial displacement of the pile-soil interface equal to 0.010 inches (254 μm), assuming both elastic compression and the Poisson Effect on the pile cross-section are ignored. Although the magnitude of radial displacement of the tapered section is depth-wise constant, the relative diameter increase is not. The relative increase is 0.14% at the top of the tapered section but 0.25% (almost twice as much) at the bottom of the tapered section, i.e. toe of the pile. Because the increase in radial pressure generated by expansion of a pre-existing cylindrical cavity is proportional to the relative increase of its diameter, this means that the relative increase in radial stress is greater at the bottom of a tapered section compared to the top of a tapered section. This, in turn, suggests that shaft resistance should be greater at the bottom of the tapered section compared to the top which is indeed what appears to be the case.

The net result is that if future research shows these initial, preliminary observations and conclusions to be correct, then this will complicate development of empirical relationships for the proposed new C_{sx} correlation coefficient for the Extended UniCone Method. This is because both taper angle and pile diameter will need to be reflected in any empirical correlations that are developed for C_{sx}.

As a final comment with regard to the shaft-resistance correlation coefficient used with the UniCone concept, neither the original (Eslami-Fellenius) nor modified (Niazi) methods include DFE typology as a variable, only soil type. The writer finds this curious as the vast majority of resistance-forecasting methodologies, both indirect and direct, that have been developed since modern foundation engineering began to evolve in the 1950s have included DFE typology as a problem variable, at least in some manner. Niazi did raise the issue in Niazi and Mayne (2015c) and provided some indication that DFE typology had an effect. However, it appears that his database was too sparse to develop a DFE-related problem parameter given the relatively large number of driven, jacked, drilled, and torqued deep-foundation alternatives that exist these days. Consequently, future research into an Extended UniCone Method should include DFE typology in general as a variable in addition to taper that has been the focus of the discussion in this monograph.

A.4.4.1.2 Toe Resistance

To complete the discussion of the UniCone concept and associated analytical methodologies, it is of interest to compare measured and calculated values of the toe-resistance correlation coefficient. To begin with, a generic toe-resistance correlation coefficient, C_{TOE}, consistent with the UniCone concept can be defined as follows:

$$C_{TOE} = \frac{r_t}{q_{E(toe)}}.$$ (A.6)

As with shaft resistance, there is no restriction as to how the unit toe resistance, r_t, which is not necessarily a ULS value, is determined, i.e. by measurement or calculation.

Note that in this case, the value of $q_{E(toe)}$ is generally taken to be an average of the q_E values over some predefined depth range relative to the pile toe as opposed to being at a single depth. The default averaging interval used for the Original (Eslami-Fellenius) UniCone Method is usually seen expressed as '+8d_{toe} to -4d_{toe}', i.e. from eight toe diameters above (+) the toe down to four toe diameters below (-) the toe. Niazi adopted the same guideline for his Modified UniCone Method.

For consistency, this guideline was also used here when evaluating the measured results as well as calculating results using the writer's original toe-resistance methodology based on traditional bearing-capacity theory as well as the Basic Manandhar-Yasufuku Toe-Resistance Method for the toe resistance of tapered piles, both of which were discussed earlier in this appendix for their use with the writer's Modified Nordlund Method.

Table 4.2 shows the measured vs. calculated (all ULS values) results for C_{TOE}. Note that both the Original (Eslami-Fellenius) and Modified (Niazi) UniCone methods yield values of C_{TOE} (C_t and C_{te} respectively) directly. On the other hand, the measured results and calculations using traditional bearing-capacity theory and the Basic Manandhar-Yasufuku Toe-Resistance Method produce values of r_t and from this C_{TOE} is calculated using Equation A.6.

Table A.2. Measured vs. Calculated UniCone Toe-Resistance Correlation Coefficients for Fellenius et al. (2000) *Monotube* Piles at JFKIA.

Pile	Measured	Toe-Resistance Correlation Coefficients			
		Calculated			
	(C_{TOE})	bearing capacity (C_{TOE})	M-Y (C_{TOE})	Eslami-Fellenius (C_t)	Niazi (C_{te})
#2	1.25	0.85	1.30	1	0.32
#3	1.59	0.79	1.19	1	0.31

The results shown in this table are more complex to interpret than they might initially appear. The underlying reason is that unlike shaft resistance due to sliding friction that always has a well-defined ULS value (or two values if peak and constant-volume/critical-state conditions are separated) that is/are mobilized at relatively small magnitudes of displacement (settlement in this case), toe resistance often does not have a well-defined ULS value as discussed in Chapter 4 and noted previously in this appendix. Furthermore, and regardless of whether there is a well-defined ULS value, toe resistance is mobilized over a settlement range that is typically an order of magnitude greater than that associated with shaft resistance although this longstanding rule of thumb is of little use when significant residual toe loads exist as with tapered piles. Thus, as has been noted several times throughout this monograph, it is not possible to meaningfully discuss toe resistance without bringing settlement into the discussion as a significant problem parameter.

With the influence of toe settlement in mind, the toe-resistance values shown in Table 4.2 are difficult to interpret. The measured toe-resistance values are clearly relatively large in magnitude and this suggests something approaching, if not at, a ULS value. On the other hand, the inferred toe settlements from the pile load tests that were performed are relatively small in magnitude. However, as noted earlier in this appendix, Fellenius et alia interpreted the load test data to indicate that toe resistance was already 100% mobilized and locked-in as a residual force at the time the static load tests began. Therefore, interpreting the measured results to reflect something approaching ULS values seems reasonable as whatever magnitude of settlement that was required to mobilize them had already occurred during driving and prior to the start of the static load tests.

On the other hand, the calculated results in Table 4.2 present a 'mixed bag' with respect to settlement. By assumption in their derivation, the results for traditional bearing-capacity theory and the Basic Manandhar-Yasufuku Toe-Bearing Method ("M-Y" in Table 4.2) are ULS resistances so are nominally consistent with the measured results in this regard. Interpreting the Original (Eslami-Fellenius) and Modified (Niazi) UniCone Method results is less clear. It appears that Eslami and Fellenius intended for their toe-resistance correlation coefficient, C_t, for the Original UniCone Method to yield something approaching a ULS value and from Table 4.2 that would appear to be more or less the case.

Interpreting Niazi's C_{te} coefficient for the Modified UniCone Method is not straightforward. The values are quite low compared to all the rest. This suggests that the parameter values relate to conditions that are much less than the ULS. In reviewing the database presented in Niazi (2014) that he used to develop his empirical relationship for C_{te}, it appears that the data are somewhat sparse and scattered, even with the $\log_{10}$ scaling he used to plot the data. This suggests that the static load tests used to construct the database involved cases where the DFEs were loaded to varying magnitudes of settlement and thus had varying degrees of toe-resistance mobilization (and perhaps residual loads as well). Unfortunately, Niazi made no attempt to include relative toe settlement, e.g. the ratio of toe settlement to toe diameter, as a variable in the relationship he developed for C_{te}. As a result, the database contains results with varying degrees of toe-resistance mobilization that undoubtedly contributes the observed scatter.

An additional issue with the database used by Niazi to develop an empirical relationship for C_{te} is that it combines DFEs of widely varying typology. Unlike with his empirical relationships for the UniCone shaft-resistance parameter, C_{se}, Niazi did not develop separate empirical relationships for toe resistance based on DFE typology. The significance of this is discussed further subsequently.

It appears that Niazi was aware, at least to some degree and for some unspecified reason(s), of the limitations of the toe-resistance forecasting component of his Modified UniCone Method, perhaps if for no other reason that the values he came up with for his C_{te} parameter were considerably less than the $C_t = 1$ recommendation that had long been part of the Original (Eslami-Fellenius) UniCone Method. For example, in Niazi and Mayne (2015c), it was emphasized that the toe-resistance component be used for <u>preliminary design</u> only but with no suggestion as to what a design professional might use for <u>final design</u>. Niazi also said that further additions to, and concomitant re-evaluation of, the database were desirable but provided no opinion or direction along those lines.

In view of these facts, it appears reasonable to conclude that the toe-resistance component of the Modified (Niazi) UniCone Method should be used with considerable caution in its current version as it reflects the limitations and uncertainties of the database used in its formulation. As a result, forecasts of toe resistance made with the current empirical equation for C_{te} are likely to be quite conservative in terms of being a ULS value or anything close to it.

To conclude this discussion of the toe-resistance component of the UniCone concept in general and how it might be approached analytically in the future for developing the writer's suggested Extended UniCone Method in particular, it would appear that Niazi's linking of the toe-resistance correlation coefficient to soil-behavior type is, in principle, an improvement over the Original (Eslami-Fellenius) UniCone Method that ignored this consideration. However, in addition to soil typology, the writer feels that two additional variables need to be included in any empirical relationship developed for the toe-resistance correlation coefficient, C_{tx}, for the writer's proposed Extended UniCone Method for use with tapered piles.

First, the dimensionless relative toe settlement, which is usually expressed as toe settlement divided by the toe diameter, should be included. As has been noted and discussed several times and in several contexts throughout this monograph in general and this appendix in particular, it is now well-established that toe settlement is an important variable for forecasting toe resistance. Independent of whether or not toe resistance ever reaches a well-defined, single-valued ULS value is the fact that toe resistance is mobilized over a range in settlements that is at least an order of magnitude greater than the settlement required to mobilize the ULS shaft resistance. This means that the state of practice should be changing to better incorporate an explicit consideration of toe settlement in the assessment of toe resistance.

Second, DFE typology should be included as a variable but for a subtly different reason than suggested earlier in this appendix for shaft resistance. The relevance of DFE typology with shaft resistance is the effect that installation has on the radial stress state acting along the shaft after installation. With regard to toe resistance, DFE typology appears to have a profound effect on the development of residual toe loads (Fellenius 2015). To the extent that residual toe loads affect the post-installation load-settlement response of the pile toe, this influence becomes relevant.

In summary, the toe resistance of DFEs appears to be a complex interaction between soil type, DFE typology, and relative toe settlement that is not adequately addressed by the traditional assumption of a single-valued ULS resistance. Regardless of which or whose resistance-forecasting methodology is used, it is clear that toe resistance requires substantial future research to better quantify it than at present.

A.4.4.2 Togliani Method

One goal of this appendix is a detailed assessment of the two *Monotube* piles that are the focus of Fellenius et al. (2000) for the purpose of assessing various embellishments and extensions of existing indirect and direct resistance-forecasting methodologies that have been proposed by the writer for use with tapered piles. In that regard, most of the material presented in this appendix has not been heretofore published.

Another goal is assessing resistance-forecasting methodologies for tapered piles that have already been developed and presented in the published literature such as the writer's Modified Nordlund Method that was originally developed circa 2000 and published in Horvath (2002). In this latter vein, it is of interest to apply the Togliani Method to the Fellenius et alia instrumented test piles to evaluate the accuracy of this method in this specific case even though earlier research by the writer that was discussed in Chapter 5 found the Togliani Method to be of questionable value in its current form. Nevertheless, the writer feels that an assessment of this method in this appendix is of value simply because the Togliani Method is the only known direct method developed to date explicitly for application to tapered piles, albeit in a very approximate manner as it utilizes on Fellenius' generic donut-on-a-stick model to account for taper effects as opposed to using a physical model more closely related to the actual wedging mechanism exhibited by tapered piles.

To begin with, it is of interest to assess the absolute accuracy of the Togliani Method. Table A.3 compares the measured and forecast results for Pile #3 which is the only one of the two discussed in Fellenius et alia for which reliable results for the entire embedded length of the pile exist.

**Table A.3. Measured vs. Forecast (Togliani Method) Results for
Fellenius et al. (2000) *Monotube* Pile #3 at JFKIA.**

| | Axial-Compressive Resistance, kilonewtons (kips) | | | |
| | Shaft | | Toe | TOTAL |
	constant-diameter section	tapered section		
measured	450 (101)	1500 (337)	550 (124)	2500 (562)
forecast	1262 (284)	1786 (401)	458 (103)	3506 (788)

As can be seen, the ULS total resistance forecast using the Togliani Method is approximately 40% greater than that measured in the static load test for Pile #3. The source of most of this relatively large discrepancy lies with the shaft contribution (especially that from the constant-diameter section) for which the forecast resistance is 56% greater than that measured (180% greater for just the constant-diameter section).

Overall, these results are consistent with the writer's earlier findings that were originally presented in Horvath (2015) and also discussed in Chapter 5. Specifically, the comparison of forecast results made using the Togliani Method consistently and significantly exceeded measured total resistances for other *Monotube* piles as well as timber piles at JFKIA.

In addition to and independent of this absolute comparison, it is of interest to compare the relative depth-wise distribution of forces within Pile #3 that is also a reflection of the development of ground resistance as a function of depth as these two quantities are mirror images of each other as can be seen in Figure A.1. This is important because if the Togliani Method can reasonably forecast the correct depth-wise attenuation of forces within the pile shaft and concomitant development of resistance along the pile shaft then it is at least possible in principle to simply adjust the empirical equations for the Togliani Method to more accurately forecast these forces and resistance. However, if the depth-wise forecasting accuracy of the Togliani Method is problematic then it legitimately calls the validity of the entire methodology in its current form into question.

Note that earlier research by the writer (Horvath 2015, reproduced in Figure 5.1 in Chapter 5) indicated that the Togliani Method did a poor job of replicating the actual depth-wise distribution of axial-compressive forces within the pile and its complement, ground resistance. However, before basing any judgment solely on that earlier work and for the sake of completeness, it is of value to perform an identical comparison specifically for the Fellenius et alia piles where the depth-wise distribution of pile forces and complementary ground resistance are known to the greatest accuracy that can reasonably be expected in routine practice.

Figure A.15 uses the same basic format and plotting nomenclature as Figure A.1 and shows the comparison between the measured and forecast results for Fellenius et alia Pile #3. Although the kink in both the load and resistance curves for the Togliani Method at the interface between the constant-diameter and tapered sections is somewhat more pronounced than that shown in Figure 5.1, it is not nearly as pronounced as the measured results. On the other hand, the writer's Modified Nordlund Method typically forecasts distributions that are in consonance with measured results as shown in Figures 5.1, A.6, and A.7.

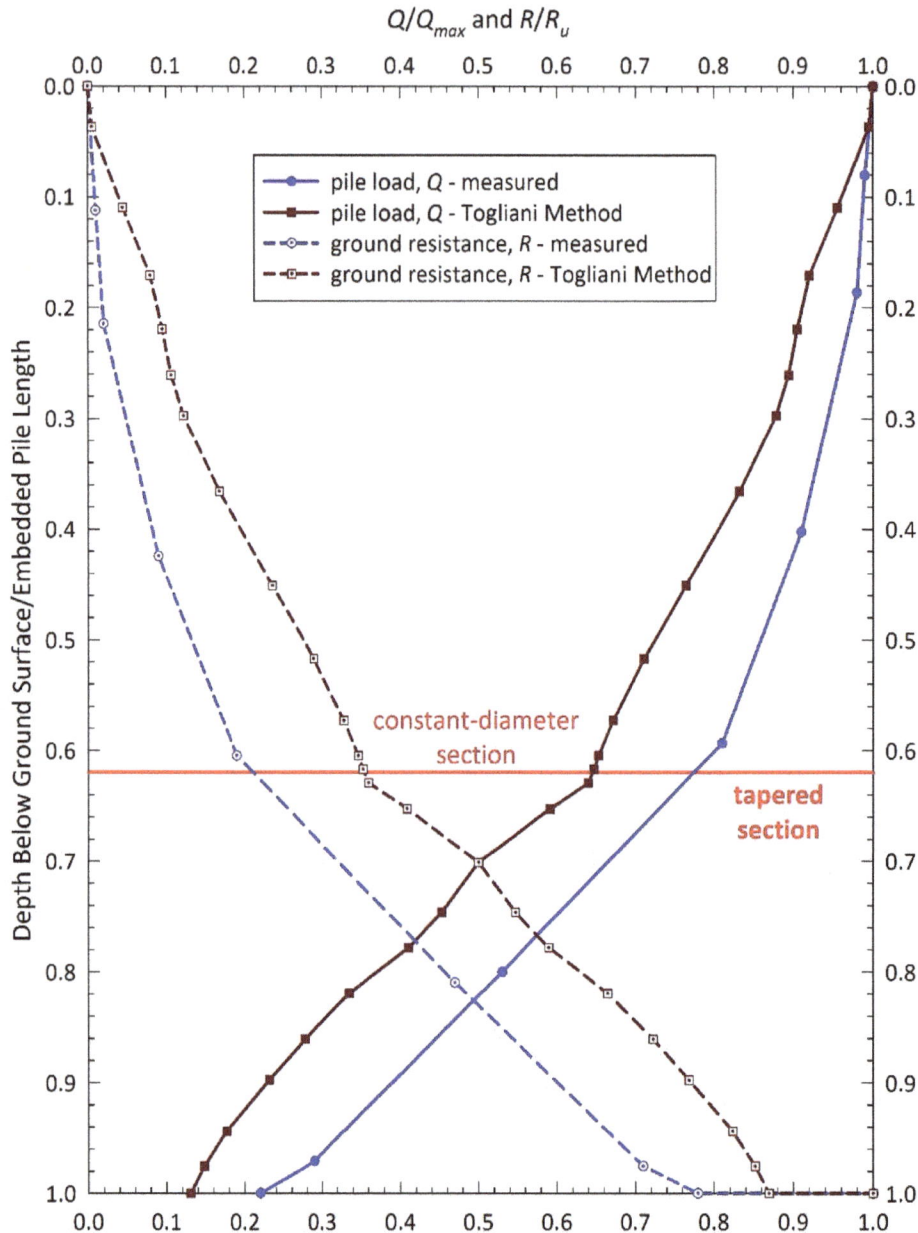

Figure A.15. Measured vs. Forecast (Togliani Method) Depth-Wise Distribution of Relative Axial-Compressive Forces and Soil Resistance for Fellenius et al. (2000) *Monotube* Pile #3 at JFKIA.

The conclusion drawn from Figure A.15 is that it is consistent with the writer's earlier findings as typified by Figure 5.1 that the Togliani Method is not only inaccurate on an absolute basis as shown in Table A.2 but is inaccurate on a relative basis in that it does not correctly apportion shaft vs. toe loads/soil resistance or constant-perimeter vs. tapered shaft loads/soil resistance. Whether these shortcomings are primarily due to Togliani's empirical relationships for shaft and toe resistance or his use of Fellenius' donut-on-a-stick model for taper or some combination of the two is unclear and is problematic to determine as these factors are all interrelated. However, with this in mind, the final aspect explored with respect to the Togliani Method is the pattern of the development of the resistance provided by the ground (that is the complement and reciprocal of axial-force attenuation within the pile) within the tapered section of the Pile #3.

Figure A. 16 shows the measured vs. forecast (using the Togliani Method) depth-wise development of resistance within the 25-foot (7620-mm) -long tapered section of Pile #3 only. As can be seen, what Togliani calls the Shaft Resistance, R_s (conceptually, this is the 'stick' component of the donut-on-a-stick model) comprises about one-third of the total calculated resistance, with the remaining two-thirds contributed by what Togliani calls the Taper Resistance, R_t, that is the 'donut' component[177].

As an aside, it is of interest to note that this forecast one-third/two-thirds allocation between shaft-resistance components within the tapered portion of Pile #3 is considerably different than the approximately one-half/one-half allocation for the almost identical *Monotube* pile at JFKIA (Pile P-5A) that was used in Chapter 6 as an exemplar for comparing forecast outcomes from resistance-forecasting methodologies. This suggests that the donut-to-stick allocation in the Togliani Method is subject to relatively substantial variation even within essentially the same subsurface conditions. Note that the only difference between Fellenius et alia's Pile #3 and the writer's Pile P-5A was the specific cone-penetrometer data used. These data are, overall, very similar as are all the cone-penetrometer data that the writer has seen for the JFKIA Central Terminal Area where both of these piles were driven (Horvath 2014b, 2015). As noted above, the piles themselves are for all intents and purposes identical as they are the same *Monotube* pile (except for pipe-wall thickness) and driven to virtually the same embedded depth.

Thus, while it is clear in this particular case of Fellenius et alia's Pile #3 that the donut or end-bearing component dominates the forecast results for the Togliani Method, this conclusion should not be taken to be a general conclusion. Based on the aforementioned results for Pile P-5A as well as numerous other calculations made by the writer for earlier studies that were summarized in Horvath (2015) in which several timber, *Monotube*, and *TAPERTUBE* piles were assessed (albeit all for predominantly coarse-grained soil conditions), it was found that the relative contributions of Togliani's R_s and R_t parameters can vary widely from pile to pile so that either or neither parameter predominates in a given applications. As might be expected, taper angle, α, plays at least some role as it affects the relative size of the donut-shaped area used to calculate R_t. The greater the taper angle, the larger the donut becomes, all other things being equal.

---

[177] Note that Togliani's parameters R_s and R_t as applied to his analytical methodology should not be confused with the generic definition of these parameters as shown in Figure 2.1 and used throughout this manuscript. There is no correlation between these Togliani and generic parameters. The notational duplication is just coincidence and Togliani's notation is retained here solely for the purposes of the current discussion.

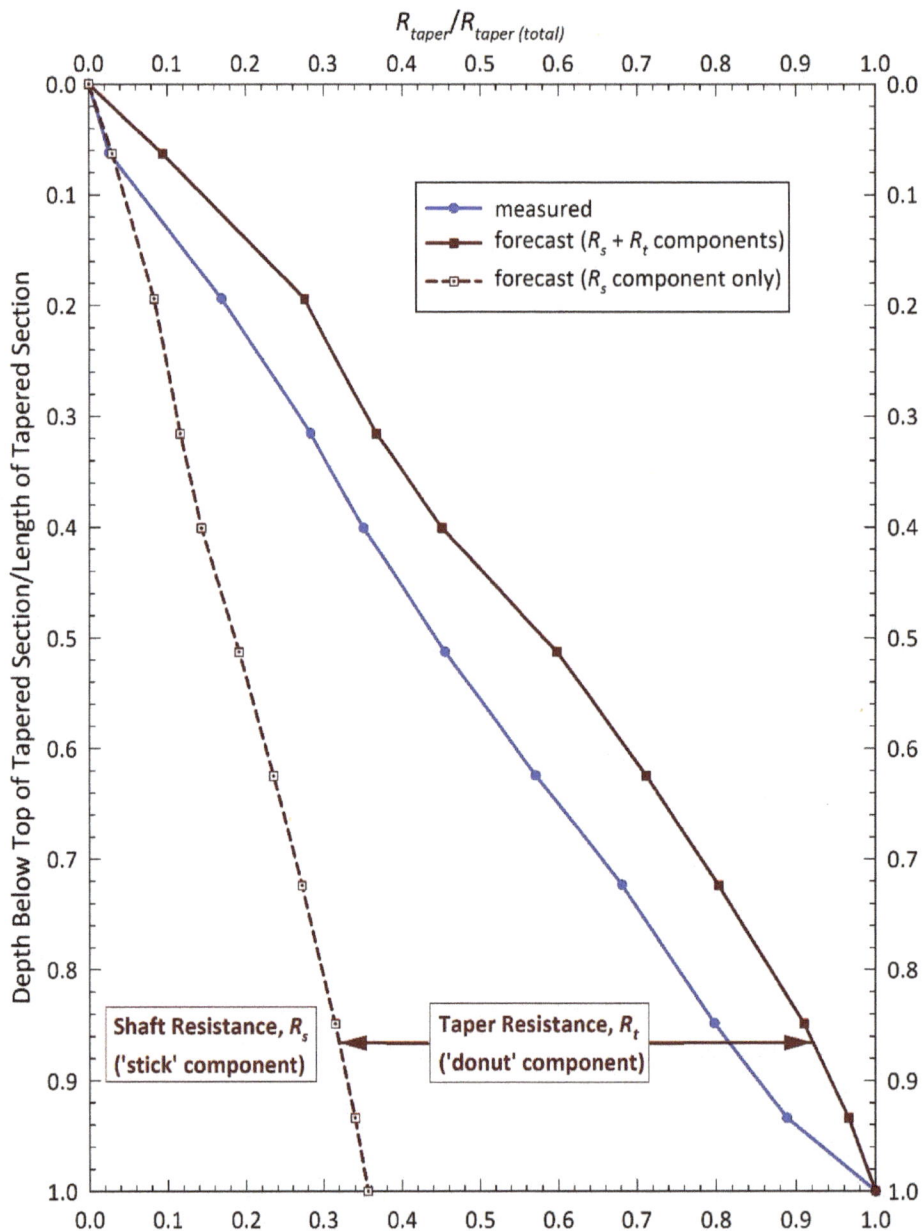

Figure A.16. Measured vs. Forecast (Togliani Method) Depth-Wise Distribution of Relative Resistance within Tapered Section for Fellenius et al. (2000) *Monotube* **Pile #3 at JFKIA.**

Appendix B

Concepts for an Improved Toe-Settlement Forecasting Methodology

---

B.1 BACKGROUND

On the one hand, the primary, fundamental premise of the Enhanced Manandhar-Yasufuku Toe-Bearing Method of relating toe resistance to toe settlement as opposed to using a fixed value of ULS toe resistance as has been done traditionally (including in the Basic Manandhar-Yasufuku Toe-Bearing Method) is correct in concept. It is thus a significant step forward and improvement to the state of foundation engineering practice over the decades-old status quo. A recurring theme throughout this monograph is that explicit quantification of load-settlement behavior should be an essential element of all deep-foundation design for axial loading, especially when it comes to toe resistance.

On the other hand, as concluded in Chapter 5 and illustrated using a case-history application in Appendix A, research to date by the writer has found the Enhanced Manandhar-Yasufuku Toe-Bearing Method to be quantitatively lacking when applied in practice to full-scale piles as it forecasts an excessively soft toe response. This behavior is typified in Figure 5.3 that illustrates the application of this method as part of the Composite JSH/M-Y Settlement Model. Additional examples of this poor agreement between forecasts and measurements for both tapered and non-tapered piles are presented in Horvath (2015) as well as in Figures A.11 and A.12.

In the writer's opinion, conclusions drawn from these observations involving application of the Enhanced Manandhar-Yasufuku Toe-Bearing Method to actual piles is much more significant and persuasive than the results obtained by Manandhar and Yasufuku from 1-g testing of small-scale model piles in a laboratory environment that were presented as the claimed 'validation' of their methodology (Manandhar and Yasufuku 2012). Reasons why the Enhanced Manandhar-Yasufuku Toe-Bearing Method may compare favorably to results in a laboratory setting but not actual field conditions include the fact that the model piles they used in the laboratory did not have a geometry that replicated actual piles and, more importantly, were not installed in a manner and in a stress field that in any way replicated actual field installation and stress conditions, including residual loads.

In particular, as discussed in Chapter 4 and Appendix A, it appears from more than 50 years of published work dating back to Nordlund (1963) that residual loads are significant for piles in general and tapered piles in particular. In the cases involving tapered piles presented by Fellenius et al. (2000) that are discussed in Appendix A, the toe resistance was interpreted to be essentially 100% mobilized at the end of driving in both cases. The cases involving tapered piles noted by Nordlund suggest that toe resistance was at least significantly mobilized at the end of driving. Such realities of actual, full-scale piles were not replicated in the laboratory testing used by Manandhar and Yasufuku.

B.2 OVERALL CRITIQUE OF CURRENT METHOD

The writer's takeaway from this discussion is that the apparent poor forecasting ability of the Enhanced Manandhar-Yasufuku Toe-Bearing Method with actual piles is not necessarily the fault of the Kondner hyperbolic model that is a key theoretical component of the method. The reasons are:

- The Kondner hyperbola, as with any mathematical model for curve-shape, simply defines the <u>qualitative</u> shape of a relationship between two variables, not the <u>quantitative</u> outcome which is entirely determined by the model parameters used.

- Fellenius (2014) indicated that what he calls the *Ratio Function* best models the load-settlement (*q-z*) behavior of the toe of a DFE. Although the Ratio Function never approaches, no less reaches, a ULS value as does the Kondner hyperbola (which Fellenius (2014) refers to as the *Chin-Kondner Hyperbolic Function*), the Ratio Function as typically quantified for use with deep foundations[178] has the same basic curve shape of a hyperbola. This suggests that a hyperbola is qualitatively an appropriate shape for the load-settlement behavior of a pile toe.

- The ULS unit toe resistance, which Manandhar and Yasufuku call q_{pcal} ($r_{t(ult)}$ in the notation used in this monograph), and the toe-resistance force, $R_{t(ult)}$, that derives from it, appear to correlate reasonably well with observed toe resistances (see Appendix A and Horvath (2015)). As will be seen later in this appendix, this ULS unit toe resistance is the theoretical limit or endpoint of the hyperbolic model in this case. This suggests that this component of the Enhanced Manandhar-Yasufuku Toe-Bearing Method is fundamentally correct and that the forecasting shortcomings lie elsewhere.

The conclusion reached from this chain of logic is that the forecasting shortcomings of the Enhanced Manandhar-Yasufuku Toe-Bearing Method as it currently exists are not the 'destination' of the hyperbolic curve but 'how it gets there'. Specifically, in the writer's opinion the problem is with the model parameters used by Manandhar and Yasufuku to quantify the <u>shape</u> of the Kondner hyperbola in this application. It appears that these model parameters were based solely on laboratory testing and other evidence that has several technical shortcomings as enumerated above and discussed in further detail later in this appendix.

The conclusion drawn by the writer is that one or more improved versions of the Enhanced Manandhar-Yasufuku Toe-Bearing Method are possible, at least in principle. The improvements could come from making alternative developmental assumptions as well as model-parameter correlation with the observed behavior of actual piles. With this in mind, it is useful to outline the generic process by which such improvements might be made, starting with the baseline of discussing in detail the original assumptions made by Manandhar and Yasufuku.

B.3 FRAMEWORK FOR METHOD IMPROVEMENTS

B.3.1 Baseline Review of the Kondner Hyperbolic Model

The Kondner Hyperbolic Model is a two-parameter mathematical relationship that is used for modeling stress-strain behavior. It has the following general form:

---

[178] This comment relates to the fact that one limiting case of the Ratio Function is a straight line, i.e. linear-elastic behavior, although all known (to the writer) published pile-related applications of the Ratio Function to date have involve curved lines.

$$\sigma = \frac{\varepsilon}{a + (b \cdot \varepsilon)} \tag{B.1}$$

where:

- a, b = model parameters (constants) that affect the shape of the curve,
- ε = normal strain, and
- σ = normal stress.

Note that there is considerable latitude as to how both the model parameters (a and b) and problem variables (stress and strain) are defined in a particular application.

The primary and enduring (in use for over 50 years) attraction of the Kondner model is that it yields a curved relationship between stress and strain that reasonably and relatively simply replicates behaviors in many geotechnical and foundation engineering applications. A detailed primer on the Kondner model for its original and still-common use as an all-purpose constitutive model for soil in finite-element (FE) analyses of geotechnical force-displacement problems can be found in Duncan and Chang (1970).

Note that in the application of the Kondner model that is the focus of this appendix, all the variable and parameter labels used subsequently in this appendix are those found in the various cited publications of Manandhar and Yasufuku. This is done intentionally so that a comparison to their version of the Kondner model is facilitated.

B.3.2 Baseline Review of the Manandhar-Yasufuku Solution Approach and Assumptions

The Kondner Hyperbolic Model as applied by Manandhar and Yasufuku to the deep-foundation toe-bearing problem is:

$$q_{cal} = \frac{S/D}{n + m(S/D)} \tag{B.2}$$

where:

- q_{cal} = unit toe resistance (equivalent to r_t in Figure 2.1) at some arbitrary level between zero and the ULS value, q_{pcal};
- S = toe settlement (equivalent to ρ_{toe} as used throughout this monograph);
- D = pile toe diameter (equivalent to d_{toe} as used throughout this monograph);
- n, m = model parameters.

In this case, the ratio S/D is dimensionless (assuming that S and D have the same units) and is expressed as a decimal (in deep-foundation applications this ratio is more commonly expressed as a percent) and functions conceptually as the strain parameter that is shown generically in Equation B.1.

The physical meanings of the two hyperbolic-model parameters, n and m, are illustrated in Figure B.1 that depicts a generic hyperbolic curve that could be used to model any force-displacement relationship. The parameter n (equivalent to the parameter a in Equation B.1) is the reciprocal of the initial tangent (to the curve) stiffness (modulus) that is often given the notation E_{ti} (*initial tangent Young's modulus*) in geotechnical applications. Thus, $n = 1/E_{ti}$ has dimensions of length squared per unit force (L^2F^{-1}). The parameter m (equivalent to the parameter b in Equation B.1) is the reciprocal of the asymptotic value of q_{cal} that the hyperbolic curve approaches but never reaches, with the same dimensions as n. Thus, m also has dimensions of length squared per unit force (L^2F^{-1}).

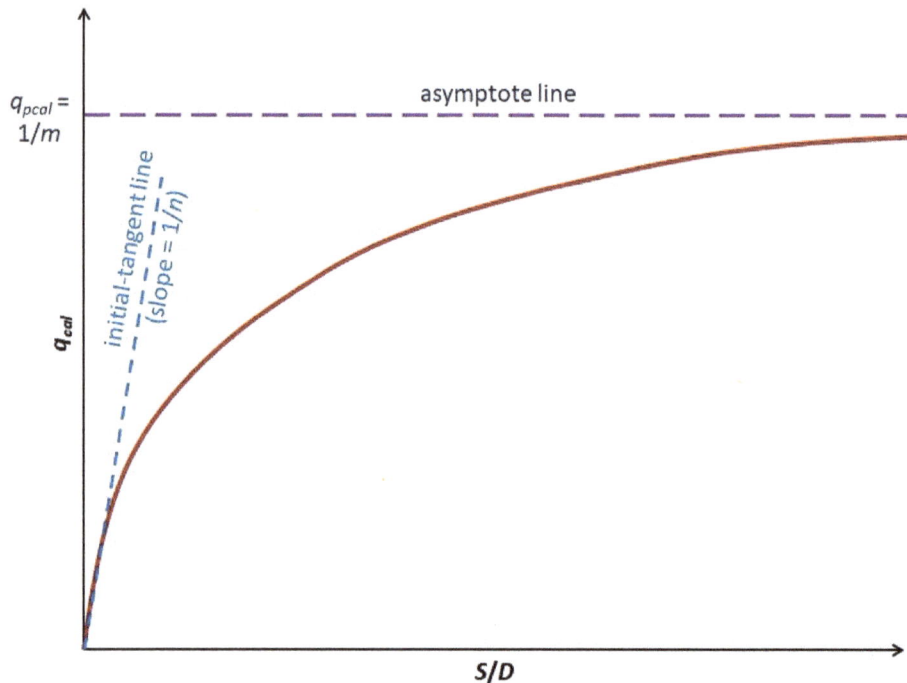

**Figure B.1. Kondner Hyperbolic Model -
Parameter Definitions Using Manandhar-Yasufuku Notation.**

As noted previously, Manandhar and Yasufuku defined this limiting, asymptotic value of q_{cal} to be the ULS unit toe resistance, q_{pcal}. It is the same parameter that is the forecast outcome of the Basic Manandhar-Yasufuku Toe-Bearing Method and is assumed to be known a priori from their bearing-capacity solution that is based on their version of spherical-cavity expansion (SCE) theory. However, as will be seen later in this appendix, it is possible in an alternative toe-bearing methodology that is developed subsequently to initially have q_{pcal} as an unknown variable that becomes part of the problem solution.

The definition of m in this application is very straightforward as it is consistent with the definition in general geotechnical problems that use a Kondner hyperbola where the limiting, asymptotic value is typically the shear strength of the soil. However, Manandhar and Yasufuku's definition of n is unique to their methodology and requires further elaboration in order to set the stage for the discussion that follows.

Manandhar and Yasufuku (2012, 2013) defined n as:

$$n = \frac{(S/D)_{ref}}{\lambda \cdot q_{pcal}}$$

(B.3)

where $(S/D)_{ref}$ is an arbitrary 'reference value' of S/D that is assumed to correlate with some assumed fractional value (defined here by the dimensionless parameter λ where $0 < \lambda < 1$) of the known ULS unit toe resistance, q_{pcal}. Stated another way, the quantity $(\lambda \cdot q_{pcal})$ reflects the portion of the ULS unit toe resistance that is assumed to be mobilized at the arbitrary reference toe-displacement (equivalent strain) level of $(S/D)_{ref}$.

Note that Manandhar and Yasufuku did not use the variable λ in their published work but, in the writer's opinion, implied its existence as explained subsequently. The writer introduced λ as a problem variable in order to make the following discussion clearer to illustrate and, hopefully, comprehend.

To begin with, the writer believes that there may be an error in Manandhar and Yasufuku's published work. This is because they clearly state in Manandhar and Yasufuku (2012/Page 5, 2013/Page 862) that $(S/D)_{ref}$ is assumed to have a value of 0.25 at one-half of q_{pcal}. Using Equation B.3, this verbal statement can, in the writer's opinion, be expressed mathematically as:

$$n = \frac{0.25}{0.5 \cdot q_{pcal}}.$$ (B.4)

This clearly implies that λ = 0.5. Yet the version of Equation B.3 that appears in their publications is as follows:

$$n = \frac{0.25}{q_{pcal}}$$ (B.5)

which clearly implies that λ = 1 which does not appear to be correct in the writer's opinion.

In June 2017, the writer initiated email contact with Prof. Manandhar (who was indicated to be the contact author for the above-referenced joint publications with Prof. Yasufuku which is logical given that the references reflect work performed for Manandhar's doctoral studies) for a clarification of this issue. While Prof. Manandhar promptly acknowledged receipt of the writer's email, there was no answer to the writer's specific query asking for clarification of this issue. However, Prof. Manandhar did state (writer's comments here and below highlighted in yellow):

"I will check it in the manuscript and will reply you [sic] *some day."*

That "some day" never came (as of the timeframe when this monograph was being finalized for publication in early 2020) as no further response was every received by the writer as of the time of publication of this monograph almost three years after than June 2017 email was sent and received.

As an aside, the above value of $(S/D)_{ref}$ = 0.25, which Manandhar and Yasufuku claimed in their 2012 paper is:

"...empirically derived for nondisplacement [sic] *piles in sands..."*

and citing Hirayama (1990) for this, means that toe resistance is 100% mobilized at a settlement magnitude that is 25% of the toe diameter. This is a substantially larger value compared to the 10% to 15% that many others, including the writer, have assumed over the years. Even Fellenius, who eschews the concept of a ULS toe resistance to begin with, notes (e.g. Fellenius 2014) that, when analytically necessary, he assumes that a toe displacement of 30 millimetres (1.2 in) constitutes toe-bearing 'failure'. For a typical pile-toe diameter of 300 millimetres (12 in), this is an S/D ratio of 10%. For a common taper-pile toe dimeter of 8 inches (200 mm), this is an S/D ratio of 15%.

In any event, it is useful to explore the implications of this apparent error, at least in a preliminary manner, by first expanding Equation B.2 in light of the preceding variable definitions and discussion in several steps to rearrange and consolidate variables:

$$q_{cal} = \frac{S/D}{\left\{\left[\frac{(S/D)_{ref}}{\lambda \cdot q_{pcal}}\right] + \left[\left(\frac{1}{q_{pcal}}\right) \cdot (S/D)\right]\right\}} , \qquad \text{(B.6a)}$$

$$q_{cal} = \frac{S/D}{\left\{\frac{\left(\left[\frac{(S/D)_{ref}}{\lambda}\right] + (S/D)\right)}{q_{pcal}}\right\}} , \qquad \text{(B.6b)}$$

$$q_{cal} = \frac{S/D}{\left\{\left[\frac{(S/D)_{ref}}{\lambda}\right] + (S/D)\right\}} q_{pcal} . \qquad \text{(B.6c)}$$

Equation B.6c is the basic form of the end result found in Manandhar and Yasufuku (2012, 2013), with $(S/D)_{ref} = 0.25$ and λ non-existent which implies a value of $\lambda = 1$ as noted previously. However, for comparative purposes in the present discussion, it is useful to rearrange Equation B.6c one more time as follows:

$$\frac{q_{cal}}{q_{pcal}} = \frac{S/D}{\left\{\left[\frac{(S/D)_{ref}}{\lambda}\right] + (S/D)\right\}} . \qquad \text{(B.7)}$$

This allows a generic, dimensionless comparison to be made for both $\lambda = 1$ as implied in Manandhar and Yasufuku's arithmetic published work and $\lambda = 0.5$ as the writer believes is correct based on Manandhar and Yasufuku's verbal published work, assuming $(S/D)_{ref} = 0.25$ in both cases. This comparison is shown graphically in Figure B.2.

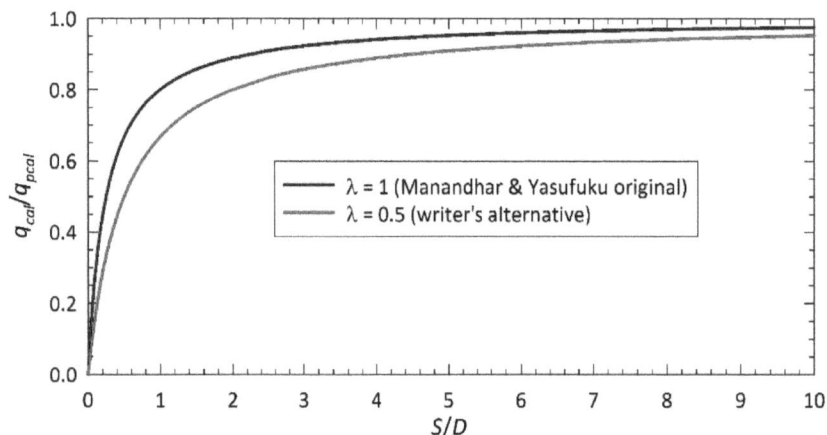

Figure B.2. λ-Parameter Comparison for Manandhar-Yasufuku Hyperbolic Model with $(S/D)_{ref} = 0.25$.

As can be seen, the difference in λ values results in a not-insignificant difference in curve shape, with the writer's $\lambda = 0.5$ curve displaying a softer load-settlement response compared to the $\lambda = 1$ curve implied in Manandhar and Yasufuku's published work. Unfortunately, this lower value of λ moves the shape of the curve in the wrong direction as the curves generated with the original, implied value of $\lambda = 1$ already consistently forecast a toe-bearing response that is much too soft in the writer's opinion. Nonetheless, there is still a lingering question concerning a possible mathematical error in Manandhar and Yasufuku's published work that can only be resolved by the original authors, presumably "some day".

B.3.3 Conceptual Development of Alternatives

B.3.3.1 Background and Overview

The writer's research to date has found that the Enhanced Manandhar-Yasufuku Toe-Bearing Method in its current formulation consistently forecasts a toe response that is much softer than measured results on full-scale tapered and constant-diameter piles installed by conventional impact driving in actual field conditions. The potential reasons for this discrepancy fall into two broad categories:

- The value of the ULS unit toe resistance, q_{pcal}, which is the same as that forecast by the Basic Manandhar-Yasufuku Toe-Bearing Method, is essentially correct but the shape of the curve leading up to failure is not.

- The value of q_{pcal} is significantly incorrect. There may or may not be issues with the shape of the curve leading up to failure as well but they are secondary in this case.

The writer's research to date suggests that the former is more likely than the latter so the primary objective of improving the outcomes of a toe-bearing methodology should focus on curve shape as opposed to the end result (q_{pcal}).

As can be seen in Figure B.2, the shape of the hyperbola is relatively sensitive to the model parameters so some thought needs to be given to a rational selection of the hyperbolic-model parameters. In this regard, it appears that more attention needs to be given to residual loads and the fact that for piles toe resistance may be substantially, if not fully, mobilized at the end of initial driving so that the toe-bearing stiffness encountered in a load test is occurring along a reload, not virgin, curve. However, good practice requires keeping an open mind so some consideration to the latter issue of the correctness of the value of q_{pcal} as currently forecast by Manandhar and Yasufuku's methodology is also appropriate.

The following sections explore the concepts for three different alternatives that can potentially be used to improve upon the Enhanced Manandhar-Yasufuku Toe-Bearing Method. There may well be additional alternatives that remain to be identified and explored in the future but this is intended to at least be a start.

B.3.3.2 Alternative #1

The simplest alternative is to work with the original solution endpoint defined by Equation B.6c but with different values of $(S/D)_{ref}$ and λ that are the problem parameters dictating the shape of the hyperbola getting to that endpoint. Note that it is really the ratio $(S/D)_{ref}/\lambda$ that is more important than the two individual components per se. However, it is useful to retain the two components individually in the governing equation (B.6c) to allow for flexibility in defining at which level, λ, of the ultimate toe resistance, q_{pcal}, the reference level of non-dimensionalized toe settlement, $(S/D)_{ref}$, is mobilized.

Figure B.3 illustrates one example of implementing this alterative. In this case, $(S/D)_{ref}$ was changed from 0.25 as used by Manandhar and Yasufuku to 0.10 and using $\lambda = 1$. These parameter values are consistent with Fellenius' aforementioned 30 millimetre (1.2 in) 'failure' criterion applied to a pile with a toe diameter of 300 millimetres (12 in). As can be seen in this figure, this results in a noticeable stiffening of the load-settlement response relative to the original results shown in Figure B.2 (these results are also shown in Figure B.3 for ease of comparison) which is, at least, a trend in the right direction.

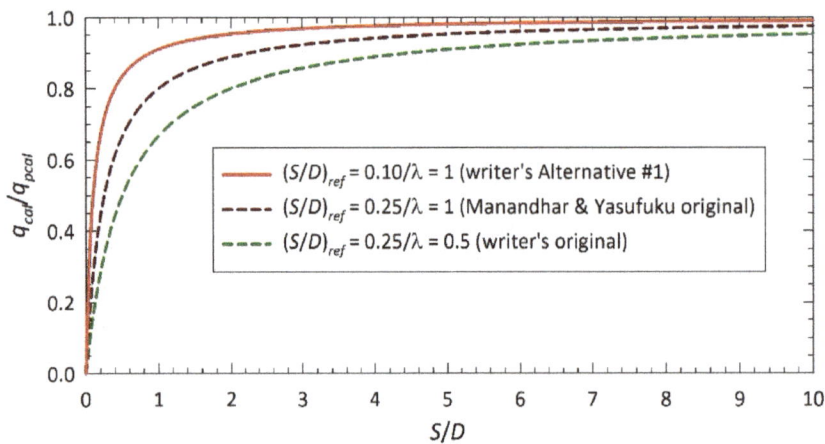

Figure B.3. Alternative #1 Model-Parameter Comparison for Manandhar-Yasufuku Hyperbolic Model.

While the results shown in Figure B.3 are of academic interest, it is important to keep in mind that any toe-bearing settlement model is only one component of an overall load-settlement methodology. Therefore, it is of more practical interest to examine the impact of different assumptions concerning the $(S/D)_{ref}/\lambda$ ratio have on overall load-settlement results.

This is done in Figure B.4 using the same *Monotube* pile as shown in Figure 5.3. The forecast outcomes were made using the Composite JSH/M-Y Settlement Model that calculates shaft resistance using the writer's Modified Nordlund Method and toe resistance using the Enhanced Manandhar-Yasufuku Toe-Bearing Method ($\lambda = 1$ was assumed in both cases). This composite settlement model is essentially an improved version of the writer's Tri-Linear Settlement Model that allows for an improved settlement-dependent contribution from toe resistance, albeit with additional computational effort. Note that $(S/D)_{ref} = 0.25$ is the original Manandhar and Yasufuku assumption and $(S/D)_{ref} = 0.10$ is an arbitrary assumption made by the writer.

As can be seen in Figure B.4, the toe-bearing component has essentially no influence for much of the load-settlement curve until the shaft resistance is assumed to be fully mobilized. At that point, toe-bearing dominates and the difference in forecast outcomes between the two $(S/D)_{ref}$ assumptions becomes apparent. It is clear that reducing the value of $(S/D)_{ref}$ from 0.25 to 0.10 results in a modest improvement in agreement between the forecast and measured results.

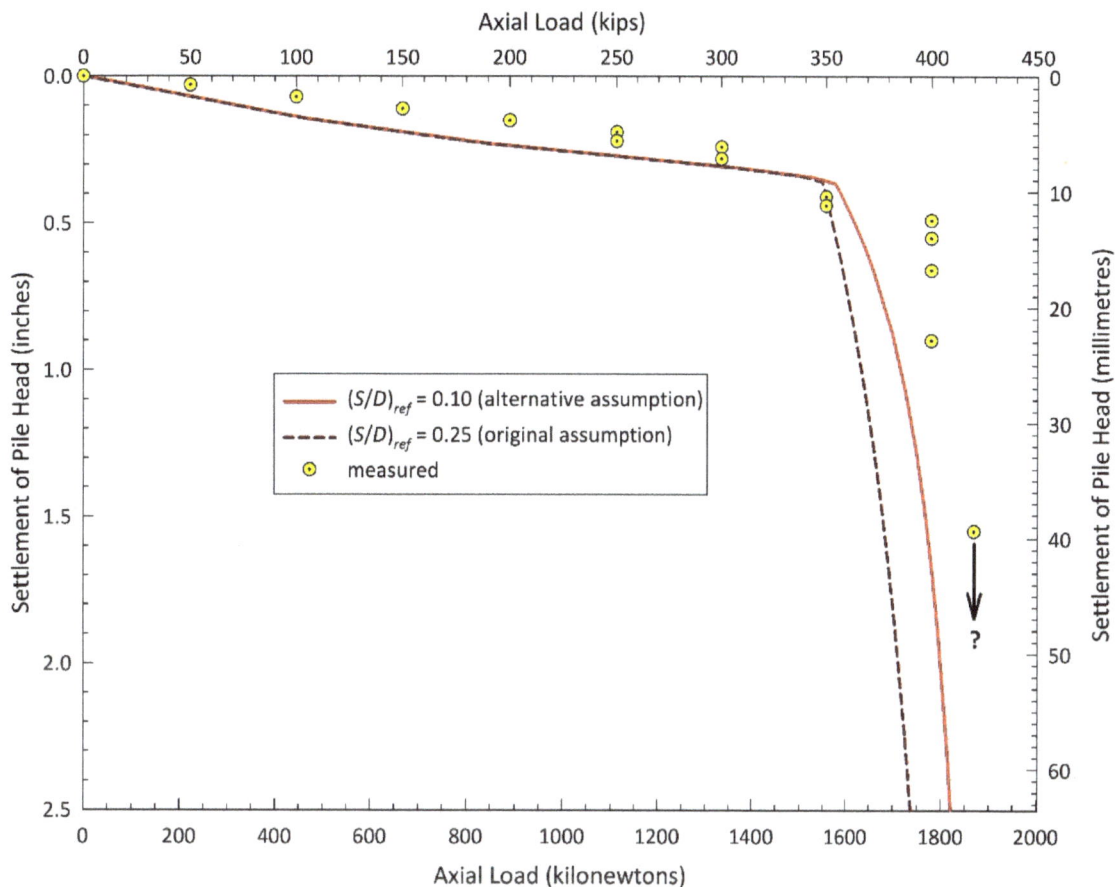

Figure B.4. Influence of $(S/D)_{ref}$ Assumption for Pile LT2-172 at JFKIA.

In conclusion, the Alternative #1 example shown in Figures B.3 and B.4 was not intended to be definitive. It was simply intended to be a proof-of-concept illustration that modifying the shape of the hyperbolic curve in the Enhanced Manandhar-Yasufuku Toe-Bearing Method by simply changing a hyperbolic-model parameter is a viable alternative. This is true whether this method is to be used alone or, more commonly, as part of another analytical model or methodology for forecasting the overall load-settlement behavior of a pile.

B.3.3.3 Alternatives #2 and #3

The second and third alternatives for improving the forecasting ability of the Enhanced Manandhar-Yasufuku Toe-Bearing Method have a common basis so are discussed together. This basis involves going back to the original equation (B.2) for the Kondner hyperbolic model as expressed using the problem parameters defined by Manandhar and Yasufuku. This equation is repeated here for ease of reference:

$$q_{cal} = \frac{S/D}{n + m(S/D)}. \tag{B.2}$$

Rearranging this equation algebraically as follows:

$$\frac{S/D}{q_{cal}} = m(S/D) + n \qquad \text{(B.8)}$$

is done so that it has the format of the generic equation:

$$y = c_1 x + c_2 \qquad \text{(B.9)}$$

where c_1 and c_2 are arbitrary constants.

This algebraic manipulation allows the two hyperbolic-model parameters, n (= c_2) and m (= c_1) to be expressed or evaluated as desired as the intercept and slope, respectively, of a straight line through datapoints defined using pairs of S/D vs. q_{cal} data that would be obtained, ideally, from the interpretation of instrumented pile load tests (note that this would require correction of the load-test results for residual forces for reasons discussed in Appendix A). These data pairs would then be plotted and interpreted as shown conceptually in Figure B.5. The graphical interpretation of the plotted data can take two different paths and this is what distinguishes between Alternative #2 and Alternative #3.

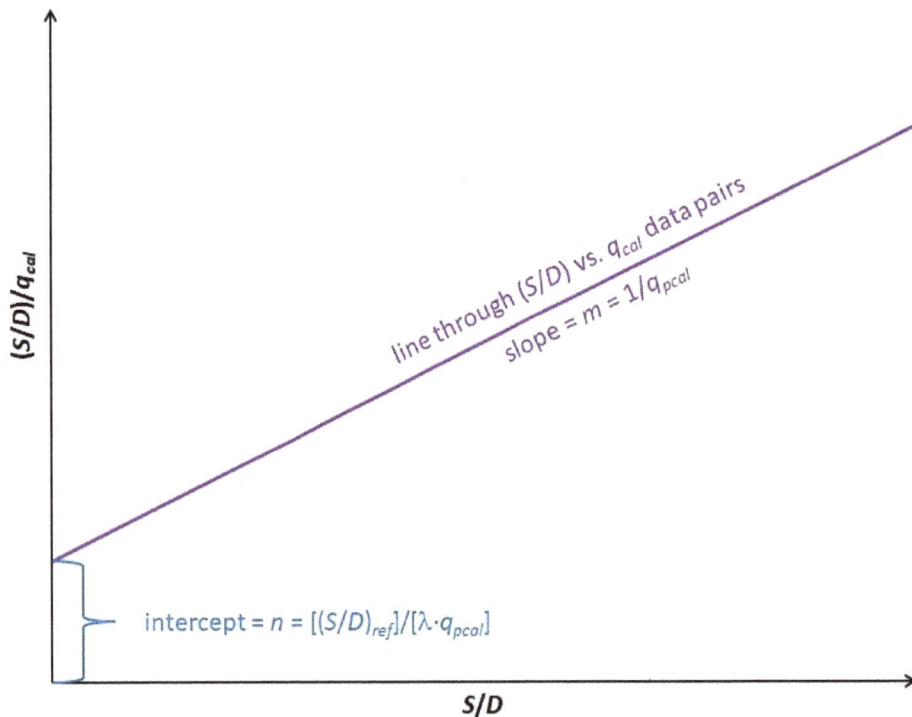

Figure B.5. Kondner Hyperbolic Model - Alternatives #2 and #3 for Parameter Evaluation.

Alternative #2 is the more restrictive of these two alternatives. It is the only alternative that is theoretically consistent with Manandhar and Yasufuku's solution for ULS unit toe resistance based on SCE that produces a value of the ULS unit toe resistance, q_{pcal}. This is because the line (shown qualitatively in Figure B.5) that one constructs through the plotted data pairs <u>must</u> be the line:

a) with slope $m = (1/q_{pcal})$ that also

b) best fits the data.

Note that this is <u>not</u> the same as a generic best-fit line that is determined solely using a mathematical algorithm and which would <u>not</u> have its slope determined a priori. In any event, this constructed line will then define the value of n which, in this case, will readily produce the 'correct' value of the $(S/D)_{ref}/\lambda$ ratio as this ratio is simply the product of n and the already-known q_{pcal}.

As an aside, a variation of this alternative would be to use an assumed known value of the ULS unit toe resistance from a solution other than that obtained using Manandhar and Yasufuku's SCE solution. This alternative value would be used in Alternative #2 in the same manner described for q_{pcal}.

Alternative #3 is more general than Alternative #2 as it assumes that q_{pcal} (and thus m) are unknown initially. The solution process begins by using a mathematically rigorous best-fit algorithm to determine the best-fit straight line through the plotted data pairs and from this a value for m and thus a value for q_{pcal}. Then, the evaluation of n and the value of the $(S/D)_{ref}/\lambda$ ratio proceeds as with Alternative #2.

There are two comments to close out both this discussion and this appendix. The more important one is that that both Alternatives #2 and #3 require substantially greater effort than Alternative #1 to implement as they require instrumented static load tests to generate the necessary data. Even then, Alternatives #2 and #3 only produce 'answers' after the fact. However, the benefit of doing this would be to develop a sufficiently robust database so that an empirical relationship for at least the $(S/D)_{ref}/\lambda$ ratio could be developed for future forecasting use. It would also provide sufficient data to either verify the accuracy of Manandhar and Yasufuku's methodology for forecasting q_{pcal} or indicate if an alternative methodology was required.

The second comment is to emphasize what was noted earlier in this section that the process for Alternatives #2 and #3 will only produce a value of n and, therefore, a unique value of only the <u>ratio</u> $(S/D)_{ref}/\lambda$. In theory, there will always be an infinite number of combinations of $(S/D)_{ref}$ and λ that can produce this singular value of the ratio. However, from a practical perspective, only the ratio is important, not the individual values of its constituent variables, as was noted earlier in this appendix.

This page intentionally left blank.

Appendix C

Theoretical Development of Shaft Resistance in the Kodikara-Moore and El Naggar-Sakr Methods

C.1 BACKGROUND

As presented originally in El Naggar and Sakr (2000) as well as in subsequent publications such as Sakr et al. (2004a), the El Naggar-Sakr Method is a resistance-only forecasting methodology of the indirect type. As discussed in Chapter 5, the novel element of this method is the analytical algorithm used for calculating the shaft resistance, as it attempts to strike a better balance between theoretical accuracy and ease of use compared to the Kodikara-Moore Method on which shaft-resistance component of the El Naggar-Sakr Method is largely based.

However, the shaft-resistance component of the El Naggar-Sakr Method has broader potential to be used as part of an integrated load-settlement methodology. All that is required is to relax one assumption made by El Naggar and Sakr and then couple the shaft-resistance algorithm to a toe-resistance methodology that allows for toe resistance to vary as a function of toe settlement, e.g. the Enhanced Manandhar-Yasufuku Toe-Resistance Method discussed in Chapter 5. It does not appear from the published literature that this opportunity to broaden the forecasting capability of the El Naggar-Sakr Method has been exploited to date.

Separate from and in addition to this is the potential to make much more substantial modifications to the shaft-resistance component of the El Naggar-Sakr Method, specifically, to allow for load-dependent degradation of the shear modulus, G, that figures prominently in the theoretical formulation of the method. Modulus degradation of ground stiffness as a result of load application and concomitant settlement is now a well-established fact for both cylindrical-cavity expansion (CCE) in all applications (Cook 2010) as well as settlement of all types of DFEs (Mayne 2007). As was noted in Chapter 5, the lack of accounting for modulus degradation is a notable shortcoming of the Kodikara-Moore Method and, as a result, the El Naggar-Sakr Method as originally formulated as well.

C.2 OVERVIEW

Because the El Naggar-Sakr Method in both its original form as well as with its potential extension and modification noted above has potential for use with tapered piles, the focus of this appendix is a detailed presentation of the development of the El Naggar-Sakr Method so that its theoretical basis, synthesis, and assumptions are clearly understood. This is necessary for anyone either desiring to use this method in its present form or contemplating making changes to the method.

This presentation is divided into two parts. First discussed are the Phases I, II, and III analytical steps from the Kodikara-Moore Method, significant portions of which form the theoretical basis of the El Naggar-Sakr Method. Then, the synthesis and simplifying assumptions made by El Naggar and Sakr to relate the theoretical elements from the Kodikara-Moore Method to their basic governing equation for shaft resistance are presented. This is intended to provide both clarity concerning the relationship between the El Naggar-

Sakr and Kodikara-Moore methods as well as a clearly defined starting point for any future extensions or modifications to the El Naggar-Sakr Method.

Throughout this presentation, the original terminology and notation used in Kodikara and Moore (1993) and El Naggar and Sakr (2000) are retained even in situations where it may deviate from the standardized terminology and notation adopted in this monograph. This is to facilitate reference back to the original works as might be desired.

C.3 SHAFT-RESISTANCE CALCULATION

C.3.1 Basic Governing Equation

The basic equation that governs the forecasting of shaft resistance in the El Naggar-Sakr Method is an algebraic relationship for the ULS unit shaft resistance, a parameter that they call "skin friction" and define using the notation τ_s. As discussed in Chapter 4, the ULS unit shaft resistance is the analytical core of any shaft-resistance calculation as integrating this unit stress over the exterior surface area of the shaft produces the desired forecast of ULS shaft resistance, R_s. The mathematical definition of τ_s was given in Chapter 5 (Equation 5.19) but is repeated here for ease of reference:

$$\tau_s = K_t \, K_s \, \tan \delta \; \sigma_v' \, . \tag{C.1}$$

Before discussing this equation, it is of interest to note that El Naggar and Sakr adopted the same analytical philosophy used by both Nordlund and Kodikara and Moore in that a single definition of ULS unit shaft resistance was adopted ($r_{s(ult)}$ in the notation of this monograph) as opposed to differentiating between the ULS unit shaft resistance due to the mechanism of sliding friction, $r_{ss(ult)}$, as occurs with a constant-perimeter DFE and ULS unit shaft resistance due to the wedging effects of taper, $r_{st(ult)}$, as was defined by the writer in Chapter 4. However, this is just an academic detail and does not impact the numerical outcomes from the El Naggar-Sakr Method one way or the other in terms of accuracy.

With regard to the other parameters in Equation C.1, recall from Chapter 5 that K_s is assumed to be the baseline-reference value of the dimensionless lateral earth pressure coefficient acting on a constant-diameter pile in the same site-specific conditions after driving. The dimensionless parameter K_t (≥ 1) represents all of the taper benefit.

As will be seen, El Naggar and Sakr assumed that K_t is a function of pile settlement due to applied load after driving. Thus, when taken together with their assumptions concerning K_s, this implies that in the El Naggar-Sakr Method all taper benefit is implied to derive solely from post-driving load application. None is the result of radial stress increase due to driving beyond what would occur with a constant-diameter pile in the same ground conditions. Stated another way, Equation C.1 implies that after driving and prior to any external load application, a tapered and constant-diameter pile of the same basic type in the same ground conditions would have the same radial stress acting along the pile shaft, as given by the quantity ($K_s \tan \delta \, \sigma'_v$). It is only after post-driving load application begins that K_t begins to increase for tapered piles from a value equal to one while it remains constant at a value equal to one for constant-diameter piles.

In any event, El Naggar and Sakr assumed that in a typical application of Equation C.1 in practice that the design professional would assume values for K_s and δ (they provided no meaningful guidance in this regard other than that K_s should be the value for a constant-perimeter pile in the same setting) and calculate σ'_v which was implied to be σ'_{vo}. This leaves only the K_t parameter and for this El Naggar and Sakr used τ_s as the link between their

Equation C.1 and the theoretical solution developed by Kodikara and Moore. Thus, before discussing the El Naggar-Sakr Method further it is necessary to discuss in detail how Kodikara and Moore evaluated τ_s, a parameter for which they used the notation τ_x.

C.3.2 Kodikara-Moore Method

C.3.2.1 Background

As discussed in Chapter 5, Kodikara and Moore (1993) assumed that in the most general case a tapered DFE went through three, successive behavioral phases of shaft-resistance development to an externally applied axial-compressive load. Although the assumed physical behavior in each of these three phases was conceptually and mathematically distinct, there are smooth, seamless transitions, both in concept and mathematics, so that a relatively smooth load-settlement curve is the calculated outcome.

The third phase of assumed post-yield, elastoplastic soil behavior is by far the most complex of the three mathematically as it involves, among other things, the evaluation of an infinite series. While this in and of itself is not insurmountable in concept, it is a pragmatic deterrent to use in routine practice unless the process is incorporated into purpose-written computer software with a user-friendly interface that hides the mathematical complexities.

In each of these three phases, the interaction between the DFE shaft and adjacent ground is mathematically defined by an assumed vertical shearing resistance, τ_x, between DFE and soil that develops along a 1-D surface defined by the interface between the DFE shaft and adjacent ground. This vertical shear stress is the key aspect of the Kodikara-Moore Method as it invokes vertical-force equilibrium as part of the overall solution methodology.

Note that because the toe response in the Kodikara-Moore Method is assumed to be linear-elastic throughout all phases of shaft response, the toe behavior is relatively trivial mathematically in the overall scheme of things. Furthermore, El Naggar and Sakr used a completely different physical model and concomitant mathematical solution for toe resistance. Consequently, toe resistance is not discussed further in this appendix.

As noted in Chapter 5, El Naggar and Sakr (2000) reasoned that for the tapered piles of interest to them, which had taper angles of approximately 1° or less, that yield of the surrounding soil would not occur under reasonable levels of pile settlement so that they could ignore Phase III of the Kodikara-Moore Method that involved assumed soil yielding around the pile shaft. Thus, El Naggar and Sakr only considered Phases I and II of the Kodikara-Moore Method when developing their own analytical methodology.

El Naggar and Sakr provided minimal details in their 2000 paper concerning the theoretical relationships that they utilized from the Kodikara-Moore Method. Furthermore, because El Naggar and Sakr dealt only with a coarse-grained soil, they used simplified versions of the theoretical relationships developed by Kodikara and Moore who had allowed for both shear-strength components, c and ϕ, of the Mohr-Coulomb failure criterion to be non-zero. Thus, in the interest of better understanding all theoretical aspects of the relationships developed by Kodikara and Moore, their original work is reproduced here before addressing the simplifications, synthesis, and assumptions made by El Naggar and Sakr.

Note that all three behavioral phases of shaft resistance in the Kodikara-Moore Method are discussed in this appendix even though El Naggar and Sakr did not use the results from Phase III. This is done for completeness as well as to fully understand what behaviors El Naggar and Sakr chose to neglect when developing their analytical methodology. It is also done so that the assumptions of Kodikara and Moore can be compared to other work involving cylindrical-cavity expansion such as Cook (2010).

C.3.2.2 Overview

Because the Kodikara-Moore Method was intended to be an integrated, one-step load-settlement forecasting methodology, their overall approach was to develop unique mathematical relationships for each of the three behavioral phases of shaft resistance between:

- DFE resistance, as reflected in the aforementioned vertical shearing resistance, τ_x, between the DFE shaft and the adjacent, surrounding ground, and

- vertical displacement (settlement) along the DFE that they defined using the notation u_p.

Although the resulting τ_x vs. u_p relationships are different for each of the three behavioral phases, they are each based on the same free-body diagram of forces applied to a typical horizontal slice through the shaft of a continuously tapered DFE with a circular cross-section and uniform taper angle α. However, as will be seen, the resulting τ_x vs. u_p relationships are, as a group, valid for the limiting $\alpha = 0$ case, i.e. a constant-diameter DFE, as well so the Kodikara-Moore Method can be used on partially tapered piles that have both tapered and constant-diameter components such as the *Monotube* and *TAPERTUBE* as well as a constant-diameter DFE although Kodikara and Moore did not note this explicitly.

The vertical axis for this phase-unifying free-body diagram was defined as the *x*-axis, not the *z*-axis (as is the typical geotechnical and foundation engineering sign convention) or the *y*-axis (as is the typical structural engineering sign convention) as discussed in Horvath (2018b). Although not stated explicitly in their paper, based on notational usage for problem parameters, the radial direction is defined as the *y*-axis direction although it arguably should have been defined as the *r*-axis as the problem is geometrically axial-symmetric (alternatively referred to as circular-symmetric).

Both vertical-force equilibrium of the free-body diagram and continuity of vertical displacements (settlements) were invoked by Kodikara and Moore in developing their model, and taper effects were incorporated in two broad ways:

- The skewed (with respect to the vertical and radial directions) orientation of the orthogonal force vectors along the shaft-ground interface due to the taper angle, α, was considered when invoking vertical-force equilibrium. Force vectors oriented parallel and perpendicular to the skewed orientation, which are thus parallel and perpendicular, respectively, to DFE shaft-ground interface, are subscripted with the notation *n*.

- In Phases II and III, the physical and concomitant mathematical model of cylindrical-cavity expansion (CCE) in the radial direction was used to approximate the apparent radial expansion of the tapered pile shaft as it settled under applied load. Note that this is slightly approximate due to the fact that the DFE-ground interface is not perfectly vertical but was apparently considered to be. This is a reasonable assumption given that another assumption that was made and that carries through all three behavioral phases in different ways is that taper angles are reasonably "small", specifically, $\alpha \leq 5°$. This is not problematic as the vast majority of tapered piles used commercially, at least in the U.S., have taper angles that are < 1°.

The next issue of general note is that Kodikara and Moore assumed that the shear strength of the ground surrounding the DFE followed the most general case of the well-known Mohr-Coulomb failure criterion with both a stress-independent component ('cohesion') and a stress-dependent component ('friction') or what was frequently called in older literature a 'c-ϕ soil'. This archaic term is now recognized to be simplistic in concept and harks back to the days when it was perceived that clay had 'cohesion'; sand and gravel had 'friction'; there were intermediate geomaterials (IGMs) that had both cohesion and friction; and there was no unifying concept in the behavior of these categories of soil type which, of course, is simply incorrect. As is now well-known, all soils can be characterized by a friction angle that is often given the notation ϕ'. Nevertheless, this c-ϕ formulation allows for both drained and undrained analyses to be performed and for the cohesion term to be viewed, as desired, as either:

- 'true' cohesion due to, say, inter-particle cementation;

- 'apparent' cohesion due to partially saturated soil behavior and matric suction; or

- the undrained shear strength in a total-stress/undrained-strength analysis.

As a final observation and comment, Kodikara and Moore neglected the Poisson Effect on the DFE shaft, i.e. the fact that the shaft-ground interface expands radially under axial-compressive loading because the Poisson ratio of the DFE material(s) is always greater than zero. This omission is not surprising as the Poisson Effect for DFEs has historically been ignored in both research and practice although from time to time researchers have noted this behavioral phenomenon and opined as to its possible effect on DFE behavior under axial loading.

The reason why the omission of the Poisson Effect is highlighted here is that the Kodikara-Moore Method is unique in that it is the first known resistance-forecasting methodology for DFEs to specifically model radial displacement of the shaft-ground interface. While the Kodikara-Moore Method links this radial displacement geometrically to tapered-DFE settlement, it is clear that the Poisson Effect, which is a mechanical as opposed to geometric phenomenon, is additive to this geometric radial displacement. Therefore, to the extent that the additional radial displacement due to the Poisson Effect may influence the calculated results, it becomes a potentially relevant issue more than in the past.

The writer developed a theoretical solution for calculating the Poisson Effect for a composite DFE cross-section composed of up to three different materials (Horvath 2010a). This solution could be incorporated into the Kodikara-Moore Method or a theoretically related method such as the El Naggar-Sakr Method so that the Poisson Effect is explicitly considered in the solution process. The one caveat is that the radial stress acting along the shaft-ground interface is assumed to be a known problem constant in the writer's solution. Because this radial stress is initially an unknown and one of the calculated outcomes of the Kodikara-Moore Method and derivative methods such as that of El Naggar and Sakr, it is clear that an iterative process would have to be included in any solution algorithm. This is a relatively minor detail and thus would not preclude consideration of the Poisson Effect.

C.3.2.3 Phase I

The fundamental assumptions of this initial behavioral phase are that:

- The overall load-settlement response of the DFE-ground system is linear-elastic.

- The response of a tapered DFE to an applied axial-compressive load is for all practical purposes the same as that of a DFE with constant diameter in that there is:
  - <u>no</u> radial displacement of the DFE-ground interface and
  - "<u>no significant increase</u>" (this is a direct quote from Kodikara and Moore (1993), with "significant" being undefined by them) in radial normal stresses acting along the DFE-ground interface so $\sigma_r = \sigma_o$.

- There is <u>no</u> relative settlement (slippage) between the exterior surface of the DFE shaft and adjacent ground so that u_p and u_g (the settlement of the ground) are the same at all times. Stated another way, the DFE shaft and adjacent ground are 'stuck together'.

Note that some of these assumptions are clearly approximate if not outright incorrect. Specifically, simple geometry as well as the Poisson Effect noted in the preceding section dictates that there <u>must</u> be some radial displacement of the DFE-ground interface when the DFE settles. In addition, there <u>must</u> be some slippage between the DFE shaft and adjacent ground as the mobilization of interface friction, which is technically called *Coulomb* or *dry friction*, physically requires relative displacement. Therefore, the collective assumptions made by Kodikara and Moore should be recognized as being approximations made for analytical simplicity and concomitant convenience as opposed to being statements of reality.

In any event, these simplistic assumptions allowed Kodikara and Moore to use the theory-of-elasticity solution for DFE settlement that was published in Randolph and Wroth (1978) and is the basis of the Randolph-Wroth Problem that is discussed at length in Chapter 5 as a settlement-only methodology that is coupled with a resistance-only methodology to create a two-step procedure for generating a load vs. settlement curve.

As noted in Chapter 5, the physical model used by Randolph and Wroth is that the ground surrounding a DFE consists of a series of imaginary, contiguous, concentric, vertical cylinders that transfer the assumed vertical shear stress, τ_x, acting along the DFE-ground interface radially by a process of vertical shearing within and between cylinders. It is important to note that perfect adhesion, i.e. no relative slippage, is assumed at all times between the DFE-ground interface as well as between all of the imaginary cylinders that constitute the physical and mathematical models of the ground around the shaft of the DFE.

Settlement within these concentric cylinders that is caused by these vertical shear stresses attenuates progressing radially away from the DFE-ground interface. At some radial distance from the DFE-ground interface, the settlement magnitude becomes effectively zero. In very broad conceptual terms, this is the same as the concept of a 'significant stressed depth' beneath an applied vertical load on the ground where at some depth settlements caused by that load become negligibly small and are assumed to be zero in magnitude for analytical convenience.

Note that the overall load-transfer mechanism that is assumed to be occurring within and between the concentric cylinders is one of vertical shearing. This is the primary reason why the shear modulus, G, as suggested by Niazi, is nowadays the preferred metric for ground stiffness as opposed to Young's modulus, E, as suggested originally by Mayne, when using the Randolph-Wroth Problem solutions, as was noted in Chapter 5. The shear modulus simply better reflects the load-transfer mechanism that is assumed to be occurring within the concentric-cylinder model. Fortuitously, the small-strain shear modulus, G_{max}, is nowadays relatively easy to determine using a sCPTu sounding or other in-situ testing methodologies so this makes using shear modulus straightforward.

Because the overall solution algorithm developed by Kodikara and Moore is a settlement-based iterative process using the aforementioned τ_x vs. u_p relationships developed for each of the three behavioral phases, the basic equation used to define behavior during Phase I is as follows:

$$u_g = \frac{\zeta \tau_x r_m}{G} \qquad (C.2)$$

where:
- ξ = a dimensionless parameter defined by the following equation where L = embedded length of the DFE and v = Poisson's ratio of the ground:

$$\zeta = \ln\left[\frac{2.5\,L(1-v)}{r_m}\right] ; \qquad (C.3)$$

- r_m = the mean (average) radius of the DFE. Because the solution algorithm requires dividing the DFE shaft into an arbitrary number of artificial segments, r_m would be the average radius of a given segment;

and all other terms have been defined previously.

There are several issues of note with respect to these two equations:

- G is assumed to be constant in magnitude with depth and under applied load. Equation C.3 is only theoretically correct for this simplistic assumption. As noted in Niazi (2014) and Niazi and Mayne (2015c), if the shear modulus varies with depth and/or is different along the shaft vs. beneath the toe, then the relationship for ξ is algebraically much more complex that that given in Equation C.3 in that it contains factors defining the shear-modulus variations.

- The magnitude of τ_x is less than the ULS value that is defined subsequently.

- The numerator in the natural-log function in Equation C.3, $[2.5\,L(1-v)]$, is the above-noted radial distance from the DFE-ground interface at which the soil settlement that is caused by DFE settlement is assumed to be zero.

Because no relative displacement between the DFE shaft and adjacent ground is assumed during Phase I, the settlement of the DFE, u_p, is the same as that of the ground, u_g. Making the appropriate parameter replacement and rearranging, Equation C.2 becomes the desired τ_x vs. u_p relationship for the DFE shaft in Phase I:

$$\tau_x = \frac{G}{\zeta r_m} u_p . \qquad (C.4)$$

Kodikara and Moore assumed that τ_x continues to increase according to Equation C.4 with increasing DFE settlement, u_p, until yielding along the planar interface between the DFE shaft and adjacent ground occurs. This yielding is assumed to be the onset of traditional relative settlement (slippage) along the DFE-ground interface as a consequence of reaching the ULS of the interface materials. The magnitude of τ_x at yield is defined as τ_o and for the baseline reference case of a DFE of constant diameter is mathematically expressed using the Mohr-Coulomb yield criterion as follows:

$$\tau_o = (\sigma_o \tan \phi_i) + c_i \qquad (C.5)$$

where:
- σ_o = the radial normal stress (lateral earth pressure) acting along the DFE-ground interface <u>after</u> DFE installation but <u>before</u> any external load application;
- ϕ_i = the Mohr-Coulomb friction angle along the DFE-ground interface (hence the subscript i). In this monograph, the notation δ is used for this angle; and
- c_i = the Mohr-Coulomb cohesion along the DFE-ground interface (hence the subscript i). This would normally be called the *adhesion*, with the notation c_a.

The σ_o parameter requires further comment:

- Normally, the notation σ_h or σ_r would be used to be unambiguous about the orientation of this stress.

- This stress is not necessarily a total stress as the notation would imply. Kodikara and Moore were notationally vague and imprecise throughout their 1993 paper in this regard, and a review of the use of this term within their paper indicates that it could be either a total or effective stress depending on the application-specific context. Thus, there should have be a more-precise indication using the appropriate notation as to whether this was intended to be a total or effective stress.

- Not only is σ_o assumed not to change in magnitude to any "significant" (a direct quote from Kodikara and Moore's 1993 paper) extent during Phase I loading, Kodikara and Moore assumed that σ_o was depth-wise uniform in magnitude throughout all three behavioral phases in the exemplar applications included in their paper. This assumption is a significant deviation from reality and the most basic concepts of elementary soil mechanics.

As noted in the overview section of this appendix, Kodikara and Moore introduced the effects of DFE taper into their analytical model by noting that the DFE-ground interface is not vertical but is rotated off-vertical by the taper angle, α. Thus, the shear stress acting along the DFE-ground interface at the ULS is τ_n, not τ_o, and the normal stress on the interface is σ_n, not σ_o, where:

$$\tau_n = (\sigma_n \tan \phi_i) + c_i \,. \tag{C.6}$$

Of course, τ_o as a function of σ_o and τ_n as a function of σ_n are related by geometry so Equation C.5 as applied to a tapered DFE with taper angle α is as follows:

$$\tau_o = [\sigma_o \tan(\phi_i + \alpha)] + c_i \left[\frac{\sec^2 \alpha}{1 - (\tan \alpha \, \tan \phi_i)} \right] \tag{C.7a}$$

$$\tau_o = [\sigma_o \tan(\phi_i + \alpha)] + c_i' \,. \tag{C.7b}$$

Note that Kodikara and Moore defined the term c_i' in Equation C.7b solely for algebraic convenience in subsequent Phase II and Phase III derivations to eliminate having to show the bracketed trigonometric terms that modify c_i in Equation C.7a. Thus, c_i' is equivalent to the terms in Equation C.7a that are highlighted in red. Note also that as would be expected, if $\alpha = 0$ in Equation C.7a then Equation C.5 for a pile with constant diameter is recovered.

As a final comment concerning Phase I, note that this analysis is valid for a constant-diameter DFE as well as a tapered DFE. Although Kodikara and Moore only considered a continuously tapered DFE, there are many tapered piles in commercial use, especially in the U.S., such as the *Monotube* and *TAPERTUBE* piles, that combine both constant-diameter and tapered components so the ability to use the Kodikara-Moore Method with constant-diameter DFEs is an important issue.

C.3.2.4 Phase II

The writer's review and interpretation of the development of the El Naggar-Sakr Method as presented in El Naggar and Sakr (2000) indicates that the shaft-resistance component of the El Naggar-Sakr Method is centered around Kodikara and Moore's behavioral Phase II.

The transition from Phase I to Phase II is assumed to occur when the value of τ_x in Equation C.4 equals the value of τ_o in Equation C.7b, i.e. yield or slippage along the DFE-ground interface initiates due to the interface shear strength reaching its maximum (ULS) theoretical value. The DFE settlement at which this transition occurs, which can be calculated by using Equation C.4 and setting $\tau_x = \tau_o$ with the latter calculated using Equation C.7b, is called $(u_p)_Y$ by Kodikara and Moore, with the subscript Y indicating yield.

The assumed behavioral characteristics that define Phase II are severalfold:

- Yield along the DFE-ground interface is the traditional assumption and classical physical mechanism of 1-D slippage or sliding of the DFE shaft relative to the adjacent ground along a planar failure surface defined by the exterior of the DFE shaft.

- Yielding along this assumed failure surface is continuously ongoing throughout this phase. In classical rigid-body mechanics, this is referred to as *kinetic friction*, i.e. the shearing resistance that develops on a contact surface between two bodies where at least one of the bodies is in motion relative to the other.

- The shearing resistance, τ_x, along this assumed interface failure surface is defined by the traditional Mohr-Coulomb failure mechanism that in the most general case has a stress-independent component (adhesion) and stress-dependent component (interface friction) as noted previously.

- Although interface shear failure is occurring along this assumed failure surface, the ground adjacent to the DFE is assumed to remain in a non-failure, linear-elastic state at all times, with ongoing ground settlement, u_g, that is independent of and less than the ongoing DFE settlement, u_p.

- Because relative slippage along the shaft-ground interface occurs, u_p becomes and remains greater in magnitude than u_g throughout this phase.

- The relative difference in settlement, $u_p - u_g$, increases throughout this phase as the applied load is increased. This occurs because although both u_p and u_g increase with increasing applied load, the relative increase in u_p is always greater due to the ongoing slippage that is assumed to be occurring along the DFE-ground interface.

- Because points on the DFE and adjacent ground no longer settle in consonance as in Phase I, there is a radial (*y*-axis) component of <u>relative</u> displacement of the DFE-ground interface called v that is mathematically defined as follows:

$$v = (u_p - u_g) \tan \alpha .$$
(C.8)

Note that v increases continuously throughout Phase II because $(u_p - u_g)$ increases continuously under increasing applied load.

- Because v represents an effective <u>radial</u> expansion of the DFE-ground interface, there is an increase in radial stress, $\Delta\sigma_r$ (the subscript 'r' was not used by Kodikara and Moore but is added here for clarity), beyond the initial value of σ_o. This, coupled with the fact that u_p and u_g are no longer in sync, means that the relationship for τ_x in Phase I is no longer valid so a new τ_x vs. u_p relationship needs to be developed.

Clearly, the key theoretical element of Phase II is developing a physical model and concomitant mathematical relationship for the relationship between the increase in radial displacement, v, and increase in radial stress, $\Delta\sigma_r$, that are each occurring as a direct result of ongoing DFE settlement, u_p. Note that for Kodikara and Moore this was conceptually the same broad challenge that Nordlund faced in developing his analytical methodology even though Nordlund did not explicitly consider pile settlement in his analytical methodology. As discussed in Chapter 5, Nordlund chose the 2-D model of a rigid wall rotating about its bottom into an infinite soil mass. However, Kodikara and Moore chose the much more realistic 3-D physical model of CCE, i.e. the case of a cavity expanding from an initial non-zero radius (as the DFE is 'wished in place' prior to the start of Phase I), that was reasonably approximate because of their assumption that the taper angle, α, would be relatively small in magnitude. Note, however, that Kodikara and Moore chose to use different theoretical solutions for CCE in their Phases II and III as they assumed different behavioral states within the ground adjacent to the DFE in each phase.

As noted above, in Phase II that is being discussed here, the ground behavior adjacent to the DFE shaft is assumed to remain linear-elastic throughout the entire phase which allows for a relatively simple CCE solution to be used. Specifically:

$$\Delta\sigma_r = K_e v$$
(C.9)

where:

$$K_e = \frac{2G}{r_m} .$$
(C.10)

The parameter K in the context of cavity mechanics is called the *tangent gradient of cavity stress to cavity radius*. In this behavioral phase (Phase II), K is subscripted with an e in this case to denote the assumed **e**lastic behavior of the expanding cylindrical cavity.

Conceptually and dimensionally, K_e is a rate-change-of-modulus or tangent-modulus gradient that relates the rate-change of stress vs. displacement in much the same way that elastic moduli relate stress and strain. The dimensions of K_e are force per length cubed, better expressed as force per length squared (a modulus) per unit length (the dimension over this modulus is changing).

As noted earlier in this discussion, the overarching objective in the development of the Kodikara-Moore Method is to develop a τ_x vs. u_p relationship for each of the three assumed behavioral phases. With this in mind, there are two ways in which τ_x in Phase II differs from that in Phase I. One is purely mechanistic, i.e. how does this shear stress physically develop. The other is conceptual, i.e. how is the parameter τ_x used.

Considering first the mechanistic issue, unlike in Phase I where the radial stress[179], σ_r, is assumed to remain constant with $\sigma_r = \sigma_o$, in Phase II σ_r is increasing continuously from the initial value of σ_o because $\Delta\sigma$ is continuously increasing with increasing v, i.e. $\sigma_r = \sigma_o + \Delta\sigma_r$. This means that the shear stress, τ_x, is increasing continuously as well due to the increased confining stress caused by the radially expanding DFE-ground interface. This is the conceptual 'heart' of cavity expansion, that an expanding cavity causes a radial-stress increase[180].

Kodikara and Moore expressed this behavior as follows (note the use of the c'_i parameter that is equivalent to what is highlighted in red in Equation C.7a):

$$\tau_x = [(\sigma_o + \Delta\sigma_r)\tan(\phi_i + \alpha)] + c'_i . \tag{C.11}$$

Note, however, that it is still the DFE settlement, u_p, that is causing the increase in τ_x in Phase II as it did in Phase I. This is because the $\Delta\sigma_r$ that is causing the increase in τ_x in Phase II (per Equation C.11) is linked to the radial displacement, v (per Equation C.9), that is, in turn, linked to u_p (per Equation C.8).

Considering next the conceptual usage of τ_x, note that the implication of τ_x in Equation C.11 is subtly different than the same notation implied in Phase I. In Phase I, τ_x increased with increasing DFE settlement up to the point of the interface ULS, τ_o, as defined by Equation C.7a. However, as used in Phase II, τ_x is <u>both</u> increasing in magnitude with u_p <u>and</u> defining the ULS (Equation C.11) at the same time. The latter is because one of the behavioral assumptions in Phase II that was stated at the outset of this presentation is that the DFE-ground interface is continuously in the ULS as relative DFE-ground slippage and ever-increasing $(u_p - u_g)$ are both continuously occurring. Consequently, Kodikara and Moore did not define and use the parameter τ_o in Phase II. As will be seen, they use another parameter, $(\tau_x)_Y$, for defining the limiting value (upper bound) of τ_x in Phase II.

The final desired Phase II relationship between τ_x and u_p, i.e. a conceptual equivalent to Equation C.4 in Phase I, is achieved by combining Equations C.2/.8/.9/.11 with the following result:

$$\tau_x = \frac{\left[K_e \cdot \tan\alpha \cdot \tan(\alpha + \phi_i) \cdot u_p\right] + [\sigma_o \cdot \tan(\alpha + \phi_i)] + c'_i}{1 + \left[\left(\frac{K_e \zeta r_m}{G}\right) \cdot \tan\alpha \cdot \tan(\alpha + \phi_i)\right]} . \tag{C.12}$$

Note the total DFE settlement that occurred during Phase I is incorporated into this equation so that it represents the <u>cumulative</u> results throughout Phase II.

---

[179] As noted previously, that Kodikara and Moore did not use the notation σ_r. This notation is introduced and used here by the writer solely for clarity in the presentation and concomitant discussion.

[180] As noted in Chapter 4, which is where the concept of cavity expansion was first discussed in an abstract manner, problems in cavity expansion are traditionally formulated so that it is a deliberate increase in pressure <u>within</u> the cavity that is the fundamental causal mechanism that is driving or causing the expansion of the cavity wall (DFE shaft-ground interface here). In other words, the cause-effect mechanism is pressure causing expansion. For example, this is what occurs in pressuremeter testing. However, in the application of cavity expansion with DFEs, the cause-effect mechanism is reversed as it is the expansion of the cavity that is causing an increase in pressure <u>outside</u> the cavity, in this case, the adjacent ground along the interface with the DFE shaft. Although the cause-effect mechanism is reversed from the traditional formulation, the mathematics are unchanged, just rearranged as the known and unknown problem parameters are reversed.

Equation C.12 is completely general in that it is also valid for constant-diameter DFEs or, more relevantly, the constant-diameter portion of a partially tapered pile such as a *Monotube* or *TAPERTUBE*. This is because if $\alpha = 0$, Equation C.12 reverts back to Equation C.5 which implies that $\tau_x = \tau_o$. This simply means that for a constant-diameter DFE or portion of a DFE, the DFE-ground shear stress, τ_x, remains at a constant, ULS magnitude, τ_o, once the ULS is reached at the end of Phase I. Furthermore, this DFE-ground shear stress is independent of any additional DFE settlement beyond the DFE settlement magnitude, $(u_p)_Y$, that is required to cause τ_o.

Note that this is the common assumption that has been made historically for constant-perimeter DFEs, i.e. once shaft resistance has been fully mobilized it remains constant in magnitude under further settlement. This assumption is generally acceptably correct for use in routine practice even though shaft resistance sometimes exhibits some slight strain-softening behavior as the soil shear strength transitions from peak to constant-volume (critical-state) conditions along the shaft.

Of greater importance and relevance to the theme of this monograph is that Equation C.12 quantitatively expresses the essence of tapered DFEs (at least in the minds of Kodikara and Moore, El Naggar et alia, and Manandhar et alia), i.e. what is referred to qualitatively as taper benefit which is here interpreted to mean the additional shaft resistance developed by a tapered DFE over and above that developed by a constant-perimeter DFE in the same soil conditions. This equation very clearly illustrates that the DFE-ground shear resistance and, therefore, the unit shaft resistance in axial compression, <u>increases</u> as DFE settlement increases and the conceptual 'cylinder' formed by the DFE-ground interface expands radially when viewed from some fixed depth below the ground surface. This is distinctly different from constant-perimeter DFEs where, as noted above, the shaft resistance 'maxes out' at some point and does not increase further with additional pile settlement.

As in Phase I, the assumed behavior in Phase II, i.e. that CCE occurs with linear-elastic behavior in the ground surrounding the DFE, has a limit. Kodikara and Moore defined the Phase II limit as being the initiation of <u>radial</u> yielding at the DFE-ground interface. Note that this is distinctly different than yielding <u>parallel to</u> (along) the DFE-ground interface (i.e. in a direction that is nominally orthogonal to the radial direction) which is assumed to be occurring continuously during Phase II. The radial stress at the initiation of radial yield, σ_Y, is given by the following equation:

$$\sigma_Y = [\sigma_o \, (1 + \sin \phi)] + (c \, \cos \phi) . \tag{C.13}$$

The various equations presented above can be combined to produce an estimate of the magnitude of pile settlement at which this radial yielding just occurs, $(u_p)_Y$:

$$(u_p)_Y =$$
$$\left\{ \frac{\sigma_o}{K_e} \cot \alpha \left[\left((1 + \sin \phi) \left(1 + \left(\frac{K_e \zeta r_m}{G} \tan(\alpha + \phi_i) \tan \alpha \right) \right) \right) - 1 \right] \right\}$$
$$+ \left\{ \frac{c}{K_e} \cos \phi \, \cot \alpha \left[1 + \left(\frac{K_e \zeta r_m}{G} \tan(\alpha + \phi_i) \tan \alpha \right) \right] \right\}$$
$$+ \left\{ \frac{r_m \zeta}{G} c_i' \right\} . \tag{C.14}$$

Kodikara and Moore noted that by substituting $(u_p)_Y$ from Equation C.14 for u_p in Equation C.12, the value of τ_x at radial yield, $(\tau_x)_Y$, can be calculated.

C.3.2.5 Phase III

Although the El Naggar-Sakr Method does not use Phase III of the Kodikara-Moore Method, it is important to summarize the basic theoretical elements of this behavioral phase so that there is a clear understanding of the pile-soil behavior that El Naggar and Sakr chose to neglect.

The basic conceptual assumption of Phase III is that a plastic zone develops within the radially yielded ground adjacent to the DFE shaft. The radius of this plastic zone, a, increases with increasing radially displacement, v, throughout this phase. Because v is geometrically linked to DFE settlement, u_p, through the taper angle, α, (as it was in Phase II as well), this means that there is a direct theoretical connection between DFE settlement and radius of the yielded zone of soil around the DFE throughout Phase III.

Although Kodikara and Moore did not note this explicitly in their 1993 paper, more-theoretical treatments of cavity expansion such as Cook (2010) illustrate that once radial yielding of the ground around a cylindrical cavity has occurred, progressing radially outward from the cavity-ground interface there is a transition within the ground from:

- a plastic zone with varying shear strength (from ϕ_{cv} to ϕ_{peak} in the problem defined and studied by Cook) to...

- a transitional elastic zone of non-linear behavior with varying shear modulus to...

- the remaining ground that extends to infinity and is in a permanent state of linear-elastic behavior, with the shear modulus equal to the small-strain value, G_{max}.

The point being made here is that the yielded/plastic zone noted by Kodikara and Moore is but one component of what is theoretically a complex system of implied ground behavior.

Of note is that fact that unlike in Phases I and II where there is a single-value, upper-bound radial stress for which each phase was valid, Kodikara and Moore did not indicate that there was an upper-bound radial stress for Phase III. The implication is that the Phase III radial stress continues to increase without limit with increasing DFE settlement, u_p, and concomitant radial displacement, v, of the DFE-ground interface. This, in turn, implies that τ_x and, therefore, shaft resistance in axial compression increases without limit as well. The plotted calculated results for the several exemplar analyses presented in Kodikara and Moore (1993) illustrate such unbounded behavior.

Note, however, that this implied unlimited-radial-stress behavior is contradicted by other published work as well as easily observed real-world behavior. Other researchers, e.g. Yu and Houlsby (1991) and Cook (2010), clearly indicate that cavity-expansion theory shows that there is <u>always</u> a single-value limiting stress, referred to as *limiting pressure* in the literature, for cavity expansion. This includes both cylindrical and spherical cavity geometries as well as for both cavity <u>creation</u>, i.e. starting from zero cavity radius, and cavity <u>expansion</u>, i.e. starting from an initial cavity radius, a_o, as in the Kodikara-Moore Method. Thus, the case used by Kodikara and Moore of CCE clearly is covered and has a limiting pressure.

Furthermore, and perhaps more conclusively because it is easily demonstrated, one need look no further than a typical expansion curve for a pressuremeter. All pressuremeter tests, if carried to a sufficient magnitude of radial expansion, will demonstrate the existence of a limiting cavity pressure. In fact, it is an essential part of pressuremeter testing.

That having been said, it is important to note that for the case of CCE that is applicable here, Yu and Houlsby (1991) showed that relatively large a/a_o ratios (a_o in this case is the same as r_m) of the order of 2 or greater are required in order to reach the limiting-pressure condition. Pressuremeter tests also indicate clearly that relatively substantial relative cavity expansion is required in order to reach a limiting pressure. For the relatively small taper angles being considered for tapered DFEs in general and tapered piles in particular, the required radial displacement, v, necessary to achieve such a/a_o ratios (note that $a = a_o + v = r_m + v$) would not even come close to being achieved for any reasonable magnitude of DFE settlement, u_p.

A simple numerical example best illustrates this point. For a taper angle, α, equal to 1°, a pile settlement of 1 inch (25 mm) produces a radial increment of CCE, Δv, equal to only about 0.02 inches (500 μm). Assuming a pile diameter of 12 inches (305 mm), this is an a/a_o ratio of 6.02/6 = 1.003 which is not even close to the value of 2 or more required to develop a limiting pressure in theory.

Thus, from a pragmatic perspective, Kodikara and Moore did not err by failing to note that the radial stress in Phase III has a limit as that limit would likely never be approached, no less reached, by any tapered DFE used in routine practice. Nevertheless, it would have been worthwhile for them to at least note, for the sake of completeness, that such a limit theoretically exists but is, in general, irrelevant for the problem of tapered DFEs.

In any event, the cavity-expansion parameter that is central to Phase III and also provides a conceptual link to Phase II is again the tangent gradient, K, but this time with the subscript p to indicate that the gradient is associated with **p**lastic soil behavior. The assumed relationship between a small (differential) increment of radial stress, $d\sigma_r$, and a corresponding small increment of radial displacement, dv, is given by the following equation that is identical in overall form and concept to that used in Phase II (Equation C.9):

$$d\sigma_r = K_p dv . \tag{C.15}$$

The dv term is defined using a relationship identical in form and concept to that used in Phase II (Equation C.8):

$$dv = \left(du_p - du_g\right) \tan \alpha \tag{C.16}$$

which, when integrated and combined with Equation C.2, yields:

$$v = \left(u_p - \frac{\tau_x \zeta r_m}{G}\right) \tan \alpha . \tag{C.17}$$

Equation C.17 is used to calculate a parameter called v_Y that is the radial cavity expansion at the end of Phase II/beginning of Phase III when yield of the ground adjacent to the DFE is initiated. This is done by:

- first calculating $(u_p)_Y$ from Equation C.14;

- then using $u_p = (u_p)_Y$ in Equation C.12 to calculate $\tau_x = (\tau_x)_Y$;

- then setting $u_p = (u_p)_Y$ and $\tau_x = (\tau_x)_Y$ in Equation C.17 to calculate v_Y.

This evaluation of v_Y allows the radial stress, σ_r, during Phase III to be calculated as it is the radial stress at the end of Phase II, σ_Y (from Equation C.13), plus the increase in radial stress, $d\sigma_r$ (from Equation C.15), that occurs during Phase III as the radial displacement increases from v_Y to v. This is expressed mathematically as follows:

$$\sigma_r = \sigma_Y + \int_{v_Y}^{v} K_p \, dv \,.$$

(C.18)

The final step is to define τ_x which is done by substituting the relationship for radial stress given by Equation C.18 into Equation C.11:

$$\tau_x = \left[\left(\sigma_Y + \int_{v_Y}^{v} K_p \, dv\right)\tan(\phi_i + \alpha)\right] + c_i' \,.$$

(C.19)

Note the continued use of the c_i' variable that was highlighted in red in Equation C.7a.

Everything up to this point has been the easy, or at least straightforward, part of Phase III. It is at this point that a decision has to be made as to how to define K_p as it by no means as simple as the definition of K_e in Phase II.

To begin with, Kodikara and Moore expressed a preference for a cavity-expansion theory that incorporated a non-associated flow-rule as they felt that an associated flow-rule was a poorer match for the dilative behavior of actual soils (they are not unique in this opinion). This left them with the following three choices, keeping in mind the circa-1990 timeframe in which their research and concomitant manuscript preparation were occurring:

- <u>Carter et al. (1986)</u>: A small-strain solution that assumes strains are infinitesimal.

- <u>Hughes et al. (1977)</u>: A large-strain solution that neglects the elastic component of displacements within the plastic zone.

- <u>Yu and Houlsby (1991)</u>: A large-strain solution that includes the elastic component of displacement within the plastic zone.

There is a considerable range in mathematical complexity of these three solutions, with the Yu and Houlsby solution the most complex by far.

Kodikara and Moore compared the calculated results from these three solutions for one exemplar problem in their 1993 paper. The calculated results began to diverge almost from the onset of Phase III behavior and this divergence increased with increasing DFE settlement. Despite the complexity of the Yu and Houlsby solution, it was selected by Kodikara and Moore for all subsequent calculated results presented in their paper.

C.3.2.6 Commentary

The variable that is the key problem parameter not only linking all three behavioral phases in the Kodikara-Moore Method but also significantly influencing the calculated results for shaft resistance is σ_o, the assumed radial stress acting along the DFE-ground interface <u>after</u> DFE installation but <u>prior</u> to the application of external loading. The writer has several observations and comments to make with regard to this parameter:

- Given its overarching importance in their analytical methodology (it is literally the starting point for forecasting τ_x that is used for calculating shaft resistance for each behavioral phase), Kodikara and Moore appear to treat it in a theoretically questionable manner. Specifically, in all of the exemplar applications illustrated in their paper they assume it is depth-wise uniform in magnitude which in and of itself violates the most basic precept of modern soil mechanics and is thus fundamentally incorrect.

- In the writer's opinion, although it has the notational form of a total stress, i.e. there is no bar or prime to suggest an effective stress, it is unclear as to whether it is intended to be interpreted as a total stress or effective stress or either depending on the context. This is important given that the exemplar applications in Kodikara and Moore's 1993 paper assume non-zero values for both c and ϕ as well as significant non-zero values for the dilation angle[181], ψ, in most cases.

- Kodikara and Moore offer no opinion, no less specific guidance, as to how this constant value of σ_o should be determined in an application of their methodology as it has to be single-valued but in any real application will involve a depth-wise variation in both total and effective horizontal stresses.

- Because it reflects post-driving conditions, it is unclear if the chosen value of σ_o should include any taper benefit in addition to the normal increase above the pre-installation lateral earth pressure at rest that would be expected, at least for (driven) piles regardless of whether they are tapered or not. Kodikara and Moore were silent on this subject so it appears reasonable to conclude that the Kodikara-Moore Method implies that taper benefit only occurs during post-installation loading, specifically, during the CCE that is assumed to occur only in Phases II and III. This overlooks the fact that the pre-installation K_o is affected by the pile-driving process as reflected in what Kulhawy termed the K_h/K_o ratio.

This final point is emphasized as it is the writer's hypothesis based on an assessment of all available information presented in this monograph that, in the more general case, taper benefit has two distinct components and thus accrues in two distinct stages, identified in the main section of this monograph as Stage I and Stage II:

- **Stage I**. Increased radial stress relative to a non-tapered (driven) pile acting along the shaft as the result of driving and prior to external load application. The assumed (by the writer) physical mechanism is that of ***enhanced cylindrical-cavity creation, CCXE***. Note that if the tapered DFE is drilled instead of driven, this relative increase in radial stress does <u>not</u> occur as it is assumed that the radial stress state around a drilled shaft (bored pile) after installation is unrelated to the geometry, i.e. it is the same whether the drilled shaft has a constant diameter or is tapered.

- **Stage II**. A further increase in radial stress as a function of DFE settlement as the result of post-installation axial-compressive load application. The assumed (by the writer) physical mechanism is that of ***cylindrical-cavity expansion, CCE***. By comparison, a non-tapered DFE of any kind develops no increase in radial stress as it is loaded.

---

[181] The <u>dilation</u> angle, ψ, should not be confused or conflated with the <u>dilatancy</u> angle, ϕ_d, that was defined in Equation 4.7. Note, however, that these two angles are related conceptually in that they broadly involve the same behavioral phenomenon of volumetric increase due to particle interlocking during shear. In both research (e.g. Cook 2010) and practice, the empirical relationship of $\phi_d = 0.8\psi$ suggested by Bolton (1986) is widely used.

The conclusion drawn by the writer is that tapered (driven) piles develop taper benefit both from the driving associated with installation and post-installation loading. Most resistance-forecasting methodologies developed to date, including the Kodikara-Moore Method, imply taper benefit from either one or the other but not both.

Furthermore, it is the writer's conclusion that the Kodikara-Moore Method is, at best, a reasonable one-step resistance-forecasting methodology but <u>only</u> for drilled shafts (bored piles) as drilled tapered DFEs can only have what the writer terms Stage II taper benefit due to post-installation axial-compressive loading and this is the only stage of taper benefit that the Kodikara-Moore Method explicitly considers.

Note that this means that the Kodikara-Moore Method is <u>not</u> acceptable for use with tapered (driven) piles unless there is a reliable way to estimate what Kodikara and Moore call σ_o, the radial stress acting along the shaft of the DFE <u>after</u> installation and <u>prior</u> to external load application. Such as estimate would have to include a consideration of not only the pre-driving K_o value but all Stage I taper benefit due to the CCXE that occurs when a tapered pile is driven into the ground. This logical conclusion drawn from the presentation and detailed discussion of the Kodikara-Moore Method sets the stage for a discussion of the El Naggar-Sakr Method that concludes this appendix after a brief discussion of modifications made to the Kodikara-Moore Method by others.

C.3.2.7 Modifications of Manandhar and Yasufuku

Manandhar and Yasufuku made one modification that impacts the shaft-resistance component of the Kodikara-Moore Method. They assumed that ϕ, and, as a result, ϕ_i, are not problem constants as assumed by Kodikara and Moore but vary with the operative stress level using Equation 4.7 and assuming that the dilatancy angle, ϕ_d, is given by Bolton's well-known empirical relationship (Bolton 1986). This relatively modest change to the Kodikara-Moore Method was implemented by simply adding an iterative step to the otherwise-unchanged overall solution algorithm presented in Kodikara and Moore (1993).

Unfortunately, in none of the publications by Manandhar, Yasufuku, and their colleagues that were reviewed in preparation of this manuscript and that are included in the Cited References and Supplemental Bibliography of this monograph was there an exemplar comparison of calculated results without and with this change to illustrate how and to what extent the results differed. Thus, the quantitative significance of this modification cannot be readily assessed although qualitatively it would appear to be a (very modest) improvement simply based on theoretical considerations.

C.3.3 El Naggar and Sakr Simplifying Assumptions and Synthesis

As noted at the beginning of this appendix, El Naggar and Sakr assumed that the unit shaft resistance, which they called "skin friction" and with the notation τ_s, for their analytical methodology is given by Equation C.1 that is repeated here for ease of reference in the current discussion:

$$\tau_s = K_t \, K_s \, \tan\delta \; \sigma_v' . \qquad\qquad \textbf{(C.20)}$$

Note that τ_s is conceptually the same as the parameter τ_x that Kodikara and Moore used as a common parameter in each of their three assumed behavioral phases.

El Naggar and Sakr made use of this conceptual equivalence between τ_s and τ_x but with two simplifying assumptions:

1. The cohesion, c, was assumed to be zero. Therefore, the resulting analytical methodology, which is referred to as the El Naggar-Sakr Method in this monograph, is effectively limited to coarse-grained soil. It appears that the reason for doing this was because the small-scale, 1-g, laboratory tests that El Naggar and Sakr used to validate their methodology involved test piles embedded only in dry sand.

2. Equation C.20 was equated with Equation C.12 that defines τ_x for Phase II in the Kodikara-Moore Method and then solved for K_t. By neglecting Phase III of the Kodikara-Moore Method, El Naggar and Sakr implicitly limit their resulting methodology to conditions where radial yield of the cylindrical cavity defined by the DFE-soil interface does not occur.

The second assumption is by far the more consequential. Unfortunately, El Naggar and Sakr did not provide a worked example or other evidence in their 2000 paper to support the validity of this assumption. Rather, they simply made the following statement in their paper, with no explicit indication of what radial displacements were measured and what constituted the metric of "large" to which these displacements were presumably compared:

> "The taper angles considered in the current study were relatively small (0.35–1.02°) and thus the large radial expansion movements required to mobilize plastic zones were not achieved during the loading tests. Only the first and second phases will be considered in analyzing the experimental results using this model."

Note that the validity of their assumption in any given application is easy enough to check and should be included as a required analytical step in any use of the El Naggar-Sakr Method in practice. Any of four parameters (u_p, v, σ, τ_s) calculated using the El Naggar-Sakr Method can be compared to values that occur at the onset of radial yield, i.e. the end of Phase II/beginning of Phase III, using the various equations presented earlier in this appendix for the Kodikara-Moore Method.

Moving on in the discussion, El Naggar and Sakr used a slightly altered form of Equation C.12 that defines Phase II behavior in the Kodikara-Moore Method to reflect their zero-cohesion assumption as well as to use parameter notations that were consistent with their methodology. The specific notational changes made to problem parameters were:

- τ_x was replaced by τ_s,

- σ_o was replaced by the product ($K_s \sigma_v$) where $\sigma_v = \sigma_{vo}$,

- ϕ_i was replaced by δ,

- u_p was replaced by U_p, and

- a dimensionless parameter called the *settlement ratio*, S_r, was defined as:

$$S_r = \frac{U_p}{d} \tag{C.21}$$

where d = the average pile diameter[182]. Note that subsequent papers that use the El Naggar-Sakr Method such as Sakr et al. (2004) use the notation Δ_r instead of S_r.

Using these notational changes, the resulting revised form of Equation C.12 is:

$$\tau_s = \frac{\left[K_e \tan\alpha \, \tan(\alpha+\delta)\, U_p\right] + \left[K_s \, \sigma_v \, \tan(\alpha+\delta)\right]}{1 + \left[\left(\frac{K_e \zeta r_m}{G}\right)\tan\alpha \, \tan(\alpha+\delta)\right]}.$$ (C.22)

As an aside, note that El Naggar and Sakr were inconsistent with the notation that they used for the vertical stress in the ground. In the basic equation (C.20) that they used to define shaft resistance in their analytical methodology, it was clearly an effective stress and was explicitly stated to be so in their 2000 paper. However, In Equation C.22 and subsequent equations, both in their 2000 paper and as presented here, they used notation consistent with total stress although it is clear that they still intended this stress to be an effective stress. Presumably, this notational conflict was simply the result of poor/sloppy editing of their 2000 paper manuscript although this does not explain why the same inconsistency appeared in subsequent papers such as Sakr et al. (2004) where the total-stress notation is used yet the parameter is explicitly stated to be an effective stress in the text of the paper.

The assumption made by the writer in this matter is that the vertical stresses in the El Naggar-Sakr Method are unambiguously intended to be effective stresses and that the use of total-stress notation in some equations was simply the result of human oversight. Note that this is distinctly different than the previously discussed stress issue in the Kodikara-Moore Method where there is ambiguity as to how these authors define σ_o. This is because Kodikara and Moore allowed for non-zero cohesion which means that the interpretation of normal stress in their analytical methodology depends on application-specific context.

Continuing on with the presentation, substituting Equation C.22 into Equation C.20 and solving for K_t yields the following:

$$K_t = A + \left(\frac{B}{\sigma_v}\right)S_r$$ (C.23)

where A and B are problem-specific constants that are defined as follows:

$$A = \frac{\tan(\alpha+\delta)\,\cot\delta}{1 + [2\zeta \tan\alpha \, \tan(\alpha+\delta)]}$$ (C.24)

$$B = \frac{4G \tan\alpha \, \tan(\alpha+\delta)\,\cot\delta}{K_s\{1 + [2\zeta \tan\alpha \, \tan(\alpha+\delta)]\}}.$$ (C.25)

It is clear from Equation C.23 that K_t is a function of pile settlement which is as expected. This is because El Naggar and Sakr used Phases I and II of the Kodikara-Moore Method and this method is an integrated, one-step load-settlement methodology.

At this point, El Naggar and Sakr made two additional assumptions to finalize their analytical methodology:

---

[182] Note that El Naggar and Sakr only studied continuously tapered piles with a uniform taper angle so the determination of d was always straightforward and unambiguous as it was simply the arithmetic mean (simple average) of the head and toe diameters.

1. The vertical effective stress ($\sigma'_v = \sigma_v$) was assumed to be depth-wise variable in the normal manner of vertical effective overburden stress, σ'_{vo}. This represents a significant improvement over the assumption of a depth-wise constant stress that was used throughout Kodikara and Moore (1993).

2. S_r (Δ_r in later papers) was assumed to be constant and equal to 0.1. This simplistic assumption, which El Naggar and Sakr attribute to Terzaghi (1942), combined with the fact that they assumed a single-valued ULS toe resistance (not discussed in this appendix) is what makes the El Naggar-Sakr Method a resistance-only forecasting methodology of the indirect type as opposed to being a one-step load-settlement methodology as is the Kodikara-Moore Method from which the El Naggar-Sakr Method is derived.

However, it appears, in principle at least, that the El Naggar-Sakr Method could be turned into a one-step methodology for forecasting load-settlement behavior by:

*   relaxing Assumption #2 (above) and using Equation C.23 in an iterative fashion with different assumed values of S_r and

*   using a toe-resistance methodology that allows for calculating toe resistance as a function of toe settlement.

The writer considered implementing these changes for the purposes of developing an improved resistance-forecasting methodology for tapered piles along the lines of the scheme presented in Chapter 7. However, the writer ultimately decided not to pursue this as there are too many 'fatal' shortcomings of the underlying El Naggar-Sakr methodology, primarily the fact that Stage I taper benefit due to pile driving is ignored. Another significant negative is the fact that the shear modulus is assumed to be constant in magnitude, both with depth and as a function of load. There are other issues but these two alone make the El Naggar-Sakr Method, either in its current, original form or in a potentially enhanced form, simply incorrect to use with tapered (driven) piles, at least in the writer's opinion.

Appendix D

Mathematical Relationships for Modeling Shear-Modulus Degradation

---

D.1 BACKGROUND

D.1.1 Introduction and Overview

Since the earliest days of modern soil mechanics, it has been recognized that the traditional moduli from the theory of elasticity that are used to define soil stiffness for analytical purposes, Young's modulus, E, and the shear modulus, G, are both <u>stress</u> and <u>stress-level</u> dependent. The former refers to the overall confining stress to which an element of soil that is of analytical interest is subjected, and the latter refers to what fraction of the maximum shear resistance (shear strength) within that element of soil is mobilized at a given confining stress and in a specific application. Taken together, stress and stress-level dependency define the *operative stress level* in some specific analytical application.

This relatively complex relationship between soil moduli and the stress state within the soil creates significant analytical complications in virtually every problem-specific application in geotechnical and foundation engineering. This is because analytical methodologies typically require single values of moduli. This is true even for sophisticated, state-of-art numerical methods such as the finite-element (FE) method (FEM). If one were to 'look under the hood' of the computer codes for such advanced analytical methodologies, for a given element in a FE mesh at a given step in the overall analysis the constitutive equations typically require single-value moduli.

Thus, the overall challenge in geotechnical and foundation engineering is to craft ways in which the analytical demand for single-valued moduli can be met while at the same time taking into account the stress and stress-level dependency of these moduli. With the aforementioned FEM example, the single-valued moduli are allowed to vary from element to element within the FE mesh and, within each element, with each cumulative analytical step in an overall simulation.

With the analytical methodologies for tapered piles that are the focus of this monograph, the approach taken to deal with modulus variation is to develop closed-form mathematical relationships called *modulus-degradation equations* that can be used in both simpler numerical-solution tools such as generic spreadsheet software as well as more-advanced numerical-solution tools such as purpose-written software using FORTRAN or some other programming language.

The discussion of such mathematical relationships is the focus of this appendix. This is because modulus-degradation relationships play an ever-increasing role in foundation engineering involving DFEs in general so it is essential to have a solid understanding of these relationships so that they can be selected appropriately and used correctly.

Although the basic concepts presented and discussed in this appendix can, in general, be equally applied to both the Young's and shear modulus, the remainder of this appendix is limited to a consideration of the shear modulus, G. Not only is this the modulus of greater interest for the purposes of this monograph, as it turns out that the vast majority of research devoted to mathematically modeling modulus degradation has focused on shear modulus. This is not surprising as this body of research has been largely devoted to the behavior of soil under dynamic loading where shear strains, γ, and shear modulus govern.

Shear strains are, of course, the natural complement to shear modulus and this appendix consequently focuses on them as well. However, as will be seen, shear modulus-degradation equations as applied to DFEs have sometimes used parameters other than shear strain in their formulation. These alternative formulations are also included in this appendix for the sake of completeness as well as relevance to the topic and goals of this monograph.

D.1.2 Stress Dependency

Dealing with the effect that overall confining stress has on shear modulus is most easily handled nowadays by using the small-strain shear modulus, G_{max}, as the baseline-reference value for any modulus-degradation model. This is because whether this parameter is measured in-situ using an sCPTu sounding or other geophysical tool, or calculated using some empirical relationship (Cook (2010), for example, used the well-known Hardin and Black (1968) equation developed over 50 years ago), the stress-dependency effects of confining stress are built into the overall process either inherently/naturally (as in the case of an sCPTu sounding) or by virtue of the parameters in an empirical equation (as with Hardin and Black (1968)). For example, the aforementioned Hardin-Black equation uses the mean effective stress non-dimensionalized and normalized to atmospheric pressure. Fahey and Carter (1993) discuss other theoretical equations with empirically derived coefficients that are similar in concept.

Therefore, the issue of stress dependency of shear modulus requires no further discussion in this appendix as it is relatively easily handled in practice. Unfortunately, the same cannot be said for stress-level dependency that requires extensive discussion.

D.1.3 Stress-Level Dependency

D.1.3.1 introduction and Overview

Analytically dealing with the stress-level dependency of the shear modulus is complicated by the fact that the stress-strain behavior of soil is non-linear except for a very limited small-strain region at the initial portion of the stress-strain curve where linear-elastic behavior can reasonably to be assumed to exist and the aforementioned G_{max} is valid as the operative shear modulus. Thus, the fundamental purpose of a modulus-degradation equation is to 'fill in the blanks' for operative values of shear modulus, $G_{operative}$ ($< G_{max}$), for strain levels that are larger in magnitude than the initial small-strain region.

The standard format for graphically portraying shear-modulus degradation, whether it is measured experimentally or forecast empirically, is to assume that shear strain is the independent variable and shear modulus is the dependent variable. Although geotechnical and foundation engineering are well-known for routinely violating the basic rules of generic plotting in math/science/engineering that call for the independent variable along the abscissa (horizontal axis) and the dependent variable along the ordinate (vertical axis), plots of modulus degradation are actually one of those cases in geotechnical engineering that follows the generic plotting norms.

Although the overall presentation format for modulus degradation is standardized, over the years some subjective variability in terms of the plotted parameters has emerged in the published literature. Consequently, there is some variation in the final form that modulus-degradation equations and associated plots take. Therefore, before presenting and discussing the modulus-degradation equations considered for use in this monograph, these subjective topics need to be discussed.

D.1.3.2 Modulus Plotting

The one element of modulus-degradation plots that is near-universal is the format used for the shear modulus along the ordinate. It is conventional to plot the dimensionless ratio $G_{operative}/G_{max}$. Because $0 \leq G_{operative} \leq G_{max}$ by definition, a linear scale that varies from 1 down to 0 is almost universally used for this ratio as it is almost always expressed in its fundamental dimensionless form as a decimal as opposed to a percent.

Note that most publications omit the 'operative' subscript for the ordinate label in plots (and thus use G/G_{max}) and only imply it as noted in Chapter 2. This subscript is retained in this monograph to avoid any ambiguity. Also reiterated from Chapter 2 is the fact that outside of the U.S. the notation G_o is generally used in lieu of G_{max}. As noted in Chapter 2, there is a specific reason totally unrelated to regional preferences for not using G_o in this context in this monograph. Specifically, it conflicts with the completely different meaning of G_o as used in Niazi's published work that is referenced extensively throughout this monograph in many different contexts.

D.1.3.3 Small-Strain Threshold

The first abscissa-related issue that is addressed is the assumed *small-strain threshold*, γ_s, that is the shear-strain magnitude that marks the upper-bound (relative to increasing shear strain) of linear-elastic behavior and concomitant applicability of the small-strain shear modulus, G_{max}. This threshold strain typically anchors one end of a modulus-degradation relationship so is essential to define with the appropriate thought and care.

Just how small is 'small' in this context is a difficult question to answer as recent published work suggests that the answer is even more complex than appears to have been thought in the past. To begin with, it is relevant to note that virtually all of the research on this subject has been done using dynamic (usually cyclic) loading for pragmatic reasons:

- It is easier to develop laboratory-testing hardware that delivers precise magnitudes of cyclic loading as opposed to static loading to measure the very small displacements associated with the linear-elastic behavior of soil.

- Applications involving dynamic loading, both transient and steady-state, became of great interest beginning in the 1960s when the earliest research into the subject of the small-strain threshold was first conducted. These applications include:
  - machine foundations and similar problems that involve cyclic loading with relatively small displacement amplitudes;
  - pavement and railway-track-system loading;
  - water-wave loading on deep-water marine foundations; and
  - seismic loading.

Thus, there is a 'leap-of-faith' element involved when directly using the modulus-degradation results obtained for dynamic loading for applications involving the static loading of DFEs. Niazi (2014), who reviewed measured modulus-degradation behavior and concomitant models as part of his doctoral research related to deep foundations, presented results of relevant research conducted by Vardanega and Bolton (2011). They indicated that there was a modest-but-noticeable difference in shear-modulus degradation between dynamic and static loading, at least for fine-grained soil that was the focus of their research, although the overall degradation trends in terms of curve shape and effect of Plasticity Index

(PI) were identical between the two forms of loading. The overall stiffer response under dynamic as opposed to static loading was attributed to what Niazi called *soil viscosity*.

A 'ballpark', order-of-magnitude value for the dynamic small-strain threshold that the writer has seen in various publications over the years is γ_s = 0.001%[183] although some researchers such as Fahey and Carter (1993) go one order-of-magnitude lower, to γ_s = 0.0001%. Niazi (2014) found that γ_s = 0.001% is reasonable but with some variation based on the PI of the soil. Specifically, the small-strain threshold tends to decrease toward the aforementioned smaller alternative, γ_s = 0.0001%, as the PI decreases toward its limiting value of zero, i.e. a 'clean' coarse-grained soil. In the other direction, the small-strain threshold increases somewhat above γ_s = 0.001% as the PI increases.

A more-detailed study of the subject of small-strain thresholds, albeit with a focus on fine-grained soils, was presented in Diaz-Rodriguez and Lopez-Molina (2008). They curated and summarized the findings presented in 27 publications on the subject that were published between 1962 and 2007. The conclusion drawn by the writer from this circa-2008 research summary is that the measured outcomes of dynamic testing of soils are governed by two broad categories of variables:

1. The mechanics of the laboratory-testing protocol that includes the specifics of the testing hardware:
   o   resonant column,
   o   torsional shear,
   o   simple shear,
   o   triaxial;
   loading frequency; number of loading cycles; and other variables.

2. Various soil properties, with PI apparently the most significant overall.

Diaz-Rodriguez and Lopez-Molina further noted that different criteria can be used to define the small-strain threshold depending on the specific aspect of soil behavior that is used to make this determination:

• stress-strain behavior (linear vs. non-linear),

• stiffness (modulus) degradation, and

• pore pressure generation and net accumulation

as these three aspects of material behavior do not, in general, exhibit changes or transitions at the same strain level.

With these different criteria in mind, Diaz-Rodriguez and Lopez-Molina defined four different thresholds of shear strain that were collectively used to defined five different behavioral regions. Of relevance to the current discussion is the smallest of these four shear-strain thresholds that they called the *linear threshold shear strain*, γ_{tl}, that is essentially the same as the small-strain threshold, γ_s, defined above.

---

[183] Note that some researchers express γ_s in its fundamental, dimensionless decimal form (usually using exponential notation because the values are orders of magnitude less than 1) and some as a percent as done here so care is required when comparing different published results. The latter (percent) format is used throughout this monograph.

The conclusion drawn by the writer is that the current state of knowledge supports using, at least on an interim basis, a uniform small-strain threshold value of γ_s = 0.0001% for a modulus-degradation model that is used with an improved resistance-forecasting methodology for tapered piles under static loading. This use of a threshold value that is at the lower end of the range of the ballpark values of γ_s = 0.0001% to 0.001% stated above is based on two considerations:

1. The aforementioned work of Vardanega and Bolton that indicates that values of γ_s trend smaller in magnitude for static loads compared to dynamic loads (static loads are of greater interest for tapered piles).

2. The fact that γ_s trends smaller in magnitude for coarse-grained soil as opposed to fine-grained soil (the former is much more common in problems involving tapered piles).

Future research related to tapered piles should address two issues:

1. A better understanding of the effect of static loading on values of γ_s.

2. A better definition of the range in the static γ_s as a function of soil plasticity. This is because there are decades of research for dynamic loading that indicate that γ_s can vary by up to an order of magnitude depending on the PI of the soil.

With regard to the second issue, note that although Niazi included a static-loading PI-dependency when developing the Niazi Method (Niazi 2014, Niazi and Mayne 2015b, Niazi and Mayne 2019), as noted in the main body of this monograph, it is the writer's opinion that Niazi's database was not sufficiently robust to support including a PI correction in his forecasting methodology. Thus, further research into this important subject is required.

D.1.3.4 Standard Strain Plotting: Actual Shear Strain

The standard plotting format for modulus degradation has historically involved using the actual values of shear strain, with the only variation being whether shear strain is expressed in its fundamental, dimensionless decimal form or as a percent. As will be seen subsequently, there are applications where deviations from this plotting convention form the basis for alternative plotting formats that are more insightful and useful in certain applications.

Shear strain in both geotechnical laboratory testing and geotechnical applications varies over several orders of magnitude so it has long been the practice to use $\log_{10}$ scaling for plotting γ along the abscissa. Typically, five log cycles are sufficient to depict the full range of interest for either measured or forecast behavior or both if the two are being compared on the same plot. With this type of plotting, the relationship between $G_{operative}/G_{max}$ and $\log_{10} \gamma$ is that of an S-shaped curve that begins at $G_{operative}/G_{max}$ = 1 and asymptotically approaches $G_{operative}/G_{max}$ = 0 with increasing values of γ.

One issue that the writer has never seen mentioned, no less discussed, relative to this application is the fact that extremely large strain ranges are involved. As a result, there are fundamental, underlying questions to be asked and concomitant issues to consider in this regard. This relates to the fact that routine civil engineering applications of engineering (a.k.a. applied, continuum, solid) mechanics are typically assumed to be 'small-strain' in nature and what is technically called *infinitesimal strain theory* is used as the mathematical approach for calculating the displacements and deformations of a body. Consequently, the fundamental derivations of both normal and shear stresses and strains are all based on:

- the original body geometry (dimensions, shape, and position in space) and

- the assumptions of relatively small displacements,

both of which greatly simplify the stress and strain derivations, the results of which are referred to as *engineering* (a.k.a. *Cauchy*) *strains* and *engineering stresses*.

This assumption of small displacements is carried over into formulating theoretical models and concomitant closed-form mathematical solutions for many basic applications in civil engineering. The flexure (bending) of beams is a classic example. *Euler-Bernoulli* (a.k.a. *simple*) *beam* theory that is based on the above assumption of the original, undeformed shape of a beam and relatively small displacements under load application is the heart of both civil engineering education and practice (Horvath 2018b).

However, as noted by the writer in Horvath (2018a), in applications where relatively large normal strains and concomitant normal stresses are expected, the assumptions of original body geometry and subsequent small displacements of same become highly questionable. The alternative is to use what are called *true* (a.k.a. *Hencky*, *logarithmic*, *natural*) *strains* and *true stresses* that take into account the changing geometry of an object as it displaces and deforms under applied loading and the internal stresses that derive from that loading[184]. This is part of what is called *finite strain theory* where the restrictive assumptions of infinitesimal strain theory are removed. Subappendix D1 contains a discussion of engineering vs. true normal strains and normal stresses that was first published in Horvath (2018a) and is presented here in an expanded and edited form to illustrate the differences.

Of course, the issue here involves shear, not normal, strains and shear stresses so the question that should be asked, but apparently has not to date, is whether the same distinction between engineering and true normal strains and normal stresses applies to shear strains and shear stresses as well. A review of the literature indicates that the distinction between engineering and true strains is not made for shear strains as it is for normal strains although it appears that such a distinction could, and arguably should, be made.

However, the literature does make a distinction between engineering shear strain and what is called *tensorial shear strain*, with the latter an outcome of the aforementioned large-strain theory. The engineering vs. tensorial shear strain issue is important as different commercial finite-element (FE) software packages that are used for load-displacement problem analyses apparently report shear strains in either one or the other formats which do not yield the same numerical results. Thus, some discussion of the nuances of shear strain appears to be warranted and this is presented in Subappendix D2.

The conclusion reached by the writer is that the issue of the interpretation and meaning of shear strain in geotechnical engineering is a subject requiring further study to better understand the frame of reference used for reported shear strains, i.e. the original geometry of a test specimen or the deformed geometry at some point in a test. The writer's interpretation and understanding of the current state of knowledge is that this is not always clearly stated or known. Furthermore, it is possible that the frame of reference used is not consistent from one publication to another.

---

[184] Similarly, *Timoshenko beam* theory is more general (and, theoretically, more accurate) than Euler-Bernoulli beam theory as it removes some of the simplifying assumptions made in the derivation of Euler-Bernoulli beam theory and considers some of the secondary effects that Euler-Bernoulli beam theory ignores.

D.1.3.5 Alternative Strain Plotting: Normalized Shear Strain

D.1.3.5.1 Background and Overview

Given the numerous and ever-increasing applications of the generic concept of modulus degradation as a function of some other problem parameter (only shear strain has been noted up to this point but other parameters are discussed later in this appendix), it is no surprise that from the earliest days of soil-dynamics research circa 1970 that there has been an interest in identifying and developing mathematical models for the relationship between shear-modulus degradation and shear strain. Two early pioneers in soil-dynamics research, Prof. Bobby O. Hardin and Prof. Vincent P. Drnevich who were both on the faculty of the University of Kentucky at this time, identified the fact that many aspects of soil behavior, including soil dynamics, could be reasonably modeled mathematically as a hyperbolic relationship (Hardin and Drnevich 1972a, 1972b). This observation was consistent with seminal research of a more-generic nature by Kondner almost a decade earlier (Kondner 1963) as well research into soil models to use with the then-new geotechnical application of the finite-element method (Duncan and Chang 1970).

In the almost 60 years since the publication of Kondner's landmark paper, there has been a substantial body of published research related to modeling modulus degradation using a hyperbolic relationship. There are two aspects of this research that are highlighted here:

1. It is analytically useful to normalize the actual shear strain, γ, to some reference value, γ_{ref}. Shear-modulus degradation is then plotted against, and mathematically correlated to, the dimensionless ratio γ/γ_{ref}. The characteristic S-shaped curve that exists with the basic shear-strain plotting is still retained. It is important to note at the outset that there is no standard or consensus as to how to define γ_{ref}. It varies from researcher to researcher and application to application so it is always essential to identify how this parameter is defined in a particular application or a particular publication.

2. It is possible to replace shear strain with another definition of strain or even another problem parameter, the latter usually related to stress. This is advantageous in certain applications, including DFE settlement which is of interest in this monograph.

The first aspect is discussed in some detail in the sections immediately following. There has been evolutionary development in the specific form of the hyperbolic relationship used over the years by various researchers. Illustrating this development is necessary as this lays the groundwork for the second aspect of relating modulus degradation to some parameter other than shear strain which is discussed later in this appendix as it has direct relevance to the theme of this monograph.

D.1.3.5.2 Hyperbolic Model: Basic Generic Equation

Fahey and Carter (1993) noted that the most basic, generic mathematical relationship for a hyperbolic model relating the degradation of the normalized shear modulus, $G_{operative}/G_{max}$, to normalized shear strain, γ/γ_{ref}, has the form shown in the equation that follows. Note that some of the variable names have been changed from the original Fahey and Carter paper to be consistent with notation used in this monograph as well as to be consistent with other results shown subsequently:

$$\frac{G_{operative}}{G_{max}} = \frac{1}{1 + \left(\frac{\gamma}{\gamma_{ref}}\right)} . \tag{D.1}$$

The first thing to note before proceeding further in the discussion is that $G_{operative}$ in this equation is a <u>secant</u> modulus, <u>not</u> a tangent modulus. This is emphasized as the distinction is very important analytically (this topic is discussed in detail later in this appendix) and most publications rarely make it clear which type of modulus is being discussed although Fahey and Carter did so in their 1993 paper as the distinction was relevant for their work (as it is to DFE applications in general).

The attraction of Equation D.1 in general is, of course, its overall simplicity as it involves only two independent problem parameters, G_{max} and γ_{ref}. As will be seen later in this appendix, a further attraction is that the latter parameter, γ_{ref}, can always be replaced as desired with another problem parameter that is either more readily definable or measurable.

However, as is usually the case, simplicity comes at a price. In this instance, the most significant issue in this regard, as Fahey and Carter pointed out, is that the hyperbola defined by Equation D.1 essentially takes forever to get to where it is going. Specifically, in this case it takes an infinite amount of strain to reach $G_{operative} = 0$.

This is an inherent, well-known behavioral issue with hyperbolas that was recognized by other, early researchers using the hyperbolic model for soil such as the aforementioned work of Duncan and Chang (1970). In that case, an arbitrary parameter defined as R_f, with $0 < R_f \leq 1$, was introduced in order to cause the soil to reach the ULS at a finite magnitude of strain. As will be seen subsequently, the way that this is typically handled for shear-modulus degradation equations is by making arithmetic alterations to Equation D.1.

D.1.3.5.3 Hyperbolic Model: Vardanega-Bolton Equation and Derivatives

The first enhancement to Equation D.1 that is discussed is the work of Vardanega and Bolton (2011). Their work was cited and discussed at length by Niazi (2014) as part of Niazi's theoretical groundwork for developing his settlement-based shear-modulus-degradation model for DFEs.

Vardanega and Bolton found that normalization of shear strain to some reference value, γ_{ref}, collapsed some of the variation due to both the aforementioned soil viscosity, i.e. data obtained in dynamic vs. static tests, and soil plasticity. For their purposes, Vardanega and Bolton defined γ_{ref} as the magnitude of shear strain that correlated with $G_{operative}/G_{max} = 0.5$. Some publications call this value of $G_{operative}$ 'G_{50}'. As will be seen, this assumption resulted in a mathematical outcome that is somewhat more general and complex than Equation D.1 but simpler than the mathematical relationship ultimately used by Niazi (2014).

The basic equation that Vardanega and Bolton proposed has the form shown in the equation that follows. As before, some of the variable names have been changed from those in the original publication to be consistent with notation used in this monograph as well as to be consistent with other results shown subsequently:

$$\frac{G_{operative}}{G_{max}} = \frac{1}{1 + \left(\frac{\gamma}{\gamma_{ref}}\right)^{\beta}} . \tag{D.2}$$

Note that the difference from Equation D.1 is the addition of the β variable as an exponent. Vardanega and Bolton found that both γ_{ref} and β varied as a function of static vs. dynamic testing used to generate the γ data. In addition, γ_{ref} varied with the PI of the soil.

To evaluate β, Vardanega and Bolton used a clever algebraic transformation that involved the judicious rearrangements of terms combined with taking the $\log_{10}$ of both sides of Equation D.2. To being with, Equation D.2 is first rearranged in two steps as follows:

$$1 + \left(\frac{\gamma}{\gamma_{ref}}\right)^{\beta} = \frac{G_{max}}{G_{operative}} \tag{D.3}$$

$$\left(\frac{\gamma}{\gamma_{ref}}\right)^{\beta} = \left[\left(\frac{G_{max}}{G_{operative}}\right) - 1\right]. \tag{D.4}$$

The $\log_{10}$ of each side is then taken:

$$\log_{10}\left[\left(\frac{\gamma}{\gamma_{ref}}\right)^{\beta}\right] = \log_{10}\left[\left(\frac{G_{max}}{G_{operative}}\right) - 1\right] \tag{D.5a}$$

$$\beta \log_{10}\left(\frac{\gamma}{\gamma_{ref}}\right) = \log_{10}\left[\left(\frac{G_{max}}{G_{operative}}\right) - 1\right] \tag{D.5b}$$

with a final rearrangement into:

$$\log_{10}\left[\left(\frac{G_{max}}{G_{operative}}\right) - 1\right] = \beta \log_{10}\left(\frac{\gamma}{\gamma_{ref}}\right). \tag{D.5c}$$

Equation D.5c is recognized as a special case of the generic equation that defines a straight line:

$$y = c_1 x + c_2. \tag{D.6}$$

In this case, $\beta = c_1$ and is thus simply the slope of this line. Note also that the intercept, c_2, of this line with the x (= $\log_{10}(\gamma/\gamma_{ref})$ in this case) = 0 axis is 0 because of the judicious choice of making $\gamma/\gamma_{ref} = 1$ when $G_{operative}/G_{max} = 0.5$ which means that its reciprocal, $G_{max}/G_{operative}$, = 2.

The way in which Equation D.5c is applied in practice is to create a new, alternative strain plot with $\log_{10}(\gamma/\gamma_{ref})$ as the abscissa and $\log_{10}[(G_{max}/G_{operative})-1]$ as the ordinate, each with a linear scale. In this format, the $G_{operative}/G_{max}$ vs. γ/γ_{ref} data have a rectilinear trend as opposed to the classical S-shaped trend using either the standard or normalized strain-plotting formats described earlier. Using the generic statistical function that is built into many software packages such as *Excel*, a best-fit straight line that goes through the origin (a requirement due to the $c_2 = 0$ assumption that is implied by the form of the original hyperbolic relationship shown in Equation D.2) is fit to the plotted data. The desired outcome of this line-fitting process is the slope, c_1, of the line which, as noted previously, is the desired β value for the hyperbolic exponent in Equation D.2.

D.1.3.5.4 Hyperbolic Model: Enhanced Generic Equation

As will be seen later in this appendix, a further enhancement of the basic relationship shown in Equation D.1 is to add an additional problem parameter to that used in the Vardanega-Bolton relationship (Equation D.2) as follows:

$$\frac{G_{operative}}{G_{max}} = \frac{1}{1 + \alpha \left(\frac{\gamma}{\gamma_{ref}}\right)^{\beta}} .$$
(D.7)

The additional variable is α[185], a constant coefficient that is evaluated empirically on a problem-specific basis using the same algebraic manipulation used by Vardanega and Bolton but with an important relaxation of one assumption.

Proceeding along the same developmental lines as with the Vardanega-Bolton relationship (Equation D.3 and following):

$$1 + \alpha \left(\frac{\gamma}{\gamma_{ref}}\right)^{\beta} = \frac{G_{max}}{G_{operative}} .$$
(D.8)

$$\alpha \left(\frac{\gamma}{\gamma_{ref}}\right)^{\beta} = \left[\left(\frac{G_{max}}{G_{operative}}\right) - 1\right] .$$
(D.9)

$$\log_{10}\left[\alpha \left(\frac{\gamma}{\gamma_{ref}}\right)^{\beta}\right] = \log_{10}\left[\left(\frac{G_{max}}{G_{operative}}\right) - 1\right]$$
(D.10a)

$$\log_{10}\alpha + \log_{10}\left[\left(\frac{\gamma}{\gamma_{ref}}\right)^{\beta}\right] = \log_{10}\left[\left(\frac{G_{max}}{G_{operative}}\right) - 1\right]$$
(D.10b)

$$\log_{10}\alpha + \left[\beta \log_{10}\left(\frac{\gamma}{\gamma_{ref}}\right)\right] = \log_{10}\left[\left(\frac{G_{max}}{G_{operative}}\right) - 1\right]$$
(D.10c)

$$\log_{10}\left[\left(\frac{G_{max}}{G_{operative}}\right) - 1\right] = \left[\beta \log_{10}\left(\frac{\gamma}{\gamma_{ref}}\right)\right] + \log_{10}\alpha .$$
(D.10d)

Equation D.10d is again recognized as the equation of a straight line (Equation D.6) with a slope $c_1 = \beta$ but in this case with an intercept, c_2, of the x (= $\log_{10}(\gamma_p/\gamma_{p\text{-}ref})$ in this case) = 0 axis that is not necessarily equal to zero as was required in the Vardanega-Bolton derivation. This means that α is not necessarily = 1 as with the Vardanega-Bolton relationship (Equation D.2) that is now seen to be a special, simplified case of the more-general relationship shown in Equation D.7. Specifically, α in this enhanced generic equation can be interpreted and evaluated as follows:

---

[185] Note that the use of α in this context is independent of the use of this parameter notation throughout the rest of this monograph to mean taper angle. There is no correlation between these two usages. There is, however, a very specific reason why this notation is used here in this context instead of an alternative letter from the Greek alphabet. This reason will become apparent later in this appendix.

$$c_2 = \log_{10} \alpha \qquad \textbf{(D.11)}$$

$$10^{c_2} = 10^{\log_{10} \alpha} = \alpha \qquad \textbf{(D.12a)}$$

which when simplified and rearranged becomes the desired final result:

$$\alpha = 10^{c_2} . \qquad \textbf{(D.12b)}$$

The way in which these results are used is identical to that outlined previously for the Vardanega-Bolton relationship:

- Equation D.10d is used to create a new, alternative strain plot with $\log_{10}(\gamma/\gamma_{ref})$ as the abscissa and $\log_{10}[(G_{max}/G_{operative})\text{-}1]$ as the ordinate, each with a linear scale.

- As previously, in this format $G_{operative}/G_{max}$ vs. γ/γ_{ref} data have a rectilinear trend as opposed to the classical S-shaped trend using the standard plotting format.

- Using generic statistical software, a best-fit straight line is fit to the plotted data but in this case with no restrictions as with the Vardanega-Bolton equation. The desired outcomes of this line-fitting process are:
  o the slope, c_1, of the line that is the desired β value for the hyperbolic exponent in Equation D.7 and
  o the intercept, c_2, of this line that is the $\log_{10}$ of the desired α constant-coefficient in Equation D.7.

D.1.3.6 Alternative Pseudo-Strain Plotting: Normalized Settlement

D.1.3.6.1 Introduction and Overview

As discussed later in this appendix, Niazi (2014) reviewed several shear-modulus-degradation equations in addition to the one proposed by Vardanega and Bolton that was presented and discussed earlier in this appendix. In the end, Niazi wound up using the Vardanega-Bolton equation but with two significant changes:

- Normalized/non-dimensionalized DFE settlement was used as what Niazi called a "pseudo-strain" in lieu of normalized shear strain.

- The modulus-degradation equation was generalized by the addition of the α parameter as outlined in the preceding section to allow for better statistical fitting to plotted data.

Each of these changes is discussed further below as they are important relative to the theme and goals of this monograph.

D.1.3.6.2 Alternative Pseudo-Strain Settlement Formulation

Degradation of elastic moduli is not unique to shear strain or even strain in general as will be seen later in this appendix. Therefore, there is no inherent, theoretical reason why the independent variable against which the dependent variable of modulus degradation is plotted cannot be changed from γ in general or γ/γ_{ref} in particular as used in Equations D.1, D.2, and D.7 for the increasingly more-general forms of a hyperbolic relationship.

Because Niazi's research as reported in detail in his 2014 doctoral dissertation focused on the load-settlement response of axially loaded DFEs, he replaced the generic shear strain, γ, with what he called a *pseudo-strain*, γ_p, for use with all types of DFEs that is defined as follows:

$$\gamma_p = w_t/d \qquad \qquad \textbf{(D.13)}$$

where:
- w_t = settlement at the head of the DFE and
- d = diameter of the DFE (which was unambiguous as he only considered constant-diameter DFEs),

both of which must have the same units.

Niazi was inconsistent in his 2014 dissertation in that he indiscriminately expressed the pseudo-strain settlement, γ_p, in both its basic, non-dimensional decimal form and as a percent. Either format is acceptable for use but, obviously, one should be consistent for clarity and to minimize confusion. To be consistent with material presented earlier in this appendix, the percent format is used in this monograph in which case the outcome of Equation D.13 is multiplied by 100.

Before proceeding further, it is obvious to note that Equation D.13 is ill-defined for any tapered pile, whether continuously tapered or partially tapered, because of the issue of how to define d. As an interim measure solely to allow for investigation and extension of the Niazi Method to tapered piles, the writer used a weighted average of diameter of a tapered pile for d in Equation D.13. However, this does not mean that this is the best alternative. Future research is needed in this regard and may prove otherwise.

In any event, Niazi went on to define what he called a *reference pseudo-strain*, $\gamma_{p\text{-}ref}$, that is equal to 0.01 = 1%. He then used the dimensionless ratio $\gamma_p/\gamma_{p\text{-}ref}$ in the same way in which the γ/γ_{ref} ratio is used in Equation D.7.

D.1.3.6.3 Modulus-Degradation Equation Generalization

Because Niazi linked $\gamma_{p\text{-}ref}$ to a fixed, pre-determined value of γ_p as opposed to a fixed, pre-determined value of $G_{operative}/G_{max}$ as was done by Vardanega and Bolton, there is no longer an assurance that the best-fit line on a $\log_{10}[(G_{max}/G_{operative})\text{-}1]$ vs. $\log_{10}(\gamma/\gamma_{ref})$ plot (actually $\log_{10}(\gamma_p/\gamma_{p\text{-}ref})$ in this case) goes through the origin as in the case of the Vardanega-Bolton relationship (Equation D.2). Consequently, the enhanced generic relationship defined in Equation D.7 must be used, rewritten here with notation consistent with Niazi's pseudo-strains[186]:

$$\frac{G_{operative}}{G_{max}} = \frac{1}{1 + \alpha \left(\dfrac{\gamma_p}{\gamma_{p-ref}}\right)^{\beta}} \cdot \qquad \textbf{(D.14)}$$

---

[186] Note that this was the reason that α was introduced and used as parameter notation in the modulus-degradation application going back to Equation D.7, i.e. to be consistent with notation that Niazi used in Niazi (2014) and subsequent publications. The notational duplication with taper angle is regrettable but unavoidable as the writer feels that it is more important to be consistent with Niazi's notation in this case.

Similarly, the end result (Equation D.10d) of manipulating this hyperbolic equation to allow for problem-specific determination of the α and β parameters has the following form when Niazi's pseudo-strain is used:

$$\log_{10}\left[\left(\frac{G_{max}}{G_{operative}}\right) - 1\right] = \left[\beta \log_{10}\left(\frac{\gamma_p}{\gamma_{p-ref}}\right)\right] + \log_{10}\alpha \ . \qquad \textbf{(D.15)}$$

As before:

- Equation D.15 is used to create a pseudo-strain plot with $\log_{10}(\gamma_p/\gamma_{p-ref})$ as the abscissa and $\log_{10}[(G_{max}/G_{operative})-1]$ as the ordinate, each with a linear scale.

- In this format, $G_{operative}/G_{max}$ vs. γ_p/γ_{p-ref} data have a rectilinear trend.

- Using generic statistical software, a best-fit straight line is fit to the plotted data. The desired outcomes of this line-fitting process are:
  - the slope, c_1, of the line that is the desired β value for the hyperbolic exponent in Equation D.14 and
  - the intercept, c_2, of this line that is the $\log_{10}$ of the desired α constant-coefficient in Equation D.14.

D.1.3.6.4 Application in the Niazi Method

Although Niazi did not include any tapered piles in the database[187] he developed for the Niazi Method as presented in Niazi (2014) and Niazi and Mayne (2015b, 2019), his method is, nevertheless, a recent and significant addition and advancement to the body of knowledge for deep foundations. This is due in large part to the Niazi Method being an integrated load-settlement methodology. As discussed in Chapter 5, such one-step procedures for simultaneously forecasting both resistance and concomitant settlement in a single, theoretically integrated manner are still quite rare in foundation engineering.

Furthermore, and of direct relevance to the theme of this monograph, the Niazi Method has potential for relatively straightforward extension to include tapered piles, also as discussed in Chapter 5. Because the Niazi Method is addressed in the main body of this monograph, the focus of the discussion in this appendix is on the α and β parameters only.

Although Niazi used Equation D.14 and its derivative outcome, Equation D.15, as the sole basis for curve-fitting the load-settlement database he created for his methodology, as noted in Horvath (2016a, 2016b), Niazi interpreted that database in several different ways. As a result, the α and β parameters in Equation D.14 were not uniquely defined by Niazi which creates some confusion when initially reviewing the results presented in Niazi (2014).

With some simplification by the writer in order to achieve conciseness in presentation, the ways in which Niazi quantified the α and β parameters in his 2014 doctoral

---

[187] In Niazi and Mayne (2019), Niazi stated that "timber" [sic] piles were included in his database. However, a review of the DFEs that comprised his database, as tabulated in Appendix E of Niazi (2014), does not indicate any DFE with a non-constant value of diameter or equivalent diameter in the case of a DFE with a non-circular cross-section. As the writer was unsuccessful in establishing communications with Prof. Niazi to clarify this point (and many other points) related to his doctoral research, the writer can only conclude that, based on the published evidence available to the writer, no tapered piles were included in Niazi's database, at least as originally published in 2014..

dissertation can be summarized as follows (note that this curation is based <u>solely</u> on the writer's interpretation of Niazi's work as Niazi did not provide such a summary in any of the published work to which the writer had access):

- A single best-fit line to the entire database of approximately 2,000 load-settlement datapoints produced single-valued parameters that are referred to herein as α_0 (= 3.634) and β_0 (= 0.942). Note that Niazi did not use these notations but simply showed the numerical values indicated with the precision indicated.

- Next, the database was segregated by DFE typology into four types of DFEs (all of which Niazi referred to as "piles"):
  - ○ auger ('augered cast in place (ACIP) pile' in U.S. terminology),
  - ○ bored ('drilled shaft' in U.S. terminology),
  - ○ driven (simply 'pile' in U.S. terminology), and
  - ○ jacked (relatively rare in U.S. practice except in underpinning work but apparently increasingly common in some parts of the world).

  Best-fit straight lines were determined separately for each of these smaller databases which produced different values of both α_0 and β_0 for each DFE type. Again, Niazi did not use the α_0 and β_0 notation. Although the variation between the four β_0 values was not very large (they were each close to 1.0), the maximum variation between α_0 values was a factor of approximately 3. This means that while all of the rectilinear relationships had approximately the same slope on a $\log_{10}[(G_{max}/G_{operative})\text{-}1]$ vs. $\log_{10}(\gamma_p/\gamma_{p\text{-}ref})$ plot, the zero-axis intercepts on such plots had not-insignificant variation. Overall, this indicates that DFE typology is a significant factor. This is not surprising given the wide variation in DFE installation/construction methodologies reflected in the four types that Niazi defined and studied. Note that this also suggests that for future extension of the Niazi Method to other types of DFEs that he did not consider that the most likely variation would be in the α parameter in Equation D.14, not the β parameter in that equation. This is because all indications to date (and this includes the previously unpublished extension of the Niazi Method by the writer to include tapered piles that is presented in the following two sections) are that the β parameter is insensitive to DFE typology and always has a value of approximately 1.

- In an effort to develop a unified equation, Niazi used the aforementioned α_0 = 3.634 and β_0 = 0.942 values for the entire database data-fit but then introduced two additional variables to each to correct for DFE typology and soil PI respectively. Thus, and with reference to Equation D.14, in this final iteration:
  - ○ $\alpha = \alpha_0 \cdot \alpha_1 \cdot \alpha_2$ and
  - ○ $\beta = \beta_0 \cdot \beta_1 \cdot \beta_2$.

  As noted in Chapter 5, in the writer's opinion, the correction factors for DFE typology (α_1 and β_1) appear to be adequately supported by the database as there are only four types or categories of DFEs considered and there appear to be a reasonable number of datapoints for each type. However, the correction factors for soil plasticity (α_2 and β_2) do not appear to be adequately supported although Niazi clearly feels otherwise (Niazi and Mayne 2019). This latter opinion of the writer derives from both the sparseness of the current databases for each DFE type for the wide range in soil PI that is possible as well as the difficulty in choosing a single PI and concomitant correction factor for cases where the DFE penetrates through multiple soil types (which is the rule rather than the exception in the real world). It is for this reason that the writer suggested that correction for soil typology <u>not</u> be applied (this simply means using $\alpha_2 = \beta_2 = 1$) unless and until such time as an expanded database as well as analytical methodology for dealing with multiple soil types for a given DFE supports such a correction.

D.1.3.6.5 Potential Extension to Tapered Piles: Rigorous Alternative

There is no reason why Equation D.14 cannot be used for tapered piles. One could simply evaluate the α and β parameters using Equation D.15 applied to some appropriate database of load-settlement data assembled for this purpose. However, as discussed in Chapter 5, a more-methodical approach is to treat tapered piles simply as an extension of Niazi's original database and concomitant analytical methodology.

What the writer termed the 'rigorous alternative' is to add additional correction factors, α_3 and β_3, to the ones discussed previously so that:

- $\alpha = \alpha_0 \cdot \alpha_1 \cdot \alpha_2 \cdot \alpha_3$ and

- $\beta = \beta_0 \cdot \beta_1 \cdot \beta_2 \cdot \beta_3$.

These additional correction factors would, as a minimum, vary with taper angle although there would be additional influences to consider such as pile material, continuity of taper (i.e. continuously tapered vs. partially tapered), and possibly toe diameter as well. Note that for a constant-perimeter DFE, $\alpha_3 = \beta_3 = 1$ so there is no loss in generality by adding these factors.

D.1.3.6.6 Potential Extension to Tapered Piles: Simplified Alternative

As noted earlier in this appendix, based on Niazi's research that included several very different types of DFEs as well as a wide range of soil conditions, the β parameter in Equation D.14 was found to be approximately equal to 1 in all cases. This was the genesis of suggesting in Chapter 5 that an alternative, simplified extension to the Niazi Method to include tapered piles could be developed based on the following simplified version of Equation D.14:

$$\frac{G_{operative}}{G_{max}} = \frac{1}{1 + \alpha \left(\frac{\gamma_p}{\gamma_{p-ref}}\right)}. \tag{D.16}$$

Note that in this case, the α parameter is evaluated as a single-value parameter for some specific database of tapered piles, not the product $(\alpha_0 \cdot \alpha_1 \cdot \alpha_2 \cdot \alpha_3)$ as in the rigorous alternative discussed in the preceding section.

D.1.3.7 Alternative Shear-Stress/Shear-Strength/Safety-Factor Plotting

D.1.3.7.1 Original Work of Fahey and Carter

Returning now to the work of Fahey and Carter (1993) and Equation D.1 that represents the simplest, most basic form of the hyperbolic model for shear-modulus degradation as a function of shear strain, the objective of these authors in this cited paper was to model certain aspects of pressuremeter (PMT) behavior. Thus, in their analytical development they wanted to ultimately replace:

- shear strain, γ, with shear stress, τ, as the correlation parameter for shear-modulus degradation. Fahey and Carter's work was an early example of extending the concepts of shear-modulus degradation that were initially developed for soil dynamics and shear strain to static and quasi-static geotechnical applications involving problem parameters other than shear strain (shear stress in this case). Note that this change to shear stress also involves, by implication, the parameters of shear strength, τ_{max}, (because it is the inherent maximum value of shear stress) as well as the traditional ASD/WSD safety factor (because it is the reciprocal of the shear stress-to-shear strength ratio); and

- shear strength, τ_{max}, with the friction angle, ϕ, of the soil as they were dealing only with coarse-grained soil[188].

As a result, the only independent problem parameters become G_{max} and ϕ.

To begin the necessary mathematical transformation of Equation D.1 (repeated here for ease of reference):

$$\frac{G_{operative}}{G_{max}} = \frac{1}{1 + \left(\dfrac{\gamma}{\gamma_{ref}}\right)} \tag{D.1}$$

recall that the definition of γ_{ref} is always arbitrary. Consequently, Fahey and Carter first defined γ_{ref} in terms of shear stress as follows:

$$\gamma_{ref} = \frac{\tau_{max}}{G_{max}}. \tag{D.17}$$

They also defined γ as follows:

$$\gamma = \frac{\tau}{G_{operative}} \tag{D.18}$$

where τ is an arbitrary value of shear stress $\leq \tau_{max}$.

They next referenced earlier work by the senior author (Fahey 1992) in which Equations D.1, D.17, and D.18 were combined and algebraically manipulated to produce a relationship between shear-modulus degradation and shear stress and strength as follows:

$$\frac{G_{operative}}{G_{max}} = 1 - \frac{\tau}{\tau_{max}}. \tag{D.19}$$

Fahey and Carter noted that Equation D.19 has the same forecasting limitations as noted earlier in this appendix for Equation D.1 that is based on shear strain as opposed to shear stress as in Equation D.19. In this case, this means that τ never reaches τ_{max} but only approaches it asymptotically with increasing shear stress. Furthermore, because there are only two independent variables in Equation D.19, the shape of the hyperbola cannot be manipulated to better match some measured behavior.

In order to overcome both of these shortcomings and in consonance with the similar approach used with shear strain that was discussed earlier in this appendix, Fahey and Carter

---

[188] Fahey and Carter used the term *effective friction angle* and notation ϕ'.

proposed an enhanced version of Equation D.19 that incorporated two additional independent, dimensionless parameters for a total of four. Note that this is conceptually identical to Equation D.7 being an enhanced version of Equation D.1. The equation they proposed as an enhancement of Equation D.19 has the following form:

$$\frac{G_{operative}}{G_{max}} = 1 - f \left(\frac{\tau}{\tau_{max}}\right)^{g}$$ (D.20)

where:
- $0 \leq f \leq 1$ and
- $0 \leq g$

and are conceptually similar to the variables α and β in Equation D.7.

The basic graphical presentation of Equation D.20 is to plot τ/τ_{max} as the independent variable along the abscissa and $G_{operative}/G_{max}$ as the dependent variable along the ordinate. However, Fahey and Carter (1993) showed that alternative plots such as the basic $G_{operative}/G_{max}$ vs. $\log_{10}\gamma$ that was discussed earlier in this appendix and τ/τ_{max} vs. γ/γ_{ref} can be created by utilizing the various equations presented above.

Based on these three plotting alternatives, an informative picture emerges with respect to the qualitative influence of the f and g variables:

- f controls the location of the endpoint of whatever curve is generated. This is particularly useful when shear stress vs. shear strain curves are generated as this allows the ULS to be reached at a finite strain level defined by the analyst. Specifically, the shear strength is fully mobilized at a normalized strain equal to $1/(1 - f)$.

- g controls the overall shape (curvature) of the curve. As g increases, the overall system stiffness, especially in the early stages of loading, increases. Another way of viewing this behavior is that as g increases, the small-strain limit that was discussed earlier in this appendix increases in magnitude.

- $f = g = 1$ is the special case of the basic, no-frills hyperbolic relationship defined by Equation D.19.

- $f = 0$ is the special case of a linear-elastic material with $G_{operative} = G_{max}$.

- $g = 0$ is the special case of a linear-elastic material with $G_{operative} = (1 - f) G_{max}$.

D.1.3.7.2 Extension to Deep Foundations by Mayne and Niazi

Approximately a decade after Fahey and Carter published their modulus-degradation model for use with the pressuremeter test (PMT), Mayne adopted it for use with deep foundations as an essential component of the settlement-only forecasting methodology that he crafted around solutions to the Randolph-Wroth Problem. For this purpose, Mayne made two parameter changes to the original Fahey-Carter relationship (Equation D.20):

- Young's modulus, E, was substituted for shear modulus. This was likely due to the fact that, in the early years of the 21st century (Mayne (2007) represents what appears to the first widespread publication of this analytical methodology), it was easier to empirically estimate the small-strain Young's modulus, E_{max}, using various in-situ testing tools than it was the small-strain shear modulus, G_{max}.

- Shear stress and shear strength were replaced by the applied load at the head of a DFE, Q, and the ultimate applied load or capacity, Q_u, respectively, where $0 < Q < Q_u$.

Because Mayne linked the Fahey-Carter model to the Randolph-Wroth Problem solutions, a further stipulation made by Mayne was that the Young's modulus values, both operative and maximum (i.e. small-strain), are those at the toe of the DFE (E_L in Figure 5.4).

Note that the issue of modulus varying along the length of an object (DFE in this case) embedded in the ground did not arise in the original Fahey-Carter derivation as they were interested in the PMT that involves a device of relatively limited length. On the other hand, DFEs have relatively significant lengths over which moduli will generally vary. Consequently, when dealing with DFEs, it becomes essential to clearly identify the way in which moduli are defined and evaluated.

The resulting revised relationship was discussed in Chapter 5 but is repeated here for ease of reference but with the applied-load notation changed to conform to that adopted for this monograph as shown in Figure 2.1 (Mayne's original notation was used in Chapter 5 for its historical relevance):

$$\frac{E_{L(operative)}}{E_{L(max)}} = 1 - f\left(\frac{Q}{Q_u}\right)^g . \tag{D.21}$$

As also discussed in Chapter 5, Mayne made relatively specific suggestions concerning the values of the f and g parameters:

- $f = 1$ and

- $g = 0.3 \pm 0.1$.

The writer has explored, to a limited degree, the sensitivity of calculated outcomes to the second assumption and it is not-insignificant, at least in some specific applications (see Figure 5.10). In various publications and presentations subsequent to his 2007 publication, Mayne himself has noted that values of g other than the default 0.3 value sometimes provide a better fit to measured results[189]. This suggests that, as a minimum, further research is warranted into the value of g to use in practice.

By the second decade of the 21st century when Niazi was conducting his doctoral research under Mayne's supervision, technology had advanced to the point where measuring G_{max} for soil subgrades using sCPTu soundings was practical on a routine basis, even for small projects. As a result, Niazi reformulated Equation D.21 in terms of shear modulus:

$$\frac{G_{L(operative)}}{G_{L(max)}} = 1 - f\left(\frac{Q}{Q_u}\right)^g . \tag{D.22}$$

It does not appear that Niazi addressed the issue of the f and g parameters at all as his research interests that culminated in development of the Niazi Method apparently quickly

---

[189] Fahey and Carter (1993) showed that $g = 0.25$ (with $f = 0.98$) provided the best match for a simple-shear test performed on a normally consolidated sand specimen.

moved away from use of the Fahey-Carter modulus-degradation model during the course of his doctoral studies at Georgia Tech. Thus, the default values of f and g noted above still represent the current state of practice when using Equation D.22.

Should further use of Equation D.22 in practice and/or research prove desirable, research into developing a rational procedure for quantifying the f and g parameters would be appear to be a worthwhile endeavor. In particular, using a value for f that is slightly less than 1 would cause the resulting hyperbolic curve to reach the ULS total resistance of a DFE at a finite magnitude of settlement as opposed to only approaching the ULS total resistance asymptotically. With regard to the g parameter, because it controls the shape (curvature) of the load-settlement curve, it would appear that this parameter needs to be matched to the stiffness of the DFE-ground system rather than using a one-size-fits-all value as at present.

D.1.3.8 Secant vs. Tangent Modulus Interpretation

D.1.3.8.1 Introduction

As noted earlier in this appendix, the various publications dealing with shear-modulus degradation often do not clearly indicate whether the modulus-degradation relationships that are presented or used produce a secant or tangent value of the operative shear modulus, $G_{operative}$. The distinction is very important as a modulus-degradation relationship must be behaviorally consistent with the analytical methodology with which it is coupled. Therefore, the purpose and focus of the present discussion is to explore the secant vs. tangent differences in some detail so that a modulus-degradation relationship that properly matches the intended application is selected for use.

D.1.3.8.2 Overview

The distinction between secant and tangent values of material properties is a recurring issue in geotechnical and foundation engineering. This is due to the fact that soil behavior is inherently a non-linear process yet virtually all analytical methodologies require a linear value of soil properties and other problem parameters, at least for some incremental step in an analysis if not for the overall analysis.

Because this issue comes up in several different ways throughout this monograph, the generic distinction between secant and tangent values of material properties and other problem parameters was addressed early on in this monograph, in Chapter 2. The discussion in this appendix is limited to application of the generic concepts specifically to modulus-degradation relationships.

First discussed in the following sections is the work of Fahey and Carter (1993). Not only is their work relevant to the present discussion (the modulus-degradation relationships they discussed form the basis for Mayne's and Niazi's work related to deep foundations) but their discussion of the distinction between secant and tangent versions of the same modulus-degradation relationship is unusually thorough compared to similar published work by others. This is followed by discussions of the work of various researchers relative to deep foundations as the distinction of secant vs. tangent is not always made clear in this work.

As will be seen, it appears that the vast majority of modulus-degradation relationships that have been used to date in geotechnical engineering in general, and for deep-foundation applications in particular, have been secant in nature. This is generally not obvious as the distinction is often not noted and the analytical methodologies in which the modulus-degradation relationships are used (for deep foundations at least) tend to result in progressive load-settlement forecasts that would generally call for a tangent-modulus approach. Thus, this usage gives the impression that a tangent modulus is being used to produce contiguous, cumulative results when, in reality, the calculated outcomes are just a series of independent secant-modulus calculations for which someone has 'connected the dots'.

D.1.3.8.3 Fahey and Carter (1993)

Fahey and Carter explicitly explored the difference between secant and tangent modulus-degradation relationships for two versions of the relationship between normalized shear modulus and normalized shear resistance as strength, not strain, was of primary relevance for their work related to the PMT. They first considered the equivalent stress-strength version (Equation D.19, repeated below for ease of reference) of the basic generic hyperbolic equation based on shear strain (Equation D.1):

$$\frac{G_{operative}}{G_{max}} = 1 - \frac{\tau}{\tau_{max}}. \tag{D.19}$$

Note that this equation produces a <u>secant</u> value of $G_{operative}$.
The related equation for degradation of the <u>tangent</u> shear modulus, $G_{t(operative)}$, is given by Fahey and Carter as:

$$\frac{G_{t(operative)}}{G_{max}} = \left(1 - \frac{\tau}{\tau_{max}}\right)^2 = \left(\frac{G_{operative}}{G_{max}}\right)^2. \tag{D.23}$$

Fahey and Carter next considered the enhanced version of Equation D.19 that includes the f and g parameters to allow for control of the endpoint and shape of the resulting curve. This equation, D.20, is also repeated here for ease of reference:

$$\frac{G_{operative}}{G_{max}} = 1 - f\left(\frac{\tau}{\tau_{max}}\right)^g. \tag{D.20}$$

Again, this equation produces a <u>secant</u> value of $G_{operative}$.
The related equation for degradation of the <u>tangent</u> shear modulus, $G_{t(operative)}$, is given by Fahey and Carter as:

$$\frac{G_{t(operative)}}{G_{max}} = \frac{\left(\frac{G_{operative}}{G_{max}}\right)^2}{\left[1 - f(1 - g)\left(\frac{\tau}{\tau_{max}}\right)^g\right]}. \tag{D.24}$$

To provide some insight into the relative difference between the secant and tangent values of $G_{operative}$ for the two different equations (basic and enhanced) considered by Fahey and Carter, Figure D.1 compares Equations D.19 to D.23 (basic, secant and tangent respectively) and Equations D.20 to D.24 (enhanced, secant and tangent respectively).

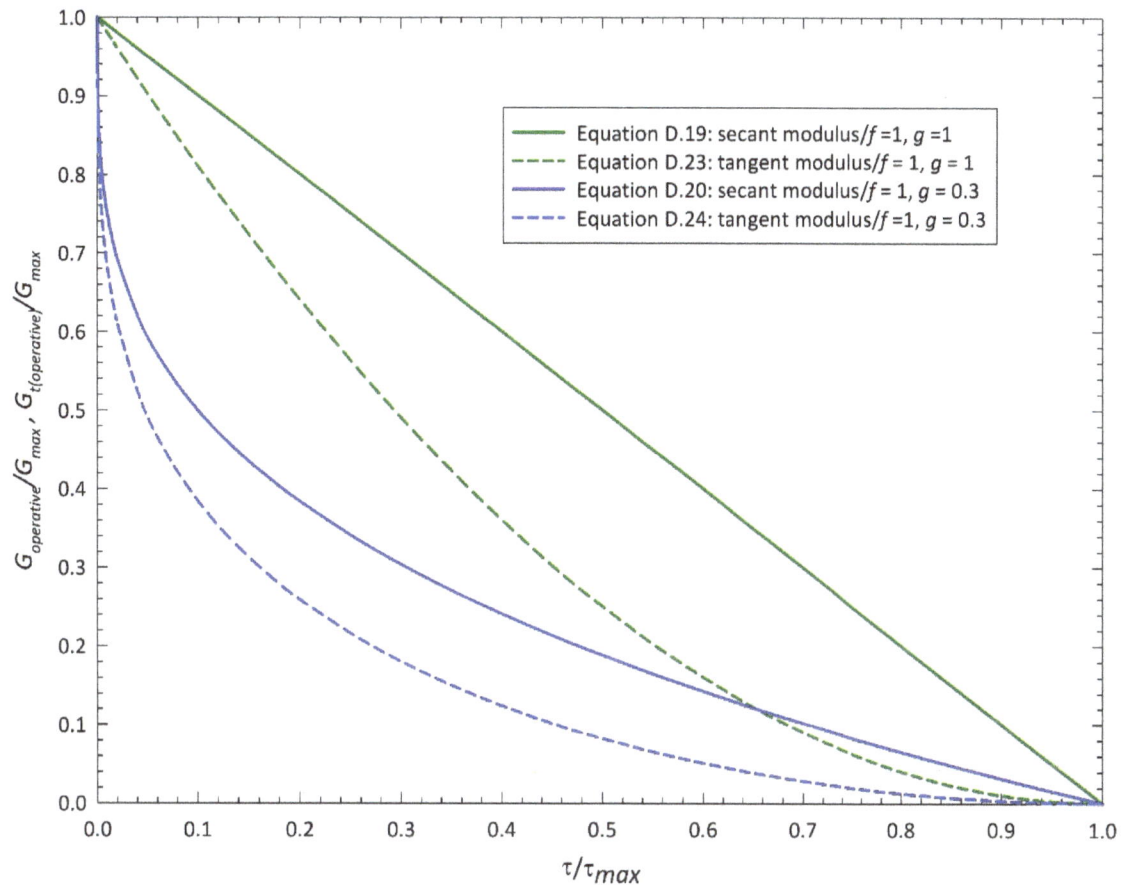

Figure D.1. Relationship between Secant and Tangent Shear Moduli for Stress-Based Hyperbolic Modulus-Degradation Model.

Note the following:

- Equations D.19 and D.23 for the basic hyperbolic model are essentially special cases of the enhanced model (Equations D.20 and D.24 respectively) with $f = g = 1$ and are so labeled in Figure D.1 in order to present a more-unified picture of results.

- Equations D.20 and D.24 for the enhanced hyperbolic model are shown using the default values of $f = 1$ and $g = 0.3$ that Mayne suggested for use with all DFEs.

The results in Figure D.1 well-illustrate the fact noted in Chapter 2 that tangent values of any parameter always decrease faster than secant values of the same parameter.

D.1.3.8.4 Deep-Foundation Applications

In order to decide whether to use a secant- or tangent-modulus approach in an improved resistance-forecasting analytical methodology for tapered piles that integrates load-settlement behavior into a one-step procedure, it is first useful to look at the generic ways this can be done and how these generic approaches have been applied in the past.

The red curve in Figure D.2 qualitatively depicts a generic load-settlement curve for a DFE subjected to axial-compressive loading. Because even the most sophisticated numerical-solution methodology requires a single-valued modulus for any arbitrary i-th stage of applied loading in an analysis, the question then becomes how to best approximate the actual curved relationship in incremental linear increments.

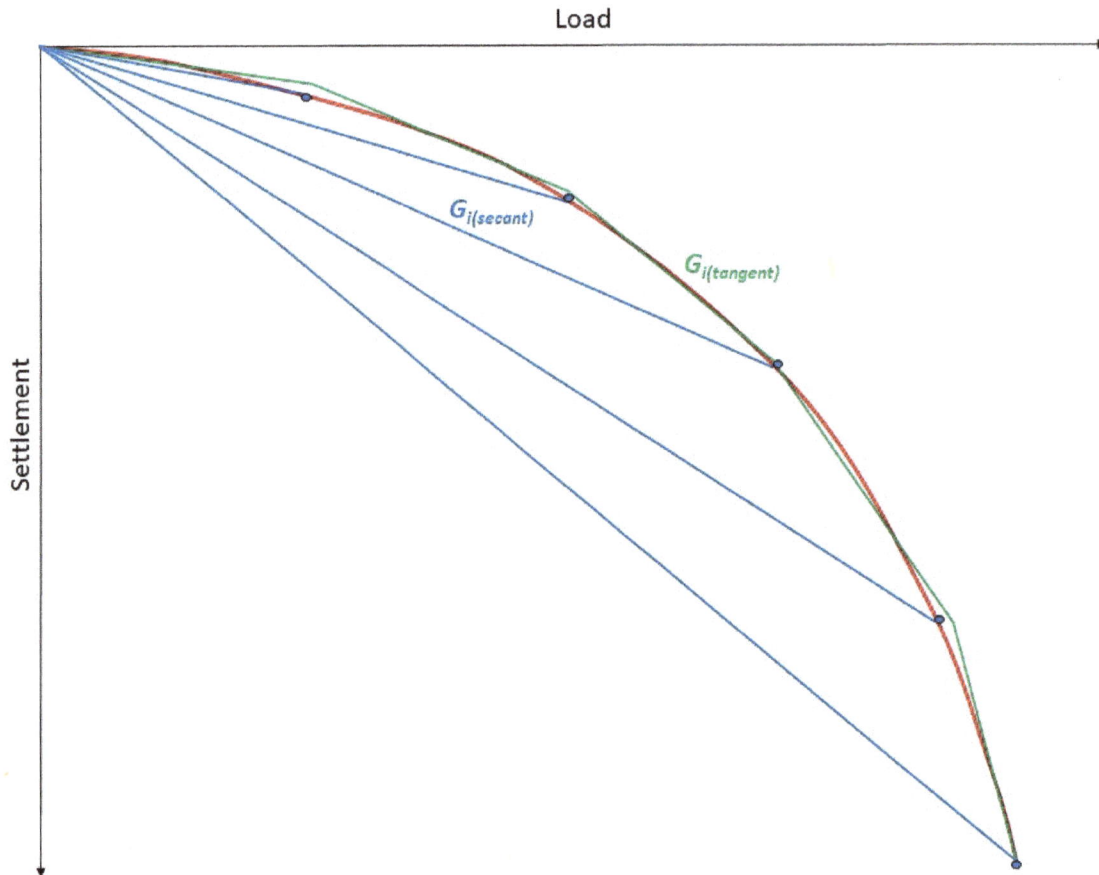

Figure D.2. Generic Application of Secant and Tangent Shear Moduli to DFE Load-Settlement.

The more-sophisticated numerical methods such as the finite-element method (FEM) approximate the actual curved relationship with a series of contiguous linear segments as shown by the green lines in Figure D.2. Use of a tangent shear modulus, $G_{i(tangent)}$, for each i-th analyses step of applied loading is appropriate in this case as that value of the tangent shear modulus represents the average operative value over that increment of applied loading.

Note that using this approach, calculated results are all cumulative and thus inherently interdependent. Also, the smaller the increments, the more closely the accumulated straight-line segments approximate the actual curved relationship.

Relevant to the present discussion, a cumulative, tangent-based approach was used in the original Kodikara-Moore Method for both the shaft- and toe-resistance components. However, Kodikara and Moore did not make full use of the potential of this approach, nor did Manandhar et alia in their relatively modest theoretical 'tweak' of the Kodikara-Moore Method. This is because all of these researchers used the same value of shear modulus for every analytical increment which defeats the purpose and concomitant benefit of performing an incremental analysis in the first place.

The only reason that Kodikara and Moore and Manandhar et alia were able to generate non-linear load-settlement outcomes (one would naturally expect a perfectly linear outcome from using the same modulus in every load step) in their various publications is that tapered-DFE settlement geometrically produces radial displacement of the shaft-ground interface. Using cylindrical-cavity expansion (CCE) theory as these researchers did, this increasing radial displacement produces increasing radial stress along the DFE shaft. This, in turn, produces increasing shaft resistance and increasing total resistance. The net result is the appearance of a non-linear load-settlement behavior as each (assumed identical) load increment encounters more ground resistance than the preceding load increment (of equal magnitude) and thus produces a smaller increment of settlement. Note, however, that if a DFE had no taper, their approach would produce a simplistic elastic-perfectly plastic load-settlement response, not a true non-linear response.

The alternative to the tangent-modulus approach is to use a secant-modulus approach as depicted by the series of blue lines and concomitant datapoints in Figure D.2. Note that this secant-modulus approach is conceptually completely different from the preceding tangent-modulus approach as each i-th calculation step is completely independent of the others (except for the fact that it uses the same equation but with a different modulus value) so the overall net results are neither contiguous nor cumulative. Rather, the net result is a series of standalone (blue) datapoints as opposed to contiguous (green) line segments as results from the tangent-modulus approach. However, if these (blue) datapoints are sufficiently closely spaced then a smooth curve can be fit to the datapoints to create a pseudo-contiguous outcome. Thus, the end result of this secant-modulus approach can mimic the inherent outcomes of a tangent-modulus approach.

Relevant to the present discussion, the analytical methodologies based on the Randolph-Wroth Problem solutions developed by Mayne (using the Fahey-Carter Modulus-Degradation Model) and Niazi (for the Niazi Method, using an empirical modulus-degradation model back-calculated from actual load-test results) are both based on using secant moduli to generate a quasi-contiguous, pseudo-continuous, load-settlement outcome.

While it may seem that both the tangent- and secant-modulus approaches will produce similar results and are thus interchangeable, this is generally not the case as it depends on the overall analytical algorithm that is being solved. For example, in the case of the Randolph-Wroth Problem where there is essentially one solution equation that is being evaluated repeatedly, the results would be expected to be similar. Mayne's choice of a secant-modulus solution approach was appropriate as in this case it is easier to set up and solve using a simple calculation tool such as a spreadsheet.

On the other hand, when solving the Kodikara-Moore Method where there are multiple behavioral phases to the solution, with the progression from one phase to the next dependent on cumulative results, only the tangent-modulus approach is correct to use. This is because the secant-modulus approach, which essentially just lumps everything into a single linear calculation, cannot properly account for the cumulative, non-linear effects of a multi-stage calculation algorithm.

A simpler type of application that comes up routinely in foundation engineering is one that calls for a single-valued problem parameter to calculate a single-valued result. Two variations of this can occur:

- A one-time-only calculation.

- An iterative calculation to converge on a single-valued result.

484

Which variation occurs in a given application depends on the theoretical content built into the solution algorithm for the application. Specifically, in some applications, the final result can depend on the final result as part of the calculation process.

An example of this is traditional shallow-foundation bearing capacity. The writer developed a solution algorithm that depends on the stress-state in the ground at the point of bearing failure. But one component of this assumed stress-state is the gross ultimate bearing pressure which is the final calculated outcome that is being sought. So, a simple iterative calculation process, in this case involving the Mohr-Coulomb friction angle, ϕ, that converges very rapidly is required (Horvath 2000a, 2000b, 2011).

An example of the simpler one-time-only calculation that is relevant to the monograph is calculating the ULS toe resistance based on spherical-cavity expansion (SCE) theory as modified specifically for use with tapered piles by Manandhar and Yasufuku (2011a, 2012, 2013). They used an empirical equation for the required single-valued shear modulus, $G_{operative}$. They assumed this would be the secant-modulus value at a strain (shear strain implied) level equal to 0.001 (= 0.1%). This strain level approximately correlates to a $G_{operative}/G_{max}$ ratio of 0.5 or 50%.

Note that this is a not-uncommon practice in geotechnical engineering, choosing a secant shear or Young's modulus to correlate to a $G_{operative}/G_{max}$ ratio of 50%. Such moduli are often given the notation G_{50} or E_{50} although Manandhar and Yasufuku did not use this notation.

In summary, the nature of the application and, more specifically, the nature of the calculation algorithm used in that application, always dictates whether a tangent- or secant-modulus approach is appropriate to use for that application. In some cases, either modulus could be used and the final choice is up to the end user. However, it is important to note that the choice is never completely arbitrary. Some thought based on an understanding of the underlying calculation algorithm is always required.

D.2 MODULUS-DEGRADATION MODEL ASSESSMENT AND SELECTION

D.2.1 Assessment Criteria

D.2.1.1 Introduction and Overview

Having addressed the choices in modulus-degradation modeling that are available, the next step in the overall process of selecting a shear-modulus degradation model or models to use in an improved resistance-forecasting methodology for tapered piles is to identify the model variables that are most significant and most impact the intended goals. This will provide a rational guideline for assessing the suitability of different models and different forms of a model for use.

Note that the process outlined here is intentionally generic and completely general so that it could be used by others if desired to critique the writer's ultimate choices or to perform their own assessments for other geotechnical and foundation engineering applications. However, it should be understood that the intended application for the purposes of this monograph is an integrated, one-step procedure for forecasting the load-settlement behavior of tapered piles. Consequently, the following discussion is biased with that in mind.

D.2.1.2 Modulus Choice

The first and most obvious consideration is the choice of modulus to use. While the shear modulus has emerged in recent years as the modulus of choice because of the relative ease in which the small-strain shear modulus can be determined in-situ nowadays using a variety of in-situ testing tools, there may be problems where Young's modulus is preferred because of the specific analytical methodology of interest. Therefore, for the sake of completeness, choice of modulus is listed here explicitly although, as stated at the beginning of this appendix, shear modulus was the only serious choice for the purposes of this monograph.

D.2.1.3 Single-Valued vs. Continuous Calculated Outcome

The next consideration is whether the chosen modulus will be used as part of a calculation process to produce a single-valued forecast of some end result such as bearing capacity or to generate a continuous end result such as a load-settlement curve. If the latter, the underlying theoretical elements of the calculation algorithm must be clearly understood to determine if it involves the repetitive calculation of a single equation (as with Mayne's solution of the Randolph-Wroth Problem) or a series of equations (as with the multi-phase Kodikara-Moore Method). This will influence the subsequent decision-making process concerning secant vs. tangent modulus.

Note that a single-valued forecast outcome does not necessarily mean that only one calculation with one value of modulus would be made to obtain it. While that is certainly true for the Basic Manandhar-Yasufuku Toe-Resistance Method noted earlier in this appendix, it is possible that some solution algorithm might involve multiple iterations, each with a different modulus value, to arrive at the final single-valued outcome. This would be identical in concept to the writer's iterative process to arrive at a forecast of the ULS gross bearing capacity using traditional bearing-capacity solutions that was also noted earlier in this appendix. While the writer's algorithm used the friction angle, ϕ, of the soil as the iteration parameter for bearing capacity, there is no reason why a similar iterative solution for bearing capacity could not be developed that includes shear modulus as an iteration parameter as well. This certainly would be the case if a SCE bearing-capacity solution is used as an SCE solution depends on both soil strength and soil stiffness.

D.2.1.4 Secant vs. Tangent Modulus

In the writer's opinion, the single most important consideration in this entire discussion is whether a secant- or tangent-modulus solution approach is used in the solution algorithm. This issue has been discussed at length earlier in this appendix so need not be repeated here. The primary thing to keep in mind in this regard is that the choice in modulus type is never trivial or arbitrary. This choice must be done with due diligence so that it is consistent with the calculation algorithm of the application.

In the writer's opinion, it is unfortunate that so many authors who use either a secant or tangent modulus do not make their choice and justification for it clear in their published work for the benefit of the reader of their work. As noted earlier in this appendix, Fahey and Carter (1993) were unusual in this regard for clearly addressing the differences and illustrating them mathematically for the benefit of the reader.

D.2.1.5 Degradation Correlation Parameter

Modulus is always the dependent variable in any modulus-degradation relationship. However, the independent variable to correlate with modulus has increasingly become a matter of choice so has to be explicitly considered for a given application.

When modulus-degradation relationships were first used, shear strain was always the correlation parameter for shear modulus as was noted earlier in this appendix. While this is still appropriate in many applications, modern alternatives fall into two broad categories:

- <u>Alternative strain definitions</u> that do not necessarily have to be a classic shear- or normal-strain parameter from the theory of elasticity but can be an arbitrarily defined strain. A relevant example is Niazi's definition of strain based on DFE head settlement.

- <u>Stress and/or strength definitions</u> such as used by Fahey and Carter for PMT modeling and as adopted by Mayne for DFE load vs. settlement. Because of the reciprocal relationship between strength and the conventional ASD definition of safety factor, it is also possible to use safety factor as the correlation parameter with modulus degradation.

D.2.1.6 Parameter Normalization and Non-Dimensionalization

Separate from, but related to, the issue of which parameters are correlated in a modulus-degradation relationship is the normalization and, where desirable, non-dimensionalization of one or both of the correlated parameters. It has long been common to simultaneously normalize and non-dimensionalize the operative value of a modulus to its small--strain value. Only more recently has it become common to do the same for the other correlation parameter, whether it is shear strain, some arbitrarily defined strain, or stress/strength/safety factor.

Note, however, that while the use of the small-strain modulus as the normalization parameter for modulus is obvious and unambiguous, the same is not true of the normalization parameter used for the other correlation parameter. When strains are involved, normalization is typically done to some 'reference' strain level that is arbitrary and not necessarily the same from one publication to another. When stress or strength are involved, normalization is typically done to some ULS value that, again, can vary depending on the particular application.

In summary, while parameter normalization and non-dimensionalization in modulus-degradation relationships has become more or less standard in recent years, the specifics of doing so are far from standard. The positive aspect of this is that it allows for customization of the normalization/non-dimensionalization process to optimally match the intended application. The negative is that there is no inherent portability of modulus-degradation relationships from one published work to another. Furthermore, one must careful to understand the assumptions made for any given modulus-degradation relationship so that the relationship is interpreted and used correctly.

D.2.2 Selection

The writer used the information presented in this appendix for developing the schema for an improved resistance-forecasting methodology for tapered piles that is presented in Chapter 7. The specifics related to shear-modulus selection are presented in that chapter.

Subappendix D1

Engineering vs. True Normal Stress and Normal Strain

D1.1 INTRODUCTION AND BACKGROUND

The purpose of this subappendix to Appendix D that deals with modulus-degradation relationships is to define and illustrate the differences between *engineering* and *true* normal stress and normal strain that are the outcomes of using *infinitesimal-strain* and *finite-strain* theories respectively. As stated in Appendix D, the reason for doing so relates to the independent parameter, usually shear strain but possibly an alternative strain definition or stress/strength/safety factor, used to model modulus degradation. As discussed in Appendix D, it is typical for shear strain or an alternative to have a range of five orders of magnitude using the traditional engineering-strain definition. This is a relatively large range and calls into question whether this behavior is better depicted using true strains.

The overall exemplar application used for this theoretical exercise is the *compressible-inclusion function* of *geofoams* which are a type or category of *cellular geosynthetics* that the writer researched extensively beginning in 1987. The compressible-inclusion function is one of two functional ways in which geofoams can reduce lateral loads on earth-retaining structures as well as vertical loads on underground conduits and structural slabs. Lateral-pressure reduction on earth-retaining structures was the subject of an earlier scholarly monograph by the writer (Horvath 2018a), and this subappendix is a revised and expanded version of an appendix in Horvath (2018a).

In the world of geosynthetics, geofoams are relatively unique in terms of both their distinctive 3-D physical nature and concomitant 3-D mechanical and geosynthetic-functional behavior (most other geosynthetics such as geogrids and geotextiles have only nominal thickness and are distinctly 2-D or planer in their mechanical and functional behavior) and diverse multi-functionality (2-D/planar geosynthetics tend to be very limited in their functionality, with most developed specifically for one intended function). Geofoam functions can be classified as either small- or large-strain in nature in reference to the operative strain levels within the geofoam product under service-load conditions.

The compressible-inclusion function is always large-strain in nature. Large-strain material and product behavior under service-load conditions is relatively rare in the world of materials that are used in engineered construction. In most applications, materials such as steel, PCC, wood, and even other types of geosynthetics are designed to have relatively small strains. This is both to prevent material failure as well as for serviceability of the structure in which products made from these materials are used. Consequently, design professionals inherently and naturally tend to think 'small strain' and use analytical methodologies that are suited to this behavior.

The inherent large-strain nature of the compressible-inclusion function raises the question of whether or not the traditional definitions of stress and strain most often used by both geotechnical and structural engineers are appropriate to use when assessing the performance of a geofoam compressible inclusion. This is because these traditional definitions are based on the original, undeformed geometry of the body being loaded and are normally considered adequate only for relatively small-strain applications.

The alternative is to use stress and strain formulations that are based on the deformed geometry of the loaded body. This is broadly similar to the analytical decision structural engineers routinely face as to whether to use a *linear analysis* for some structural system that is based on the undeformed geometry of the system or a *nonlinear analysis* that is based on the deformed geometry of the system. The benefit of the latter is that it considers what is commonly known as the *P-Δ effect* and the additional forces and moments associated with it.

The objective of this subappendix is to explore some basic concepts of evaluating stress and strain using the traditional *small-strain approach* that produces engineering stresses and strains vs. the less-common *large-strain approach* that produces true stresses and strains. In the discussion that follows, only normal stress and normal strain are considered as these are the conditions that predominate when loading a geofoam compressible inclusion. Also, the problem of a small-scale test specimen of simple geometry, e.g. a cube or a right-circular cylinder with a solid cross-section and uniform diameter, subjected to uniaxial loading will be used for simplicity to illustrate the basic concepts presented. However, the discussion is completely general and can easily be extended to any type of body with a complex geometry and loading.

Before proceeding further, it is of interest to note and comment on the genesis of the writer's perceived need for this subappendix. Having spent more than three decades in academic instruction, it is clear to the writer that civil engineering students, in the U.S. at least, get surprisingly little (if any) formal education on the topic of large-strain theory and true stresses and strains, even on the graduate level. While one might argue that this omission is reasonable given the fact that the vast majority of civil engineering designs involve small-strain material behavior, the writer feels that this argument is not persuasive. One needs to look no further than the basic behavior of mild or structural steel in tension that is probably the first topic in engineering materials taught to undergraduate civil engineering students. This forms a lasting impression that a student carries with them for decades of a professional career. Yet, for reasons discussed in detail later in this subappendix, that lasting impression is grossly incorrect due to the way that it is usually presented.

D1.2 OVERVIEW

The writer is only aware of the traditional small-strain formulation for normal stress and normal strain being used to assess geofoam behavior in compressible-inclusion applications, including in the writer's own extensive publication on the subject that goes back to 1990. However, there are a number of arguments that can be made for at least considering using the alternative large-strain formulation, especially for strain:

- The simple, basic fact is that geofoam compressible inclusions are inherently a large-strain problem where engineering compressive strains of the order of tens of percent under service loads are possible. This compares to the well-known lightweight-fill function of geofoam that is small-strain in nature, with engineering compressive strains typically held to 1% or less under service-load conditions.

- Polymeric materials in general are often cited as being a broad class of materials for which large-strain formulations are preferable because of the ability of polymeric materials, as a group, to undergo large strains prior to material rupture, especially in tension.

- Applications in which loads are applied in stages and thus build on each other are also cited as being more-rationally analyzed using large-strain formulations. This would certainly apply to many geofoam compressible-inclusion applications where there would be immediate, permanent compression of the geofoam under permanent, long-term gravity loading. In many cases this could be followed by cyclic applications of a surface surcharge throughout the design life of the structure and/or seismic loading at some random time in the future.

As noted earlier in this subappendix, it is the writer's first-hand experience as both a student and former academician that civil engineers, at least in the U.S., do not receive much, if any, formal education about large-strain formulations of stress and strain. Therefore, this subappendix is intended to be a brief primer and overview of the subject. However, before doing so, some cautions are noted:

- Because of the Web, there is no shortage of websites that can be found that address this subject in varying detail and with varying clarity of presentation.

- There is no consistency of notation for either normal stress or normal strain among the various websites. In particular, the two symbols, ε and e, used for engineering normal strain and true strain are often switched from one website to the other which can be particularly confusing and misleading. Furthermore, the writer has not observed that there is a clear 'best' notation or at least a notation that appears to be used more often. Therefore, the notation used in this subappendix, while arbitrary, is at least consistent and, where possible, is consistent with the notation used throughout this monograph.

- Every such website that the writer has seen shows both the traditional small-strain and alternative large-strain derivations and final results based on the traditional structural engineering sign conventions. This means that tensile forces and concomitant tensile normal stresses are positive, and the elongation strains they produce in a body are also positive. As is well known, the typical geotechnical engineering sign conventions are the opposite of this. The derivations and results shown subsequently in this subappendix reflect the geotechnical engineering sign conventions as they are more relevant to geotechnical engineering in general and the subject of geofoam compressible inclusions in particular[190].

- Every website the writer has seen compares the traditional small-strain and alternative large-strain results only for tension. As the writer has learned through personal experience and confusion, this is misleading. This is because although the traditional small-strain theory produces the same results in tension and compression for the same body subjected to the same load that only differs in sign, this is not true for large-strain theory. The outcomes for large-strain theory are not symmetrical in tension and compression. Therefore, to fully and correctly understand large-strain theory, results for both tension and compression need to be shown.

- Many websites show a simple equation to convert stresses obtained in one formulation to the other, and illustrate the conversion using an example of a body with simple geometry subjected to uniaxial loading. What very few of these websites make clear is that this simple conversion relationship is based on two key assumptions:

---

[190] A detailed presentation and discussion of the different sign conventions used in geotechnical and structural engineering, and the conflicts and confusion that can result when these two specializations in civil engineering come together in foundation engineering, are presented in Horvath (2018b).

1. The body maintains a <u>constant volume</u> as it deforms under load application.
2. The cross-sectional dimensions of the body are the same from one end to the other at any given stage of load application, i.e. there are <u>no end effects</u> from whatever hardware is used for load application.

The former assumption is nowhere near to being correct for geofoam materials where the gas-filled voids of the cellular structure (which may account for as much as 99% of the total volume of the material initially) are literally crushed when loaded in compression, with the result that the geofoam material undergoes significant volume reduction during loading. The latter assumption is not met in many laboratory testing scenarios of small-scale geofoam test specimens as friction between the upper and lower surfaces of the specimen and platens that are part of the testing apparatus tends to cause the middle of a test specimen to deform (either bulge or neck depending on the specific material being tested) differently than ends of the test specimen. Therefore, this simple correlation between small and large strain-based stresses that is typically found on the Web is clearly <u>not</u> applicable to geofoam materials.

Although not noted in Horvath (2018a), Item #1 above is also relevant for soil specimens. Unless the undrained behavior of a saturated soil is being replicated, a test specimen of soil will <u>always</u> undergo volume change during loading. Furthermore, and with reference to Item #2 above, regardless of the drainage conditions, unavoidable end effects will <u>always</u> result in a non-uniform specimen geometry (a barrel shape under the usual compressive loading is quite common) as load application progresses during a test.

In consideration of all the bulleted items presented above, a zero-based presentation and discussion of the theoretical aspects of engineering vs. true normal stresses and strains is required and is presented in the remainder of this subappendix. Because the difference in strain tends to be more significant than the difference in stress in terms of the final outcomes, strain is discussed first.

D1.3 NORMAL STRAIN

As noted previously, civil engineers, at least in the U.S., are educated primarily, if not exclusively, to use what is called engineering, a.k.a. *Cauchy*, strain that is based on small-strain, a.k.a. infinitesimal-strain, theory. Strain formulations based on this approach are based on the dimensional changes of a body relative to the body's original dimensions. For convenience in the current discussion, this body is assumed to be some laboratory test specimen of arbitrary, but simple, geometric shape and dimensions.

Using the geotechnical sign convention of compression positive, *engineering normal strain*, ε, due to uniaxial loading in either compression or tension is defined as:

$$\varepsilon = -\left(\frac{\Delta L_i}{L_o}\right) = \frac{-(L_i - L_o)}{L_o} = \frac{L_o - L_i}{L_o} \tag{D1.1}$$

where:
- L_i = length of the test specimen at an arbitrary point i during loading,
- L_o = original length of the test specimen, and
- $\Delta L_i = (L_i - L_o)$ = change in length of the test specimen at an arbitrary point i during loading relative to its original length. Note that the sign of ΔL_i in this and all subsequent equations presented in this appendix is important, with compression negative and tension positive.

Equation D1.1 produces results in dimensionless decimal form that is the basic, generic standard for strain. It is common in civil engineering to express strains in percent which simply requires the result obtained from Equation D1.1 to be multiplied by 100.

Conceptually, engineering normal strain can be viewed or interpreted as the <u>average</u> or <u>secant</u> strain over some range that begins at the origin, i.e. the zero stress/zero strain state. Consequently, engineering strain is the <u>total</u>, overall strain that occurs within some body due to some applied load. This is considered to be acceptable for use for 'small' strains although what the limiting strain value is for 'small' does not seem to be universally agreed upon. A graphic is presented subsequently to provide some insight into this but for now it is noted that various websites suggest that strain levels as small as 0.05 (5%) are the limit of 'acceptable' accuracy for engineering strains. While this may cover many structural, geotechnical, and foundation engineering applications, this is well below the typical operational strain range of compressible-inclusion applications involving geofoam.

The alternative to engineering strain is most commonly called true strain although alternative terminology exists (*Hencky, logarithmic, natural*). True strain is defined as the <u>incremental</u> strain at some arbitrary point in the loading sequence <u>relative to the specimen length at that point</u> (as opposed to the specimen length at the start of the test) so can be viewed as a <u>tangent</u> strain at some point on a stress-strain curve. Thus, in the same way that a non-linear stress-strain curve can be interpreted to have secant vs. tangent Young's or shear moduli as is discussed in Appendix D, a body subjected to an external load can be interpreted to have secant (engineering) vs. tangent (true) normal strains.

Using the geotechnical sign convention of compression positive, *true normal strain, e,* due to uniaxial loading in either compression or tension is defined as:

$$de = -\left(\frac{dL_i}{L_i}\right) \tag{D1.2}$$

where:
- d = the differentiation operator and
- L_i = length of the test specimen at an arbitrary point i during loading.

The aggregate or total strain that occurs between the start of load application when the test specimen has an initial length of L_o and load-point i is obtained by simply summing the strains that have occurred which, in this case, is mathematically accomplished by integrating both sides of Equation D1.2:

$$e = \int de = \int_{L_o}^{L_i} -\left(\frac{dL_i}{L_i}\right) = -\ln\left(\frac{L_i}{L_o}\right). \tag{D1.3}$$

As with engineering normal strain, sign is important with true normal strain as an indicator of tension or compression. In this case, the sign is dictated by the natural-log function, i.e. whether it returns a positive (tension) or negative (compression) result depending on whether the L_i/L_o ratio in the function argument is greater or less than one respectively. Note, however, that the sign of the resulting true normal strain, *e*, is the opposite of the sign returned by the natural-log function when using the geotechnical sign convention as is the case in this subappendix.

Also, as with engineering normal strain, Equation D1.3 produces a result in the basic, dimensionless decimal format. As with engineering normal strain, converting this result to a percentage is trivial.

The correlation between engineering normal strain, ε, and true normal strain, e, is obtained by expanding the final term in Equation D1.3 using relationships defined in Equation D1.1:

$$e = -\ln\left(\frac{L_i}{L_o}\right) = -\ln\left(\frac{\Delta L_i + L_o}{L_o}\right) = -\ln\left[\left(\frac{\Delta L_i}{L_o}\right) + \left(\frac{L_o}{L_o}\right)\right]$$
$$= -\ln[(-\varepsilon) + 1]$$

(D1.4a)

that simplifies to the desired final result:

$$e = -\ln(1 - \varepsilon).$$

(D1.4b)

Before proceeding further, it is of interest to explore the numerical relationship between engineering and true normal strain to gain insight into:

- when engineering strain is reasonably accurate and

- the generic implications of true normal strain that are likely to be unfamiliar to most civil engineers (a fact to which the writer can personally attest) and counter-intuitive relative to the well-established engineering strains.

Figure D1.1 shows true strain as the dependent variable relative to engineering strain as the independent variable. The engineering strain trendline is also shown so that the difference between the two strain definitions is clear. Note that all strains are expressed as a percent as is typically done in civil engineering and that tensile strains are shown to relatively large values as some polymeric materials can withstand substantial tensile strain before material rupture.

There are several items of note in this figure:

- Unlike the trend in engineering normal strain that is symmetrical about the origin (i.e. compressive and tensile strains are mirror images of each other), true normal strain is asymmetrical with respect to the origin as was noted earlier in this subappendix. This illustrates the point made earlier that websites that only show derivation of tensile true normal strains present a misleading behavioral picture.

- There is only a relatively small range in normal strain centered at the origin over which engineering and true normal strain yield essentially the same results.

- The deviation between engineering and true normal strain increases with increasing strain although the deviation is not symmetrical in either magnitude or 'sense', i.e. which is greater than the other.

- Because true normal strain is calculated based on the instantaneous strain at some point, it tends to be larger than engineering strain in compression and smaller in tension[191]. This is because in compression, true normal strain in calculated relative to an ever-<u>decreasing</u> length dimension whereas in tension the calculation is relative to an ever-<u>increasing</u> length dimension.

---

[191] This corrects a misstatement that appeared in Appendix D of Horvath (2018a).

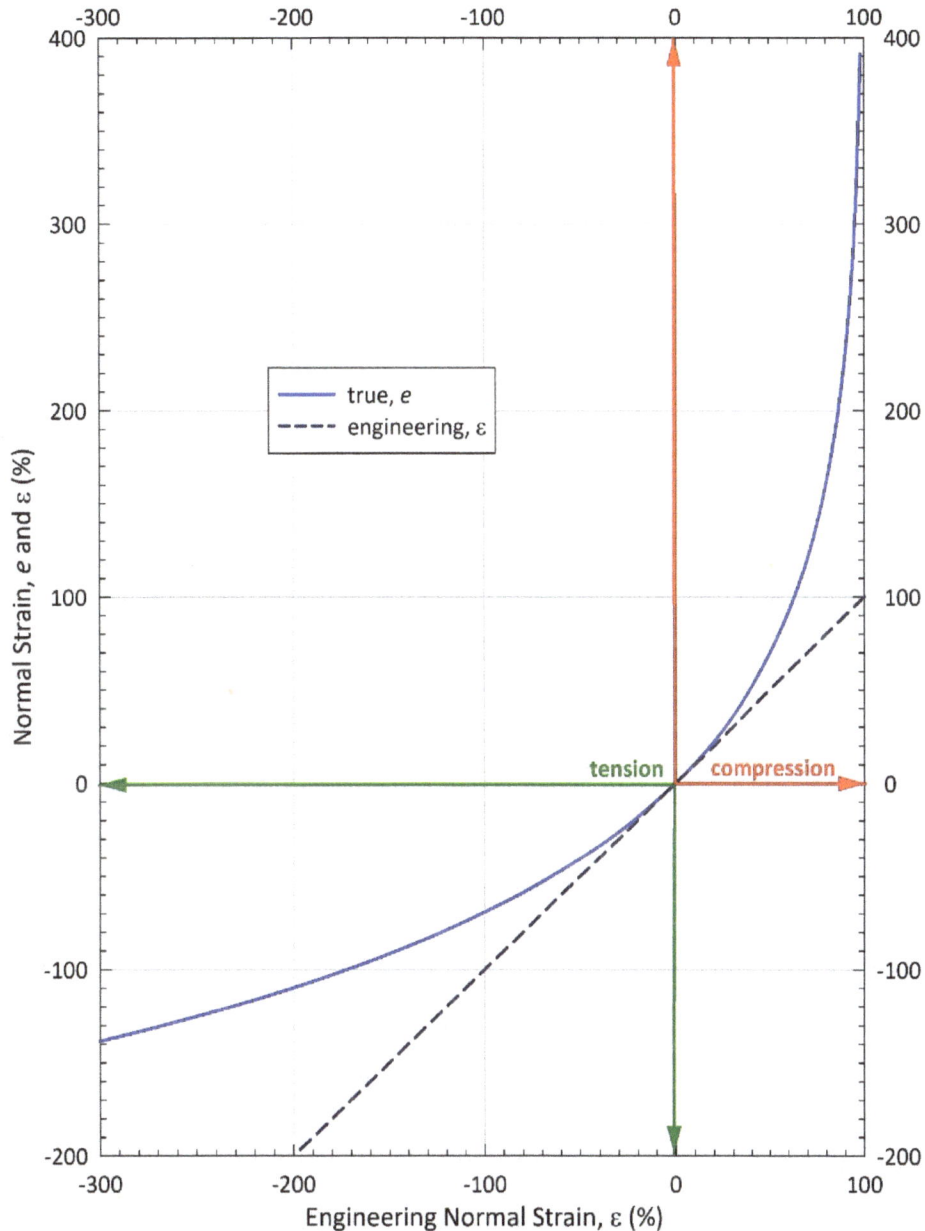

Figure D1.1. Relationship Between Engineering Normal Strain and True Normal Strain (Large Strain Range).

- In particular and for the same reasons, true normal strain in compression is non-intuitive in that it can be greater than 100%. In fact, true normal strain in compression trends toward infinity as engineering normal strain approaches 100%.

Figure D1.2 presents the same results as Figure D1.1 but over a much smaller range in normal strain that would be the maximum encountered in most civil engineering applications, even geofoam compressible inclusions. Also shown in Figure D1.2 using a shaded yellow box is the suggested ±5% strain range for engineering normal strain validity that is found in the published literature as well as on websites that was noted earlier in this subappendix. As can be seen, this strain range is certainly reasonable although somewhat conservative. Depending on the relative error between engineering and true normal strain that is considered to be acceptable, this range can be extended to at least ±10% if not somewhat greater.

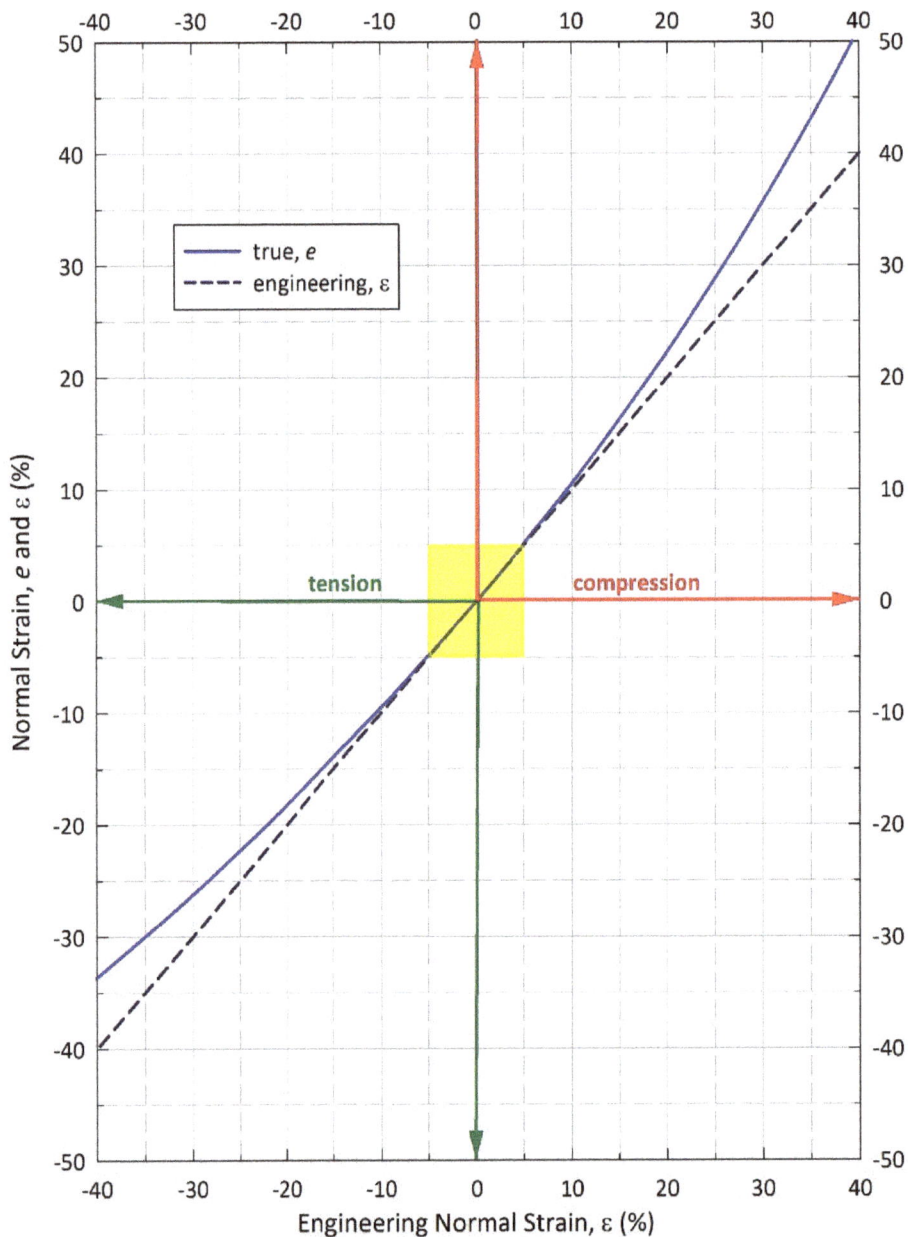

Figure D1.2. Relationship Between Engineering Normal Strain and True Normal Strain (Limited Strain Range).

D1.4 NORMAL STRESS

The difference between *engineering normal stress* and *true normal stress* is broadly similar to that discussed for normal strain in that engineering normal stress is based on the original cross-sectional area of a test specimen whereas true normal stress is based on the cross-sectional area at the specific point in the test for which the stress calculation is made. It is important to note that unless certain simplifying assumptions related to both volumetric changes (none) and dimensional changes (uniform cross-sectional area throughout the test specimen) under loading that were discussed earlier in this subappendix are valid throughout a test, true-stress calculations require on-specimen displacement measurements that would allow accurate calculation of the test-specimen cross-sectional area to be made. This requirement is certainly applicable to geofoam materials.

Nevertheless, it is of interest to view the theoretical relationship between true and engineering normal stresses for the idealized case of no volume change and uniform specimen deformation. This is most easily viewed as the dimensionless ratio of true normal stress to engineering normal stress and is shown in Figure D1.3 for the same extended range of engineering normal strain used in Figure D1.1.

Note that the relationship between the two different definitions of normal stresses is not unique as it was for normal strains (Figure D1.1) as the results are affected by the Poisson ratio, ν, as would be expected. This is because even under the simple case of uniaxial loading with no net volume change, a Poisson ratio other than zero will cause normal strains in the direction transverse to the direction of loading and thus result in changes in cross-sectional area of the test specimen. Consequently, results are shown in Figure D1.3 are for four values of ν that includes both positive and negative values.

As an aside, showing results for a negative Poisson ratio is not just an academic exercise. There are some materials, block-molded expanded polystyrene[192] (EPS-block) being one of them, that exhibit the phenomenon of *necking* at large compressive strains and thus have a negative Poisson ratio at large compressive strain. This behavior is potentially important in real-world applications as EPS-block is the most common geofoam material in the world for approximately 50 years now (Horvath 2018a).

As was done for normal strains, Figure D1.4 shows the same results as Figure D1.3 but over a much more limited range of engineering normal strains. Also shown in Figure D1.4 with yellow shading is the ±5% range in normal strains over which engineering and true normal strains are essentially identical. Note that no such condition exists for normal stresses, i.e. for any strain level other than zero and for any Poisson ratio other than zero, engineering and true normal stresses will always differ noticeably. Whether that difference is considered significant or not is a separate, subjective issue that is left for the individual design professional to decide.

---

[192] *Polystyrene* is called *polystyrol* in some countries such as Japan.

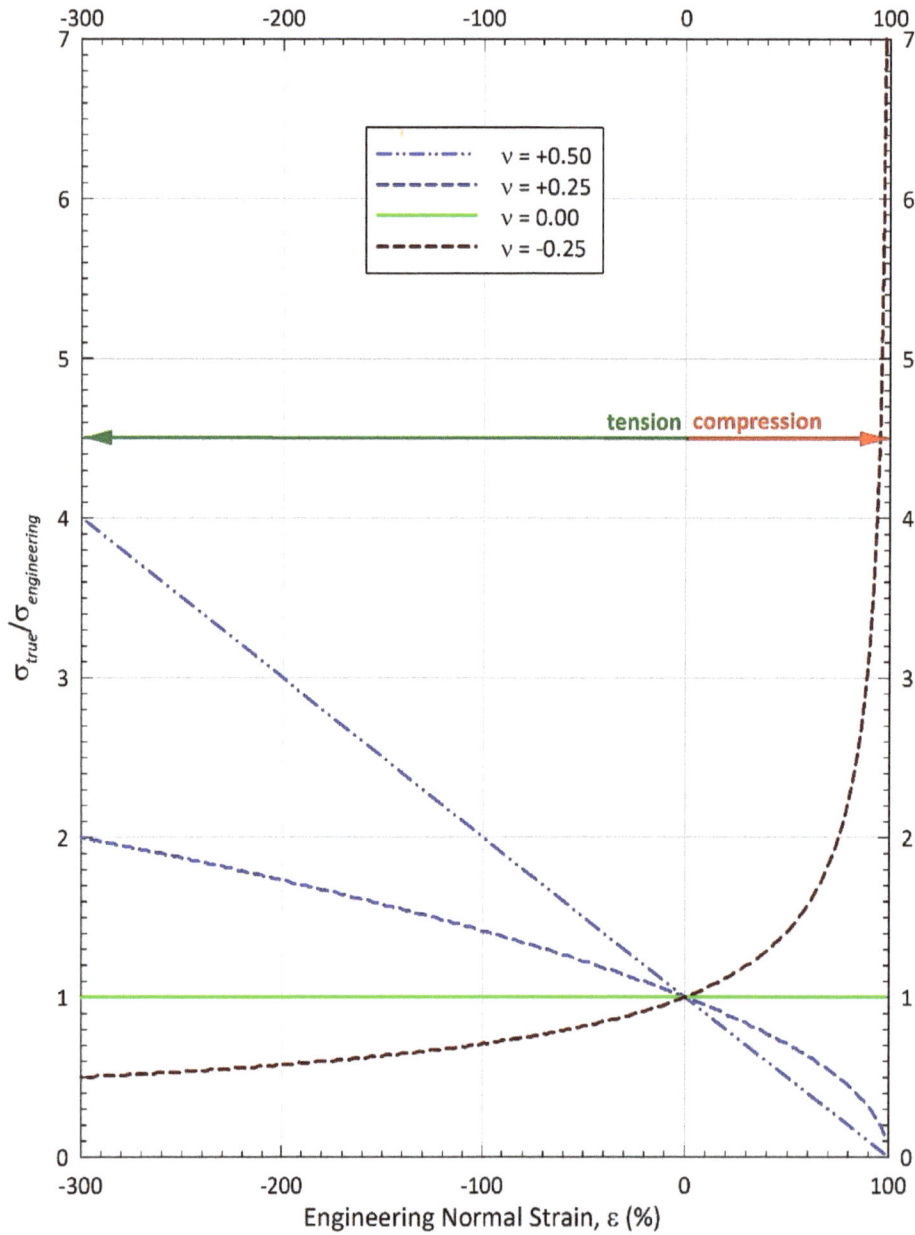

Figure D1.3. Variation between Engineering and True Normal Stress as a Function of Engineering Normal Strain (Large Strain Range).

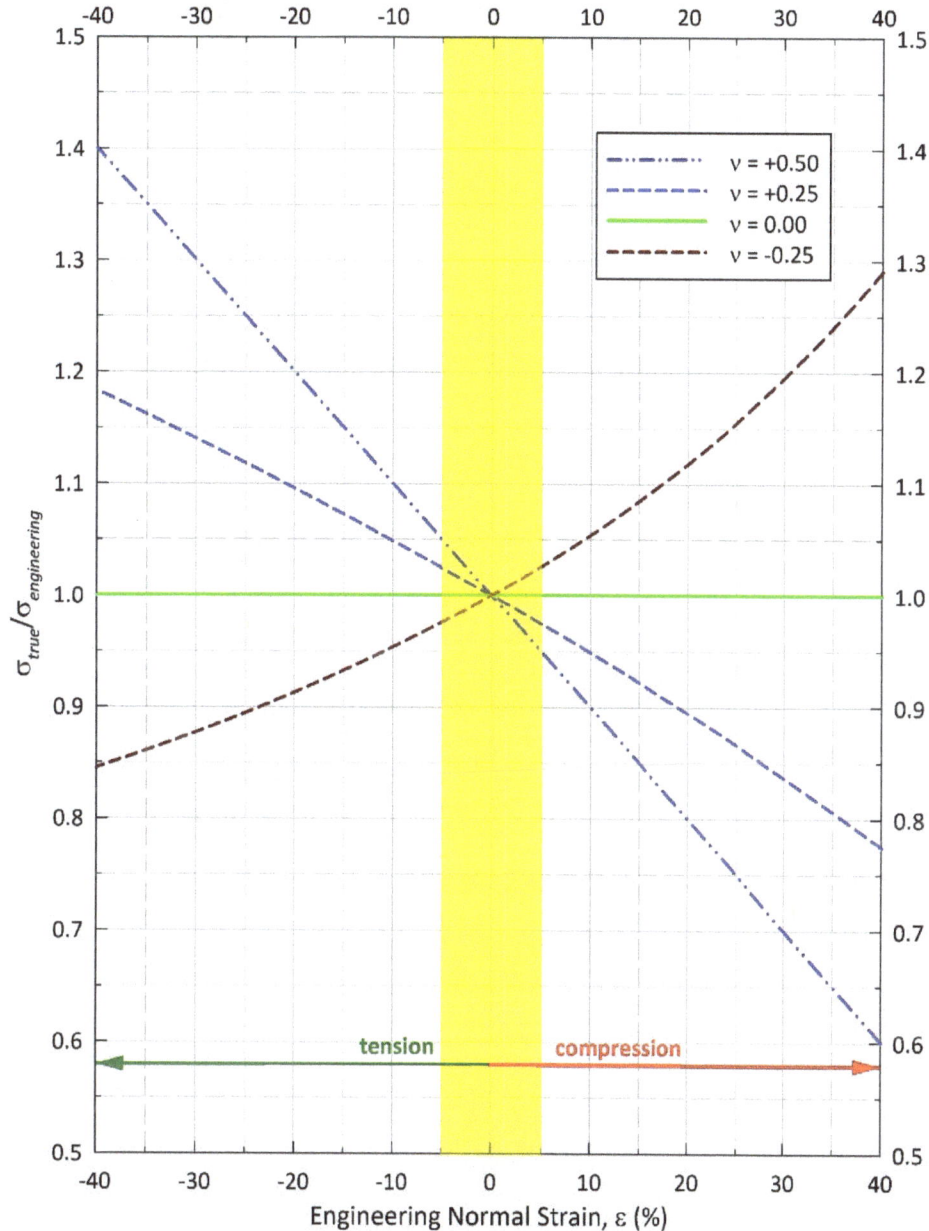

Figure D1.4. Variation between Engineering and True Normal Stress as a Function of Engineering Normal Strain (Limited Strain Range).

D1.5 COMBINED STRESS-STRAIN EFFECTS

D1.5.1 Overview

It is beyond the intent and scope of this subappendix to perform a detailed comparison between stress-strain tests interpreted using engineering vs. true normal stress-strain methodologies. However, it is useful to present some generalities and illustrate them with some simple examples.

To begin with, to summarize and synthesize the key points made in this subappendix up to this point:

- With regard to normal strain, as shown in Figures D1.1 and D1.2, theoretically the two methods agree only at the origin and diverge from there in both compression and tension. True normal strains are larger in compression but smaller in tension compared to engineering normal strains.

- With regard to normal stresses, again, the two methods yield the same results only at the origin and then diverge with increasing strain. However, the relative divergence is opposite that observed for normal strain in both compression and tension. Specifically, in compression, engineering normal stresses are larger in magnitude than true normal stresses provided that the Poisson ratio is non-negative. While this might seem to always be the case, it is not. Some closed-cell geofoam materials such as EPS-block have actually been observed to 'neck' at large <u>compressive</u> strains which means that the Poisson ratio is negative at that point which is certainly counter-intuitive. In such cases, the true normal stresses would therefore be larger than the engineering normal stresses. On the other hand, under tensile loading true normal stresses are always larger than engineering normal stresses, sometimes substantially so due to necking of test specimens.

- The combined outcome of these trends is that, overall, normal stress vs. normal strain behavior will tend to appear stronger/stiffer in tension and weaker/softer in compression when viewed from the perspective of true normal stresses and strains as opposed to the traditional engineering normal stresses and strains. The important exception to this overall trend is that geofoam materials that neck in compression will tend to appear stronger but still softer/less-stiff when viewed using true stress-strain.

There are some general observations to add to this:

- The differences in the interpreted outcomes are both qualitative (conceptual) and quantitative and can be significant. As will be seen subsequently using the classical example of steel, the qualitative differences can often be more important than the quantitative. This is because the former provides broad, visual conceptual depictions of what is happening within a material under some specific type of loading. In general, the de-facto standard engineering stress-strain interpretation can provide surprisingly misleading insight into actual material behavior with increasing strain leading up to and culminating in material failure. On the other hand, the true stress-strain interpretation <u>always</u> provides exactly what its name says, a <u>true</u> behavioral picture.

- To be consistent, only one method or the other should be used to interpret both stress and strain data. This is especially true if the objective is to have a correct interpretation of material behavior both qualitatively and quantitatively. That having been said, there are exceptions to this that have evolved over the years that are noted subsequently. Whether such exceptions are right or wrong is a subjective issue that is not addressed here.

In conclusion up to this point, at relatively small strain levels (again, how small is 'small' is not universally defined or accepted and is actually something that varies with the specific material and loading, compressive or extensive/tensile) the two methods yield essentially the same results. Thus, it is not surprising that engineering normal strain and engineering normal stress, which are far simpler to both teach and grasp than true normal strain and true normal stress, have always and continue to be the backbone of civil engineering education, practice, and research.

That having been said, as illustrated in the following examples, this blind devotion to the engineering interpretation of stress and strain provides highly misleading qualitative, conceptual insight into general material behavior in some cases. Such misleading insights can be detrimental in the education process of civil engineers as first impressions are often lifelong impressions (the writer says this from personal experience). Therefore, in the writer's opinion, a strong argument can be made for introducing the concept of true normal stress and true normal strain even at the elementary undergraduate level so that the correct impressions of material behavior are learned from the start of the professional-engineering educational process.

D1.5.2 Examples

D1.5.2.1 Steel

How a laboratory test specimen of generic *mild*, a.k.a. *plain-carbon* or *low-carbon*, *steel* behaves in uniaxial extension (tension) is the most common example used in websites to illustrate the qualitative difference in the same test results interpreted using engineering vs. true normal stress and strain. The stress-strain curve seen by every civil engineering student that is based on the engineering definitions of normal stress and normal strain shows a relatively complex post-yield behavior in the plastic[193] range of work-hardening followed by strain-softening to the point of physical rupture. This gives the clear impression that steel:

- Possesses two yield points, one at the transition from elastic to plastic behavior followed by a second within the plastic zone where behavior transitions from work-hardening to strain-softening.

- Continues to gain strength after the first yield point.

- For some reason, <u>loses</u> strength at relatively large strains beyond the second yield point and continues to lose strength until it ruptures.

The reality, of course, is completely the opposite and can only be seen using the true definitions of both normal stress and normal strain. Steel actually has only one yield point where it transitions from elastic to plastic behavior. Furthermore, the post-yield, plastic region is all work-hardening in nature although the slope of the stress-strain curve within the plastic region is variable up to the point of rupture due to the development of significant necking in the test specimen just prior to rupture. Note that this necking causes the ratio of $\sigma_{true}/\sigma_{engineering}$ to be even greater than that implied in Figure D1.4 (the $\nu = +0.25$ curve is reasonably close for steel) as the assumption implied in this figure is that the cross-sectional area of the test specimen is uniform from one end to the other at all stages of loading. Thus, not only does steel <u>not</u> lose strength prior to rupture as it transitions through is plastic behavior, the actual rupture strength is substantially <u>greater</u> than that interpreted using the conventional engineering interpretation of normal stress and strain.

---

[193] 'Plastic' is used here in its historical engineering mechanics definition of inelastic or non-recoverable strain.

This significant difference in implied material behavior between the engineering and true definitions of normal stress and normal strain perfectly illustrates the point made earlier in this subappendix that there can be significant <u>qualitative</u> differences of a conceptual nature between the two definition, not just <u>quantitative</u> as illustrated in some of the figures presented in this subappendix.

D1.5.2.2 EPS Geofoam in Compressible-Inclusion Applications

Focusing now on how the difference between engineering and true normal stress-strain affect geofoam materials used in compressible-inclusion applications, it is useful to first review the generic behavior of both 'normal' EPS-block and *resilient block-molded EPS*[194] (R-EPS) when laboratory test specimens (typically 2-inch (50 mm) cubes) are subjected to uniaxial compression and then interpreted and plotted using the conventional engineering interpretation of both normal stress and normal strain. Note that R-EPS is the generic geofoam material of choice for most compressible-inclusion applications.

The comparison between these two geofoam materials is shown in Figure D1.5 that was originally published in Horvath (2010b) and subsequently republished in Horvath (2018a). Stresses and strains shown in this figure are based on the traditional engineering definition.

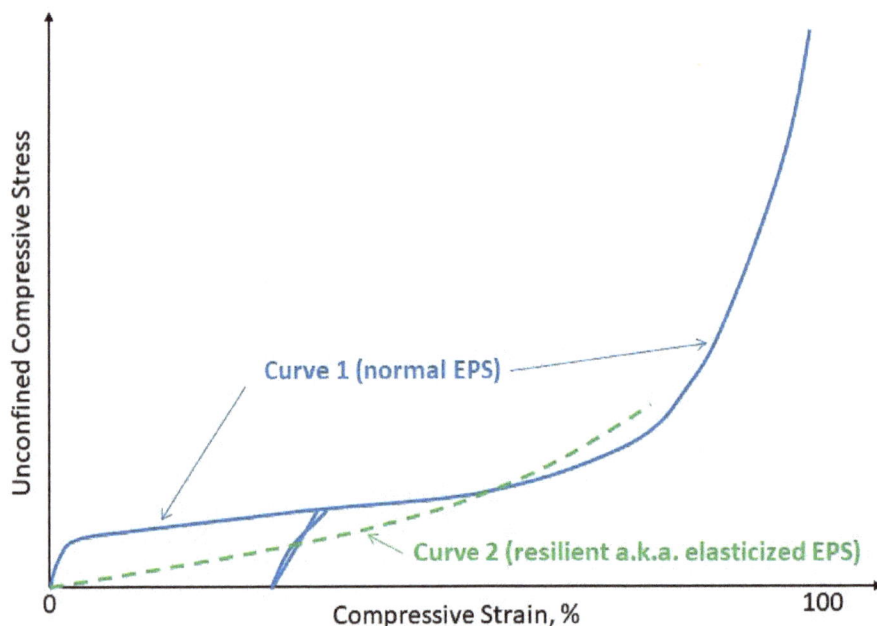

Figure D1.5. Generic Engineering Stress-Strain Behavior of Compressible-Inclusion Materials [Horvath (2010b) as republished in Horvath (2018a)].

---

[194] Resilient block-molded EPS was called *elasticized* EPS when originally developed in Western Europe in the late 20th century. The term 'resilient' was later introduced into practice by the writer and is now preferred as it more accurately describes the behavior of this material. It is normal EPS-block that has been subjected to an additional manufacturing step of post-molding processing to a) make it substantially more compressible under relatively low normal stresses and b) significantly extend the strain range of essentially linear-elastic behavior (Horvath 1995). The latter aspect is what makes R-EPS particularly attractive for applications involving cyclic loading.

To begin with, if normal strains were recalculated using the true-strain definition and plotted using the same axis scales shown in Figure D1.5, both material curves would shift significantly to the right, with a true normal strain approaching 400%, not 100% as in this figure, as this is the asymptotic limit approached by Curve 1 for normal EPS-block[195]. Furthermore, the compressive strain level at which EPS-block begins to stiffen appreciably in Figure D1.5, which is typically an engineering compressive strain of approximately 70%, would be approximately 120% true compressive strain.

The effect on normal stresses is less certain. Both EPS-block and R-EPS have very small positive values of the Poisson ratio under relatively low strain levels, so small that in routine practice a value of zero is commonly assumed (Horvath 1995; Stark et al. 2004a, 2004b). Consequently, per the results shown in Figure D1.3, the true compressive stresses would not be expected to differ much from the engineering compressive stresses depicted in Figure D1.5 (labeled "Unconfined compressive stress" in the figure), at least under relatively small strain levels. At higher stress levels, some specimen necking might develop which would tend to make the true stress levels <u>larger</u> in magnitude than the stresses shown (see the $\nu = -0.25$ curve in Figure D1.3 for a sense of this). Unfortunately, the necking of both EPS-block and R-EPS have only been incidentally observed and qualitatively noted during laboratory testing, not specifically studied and quantified by the necessary on-specimen measurements. Consequently, the extent to which the normal stresses shown in Figure D1.5 would be affected cannot be stated with certainty at this time. However, if necking occurred then the compressive stresses would be somewhat larger than those shown in Figure D1.5 at higher strain levels.

In any event, it is likely that the overall true stress-strain curves would have the same basic shape as the curves shown in Figure D1.5 so the dramatic qualitative difference noted and discussed previously for steel under uniaxial tensile loading would not occur. However, the slopes of the curves for both EPS-block and R-EPS interpreted using true normal stress and true normal strain would likely be somewhat flatter than those with the conventional engineering stress-strain interpretation. This would mean that the calculated Young's modulus would be somewhat smaller and the overall apparent stiffness of a commercial product consisting of either material would be somewhat smaller as well. Furthermore, the nominally elastic range of R-EPS in which its signature resilient behavior occurs would be significantly greater when interpreted using true stresses and strains. Collectively, these two behaviors are potentially significant as all compressible-inclusion functional applications are analytically stiffness-based and thus require an accurate estimate of the stiffness of the geofoam product used under design loading.

D1.5.2.3 Hybrid Applications in Geotechnical Engineering

As noted earlier in this subappendix, there are cases where engineering vs. true compressive stress-strain behavior is not implemented in a consistent manner. The example with which the writer, and, most likely, most readers of this monograph, is most familiar involves geotechnical engineering and the triaxial testing of soil specimens.

For decades, when reducing triaxial test data true compressive stresses are calculated. Specifically, it is the rule, not the exception, in the writer's personal experience for test data to be reduced assuming that the test specimen deforms either as a perfect right-circular cylinder or, more commonly, in some 'barrel' shape that is usually defined by some assumed curve, such as a parabola, that has its largest diameter at mid-height of the test

---

[195] This statement corrects a misstatement that appeared in Appendix D of Horvath (2018a).

specimen. However, these empirical corrections and concomitant estimates of true normal stresses are then plotted against engineering normal strains that results in a conceptually inconsistent hybrid plot.

This, then, raises the question as to whether true normal strains should be used instead. It would certainly result in plots that were more-rigorously correct and perhaps more insightful in terms of true material behavior. However, realistically, the difference is neither large (with reference to Figure D1.2, most triaxial tests are run to a maximum engineering normal strain of approximately 20%) nor significant.

Thus, in conclusion, while the differences between engineering and true normal stresses are always of some significance (see Figure D1.4) thus justifying the specimen-area correction in triaxial tests, the differences between engineering and true normal strains over the same range are of much less significance (see Figure D1.2) so can reasonably be ignored.

D1.6 CONCLUSIONS

Although the issue of engineering vs. true normal stress and strain has only been covered in a very introductory and preliminary manner in this subappendix, available information suggests that it is a topic worth pursuing further for the compressible-inclusion function of geofoams. The use of true compressive normal stresses and strains may provide better insight into the behavior of compressible inclusions that are usually subjected to relatively complex loading such as:

- gravity loading under staged construction;

- gravity loading followed by application of a surface surcharge, especially if it is cyclic in nature; and

- gravity loading followed by cyclic seismic loading.

This is because the behavior of compressible-inclusion applications centers around material stiffness and, as noted earlier in this subappendix, the operational stiffness of geofoam materials at relatively large strains is potentially significantly different depending on whether it is calculated using engineering vs. true stresses and strains.

Looking at the broader picture, there are other geotechnical and foundation engineering applications that might benefit from a zero-based reassessment into the potential use of true normal stresses and strains. This includes applications where relatively large strains of polymeric materials might occur as well as applications were staged loading, e.g. permanent gravity loading followed by cyclic loading from various sources, is involved. This is because the behavior under cyclic live loading is not independent but 'builds on' the stresses and strains that have already occurred under gravity loading.

Subappendix D2

Shear-Strain Fundamentals

D2.1 INTRODUCTION AND OVERVIEW

The goal of this subappendix is to illustrate two issues related to shear strain:

- How engineering shear strain and tensorial shear strain are related qualitatively but differ quantitatively.

- The small-strain, a.k.a. infinitesimal-strain, assumptions that are made to arrive at the strain-displacement relationships for shear strain with which civil engineers are most familiar.

It is important to note that this subappendix is intended to be just a brief primer on these topics so that assumptions inherent in shear-modulus degradation and the analytical equations used to model this behavior are clearly known and understood.

D2.2 THEORY

D2.2.1 Free-Body Diagram of Displacements

Figure D2.1 shows a generic 2-D free-body diagram of displacements that is the most general case for defining shear strains in that shear distortion occurs in both the x- and y-axis directions. The original object of dimensions dx by dy is shown with green dashed lines. The distorted object (exaggerated for clarity and labeling purposes) is shown with dark-red solid lines along with the incremental displacements (elongations in this case) of each pair of parallel sides. The angular rotations, α and β, of the sides relative to the original position of the sides are also shown[196].

D2.2.2 Engineering vs. Tensorial Shear Strain

The qualitative concepts of normal strain (the focus of Subappendix D1) and shear strain (the focus of this subappendix) are universal in terms of the nature of the distortion or deformation to some body or object that each defines. However, the notation used to define strain (and stress for that matter) is not universal in that there are several different notational systems that are used. Only two of these systems, *engineering notation* and *tensor notation*, are discussed in this subappendix as they are the ones most often used in civil engineering applications. In addition, they are the ones that have different definitions of shear strain and thus the potential for conflicting analytical outcomes for the same problem.

---

[196] Note that the α and β parameter notations used in Figure D2.1 and in subsequent equations in this subappendix are unrelated to any other use of these parameter notations elsewhere in this monograph.

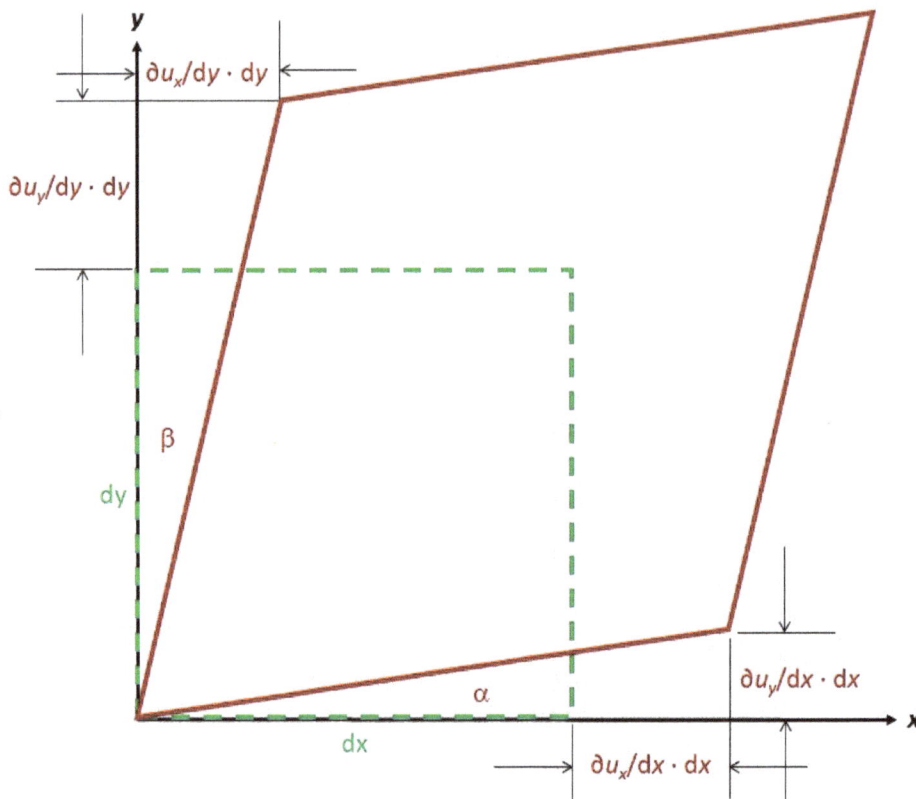

Figure D2.1. Free-Body Diagram of Displacements for Shear-Strain Definition Purposes.

There are several ways in which the engineering and tensor notational systems differ. For example, engineering notation is always letter-based so that the usual subscripts x, y, and z are essential for indicating directions of coordinate axes, strains, and stresses. On the other hand, tensor notation is an indicial system so numerical subscripts $(1, 2, 3)$ are essential for directional indication whereas letters are relatively less important and, in some cases, have no inherent meaning in and of themselves. For example, in tensor notation the coordinate axes would be referred to not as the usual x-y-z but as a_i where $i = 1, 2, 3$ where the letter a could be replaced by any other letter.

Other notational differences include the way that strains and stresses are defined. In tensor notation, the notation ε is used for <u>all</u> strain components, normal and shear, whereas engineering notation uses separate notation, ε and γ, for normal and shear strains respectively[197]. Likewise, tensor notation uses σ for all stress components vs. the familiar σ and τ used in engineering notation. Therefore, in tensor notation ε_{11} would be a normal strain whereas ε_{12} would be a shear strain. Similarly, σ_{11} would be a normal stress whereas σ_{12} would be a shear stress.

However, the most significant difference, and the one most likely to cause confusion in practice, especially when using commercial software (some software packages use

---

[197] Some reference material found on websites states that engineering notation sometimes uses the notation ε for both normal and shear strains, just like tensor notation, with the only difference being that engineering notation uses letter (x, y, z) subscripts whereas tensor notation uses numerical $(1, 2, 3)$ subscripts. This may be true but the writer has never seen this done in personal experience dating back to the late 1960s in U.S. education, practice, or research.

engineering notation whereas other packages use tensorial notation), is how the two notational systems define shear strains. With reference to Figure D2.1, the engineering interpretation of the shear strain experienced by the deformed object is the <u>total</u> angular distortion of the body, i.e. $\alpha + \beta$. Thus, the shear strain, γ_{xy}, can be fundamentally defined as:

$$\gamma_{xy} = \alpha + \beta. \tag{D2.1}$$

However, tensor notation treats the two angular contributions as two separate components of shear strain, e.g. ε_{12} and ε_{21}. Thus, in tensor notation, Equation D2.1 is replaced by:

$$\varepsilon_{12} + \varepsilon_{21} = \alpha + \beta. \tag{D2.2}$$

The way that the difference between Equations D2.1 and D2.2 is often stated is that engineering shear strain is twice the corresponding tensorial shear strain or that tensorial shear strain is one-half of the corresponding engineering shear strain.

D2.2.3 Small-Deformation Derivation of Shear Strain in Engineering Notation

As defined in the preceding section, regardless of the notational system used shear strain is fundamentally an angular distortion in radians. However, while angular distortion is intuitive and easy to visualize, it can be difficult to measure so an alternative definition based on displacements and trigonometry is more commonly used.

With reference to Figure D2.1, this alternative definition is based on the trigonometric tangents to the two components of angular distortion, α and β:

$$\tan \alpha = \frac{\frac{\partial u_y}{\partial x} dx}{dx + \frac{\partial u_x}{\partial x} dx} = \frac{\frac{\partial u_y}{\partial x}}{1 + \frac{\partial u_x}{\partial x}} \tag{D2.3a}$$

$$\tan \beta = \frac{\frac{\partial u_x}{\partial y} dy}{dy + \frac{\partial u_y}{\partial y} dy} = \frac{\frac{\partial u_x}{\partial y}}{1 + \frac{\partial u_y}{\partial y}}. \tag{D2.3b}$$

At this point, small-strain theory is invoked to simplify Equations D2.3a and D2.3b. Specifically, it is assumed that:

$$\frac{\partial u_x}{\partial x} \ll 1 \tag{D2.4a}$$

$$\frac{\partial u_y}{\partial y} \ll 1 \tag{D2.4b}$$

in which case Equations D2.3a and D2.3b simplify to:

$$\tan \alpha = \frac{\partial u_y}{\partial x} \tag{D2.5a}$$

$$\tan \beta = \frac{\partial u_x}{\partial y}.$$ (D2.5b)

The next small-strain-theory assumption is that the two angular distortions are each sufficiently small in magnitude so that Equations D2.5a and D2.5b simplify further to:

$$\tan \alpha = \frac{\sin \alpha}{\cos \alpha} \cong \frac{\alpha}{1} = \alpha \cong \frac{\partial u_y}{\partial x}$$ (D2.6a)

$$\tan \beta = \frac{\sin \beta}{\cos \beta} \cong \frac{\beta}{1} = \beta \cong \frac{\partial u_x}{\partial y}.$$ (D2.6b)

The desired final result of replacing angular changes with changes in linear dimensions is obtained by substituting the final results of Equations D2.6a and D2.6b into Equation D2.1:

$$\gamma_{xy} = \alpha + \beta = \frac{\partial u_y}{\partial x} + \frac{\partial u_x}{\partial y}.$$ (D2.7)

The relevant point being made here is that the usual definition of shear strain (Equation D2.7) that civil engineers are familiar with and assume is 'correct' is, theoretically, not exact as it is a small-strain <u>approximation</u> that relies on the validity of two assumptions as stated in Equations D2.4a and D2.4b and Equations D2.6a and D2.6b. The writer is not aware of any published work that explores the range of validity of the small-strain formulation of shear strain that is reflected in Equation D2.7 although such a comparison has, presumably, been done by someone at some point.

Appendix E

New Structural Model for Deep-Foundation Elements with Composite (Multicomponent) Cross-Sections Under Axial Load

---

E.1 INTRODUCTION AND ACKNOWLEDGMENT

The basic elements of this appendix were originally developed by the writer in 2009 and published in Horvath (2010a). They are presented here in an expanded and updated form in view of their overall relevance to the theme of this monograph.

The author acknowledges and thanks Mr. Regis J. Colasanti, M.S., P.E., M.ASCE for performing the *ANSYS* analyses that were used to validate both the basic theory-of-linear-elasticity closed-form solution that was the theoretical basis for the writer's 3-D analytical model presented originally in Horvath (2010a) as well as for providing critical peer review and discussion of the author's theoretical work. At the time the writer's original research was conducted circa 2009, Mr. Colasanti was employed by URS Washington Division in Denver, CO, a company that has since been acquired by AECOM. The writer acknowledges with the thanks the computer hardware and software resources that URS Washington Division made available to Mr. Colasanti for performance of the work presented in Horvath (2010a) and republished herein.

E.2 BACKGROUND

The writer's original interest in, and ultimate motivation for, conducting the research presented in Horvath (2010a) and this appendix evolved in the early years of the 21st century from a series of discussions with a professional colleague who was actively involved with micropiles through a U.S-based industry organization with international membership and one of its technical committees. At that time, a 'hot topic' tasked to this committee was whether or not the permanent casing of micropiles provided radial confinement to the Portland-cement grout (PCG) core. The significance of the answer to this question is that confinement could justify the use in routine practice of compressive strengths for the PCG core greater than those obtained from the usual unconfined-compression test. There were obvious economic consequences to this in that a given micropile could support a larger load or a smaller micropile could be used to support a given load.

The genesis of this PCG-core-confinement question is based on the well-known fact that the compressive strength of Portland-cement concrete (PCC) increases with increasing confinement. However, this is rarely taken into account in routine practice so traditional PCC design has always been, and still is, based on the unconfined compressive strength, f'_c. This is true in general and true specifically with traditional steel shell or pipe piles that are filled with fluid Portland-cement concrete (PCC) after driving. Thus, this includes many of the tapered piles, past and present, that are discussed in Chapter 3.

In any event, the rapidly growing use of micropiles in the U.S. that began toward the end of the 20th century provided the heavy-construction industry in the U.S. with impetus and incentive to take a closer look at how the various micropile components (casing, PCG core, and reinforcing bar (rebar)) interact with each other structurally. One deliverable of this industry-funded research was a technical report (Fuller et al. 2003), a copy of which was provided to the writer by the aforementioned professional colleague.

E.3 OVERVIEW

As might be imagined, there was, apparently, no shortage of professional opinion on the issue of PCG confinement in the years leading up to the Fuller et alia report. The writer had no vested interest in the matter and thus was not involved in any of these discussions.

However, upon learning about the discourse on the subject, it reminded the writer of similar passionate and often-heated discussions, both in person at conferences and in published works, years earlier during the early development soil nailing. In that case, the issue was whether the passive (i.e. initially unstressed) steel nails deformed in tension or flexure where they crossed a hypothetical 'failure plane' in some 2-D analysis. The behavioral assumption influenced how stresses in the nails were to be calculated and was thus a significant design issue.

Recalling that objective research ultimately answered the question for soil nailing, the writer was of the opinion that the best way to resolve the issue of micropile PCG confinement was to perform objective research to model the problem structurally and simply examine the outcomes. To the writer at least, this seemed to be a more fruitful approach as opposed to engaging in continued oral and verbal discourse ad infinitum.

The use of DFEs with a composite cross-section consisting of two or more different materials is certainly not new. As discussed in Chapter 3, this dates back to at least the latter part of the 19th century when PCC began to emerge as an alternative material to timber in piles. However, historically the interest in composite cross-sections was only in apportioning axial loads and concomitant normal stresses between or among the different material components. The issue of radial stresses was not of any practical value or, consequently, interest. As a result, a simple 1-D model that is discussed in detail later in this appendix has long been used for forecasting axial normal stresses in composite piles.

Because this simple 1-D model does not produce estimates of radial stresses in a composite DFE, it was apparent to the writer that a true 3-D model was required. Of course, there are any number of commercially available programs such as the aforementioned *ANSYS* that can be used for this purpose. However, the writer felt that it was desirable to develop an analytical methodology that was theoretically rigorous yet easier to implement in practice without resorting to a finite-element analysis (FEA) of a continuum. The ability to solve such a simpler 3-D model with generic spreadsheet software such as *Excel* was also judged to be highly desirable (see the discussion of the 'solution by spreadsheet' concept in Chapter 7).

The remainder of this appendix is devoted to, first, a review of the traditional 1-D analytical model used with composite DFEs as this model is used for comparative purposes subsequently. This is followed by:

- A detailed outline of development of the writer's 3-D analytical model.

- The validation process used to verify the accuracy of the outcomes from this model.

- The results of exemplar analyses showing application to micropiles as well as a traditional PCC-filled steel pipe pile as would be representative of *Monotube-* and *TAPERTUBE*-brand tapered piles.

E.4 TRADITIONAL ONE-DIMENSIONAL (1-D) ANALYTICAL MODEL

E.4.1 Concept

The structural analysis or design of any type of DFE with a composite cross-section that is subjected to applied axial loads at the head requires that the axial stresses caused by the applied axial force be apportioned to the components comprising the cross-section. Historically, this stress allocation has been done using a simple, approximate 1-D analytical model that can be visualized as a system of independent axial springs oriented and acting in parallel with the longitudinal axis of the DFE. Each material component of the cross-section is assumed to be represented by its own spring. The springs are usually assumed to be linear in behavior so that the equivalent spring stiffness of each material is simply AE per unit length, L, of pile where A and E are the cross-sectional area and Young's modulus respectively of each material. Note that where appropriate for cementitious components such as PCC or PCG, the time-dependency of, and other factors influencing, E should be taken into account (York et al. 1974, Fellenius et al. 2000).

To develop a closed-form solution for this model, it is further assumed that the axial strains of the different materials comprising the cross-section, i.e. displacements of the different springs, are the same. This is a reasonable assumption in that it reflects reality in practice. It is then straightforward to develop a simple algebraic equation to apportion an applied force and calculate the respective axial normal stresses within each material in the cross-section as, in simple terms, this is the basic problem from elementary physics of two or more springs acting in parallel.

Relevant to the contents of this appendix, it is worth noting that this 1-D model is conceptually and mathematically identical to assuming that each material in the cross-section exhibits linear-elastic behavior under uniaxial stress conditions. This means that the axial stress in a material is dependent solely on the axial strain and Young's modulus of that component.

E.4.2 Commentary

There is are subtle implications with this 1-D model that, in the writer's experience, are rarely appreciated and considered in practical applications of this model. First and foremost is the fact that it is typically assumed that all the cross-sectional components are initially stress-free and thus develop axial-normal stresses in consonance as an axial load is applied to the head of the DFE. As discussed elsewhere in this monograph, this is, in general, never the case for most types of DFEs due to residual loads from installation. Relevant to the theme of this monograph is that tapered piles tend to have the most significant residual loads of any type of DFE. This means that the assumption of initially stress-free components of a composite tapered-pile cross-section will generally be incorrect and potentially seriously so.

Second, the 1-D model and its results are theoretically equivalent to assuming that the different materials comprising the cross-section of the DFE do not interact with each other than having a common axial strain. Stated another way, the Poisson Effect on:

- radial and tangential normal stresses,

- radial displacements, and

- radial interaction between and among the components of the DFE are neglected under an applied external axial load even though the different system components are in intimate contact with each other, at least after initial construction and prior to external load application. Furthermore, because the Poisson Effect is completely neglected means that any radial interaction with the surrounding ground, either in terms of cause or effect, is also neglected.

In view of these two issues, it should be recognized that this traditional 1-D model is <u>never</u> rigorously correct in practical applications and is, at best, an <u>approximation</u> of reality due to the simple fact that initially stress-free, purely uniaxial stress conditions never exist in any composite DFE. That having been said, it should also be recognized that all indications are that this 1-D model has produced acceptable results in routine practice for well over a century now. There is no known structural failure of a DFE that can be directly attributed to the use of this model although this is a tenuous defense, at best, of the status quo, especially in view of the significant impact of residual loads as illustrated in Chapter 7.

E.5 THREE-DIMENSIONAL (3-D) ANALYTICAL MODEL

E.5.1 Introduction and Overview

Although the traditional 1-D model might be considered adequate for use in practice for its intended purpose of estimating axial stresses in a composite cross-section, it is inherently inadequate for the purpose of evaluating confinement of the inner core of a composite cross-section. This is because the assumptions made during theoretical development of the 1-D model neglect all radial and tangential stress-strain behavior.

With this mind, it is of interest to investigate the issue of 3-D effects between and among not only the components of composite DFEs but with the adjacent ground as well in a manner that is much more rigorous than considered to date. The scope of this appendix is to report on a 3-D analytical model for this purpose that was developed by the writer circa 2009. This model is theoretically rigorous yet intentionally easy to implement in practice using nothing more than commercially available spreadsheet software.

The analytical work to date involving this new 3-D model has been limited to no more than three components. However, the basic theory and methodology used for this new model are completely general so there is no conceptual reason why the number of components could not be increased without limit beyond three. Directions for how to accomplish this are also given in this appendix.

E.5.2 Concept and Theoretical Basis

The 3-D model presented in this appendix is based on a theory-of-linear-elasticity boundary-value problem and concomitant closed-form solution published in Poulos and Davis (1974). The particular problem used can be found on Page 297/Section 15.1 and is titled "Thick-Wall Cylinder in Triaxial Stress Field".

Figure E.1 illustrates the basic elements of the problem as presented by Poulos and Davis. Note that the normal stresses that are applied to the internal and external cylinder walls as well as the cylinder ends are assumed to be positive in sign as shown in this figure. This is consistent with the usual geotechnical sign convention of compression positive (Horvath 2018b). The problem formulation and resulting solution presented by Poulos and Davis also uses the geotechnical sign convention.

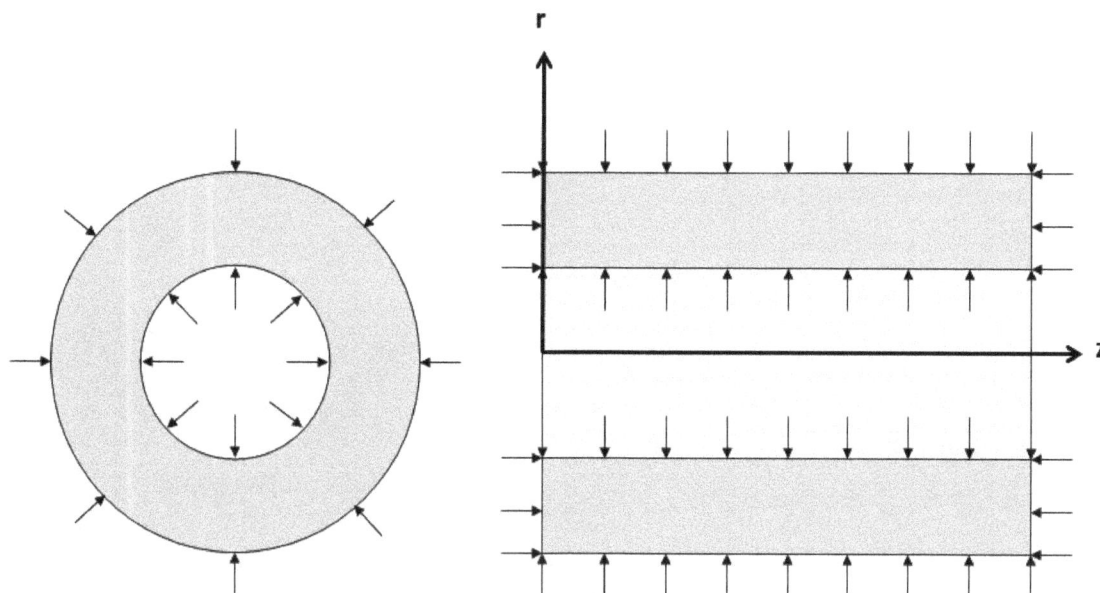

Figure E.1. Geometry of "Thick-Wall Cylinder in Triaxial Stress Field" Problem.

Before proceeding further, it is important to note that Equation 15.1f of the closed-form solution to this problem that is presented in Poulos and Davis (1974), which is for radial displacement of the cylinder, contains a typographical error. Specifically, the positive sign that begins the second line of this equation should be a negative sign. This typo was found by the writer to have a significant impact on calculated results for both the basic solution to the boundary-value problem shown in Figure E.1 as well as the new 3-D model presented in this appendix. This error was only discovered after some considerable effort by the writer during the validation phase (discussed subsequently) of the writer's new 3-D model.

Turning now to the theoretical basis of the writer's new 3-D model, the well-known principle and technique of superposition that is valid for all linear-elastic systems was employed as the basic modeling strategy. Specifically, the basic problem shown in Figure E.1 was used repeatedly to represent each component of a multi-component, composite cross-section by 'nesting' as many thick-walled cylinders as necessary, one within the other in assumed perfect contact, to produce a two- or three-component physical model that matches the actual cross-section being modeled. For example, a three-component cross-section requires three nested cylinders that overall results in a solid cross-section. Note that the innermost cylinder always has a zero inner radius and thus zero internal normal stress as the solution equations allow for this assumption[198].

Note that this modeling strategy can easily be modified to meet other modeling requirements. For example, the writer's model could easily be extended beyond three components simply by nesting the basic problem (Figure E.1) as many additional times as desired. In addition, although the innermost component in both the two- and three-component versions of the writer's model is solid-cylindrical in geometry, it does not have to

---

[198] The resulting simplified equations were compared to those for a solid right-circular cylinder based on other solutions found in Poulos and Davis (1974) and Timoshenko and Goodier (1970) and were found to be the same. This demonstrated that the problem of a solid right-circular cylinder subjected to triaxial externally applied stresses is essentially a special case of the more-general problem shown in Figure E.1.

be so. The innermost cylinder could be assumed to have a non-zero inner radius in order to model hollow rebar for example. The assumed stresses acting on that inner void could vary as well, for example, to simulate grout injection.

E.5.3 Equilibrium and Compatibility Assumptions

The following additional equilibrium and compatibility assumptions were made by the writer when developing this new 3-D model:

- The sum of the axial forces carried by each component equals the total axial force applied to the overall cross-section. Note that these forces can be either compressive or tensile as desired.

- The longitudinal (axial or z-axis) strains are the same in each component.

- The radial (r-axis) displacements are continuous and compatible at each interface between components.

- The radial stresses are continuous and compatible at each interface between components.

- The external applied stress on the outermost cylinder that represents the radial stress in the ground adjacent to the DFE is assumed to be a known, constant input variable.

There are a number of observations and comments to make concerning these assumptions:

- The assumptions concerning forces and strains parallel to the longitudinal (z-axis) are the same as those made for the traditional 1-D model discussed previously. This includes that all structural components start from a stress-free condition. In view of what is now known about residual loads in DFEs, a revised solution that removes this assumption appears to be something worth doing in the future.

- The assumption of radial (r-axis) displacement and stress continuity at component interfaces implies that no gap develops at an interface even if a tensile radial normal stress develops. This was felt to be reasonable as the combinations of cementitious and metallic materials that are typically used for deep foundations would be expected to have some tensile interface strength due to interface adhesion.

- The assumption of radial (r-axis) displacement and stress compatibility at component interfaces implies that the magnitudes and signs of these parameters are the same in each component on either side of the interface. This was felt to be reasonable simply based on physical consideration of the aforementioned interface adhesion.

- The assumption of the radial stress in the ground being a known input parameter was judged to be reasonable for constant-perimeter DFEs that were the focus of the writer's original research circa 2009. However, the model could be generalized for tapered DFEs by leaving this stress as an unknown variable and linking its magnitude to radial displacement of the pile-soil interface based on cylindrical-cavity mechanics.

The net result of the above-described assumptions is that the number of unknown variables in the general, closed-form solution presented in Poulos and Davis is reduced to:

1. overall axial normal strain,

2. axial normal stress within each cylinder/cross-section component,

3. radial normal stress between each adjoining pair of cylinders/cross-section components,

4. radial displacement at the interface between each adjoining pair of cylinders/cross-section components, and

5. radial displacement of the exterior of the outermost cylinder/cross-section component that represents the interface between the DFE shaft and surrounding ground.

Note that the first two unknowns are the same as those in the 1-D model discussed previously. The remaining three unknowns related to radial normal stresses and displacements represent the additions included in the 3-D model.

Note also that tangential normal stresses within each cylinder/cross-section component components are initially unknown as well and could be made part of the solution if desired. However, these stresses were judged not to be of importance and were not made part of the writer's research that was presented in Horvath (2010a).

E.5.4 Problem Solution

The unknown stress, strain, and displacement variables were assembled into a system of simultaneous linear equations that were solved using commercially available spreadsheet software (*Excel*) to illustrate the relative simplicity of the overall methodology. Other solution alternatives such as purpose-written software using *FORTRAN* or other programming language could obviously be used[199]. Separate spreadsheet versions for two- and three-component piles, named *HEX2* and *HEX3* respectively, were developed by the writer and used for this study.

E.5.5 Solution Validation

E.5.5.1 Introduction and Overview

Before presenting example applications of both the two- and three-component versions of this new 3-D model, it is necessary to present the results of the validation process that the writer went through in 2009 with the assistance of Mr. Colasanti. Validation is always an essential component of developing a new analytical model and can take on many forms. In this case, two broad strategies were used:

1. The solution components (equations for stresses and displacements in this case) for the basic problem shown in Figure E.1 as presented in Poulos and Davis (1974) were corroborated using other sources and methodologies.

---

[199] Mr. Colasanti wrote a *FORTRAN* code named *Micropile* in 2009 to illustrate this alternative.

2. The results of the 3-D model as implemented in the aforementioned *HEX2* and *HEX3* programs were checked against calculated results obtained using a completely different analytical methodology.

The details of these efforts are presented in the following two sections.

E.5.5.2 Original Theory-of-Elasticity Problem

The (corrected) basic elastic solution published in Poulos and Davis for the problem shown in Figure E.1 was validated in two completely different ways. First, the basic elements of this solution were sought in other published work. Timoshenko and Goodier (1970, Pages 68-71) was the specific one used in this case. Note that Timoshenko and Goodier used structural-engineering notation of tension positive in their solution equations whereas all solution equations in Poulos and Davis use the geotechnical notation of compression positive that the writer used. It was actually through the material published in Timoshenko and Goodier that the aforementioned typographical error in Poulos and Davis was confirmed.

Note that another, indirect validation along these same lines is the fact that the corrected solution in Poulos and Davis for the special case of zero interior radius and applied stress matched the solution for a solid, right-circular cylinder as published in both Poulos and Davis (1974) and Timoshenko and Goodier (1970) as noted previously.

The second validation methodology used was to formulate an arbitrary problem based on Figure E.1; calculate results using the corrected (by the writer) equations published in Poulos and Davis; and then compare these results to those obtained using a finite-element (FE) continuum model solved numerically using *ANSYS* (Version 11.0). Obviously, the equations contained in *ANSYS* are those of a general linear-elastic continuum and are thus independent of the specific solution for the problem shown in Figure E.1 so provide a truly independent analytical model as is appropriate.

The following were the assumed problem parameters (refer to Figure E.1 for visualization). SI units were primary by the writer's choice:

*   cylinder height = 127 mm (5.00 in)

*   inside radius of cylinder = 22200 μm (0.875 in)

*   outside radius of cylinder = 757400 μm (2.982 in)

*   axial (*z*-axis) normal stress = 520 MPa (75 ksi)

*   internal radial (*r*-axis) normal stress = 690 MPa (100 ksi)

*   external radial (*r*-axis) normal stress = 345 MPa (50 ksi)

*   Young's modulus of cylinder material = 17 GPa (2500 ksi)

*   Poisson's ratio of cylinder material = 0.35.

The calculated results are summarized in Table E.1 and confirm the validity of the corrected solution in Poulos and Davis that was used by the writer as the basis for the new 3-D model.

Table E.1. Results of Validation Analyses of "Thick-Wall Cylinder in Triaxial Stress Field" Problem Used as Basis for New 3-D Model.

Analytical Method	Displacements, nm (in)		
	axial	radial	
		inside	outside
corrected closed-form solution in Poulos and Davis (1974)	2199510 (0.086595)	628269 (0.024735)	96076 (0.0037825)
finite-element numerical solution using *ANSYS*	2200300 (0.086626)	626872 (0.024680)	94412 (0.0037170)

Notes:
1. Axial displacements defined as compression positive.
2. Radial displacements defined as outward positive.
3. Signs of *ANSYS* results adjusted as necessary to be consistent with sign conventions given in Notes 1 and 2.

E.5.5.3 New 3-D Model

As discussed subsequently, several example problems were analyzed using the *HEX2* and *HEX3* spreadsheet programs to investigate the 3-D model for both micropiles and traditional (driven) piles. These problems were also analyzed using a FE continuum model that was solved using *ANSYS* (Version 11.0) to provide validation of the 3-D model. The *ANSYS* analyses utilized an axisymmetric mesh consisting of 400 elements.

There was excellent agreement between the spreadsheet and FE results in all cases. The results of only one such case are shown here in Table E.2 as an exemplar validation analysis. The material properties assumed for the different micropile components are presented later in this appendix.

The specific problem presented in Table E.2 is a three-component micropile that was studied in Fuller et al. (2003) and is referred to herein as FMSM Sample #3. For validation purposes, the writer assumed an arbitrary externally applied radial stress that Fuller et alia did not consider. Further explanation of the composition of FMSM Sample #3 is given in the following sections but, as can be seen in Table E.2, the excellent comparison between the 3-D model and *ANSYS* FE results provides validation of the new 3-D model.

E.5.6 Example Applications

E.5.6.1 Introduction and Overview

As noted in the preceding section, several example applications were analyzed to provide an overall sense of typical results obtained using the 3-D model as well as to provide a comparison to results obtained using the traditional 1-D model. Although the cases analyzed were not intended to be exhaustive and cover all reasonable combinations of materials and geometries that might be encountered in practice, they were chosen to be typical and thus representative. Particular emphasis was placed on analyzing micropiles as this is what had motivated the writer to pursue development of a 3-D model in the first place as was noted at the beginning of this appendix.

Table E.2. Validation of 3-D Model for Micropile FMSM Sample #3 with External Stress.

Analytical method	Internal stresses, kPa (psi)					Axial strain, x 10⁻⁶	Radial displacements, nm (in)		
	axial			radial			rebar-core interface	core-shell interface	shell-ground interface
	rebar	core	shell	rebar-core interface	core-shell interface				
3-D model	277770 (40257)	24310 (3523)	277210 (40175)	600 (87)	1100 (159)	1363	9040 (0.000356)	33810 (0.001331)	37080 (0.001460)
ANSYS	277770 (40257)	24310 (3523)	277125 (40177)	607 (88)	1100 (159)	1363	9040 (.000356)	33810 (0.001331)	37080 (0.001460)

Notes:
1. Axial normal stresses defined as compression positive.
2. Radial-interface normal stresses defined as compression positive.
3. Axial strain defined as compression positive.
4. Radial displacements defined as outward positive.
5. Signs of *ANSYS* results adjusted as necessary to be consistent with sign conventions given in Notes 1 through 4.

Although the Fuller et al. (2003) study cited earlier in this appendix was for purposes completely unrelated to the writer's 3-D model development, it was convenient to select two micropile sections, one each with two and three components, from the Fuller et alia report as they were presumed by the writer to be representative of what might be encountered in practice, at least in the early years of the 21st century when the Fuller et alia study was commissioned and conducted. In addition, the Fuller et alia report provided useful data concerning the relevant dimensions and engineering properties of micropile components.

In addition to the two micropile examples analyzed, the writer analyzed one steel pipe pile with a PCC core. Although not intended at the time the writer performed these analyses in 2009, the example chosen is broadly similar to the *Monotube-* and *TAPERTUBE*-type tapered piles that figure prominently in this monograph.

Table E.3 summarizes the material properties assumed for the micropile and pile components in the analyses shown subsequently. Note that these material properties were also used for the exemplar verification analysis shown previously in Table E.2.

Table E.3. Assumed Physical Properties of Deep-Foundation Element Materials Used for Example Analyses.

Material	Elastic Parameters	
	Young's modulus, MN/m² (ksi)	Poisson's ratio
Portland-cement concrete (PCC)	20700 (3000)	0.15
Portland-cement grout (PCG)	17000 (2500)	0.35
steel	205000 (29700)	0.30

Note that no adjustment was made to the PCG modulus, which was developed from data presented in Fuller at al. (2003), to account for long-term behavior. Because the PCC modulus was simply assumed by the writer, the issue of time-dependency is not relevant. However, had the PCC modulus been for a specific material then the issue of time dependency would be an important consideration (York at al. 1974).

E.5.6.2 Micropiles

Two micropile cross-sections, one designated FMSM Sample #7 and comprised of two components (steel shell with PCG core) and the other designated FMSM Sample #3 with three components (steel shell with PCG core and a single steel rebar in the center of the core), were analyzed using data presented by Fuller et al. (2003). Note that the three-component micropile, FNSM Sample #3, was used earlier in this appendix as the verification exemplar. In both cases, the Schedule 40 steel shell had an outside diameter of 168 mm (6.625 in) and an inside diameter of 152 mm (5.964 in), and the PCG used had a nominal unconfined compressive strength, f'_c, = 35 MPa (5000 psi). For FMSM Sample #3, the Grade 150 rebar was 44 mm (1.75 in) in diameter. This correlates to present-day size #14 rebar using Imperial-unit designations and size #43 rebar using metric designations.

Analyses were performed using *HEX2* and *HEX3* as appropriate for each section both without and with an arbitrary externally applied radial stress to assess the effect of this variable on the calculated results[200]. In addition, results for the traditional 1-D model were also calculated for comparison (this 1-D calculation is also incorporated into the *HEX2* and *HEX3* software). An arbitrary axial-compressive force of 2000 kN (450 kips) was applied in all cases. The results are presented in Tables E.4 and E.5 for the two- and three-component sections respectively. Conclusions drawn from these analyses are presented at the end of this appendix.

Table E.4. Analysis Results for Two-Component Micropile (FMSM Sample #7).

Model	External radial stress, kPa (psf)	Internal stresses, kPa (psi)			Axial strain, x 10⁻⁶	Radial displacements, nm (in)	
		axial		radial		core-shell interface	shell-ground interface
		core	shell	core-shell interface			
3-D	0	30260 (4385)	345770 (50111)	1260 (182)	1703	42900 (0.001689)	46990 (0.001850)
	240 (5000)	30330 (4394)	345500 (50072)	1420 (206)	1700	42550 (0.001675)	46630 (0.001836)
1-D	any	29410 (4262)	349380 (50635)	0	1705	0	0

Note: Sign conventions defined in Table E.2.

---

[200] As noted earlier in this appendix, the same problems were also analyzed using *ANSYS* but the results are not shown as they essentially were the same as those obtained using the spreadsheet software.

518

Table E.5. Analysis Results for Three-Component Micropile (FMSM Sample #3).

Model	External radial stress, kPa (psf)	Internal stresses, kPa (psi)					Axial strain, x 10⁻⁶	Radial displacements, nm (in)		
		axial			radial			rebar-core interface	core-shell interface	shell-ground interface
		rebar	core	shell	rebar-core interface	core-shell interface				
3-D	0	280000 (40579)	24230 (3512)	277370 (40199)	390 (57)	930 (135)	1365	9070 (0.000357)	34140 (0.001344)	37440 (0.001474)
	240 (5000)	277770 (40257)	24310 (3523)	277210 (40175)	600 (87)	1100 (159)	1363	9040 (0.000356)	33810 (0.001331)	37080 (0.001460)
1-D	any	279980 (40577)	23570 (3416)	279980 (40577)	0	0	1366	0	0	0

Note: Sign conventions defined in Table E.2.

E.5.6.3 Steel Pipe Pile

A steel pipe pile filled with PCC was also analyzed to illustrate results for this traditional type of DFE that can be installed in a number of ways, including impact driving. As it turns out, this example is quite relevant for the purposes of this monograph as it is broadly similar to both *Monotube*- and *TAPERTUBE*-brand tapered piles. The pile was assumed to have a 457 mm (18 in) outside diameter with a wall thickness of 13 mm (0.5 in). The assumed elastic parameters for the steel and PCC core are given in Table E.3. The calculated results are given in Table E.6 for an arbitrary axial-compressive force of 6000 kN (1,350 kips). Note that the radial core-shell interface stresses are tensile in this case. Conclusions drawn from these analyses are presented at the end of this appendix.

Table E.6. Analysis Results for Steel Pipe Pile.

Model	External radial stress, kPa (psf)	Internal stresses, kPa (psi)			Axial strain, x 10⁻⁶	Radial displacements, nm (in)	
		axial		radial		core-shell interface	shell-ground interface
		core	shell	core-shell interface			
3-D	0	18170 (2633)	188840 (27368)	-1120 (-164)	894	38460 (0.001514)	42370 (0.001668)
	240 (5000)	18170 (2633)	188830 (27367)	-959 (-139)	892	36960 (0.001455)	40870 (0.001609)
1-D	any	18660 (2705)	184760 (26777)	0	902	0	0

Note: Sign conventions defined in Table E.2.

E.6 SUMMARY

From the example problems presented in this appendix as well as additional, undocumented analyses performed by the writer (one of the attractions of a spreadsheet solution is that it is very easy to quickly and efficiently play 'what-if' games with input variables) several broad trends are observed:

- The calculated results are qualitatively very similar for both the micropile and steel pipe pile sections analyzed.

- The axial normal stresses within components obtained using the traditional, approximate 1-D model are very close in magnitude to those obtained using the exact 3-D model.

- The radial normal stresses at the interfaces between components may be either compressive or tensile depending on the particular problem. The nature of the pile components in general, and their relative Poisson's ratios in particular, are the key variables influencing these results. However, regardless of sign, these stresses are always relatively small in magnitude compared to the axial stresses within components.

- The radial normal stress applied to the exterior of any type of DFE has very little influence on the calculated results for the range in stresses studied.

E.7 CONCLUSIONS

Addressing first the specific issue of 3-D vs. 1-D analysis of composite cross-sections, the overall conclusion drawn from analyses performed by the writer (including many not included in this appendix) using the 3-D analytical model presented in this appendix is that for all intents and purposes the components of a composite DFE cross-section under axial load can be adequately analyzed structurally as having no significant radial interaction with regard to both stresses and strains, either between and among the components or with the outside ground. Therefore, the continued use of the traditional 1-D model that ignores all radial effects would appear to be justified in routine practice.

Nevertheless, in some cases it might be appropriate to consider 3-D effects. This might occur:

- with a DFE of unusual or novel size, geometry, materials, or construction;

- when some advanced analytical method for geotechnical resistance considers radial displacements of the interface between the DFE and surrounding ground; or

- for more-exact interpretation of strain-gauge measurements within an instrumented pile.

In such cases, the methodology presented in this appendix should prove useful.

Note that one of the exceptions cited includes tapered piles as modeled using cylindrical-cavity mechanics, both in expansion and contraction. This means that the issue of 3-D effects should at least be raised and discussed in this monograph. which was the purpose of including this appendix.

However, the larger issue relative to stress-calculations in a composite DFE cross-section that transcends the issue of 1-D vs. 3-D analysis is that of residual loads. As discussed throughout this monograph and illustrated specifically in Chapter 7 using case-history data, a composite (driven) pile consisting of a steel pipe that is later filled with fluid PCC after driving can have post-installation axial normal stresses within the pile components that are significantly different from that forecast using the traditional 1-D analytical methodology. It would be expected that the 3-D analytical methodology presented in this appendix would be similarly in error.

The point being made here is that the time has clearly come for design professionals involved with DFEs of all types to question and reconsider the continued blind faith in the 1-D analytical model as that model is currently formulated, i.e. assuming all system components are initially stress-free. This is clearly not the case, at least when (driven) piles are concerned and most likely in other cases, such as jacked piles and possibly micropiles, when one DFE component (typically a steel pipe or casing) is installed prior to installing the other component or components (typically fluid PCC but possibly PCG as well). Note that this same comment applies to the writer's 3-D model as it was originally formulated as it, too, assumes a stress-free initial condition.

However, this does not mean that either the 1-D or 3-D models should be abandoned. It simply means that the analytical methodologies built around these models need to be revised to allow for an initial stress state in the component that is installed initially such as the steel pipe or casing. Assuming that all structural materials in the final DFE cross-section remain within linear-elastic limits, this should be a relatively easy and straightforward change to make while still maintaining the analytical simplicity of these methods.

Note that the solution for the 3-D model will need to be revised to accommodate this change. This is because while the steel pipe or casing that is installed initially will have axial-compressive normal and tangential stresses, the interior radial normal stress will still be zero initially as the pipe or casing is hollow when first installed.

Appendix F

Sabry Model for Forecasting K_h

F.1 INTRODUCTION

Sabry and Hanna (2009) contains a brief overview and summary of a research effort to develop a forecasting methodology for the post-driving, pre-load-application coefficient of lateral earth pressure, K_h, acting along the shaft of a (driven) pile in coarse-grained soil. These authors apparently saw this parameter as the key to improving indirect resistance forecasting for piles subjected to axial-compressive loads. In this sense, their thinking is in consonance with the writer's position as expressed throughout this monograph in general, and in Chapter 7 in particular, that any improved forecasting methodology for tapered piles requires an accurate depth-wise estimate of K_h as a starting point.

Although the overall problem that Sabry and Hanna considered was simplistic to the point of being non-existent in the real world, their work is nonetheless significant with respect to the goals of this manuscript. This is because efforts to define the depth-wise variation of K_h along the shaft of a pile prior to external load application appear to be virtually non-existent in the published literature. Consequently, some discussion of what is herein called the *Sabry Model* for K_h is presented in this appendix (the reason for crediting this model solely to Sabry is explained in the following section).

F.2 BACKGROUND

Before proceeding further, it is relevant to note that the 2009 paper in which the Sabry Model was presented apparently had severe length limitations dictated by the conference proceedings in which it was published (in the writer's experience, this is the rule rather than the exception in conference publications). This means that many technical details that one would ordinarily expect in a scholarly paper of this nature were likely omitted for conciseness. Unfortunately (and curiously), these authors did not reference any other published work such as an academic research report[201] that one would have expected to be produced as part of the research study that produced the contents of the 2009 conference paper. In the writer's experience (including personal), it is typical for a scholarly publication such as conference paper that generally has significant length limitations to be only a limited-scope summary of a much larger research study that is documented fully in an academic research report, master thesis, or doctoral dissertation.

With this in mind and after a Web search by the writer, it turns out the 2009 Sabry and Hanna paper was indeed the briefest of summaries of Dr. Mohab Sabry's 2005 doctoral dissertation (Sabry 2005) at Concordia University in Canada for work performed under the supervision of Prof. Hanna. However, in keeping with the writer's self-imposed guideline of

---

[201] The second author, Prof. Adel M. Hanna, is a professor of civil engineering at Concordia University in Montreal, QC, Canada with an extensive publication record in geotechnical and foundation engineering. Thus, in the writer's experience, this suggested that the contents of Sabry and Hanna (2009) were likely the outcomes of academic, as opposed to industry, research.

crediting the outcomes of doctoral research solely to the person who was awarded the doctorate for that research (as was done earlier in this monograph with the Niazi Method for example), the writer credited the model for forecasting the depth-wise distribution and magnitude of K_h solely to Sabry.

Fortunately, a complete copy of Sabry (2005) is available online (at least as of August 2019) and was downloaded and reviewed in its entirety by the writer during development of this monograph. Thus, the discussion of the Sabry Model presented in this appendix is based on the totality of the work published in Sabry (2005), not just the very limited summary that was published in Sabry and Hanna (2009). For simplicity in the remainder of this appendix, reference is only made to Sabry (2005) as it does not appear that any material presented in Sabry and Hanna (2009) was not part of Sabry's earlier published work.

F.3 SABRY MODEL

F.3.1 Introduction

In order to facilitate reference to the original published work as may be desired, the notation used in Sabry (2005) is retained throughout this appendix although the equivalency with notation as standardized in this monograph is noted where appropriate. Of particular note in this regard is the fact that Sabry used the notation K_s for what the writer calls K_h. As noted in Chapter 7, K_s as defined and used by Sabry was not explicitly assumed to be the same as the parameter K_s used in the several works by El Naggar and his students (or vice versa, as best as can be determined by the writer). Rather, it is just a notational coincidence. That having been said, they are essentially the same parameter in terms of their function in the Sabry Model for K_s (K_h) and the El Naggar-Sakr Method for resistance forecasting.

F.3.2 Background

The overall objective of Sabry's doctoral research was to develop an improved resistance-only forecasting methodology for (driven) piles in coarse-grained soil subjected to axial-compressive loading. The overall resistance model he developed was of the indirect type with a traditional conceptual structure, i.e. it assumed that the single-valued ULS total resistance is simply the sum of a single-valued ULS shaft resistance due to sliding friction and a single-valued ULS toe resistance. Although all of the new and novel research by Sabry focused on shaft resistance, he did include his own version of a toe-resistance model (albeit based on a simplified version of traditional bearing-capacity theory). This toe-resistance model is discussed briefly in this appendix only because it is relevant to the ULS total resistance that Sabry used to 'prove' the validity of his overall resistance-forecasting method.

Sabry actually developed two versions of his shaft-resistance model, only one of which was based on the post-driving, pre-load application coefficient of lateral earth pressure, K_s, acting on the shaft. In general, the two different shaft models produce different results. However, both model versions were based on the same body of research that was simply interpreted in two different ways so the results were not grossly different.

F.3.3 Overview

Sabry's research related to shaft resistance was based entirely on numerical-model simulations made using the finite-element method (FEM). He used a well-known (to geotechnical engineers) commercial software package named *PLAXIS*. A single, basic problem setup that is discussed in the following section was used. The only problem variables were the diameter and length of the assumed pile, and the strength and stiffness of the assumed coarse-grained soil.

By comparison, Sabry's research related to toe resistance appears to have been treated as an afterthought necessary to complete his overall resistance-forecasting method. His toe-resistance model does not appear to be related to the finite-element (FE) analyses but is based solely on an assumed failure mechanism that is essentially a simplistic variation of traditional shallow-foundation bearing capacity where the vertical effective stress at the pile toe controls the calculated outcome. Thus, the modern concept that radial, not vertical, stresses at the toe govern toe resistance, and that toe resistance is much more settlement dependent than shaft resistance, were both ignored.

F.3.4 Finite-Element Model

Sabry only analyzed a pile that had the geometry of a solid, right-circular cylinder with a flat end (pile toe) and the linear-elastic material properties of PCC. The circular cross-sectional geometry of the assumed pile allowed for an axisymmetric FE mesh to be used for computational simplicity and efficiency.

The diameter and length of the cylinder were varied for each of three different assumed consistencies ("loose", "medium", "dense") of a dry sand. There was no provision for a groundwater table or layered soil systems. The soil was modeled as an elastoplastic material with a Mohr-Coulomb failure criterion. By implication of the assumed soil properties, the soil was always normally consolidated at the start of each simulation.

F.3.5 Numerical Simulation

Each combination of pile diameter, pile length, and soil consistency was subjected to the same three-stage simulation:

1. A cylindrical cavity was <u>expanded</u> (<u>not</u> created) within the FE mesh to the final diameter and length of the pile being simulated. Note that this expansion did not start from a zero-volume condition but from an initial small-diameter cylindrical void that was assumed to be pre-existing. This initial assumption was apparently done not for any theoretical reason but solely to simplify the numerical simulation in terms of the FE mesh and nodes. The fact that a pre-existing hole in the ground does not exist when actual piles are driven, and thus needs to be created, was overlooked.

2. The expanded cylindrical void was 'filled' with solid structural material to simulate the PCC pile. Thus, there were no residual loads within the pile shaft as would exist with an actual (driven) pile.

3. The completed pile was subjected to an axial-compressive load to what was assumed to be the geotechnical ULS.

F.3.6 Key Outcomes

F.3.6.1 Shaft and Shaft Model for K_s (K_h)

Of greatest interest for the purposes of this appendix and monograph are the results from the first stage of Sabry's three-stage FE simulation as outlined above. The calculated results at the end of the first stage are intended to simulate conditions around a pile after it is driven into the ground.

Because of the range in simulated pile diameters, pile lengths, and soil consistency in Sabry's work, there was, not surprisingly, substantial variation in the calculated magnitudes of K_s. However, there was consistency in the relative depth-wise variation of K_s.

Overall, the depth-wise variation in K_s was highly non-uniform, with distinct peaks relatively close to the ground surface and another peak just above the pile toe. While the non-uniform depth-wise variation in and of itself was not inherently questionable, the reported values of K_s were. Peak values of the order of 80 were reported, and even for the loosest soil condition analyzed the <u>minimum</u> value was greater than 5. Given that the soil in all cases was initially assumed to be normally consolidated and modeled as an elastoplastic material, this means that K_o can be calculated using the well-known theoretical relationship:

$$K_o = \frac{\nu}{1 - \nu}. \tag{F.1}$$

This yields a value of K_o of the order of 0.4 which, when combined with the above-cited values of K_s, yield values of the K_h/K_o ratio of the order of 10 to 200. These values exceed those reported in the historical literature by one to two orders of magnitude which is simply implausible in the writer's opinion.

Because the depth-wise distribution of K_s varied in a complex manner in all analyzed combinations of pile diameter and length and soil consistency and stiffness, Sabry went through a multi-step process of first coming up with a generic, simplified geometric pattern for the depth-wise variation that was defined by several behavioral zones. This was followed by a complex algebraic equation defining these zones, with the net outcome being an average value of K_s for a given combination of problem variables. The net result was a single plot (Figure 4.9 in Sabry (2005)) with L/D, the length-to-diameter ratio of the pile, as the independent variable plotted as the abscissa and the average value of K_s as the dependent variable plotted as the ordinate. Although these average values of K_s have a somewhat smaller range in magnitude than the minimum and maximum values noted above, they are still much too large in the writer's opinion. Furthermore, in view of the fact that these results are for normally consolidated sand, when these K_s values are divided by K_o of the order of 0.4 to produce K_h/K_o ratios they are much too large as well.

F.3.6.2 Toe Model

As noted previously, Sabry developed a toe-resistance model that was intended to be used with either one of his two shaft-resistance models, of which the one based on K_s as described in the preceding section was one. Sabry's toe-resistance model is based on the traditional bearing-capacity approach using vertical effective stresses at toe level and a simplified failure mechanism that he postulated. The resulting bearing-capacity factor, which he refers to both as N_q and N^*_q in the same figure (4.7 in Sabry (2005)), is only 10% to 20% of the magnitude of the same parameter in other traditional bearing-capacity solutions, a summary of which is presented in Figure 2.2 of Sabry (2005). As a result, calculated single-valued ULS toe resistances from Sabry's toe-resistance model are much smaller than comparable values calculated using other traditional solutions.

F.3.7 Resistance-Forecasting Method Assessment

Sabry combined his shaft-resistance models plus toe-resistance model into two different versions of a resistance-only forecasting methodology. He then compared each version of his overall methodology to the same database of 21 piles in the published literature despite the fact that the simplistic assumptions on which his overall methodology is based (a homogeneous, isotropic, normally consolidated sand with no groundwater) were likely met by none of these 21 actual piles.

Overall, the forecast outcomes from the two versions of Sabry's methodology, which differed only in the forecast shaft resistance, tended to produce results that were in the range of too high to reasonably accurate. Unfortunately, Sabry did not show the separate calculated shaft and toe resistances for his method. This would have been useful, as based on the preceding discussion of Sabry's shaft and toe models, it is likely that the calculated shaft resistances were much too large but compensated, to some degree, by calculated toe resistances that were much too small. However, this did not stop Sabry from claiming that his overall analytical methodology is:

"...superior to the methods presented in the literature..."

which the writer interprets to mean superior to any other resistance-only forecasting methodology published prior to 2005. This is, in the writer's opinion, an astoundingly bold statement given the fact that, by Sabry's own statement on Page 163 of Sabry (2005), his overall methodology is considered correct to use for a narrow set of conditions (dry, normally consolidated sand) that will likely never be encountered in the real world.

F.4 COMMENTS AND CRITIQUE

In the writer's opinion, one of the most pressing R&D needs in foundation engineering is to be able to estimate the post-driving, pre-external-load application K_h/K_o ratio for (driven) piles in general and tapered piles in particular in a rational, site- and application-specific manner. Evaluation of this ratio is the crucial first step of an improved indirect resistance-forecasting methodology for tapered piles subjected to axial loading for three reasons:

1. It allows the Stage I taper benefit due to the enhanced cylindrical-cavity creation (CCXE) that results from tapered-pile installation to be determined by virtue of the difference in K_h/K_o ratios between the tapered and constant-diameter sections of the same pile.

2. It provides a well-defined starting point for the additional, Stage II taper benefit due to the cylindrical-cavity expansion (CCE) that occurs within the tapered section when the pile is externally loaded in axial compression.

3. Conversely, it provides an equally well-defined starting point for the cylindrical-cavity contraction (CCC) within the tapered section when the pile is subjected to uplift loading.

In principle and concept, Sabry's research represents a step toward achieving this goal of being able to forecast the post-driving K_h/K_o ratio and concomitant Stage I taper benefit for tapered piles. Unfortunately, the outcomes of Sabry's FE analyses in terms of K_s (= K_h) values are unrealistically much too high to the point of having absolutely no practical value. The underlying cause of this is likely a complex combination of several factors:

*   The process of simply statically expanding an existing cylindrical cavity in the ground to simulate pile driving is too simplistic and bears no resemblance to the mechanics of actual pile driving.

*   The simple elastoplastic soil model used may not properly model the substantial soil volume change that occurs in the vicinity of pile driving.

*   The inability to include soil-particle crushing that may occur during pile driving.

*   The suitability of the FEM for modeling large-deformation problems as opposed to, say, the finite-difference method (FDM) that can be better suited to problems involving large mesh deformations.

The issues of an appropriate constitutive model for soil and an appropriate large-deformation numerical tool in which to use that model appear to be key. These issues are addressed in some detail in Dijkstra et al. (2011).

In conclusion, the ability to accurately forecast the depth-wise K_h/K_o ratio on a site- and application-specific basis (in any real-world scenario, K_o typically varies with depth and it appears that K_h does as well so certainly the ratio of these parameters will vary with depth) remains an elusive goal. This means that the ability to distinguish between different K_h/K_o ratios for the tapered and constant-diameter sections of a pile remains an elusive goal at present and for the foreseeable future.

Appendix G

Bearing-Capacity Models for Toe Resistance

G.1 BACKGROUND

Bearing capacity was one of the first theoretical developments of modern soil mechanics in the 1920s. Terzaghi's bearing-capacity model was a clever, well-thought-out extension of Prandtl's classical problem in the theory of plasticity for the indentation of a flat, rigid punch into a sheet of metal. The three-factor (N_c, N_q, N_γ) equation that Terzaghi developed to account for soil weight within the assumed failure zone beneath foundation level and foundation embedment below the ground surface that are both unique and essential to the shallow-foundation application of the punch problem was copied by numerous subsequent researchers who put their own interpretation on the same basic problem.

The common element of all traditional bearing-capacity solutions that is most relevant to this monograph is that they all assume that the stress-state governing bearing capacity is the vertical overburden stress, either effective or total as appropriate, at foundation level (toe level in the case of DFEs). In simplistic terms, it is assumed that vertical stress alone is providing confinement (and thus shear strength when effective-stress strength is assumed) to the soil within and along the assumed failure surface.

In the years that followed Terzaghi's original, seminal publications on the subject, literally volumes have been written about traditional bearing-capacity solutions. It was soon recognized that solutions that worked reasonably well with shallow foundations (Horvath 2000a, 2000b) either required modification when used with deep foundations (Kulhawy 1984) or replacement entirely by solutions that were tailored to DFEs (Nordlund 1963). Nevertheless, the basic bearing-capacity concepts that have existed now for almost a century continue to be used in both routine practice and research.

However, research dating back several decades now indicates that:

- The failure pattern observed beneath the toe of piles, at least, does not match that assumed in traditional bearing-capacity solutions (Vesic 1977).

- Cavity mechanics in the form of either cavity expansion (Vesic 1972, 1977) or cavity creation (Salgado and Prezzi 2007) provides a closer match to the failure mechanism that develops beneath the toe of a DFE.

- A significant implication of both spherical and cylindrical cavity mechanics is that radial, not vertical, stresses at the DFE toe play a significant role that cannot be overlooked or ignored. This places significant emphasis on the need to estimate the pre-driving depth-wise distribution of K_o as part of the site-characterization process.

G.2 OVERVIEW

Although cavity solutions for DFE ULS toe resistance, especially Vesic's spherical-cavity expansion (SCE), have been around for several decades now, they have seen limited use in both practice and research because of the difficulty in quantifying the stiffness-related soil properties necessary to use cavity solutions. Recent developments in site characterization based on cone-penetrometer soundings have overcome these historical difficulties, at least for Vesic's SCE theory. The primary purpose of this appendix is to review the current state of knowledge for parameter assessment for Vesic's SCE in order to promote greater use of this analytical methodology in both routine practice as well as future research.

G.3 VESIC SPHERICAL-CAVITY-EXPANSION THEORY

G.3.1 Introduction and Overview

In the following presentation to illustrate Vesic's SCE model, the notation used in various cited publications has been changed in most cases to be consistent with the standardized notation used in this monograph. Exceptions are made in cases of parameters such as bearing-capacity factors that are unique to the original methodology and publication. In such cases, the original notation is retained.

To begin with, Vesic defined what is traditionally called the *gross ultimate bearing pressure*, q_{ult} (he used the notation q_o) but in this monograph is termed the *ULS unit toe resistance*, $r_{t(ult)}$, as:

$$r_{t(ult)} = cN_c^* + \sigma'_{mean}N_\sigma \qquad (G.1)$$

for the general case of a soil where both Mohr-Coulomb strength parameters (cohesion, c, and friction angle, ϕ) are non-zero.

Note that while Equation G.1 has the general form of the traditional three-factor (N_c, N_q, N_γ) bearing-capacity equation[202], there are two important differences. First, the two bearing-capacity factors, N_c^* and N_σ, are not only unique to Vesic's solution but also contain parameters related to soil <u>stiffness</u> and not just soil <u>strength</u> (as would also be expected). The inclusion of soil stiffness is one issue that distinguishes solutions based on cavity mechanics in general and, unfortunately, makes these solutions difficult to use in real applications as was noted in Chapter 4 and elsewhere throughout this monograph.

The algebraic equations for these two bearing-capacity factors can be found in Appendix A of Vesic (1977). However, it is noted that the equation for N_σ (Equation A-1 in Vesic (1977)) is typeset in a confusing manner and, more importantly, appears to contain a typographical error. For this reason, the writer used the version published in Mayne (2006a) that is not only unambiguous in its format but also contains a simpler alternative trigonometric formulation of the tangent-squared trigonometric expression used by Vesic.

The second difference of note in Equation G.1 is the use of the mean effective stress, σ'_{mean}, as opposed to the vertical stress that is used with the traditional bearing-capacity solution. In this case, σ'_{mean} is defined as:

$$\sigma'_{mean} = \left(\frac{1 + 2K_o}{3}\right) \sigma'_{vo} . \qquad (G.2)$$

---

[202] Vesic's N_c^* parameter is comparable to the N_c 'soil strength' term of traditional bearing-capacity solutions. The N_σ parameter is comparable to the traditional N_q 'embedment' term. There is no N_γ 'soil weight' term in Vesic's SCE model but this term is typically of negligible magnitude for deep-foundation problems and more or less cancels out the weight of the DFE in any case.

The significant influence of the horizontal effective overburden stress ($\sigma'_{ho} = K_o \cdot \sigma'_{vo}$) on DFE toe bearing that is implied by Equation G.2 is consistent with the comments made earlier in this appendix as well as elsewhere throughout this monograph concerning the increased appreciation for the literally head-to-toe influence of radial stresses on the axial-compressive resistance of all types of DFEs in general and tapered piles in particular. This influence also underscores the importance of including a determination of the pre-installation profile of K_o at a site, a point made throughout this monograph with respect to determining the post-driving, pre-applied-load K_h ($= K_h/K_o \cdot K_o$) profile.

G.3.2 Application

Considering now the central discussion of how to use Equation G.1 with DFEs, in any application the DFE toe could have either one of the two following conditions:

- Bearing on or in coarse-grained soil where a drained-strength analysis with $c = 0$ and $\phi > 0$ is assumed for the Mohr-Coulomb strength parameters. In this case, only the N_σ term would have to be evaluated.

- Bearing on or in fine-grained soil where an undrained-strength analysis with $c > 0$ and $\phi = 0$ is assumed for the Mohr-Coulomb strength parameters. In this case, only the N_c^* term would have to be evaluated. Note that the N_σ term would still exist but always has a value of one in this case so does not require explicit evaluation.

Only the former (coarse-grained soil) case is addressed here in detail as it is by far the more common case in practice when tapered piles are involved. However, for the sake of completeness, it is noted that should the latter (fine-grained soil) case be encountered, Appendix A in Vesic (1977) contains the requisite equation (A-4) for evaluating the N_c^* parameter. This parameter only depends on the *reduced rigidity index*, I_{rr}, of the soil. However, for undrained-strength conditions, I_{rr} is the same as the *rigidity index*, I_r. Recent research has shown that I_r can be expressed as:

$$I_r = \frac{G_{operative}}{s_u} \tag{G.3}$$

where $G_{operative}$ = an operative value of the shear modulus and s_u = undrained shear strength.

Equation G.3 is deceptively simple as it has historically proven difficult to evaluate accurately. Thus, current research has focused on developing empirical relationships based on data obtained in CPTu soundings (Agaiby and Mayne 2018, Mayne 2019b). However, this turns out to be relatively unimportant in this case as, from a practical perspective, the tabulated values of N_c^* in Appendix A of Vesic (1977) for the undrained ($\phi = 0$) strength condition indicate that the value of N_c^* is relatively insensitive to the value of I_r. Specifically, $N_c^* \cong 10$ for a wide range in I_r values and is thus not much different from the $N_c = 9$ value that is often used for deep foundations with the traditional bearing-capacity equation.

Considering now the more-common $\phi > 0$, $c = 0$ case, it is useful to start by first stating the equation for N_σ so that the soil properties that are required to quantify this parameter in any application are clearly identified. Note that the version of the equation that appeared in Mayne (2006a) is used here as it corrects the aforementioned deficiencies in the version (Equation A-1) published in Vesic (1977):

$$N_\sigma = \frac{3}{3 - \sin\phi} \cdot e^{\left[\left(\frac{\pi}{2} - \phi\right)\tan\phi\right]} \cdot \frac{1 + \sin\phi}{1 - \sin\phi} \cdot I_{rr}^{\left[\frac{4\sin\phi}{3(1+\sin\phi)}\right]} .\tag{G.4}$$

The value of ϕ to be used in this equation is generally taken to be the operative peak value as obtained in a triaxial test. There are any number of empirical relationships based on CPTu data that can be used for this purpose.

However, evaluating I_{rr}, which is not the same as I_r in this case, has historically been problematic as it requires estimating operative values of shear modulus as well as volume change. In the writer's opinion and based on the writer's personal experience, this difficulty in quantifying I_{rr} has been the singles biggest factor that has limited use of Vesic's SCE solution in foundation engineering applications for the past five decades.

However, Mayne (2006a) presents the details of a clever back-door approach to evaluating I_{rr} for sands. Specifically, Mayne (2006a) contains an empirical equation for 'normal' sands (soil-particle mineralogy appears to play a role that affects the value of the constant coefficients shown below) as follows:

$$I_{rr} \cong \left[\frac{\left(\frac{G_{max}}{\sigma'_{vo}}\right)}{85}\right]^{2.21} .\tag{G.5}$$

The small-strain shear modulus, G_{max}, in this equation should always be obtained from an sCPTu sounding or other in-situ testing methodology to account for any site-specific structure (aging, cementation). Alternatively, it can reasonably be estimated in some cases using empirical relationships that are based on 'well-behaved' Holocene soils.

In conclusion, by using Equation G.5 together with an estimate of the operative peak value of ϕ from a CPTu or sCPTu sounding it then becomes a simple exercise in arithmetic to evaluate Equation G.4 and Equation G.1.

G.4 NTH LIMIT PLASTICITY THEORY

G.4.1 Introduction and Overview

For the sake of completeness, it is worth mentioning and briefly discussing recent, state-of-art research related to cone penetrometers, as reported in numerous publications by Mayne and his collaborators (Mayne 2019a, 2019b are just two examples), that has made use of another bearing-capacity solution named *Limit Plasticity Theory* that was developed beginning in the early 1970s at the former Norwegian Institute of Technology (Norges Tekniske Høgskole, NTH)[203]. This is a 'new-old' solution, dating back to at least the early 1970s (Janbu and Senneset 1974) but only relatively recently getting renewed and more-widespread attention by a more-global audience (Mayne 2019a).

---

[203] The NTH is now part of the Norwegian University of Science and Technology (Norges Teknisk-Naturvitenskapelige Universitet, NTNU) where work on the Limit Plasticity Theory has continued. In English-language publications, the Norwegian-language acronyms NTH and/or NTNU, not their English-language equivalents (NIT and NUST), are used to refer to the Limit Plasticity Theory. In this monograph, the term 'NTH Limit Plasticity Theory' is used as this appears to predominate in the published literature and also reflects the heritage of this analytical methodology with the NTH.

There are several unique aspects of the NTH Limit Plasticity Theory compared to other bearing-capacity solutions, including Vesic's SCE solution:

- It was developed specifically and, to date, solely for use to interpret data from CPTu soundings.

- It treats soil shear strength on a fundamental, effective-stress basis, including for fine-grained soils under undrained loading such as occurs with cone-penetrometer soundings. This dispenses with the traditional approach of analyzing the undrained strength of fine-grained soils using the well-known total-stress concept where $\phi = 0°$. Thus, all soils are presumed to have $\phi > 0°$ although the assumed failure envelope does not have to go through the origin. In such cases, there is an apparent, effective-stress cohesion intercept, c', although this parameter is not used. Rather, a new parameter called the *attraction*, a, is used where $c' = a \cdot \tan \phi$. Note that subsequent publications by others such as Mayne and his collaborators sometimes use notational variations such as a' for a and ϕ' for ϕ.

- It dispenses with the usual three-factor format of traditional bearing-capacity solutions and is based on two bearing-capacity factors, labeled N_q and N_u.

- Although it nominally only uses vertical overburden stresses in its formulation, by virtue of the fact that it correlates to cone tip resistance there is at least an implied influence of radial stresses by virtue of the well-established fact that cone tip resistance is strongly influenced by horizontal overburden stresses.

G.4.2 Original Solution

Although NTH Limit Plasticity Theory was not developed for use as a general-purpose bearing-capacity methodology, it is, in principle, possible, as well as theoretically defensible, to scale-up its intended use with a cone penetrometer for use to calculate the ULS toe resistance of DFEs in general and tapered piles in particular. With this in mind, the basic elements of the NTH Limit Plasticity solution are outlined in this section.

The basic governing equation of the NTH Limit Plasticity Theory is:

$$q_{net} = q_t - \sigma_{vo} = \left[(N_q - 1)(\sigma'_{vo} + a)\right] - [N_u \cdot \Delta u_2] \qquad \textbf{(G.6)}$$

where:
- q_{net} = net cone tip resistance,
- q_t = corrected cone tip resistance,
- $N_q = \tan^2(45° + \phi/2) \cdot e^{[(\pi - 2\beta) \cdot \tan \phi]}$
- $\beta \overset{\text{def}}{=}$ *angle of plastification* (expressed in radians in the above equation)
- $a \overset{\text{def}}{=}$ *attraction* = $c'/\tan \phi$
- $N_u = 6 \cdot \tan \phi \cdot (1 + \tan \phi)$
- $\Delta u_2 = u_2 - u_o$.

Note that in coarse-grained soils both a and Δu_2 are zero so Equation G.6 simplifies substantially. Note also the parameter β as used here is completely unrelated to the use of this parameter notation elsewhere in this monograph.

The most significant difficulty to evaluating Equation G.6 in any application is that it requires quantification of a novel problem parameter called the angle of plastification that has the notation β. As illustrated in Figure 7 in Senneset et al. (1989), this parameter relates to the geometry of the assumed failure surface beneath the tip of a cone penetrometer (the assumed NTH failure surface shown in this figure appears to be a gross approximation and geometric simplification of the geometrically complex SCE failure mechanism shown in Vesic (1977)). Figure 7 in Senneset et al. (1989) shows the plausible range in β values to be from +15° to -30° while Mayne (2019a) extended this to +30° to -40°. Thus, the potential range in q_{net} for a given set of remaining parameters (ϕ, etc.) can be significant.

Most of the published body of research to date that involves the NTH Limit Plasticity Theory, including recent work by Mayne et alia, relates to cone-penetrometer soundings in fine-grained soil where an assumption of $\beta = 0$ together with $a = 0$ appears to have yielded satisfactory outcomes. Limited information (Figure 7 in Senneset et al. (1989)) suggests that for coarse-grained soils the value of β tends to be slightly negative, with values in the range of approximately 0° to -15° for values of ϕ likely to be encountered in practice.

G.4.3 Potential Deep-Foundation Application

Examination of Equation G.6 suggests that a slightly modified version of this equation would be more appropriate for the ULS toe resistance of DFEs. This is because q_t, not q_{net}, better correlates with the ULS unit toe resistance, $r_{t(ult)}$, for DFEs that was used, for example, in Equation G.1 for the Vesic SCE solution. With this in mind, Equation G.6 becomes:

$$q_t = r_{t(ult)} = \left[(N_q - 1)(\sigma'_{vo} + a)\right] - \left[N_u \cdot \Delta u_2\right] + \sigma_{vo} . \qquad \textbf{(G.7)}$$

Because Equation G.7 is theoretically rigorous, any resistance-forecasting methodology for DFEs that uses this equation to forecast the ULS toe resistance should include explicit consideration of the effective weight of the DFE. Note that doing so will generally more or less cancel out the effect of adding σ_{vo} to the right-hand side of Equation G.7.

G.4.4 Exemplar Tapered-Pile Application

It is of interest to examine, at least in a very preliminary manner, the application of the NTH Limit Plasticity solution to forecasting the ULS toe resistance, $R_{t(ult)}$, of a DFE, specifically, a tapered pile. The exemplar *Monotube* pile at JFKIA (P-5A) that was studied in Chapter 6 is used for this purpose.

In addition to the subsurface and pile details presented in Chapter 6, the following specific soil properties were used to evaluate Equation G.7 for this case-history application:

- $\phi = 38.2°$ [averaged over $+1d_t$ to $-3d_t$] $\rightarrow \tan \phi = 0.79$[204]

- $\sigma_{vo} = 7,060$ lb/ft² (338 kPa)

- $\sigma'_{vo} = 3,570$ lb/ft² (171 kPa)

- $c = a = \Delta u_2 = 0$.

---

[204] The tangent of ϕ is shown for convenience for referencing Senneset et al. (1989) as the stress-dependent Mohr-Coulomb strength parameter, ϕ, is shown in that form in that publication.

Figure G.1 presents the outcomes of this exercise. The blue curved line shows the forecast value of the ULS toe resistance, $R_{t(ult)}$, over the full range of β values shown in Mayne (2019a) which, as noted above, is a significantly larger range than shown in Senneset (1989) and covers the full range in soil conditions, not just coarse-grained soils as in this particular case. This was done intentionally to illustrate the sensitivity of the forecast outcome to the assumed value of the angle of plastification, β.

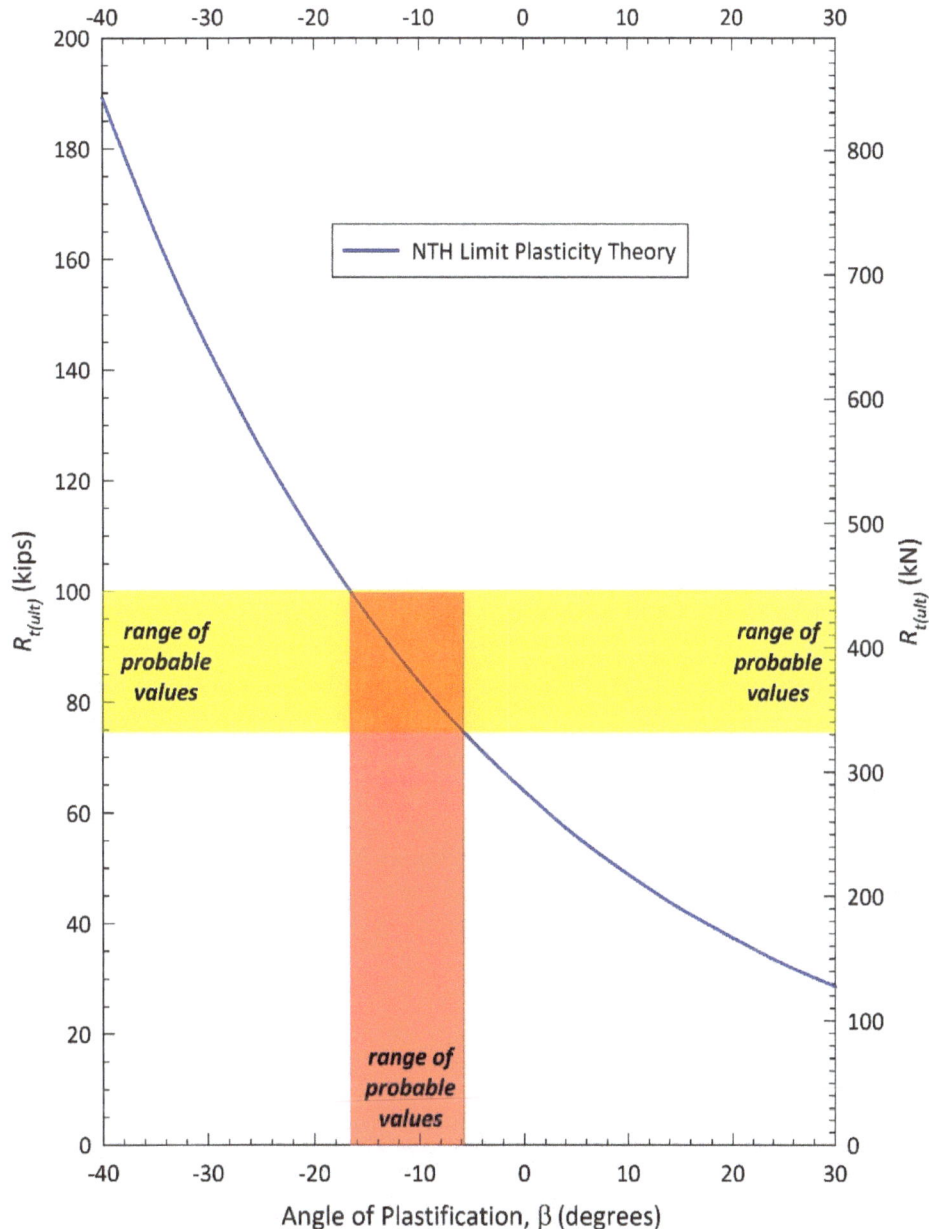

Figure G.1. Calculated ULS Toe Resistance Using NTH Limit Plasticity Theory - JFKIA Pile P-5A.

As can be seen, the range of forecast $R_{t(ult)}$ values covers almost a full order of magnitude of variation. The yellow shaded area shows the range of likely <u>actual</u> values of

534

$R_{t(ult)}$ based on a curated combination of forecasts made using various analytical methodologies that are summarized in Table 6.1 as well as the field-measured results from two almost-identical *Monotube* piles at JFKIA that are discussed in Appendix A. These two piles showed the ULS toe resistance as being approximately 20% of $Q_{t(max)}$, the maximum applied load at the head of the pile in an instrumented static load test after correction for residual loads. The red shaded area indicates the interpreted (by the writer) range of β values over which the forecast (NTH) and likely range of 'true' values intersect. This overlap defines what the writer considers to be the range in <u>plausible</u> β values for this pile. This range in β values is from approximately -6° to -16°, with the most likely value closer to -16° as this is closer to the two aforementioned field-measured results. This is in very good agreement with what one would expect based on Figure 7 in Senneset et al. (1989) and tan φ = 0.79.

G.4.5 Closing Comments

The writer has investigated several different analytical methodologies for the ULS toe resistance of tapered piles over the past 20 years. They range from traditional bearing-capacity solutions (Hansen) with correction (Vesic) for soil compressibility/rigidity to spherical-cavity expansion (both Vesic and Manandhar-Yasufuku) and now to non-traditional bearing-capacity solutions (NTH Limit Plasticity).

Based on applying these methodologies to dozens of tapered piles bearing in coarse-grained soil for which load-test data are available, the overarching takeaway is that each of these methodologies tends to be very sensitive in its calculated outcomes to relatively small changes on one or more soil properties or problem parameters. This behavior is consistent with the writer's experiences with the bearing capacity of shallow foundations in coarse-grained soil (Horvath 2000a, 2000b) where relatively small changes in φ tend to have an outsized influence on the forecast outcome.

As a result, the writer had found that for any given pile any one of these analytical methods for toe resistance will be the most accurate. Unfortunately, to date the writer has not found any one of these methods in particular to be the <u>consistently</u> most accurate. Therefore, it is not unreasonable to calculate the ULS toe resistance using several different analytical methods, compare the outcomes, and then make a decision as to the value of ULS toe resistance to use to calculate the ULS total resistance.

With this in mind, it appears reasonable to make the NTH Limit Plasticity solution one more 'tool in the toolbox' for calculating the ULS toe resistance of tapered piles and perhaps DFE's in general. However, this will require research into developing reliable correlations between φ or tan φ and β to reduce the current state of uncertainty when using this solution methodology.

Appendix H

Norwegian Geotechnical Institute Holmen Island Test Site - 1969 Test-Pile Program

---

H.1 PROLOGUE

The paper by Gregersen at al. (1973), which is based on research conducted under the auspices of the Norwegian Geotechnical institute (NGI) in 1969, is cited multiple times throughout this monograph for its seminal insight into topics that have particular relevance to the theme and goals of this monograph. Specifically, the research reported in the Gregersen et alia paper is one of the very rare (in the published literature at least) case histories where instrumented, full-scale tapered[205] and constant-diameter piles were driven into coarse-grained soil conditions and statically load tested to magnitudes of head settlement sufficient to define the geotechnical ULS in both compression and uplift.

The aspect of Gregersen et alia's work that is noted most often throughout this monograph is something that was highlighted in Fellenius et al. (2000) which is where the writer first learned of the Gregersen et alia paper. This is the fact that the tapered piles (both fully and partially tapered piles were driven and load tested) exhibited substantially greater residual loads than essentially identical constant-perimeter piles, with the increase in residual load most noticeable within the tapered portion of the partially tapered pile. This is perhaps all the more surprisingly given the fact that the taper angle, α, was only 0.29° or approximately that of typical timber foundation piles used in both Norway and the U.S. circa 1970.

Note that the use of precast-PCC *Brynildsen* piles (discussed in Chapter 3) in the Gregersen at alia work made the measurement of axial forces in the piles during all stages of construction much easier than in the efforts of Fellenius et al. (2000) who were dealing with partially tapered piles with a composite cross-section (steel pipe filled with fluid PCC after driving). The *Brynildsen* piles were instrumented with vibrating-wire strain gauges during pile manufacture so axial-force measurements within the piles could be made directly:

- before driving;

- after driving but before external load application; and

- during all phases of external load application, both in compression and uplift.

Furthermore, because the piles were extracted after the test program was completed, it was possible to exhume the instrumentation and check the zero readings of the gauges. As a result, there was a relatively high degree of confidence in the veracity of the measured results during all stages of the test program, and the highly subjective indirect interpretation of residual loads that had to be performed by Fellenius et alia for the *Monotube* piles used in their instrumented pile load tests was avoided as residual loads were measured directly.

---

[205] Gregersen et alia refer to tapered piles as *conical piles* as was noted in Chapter 1.

H.2 BACKGROUND

The test piles studied by Gregersen et alia in 1969 and reported in their 1973 conference paper were installed on Holmen Island that is located in the Drammensfjord and part of the city of Drammen, Norway that is bisected by the fjord. For many years, the NGI's 'sand' research site was located on Holmen Island although NGI has since moved their research activities related to coarse-grained soils elsewhere in Norway (Paul Mayne, personal communication, 2020).

Because the testing reported in Gregersen et alia occurred relatively early in NGI's use of the Holmen Island test site, the site-characterization information presented in their 1973 was understandably relatively Spartan by today's standards. Much more site investigation was conducted in subsequent years as can be seen in the survey paper by Lunne et al. (2003) that presents a curation of what was by that time an extensive accumulation of information from both mainstream and experimental site-characterization geotechnologies. Of particular note in the Lunne et alia paper is that the work of Gregersen et alia was apparently revisited more than once by various NGI researchers in later years.

In early 2020, the writer initiated a discussion concerning the sCPTu-based site characterization of the NGI Holmen Island site with Prof. Paul Mayne of Georgia Tech. As a result, Prof. Mayne unexpectedly and generously provided the writer with the raw, digitized (in *Excel* spreadsheet format) data for an sCPTu sounding that had been performed at the Holmen Island site at some time after 1969. This provided the writer with the necessary raw data for an independent (by the writer) state-of-art site-characterization assessment that could form the basis for a critical, independent (by the writer) assessment of the 1969 test piles. The writer's site characterization and pile assessment are presented in this appendix.

H.3 OVERVIEW

The writer originally intended to use the static load tests performed in 1969 at the NGI Holmen Island test site and as reported in Gregersen et al. (1973) only for exemplar illustrations of resistance efficiency and taper benefit as discussed in Chapter 8 as these piles offer a unique example of essentially identical tapered and non-tapered piles loaded to the geotechnical ULS in both compression and uplift. However, access to sCPTu data for the Holmen Island site presented an opportunity for the writer to independently and much more extensively parse the pile load-test results for data of interest such as K_h/K_o ratios.

The remainder of this appendix is devoted to a detailed assessment of the Holmen Island test site and the instrumented test piles that were driven and load tested there in 1969. However, it is essential that both the Gregersen et al. (1973) and Lunne et al. (2003) papers be read first as these present essential factual information that it is neither efficient nor practical to reproduce here. Both papers were readily available online at the time this monograph was finalized in early 2020 so access to these papers should not be a problem.

First presented in this appendix are key aspects of site characterization. The *Excel* spreadsheet that the writer developed for this purpose uses CPTu and sCPTu data to yield several dozen different soil properties and parameters. For many of these soil properties and parameters, multiple values are calculated using different empirical correlations so that the results can be compared and averaged as appropriate to increase the level of confidence in the calculated outcomes.

It is neither feasible nor relevant to present all of these calculated site-characterization results. Only those deemed of interest to the goals of this monograph are shown in this appendix.

This selective site characterization is followed by an assessment of the static load-test results (both in compression and uplift) for what were essentially four separate piles although the unique nature of *Brynildsen* piles (screwed connection of precast segments that were each 8 metres (26.2 ft) long) allowed two of the pile segments to be re-used by simply adding an additional segment to each and then driving them farther into the ground so that at the conclusion of the testing there were really only two installed piles. The fifth uninstrumented pile (referred to as Pile E in Gregersen et al. (1973)) that was driven as part of the 1969 study was not included in the writer's assessment as it was felt not to provide any information of relevance and value to the assessment presented in this appendix.

An important item of note with respect to the piles studied by Gregersen et alia is that the instrumented piles had slightly modified cross-sectional geometries. Normal *Brynildsen* piles are circular in cross-section. However, as illustrated in Gregersen et al. (1973), their instrumented *Brynildsen* test piles were cast with slightly flattened faces on opposing sides running the full length of the pile segments. This was to accommodate lateral-earth-pressure measurement cells at several depths along the pile shaft.

Unfortunately, the details illustrated in Gregersen et al. (1973) do not contain dimensions that would allow for calculation of cross-sectional and circumferential areas of the altered geometrical shapes. Thus, the writer could not independently calculate any stress-related outcomes for these piles and all interpreted results shown later in this appendix are based on the published values of stress-related parameters in Gregersen et al. (1973).

As a final general comment, SI units are primary throughout this appendix. This reflects the units used in the Gregersen et alia paper, the sCPTu sounding used by the writer, and the unit-preference in the writer's site-characterization spreadsheet.

H.4 SITE CHARACTERIZATION

H.4.1 Overview

Lunne et al. (2003) provide a detailed interpretation of the Holocene Epoch geology that resulted in the formation of Holmen Island. Within the depths of primary interest to the present discussion, which range from the current ground surface down to approximately 20 metres (65 ft), there is a relatively thin surficial stratum of fill (approximately 3 metres (10 ft) thick at the location of the sCPTu sounding) underlain by a predominantly sand stratum deposited within the fluvial environment of the Drammen River. Lunne et alia opine that these coarse-grained soils reflect Holocene Epoch transport and redeposition of Pleistocene Epoch deposits from farther inland along the course of the Drammen River.

Based on a review of the u_2 data from the sCPTu sounding that was provided to the writer, the groundwater table at the time of this sounding was estimated to be 1200 millimetres (4 ft) below the ground surface. This is in broad agreement with the relatively shallow groundwater level depicted graphically in Gregersen et al. (1973). Furthermore, the groundwater conditions appear to be hydrostatic, at least to the limits of the depth explored.

H.4.2 Stress History and Current Stress-State

Qualitatively, the stress history of the natural soils that comprise Holmen Island are complicated by juxtaposing natural occurrences in the post-Pleistocene era. Specifically, the area has been subjected to both isostatic rebound as well as global sea-level rise (Drammensfjord connects to the North Sea through a series of intermediate named bodies of water such as the Skagerrak strait). Thus, both the island ground surface and surrounding water level are rising and not inherently in consonance. Gregersen et alia opined that the coarse-grained soils that governed the load-bearing behavior of the test piles installed in 1969 were normally consolidated under the then-current level of filling.

The writer's assessment of the sCPTu data confirms this. Figure H.1a shows the calculated vertical effective overburden stress, σ'_{vo}, and yield stress, σ'_{vm}, to the depth limit of the sCPTu sounding (almost 24 metres/80 feet). The latter soil property was determined using empirical equations based on a parameter called m' that is dependent on I_{cn} that reflects the Normalized Soil Behavior Type. The writer used Mayne (2019b) as the specific reference although Mayne (2017) is more accessible.

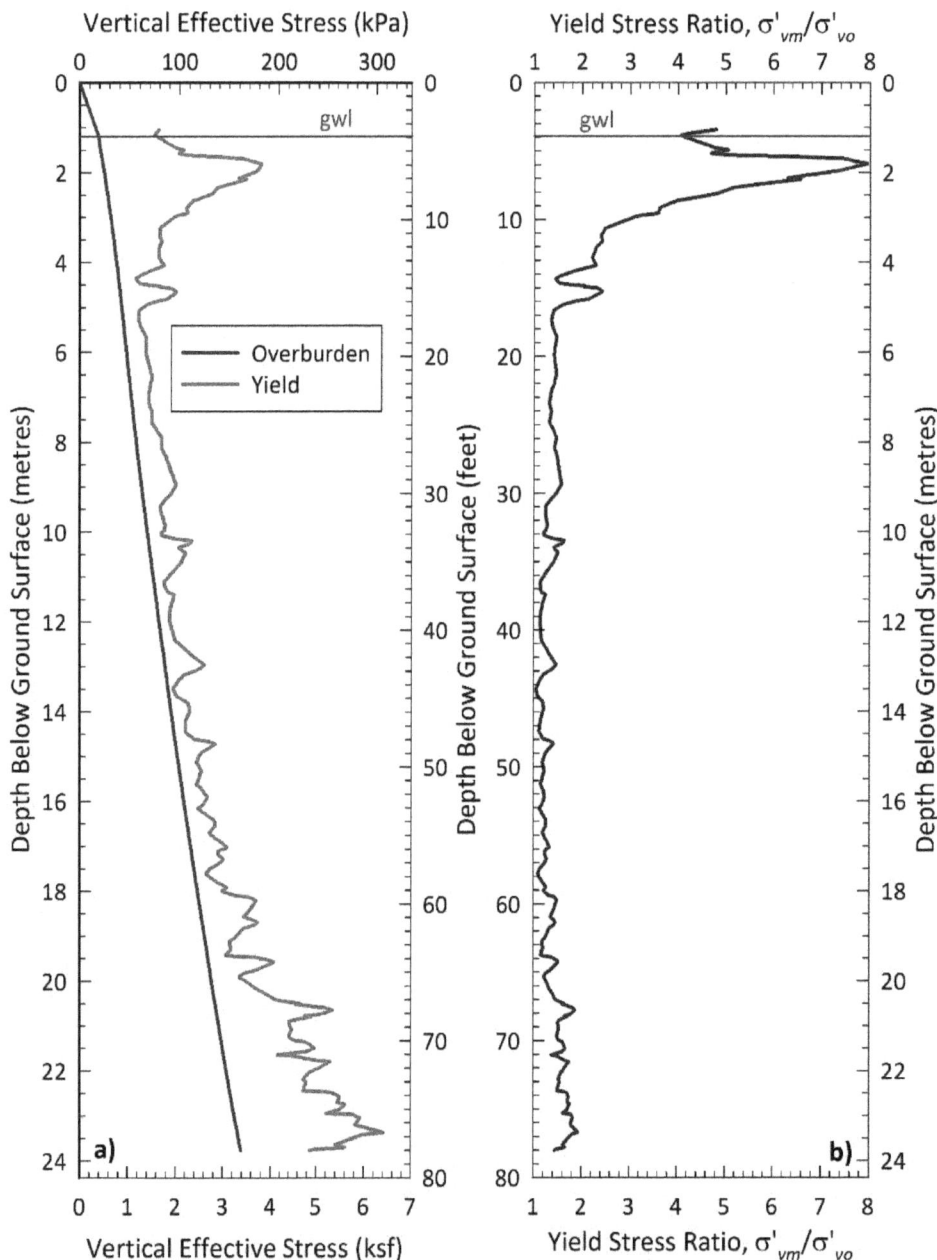

Figure H.1. Vertical Effective Stresses Based on sCPTu Sounding - Holmen Island, Norway.

It is clear that, except for the upper few metres, the yield and overburden stresses are very close in magnitude as would be expected for nominally normally-consolidated conditions. This is reflected in the calculated Yield Stress Ratio (*YSR*), $\sigma'_{vm}/\sigma'_{vo}$, that is shown in Figure H.1b as it is very close to a value of 1 for much of the explored depth.

The stress-state parameter of greatest interest and use for the purposes of this appendix is the coefficient of lateral earth pressure at rest, K_o, prior to pile driving. As has been noted many times throughout this monograph, this soil property is now recognized as being a crucial soil property for <u>both</u> the shaft <u>and</u> toe resistances of all types of DFEs. This is because it defines the starting point for radial stress changes that occur during DFE installation and post-installation external axial-load application as well.

Figure H.2 shows the calculated depth-wise variation in K_o that reflects the arithmetic mean (simple average) of the empirical relationship (applicable to coarse-grained soils only as at this site) given in Mayne (2006b, 2019b) and the theoretical relationship based on *YSR* and ϕ (applicable for all soils) given in Mayne (2007). As shown in Figure A.2 in Appendix A for the JFKIA CTA that has a predominantly coarse-grained-soil conditions, these two methods yield very similar results.

The results shown in this figure are consistent with those shown in Figure H.1 in that the uppermost 3± metres (10± ft) exhibits some widely varying overconsolidation, perhaps from human activity, and then exhibits conditions reflective of sand that is essentially normally consolidated.

H.4.3 Soil Consistency

The consistency of coarse-grained soils is often of interest in geotechnical and foundation engineering applications for any number of reasons. Historically, relative density, D_r, was the primary soil property used for this purpose but in recent years the soil property of state parameter, ψ, from Critical-State Soil Mechanics has been preferred.

Figure H.3 shows both of these soil properties as interpreted for the Holmen Island site. Note that the plot for state parameter is shown as a mirror image of the plotting format used in many publications. This is done intentionally so that the depth-wise trend in ψ is qualitatively the same as that for D_r and thus facilitates visual comparison.

Note also that, theoretically, the dividing line between contractive and dilative behavior for state parameter is $\psi = 0$. However, it has become common in the published literature to conservatively use a value of $\psi = -0.05$ (i.e. slightly dilative) as the transition from dilative to contractive behavior. This is likely because of the potentially negative consequences, e.g. risk of seismic liquefaction, that are typically associated with contractive coarse-grained soils, and the fact that there is likely some uncertainty in the empirical correlation with cone-penetrometer data that produces as estimate of ψ.

In the writer's opinion, in reality one would not expect soil behavior to change between dilative and contractive suddenly but over some range. Consequently, it seems reasonable to define that range as occurring between $\psi = 0$ and $\psi = -0.05$. This is shown as the yellow shaded zone in Figure H.3.

The results shown in Figure H.3 are broadly in consonance with those shown in Figures H.1 and H.2 in that they indicate that they uppermost 3± metres (10± ft) of soil is noticeably denser than the soils below that depth although there is a noticeable uptick again in soil density below a depth of approximately 20 metres (65 feet). However, as will be seen, the four piles of interest from the 1969 NGI test-pile program on Holmen Island were only driven to depths of 8 and 16 metres (26 and 52 ft) so their behavior was largely dominated by relatively loose, contractive soil conditions.

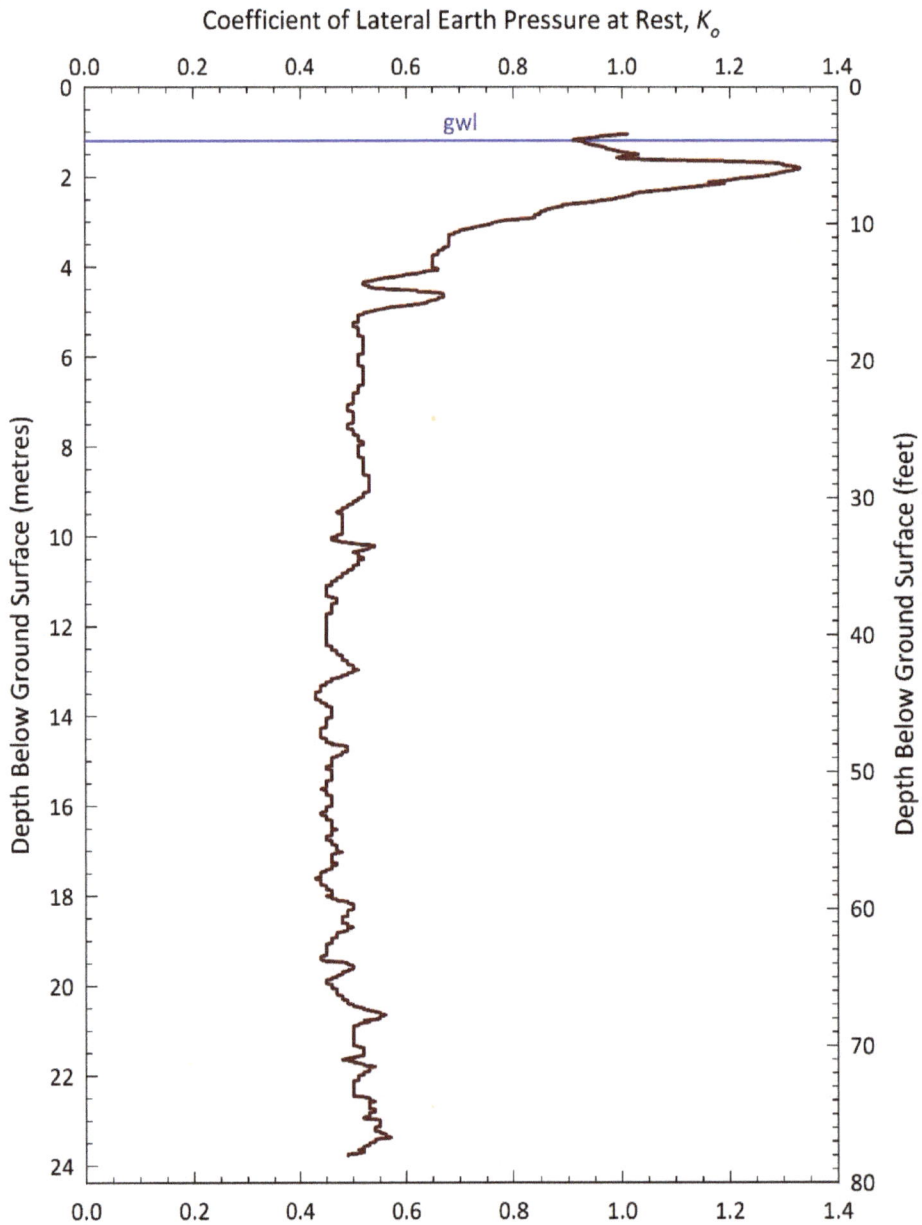

Figure H.2. K_o Profile Based on sCPTu Sounding - Holmen Island, Norway.

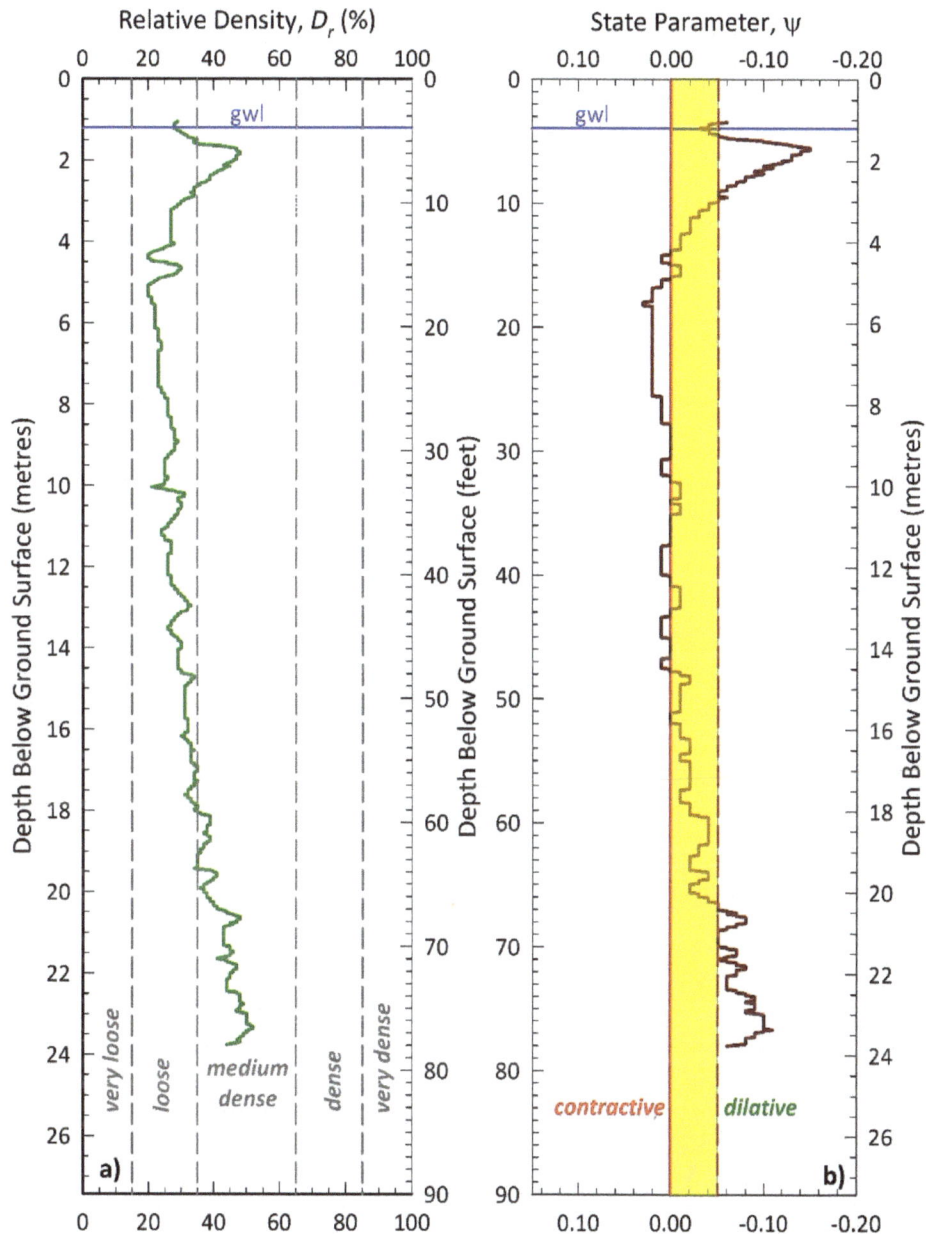

Figure H.3. Soil Consistency Based on sCPTu Sounding - Holmen Island, Norway.

H.4.4 Soil Shear Strength

The shear strength of coarse-grained soils is defined by the stress-dependent Mohr-Coulomb strength parameter, ϕ. The current procedure used by the writer for evaluating either CPTu or sCPTu data for the various versions of this parameter involves a three-step process crafted around several empirical and theoretical relationships:

1. Most cone-penetrometer research has focused on empirical correlations for ϕ_{peak} so the first step is to determine the depth-wise variation in this parameter. The writer used the

arithmetic mean of five different empirical correlations published in Mayne (2014, 2019a) for the results are shown in Figure H.4a. Although not shown, the variation in results between and among the five different correlations was never more than a few tenths of one degree which the writer considers to be overall excellent agreement given that each of these empirical correlations is its own statistical 'best fit' of some database.

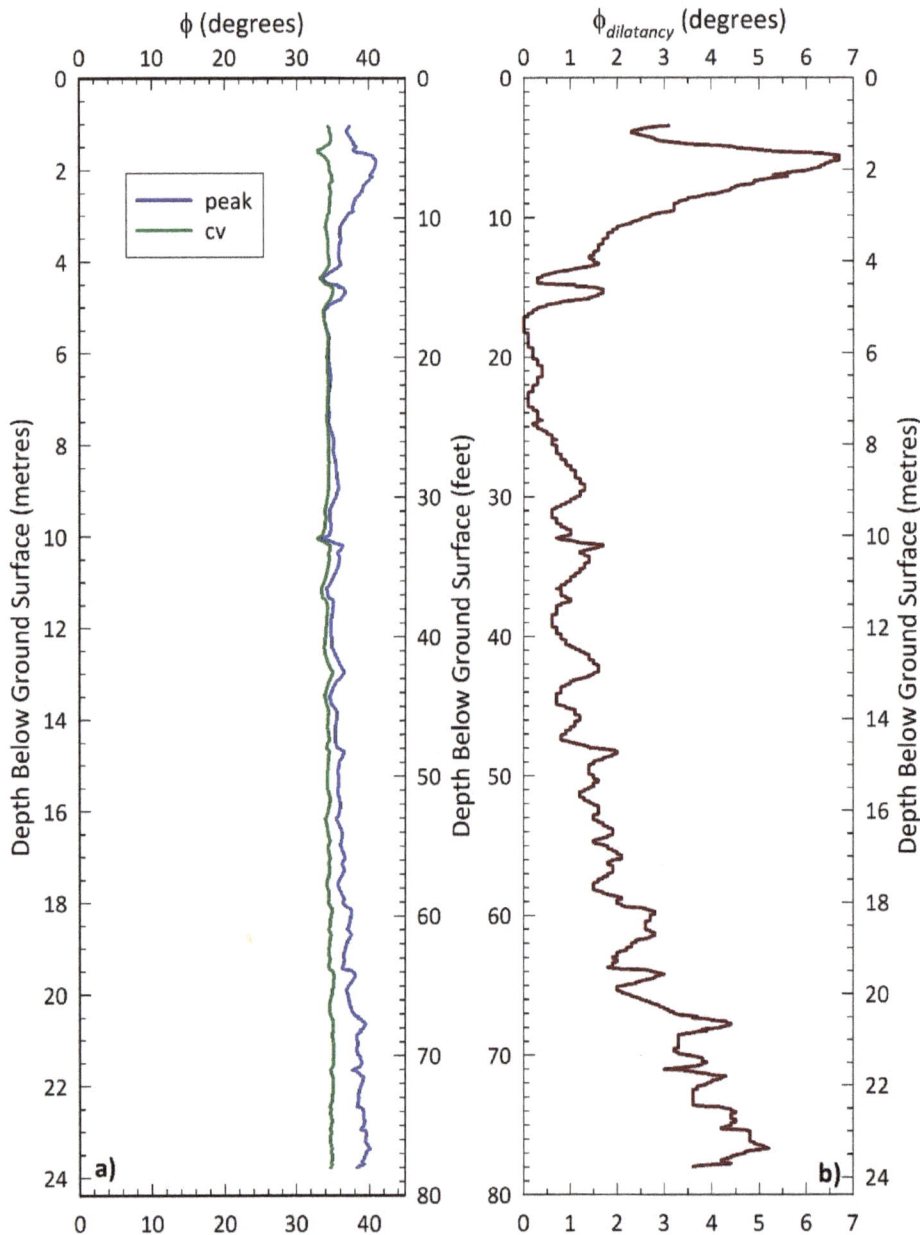

Figure H.4. Shear-Strength Parameters Based on sCPTu Sounding - Holmen Island, Norway.

2. Next calculated is $\phi_{dilatancy}$. Two different methods were used for this. One was an empirical relationship based on the state parameter, ψ, published in Robertson (2012) and the other was based on Bolton's well-known relationship based on relative density, D_r. Both

ψ and D_r were shown previously in Figure H.3. In this case, these two different methods yielded very similar results. The arithmetic mean of the values of these two methods is shown in Figure H.4b.

3. Finally, ϕ_{cv}, the constant-volume or critical-state value, is calculated using the theoretical relationship given in Equation 4.7. The results are shown in Figure H.4a.

It is clear from the results shown in Figure H.4 that, except for the uppermost 3 metres (10 ft), the soil shear strength is very close to the constant-volume value. This is to be expected based on the soil-consistency metrics shown in Figure H.3.

H.4.5 Soil Stiffness

As has been noted throughout this monograph, knowledge of soil stiffness has historically not been a technical requirement for assessing DFE behavior under axial loading as the vast majority of resistance-forecasting methodologies in widespread use since the dawn of modern foundation engineering in the 1950s have been strength-based. This is apparent in the review of existing resistance-forecasting methodologies for tapered piles presented in Chapter 4. However, a central tenet of this monograph is that explicit assessment of settlement should be a part of routine practice for DFE resistance forecasting going forward in the future. Toward that end, it is of interest to look at soil stiffness at the Holmen Island site.

As has been noted elsewhere in this monograph, the small-strain shear modulus, G_{max}, has evolved as the preferred soil property for defining the baseline soil stiffness in many geotechnical and foundation engineering applications, with DFE resistance forecasting among them. The preferred approach to evaluating G_{max} is to perform site-specific, non-invasive testing. The sCPTu is the most-common tool in use for this nowadays, at least in soil conditions that are conducive to performing a cone-penetration sounding.

Figure H.5a shows the depth-wise variation of the interval shear-wave velocity, V_s, that was obtained in the sCPTu sounding for the Holmen Island site. Datapoints were obtained on a nominal 1-metre (3.3 ft) vertical spacing that is a typical lower-bound in current routine practice (1.5-m/5-ft spacing is a typical upper-bound). Note that the values shown in this figure were calculated by the writer using the raw data provided to the writer by Mayne.

Note also that the values are toward the low end of the range that the writer has seen for coarse-grained soils (even for the relatively loose sands within the Upper Glacial Aquifer bearing stratum at JFKIA the V_s values are at least 200 m/s (650 ft/s) and usually somewhat greater) and this is likely due to the relatively loose condition of the soils at this site as is evident from Figure H.3.

G_{max} is calculated from V_s using the theoretical relationship correlating these two parameters with soil density (<u>not</u> unit weight!). In the writer's experience, most people and organizations tend to simply 'guesstimate' a single-valued, one-size-fits-all value for soil density for these purposes but the writer prefers to use a more-exact, site- and data-specific approach for this.

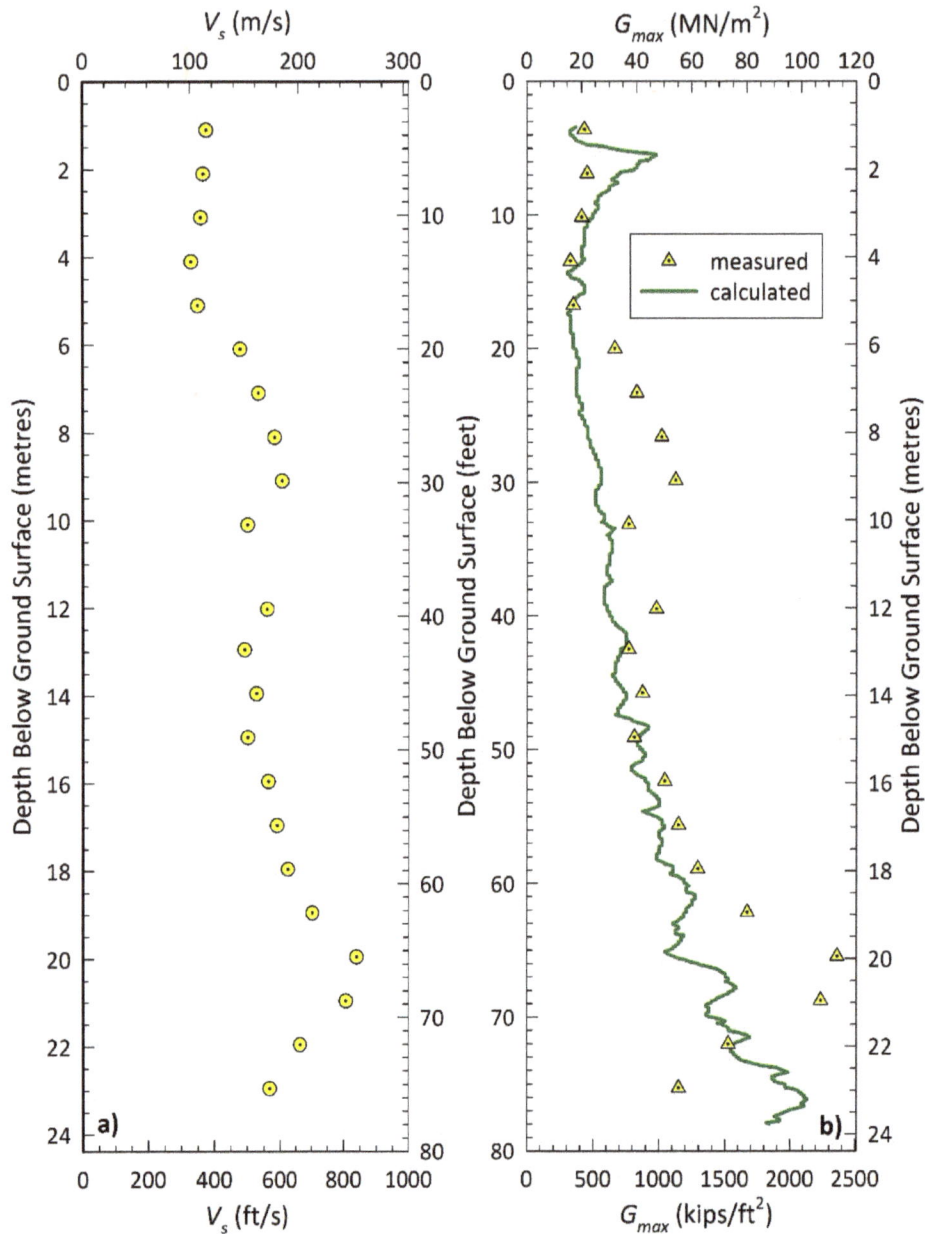

Figure H.5. Small-Strain Soil Stiffness Based on sCPTu Sounding - Holmen Island, Norway.

Specifically, the writer calculated soil density for each increment of shear-wave testing based on the arithmetic mean[206] of the soil unit weights for that increment of depth.

---

[206] As discussed in Horvath (2016b), there is a school of thought that the <u>median</u> value of some number of pieces of data measured in a cone-penetrometer sounding is better to use in a calculation requiring some 'average' value than the <u>mean</u> value of those pieces of data. This is because the median value 'smooths out' the variations from one set of cone-penetrometer data to the next that can sometimes occur. In the Holmen Island case, such variations were relatively minor and the mean and median values of various soil properties and parameters were essentially identical. Therefore, the mean value was used in this case.

These soil unit weights were, in turn, calculated using several empirical relationships for CPTu/sCPTu soundings published in the literature. Specifically, the writer used relationships published by Mayne[207] (2014, both the "approximate" and "exact" equations) and Robertson and Cabal (2010, 2012). In this case, each of these methods yielded very similar results except for the very loose shallow soils where there was greater discrepancy between and among the results of the various empirical relationships. Nevertheless, the writer used the mean value produced by these different equations for each depth-wise dataset.

As an aside, it should be noted that the vertical effective overburden stresses shown in Figure H.1a were calculated using these site-specific soil unit weights. This is noted as, in the writer's experience, vertical effective overburden stress profiles against which yield stress and other soil properties and parameters are compared such as in Figure H.1a are typically based on a generic, single-valued guesstimate of unit weight made beforehand.

The calculated values of G_{max} based on the actual V_s values are shown as individual datapoints in Figure H.5b. Also shown in this figure as a continuous line are calculated results based on an empirical methodology presented in Robertson (2012) that only requires CPTu data. While it is always preferable to determine G_{max} values based on actual V_s measurements (as was noted earlier in this monograph), there are situations where this might not be feasible, or it may be desirable on a project to supplement values obtained using sCPTu soundings with empirically derived values based on CPTu soundings. In this case, the comparison between G_{max} values was done for academic interest.

Note, however, that the use of empirical relationships to determine G_{max} profiles should always be done with great caution and site-specific comparison to results obtained from actual sCPTu measurements. To begin with, as Robertson has noted in various publications, such relationships for G_{max} typically assume behaviorally 'young' soils that are 'well behaved', i.e. lack 'structure' due to aging, cementation, etc. Even at that, some deviation should be expected.

For example, Robertson and Cabal (2015) suggest a variation of the order of approximately 2.5 between the upper and lower bounds of calculated G_{max} values even for soils meeting the behaviorally young/well-behaved criteria. So, while the writer has found that the empirical methodology for G_{max} given in Robertson (2012) and used here often produces excellent results, there can be deviations even at the same site although the deviations typically are due to the effects of structure (aging, cementation) (Horvath 2016b).

The deviations in G_{max} values apparent in Figure H.5b were surprising to the writer as the soils at the Holmen Island site can reasonably be assumed to be both young and well behaved. As noted above, in such cases the writer has consistently found good agreement between G_{max} values based on actual V_s measurements and those based on empirical assessment of the basic cone-penetrometer data.

It is not obvious to the writer what the root cause of these deviations at the Holmen Island site might be although the evaluation of other metrics (not shown here but discussed at length in Horvath (2016b) for Pleistocene coarse-grained soils at JFKIA) suggests that there might be some sort of structure of undetermined origin between the depths of approximately 5 and 10 metres (16 and 32 ft) and again toward the bottom of the sCPTu sounding. In any event, the results shown in Figure H.5b demonstrate clearly and conclusively why it is <u>always</u> good practice to include sCPTu soundings in <u>every</u> site-characterization program whenever the cone penetrometer is used for this purpose and not to rely solely on empirical correlations based on CPTu data.

---

[207] An alternative formulation of the Mayne (2014) "approximate" equation was published in Mayne and Sharp (2019) but appears to contain an error.

H.5 BEHAVIOR UNDER AXIAL LOADING

H.5.1 Background

The 1969 test-pile program at Holmen Island involved driving two heavily instrumented piles approximately 5 metres (16 ft) apart, each with a final installed length of 16 metres (52 ft)[208]. However, each pile was driven in two installments by making use of the sectional nature of *Brynildsen* piles. Specifically, two 8-metre (26 ft) long sections were first independently driven approximately 5 metres (16 ft) apart and then each was load tested in compression (in multiple cycles of loading and unloading before final loading) and then uplift. The magnitudes of settlement and uplift for both piles were sufficient to state unambiguously that the geotechnical ULS was achieved in both modes of load application.

One of the two initially installed sections, designated Pile A, had a constant perimeter with a nominal diameter[209] of 280 millimetres (11.0 in). The other initially installed section, designated Pile C, was continuously tapered (called "conical" by Gregersen et alia), with a nominal toe diameter of 200 millimetres (7.9 in) and a nominal head diameter of 280 millimetres (11.0 in). The taper angle, α, was thus 0.29°.

After all testing was completed on Piles A and C, an identical constant-perimeter extension that was 8-metres (26 ft) long and nominally 280 millimetres (11.0 in) in diameter was attached to each pile. These extensions were designed 'D' and B' respectively. Each extended pile (designated Pile D/A and Pile B/C) was then driven an additional 8 metres (26 ft) for a total, final embedded length of 16 metres (52 feet).

Both the pile with the constant perimeter from head to toe (Pile D/A) and the partially tapered pile (Pile B/C) were subjected to a similar protocol of compression testing in multiple cycles followed by uplift testing, each to the geotechnical ULS. At that point, the load testing was complete and the piles were extracted.

Gregersen et alia treated Piles A, D/A, C, and B/C as four separate piles in their paper for the purposes of presenting measured and interpreted results, and a similar treatment is applied in this appendix. However, because these four piles were, in reality, only two piles that were each installed and load-tested in two distinct increments, it is obvious that there is the unanswered question as to the effect that the first series of loading, i.e. on the initial 8-metre (26 ft) installed sections, had on the second series of loading on the full-length (16 m/52 ft) piles. Because this question is unanswerable, there remains an open question of whether the load-test behavior of Piles D/A and B/C would have been the same had these piles been driven their full length initially and not been subjected to the intermediate testing.

H.5.2 Overview

Gregersen et alia presented the load-testing results for each of the four piles in three different types of plots:

- Load vs. displacement, with both the compression and uplift load tests shown on the same figure. The multiple cycles of compression loading and unloading were shown in the

---

[208] As noted earlier in this appendix, there was a fifth, uninstrumented pile, designated Pile E, that is not considered here.

[209] As noted earlier in this appendix, the instrumented-pile sections all were atypical *Brynildsen* piles in that they were not perfectly circular in cross-section but had diametrically opposing flattened faces to accommodate geotechnical instrumentation (earth-pressure cells).

correct cumulative format. For compression loading, there were separate plots of toe, shaft, and total loading. As a result, these figures are visually very 'busy' and some time is required to study and understand them.

- The depth-wise variation of instrument-measured, axial-compressive force within the pile shafts at three different load-level conditions:
  1. After driving but prior to external load application to show residual loads.
  2. At one-half of the interpreted geotechnical ULS (the geotechnical ULS is called "Q_{90}" by Gregersen et alia and this parameter is defined subsequently).
  3. At the interpreted geotechnical ULS, Q_{90}.

- The calculated depth-wise variation of unit shaft resistance (referred to as "skin friction" by Gregersen et alia) at the same three different load-level conditions listed above.

The writer's assessment of Piles A, D/A, C, and B/C that comprises the remainder of this appendix was based on the writer's independent interpretation and replotting of these results combined with the writer's independent site-characterization assessment of the sCPTu sounding. This assessment is divided into three primary sections for the purposes of presentation and discussion:

1. Post-installation/pre-external load application residual loads.

2. Compression load tests.

3. Uplift load tests.

Overall pile behavior is addressed for all three cases and the discussion of the first two cases is further divided into shaft and toe resistances as well.

Because Gregersen et alia used non-SI metric units for their various plots of applied load, axial forces, and unit shaft resistance, the writer converted their units to SI units as well as Imperial units for plotting and discussion purposes as follows:

- Gregersen et alia used a load (force) unit that was actually not a true force unit but mass. Specifically, these authors used "tons" (colloquially called a 'metric ton' but more technically, tonnes or t) that the writer converted to true SI and Imperial force units using the relationship 1 t = 1000 kg = 9.81 kN = 2.2 kips.

- Similarly, Gregersen et alia used a stress unit that was actually not a true stress unit but mass per unit area. Specifically, these authors used "tons/m²" (more technically, tonnes/m² or t/m²) that the writer converted to true SI and Imperial stress units using the relationship 1 t/m² = 9.81 kPa = 205 psf.

One of the writer's novel additions to the present discussion that was not included in the Gregersen et alia paper is the interpretation of unit shaft resistance in terms of K_h and the K_h/K_o ratio. These parameters, especially the K_h/K_o ratio, have been used throughout this monograph and figured prominently in the writer's back-calculated assessment in Appendix A of the *Monotube* piles at JFKIA that were presented in Fellenius et al. (2000).

Determination of K_h and the K_h/K_o ratio is straightforward if the unit shaft resistance, r_s, is known (as it is for the Gregersen et alia piles) by using the following equation that is based on material presented in Chapter 4:

$$r_s = K_h \cdot \tan \delta \cdot \sigma'_{vo} = \left(\frac{K_h}{K_o}\right) \cdot K_o \cdot \tan \delta \cdot \sigma'_{vo} . \qquad \textbf{(H.1)}$$

For the purposes of this appendix, the values of δ, σ'_{vo}, and K_o necessary to solve Equation H.1 for K_h came from the writer's assessment of the sCPTu data for the Holmen Island site. Note that in this process, δ is not determined directly. Rather, ϕ is calculated. A δ/ϕ ratio equal to 0.9 was assumed by the writer based on recommendations in Kulhawy (1984) for the interface friction between sand and smooth, formed PCC.

In some cases presented subsequently, the writer made an independent assessment of the geotechnical ULS and did not use Gregersen et alia's values for the geotechnical ULS that they called "bearing capacity" and for which they used the previously noted notation Q_{90}. They defined this parameter as follows (quoting exactly from their paper):

> "The bearing capacity Q_{90} is defined as that load at which the settlement is twice as large as the settlement at 90% of the load."

This is a unique definition of the geotechnical ULS for DFEs in the writer's experience and is perhaps better understood using the graphical depiction shown in Figure H.6. Note that Gregersen et alia also used the Q_{90} definition for the uplift load tests so it would appear better to replace the word "settlement" in the above verbal definition with 'displacement' for generality as is done in Figure H.6.

As an aside, using this definition of Q_{90}, the load corresponding to $0.5 \cdot Q_{90}$ is considered to be the "working load" or 'allowable load' in the context of ASD. Furthermore, although this interpretative procedure focuses on loads, it does, indirectly, provide an estimate of pile settlement at working/allowable load, i.e. $0.5 \cdot \Delta_{90}$.

For example, for the two 8-metre (26-ft) long piles (A and C), the $0.5 \cdot \Delta_{90}$ settlement was approximately 6 to 7 millimetres ($\sim \frac{1}{4}$ inch). For the two 16-metre (52-ft) long piles, (D/A and B/C), the $0.5 \cdot \Delta_{90}$ settlement was approximately 11 millimetres ($\sim \frac{1}{2}$ inch). Such settlement magnitudes are quantitatively consistent with the traditional ASD assumption that pile settlements at the working/allowable load are 'small' in magnitude and thus negligible, although historically there was generally no attempt to numerically define exactly what constitutes 'small'.

H.5.3 Post-Installation Residual Loads

H.5.3.1 Overview

Although the axial load at the head of a DFE is, by definition, zero after installation, this does not mean that the DFE shaft and toe are force-free or that the shaft-ground interface are stress-free as is typically at least implied (even if not assumed explicitly) by virtue of the analytical assumptions and resistance-forecasting methodologies used in both routine practice and research. This fact is particularly important for the Holmen Island test piles as all four piles under consideration in this appendix exhibited significant residual loads after driving. As will be see, these residual loads had a profound influence on the response of these piles in the compressive load test that followed installation. Consequently, it is essential to discuss these residual loads in some detail.

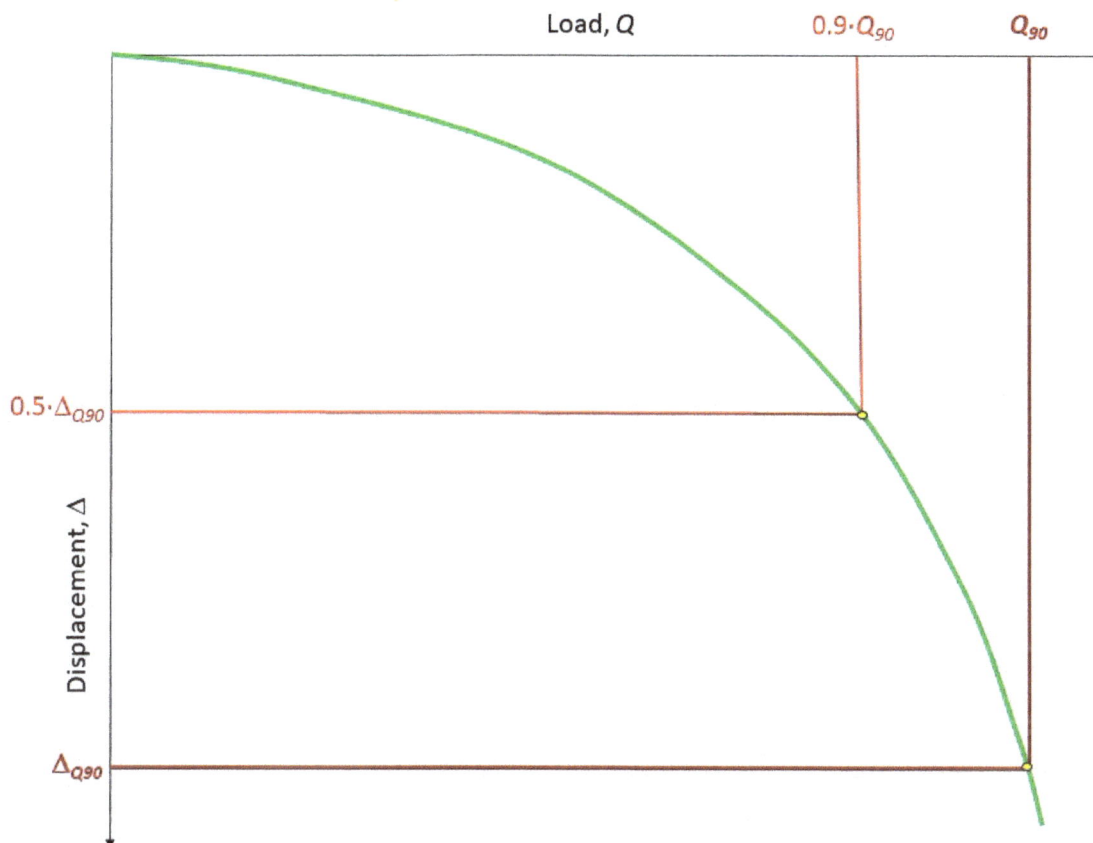

Figure H.6. Gregersen et alia Definition of Geotechnical ULS, Q_{90}, Based on Static Load Test.

The significance of residual loads for the Holmen Island test piles is clearly shown by Gregersen et al. (1973) in their Figure 6 (that shows instrument-measured axial-compressive load, Q, vs. depth) and Figure 7 (that shows calculated unit shaft resistance, r_s, vs. depth) by the curves labeled "I" in each figure. In all four cases, the neutral plane (defined in Figure 2.1 of this monograph) occurred at a depth that was approximately 60% of the installed pile length and was thus relatively independent of pile geometry, i.e. taper vs. constant perimeter. This relative depth is noticeably shallower than for the two *Monotube* piles at JFKIA that are discussed in Appendix A. For those piles, the depth of the neutral plane was approximately 80% of the installed pile length.

Although the relative depth of the neutral plane was relatively uniform for all four of the Holmen Island piles, the magnitudes of the residual load and concomitant unit shaft resistance for the tapered piles were significantly greater than those for the corresponding constant-perimeter piles. Because all four piles were, at most, modest variations of the same basic pile type and were driven in the same soil conditions during the same timeframe with (presumably) the same pile-driving equipment, the magnitude difference in residual loads can only be attributed to taper, a point that was highlighted by Fellenius (2002a).

A crucial (based on the current state of knowledge) piece of information that was not provided in Gregersen et al. (1973) was the time between driving a pile and taking measurements of the residual loads. As it now well-known and has been discussed throughout this monograph, it is now appreciated that significant increases in resistance tend to develop in piles driven in coarse-grained soil. This is especially true for tapered piles. An intriguing question, and one that has never been studied to date to the best of the writer's knowledge, is whether or not residual loads change with time after driving and before any external loads are applied.

H.5.3.2 Shaft Loads

H.5.3.2.1 Pile A

This is the intermediate-length (8 m/26 ft) pile with a constant perimeter that ultimately comprised the lower half of Pile D/A that had a constant perimeter from head to toe. Thus, all of the interpreted values of r_s (unit shaft resistance) for Pile A that are presented by Gregersen et alia in their Figure 7 are, conceptually at least, r_{ss} (the traditional unit shaft resistance due to sliding friction) in the nomenclature adopted for this monograph.

Figure H.7 shows the depth-wise variation of the instrument-measured axial-compressive forces within the pile shaft due to residual loads. As an aside, the axis scales chosen for this figure may appear to be poor choices for the results shown. However, the scale choices were intentional as the writer felt that it was useful to use the same scales for all similar figures shown in this appendix, and subsequent piles have considerably larger residual loads. The writer is a firm believer in the power of visual presentation as a complement and enhancement technique when presenting information, even with highly technical results that are essentially numerical in nature. In this case, use of the same axis scales facilitates comparison between and among the different results that will be shown.

Note that the non-zero force at the toe of the pile in this and subsequent figures is the residual toe load that is discussed separately later in this appendix. As can be seen, the neutral plane in this case occurs at a depth of approximately 4.5 metres (15 ft).

Figure H.8 shows the depth-wise variations of K_h and the K_h/K_o ratio that were calculated by the writer for this pile using Equation H.1 and the unit shaft resistances scaled from figures shown in Gregersen et alia. There are no results shown between the ground surface and a depth of 1 metre (3.3 ft) because the cone-penetrometer readings necessary for use of Equation H.1 did not begin until that depth.

Note that negative values of K_h and the K_h/K_o ratio indicate areas where a drag load as defined in Figure 2.1 is occurring along the pile shaft, i.e. the unit shaft resistance is acting in reverse (i.e. <u>downward</u>) to cause an <u>increase</u> in load within the pile shaft as opposed to acting <u>upward</u> and providing resistance and a concomitant <u>decrease</u> in load within the pile shaft. As is typical with residual loads, it is the uppermost portion of the shaft above the neutral plane that is subjected to such drag-load conditions.

Note that some authors refer to drag loads that are due to residual loads as *negative unit shaft resistance* or *negative skin friction* and use the notation $-r_s$. However, the writer prefers to call such unit stresses caused by residual loads *unit drag loads* and use the notation q_n shown in Figure 2.1. This is because regardless of whether such downward-acting stresses on the shaft of a DFE occur due to residual loads or downdrag due to soil settlement (the classical cause of drag load), the effect on the DFE is the same. So, in the writer's opinion, it makes sense to use a single, consistent notation (q_n) and terminology for these unit stresses.

In any event, the relatively small (< 1) positive values of the K_h/K_o ratio over most of the pile length indicate that shaft resistance was nowhere near fully mobilized in either direction by the residual loads except very close to the pile head. Based on tabulated results given in Kulhawy (1984), a K_h/K_o ratio in the range of perhaps 1 to 1.5 would be expected for a (driven) pile of relatively modest constant diameter as in this case.

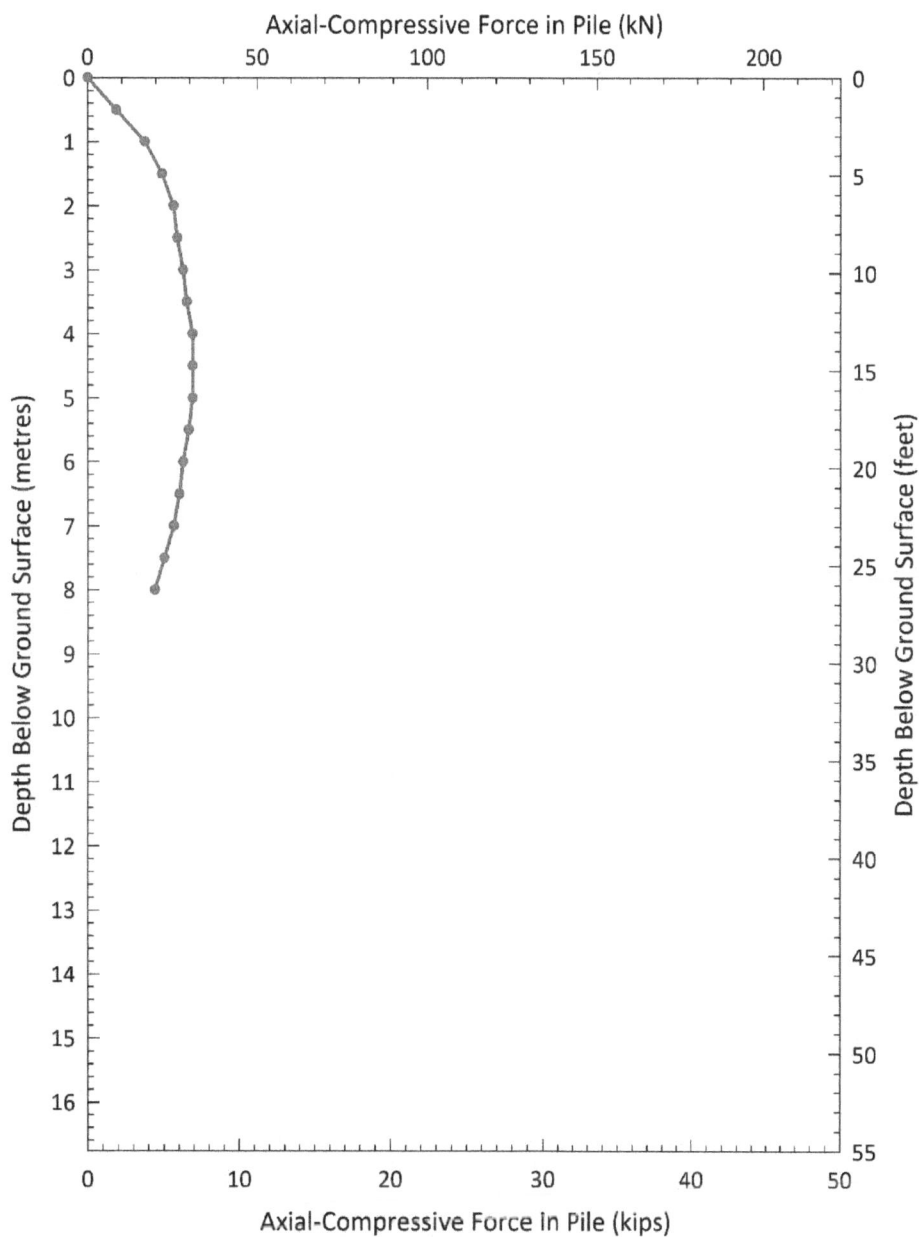

Figure H.7. Holmen Island Pile A - Measured Axial-Compressive Forces within Pile Shaft Due to Residual Loads.

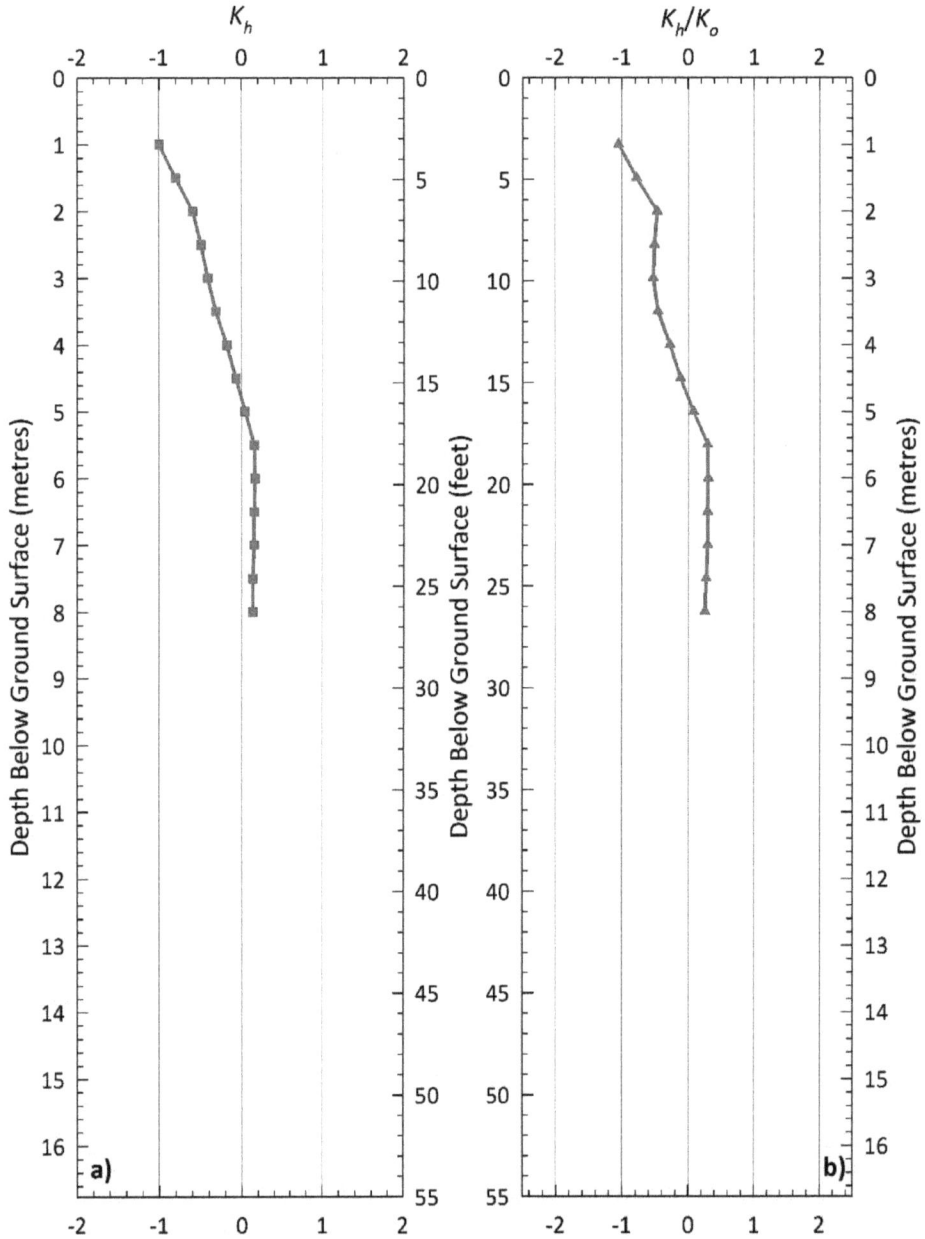

Figure H.8. Holmen Island Pile A - Calculated K_h and K_h/K_o Along Shaft-Soil Interface Due to Residual Loads.

H.5.3.2.2 Pile C

This is the intermediate-length (8 m/26 ft) pile with a continuous taper angle, α, = 0.29° over its entire length that ultimately comprised the lower half of partially-tapered Pile B/C. Thus, all of the interpreted values of r_s for Pile C that are presented by Gregersen et alia in their Figure 7 are r_{st} (unit shaft resistance due to taper-induced wedging) in the nomenclature adopted for this monograph.

Even though the taper angle is relatively modest and typical of that of many timber piles, not manufactured piles such as the *Monotube* or *TAPERTUBE*, Pile C nonetheless provides an interesting tapered vs. non-tapered comparison to the previously discussed Pile A. Such apples-to-apples comparisons of essentially identical piles driven under and in the same conditions are extremely rare in the published literature.

Figure H.9 shows the depth-wise variation of instrument-measured axial-compressive forces within the pile shaft due to residual loads, with the previously shown (Figure H.7) results for constant-perimeter Pile A included for comparison. As can be seen, the neutral plane for Pile C again occurs at a depth of approximately 4.5 metres (15 ft), the same as that for Pile A. However, the maximum force in the pile from the residual loads is substantially greater (by a factor of approximately 2.4) for Pile C compared to Pile A. This clearly demonstrates and supports key points made in the main text of this monograph:

- Taper benefit due to the shaft-resistance mechanism of taper-induced wedging and as modeled using cylindrical-cavity mechanics occurs during driving and not just as a result of post-driving external load application as assumed by Kodikara and Moore, El Naggar et alia, and Manandhar et alia.

- The wedging that occurs during the driving of tapered piles is capable of retaining and locking in a much greater proportion of pile-driving energy compared to the traditional shaft-resistance mechanism of sliding friction. This reinforces the point made in this monograph that 1-D wave mechanics as currently used for wave-equation software as well as dynamic measurements made on piles in the field is inadequate to properly model what occurs when a stress wave travels along a pile during driving. Consequently, 1-D wave mechanics either needs to be replaced for this purpose or modified in some manner to better capture the kinematics of what occurs when tapered piles are driven.

The genesis of the significantly greater residual loads in Pile C relative to Pile A can be seen in Figure H.10 that shows the calculated depth-wise variations of K_h and the K_h/K_o ratio for Pile C along with those shown previously for Pile A (Figure H.8) for comparison. The locked-in radial stresses around Pile C are substantially larger than those for Pile A.

The substantially larger inferred radial stresses for the tapered pile (Pile C) compared to those for the constant-perimeter pile (Pile A) can only be attributed to the unique wedging mechanism of tapered piles. Furthermore, this wedging occurs regardless of the final relative pile-soil movement at the end of pile driving, i.e. whether the unit shaft resistance acts in a drag-load or resistance mode.

As for the magnitudes of the inferred radial stresses for the tapered pile (Pile C), the peak values of the calculated K_h/K_o ratios are of the order of 2. Based on research by the writer than dates back to circa 2000, for timber piles with average taper angles that were essentially the same as that for Pile C in soil conditions at JFKIA that were broadly similar to those at Holmen Island, the maximum value of the K_h/K_o ratio is of the order of 4.

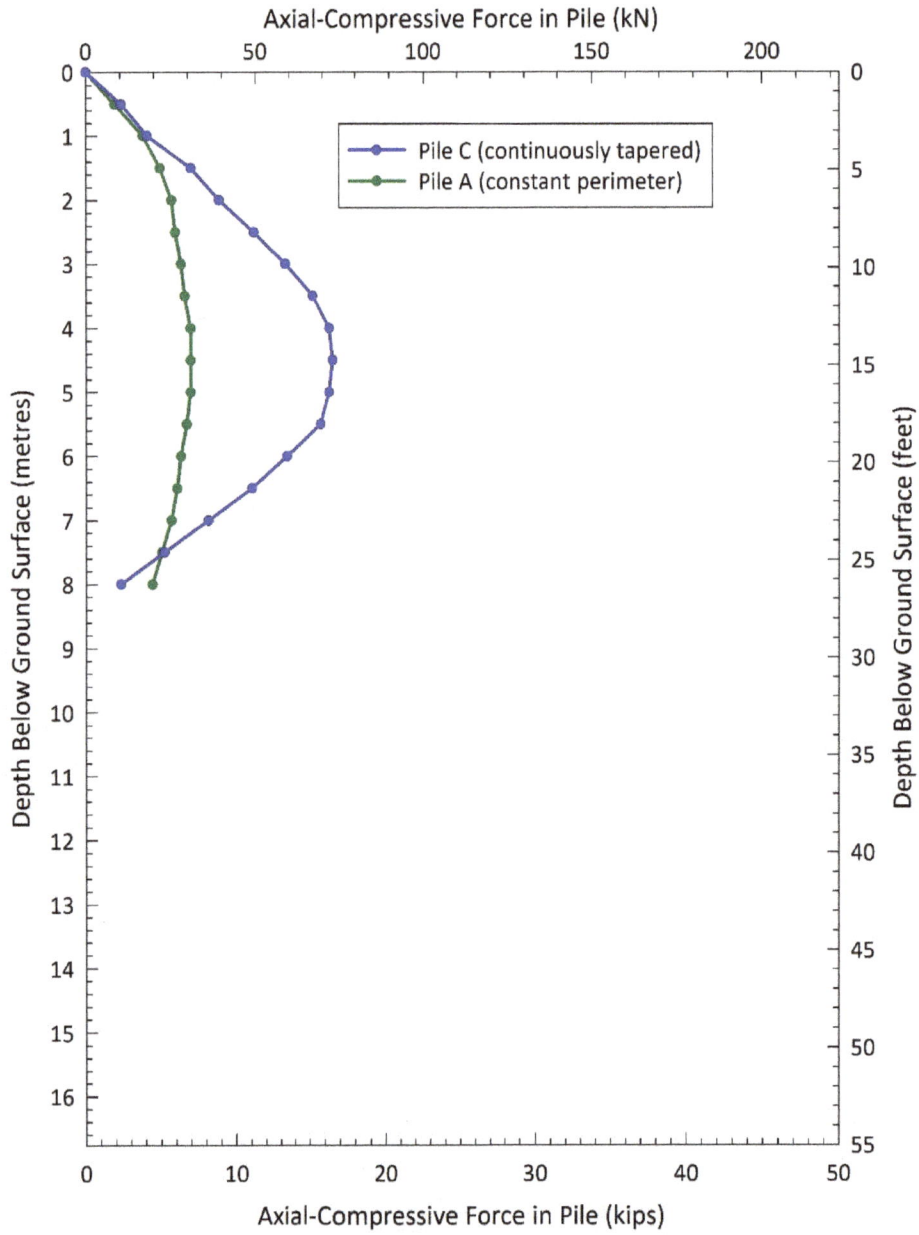

Figure H.9. Holmen Island Pile C vs. Pile A - Measured Axial-Compressive Forces within Pile Shaft Due to Residual Loads.

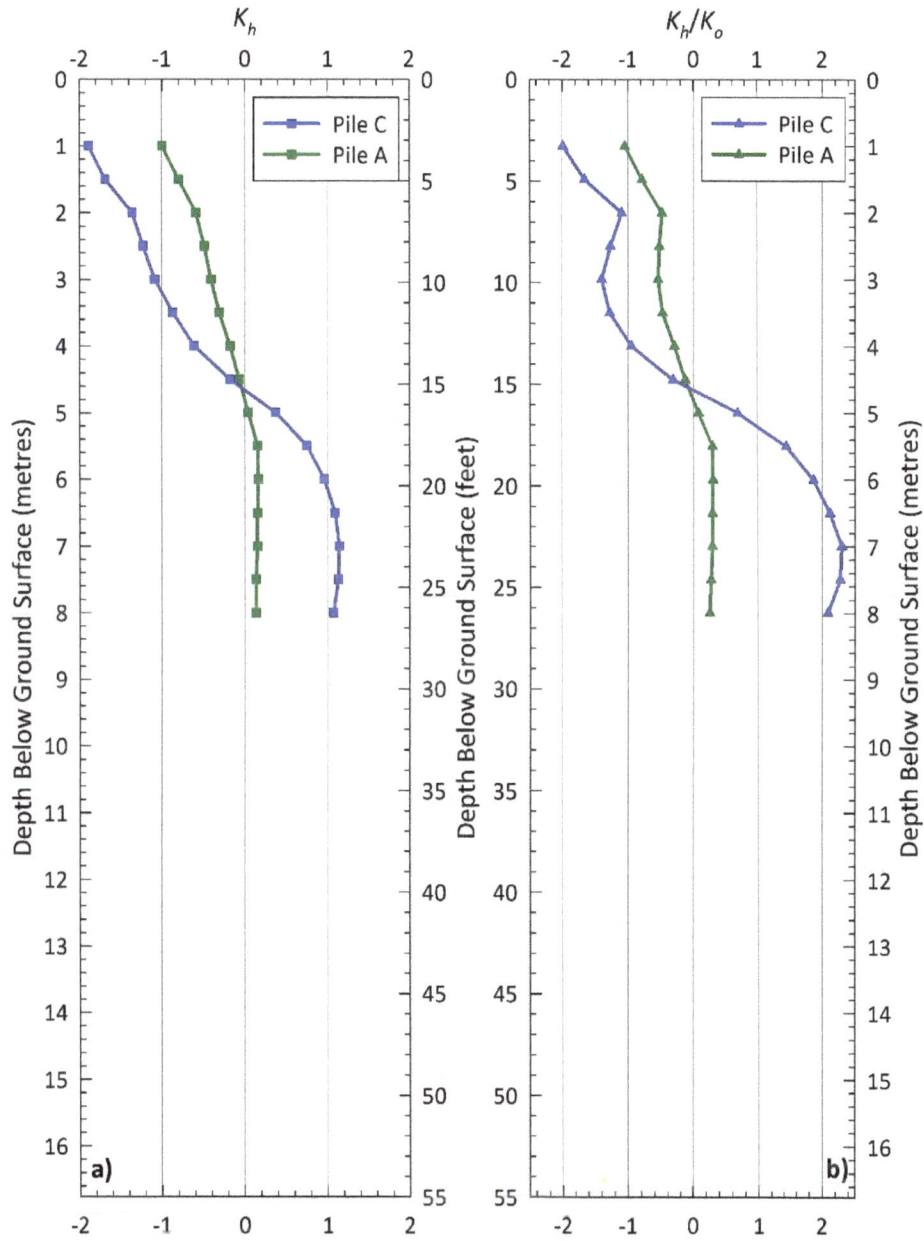

Figure H.10. Holmen Island Pile C vs. Pile A - Calculated K_h and K_h/K_o along Shaft-Soil Interface Due to Residual Loads.

H.5.3.2.3 Pile D/A

This is the final 16-metre (52-ft) -long pile with a constant perimeter from head to toe that was created by attaching an 8-metre (26-ft) -long constant-perimeter extension (referred to as Segment D) to Pile A after the latter was load-tested in compression and uplift, and then driving the two combined segments an additional 8 metres (26 ft) into the ground. Thus, all of the interpreted values of r_s for Pile D/A that are presented by Gregersen et alia in their Figure 7 are r_{ss} in the nomenclature adopted for this monograph.

Figure H.11 shows the depth-wise variation of instrument-measured axial-compressive forces within the Pile D/A shaft due to residual loads, with the previously shown (Figure H.7) results for Pile A included for comparison to show the increase that resulted from attaching the extension and then driving the extended pile farther into the ground.

As can be seen, the neutral plane for Pile D/A occurs at a depth of approximately 8.5 metres (28 ft) and there is a substantial (more than five-fold) increase in the maximum axial-compressive force within the pile shaft relative to that which existed after the first pile segment (Pile A) was driven.

Because Pile A was subjected to substantial vertical displacement (approximately 85 millimetres (3.3 in) of settlement followed by approximately 20 millimetres (0.8 in) of uplift) during post-driving load testing, it assumed that all of the effects of residual load that existed for Pile A after driving and as shown previously in Figure H.7 had been 'erased' by this post-driving load testing. Consequently, it seems reasonable to assume that all of the residual load for Pile D/A shown in Figure H.11 reflects developments during the final 8 metres (26 ft) of driving.

Figure H.12 shows the depth-wise variations of the calculated values of K_h and the K_h/K_o ratio for Pile D/A along with those shown previously for Pile A (Figure H.8) for comparison. The writer cautions about reading too much into both the K_h and K_h/K_o ratio values for the upper portion of Pile D/A that are shown in this figure. The reason is that they are based on applying Equation H.1 using the interpreted soil properties from the sCPTu sounding that is assumed to reflect virgin site conditions prior to any pile driving in 1969.

It is a virtual certainty that the driving and subsequent static load testing of Pile A significantly densified and stiffened the soils within the uppermost 8 metres (26 ft) of this site. Consequently, when the remaining 8 metres (26 ft) of Pile D/A was driven the ϕ values would not have been those shown in Figure H.4 but something larger. This, in turn, affects the δ values used in Equation H.1 as the δ/ϕ ratio is assumed to be fixed at a value of 0.9.

Unfortunately, it is not possible to rationally and reliably estimate what the increase in ϕ might have been for the uppermost 8 metres (26 ft) at the Holmen Island site. The well-known Bolton equation cannot be used as it requires a value for relative density, D_r, and this soil property would have increased as well from the values shown in Figure H.3 as the result of driving and load testing Pile A. It is of interest to observe and note that Nordlund (1963) was prescient in this regard when he allowed for the δ/ϕ ratio in his resistance-forecasting methodology to have values greater than 1 to account for densification that accompanies pile driving.

In any event, the point being made here is that the <u>actual</u> values of K_h and the K_h/K_o ratio for the uppermost 8 metres (26 ft) or so of Pile D/A are likely somewhat smaller in magnitude than the values shown in Figure H.12.

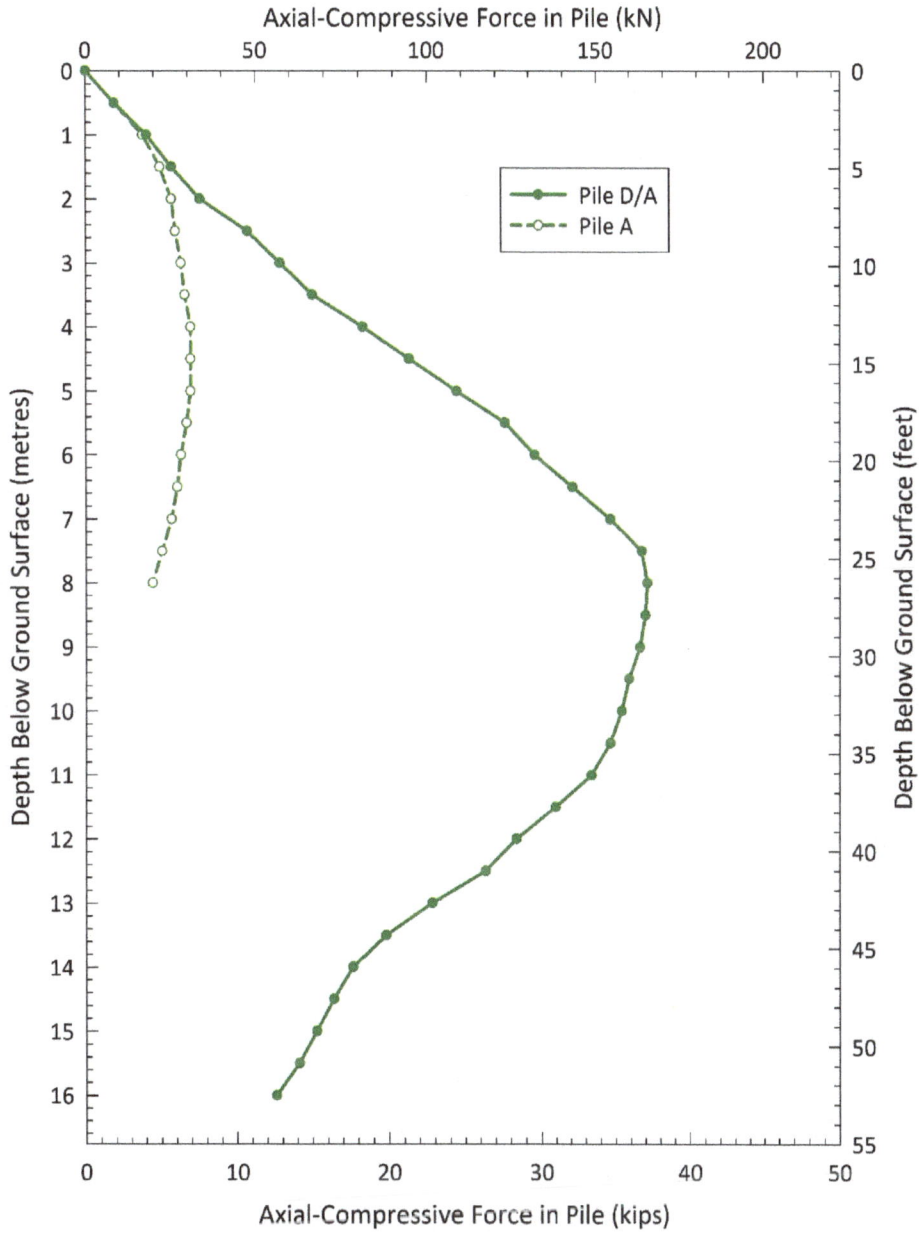

Figure H.11. Holmen Island Pile D/A vs. Pile A - Measured Axial-Compressive Forces within Pile Shaft Due to Residual Loads.

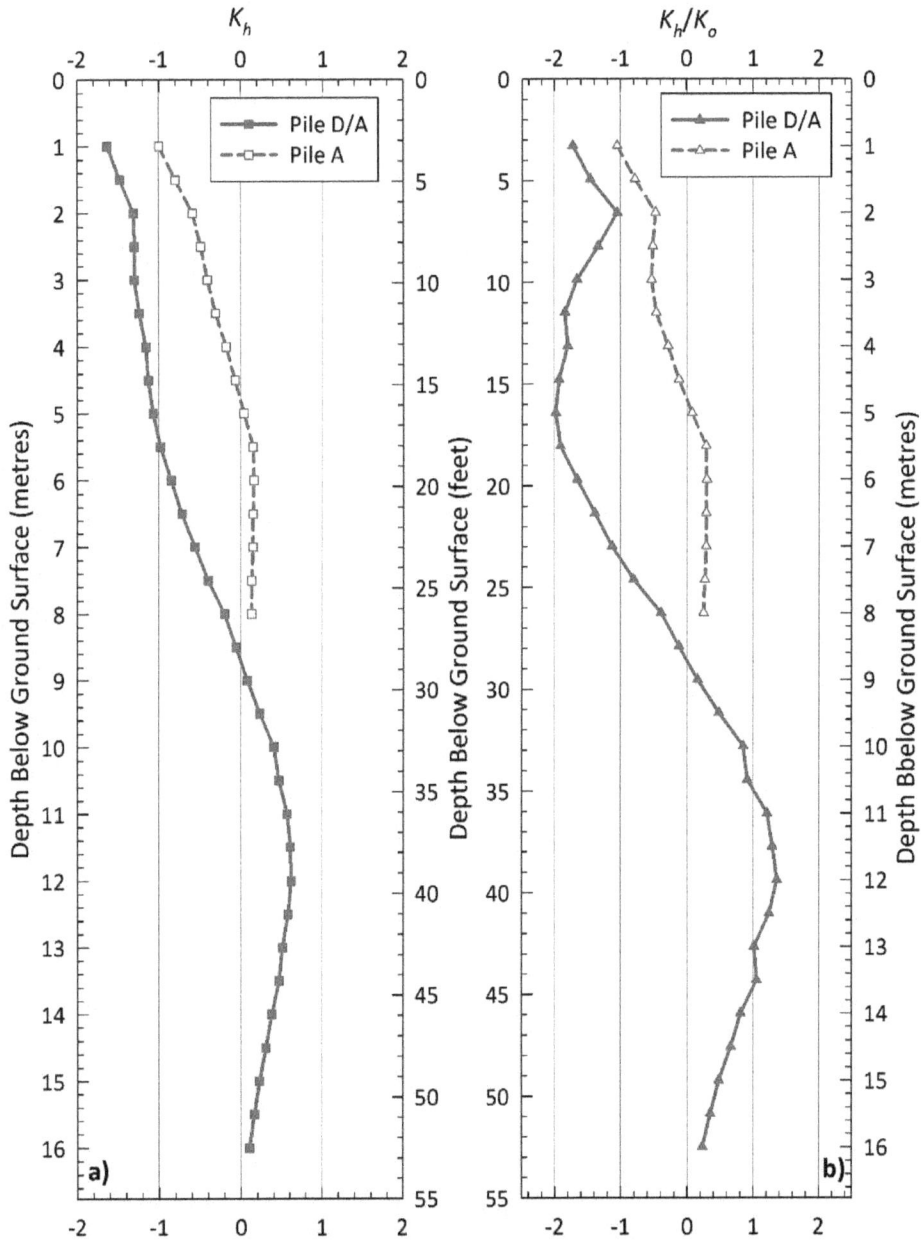

Figure H.12. Holmen Island Pile D/A vs. Pile A - Calculated K_h and K_h/K_o along Shaft-Soil Interface Due to Residual Loads.

H.5.3.2.4 Pile B/C

This is the final 16-metre (52-ft) -long partially tapered pile that was created by attaching an 8-metre (26-ft) -long constant-perimeter extension (referred to as Segment B) to the continuously tapered Pile C after the latter was load-tested in compression and uplift, and then driving the two combined segments an additional 8 metres (26 ft) into the ground. Thus, the interpreted values of r_s for Pile D/A that are presented by Gregersen et alia in their Figure 7 are a combination of r_{ss} (upper 8 m/26ft) and r_{st} (lower 8 m/26ft) in the nomenclature adopted for this monograph.

Figure H.13 shows the depth-wise variation of instrument-measured axial-compressive forces within the Pile B/C shaft due to residual loads, with the previously shown (Figure H.9) results for Pile C included for comparison. Note that the transition point from constant-perimeter to tapered for Pile B/C is noted in this figure.

As can be seen, the neutral plane for Pile B/C occurs at a depth of approximately 10 metres (33 ft) and there is a substantial (approximately three-fold) increase in the maximum axial-compressive force within the pile shaft relative to that which existed after the first pile segment (Pile C) was driven. Note, however, that because Pile C was continuously tapered from head to toe that it actually had larger residual loads within the uppermost 4.5 metres (15 ft) compared to Pile B/C that has a constant perimeter within that depth. This is additional evidence that pile taper has the ability to lock-in greater residual loads compared to non-tapered piles that are otherwise identical and driven in the same ground conditions.

Because Pile C was subjected to substantial vertical displacement (approximately 70 millimetres (2.8 in) of settlement followed by almost 25 millimetres (1 in) of uplift) during post-driving load testing, it assumed that all of the effects of residual load that existed for Pile C after driving had been erased by this load testing. Consequently, it seems reasonable to assume that all of the residual load for Pile B/C shown in Figure H.13 reflects developments during the final 8 metres (26 ft) of driving.

In the same way and for the same reasons that it was of interest to compare the instrument-measured axial-compressive forces from residual loads for the two 8-metre (26-ft) -long piles (Figure H.9), it is of interest to do this as well for the two full-length (16 m/26ft) piles, B/C and D/A. This is done in Figure H.14 that is similar to that shown in Figure 2a in Fellenius (2002a) although the writer performed an independent assessment of pile forces so the writer's results do not necessarily agree with those shown by Fellenius.

The tendency for pile taper to cause larger residual loads is quite apparent in this figure. Both piles exhibit similar residual loads for the upper 8 metres (26 ft) where both piles have a constant-perimeter cross-section. However, within the lower 8 metres (26 ft) where only Pile B/C is tapered there is a rapid increase in residual loads that causes the neutral plane to be noticeably deeper (approximately 10 metres (33 ft) below the ground surface) compared to that for Pile D/A (approximately 8.5 metres (28 ft) below the ground surface).

Figure H.15 shows the depth-wise variations of the calculated values of K_h and the K_h/K_o ratio for Pile B/C along with those shown previously for Pile C (Figure H.10) for comparison. Once again, the writer cautions about reading too much into both the K_h and K_h/K_o ratio values for the upper 8-metre (26-ft) portion of Pile B/C that are shown in this figure for the same reasons given previously for Pile D/A vs. Pile A.

As was done in Figure H.14 with the instrument-measured axial-compressive pile loads, it is of interest to compare K_h and the K_h/K_o ratio between Pile B/C and Pile D/A, i.e. partially tapered vs. constant-perimeter. This is done in Figure H.16.

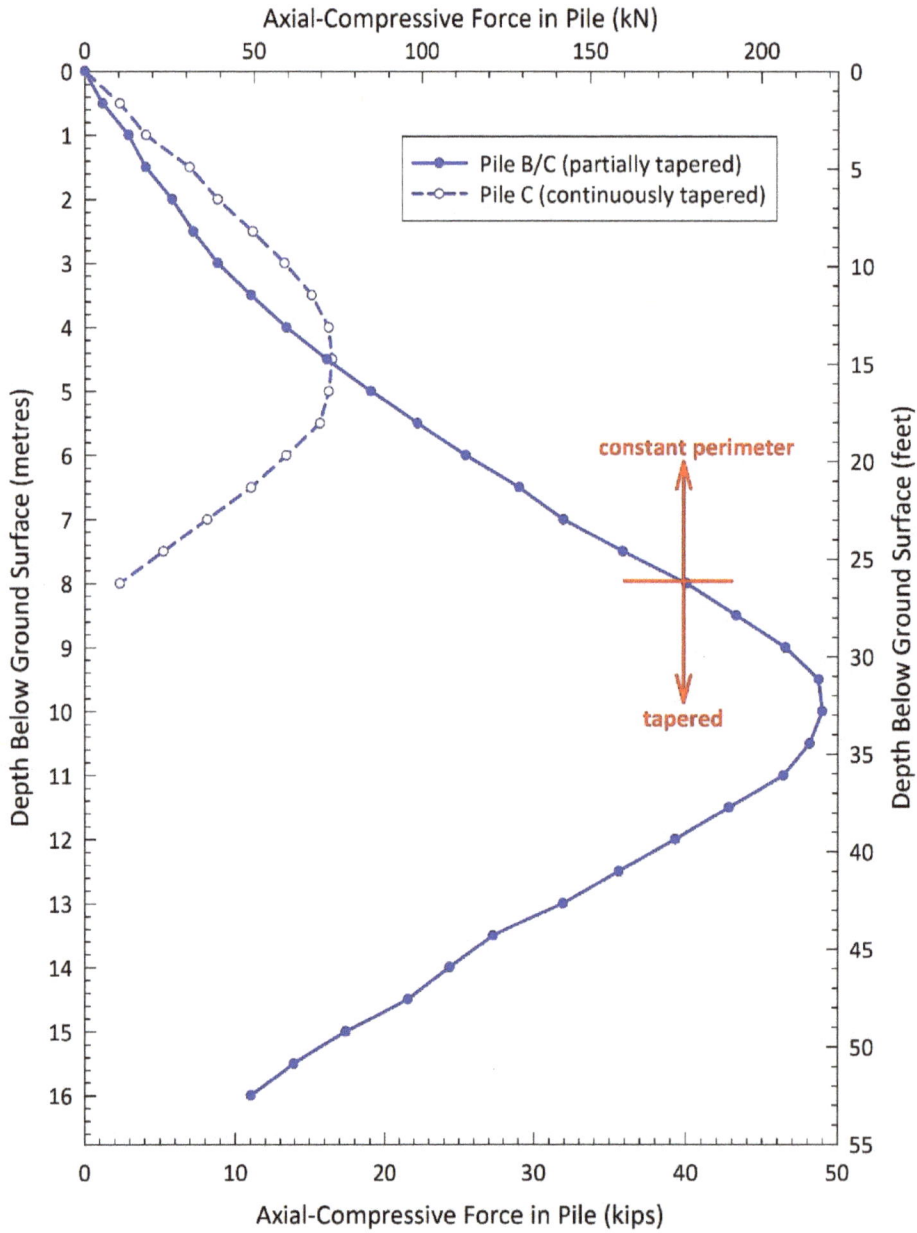

Figure H.13. Holmen Island Pile B/C vs. Pile C - Measured Axial-Compressive Forces within Pile Shaft Due to Residual Loads.

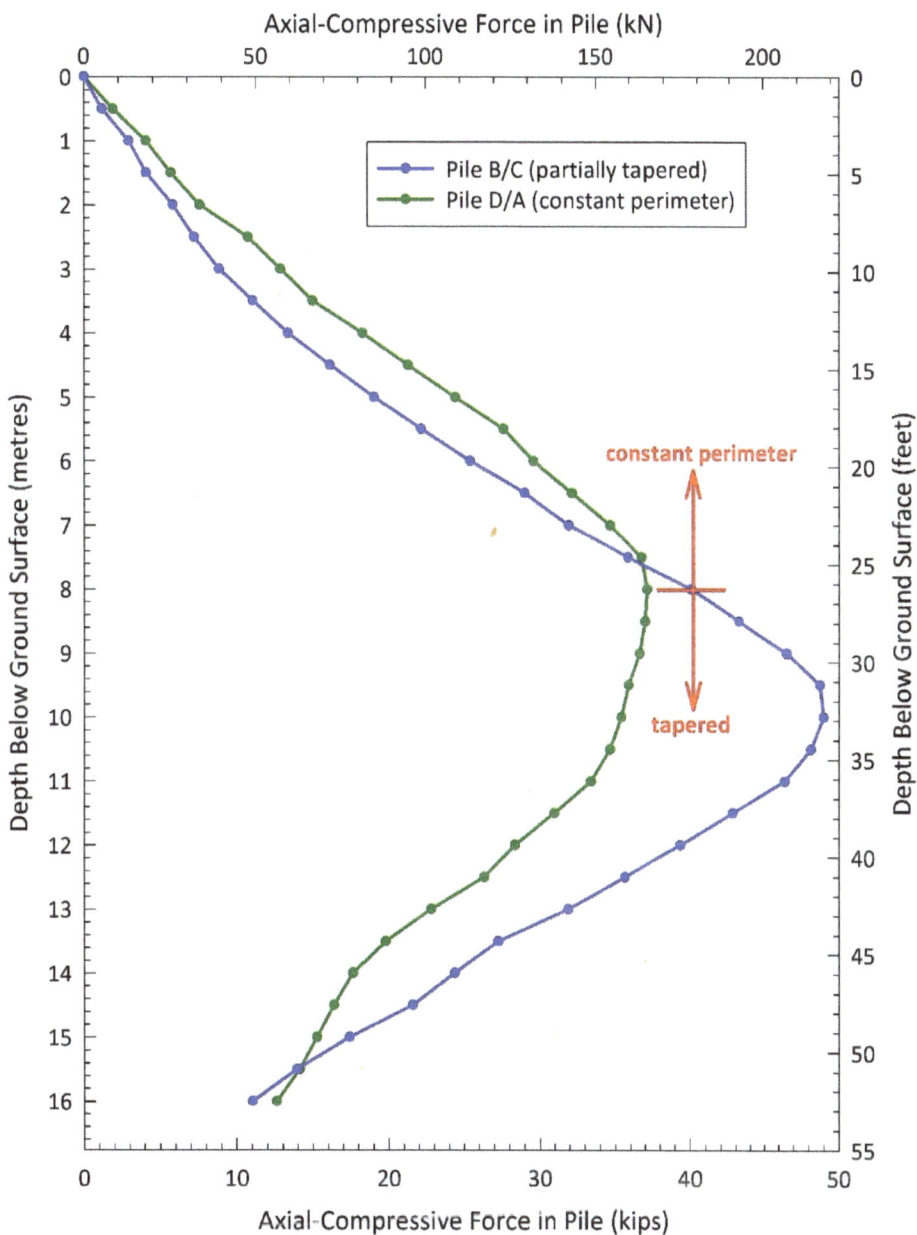

Figure H.14. Holmen Island Pile B/C vs. Pile D/A - Measured Axial-Compressive Forces within Pile Shaft Due to Residual Loads.

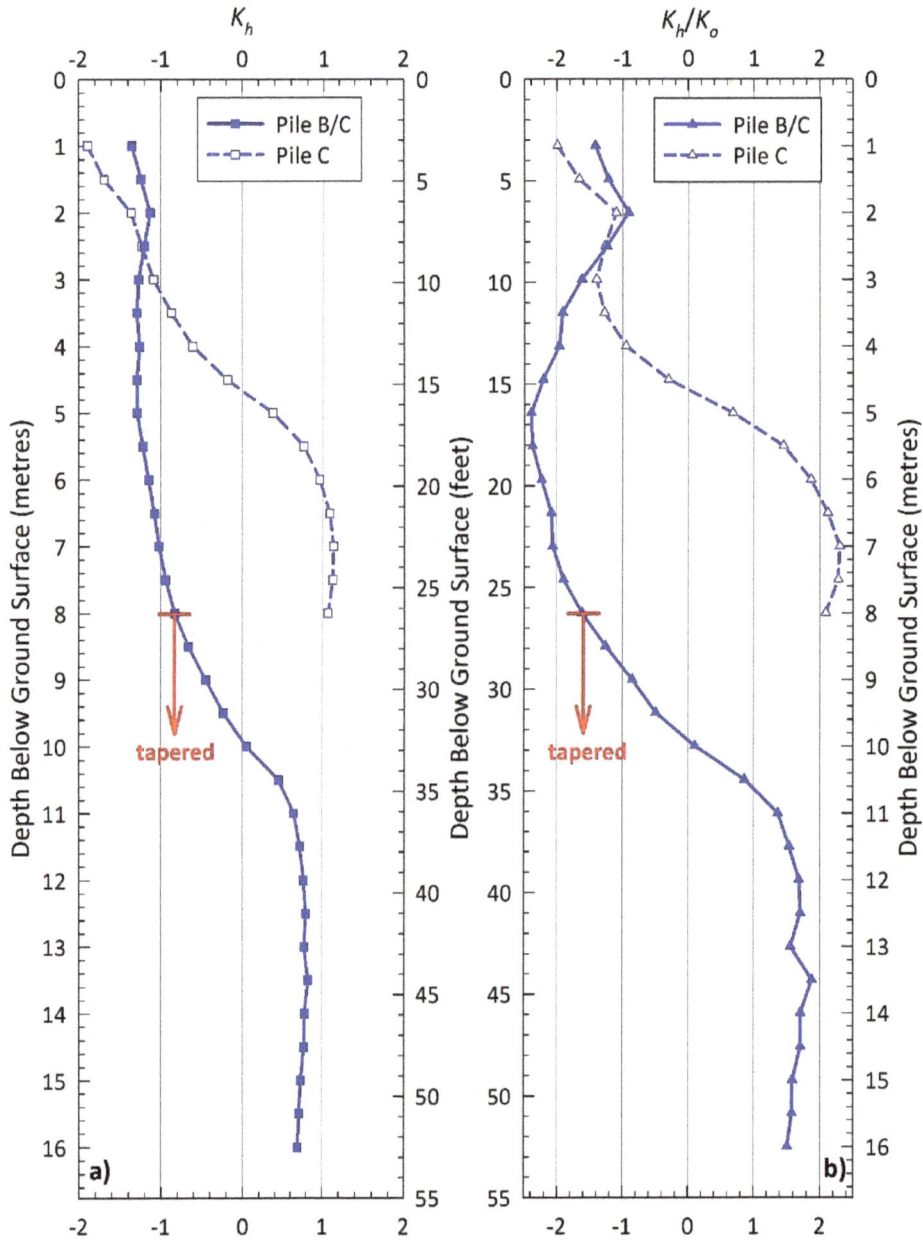

Figure H.15. Holmen Island Pile B/C vs Pile C - Calculated K_h and K_h/K_o along Shaft-Soil Interface Due to Residual Loads.

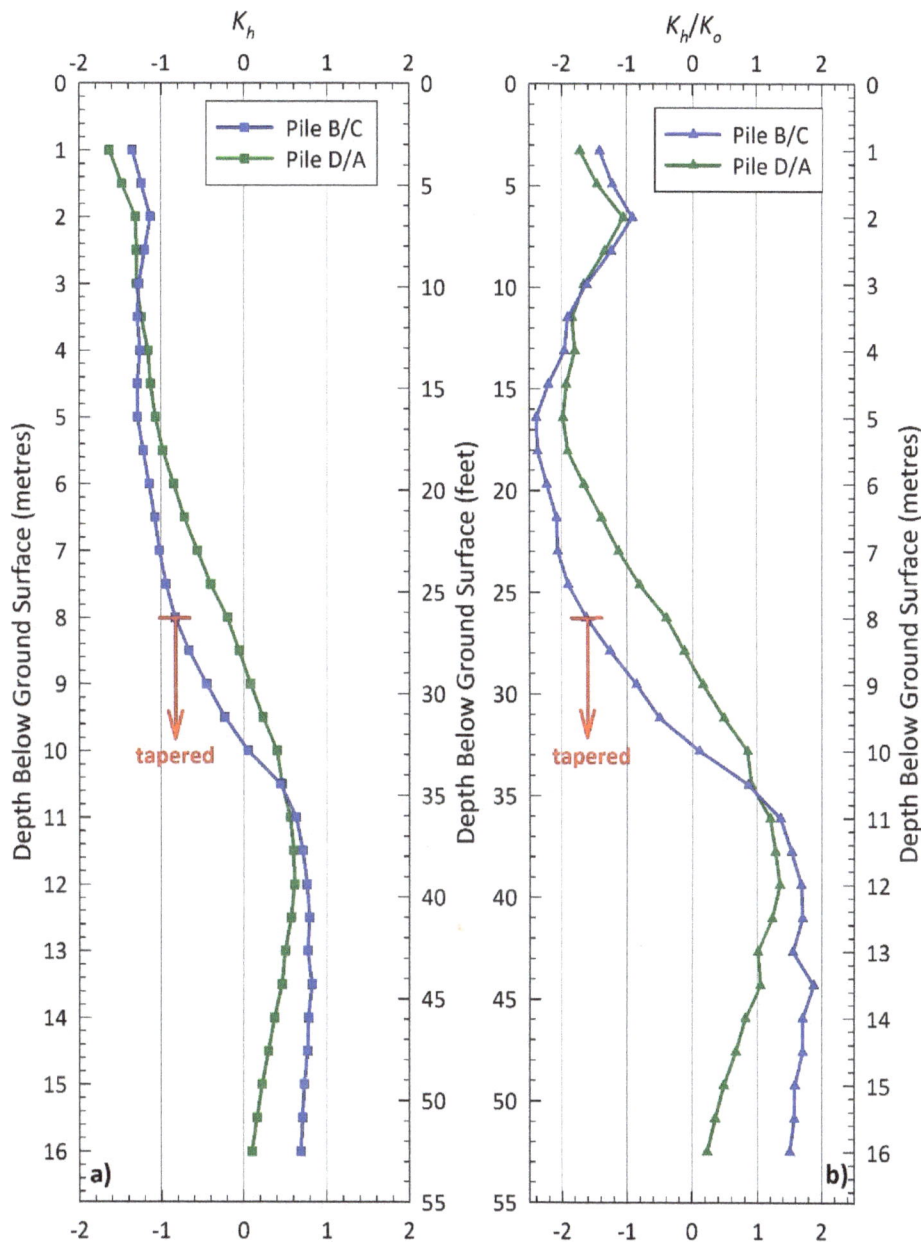

Figure H.16. Holmen Island Pile B/C vs. Pile D/A - Comparison of Calculated K_h and K_h/K_o along Shaft-Soil Interface Due to Residual Loads.

It is clear that even a partially tapered pile will tend to lock-in larger radial stresses around the pile shaft compared to a comparable pile of constant perimeter. This is reflected in Figure H.16 in two ways:

- K_h and K_h/K_o are larger within the upper 8 metres (26 ft) of depth for Pile B/C (the partially tapered pile) even though that segment of the pile has a constant perimeter identical to Pile D/A (the constant-perimeter pile). This reflects the lasting benefit of the tapered portion of Pile B/C having passed through that upper zone during the initial phase of driving followed by load testing.

- K_h and K_h/K_o are larger within the lower 8 metres (26 ft) of depth for Pile B/C (the partially tapered pile) as a result of that being the final location of the tapered segment. Note that this occurs even though the average diameter of Pile B/C within that zone is smaller than that of Pile D/A (240 mm/9.4 in vs. 280 mm/11 in). This reflects the significance of the wedging mechanism in increasing radial stresses around a pile shaft, even when the taper angle is very modest as it is in this case.

H.5.3.3 Toe Loads

H.5.3.3.1 Background

Residual toe loads are of particular interest in the present context for a number of important reasons:

- Fellenius (2015) showed how the presence of residual toe loads can significantly alter the shape of the load-settlement curve in a conventional static load test so that plunging failure is interpreted when, in reality, such behavior is not occurring as can be demonstrated by creating a corrected, 'true' load-settlement curve that incorporates the preloading caused by residual loads, especially at the toe.

- In the same 2015 paper, Fellenius showed that methods for interpreting a load-test load-settlement curve to define a single-valued total geotechnical ULS for a DFE can be significantly impacted by residual toe loads. As a result, such interpretive methodologies, when applied to a typical 'raw' load-settlement curve as determined by a static load test, tend to forecast a value for the geotechnical ULS that is greater than a value that would be obtained had the load-settlement curve been corrected for residual loads. Thus, the presence of residual toe loads tends to result in an interpretation error of an unconservative and, consequently, potentially unsafe nature when it comes to interpreting a single-valued geotechnical ULS. This is obviously a serious and widespread problem in both routine practice and research as the vast majority of analytical methodologies in common, widespread use to interpret static load tests for a geotechnical ULS value are affected by this. This includes:
  - the Davisson Offset Method (called the "Offset Limit" method in Fellenius (2015)) that dominates U.S. practice,
  - a method based on a fixed magnitude (e.g. 30 mm/1.2 in) of toe settlement that is apparently widely used in Europe, and
  - the Q_{90} method used by Gregersen et al. (1973) that was defined earlier in this appendix
  to name but a few.

- Residual toe loads tend to be significant for piles, especially in coarse-grained soil and even more so for tapered piles. This was noted by Nordlund as far back as 1963. As discussed in Appendix A, the two *Monotube* piles in coarse-grained soil at JFKIA that were studied by Fellenius et al. (2000) were estimated to have 100% of the ULS toe resistance mobilized by residual loads at the end of driving. The conclusion drawn from all of these factors suggests that residual toe loads for the Holmen Island piles could be significant.

H.5.3.3.2 Overview

The behavior of the Holmen Island piles under the compressive load tests to the geotechnical ULS is discussed later in this appendix in a separate section. However, in view of the preceding discussion that provides background information concerning the importance of residual toe loads and the fact that they can be of relatively substantial magnitude, especially for tapered piles in coarse-grained soil, it is desirable to note and use the maximum measured toe resistances for the Holmen Island piles in the present discussion of residual loads.

Note, however, that these maximum values of toe resistance are only stated here without comment. A more-complete discussion of their determination is presented later in this appendix in the context of discussing the measured and interpreted outcomes of the compressive load tests.

H.5.3.3.3 Results

Table H.1 presents a summary of the residual toe loads that were determined independently by the writer using the instrument-measured axial-compressive loads in the piles presented in the Gregersen et alia paper. Note that these loads were actually shown in graphical form earlier in this appendix in the several figures (e.g. Figure H.14) depicting the depth-wise variation of axial-compressive loads in the piles. These toe loads are considered to be exceptionally accurate as there was an electronic load cell embedded in the toe of each pile specifically for this purpose. Also shown in this table are the ULS toe loads and the ratio of residual-to-ULS toe loads expressed as a percent.

Table H.1. Residual Toe Loads for Holmen Island 1969 Test Piles.

Load/Resistance Case	Piles			
	Intermediate length (8 m/26 ft)		Full length (16 m/52 ft)	
	A	C (tapered)	D/A	B/C (part tapered)
Residual toe load, kN (kips)	20 (4.5)	10 (2.2)	56 (13)	49 (11)
ULS toe resistance, kN (kips)	72 (16)	44 (9.9)	140 (31)	83 (19)
Residual/ULS, %	28	23	40	59

While none of the residual-to-ULS ratios approaches 100% as observed or at least inferred for other tapered piles such as the *Monotube* piles at JFKIA that are discussed in Appendix A, the percentages are still significant, especially for the full-length, partially tapered Pile B/C (59%). Another relevant issue to point out is that while the toe loads for the two tapered piles are each less than those of the comparable constant-perimeter pile under both residual and ULS conditions, it should be noted that the toe cross-sectional areas of the two tapered piles were each approximately <u>one-half</u> that of the constant-perimeter piles. Thus, the unit toe resistance (r_t in Figure 2.1, what is often colloquially called the *bearing stress* or *bearing pressure*), of the two tapered piles were each either comparable to or even greater than the corresponding unit toe resistance of the constant-perimeter piles under both the residual and maximum load cases. This is particularly true for the full-length, partially tapered Pile B/C relative to Pile D/A. This observation offers some support to the hypothesis put forth by Manandhar and Yasufuku that taper affects the toe resistance of piles.

H.5.4 Compressive Load Tests

H.5.4.1 Overview

All four piles were subjected to conventional static load testing after driving. It appears from the information given in Gregersen et al. (1973), who referred to these as "push tests", that a fairly conventional load-application system consisting of:

- a reaction frame supported on four reaction piles acting in tension,

- a hydraulic jack for load application to the head of the test pile,

- an electronic load cell for load measurement at the head of the test pile, and

- two mechanical dial gauges supported on independent support beams to measure the vertical displacement of the pile head

was used.

The compressive-loading protocol for each pile consisted of seven consecutive unload-reload cycles, with each cycle having multiple load steps and a maximum load that increased with each cycle so that the total, cumulative pile-head settlement at the end of the final cycle was in the range of 65 to 85 millimetres (2.6 to 3.3 in). Although the overall total test durations were not stated, based on the information given it appears that all compressive loading of one pile was completed within one workday. Given the number of load cycles and the number of steps for each cycle, this suggests fairly rapid load application.

The focus of the presentation and discussion in the following sections is on the results for the geotechnical ULS in axial compression for each pile. Unfortunately, Gregersen et alia only provided results for their interpretation of the geotechnical ULS, Q_{90}. Therefore, before discussing the specific results for the shaft and toe resistance as was done for the residual loads, the overall total load-settlement response is discussed separately for each pile so that Q_{90} as determined by Gregersen et alia can be understood in the context of the writer's independent assessment of the actual maximum resistance of the pile.

As a final but potentially very significant comment, a crucial (based on the current state of knowledge) piece of information that was not provided in Gregersen et al. (1973) was any indication of the time lapse between driving and load testing a pile. This was noted earlier in this appendix in the context of the potential effect on measured residual loads but is mentioned again here for the almost-certain effect on the load-settlement response in axial compression. It is always dangerous to speculate but given the fact that in 1969 when the Holmen Island test piles were driven and load tested that the temporal effects on pile resistance in coarse-grained soil was not appreciated to the extent that it is today, it is likely that the time between pile driving and load testing was minimal and may well have been only a matter of days. If for no other reason, this is because the four piles under discussion were, in reality, only two piles that were each driven and load tested in two segments. Thus, it seems likely that there was a desire to move the work along so that there would not be lengthy time lapses in between driving and testing the segments.

H.5.4.2 Total Resistance

H.5.4.2.1 Overview

Shaft resistance dominated the qualitative shape of the overall load-settlement curve in compression for all four piles under discussion. This is because under load levels that can be taken to be the geotechnical ULS, shaft resistance accounted for 80% or more of the total measured resistance. This is not surprising given the relatively modest toe diameters of the piles and, more importantly, the very loose soil conditions that existed at this site.

However, the qualitative shape of the toe-load vs. settlement curve differed from that of the shaft for three of the four piles although not enough to influence the overall load-settlement behavior to any significant degree. Consequently, it was difficult to define consistent behavioral trends and thus draw broad conclusions about the qualitative nature of the development of shaft, toe, and total resistance for the four piles as a group. Thus, the relative behaviors of each pile are each discussed in some detail.

H.5.4.2.2 Pile A

This is the intermediate-length (8 m/26 ft) pile with a constant perimeter that ultimately comprised the lower half of Pile D/A that had a constant perimeter from head to toe.

The overall post-yield trend of the total load vs. pile-head settlement for this pile was that of modest work hardening, i.e. a very slight increase in total resistance with increasing settlement, at least up to the maximum cumulative settlement at the end of compressive load testing that was approximately 85 millimetres (3.3 in). It appears that this work-hardening behavior was due solely to the shaft resistance that exhibited the same work-hardening trend as the toe resistance exhibited modest strain softening, i.e. a very slight decrease in magnitude after reaching a peak at approximately 40 millimetres (1.6 in) of pile-head settlement[210].

Gregersen et alia interpreted the geotechnical ULS total resistance for this pile (Q_{90} in their terminology) to be 263 kilonewtons (59 k). This occurred at a pile-head settlement of approximately 15 millimetres (0.6 in). This compares to the total resistance of 295 kilonewtons (66 k) that occurred at the cumulative maximum pile-head settlement of approximately 85 millimetres (3.3 in).

H.5.4.2.3 Pile C

This is the intermediate-length (8 m/26 ft) pile with a continuous taper angle, α, = 0.29° over its entire length that ultimately comprised the lower half of Pile B/C.

The overall post-yield trend of this pile was similar to that of Pile A. Specifically, the total load vs. pile-head settlement for this pile exhibited very slight work hardening with increasing settlement, at least up to the maximum cumulative settlement at the conclusion of compressive load testing that was approximately 70 millimetres (2.8 in). It appears that this

---

[210] The toe settlement at peak toe resistance was likely not much less than the head settlement, not only for this but for the other three piles under discussion as well. Calculations by the writer indicate that the elastic compression of the pile shafts was likely very small in magnitude. This is supported by the very small rebound of the pile-head settlement when all applied load was removed during the unload-reload cycles (see Figure 5 in Gregersen et al. (1973)).

behavior was due solely to the shaft resistance that exhibited the same trend as in this case the toe resistance remained more or less constant in magnitude after reaching a peak at approximately 30 millimetres (1.2 in) of head settlement.

Gregersen et alia interpreted the geotechnical ULS total resistance for this pile (Q_{90}) to be 278 kilonewtons (63 k). This occurred at a pile-head settlement of approximately 13 millimetres (0.5 in). This compares to the total resistance of 310 kilonewtons (70 k) that occurred at the cumulative maximum pile-head settlement of approximately 70 millimetres (2.8 in).

H.5.4.2.4 Pile D/A

This is the final 16-metre (52-ft) -long pile with a constant perimeter from head to toe that was created by attaching an 8-metre (26-ft) -long constant-perimeter extension (referred to as Segment D) to Pile A after the latter was load-tested in compression and uplift, and then driving the two combined segments an additional 8 metres (26 ft) into the ground.

The overall post-yield trend for this pile was unique among the four piles under discussion in that the total load vs. pile-head settlement for this pile reached a peak value at a pile-head settlement of approximately 40 millimetres (1.6 in) and then exhibited very slight strain-softening behavior up to the maximum cumulative settlement at the end of compression load testing (approximately 65 millimetres (2.6 in)). This behavior was entirely due to the shaft resistance that exhibited much more noticeable strain-softening behavior that was sufficient to counteract the effects of toe resistance that exhibited the most significant work-hardening behavior of all four piles in this regard.

Overall, this pile exhibited what might be considered 'textbook' behavior of a DFE in axial compression, not surprising given the fact that this was a 'textbook' constant-perimeter pile from head to toe. Specifically, shaft resistance mobilizes the peak shear strength of the soil in interface shear then decreases as the shear strength decreases to constant-volume (critical-state) conditions. On the other hand, toe resistance never reaches a well-defined, single-valued peak but continues to develop additional resistance with increasing settlement. This behavior, with a lack of a well-defined ULS value as traditional bearing-capacity theories would forecast, is what Fellenius has argued for many years reflects the true behavior of the toe resistance of DFEs.

In any event, Gregersen et alia interpreted the geotechnical ULS total resistance for this pile (Q_{90}) to be 482 kilonewtons (108 k). This occurred at a pile-head settlement of approximately 22 millimetres (0.9 in). This compares to a peak total resistance of 510 kilonewtons (115 k) that occurred at approximately 40 millimetres (1.6 in) of pile-head settlement and the post-peak final total resistance of 495 kilonewtons (112 k) that occurred at the end of compressive load testing with a cumulative pile-head settlement of approximately 65 millimetres (2.6 in).

H.5.4.2.5 Pile B/C

This is the final 16-metre (52-ft) -long partially tapered pile that was created by attaching an 8-metre (26-ft) -long constant-perimeter extension (referred to as Segment B) to the continuously tapered Pile C after the latter was load-tested in compression and uplift, and then driving the two combined segments an additional 8 metres (26 ft) into the ground.

The overall post-yield trend of the total load vs. pile-head settlement for this pile was that of noticeable work hardening, at least up to the maximum cumulative settlement at the end of compressive load testing that was approximately 55 millimetres (2.2 in). This work-hardening behavior was due to both the shaft and toe resistances as both exhibited work-hardening (the shaft much more so than the toe, however).

As an aside, the noticeable work-hardening behavior of the shaft resistance of this pile combined with the modest work-hardening behavior of the shaft resistance for Pile C (the continuously tapered pile) tentatively answers a question raised by the writer in Horvath (2002). Specifically, while it has long been established that the traditional shaft-resistance mechanism of sliding friction often exhibits the behavior observed with Pile D/A as discussed above, i.e. peak resistance followed by some strain softening to a critical-state value, the qualitative behavior of the shaft-resistance mechanism of wedging is an unknown as noted by the writer in Horvath (2002). The observed behavior for the two tapered piles under discussion (Pile B/C and Pile C) suggests that the shaft-resistance mechanism of wedging can continue to exhibit work-hardening behavior even at magnitudes of pile-head settlement that can reasonably be defined as the geotechnical ULS. This is consistent with cylindrical-cavity expansion (CCE) theory as well as observed behavior in PMT testing that indicate that relatively substantial cavity expansion is required before the so-called limiting pressure (maximum radial stress acting along the shaft-soil interface) is reached.

In any event, Gregersen et alia interpreted the total geotechnical ULS resistance (Q_{90}) for Pile B/C to be 467 kilonewtons (105 k). This occurred at a pile-head settlement of approximately 23 millimetres (0.9 in). This compares to a peak total resistance of 508 kilonewtons (114 k) that occurred at the end of compressive load testing with a cumulative pile-head settlement of approximately 55 millimetres (2.2 in).

H.5.4.2.6 Comments

As this was the first time that the writer had encountered the Q_{90} method for interpreting the load-settlement curve from a static load test in order to quantify a single-valued geotechnical ULS, it is of interest to make several observations concerning the outcomes for the 1969 Holmen Island test piles with respect to this interpretive methodology. Note that the writer did not independently determine the Q_{90}-method results but used the values scaled from Figure 5 in Gregersen et al. (1973):

- For each of the four piles, the Q_{90} method yielded a result that was approximately 90% of the actual maximum total resistance reached in the load test. In each case, these maximum values could be reasonably interpreted as being a more or less 'true' value of the geotechnical ULS. Note, however, that this numerical coincidence is <u>not</u> the reason for the "90" in Q_{90}. The reason for this nomenclature lies in the definition of this term as given earlier in this appendix and illustrated graphically in Figure H.6.

- The magnitude of pile-head settlement that correlates with Q_{90} varied considerably between the intermediate length (8 m/26 ft) and full-length (16 m/52 ft) piles. This is not surprising as, all other things being equal, a longer pile and the concomitant larger applied loads will produce larger values of elastic compression of the shaft. Thus, the overall pile-soil system will have a softer load response. Because Q_{90} is dependent on the stiffness of the overall pile-soil system, the softer the system, the larger the settlement associated with Q_{90}.

- Although the writer did not formally apply the Davisson Offset Method to these four piles due to a lack of published information concerning the axial stiffness of the piles, it was at least possible to make an approximate, order-of-magnitude assessment of the outcomes from Davisson's widely used (in the U.S. at least) method. For all four piles, the Davisson method yielded a single-valued estimate for the geotechnical ULS that was noticeably less than the value from the Q_{90} method and at a substantially smaller magnitude of pile-head settlement. This is not surprising as the Davisson method is well-known for being a very conservative methodology in most cases for evaluating the geotechnical ULS from the load-settlement results of a conventional static load test.

H.5.4.3 Shaft Resistance

H.5.4.3.1 Overview

The same type of figures showing the depth-wise, instrument-measured, axial-compressive forces within the pile shaft and calculated (by the writer) K_h and K_h/K_o ratio values acting along the shaft-soil interface that were presented for the residual-load case are presented in the following sections for the compressive load testing. Note that all results presented subsequently are based on the calculated (by Gregersen et alia) Q_{90} resistance, <u>not</u> the maximum resistance determined by the writer, as Gregersen et alia presented detailed results only for their interpretation of the geotechnical ULS (Q_{90}). Thus, the presented results should not be interpreted as reflecting the test maximums although they are not far from them as discussed in the preceding discussion of total resistance.

Overall and as expected, the largest changes (from the residual-load case) in both axial-compressive forces within the pile shaft and, in particular, the radial stresses (as defined by K_h and the K_h/K_o ratio) acting along the shaft-soil interface occur within the upper roughly 50% of each pile that was above the neutral plane at the end of driving. This is because the external applied loads during the compressive load tests and the concomitant pile settlement relative to the surrounding soil first caused the drag loads to reduce to zero and then reverse direction to produce resistance. With reference to Figure 2.1, as the pile settled relative to the adjacent ground, the unit drag load, q_n, first reduced to zero and then increased as unit shaft resistance, r_{ss} or r_{st}, depending on the shaft geometry.

In the lower portions of the piles where the axial-compressive forces within the pile shaft and associated shaft resistance was already partially mobilized, the increases in both piles forces and radial stresses were much more modest. This was especially true for the two tapered piles (C and B/C) where the increases were surprisingly small for the lower roughly 25% of the pile lengths. This indicates that the residual loads had already mobilized the bulk of the available pile resistance in that area for the tapered piles.

This last observation is of great interest and relevance to the focus of this monograph. It clearly implies that for tapered piles the bulk of the taper benefit is achieved and locked-in during driving. This further implies that subsequent external load application, even to levels approaching the geotechnical ULS for the pile as a whole, produce relatively little additional taper benefit, at least within the lower portion of the tapered section. Clearly, this calls into serious question the primary assumption by Kodikara and Moore (and El Naggar et alia and Manandhar et alia as well) that taper benefit derives <u>solely</u> from post-driving external load application (and only a portion of that load application at that).

H.5.4.3.2 Pile A

This is the intermediate-length (8 m/26 ft) pile with a constant perimeter that ultimately comprised the lower half of Pile D/A that had a constant perimeter from head to toe. The Q_{90}-based ULS resistance for this pile was determined to be 263 kilonewtons (59 k), of which 76% (201 kN/45 k) was due to shaft resistance.

Figure H.17 shows the depth-wise variation of the instrument-measured, axial-compressive forces within the pile shaft at the Q_{90} load level in compressive load testing. As with similar figures shown for the residual-load case, the force at the toe of the pile represents the toe load that is discussed separately later in this appendix. Also shown in this figure are the results for this pile under residual loads so that the changes that occurred as a result of external load applications can be readily visualized.

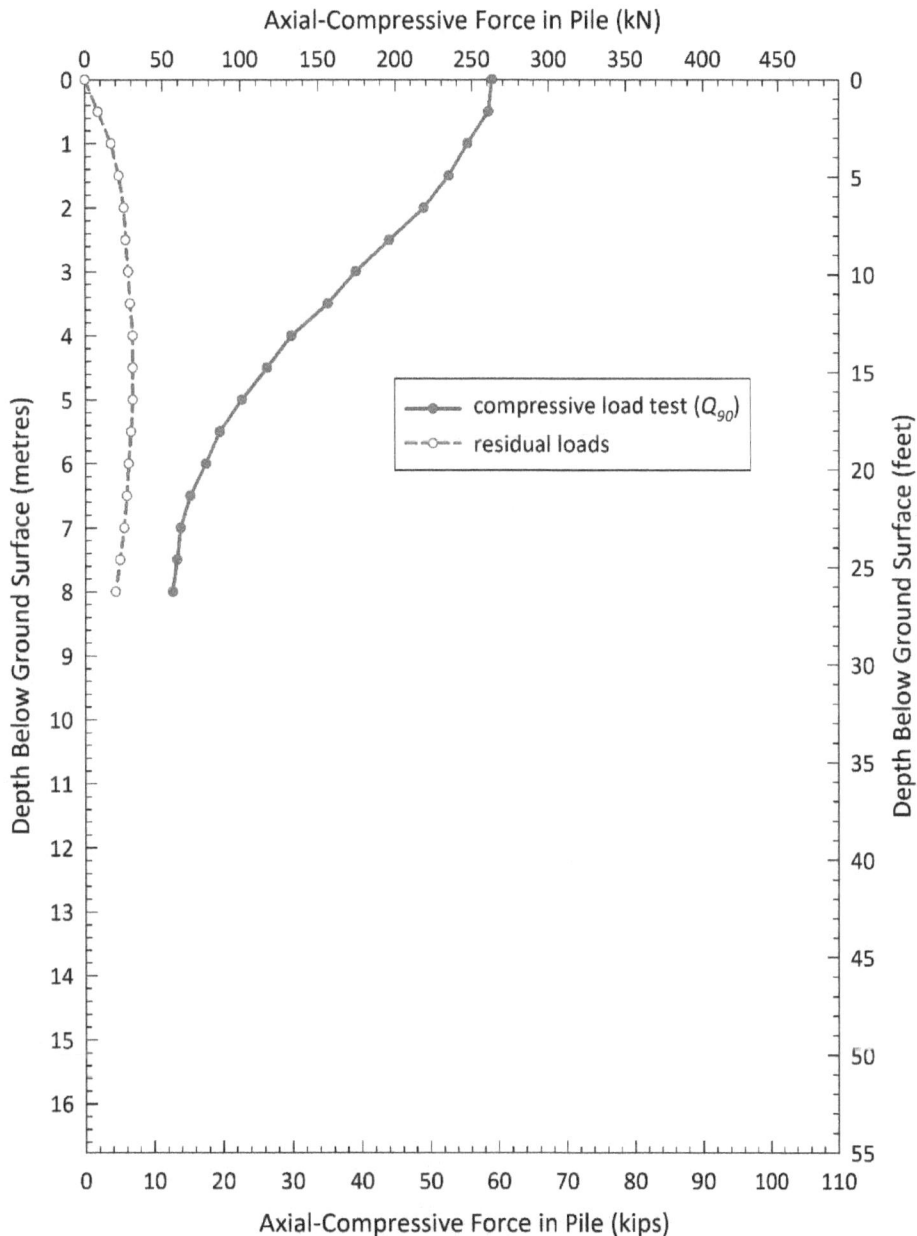

Figure H.17. Holmen Island Pile A - Measured Axial-Compressive Forces within Pile Shaft at Q_{90} Load Level in Compressive Load Test.

Figure H.18 shows the depth-wise variations of K_h and the K_h/K_o ratio that were calculated by the writer for this pile. Also shown in this figure are the results for this pile under residual loads so that the changes that occurred, especially the transition from drag load to resistance above the neutral plane that existed after driving at a depth of approximately 4.5 metres (15 ft), as a result of external load applications can be readily visualized.

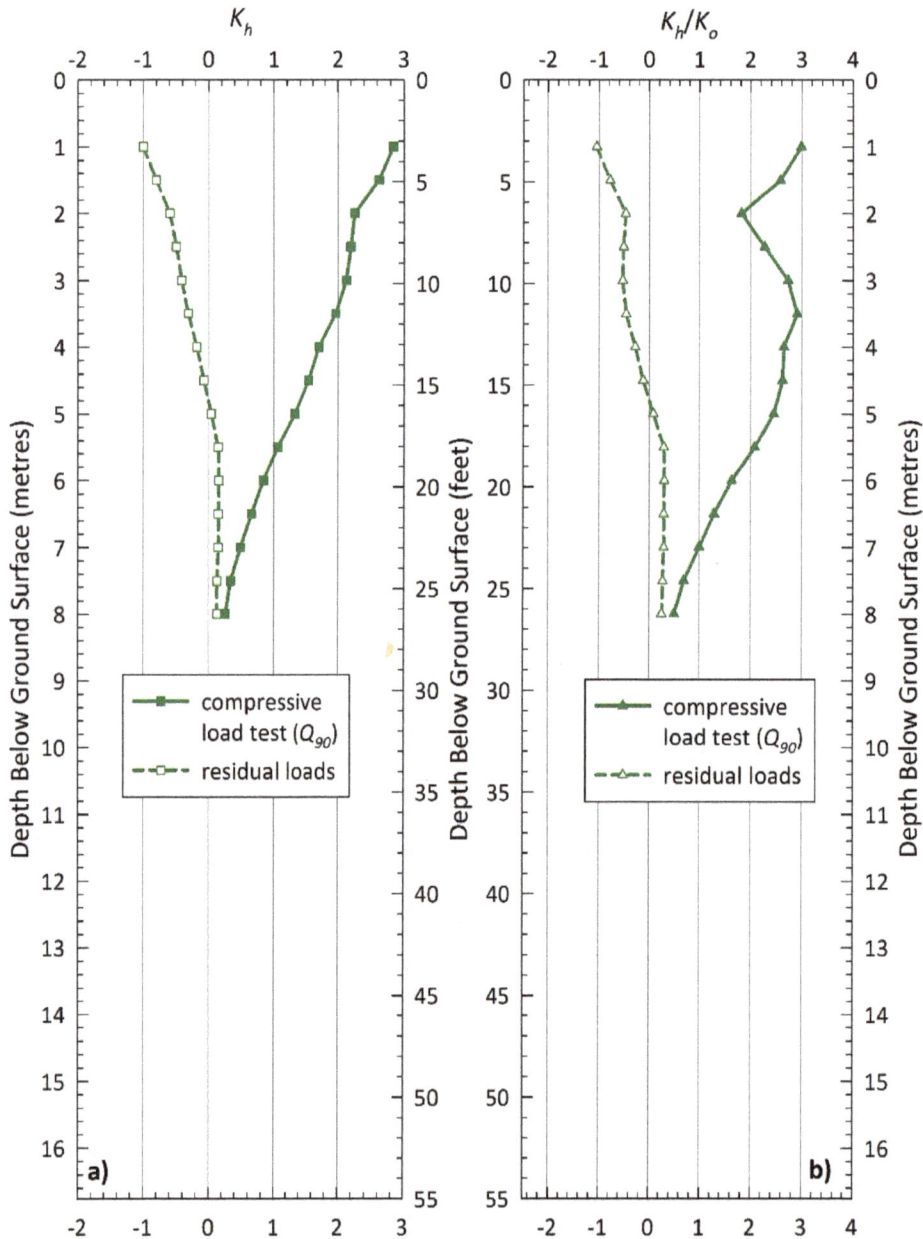

Figure H.18. Holmen Island Pile A - Calculated K_h and K_h/K_o along Shaft-Soil Interface at Q_{90} Load Level in Compressive Load Test.

While the large swing in both the sense and magnitude of the shaft-soil interface stresses along the upper portion of the pile was expected as the pile displaced downward relative to the surrounding ground during the compressive load test, it is somewhat surprising that toward the toe of the pile the trend was that of much more modest increases in the unit shaft resistance. This is evidenced by the depth-wise decrease in the difference in both the K_h and K_h/K_o values between that which existed under the post-driving residual loads and under the geotechnical ULS (based on Q_{90}) when the pile-head settlement was approximately 15 millimetres (0.6 in) and the toe settlement was likely not much less than that.

H.5.4.3.3 Pile C

This is the intermediate-length (8 m/26 ft) pile with a continuous taper angle, α, = 0.29° over its entire length that ultimately comprised the lower half of Pile B/C. The Q_{90}-based ULS resistance for this pile was determined to be 278 kilonewtons (63 k), of which 88% (244 kN/55 k) was due to shaft resistance.

Figure H.19 shows the depth-wise variation of the instrument-measured, axial-compressive forces within the pile shaft at the Q_{90} load level in compressive load testing. Also shown in this figure are the results for this pile under residual loads so that the changes that occurred as a result of external load applications can be readily visualized.

Figure H.20 shows the depth-wise variations of K_h and the K_h/K_o ratio that were calculated by the writer for this pile. Also shown in this figure are the results for this pile under residual loads so that the changes that occurred, especially the transition from drag load to resistance above the neutral plane that existed after driving at a depth of approximately 4.5 metres (15 ft), as a result of external load applications can be readily visualized.

Overall, the same trend observed for Pile A (Figure H.18) occurs for Pile C. Specifically, there is a large swing in both the sense and magnitude of the shaft-soil interface stresses along the upper portion of the pile as expected as the pile displaced downward relative to the surrounding soil during the compressive load test. However, toward the toe of the pile there was a rapid decrease in the increase in the unit shaft resistance so that just above the toe of the pile the unit shaft resistance is not much changed from that which existed after driving under residual loads only.

As can be seen in the comparison shown in Figure H.21, tapered Pile C had a modestly (6%) larger value of the Q_{90}-level geotechnical resistance (278 kN/63 k vs. 263 kN/59 k) compared to the comparable constant-perimeter Pile A discussed previously but developed a significantly larger portion (88%) of its overall axial-compressive resistance from shaft resistance compared to the comparable Pile A (76%). In terms of absolute values, shaft resistance was 244 kN (55 k) for Pile C vs. 210 kN (45 k) for Pile A or 21% larger. This is despite the fact that Pile C had only 86% of the shaft circumferential area and 73% of the total volume of Pile A. These results are not surprising given the fact that Pile C is tapered while Pile A is not, and reflect what is broadly and qualitatively referred to as taper benefit.

The genesis of this difference in shaft resistance between Pile C and Pile A can be readily seen in Figure H.22 that compares the radial stresses (expressed using the K_h and K_h/K_o lateral earth pressures parameters) acting along the shaft-soil interface of each pile. The results shown in this figure clearly indicate that substantially larger radial stresses exist over approximately 75% of the length of Pile C compared to Pile A. More importantly, this difference increases with depth and concomitant increasing vertical effective overburden stresses.

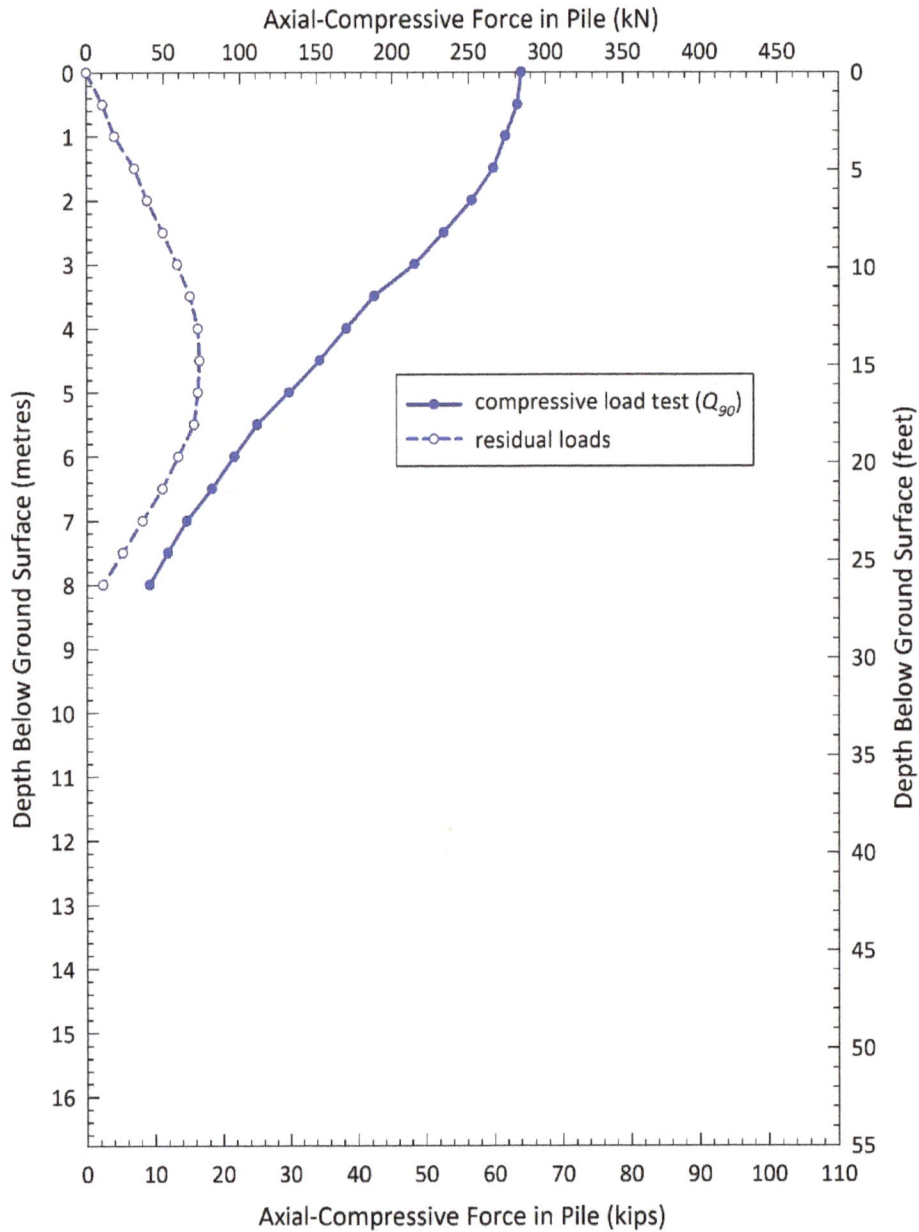

Figure H.19. Holmen Island Pile C - Measured Axial-Compressive Forces within Pile Shaft at Q_{90} Load Level in Compressive Load Test.

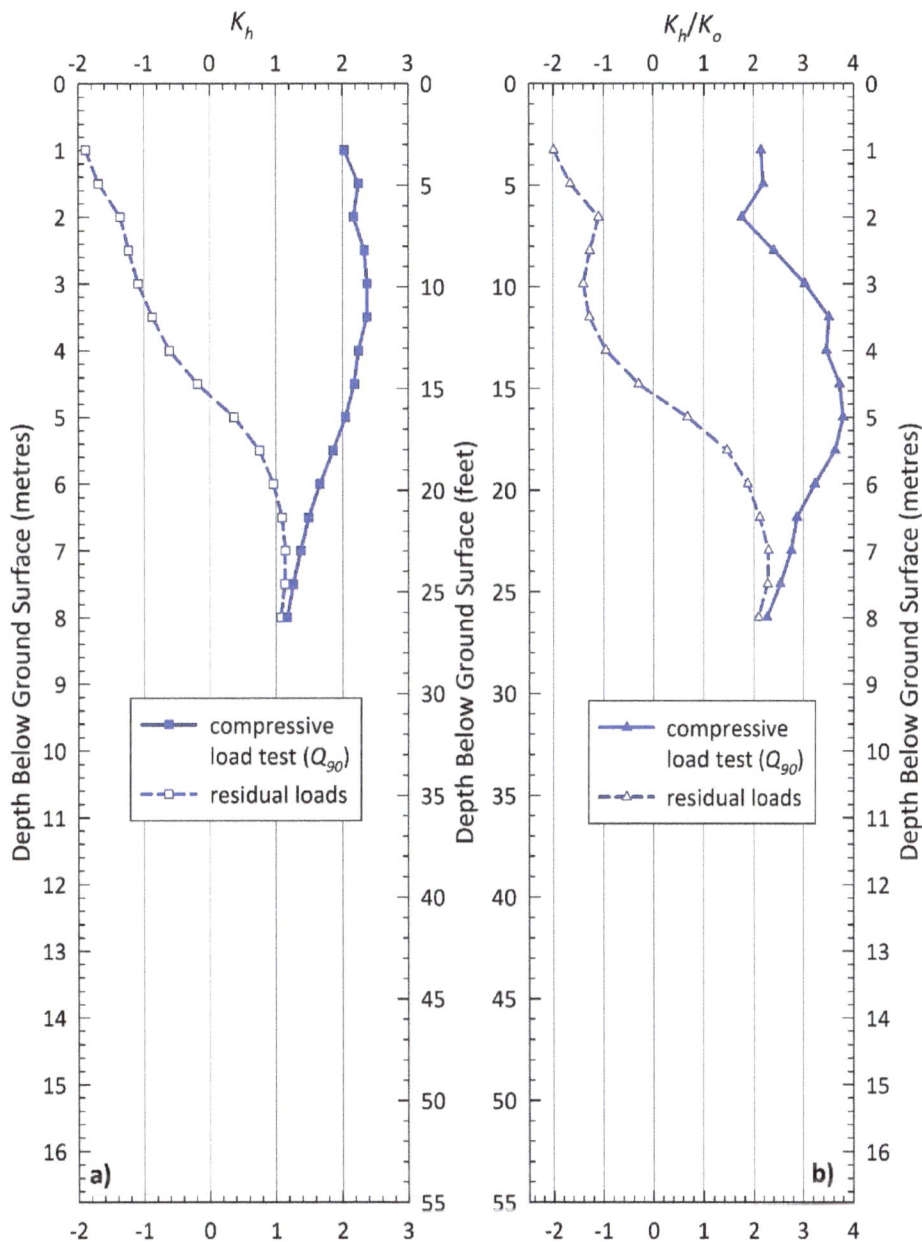

Figure H.20. Holmen Island Pile C - Calculated K_h and K_h/K_o along Shaft-Soil Interface at Q_{90} Load Level in Compressive Load Test.

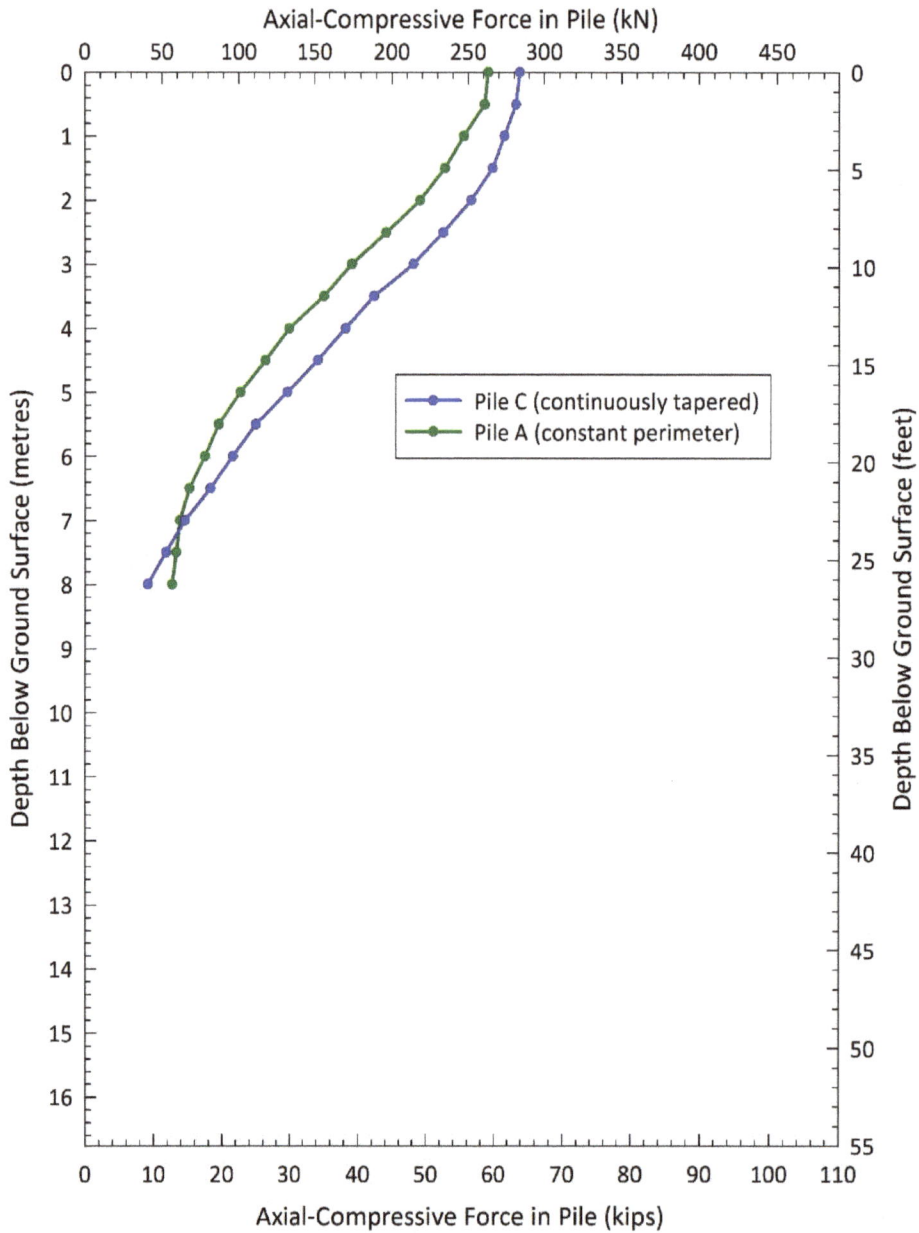

Figure H.21. Holmen Island Pile C vs. Pile A - Measured Axial-Compressive Forces within Pile Shaft at Q_{90} Load Level in Compressive Load Test.

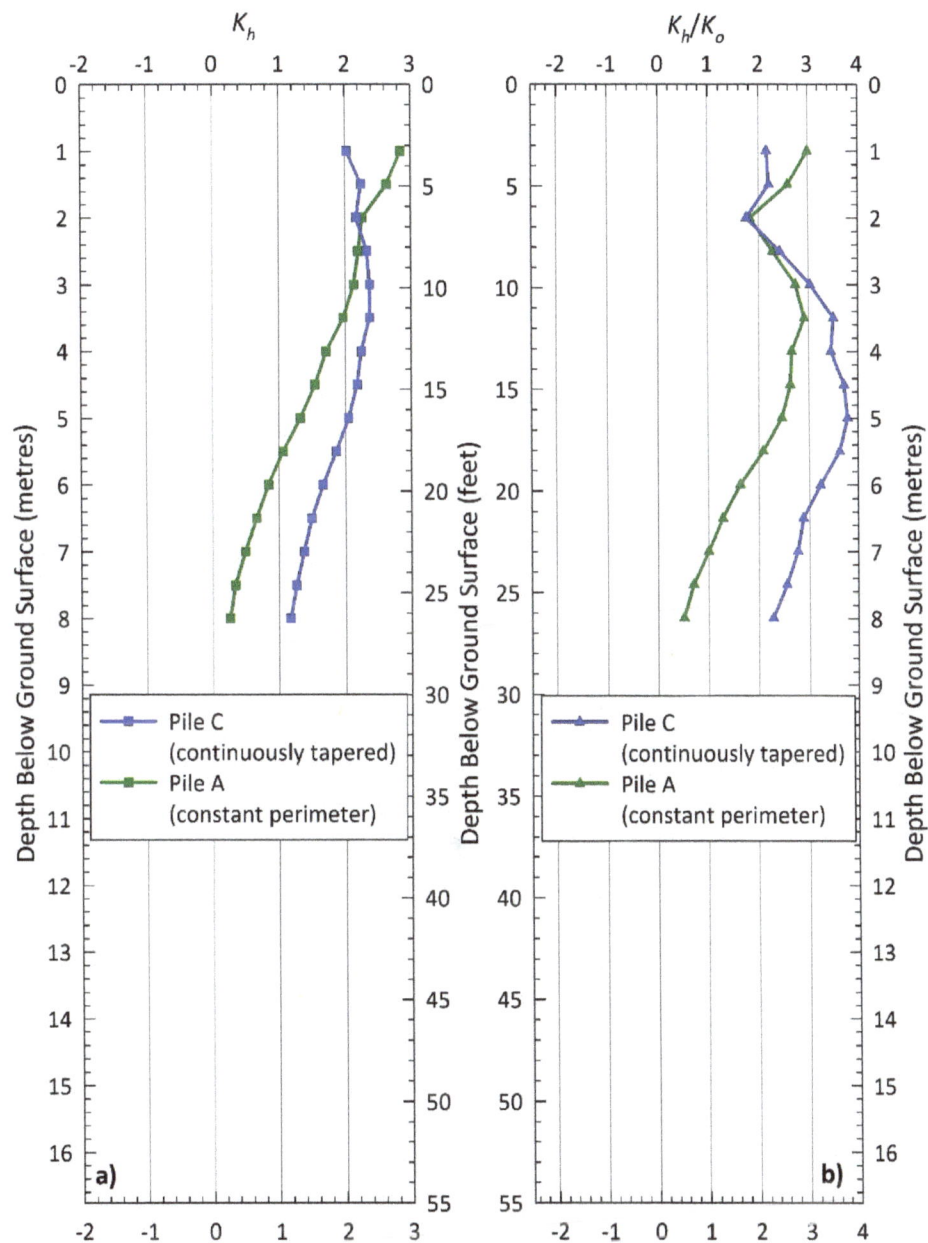

Figure H.22. Holmen Island Pile C vs. Pile A - Calculated K_h and K_h/K_o along Shaft-Soil Interface at Q_{90} Load Level in Compressive Load Test.

H.5.4.3.4 Pile D/A

This is the final 16-metre (52-ft) -long pile with a constant perimeter from head to toe that was created by attaching an 8-metre (26-ft) -long constant-perimeter extension (referred to as Segment D) to Pile A after the latter was load-tested in compression and uplift, and then driving the two combined segments an additional 8 metres (26 ft) into the ground. The Q_{90}-based ULS resistance for this pile was determined to be 482 kilonewtons (108 k), of which 77% (373 kN/84 k) was due to shaft resistance. Note that the ratio of shaft-to-total resistance for this pile is essentially the same as that for its initial component (Pile A).

Figure H.23 shows the depth-wise variation of the instrument-measured, axial-compressive forces within the pile shaft at the Q_{90} load level in compressive load testing. Note that the zero-depth load shown (466 kN/105 k) is slightly lower than the Q_{90} value (482 kN/108 k) stated in the preceding paragraph (they should obviously be the same). This reflects an inconsistency in the original Gregersen et al. (1973) paper. The larger value was scaled from the overall load-settlement plot (Figure 5 in the original paper) in which the Q_{90} point was noted explicitly. The smaller value was scaled from the load vs. depth plot (Figure 6 in the original paper). Because Figure H.23 is intended to replicate the load vs. depth plot, it was appropriate to use the smaller value as the starting point that is Q_{90} by definition. Also shown in Figure H.23 are the results for this pile under residual loads so that the changes that occurred as a result of external load applications can be readily visualized.

It is of interest to note that the results shown in this figure indicate a surprisingly low reduction in the axial force within the pile shaft over almost the first half (8 m/26ft) of the pile. This indicates that the unit shaft resistance, r_{ss} in this case, was relatively small in magnitude along a substantial portion of the shaft even though the pile was loaded to nominally the geotechnical ULS.

The source of this surprising (to the writer at least) behavior can be seen in Figure H.24 that shows the depth-wise variations of K_h and the K_h/K_o ratio that were calculated by the writer for this pile. Also shown in this figure are the results for this pile under residual loads so that the changes that occurred, especially the transition from drag load to resistance above the neutral plane that existed after driving at a depth of approximately 8.5 metres (28 ft), as a result of external load applications can be readily visualized.

The results shown in Figure H.24 are both interesting and illustrative. To begin with, they clearly indicate that the load applied during the compressive load test produced resistance throughout the entire length of the pile shaft as would be expected as the entire pile settled under external load application. However, between the ground surface and the neutral plane that existed after driving (approximately halfway down the pile shaft in the case of Pile D/A), this development of resistance had to first overcome the drag load that existed due to the residual loads. Thus, the resistance that developed during the compressive load test first had to reduce the relatively sizable drag load to zero before there could be a net increase in resistance.

As is well-known, in order for sliding friction to develop along a planar interface between soil and a non-soil material (a pile shaft in this case) there must always be relative displacement between the two dissimilar materials. However, the soil adjacent to a pile (or DFE in general) shaft is not fixed in space as is often simplistically assumed or at least implied so that all pile settlement translates into relative shaft-soil displacement (slip) that translates into shear stress and thus resistance as in the situation under current discussion. In reality, the soil adjacent to the pile is settling as well but it is not perfectly adhered to the pile shaft as assumed in the Randolph-Wroth Problem discussed in Chapter 5. The reality is somewhere in between, i.e. both the DFE and soil settle but the former more than the latter.

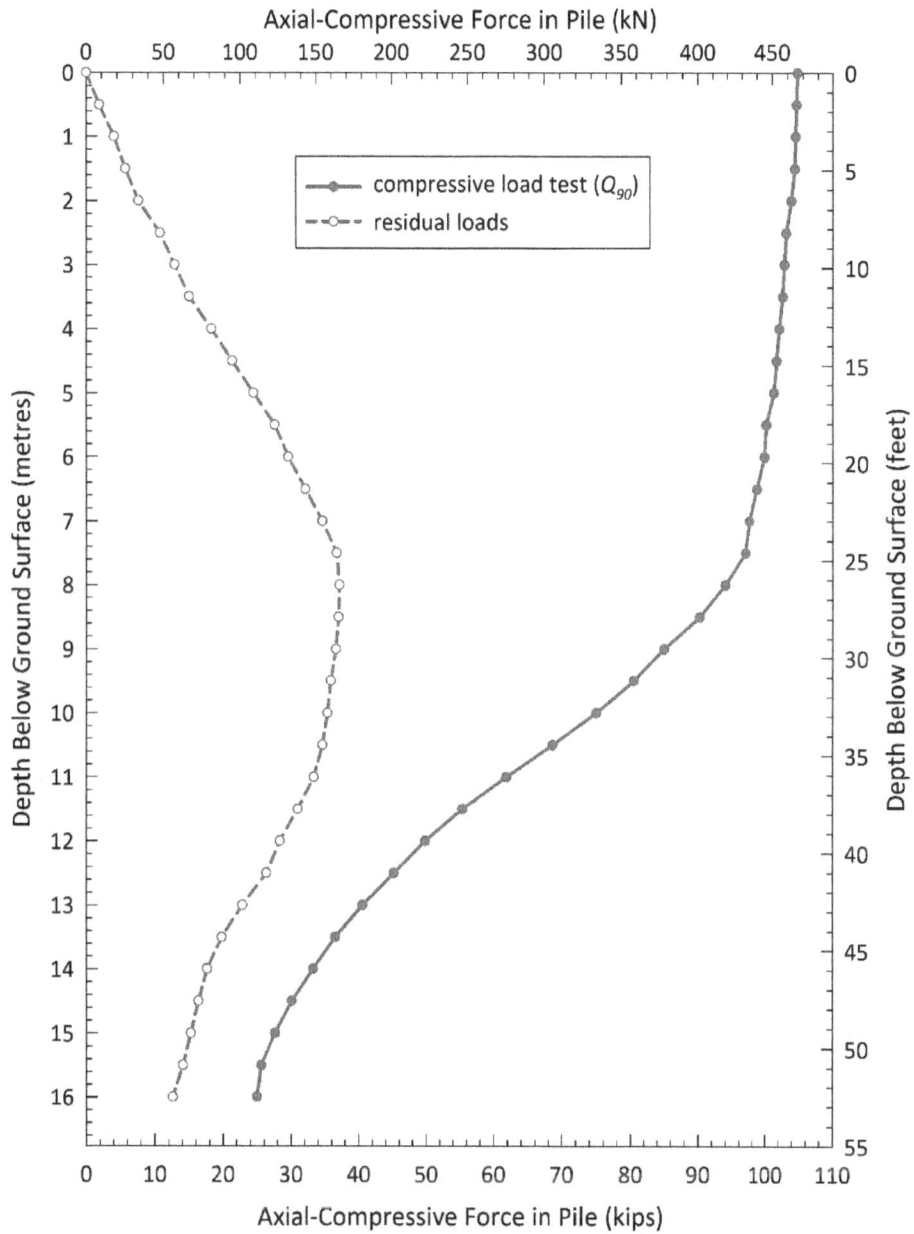

Figure H.23. Holmen Island Pile D/A - Measured Axial-Compressive Forces within Pile Shaft at Q_{90} Load Level in Compressive Load Test.

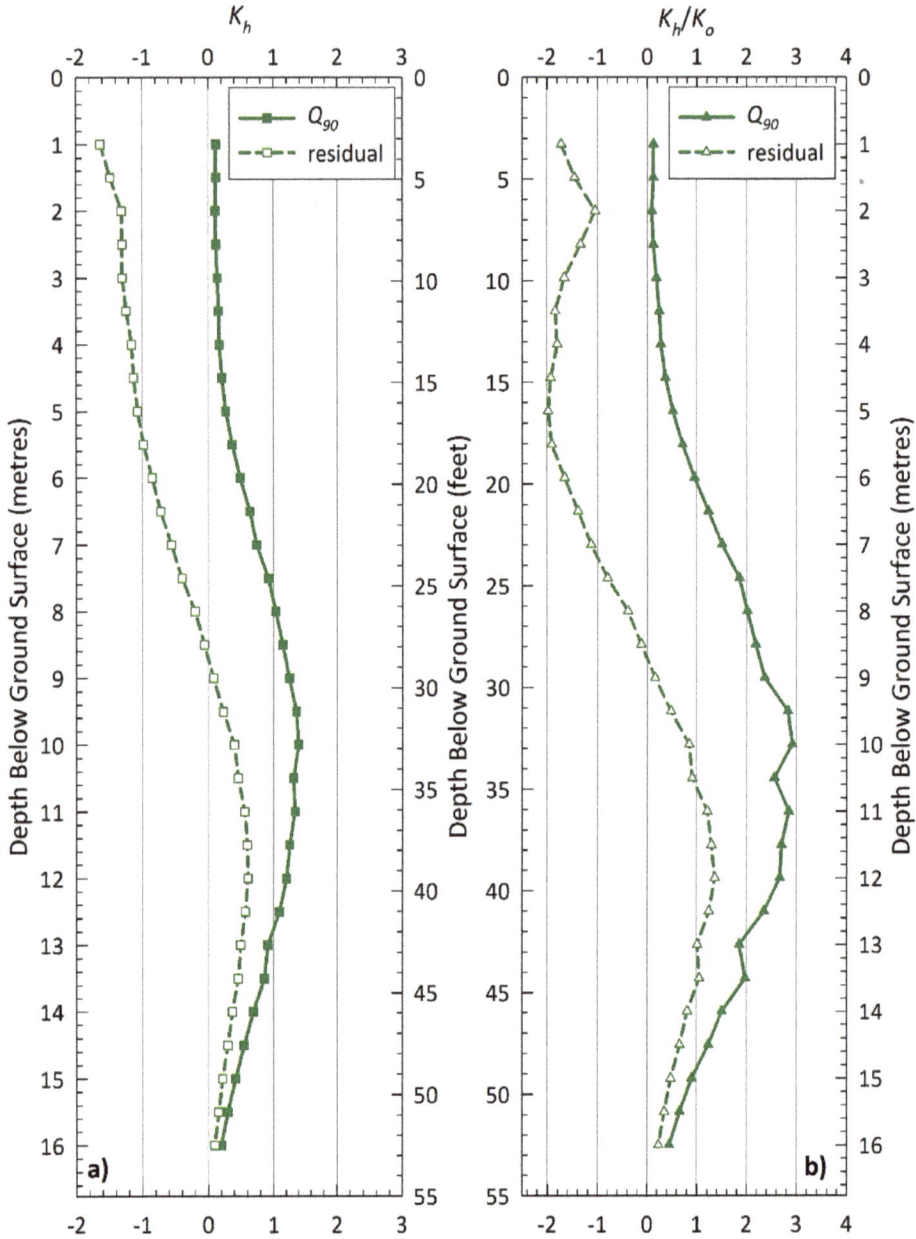

Figure H.24. Holmen Island Pile D/A - Calculated K_h and K_h/K_o along Shaft-Soil Interface at Q_{90} Load Level in Compressive Load Test.

The point being made here is that as a pile (Pile D/A in this specific case) settles under externally applied axial-compressive load, the necessary pile-soil slip to first overcome the drag load and then develop resistance occurs much more slowly than might be imagined as at least some of the pile settlement that occurs due to the applied load simply drags the adjacent soil down with the pile. As a result, the net magnitude of pile-soil slip is always something less than the gross settlement of the pile shaft.

Now, how much pile settlement needs to occur before sufficient pile-soil slip and concomitant net resistance develops likely depends on many factors including:

- The magnitude of drag load to be overcome.

- The compressibility of the pile shaft (which affects elastic compression).

- The shear stiffness of the soil surrounding the pile shaft as this influences the magnitude of ground settlement adjacent to the pile shaft.

- The magnitude of pile settlement from the externally applied load.

In the case of Pile D/A, along the upper portion of the shaft there was so much drag load to reverse combined with the fact that the pile was relatively long and compressible and the adjacent ground relatively soft (as indicated by the relatively small G_{max} values in Figure H.5b) that by the time the pile-head settlement reached the criterion for Q_{90} the upper portion of the pile had barely started to develop a net resistance. Note that by comparison, as can be seen in Figure H.18, the shorter and stiffer Pile A was able to both recover from the residual load and develop relatively significant net resistance by the time that the Q_{90} criterion was reached.

Thus, it appears that in some cases, with Pile D/A being one, it is possible that the magnitude of pile settlement necessary to first overcome the residual load and then produce a net resistance may be surprisingly large. In fact, it may be larger than the approximately 22 millimetres (0.9 in) of pile-head settlement that is associated with the Q_{90} load of Pile D/A.

H.5.4.3.5 Pile B/C

This is the final 16-metre (52-ft) -long partially tapered pile that was created by attaching an 8-metre (26-ft) -long constant-perimeter extension (referred to as Segment B) to the continuously tapered Pile C after the latter was load-tested in compression and uplift, and then driving the two combined segments an additional 8 metres (26 ft) into the ground. The Q_{90}-based geotechnical ULS for this pile was determined to be 467 kilonewtons (105 k), of which 88% (410 kN/92 k) was due to shaft resistance. Note that the ratio of shaft-to-total resistance for this partially tapered pile is identical to that of its initial continuously tapered component (Pile C).

Figure H.25 shows the depth-wise variation of the instrument-measured, axial-compressive forces within the pile shaft at the Q_{90} load level in compressive load testing. Also shown in this figure are the results for this pile under residual loads so that the changes that occurred as a result of external load applications can be readily visualized.

Of note in this figure is the near-verticality of the Q_{90} curve in this figure for the first several metres of depth that implies virtually no resistance was developed within that depth range for the same reasons discussed at length above for constant-perimeter Pile D/A. However, this behavior is even more pronounced and dramatic for Pile B/C than it was for Pile D/A as can be seen in the comparison between Pile B/C and Pile D/A shown in Figure H.26.

The root cause of the observed results for Pile B/C shown in Figures H.25 and H.26 is best illustrated using radial stresses acting along the shaft-soil interface as expressed using K_h and the K_h/K_o ratio. This is shown in Figure H.27 where these parameters are shown for Pile B/C under both residual loads (as shown earlier in Figure H.15) and at the Q_{90} load level that corresponds to the results shown in Figures H.25 and H.26.

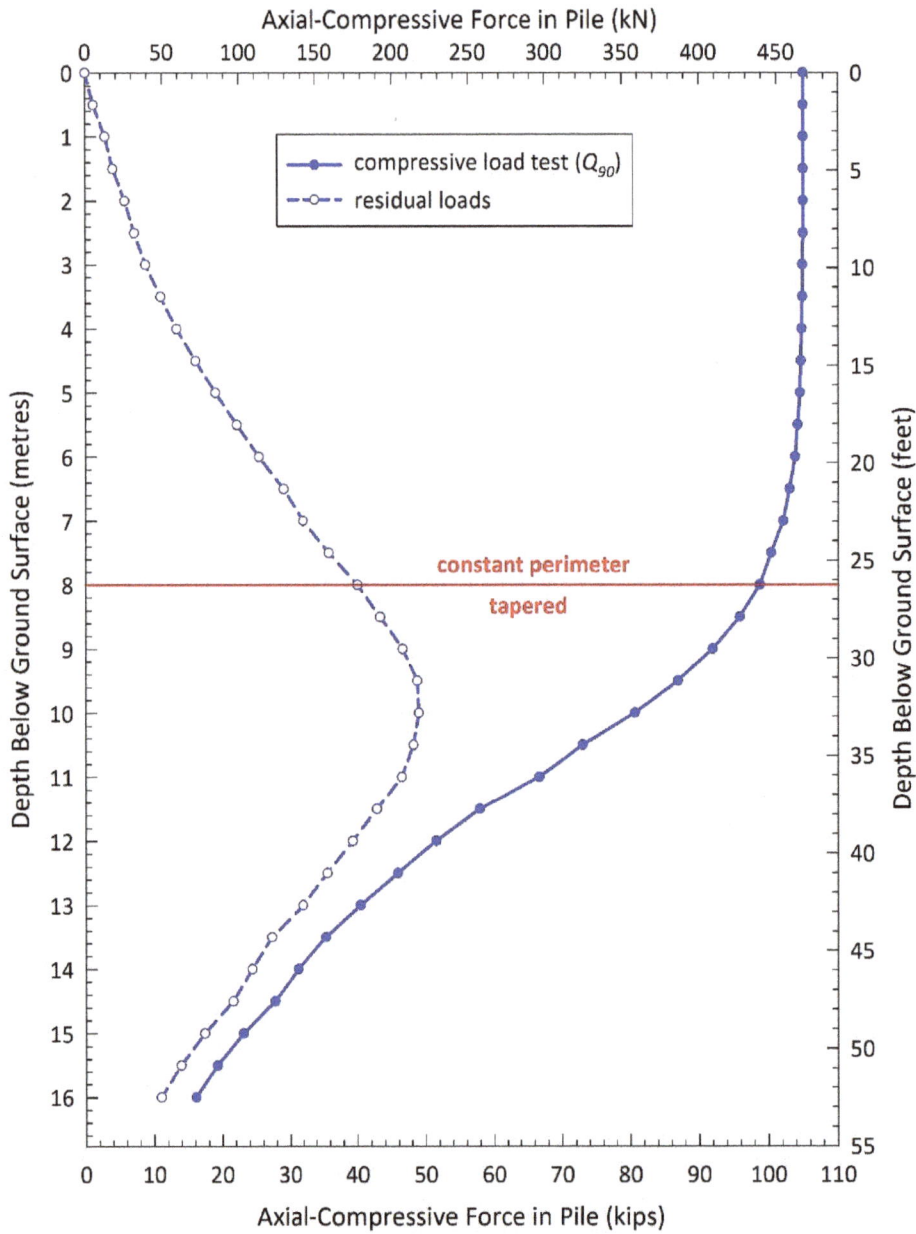

Figure H.25. Holmen Island Pile B/C - Measured Axial-Compressive Forces within Pile Shaft at Q_{90} Load Level in Compressive Load Test.

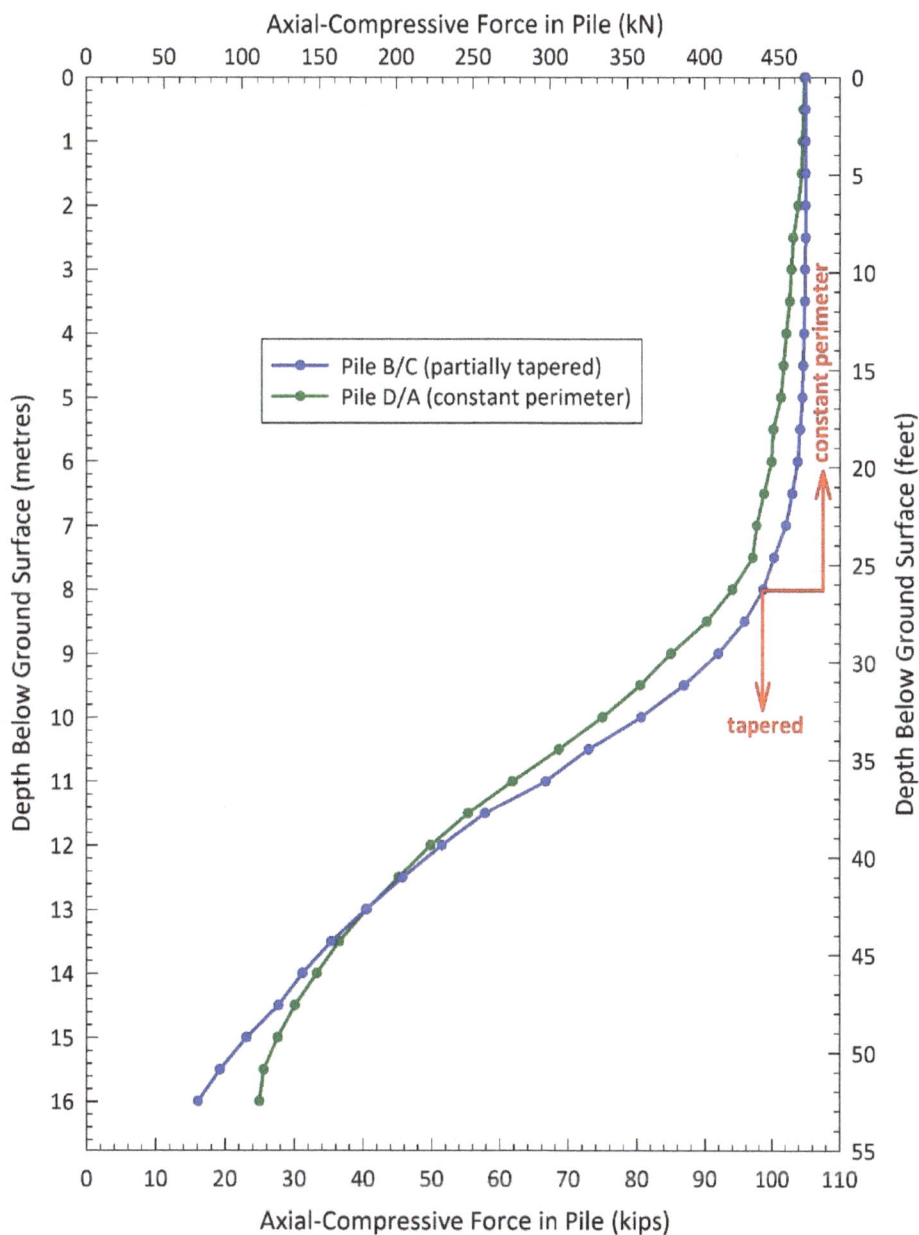

Figure H.26. Holmen Island Pile B/C vs. Pile D/A - Measured Axial-Compressive Forces within Pile Shaft at Q_{90} Load Level in Compressive Load Test.

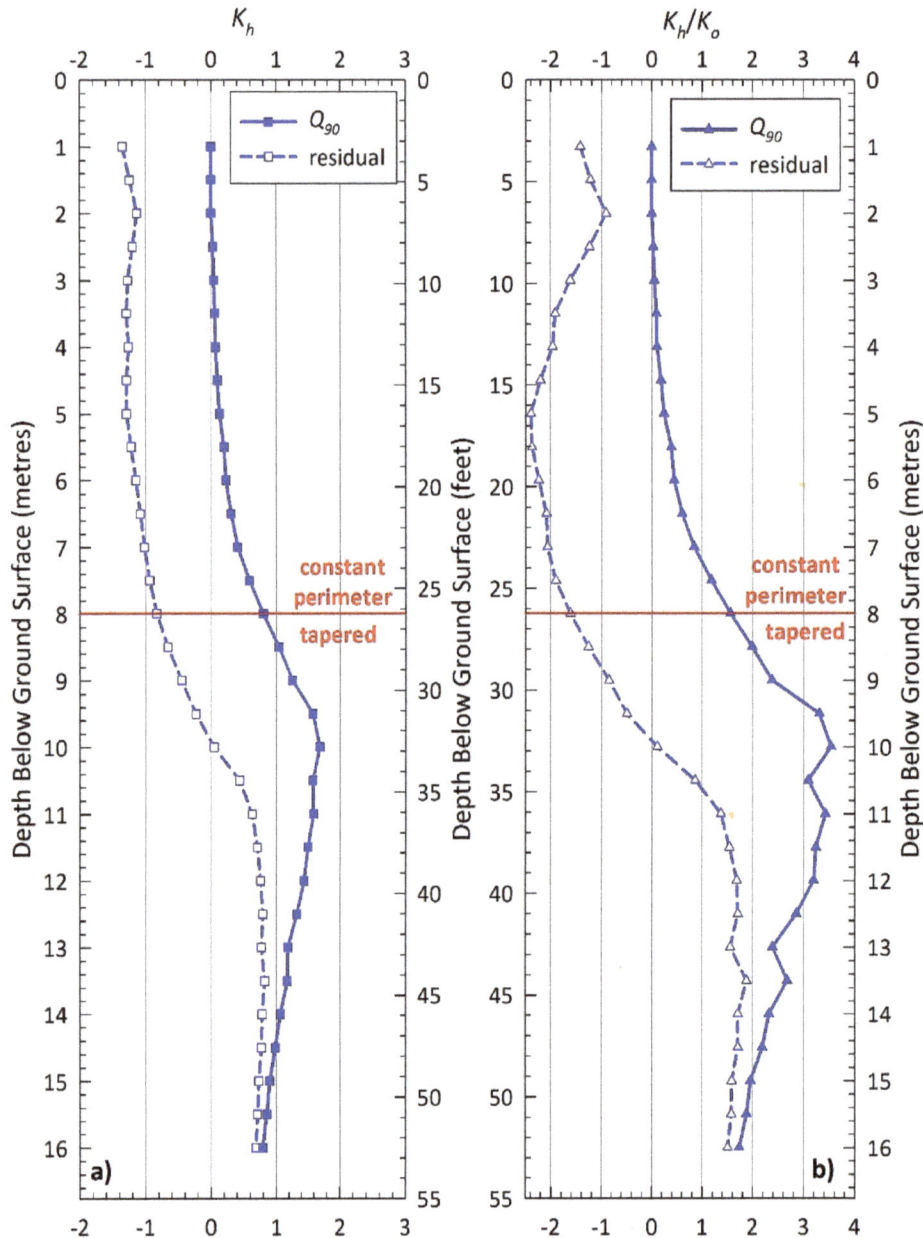

Figure H.27. Holmen Island Pile B/C - Calculated K_h and K_h/K_o along Shaft-Soil Interface at Q_{90} Load Level in Compressive Load Test.

As can be seen, the compressive load test caused a <u>gross</u> increase in resistance along the entire length of the pile as expected. However, this resistance first had to reverse the very substantial drag load due to residual loads that existed along the shaft after driving. As can be seen, the post-driving neutral plane was more than halfway down the pile shaft and extended a quarter of the way along the tapered section, at a depth of approximately 10 metres (33 ft). Thus, by the time the Q_{90} load level was reached in the compressive load test, the drag load from post-driving residual loads had just barely been neutralized along the uppermost several metres of the pile shaft so that the <u>net</u> resistance was only at or close to zero down to a depth of approximately 3 metres (10 ft). It was only due to the superior net resistance provided within the lower half of the pile that Pile B/C had approximately the same Q_{90} resistance as the constant-perimeter Pile D/A as can be seen in Figure H.26.

Figure H.28 compares the K_h and the K_h/K_o ratio values between Pile B/C (partially tapered) and Pile D/A (constant perimeter) at their respective Q_{90} load levels. Note that the dividing line between the constant-perimeter and tapered sections applied to Pile B/C only.

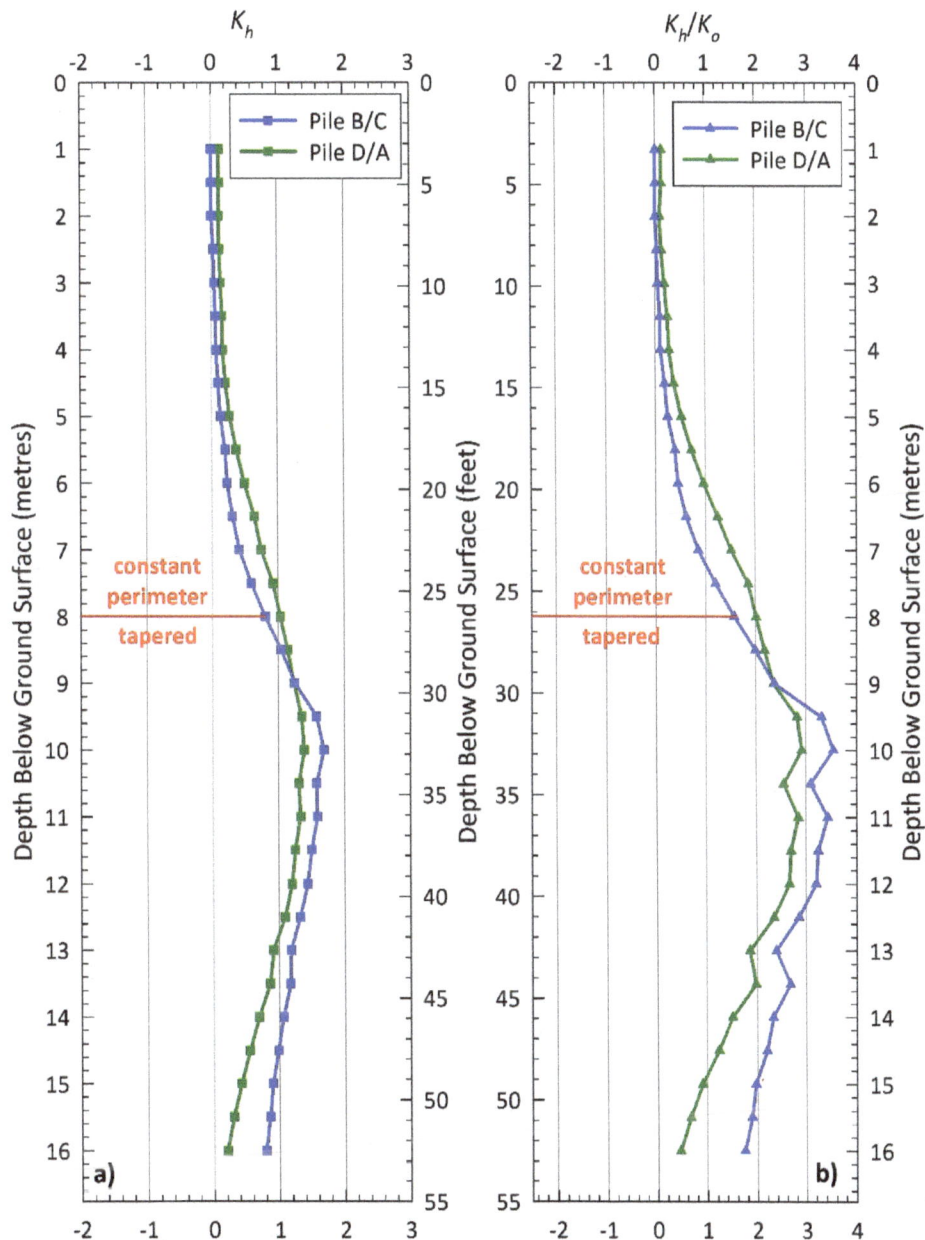

Figure H.28. Holmen Island Pile B/C vs. Pile D/A - Calculated K_h and K_h/K_o along Shaft-Soil Interface at Q_{90} Load Level in Compressive Load Test.

586

As can be seen in this figure, Pile B/C has larger lateral earth pressure coefficients within the all-important lower half of the piles where vertical effective overburden stresses are the largest in magnitude. This translates into larger radial stresses and larger unit shaft resistances so that Pile B/C, which overall has a smaller pile volume and smaller shaft circumferential area compared to Pile D/A plus a smaller toe resistance (see Figure H.26), still has approximately the same total Q_{90} resistance as can be seen in Figure H.26.

H.5.4.3.6 Intermediate (Allowable/Working) Load Case

Gregersen et alia also presented plots of the instrument-measured axial-compressive forces within the pile shaft and calculated unit shaft resistances for the case of an external applied load that was one-half of the Q_{90} value for each pile. This load magnitude was of interest to them as it represented what would have been taken to be the maximum 'allowable' or 'working' pile capacity at the time when traditional ASD/WSD design methodologies were still in widespread use globally. This corresponds to what nowadays would be called the service-load capacity.

Detailed plots for this intermediate load case are not presented in this appendix as the residual-load case and Q_{90} applied-load case that were presented previously are of greater interest as they bracket the behaviors for each of these four piles. However, there was one very interesting aspect of the one-half Q_{90} results that is worth mention and discussion. This is the fact that some of the effects of the residual loads had not been fully reversed under this intermediate applied-load level. Specifically, some of the drag load still existed within the upper portions of the piles. The residual drag load was very small for the intermediate-length (8 m/26 ft) piles but much more significant for the full-length (16 m/52 ft) piles.

In the writer's opinion, the consequences of greatest practical interest from this observed behavior under this intermediate applied-load level that conceptually reflected the maximum service-load level of a production pile are:

- The neutral plane still remained some distance below the top of the pile. The location of this neutral plane varied from a depth of approximately 1 metre (3.3 ft) for Pile A and 2 metres (6.6 ft) for Pile C to 6 metres (19.7 ft) for Pile D/A and 8 metres (26 ft) for Pile B/C. While these reflect a substantial reduction in residual loads for the intermediate-length Piles A and C, there was proportionately much less reduction in residual loads for the full-length Piles A/D and B/C.

- Because there was still some drag load in each of the four piles, the maximum axial-compressive force in each pile was <u>not</u> at the pile head but at some depth below the pile head that was commensurate with the depth of the neutral plane. Thus, the maximum axial-compressive force within a pile could <u>not</u> be simplistically assumed to be the applied load as would generally be the case in a compressive load test. Rather, the maximum force was greater than the applied load and occurred at a depth well below the head of the pile.

- It appears that the drag-load effects of the residual loads were only barely overcome for the two full-length piles (D/A and B/C) by the time that the Q_{90} applied load was reached. Stated another way, it essentially required reaching the nominal geotechnical ULS for the entire pile in order to overcome all residual-load effects in these two piles. As a result, very little net resistance ultimately developed along the upper few metres of both Pile D/A and Pile B/C. This accounts for the surprisingly small values of K_h and the K_h/K_o ratio that were calculated for the upper portions of these piles as can be seen in Figure H.28. These anomalously small values were noted earlier in the detailed discussions of these two piles.

Because of this surprising and interesting observed behavior under the one-half Q_{90} applied compressive loading, it is of interest to explore this behavior on a more-detailed and granular level for at least one of the four piles. Pile D/A was chosen for this as it is both a generic, constant-perimeter pile from head to toe and exhibited significant drag-load effects throughout the entire compressive load-testing process.

Figure H.29 shows the depth-wise variation of the instrument-measured, axial-compressive forces within the pile shaft of Pile D/A at the one-half Q_{90} load level in compressive load testing. Also shown in this figure for visual comparison are the results for this pile under both residual loads and the full Q_{90} load level in compression load testing that were shown previously in Figure H.23.

As can be seen in Figure H.29, the applied load at the head of the pile at the one-half Q_{90} load level is 233 kilonewtons (52 k). However, the <u>maximum</u> axial-compressive force in the pile is actually 281 kilonewtons (63 k) at a depth of approximately 7 metres (23 ft), almost halfway down the pile shaft. This actual maximum is <u>21% larger</u> than the applied load which is not an insignificant difference. In normal practice where the applied load at the head of a constant-perimeter pile is presumed to be the maximum axial force for structural-analysis or structural-design purposes, this routine assumption could, as a minimum, lead to a significant underestimation of normal stresses acting within the pile cross-section.

Figure H.30 shows the depth-wise variation of the calculated (by the writer) K_h and K_h/K_o parameters acting along the shaft-soil interface at the one-half Q_{90} load level in compressive load testing. Also shown in this figure for visual comparison are the results for this pile under both residual loads and the full Q_{90} load level in compression load testing that were shown previously in Figure H.24.

As can be seen in this figure, under the nominal service load of one-half Q_{90}, there was still a considerable amount of drag load along the upper portion of the pile shaft that remained to be overcome by additional pile settlement under further load application.

H.5.4.3.7 Comments

Despite the fact that there are geometric and geotechnical differences between the *Monotube* piles at JFKIA that were the focus of the paper by Fellenius et al. (2000) and evaluated independently in detail by the writer in Appendix A and the modified *Brynildsen* piles that are the focus of this appendix, the fact that each group of piles included partially tapered piles of comparable lengths and identical toe diameters in predominantly coarse-grained soil conditions does render some broad comparisons and resulting comments valid.

The responses of Holmen Island Piles D/A and B/C, wherein they displayed very small values of shaft resistance along a significant portion of the pile shaft even at the nominal geotechnical ULS, was quite surprising to the writer. These results differ substantially from the two *Monotube* piles at JFKIA. Although these *Monotube* were loaded to only the cusp of the geotechnical ULS, they displayed substantial resistance along the entire length of the pile shaft (after allowing for the fact that one of these piles, Pile #2, had virtually no resistance for a few metres below the head of the pile due to installation issues that are discussed in Appendix A). This can be seen in Figure A.4 in Appendix A that shows the K_h and K_h/K_o parameters calculated by the writer at the maximum applied load levels for these two piles.

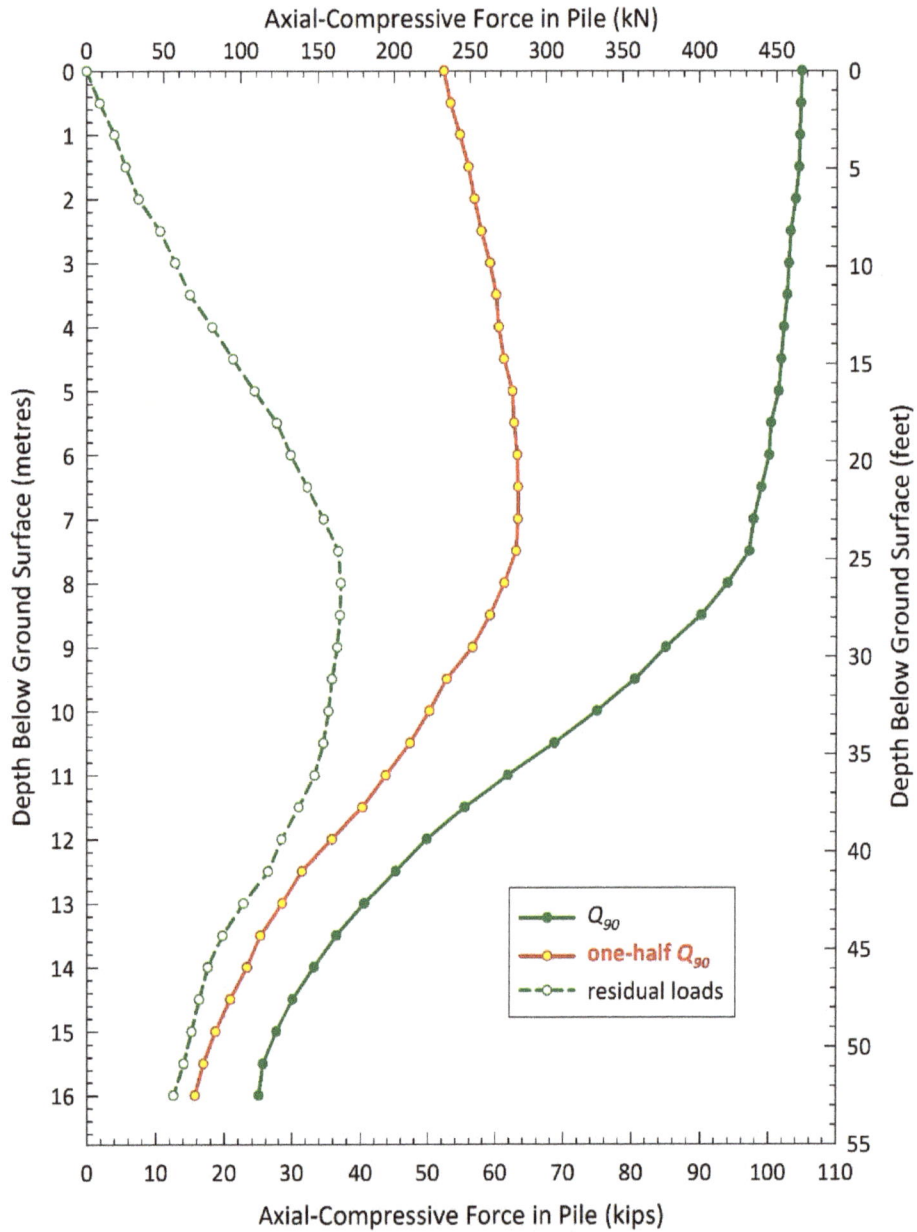

Figure H.29. Holmen Island Pile D/A - Measured Axial-Compressive Forces within Pile Shaft at End of Driving and During Compressive Load Test.

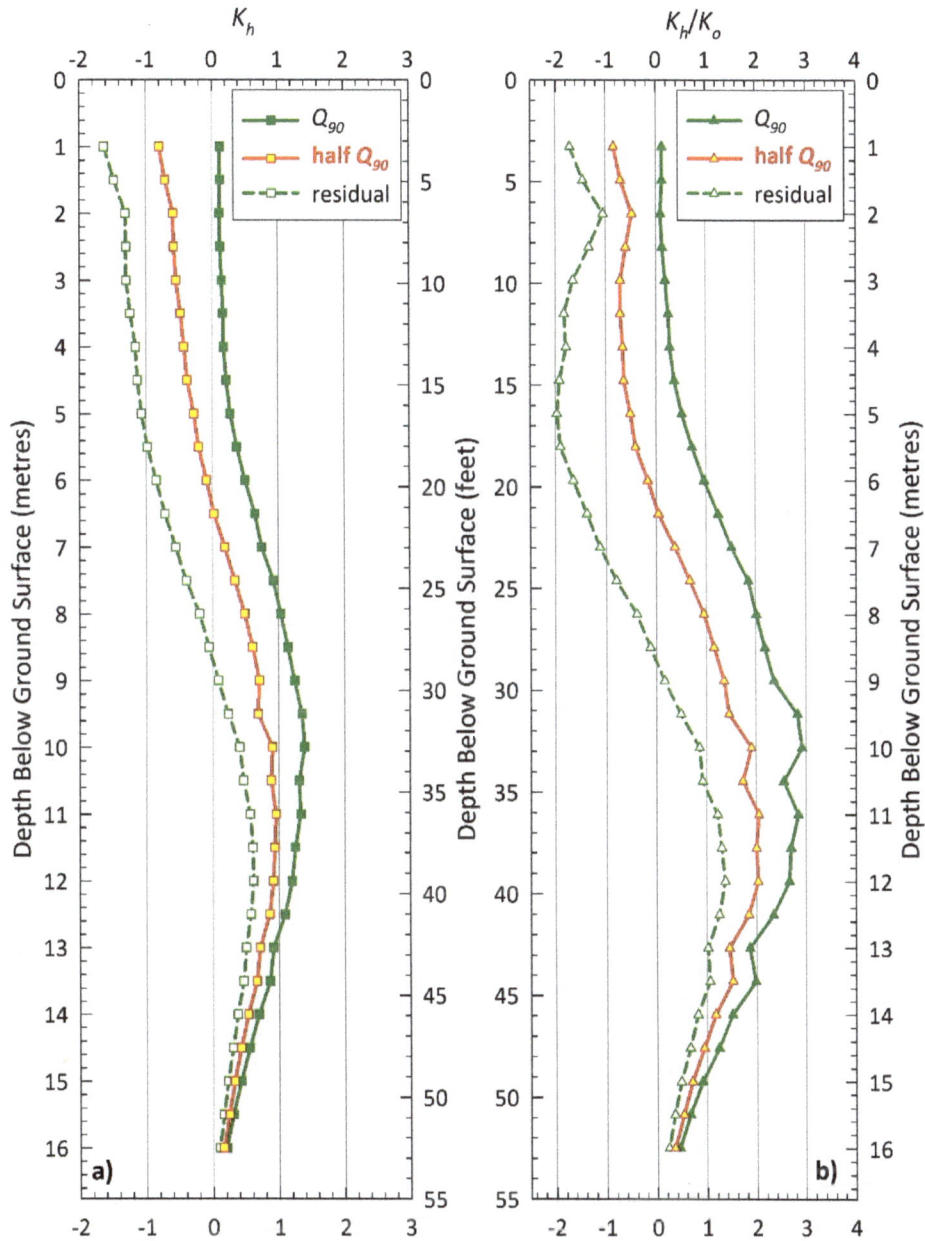

Figure H.30. Holmen Island Pile D/A - Calculated K_h and K_h/K_o along Shaft-Soil Interface at End of Driving and During Compressive Load Test.

As a consequence of this very different resistance behavior along the upper portion of the shaft, it is of interest to note that partially tapered Pile B/C at Holmen Island had an axial-compressive-resistance allocation at the nominal geotechnical ULS (Q_{90}) load level of 6-79-15 (percentages of constant-perimeter shaft/tapered shaft/toe resistance respectively) while the two partially tapered *Monotube* piles at JFKIA had a 20-60-20 resistance distribution. Thus, although the two different groups of piles had similar proportions of toe resistance (15% vs. 20%), Holmen Island Pile B/C derived 93% of its overall shaft resistance from the tapered portion of the pile while the two *Monotube* piles at JFKIA derived 75% of their overall shaft resistance from the tapered portion.

Another insightful comparison between the four Holmen Island piles and the two JFKIA *Monotube* piles is shown in Figure H.31. This shows the depth-wise variation of the post-driving residual loads for each of these six piles. Note that in order to allow apples-to-apples comparison, these loads have been normalized to either the pile-specific Q_{90} load (for the Holmen Island piles) or the pile-specific maximum applied load (for the JFKIA piles) so that these loads are presented as a fraction of more or less the maximum geotechnical resistance ultimately displayed by that pile. In addition, the pile lengths have also been normalized (to the total pile length for each pile) so that relative-depth comparisons are valid.

The primary takeaway of note in this figure is that the residual loads were, relatively speaking, substantially larger for the Holmen Island piles, especially along the upper portion of these piles. Because the drag-load component of residual loads must be overcome by the resistance that results from external load application during the post-installation static load test, much more of the relative pile-soil settlement that occurred during the compressive load testing was 'used up' negating the drag load on the Holmen Island piles compared to the JFKIA piles.

The conclusion reached from this comparison between the Holmen Island and JFKIA test piles is that it appears that both the relative magnitude and depth-wise distribution of residual loads plays a significant role in the post-driving response of a DFE to an externally applied axial load. Because compressive load testing was performed on the Holmen Island piles prior to uplift load testing (which is still to be presented and discussed in this appendix), at this point the effect of residual loads can only conclusively be noted for post-installation compressive loading. However, it appears logical that residual loads would affect post-installation uplift loading as well, at least in cases where such loading was the first externally applied load after DFE installation.

To the best of the writer's knowledge based on the known (to the writer) published literature, the linkage between residual loads and post-driving externally applied axial loads has not been noted, no less researched, to date. Thus, it is impossible to opine as to whether the Holmen Island test-pile results are a behavioral outlier due to a synergistic combination of:

- the extremely loose initial state of the site soils,

- relatively low values of small-strain shear stiffness, and

- the atypical manner in which the piles were installed

or if they are exemplars of more-widespread behavior that has simply never been recognized to date. Regardless of the answer to this question, this is certainly a topic that merits substantial basic research in the future.

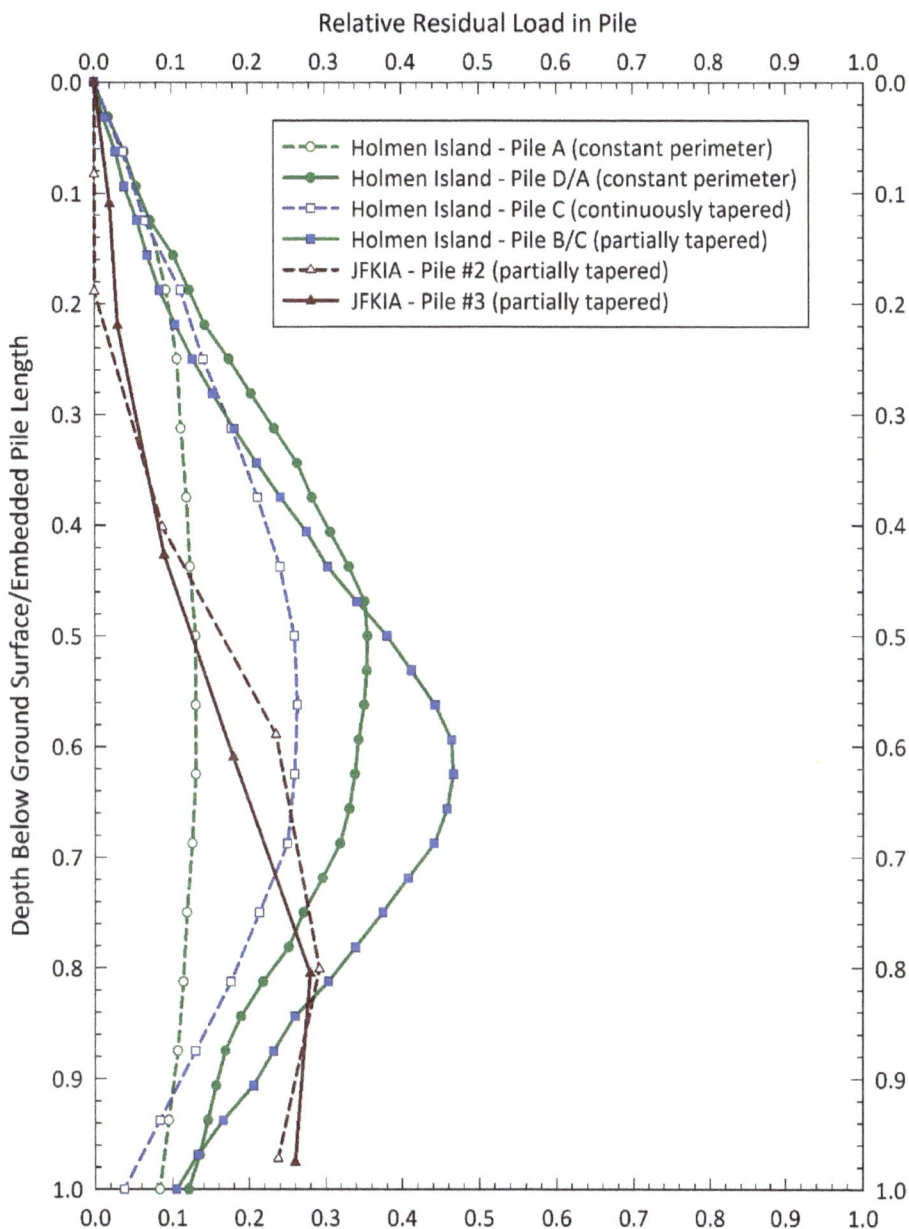

Figure H.31. Relative Depth-Wise Variation of Residual Loads in Holmen Island *Brynildsen* Piles vs. JFKIA *Monotube* Piles.

H.5.4.4 Toe Resistance

H.5.4.4.1 Background

As noted earlier in this appendix in the presentation and discussion of residual toe loads after driving, a maximum toe resistance for each of the four Holmen Island piles was independently determined by the writer based on the information presented in Gregersen et al. (1973). These maximum values were presented and used in Table H.1. The reason for determining the maximum toe resistances recorded in the compressive load tests is because the toe resistances corresponding to Gregersen et alia's Q_{90} load level were not the maximum values recorded in the compressive load tests. The two intermediate-length piles, A and C, both exhibited maximum toe resistances at settlement levels somewhat greater than that corresponding to the Q_{90} load level. These resistances then remained more or less constant for the rest of the load test.

However, the two full-length piles, D/A and B/C, exhibited the behavior long noted by Fellenius. Specifically, the fact that toe resistance often continues to increase, albeit at a much-reduced rate, with increasing DFE settlement of the order of tens of millimetres or inches. This magnitude of settlement is typically well beyond that normally taken to define the geotechnical ULS.

This was certainly true for the Holmen Island piles with respect to the settlement magnitudes that corresponded to the Q_{90} load levels as determined by Gregersen et alia. Therefore, the maximum toe resistances noted for these piles in Table H.1 were simply the values at the maximum head-settlement levels reached in the compressive load tests which was roughly 60 millimetres (2.5 in) or 30% of the toe diameter in each case. There is every indication in each of these load tests that had additional pile settlement occurred that the toe resistance would have increased further beyond the magnitudes shown in Table H.1.

H.5.4.4.2 Results

Table H.2 presents a summary of the toe loads and resistances under four load cases:

1. residual loads,

2. one-half of the Q_{90} load,

3. Q_{90} load, and

4. maximum.

Note that the results for Cases 1 and 4 were shown previously in Table H.1.

Table H.2. Holmen Island 1969 Test Piles - Toe-Resistance Summary.

Load Case	Piles			
	Intermediate length (8 m/26 ft)		Full length (16 m/52 ft)	
	A	C (tapered)	D/A	B/C (partially tapered)
1. Residual, kN (kips)	20 (4.5)	10 (2.2)	56 (13)	49 (11)
2. One-half Q_{90}, kN (kips)	33 (7.4)	23 (5.2)	71 (16)	62 (14)
3. Q_{90}, kN (kips)	57 (13)	41 (9.2)	111 (25)	72 (16)
4. Maximum, kN (kips)	72 (16)	44 (10)	140 (31)	83 (19)

As an aside, it is again pointed out that the toe cross-sectional areas of the two tapered piles, C and B/C, were each approximately one-half that of the constant-perimeter piles, A and D/A. Thus, the unit toe resistances, a.k.a. 'bearing stress' or 'bearing pressure', of the two tapered piles were each either comparable to or greater than the corresponding unit toe resistances of the constant-perimeter piles under all four load cases shown in Table H.2. This is particularly true for the full-length, partially tapered Pile B/C relative to the full-length, constant-perimeter Pile D/A. This observation offers some support to the hypothesis put forth by Manandhar and Yasufuku that taper affects the toe resistance of piles.

Another metric of toe resistance that is of interest to calculate and compare among the four piles is the rate at which toe resistance (the dependent variable) develops as a function of the externally applied compressive load on the pile (the independent variable). In this case, the abscissa is the applied load divided by the Q_{90} load, expressed as a percent. The Q_{90} load was chosen for the reference load as it is consistently defined for all piles and reflects the geotechnical ULS as interpreted by Gregersen et alia. Note that the residual load is the 0% point using this definition.

The ordinate in this case is the toe resistance at a given stage of load application (here, Cases 1, 2, and 3 noted above are used) divided by the maximum resistance (Case 4), also expressed as a percent. The maximum toe resistance was chosen for the reference load in order to provide some insight into what portion of the maximum measured toe resistance was mobilized at the Q_{90} load level that represents the assumed geotechnical ULS, at least per Gregersen et alia. For comparative purposes, this is best done graphically and this is shown in Figure H.32.

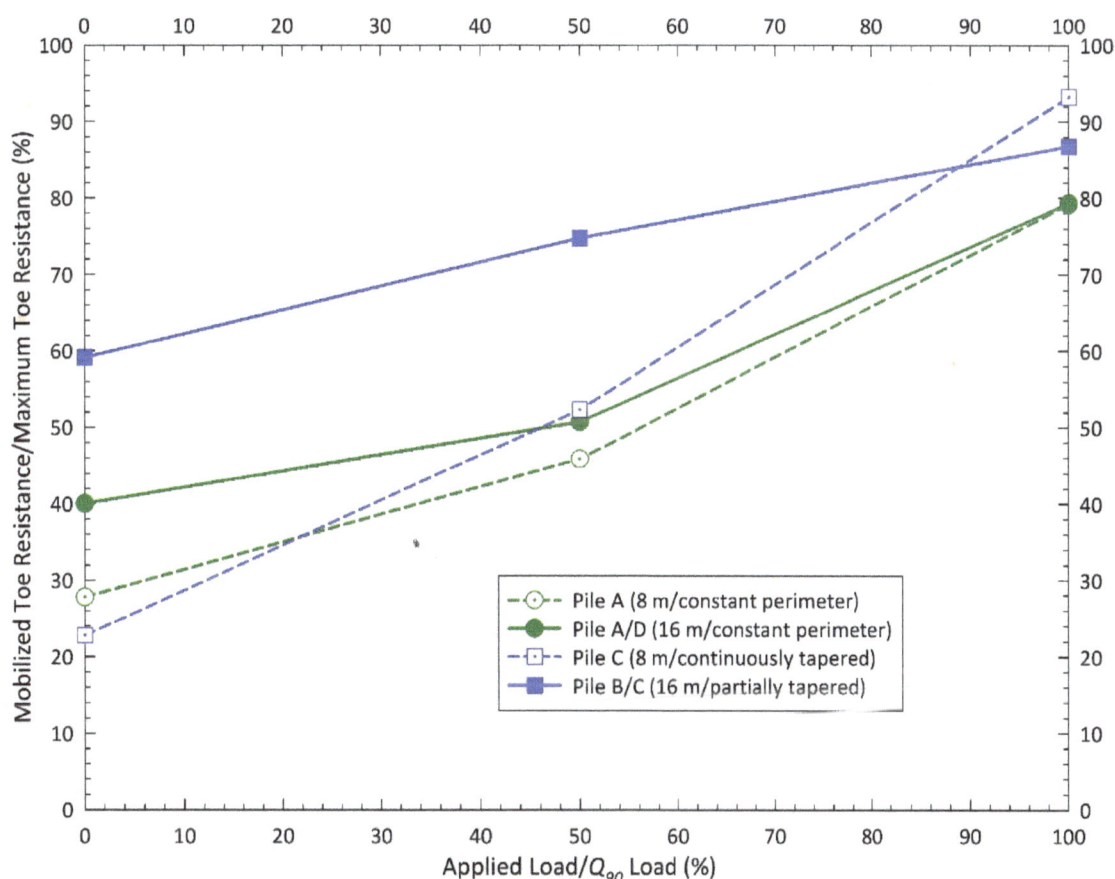

Figure H.32. Holmen Island 1969 Test Piles - Rate of Toe-Resistance Mobilization.

As can be seen, all four piles had a relatively significant portion (approximately 25-60%) of the eventual maximum toe resistance mobilized by post-driving residual loads. Furthermore, all four piles had mobilized a substantial portion (approximately 80-95%) of the eventual maximum measured toe resistance by the time the Q_{90} load level was reached. The two tapered piles (C and B/C) stand out in this regard as they averaged 90% mobilization vs. 80% average for the two constant-perimeter piles (A and D/A). As noted earlier in this appendix, this lends some credence to the hypothesis by Manandhar et alia that toe resistance is influenced by pile taper. However, in the writer's opinion, the physical mechanism for this proposed by Manandhar et alia is unproven. Consequently, if the effect of taper on toe resistance does indeed exist, the physical mechanism by which this happens remains to be determined by future research.

H.5.5 Uplift Load Tests

H.5.5.1 Overview

Each of the four piles was subjected to uplift load testing (Gregersen et al. (1973) referred to these as "pull out tests") after completion of the complex sequence of compressive load testing. Unlike the compressive load test procedure that was outlined in detail by Gregersen et alia, nothing was said about the procedure followed for the uplift load tests. However, the load-displacement curves shown in the Gregersen et alia paper imply that uplift loading was applied in one continuous (presumably relatively rapid) process until the test was terminated and the load removed.

The two intermediate-length (8 m/26 ft) piles, A and C, were each subjected to approximately 20 millimetres (0.8 in) of upward pile-head displacement before the test was terminated. The two full-length (16 m/52 ft) piles, D/A and B/C, were each subjected to approximately 30 millimetres (1.2 in) of upward pile-head displacement before the test was terminated.

As discussed subsequently, the maximum upward displacement reached in each of the four uplift load tests was approximately twice the upward displacement at which the Q_{90} geotechnical ULS load in uplift was reached. This compares to the compressive load tests where the tests were carried to settlement magnitudes that ranged from three to five times the settlement at which the Q_{90} geotechnical ULS load in compression was reached.

H.5.5.2 Resistance

H.5.5.2.1 Overview

It is assumed that because of the coarse-grained soil conditions at the Holmen Island site that no suction effects developed during the course of the uplift load tests. Therefore, all resistance measured in these tests and discussed herein are assumed to be shaft resistance only.

Gregersen et alia presented substantially fewer graphical outcomes in their paper for the uplift load tests compared to the compressive load tests. They only provided the overall load-displacement curves (superimposed on the load-settlement curves for the compressive load tests), each interpreted for the Q_{90} geotechnical ULS load in uplift. No detailed results showing calculated unit shaft resistances that could be interpreted for the K_h and K_h/K_o parameters were provided. Consequently, due to the complete lack of necessary data, detailed assessments with multiple figures presenting and comparing results were not done by the writer for the uplift load tests as they were for the compressive load tests reported earlier in this appendix. Rather, only a tabulated summary of results is presented in the following section. This is followed by a commentary section concerning the uplift load tests and observed behavior to close out this appendix.

H.5.5.2.2 Results

Table H.3 summarizes three sets of measured outcomes for the four Holmen Island piles:

1. The Q_{90} geotechnical ULS in uplift as determined by Gregersen et alia.

2. The maximum uplift resistance at the end of the uplift load test, as scaled by the writer from the results presented in Figure 5 in Gregersen et al. (1973).

3. The shaft-only component of the Q_{90} geotechnical ULS in <u>compression</u> as determined by Gregersen et alia.

The reason for including Item 3 is to provide direct comparison with at least Item 1 if not Item 2 as well. This is because one of the issues that has been debated for decades now is whether (or not) for a constant-perimeter DFE the shaft resistance in compression is the same as that in uplift. There are published results that claim to support both sides of this argument and, as discussed subsequently, Gregersen et alia raised and commented on this issue in their paper.

Table H.3. Holmen Island 1969 Test Piles - Uplift Load Tests Resistance Summary.

Resistance Case	Piles			
	Intermediate length (8 m/26 ft)		Full length (16 m/52 ft)	
	A	C (tapered)	D/A	B/C (part tapered)
1. Uplift Q_{90}, kN (kips)	93 (21)	120 (27)	265 (60)	239 (54)
2. Uplift maximum, kN (kips)	101 (23)	127 (29)	273 (61)	239 (54)
3. Compressive shaft Q_{90}, kN (kips)	140 (32)	207 (47)	266 (60)	320 (72)

With respect to the results presented in this table, there are three broad questions to ask:

1. For a given pile, how do the Q_{90} and maximum resistances in uplift compare (this is an indication of the behavioral trend of the load-displacement curve, i.e. work hardening, strain softening, or constant)?

2. For piles of comparable length, how do the results compare for the tapered vs. constant-perimeter piles in uplift?

3. For a given pile, how do the results compare in uplift vs. compression (shaft resistance only)?

With respect to Question #1, three of the four piles exhibited <u>very</u> slight work-hardening tendencies, i.e. a further increase in uplift resistance, after reaching the Q_{90} point and up to the maximum uplift load applied. Only the full-length, partially tapered pile (Pile B/C) plateaued resistance-wise after reaching the Q_{90} point.

With respect to Question #2, the tapered piles performed <u>much</u> better than might have been expected, or at least expected based on the traditional assumption that tapered piles have negligible uplift resistance. The shorter (continuously) tapered pile (Pile C) actually <u>outperformed</u> its constant-perimeter counterpart (Pile A) and by a relatively substantial margin at that (25-30% <u>more</u> uplift resistance depending on whether the Q_{90} or maximum load level is used as the metric). This is more than a little surprising in the writer's opinion given the fact that the tapered pile only had approximately 85% of the circumferential shaft area of the constant-perimeter pile.

The full-length (partially) tapered pile (Pile B/C) performed very well compared to its full-length constant-perimeter counterpart (Pile D/A), with approximately 90% of the uplift resistance. Note that the tapered pile only had approximately 90% of the circumferential shaft area of the constant-perimeter pile in this case.

Finally, with respect to Question #3, the results are a mixed bag. First considering the constant-perimeter piles, the full-length pile (Pile D/A) had essentially the same shaft resistance in uplift and compression as Fellenius and others have opined should always be the case. However, the shorter pile (Pile A) displayed substantially less (approximately two-thirds) shaft resistance in uplift compared to compression which is a significant difference and one not readily explained.

Both tapered piles exhibited substantially less shaft resistance in uplift compared to compression but this outcome is very deceptive in its implications. To begin with, due to the wedging mechanism and concomitant taper benefit, both piles derived a <u>much</u> larger percentage of their ULS total resistance in compression from the shaft compared to the constant-perimeter piles so the compressive shaft Q_{90} values were relatively very high to begin with for the tapered piles. As a result, this masks the fact that the two tapered piles actually performed very well in uplift compared to their constant-perimeter counterparts as discussed above in the answer to Question #2.

Furthermore, based on the current state of knowledge as expressed in this monograph, a tapered pile would not be expected to have the same shaft resistance in compression and uplift. This is because in compression a tapered pile develops additional taper benefit due to Stage II wedging as modeled using cylindrical-cavity <u>expansion</u> (CCE). On the other hand, in uplift a tapered pile undergoes unwedging as modeled using cylindrical-cavity <u>contraction</u> (CCC) so some reduction in unit shaft resistance is bound to occur in uplift vs. compression.

H.5.5.3 Comments

The primary takeaway from the uplift load tests performed on the Holmen Island test piles in 1969 is that they clearly dispelled the widespread, common (mis)conception that a tapered pile has much-reduced uplift resistance relative to a comparable constant-perimeter pile in the same soil conditions. The writer presented a theory-based argument against this misconception in Chapter 4. However, as always, it is essential to support any theoretical hypothesis with ground-truth. The writer feels that the material presented in this appendix does this satisfactorily with respect to tapered piles and uplift resistance.

CITED REFERENCES AND SUPPLEMENTAL BIBLIOGRAPHY

Agaiby, S. S. and Mayne. P. W. (2018). "Evaluating Undrained Rigidity Index of Clays from Piezocone Data". Proceedings of CPT18-Delft - The Fourth International Symposium on Cone Penetration Testing.

Alawneh, A. S. and Husein Malkawi, A. I. (2000). "Estimation of Post-Driving Residual Stresses Along Driven Piles in Sand". *Geotechnical Testing Journal*, Vol. 23, No. 3, pp. 313-326.

Allen, T. A. (2005). *Development of the WSDOT Pile Driving Formula and Its Calibration for Load and Resistance Factor Design (LRFD)*. Report WA-RD 610.1, Washington State Department of Transportation, Olympia, WA.

Bakholdin, B. V. (1971). "Bearing Capacity of Pyramidal Piles". Proceedings of the 4th Conference on Soil Mechanics and Foundation Engineering, Budapest, Hungary, pp. 507–510.

Bakholdin, B. V. and Igon'kin, I. T. (1978). "Investigation of the Bearing Capacity of Pyramidal Piles". *Soil Mechanics and Foundation Engineering*, Vol. 15, No. 3, pp. 165-170.

Baligh, M. M. (1976). "Cavity Expansion in Sands with Curved Envelopes". *Journal of the Geotechnical Engineering Division*, American Society of Civil Engineers, Vol. 102, No. 11, pp. 1131-1146.

Been, K. and Jefferies, M. (1985). "A State Parameter for Sands". *Géotechnique*, Vol. 35, No. 2, pp. 99-112.

Bolton, M. D. (1986). "The Strength and Dilatancy of Sands". *Géotechnique*, Vol. 36, No. 1, pp. 65-78.

Brand, A. (1997). "Game, Set and Match". *Civil Engineering*, American Society of Civil Engineers, May, pp. 2A-8A.

Briaud, J. L. and Tucker, L. (1984). "Piles in Sand: A Method Including Residual Stresses". *Journal of Geotechnical Engineering*, Vol. 100, No. 11, pp. 1666-1680.

Buckland, B. (2005). "An Introduction into the Production and Specification of Steel Pipe". *PileDriver*, Pile Driving Contractors Association, 1st Quarter, pp. 20-26.

Bulletin No. 113 (1909). American Railway Engineering and Maintenance of Way Association.

Carroll, T. and Hu, S. (2015). "Using Cone Penetration Testing to Predict the Capacity and Drivability of Tapered Piles in Sands". Proceedings of the International Foundations Congress and Equipment Expo 2015, San Antonio, TX.

Carter, J. P., Booker, J. R., and Yeung, S. K. (1986). "Cavity Expansion in Cohesive-Frictional Soils". *Géotechnique*, Vol. 36, No. 3, pp. 349-353.

Chellis, R. D. (1961). *Pile Foundations*, 2nd edition. McGraw-Hill.

Chin, F. K. (1970). "Estimation of the Ultimate Load of Piles Not Carried to Failure". Proceedings of the 2nd Southeast Asian Conference on Soil Engineering, pp. 81-90.

Chin, F. K. (1972). "The Inverse Slope as a Prediction of Ultimate Bearing Capacity of Piles". Proceedings of the 3rd Southeast Asian Conference on Soil Engineering, pp. 83-91.

Concrete (1910). Concrete Publishing Company, Detroit, MI, Vol. X, No. 1.

Cook, B. R. (2010). *Numerical Solution of Cylindrical Cavity Expansion in Sands: Effects of Failure Criteria and Flow Rules.* Thesis submitted to Washington State University, Department of Civil and Environmental Engineering, Pullman, WA in partial fulfillment of the requirements for the degree of Master of Science.

Costa, L. M. and Lopes, F. R. (2001). "Prediction of Residual Driving Stresses in Piles". *Canadian Geotechnical Journal*, Vol. 38, No. 2, pp. 401-421.

D'Appolonia, E. and Hribar, J. A. (1963). "Load Transfer in a Step-Taper Pile". *Journal of the Soil Mechanics and Foundations Division*, American Society of Civil Engineers, Vol. 89, No. SM6, pp. 57-80.

Damen, R. and Denes, D. (2017). "Improving Site Specific Modified Driving Formulae Using High Frequency Displacement Monitoring". Proceedings of the 20th New Zealand Geotechnical Society Geotechnical Symposium.

Darrag, A. A. (1987). *Capacity of Driven Piles in Cohesionless Soils Including Residual Stresses.* Dissertation submitted to Purdue University, School of Civil Engineering, West Lafayette, IN in partial fulfillment of the requirements for the degree of Doctor of Philosophy.

Darrag, A. A. and Lovell, C. W. (1989). "A Simplified Procedure for Predicting Residual Stresses for Piles". Proceedings of the 12th International Conference on Soil Mechanics and Foundation Engineering, Vol. 2, pp. 1127-1130.

Díaz-Rodríguez, J. A. and López-Molina, J. A. (2008). "Strain Thresholds in Soil Dynamics". Proceedings of the 14th World Conference on Earthquake Engineering, Beijing, China.

Dijkstra, J., Broere, W., and Heeres, O. M. (2011). "Numerical Simulation of Pile Installation". *Computers and Geotechnics*, Vol. 38. pp. 612-622.

Duncan, J. M and Chang, C.-Y. (1970). "Nonlinear Analysis of Stress and Strain in Soils". *Journal of the Soil Mechanics and Foundations Division*, American Society of Civil Engineers, Vol. 96, No. SM5, pp. 1629-1653.

Duncan, J. M and Seed, R. B. (1986). "Compaction-Induced Earth Pressures Under K_o-Conditions". *Journal of Geotechnical Engineering*, Vol. 112, No. 1, pp. 1-22.

Duncan, J. M, Williams, G. W., Sehn, A. L., and Seed, R. B. (1991). "Compaction-Induced Earth Pressures Under K_o-Conditions". *Journal of Geotechnical Engineering*, Vol. 117, No. 12, pp. 1833-1847.

El Naggar, M. H. and Sakr, M. (1999). "Centrifuge Testing of Tapered Piles, Axial Behaviour". Proceedings of the 52nd Annual Canadian Geotechnical Conference, pp. 143-150.

El Naggar, M. H. and Sakr, M. (2000). "Evaluation of Axial Performance of Tapered Piles from Centrifuge Tests". *Canadian Geotechnical Journal*, Vol. 37, No. 6, pp. 1295-1308.

El Naggar, M. H. and Wei, J. Q. (1999a). "Response of Tapered Piles Subjected to Lateral Loading". *Canadian Geotechnical Journal*, Vol. 36, No. 1, pp. 52-71.

El Naggar, M. H. and Wei, J. Q. (1999b). "Axial Capacity of Tapered Piles Established from Model Tests". *Canadian Geotechnical Journal*, Vol. 36, No. 6, pp. 1185-1194.

El Naggar, M. H. and Wei, J. Q. (2000a). "Uplift Behaviour of Tapered Piles Established from Model Tests". *Canadian Geotechnical Journal*, Vol. 37, No. 1, pp. 56-74.

El Naggar, M. H. and Wei, J. Q. (2000b). "Evaluation of Axial Performance of Tapered Piles from Centrifuge Tests". *Canadian Geotechnical Journal*, Vol. 37, No. 6, pp. 1295-1308.

Eslami, A. and Fellenius, B. H. (1997). "Pile Capacity by Direct CPT and CPTu Methods Applied to 102 Case Histories". *Canadian Geotechnical Journal*, Vol. 34, No. 6, pp. 880-898.

Fahey, M. (1992). "Shear Modulus of Cohesionless Soil: Variation with Stress and Strain Level". *Canadian Geotechnical Journal*, Vol. 29, No. 1, pp. 157-161.

Fahey, M. and Carter, J. P. (1993). "A Finite Element Study of the Pressuremeter Test in Sand Using a Nonlinear Elastic Plastic Model". *Canadian Geotechnical Journal*, Vol. 30, No. 2, pp. 348-362.

Fahmy, A. (2015). *Monotonic and Cyclic Performance of Spun-Cast Ductile Iron Helical Tapered Piles*. Dissertation submitted to Western University, Department of Civil and Environmental Engineering, London, ON, Canada in partial fulfillment of the requirements for the degree of Doctor of Philosophy.

Fahmy, A. and El Naggar, M. H. (2016a). "Uplift Performance of Spun-Cast Ductile Iron Piles". Proceedings of the American Society of Civil Engineers Geotechnical and Structural Engineering Congress, Phoenix, AZ.

Fahmy, A. and El Naggar, M. H. (2016b). "Cyclic Axial Performance of Helical-Tapered Piles in Sand". *DFI Journal - The Journal of the Deep Foundations Institute*, Vol. 10, No. 3, pp. 98-110.

Fahmy, A. and El Naggar, M. H. (2017a). "Uplift and Lateral Performance of Tapered Helical Piles in Sand". Proceedings of CSEE'17 - The 2nd World Congress on Civil, Structural, and Environmental Engineering, Barcelona, Spain.

Fahmy, A. and El Naggar, M. H. (2017b). "Axial Performance of Helical Tapered Piles in Sand". *Geotechnical and Geological Engineering*, Vol. 35, No. 4, pp. 1549-1576.

Fahmy, A. and El Naggar, M. H. (2017c). "Axial and Lateral Performance of Spun-Cast Ductile Iron Helical Tapered Piles in Clay". *Proceedings of the Institution of Civil Engineers - Geotechnical Engineering*, Vol. 170, Issue 6, pp. 503-516.

Fellenius, B. H. (1984). "Negative Skin Friction and Settlement of Piles". Proceedings of the Second International Seminar on Pile Foundations, Nanyang Technological Institute, Singapore.

Fellenius, B. H. (1988). "Unified Design of Piles and Pile Groups". *Transportation Research Record 1169*, Transportation Research Board, Washington, DC, pp. 75-82.

Fellenius, B. H. (1990). *Guidelines for the Interpretation and Analysis of the Static Loading Test.* Deep Foundations Institute, Sparta, NJ.

Fellenius, B. H. (1997). Discussion of "Piles Subjected to Negative Friction: A Procedure for Design" by H. G. Poulos, *Geotechnical Engineering*, Vol. 28, No. 2, pp. 277-281.

Fellenius, B. H. (1998). "Recent Advances in the Design of Piles for Axial Loads, Dragloads, Downdrag, and Settlement". Proceedings of the Seminar on Urban Geotechnology and Rehabilitation, American Society of Civil Engineers - Metropolitan Section, New York, NY.

Fellenius, B. H. (1999a). "Bearing Capacity of Footings and Piles - A Delusion?". Presentation at Deep Foundations Institute annual meeting.

Fellenius, B. H. (1999b). "On the Preparation of a Piling Paper". Presentation at Deep Foundations Institute annual meeting.

Fellenius, B. H. (2000). Discussion of "Downdrag Settlement a Single Floating Pile" [sic] by W. H. Ting, *Geotechnical Engineering*.

Fellenius, B. H. (2001). "What Capacity Value to Choose from the Results a Static Loading Test" [sic]. *Fulcrum*, Deep Foundations Institute, Winter 2001, pp. 19-22.

Fellenius, B. H. (2002a). "Determining the True Distributions of Load in Instrumented Piles". Proceedings of the International Deep Foundation Congress, American Society of Civil Engineers, pp. 1455-1470.

Fellenius, B. H. (2002b). "Determining the Resistance Distribution in Piles; Part 1. Notes on Shift of No-Load Reading and Residual Load". *Geotechnical News*, Vol. 20, No. 2, pp 35-38.

Fellenius, B. H. (2002c). "Determining the Resistance Distribution in Piles; Part 2. Method for Determining the Residual Load". *Geotechnical News*, Vol. 20, No. 3, pp. 25-29.

Fellenius, B. H. (2006). "Results from Long-Term Measurement in Piles of Drag Load and Downdrag". *Canadian Geotechnical Journal*, Vol. 43, No. 4, pp. 409-430.

Fellenius, B. H. (2014). "Analysis of Results from Routine Static Load Tests with Emphasis on the Bidirectional Test". Proceedings of the 17th Congress of the Brasiliero de Mecanica dos Solos e Egenharia.

Fellenius, B. H. (2015). "Static Tests on Instrumented Piles Affected by Residual Load". *The Journal of the Deep Foundations Institute*, Vol. 9. No. 1, pp. 11-20.

Fellenius, B. H. (2016). "Fallacies in Piled Foundation Design". Proceedings of Geotec Hanoi 2016 - Geotechnics for Sustainable Infrastructure Development, pp. 41-46.

Fellenius, B. H. (2017). "Best Practice for Performing Static Loading Tests; Examples of Test Results with Relevance to Design". 3rd Bolivian International Conference on Deep Foundations.

Fellenius, B. H. (2018). "Pitfalls and Fallacies in Geotechnical Design". *Innovations in Geotechnical Engineering*, American Society of Civil Engineers, pp. 299-316.

Fellenius, B. H. (2019). *Basics of Foundation Design; Electronic Edition (January 2019)*.

Fellenius, B. H. (2020). *Basics of Foundation Design; Electronic Edition (January 2020)*.

Fellenius, B. H. and Altaee, A. A. (1995). "Critical Depth: How It Came into Being and Why It Does Not Exist". *Proceedings of the Institution of Civil Engineers - Geotechnical Engineering*, Vol. 113, Issue 2, pp. 107-111.

Fellenius, B. H., Brusey, W. G., and Pepe, F. (2000). "Soil Set-Up, Variable Concrete Modulus, and Residual Load for Tapered Instrumented Piles in Sand". Proceedings of the Specialty Conference on Performance Confirmation of Constructed Geotechnical Facilities, American Society of Civil Engineers.

Flynn, K. N., McCabe, B. A., and Egan, D. (2012). "Residual Load Development in Cast in Place Piles - A Review and a New Case History". Proceedings of IS Kanazawa - 9th International Conference on Testing and Design Methods for Deep Foundations, Kanazawa, Japan, pp. 765-773.

Fragaszy, R. J., Argo, D. and Higgins, J. D. (1989). "Comparison of Formula Predictions with Pile Load Tests". *Transportation Research Record 1219*, Transportation Research Board, Washington, DC, pp. 1-12.

Fuller, Mossbarger, Scott and May Engineers, Inc. (2003). *Grout Confinement Influence on Strain Compatibility in Micropiles; ADSC-DFI Micropile Research; Micropile Strain Compatibility*. Report issued in digital format to ADSC - The International Association of Foundation Drilling and Deep Foundations Institute.

Ghasemi, M. (2006). *Experimental Investigation of Bearing Capacity of Pyramidal Piles in Sand*. Thesis submitted to Yazd University in Iran in partial fulfillment of the requirements for the degree of Master of Science.

Ghazavi, M. (2000). "Theoretical and Experimental Aspects of Tapered Piles Subjected to Static Loads". Proceedings of the 5th International Conference on Civil Engineering, Mashhad, Iran, Vol. 1, pp. 84-92.

Ghazavi, M. and Ahmadi. H. A. (2008). "Time-Dependent Bearing Capacity Increase of Uniformly Driven Tapered Piles- Field Load Test". Paper No. 1.66 in the proceedings of the 6th International Conference on Case Histories in Geotechnical Engineering, Arlington, VA.

Ghazavi, M. and Etaati, M.A. (2001a). "Analysis of Tapered Piles under Axial Loading Using Finite Element Method". Proceedings of 2nd International Conference on Tall Buildings, Iran University of Science and Technology, Tehran, Iran, pp. 443-449.

Ghazavi, M. and Etaati, M. A. (2001b). "Finite Element Analysis of Tapered Piles Subjected to Lateral Loading". Proceedings of the 14th Southeast Asian Geotechnical Conference, Vol. 2, pp. 1005-1008.

Ghazavi, M. and Kalantari, A. B. (2008). "Experimental Study of Taper Piles with Different Taper Angles". Proceedings of the 27th International Conference on Offshore Mechanics and Arctic Engineering, American Society of Mechanical Engineers, Vol. 2.

Ghazavi, M. and Lavasan, A. A. (2006). "Bearing Capacity of Tapered and Step-tapered [sic] Piles Subjected to Axial Compressive Loading". Proceedings of the 7th International Conference on Coastal, Ports & Marine Structures.

Ghazavi, M., Williams, D. J., and Morris, P. H. (1997a). "Analysis of Statically Loaded Tapered Piles in Layered Media". Proceedings of the 2nd International Symposium on Structures and Foundations in Civil Engineering, Hong Kong, pp. 163-169.

Ghazavi, M., Williams, D.J., and Morris, P.H. (1997b). "Numerical Analysis of Dynamically Loaded Tapered Piles". Proceedings of 2nd International Symposium on Structures and Foundations in Civil Engineering, Hong Kong, pp. 170-176.

Ghazavi, M., Williams, D. J., and Wong, K.Y. (1996). Analysis of a Tapered Pile during Pile Driving". Proceedings of 2nd International Conference on Multi-Purpose High-Rise Towers and Tall Buildings, Singapore, pp. 87-94.

Gibbs, H. J. and Holtz, W. G. (1956). "Research on Determining the Density of Sands by Spoon Penetration Testing". Proceedings of the Fourth International Conference on Soil Mechanics and Foundation Engineering.

Gilbreth, F. B. (1906). *The Making and Driving of Corrugated Concrete Piles*. Bulletin No. 7, Association of American Portland Cement Manufacturers, Philadelphia, PA.

Glanville, W. H., Grime, G., Fox, E. N., and Davies, W. W. (1938). *An Investigation of the Stresses in Reinforced Concrete Piles During Driving*. Technical Paper No. 20, Department of Scientific and Industrial Research, Building Research Station, U.K.

Goble, G. G. and Hery, P. (1984). "Influence of Residual Force on Pile Driveability [sic]". Proceedings of the 2nd International Conference on Stress-Wave Theory on Piles, Stockholm, Sweden, pp. 154–61.

Gotman, A. L. (1987). "Analysis of Tapered Piles under Combined Action of Vertical, Horizontal, and Flexural Loads". *Soil Mechanics and Foundation Engineering*, Vol. 24, No. 1, pp. 7-12.

Gotman, A. L. (2000). "Finite-Element Analysis of Tapered Piles under Combined Vertical and Horizontal Loadings". *Soil Mechanics and Foundation Engineering*, Vol. 37, No. 1, pp. 5-12.

Gow, C. R. (1917). "History and Present Status of the Concrete Pile Industry". *Journal of the Boston Society of Civil Engineers*, Vol. IV, No. 4, pp. 143-200.

Gregersen, O. S., Aas, G., and DiBiagio, E. (1973). "Load Tests on Friction Piles in Loose Sand". Proceedings of the 8th International Conference on Soil Mechanics and Foundation Engineering, Vol. 2, pp. 109-117.

Hanna, T. H. (1987). "Ground Anchorages: Ultimate Load Estimation by the Chin Method". *Proceedings of the Institution of Civil Engineers*, Vol. 82, Issue 3, Part 1, pp. 601-605.

Hannigan, P. J., Goble, G. G., Thendean, G., Likins, G. E. and Rausche, F. (1998a). *Design and Construction of Driven Pile Foundations - Volume I*. Report FHWA-HI-97-013, U.S. Department of Transportation, Federal Highway Administration, Washington, DC.

Hannigan, P. J., Goble, G. G., Thendean, G., Likins, G. E. and Rausche, F. (1998b). *Design and Construction of Driven Pile Foundations - Volume II*. Report FHWA-HI-97-014, U.S. Department of Transportation, Federal Highway Administration, Washington, DC.

Hannigan, P. J., Rausche, F., Likins, G. E., Robinson, B. R., and Becker, M. L. (2016a). *Geotechnical Engineering Circular No. 12; Design and Construction of Driven Pile Foundations - Volume I*. Report FHWA-NHI-16-009, U.S. Department of Transportation, Federal Highway Administration, Washington, DC.

Hannigan, P. J., Rausche, F., Likins, G. E., Robinson, B. R., and Becker, M. L. (2016b). *Geotechnical Engineering Circular No. 12; Design and Construction of Driven Pile Foundations - Volume II*. Report FHWA-NHI-16-010, U.S. Department of Transportation, Federal Highway Administration, Washington, DC.

Hannigan, P. J., Rausche, F., Likins, G. E., Robinson, B. R., and Becker, M. L. (2016c). *Geotechnical Engineering Circular No. 12; Design and Construction of Driven Pile Foundations - Comprehensive Design Examples*. Report FHWA-NHI-16-064, U.S. Department of Transportation, Federal Highway Administration, Washington, DC.

Hardin, B. O. and Black, W. L. (1968). "Shear Modulus and Damping of Soils". *Journal of the Soil Mechanics and Foundations Division*, American Society of Civil Engineers, Vol. 94, No. 2, pp. 353-369.

Hardin, B. O. and Drnevich, V. P. (1972a). "Shear Modulus and Damping in Soils: Measurement and Parameter Effects". *Journal of the Soil Mechanics and Foundations Division*, American Society of Civil Engineers, Vol. 98, No. SM6, pp. 603-624.

Hardin, B. O. and Drnevich, V. P. (1972b). "Shear Modulus and Damping in Soils: Design Equations and Curves". *Journal of the Soil Mechanics and Foundations Division*, American Society of Civil Engineers, Vol. 98. No. SM7, pp. 667-692.

Hataf, N. and Nasrollahzadeh, E. (2019). "Experimental and Numerical Study on the Bearing Capacity of Single and Groups of Tapered and Cylindrical Piles in Sand". *International Journal of Geotechnical Engineering*.

Hataf, N. and Shafagat, A. (2015). "Numerical Comparison of Bearing Capacity of Tapered Pile Groups Using 3D FEM". *Geomechanics and Engineering*, Vol. 9, No. 5, pp. 547-567.

Hery, P. (1983). *Residual Stress Analysis in WEAP*. Thesis submitted to University of Colorado, Department of Civil Engineering and Architectural Engineering, Boulder, CO in partial fulfillment of the requirements for the degree of Master of Science.

Hirayama, H. (1990). "Load-Settlement Analysis for Bored Piles Using Hyperbolic Transfer Functions". *Soils and Foundations*, Vol. 30, No. 1, pp. 55–64.

Holloway, D. M., Clough, G. W., and Vesic, A. S. (1975). *The Mechanics of Pile-Soil Interaction in Cohesionless Soil*. Report S-75-5, U.S. Army Engineer Waterways Experiment Station, Vicksburg, MS.

Holloway, D. M., Clough, G. W., and Vesic, A. S. (1978). "The Effects of Residual Driving Stresses on Pile Performance under Axial Loads". Proceedings of the 10th Offshore Technology Conference, pp. 2225-2236.

Horvath, J. S. (1989). "A Prediction of Deep Foundation Capacity and Settlement". *Predicted and Observed Axial Behavior of Piles: Results of a Pile Prediction Symposium*, American Society of Civil Engineers.

Horvath, J. S. (1994). "Estimation of Spread-Footing Settlement on a Sand Subgrade". *Predicted and Measured Behavior of Five Spread Footings on Sand*, American Society of Civil Engineers.

Horvath, J. S. (1995). *Geofoam Geosynthetic*. Published by Horvath Engineering, P.C., Scarsdale, NY.

Horvath, J. S. (2000a). *Coupled Site Characterization and Foundation Analysis Research Project: Rational Selection of ϕ for Drained-Strength Bearing Capacity Analysis*. Research Report CE/GE-00-1, Manhattan College, School of Engineering, Civil Engineering Department, Geotechnical Engineering Program, Bronx, NY.

Horvath, J. S. (2000b). *Coupled Site Characterization and Foundation Analysis Research Project: Further Research into the Rational Selection of ϕ for Bearing Capacity Analysis under Drained-Strength Conditions*. Research Report CE/GE-00-3, Manhattan College, School of Engineering, Civil Engineering Department, Geotechnical Engineering Program, Bronx, NY.

Horvath, J. S. (2001). *Geomaterials Research Project; Geofoam and Geocomb Geosynthetics: A Bibliography Through the Second Millennium A.D.* Research Report CGT-2001-1, Manhattan College, School of Engineering, Center for Geotechnology, Bronx, NY.

Horvath, J. S. (2002). *Integrated Site Characterization and Foundation Analysis Research Project; Static Analysis of Axial Capacity of Driven Piles in Coarse-Grained Soil*. Research Report CGT-2002-1, Manhattan College, School of Engineering, Center for Geotechnology, Bronx, NY.

Horvath, J. S. (2003a). *Integrated Site Characterization and Foundation Analysis Research Project: Updated Site-Characterization Algorithm for Coarse-Grained Soils*. Research Report CGT-2003-2, Manhattan College, School of Engineering, Center for Geotechnology, Bronx, NY.

Horvath, J. S. (2003b). "Tapered Driven Piles: New Directions for an Old Concept". Proceedings of IeC GEO3 - International e-Conference on Modern Trends in Foundation Engineering: Geotechnical Challenges and Solutions.

Horvath, J. S. (2003c). "A New Analytical Method for the Axial-Compressive Static Capacity of Tapered Driven Piles in Coarse-Grained Soil". Proceedings of IeC GEO3 - International e-Conference on Modern Trends in Foundation Engineering: Geotechnical Challenges and Solutions.

Horvath, J. S. (2004). *Integrated Site Characterization and Foundation Analysis Research Project: A Technical Note re Effect of K_{onc} Assumption on Site-Characterization Algorithm for Coarse-Grained Soil*. Research Report CGT-2004-2, Manhattan College, School of Engineering, Center for Geotechnology, Bronx, NY.

Horvath, J. S. (2010a). "New Structural Model for Multicomponent Pile Cross Sections under Axial Load". *Journal of Geotechnical and Geoenvironmental Engineering*, Vol. 136, No. 6.

Horvath, J. S. (2010b). "Lateral Pressure Reduction on Earth-Retaining Structures Using Geofoams: Correcting Some Misunderstandings." Proceedings of ER2010 - Earth Retention Conference 3, American Society of Civil Engineers.

Horvath, J. S. (2011). "Improved Geotechnical Analysis through Better Integration and Dynamic Interaction between Site Characterization and Analytical Theory". *Geo-Frontiers 2011: Advances in Geotechnical Engineering*, American Society of Civil Engineers, pp. 2335-2344.

Horvath, J. S. (2014a). *Some Observations on the Behavior of Cantilever Sheet-Pile Walls: A White Paper*. Published by John S. Horvath Consulting Engineer, Scarsdale, NY.

Horvath, J. S. (2014b). *John F. Kennedy International Airport: A Seven-Decade Case Study of the Evolution of Geotechnical and Foundation Engineering Design and Construction Practice*. White paper published by John S. Horvath Consulting Engineer, Scarsdale, NY.

Horvath, J. S. (2015). *New Developments in Site Characterization and Pile-Resistance Calculation: A White Paper*. Published by John S. Horvath Consulting Engineer, Scarsdale, NY.

Horvath, J. S. (2016a). *Addendum #1 to New Developments in Site Characterization and Pile-Resistance Calculation: A White Paper*. Published by John S. Horvath Consulting Engineer, Scarsdale, NY.

Horvath, J. S. (2016b). *Addendum #2 to New Developments in Site Characterization and Pile-Resistance Calculation: A White Paper*. Published by John S. Horvath Consulting Engineer, Scarsdale, NY.

Horvath, J. S. (2018a). *Lateral Pressure Reduction on Earth-Retaining Structures Using Geofoam*. Published by John S. Horvath Consulting Engineer, Scarsdale, NY.

Horvath, J. S. (2018b). *Subgrade Modeling and Models in Foundation Engineering*. Published by John S. Horvath Consulting Engineer, Scarsdale, NY.

Horvath, J. S. and Trochalides, T. (2004). "A Half Century of Tapered Pile Usage at the John F. Kennedy International Airport". Proceedings of Fifth Case History Conference on Geotechnical Engineering, New York, NY.

Horvath, J. S., Trochalides, T., Burns, A., and Merjan, S. (2004a). "Axial-Compressive Capacities of a New Type of Tapered Steel Pipe Pile at the John F. Kennedy International Airport". Proceedings of Fifth Case History Conference on Geotechnical Engineering, New York, NY.

Horvath, J. S., Trochalides, T., Burns, A., and Merjan, S. (2004b). "A New Type of Tapered Steel Pipe Pile for Transportation Applications". Proceedings of Geo-Trans 2004, American Society of Civil Engineers.

Hughes, J. M. O., Wroth, C. P., and Windle, D. (1977). "Pressuremeter Tests in Sands". *Géotechnique*, Vol. 27, No. 4, pp. 455-477.

Hunter, A. H. and Davisson, M. T. (1969). "Measurements of Pile Load Transfer". *Performance of Deep Foundations - Special Technical Publication 444*, American Society for Testing and Materials, pp. 106-117.

Isaacs, D. V. (1931). "Reinforced Concrete Pile Formulae". *Transactions of the Institution of Engineers Australia/The Journal of the Institution of Engineers Australia*, Vol. 3, No. 9, pp. 305-323.

Janbu, N., and Senneset. K. (1974). "Effective Stress Interpretation of In Situ Static Penetration Tests". Proceedings of the 1st European Symposium on Penetration Testing, Vol. 2, pp. 181-193.

Kerisel, J. (1964). "Deep Foundations - Basic Experimental Facts". Proceedings of the North American Conference on Deep Foundations, Mexico City, Mexico, pp. 5-44.

Khan, M. K., El Naggar, M. H., and Elkasabgy, M. (2008). "Compression Testing and Analysis of Drilled Concrete Tapered Piles in Cohesive-Frictional Soil". *Canadian Geotechnical Journal*, Vol. 45, No. 3, pp. 377-392.

Kodikara, J. K. and Moore, I. D. (1993). "Axial Response of Tapered Piles in Cohesive Frictional Ground". *Journal of Geotechnical Engineering*, Vol. 119, No. 4, pp. 675-693.

Kondner, R. L. (1963). "Hyperbolic Stress-Strain Response of Cohesive Soils". *Journal of the Soil Mechanics and Foundations Division*, American Society of Civil Engineers, Vol. 89, No. SM1, pp. 115-143.

Krinitzsky, E. L. (1970). *Radiography in the Earth Sciences and Soil Mechanics*. Plenum Press.

Kulhawy, F. H. (1984). "Limiting Tip and Side Resistance: Fact or Fallacy?". *Analysis and Design of Pile Foundations*, American Society of Civil Engineers, pp. 80-98.

Kulhawy, F. H. (1991). "Drilled Shaft Foundations". Chapter 14 in *Foundation Engineering Handbook*, 2nd edition, Van Nostrand Reinhold, pp. 537-552.

Kulhawy, F. H. and Mayne, P. W. (1990). *Manual on Estimating Soil Properties for Foundation Design*. Report EL-6800, Research Project 1493-6, Electric Power Research Institute, Palo Alto, CA.

Kurian, N. P. and Srinivas, M. S. (1994). "Comparative Behaviour of Uniform and Tapered Piles". Proceedings of the 3rd International Conference on Deep Foundation Practice, Singapore, pp. 155-159.

Kurian, N. P. and Srinivas, M. S. (1995). "Studies on the Behaviour of Axially Loaded Tapered Piles by the Finite Element Method". *International Journal for Numerical and Analytical Methods in Geomechanics*, Vol. 19, No. 12, pp. 869-888.

Ladanyi, B., and Guichaoua, A. (1985). "Bearing Capacity and Settlement of Shaped Piles in Permafrost". Proceedings of the 11th International Conference on Soil Mechanics and Foundation Engineering, Vol. 4, pp. 1421–1427.

Lancellotta, R. (1995). *Geotechnical Engineering*. A.A. Balkema.

Leonards, G. A. (1970). Summary and review of Part II of *Pile Foundations; Highway Research Record No. 333*, Highway Research Board, Washington, DC, pp. 55-59.

Likins, G., Fellenius, B. H., and Holtz, R. D. (2012). "Pile Driving Formulas". *Piledriver*, Pile Driving Contractors Association, Vol. 9, No. 2, pp. 60-67.

Liu, J. and Zhang, M. (2012). "Measurement of Residual Force Locked in Open-ended Pipe Pile Using FBG-Based Sensors". *Electronic Journal of Geotechnical Engineering*, Vol. 17, Bundle O, pp. 2145-2154.

Long, J. H., Hendrix, J., and Jaromin, D. (2009). *Comparison of Five Different Methods for Determining Pile Bearing Capacities*. Report 0092-07-04, Wisconsin Highway Research Program, Wisconsin Department of Transportation, Madison, WI.

Lowery, L. L., Hirsch, T. J., Edwards, T. C., Coyle, H. M., and Samson, Jr., C. H. (1969). *Pile Driving Analysis; State of the Art*. Research Report 33-13 (Final), Texas A&M University, Texas Transportation Institute, College Station, TX.

Lunne, T., Long, M., and Forsberg, C. F (2003). "Characterisation and Engineering Properties of Holmen, Drammen Sand". In *Characterisation and Engineering Properties of Natural Soils*, Vol. 2, A. A. Balkema, pp. 1121-1148.

Manandhar, S. (2010). *Bearing Capacity of Tapered Piles in Sands*. Dissertation presented to Kyushu University, Department of Civil Engineering, Kyushu, Japan in partial fulfillment of the requirements for the degree of Doctor of Philosophy.

Manandhar, S. and Yasufuku, N. (2011a). "End Bearing Capacity of Tapered Piles in Sands Using Cavity Expansion Theory". *Memoirs of the Faculty of Engineering*, Kyushu University, Vol. 71, No. 4, pp. 77–99.

Manandhar, S. and Yasufuku, N. (2011b). "Evaluation of Skin Friction of Tapered Piles in Sands based on Cavity Expansion Theory [sic]". *Memoirs of the Faculty of Engineering*, Kyushu University, Vol. 71, No. 4, pp. 101-126.

Manandhar, S. and Yasufuku, N. (2012). "Analytical Model for the End-Bearing Capacity of Tapered Piles Using Cavity Expansion Theory". *Advances in Civil Engineering*, Hindawi Publishing Corp., Vol. 2012.

Manandhar, S. and Yasufuku, N. (2013). "Vertical Bearing Capacity of Tapered Piles in Sands Using Cavity Expansion Theory". *Soils and Foundations*, Vol. 53, No. 6, pp. 853-867.

Manandhar, S., Suetsugu, D., and Yasufuku, N. (2013). "Extended Model of Cavity Expansion Theory for Evaluating Skin Friction of Tapered Piles in Sand". Presentation at the 7th Geo[3] T[2], North Carolina Department of Transportation, Cary, NC.

Manandhar, S., Yasufuku, N., and Omine, K. (2010). "Tapering Effects of Piles in Cohesionless Soil". Proceedings of the 4th Japan–China Geotechnical Symposium on Recent Developments of Geotechnical Engineering, Okinawa, Japan, pp.477–482.

Manandhar, S., Yasufuku, N., and Shomura, K. (2009). "Skin Friction of Taper-Shaped Piles in Sands". Proceedings of the 28th International Conference on Ocean, Offshore and Arctic Engineering, American Society of Mechanical Engineering, pp. 93-102.

Manandhar, S., Yasufuku, N., Omine, K., and Kobayashi, T. (2010). "Response of Tapered Piles in Cohesionless Soil Based on Model Tests". *Journal of Nepal Geological Society*, Vol. 40, pp. 85-92.

Manandhar, S., Yasufuku, N., Omine K., and Qiang, L. (2009). "Mobilized Mechanism of Skin Friction of Tapered Piles in Sand". Proceedings of the International Joint Symposium on Geo-Disaster Prevention and Geoenvironment in Asia, Fukuoka, Japan, pp.171–178.

Manandhar, S., Yasufuku, N., Omine, K., and Taizo, K. (2010). "Response of Tapered Piles in Cohesionless Soil Based on Model Tests". *Journal of the Nepal Geological Society*, Vol. 40, pp. 85–92.

Mansur, C. L. and Kaufman, R. I. (1958). "Pile Tests, Low Sill Structure, Old River, Louisiana". *Transactions of the American Society of Civil Engineers*, Vol. 123, pp. 715-744.

Mathias, D. and Cribbs, M. (1998). *Driven 1.0: A Microsoft Windows™ Based Program for Determining Ultimate Vertical Static Pile Capacity*. Report FHWA-SA-98-074, U. S. Department of Transportation, Federal Highway Administration, Office of Technology Applications, Washington, DC.

Mayne, P. W. (2006a). "The 2006 James K. Mitchell Lecture: Undisturbed Sand Strength from Seismic Cone Tests". Presented at the GeoShanghai Conference, China and published in *Geomechanics and GeoEngineering: An International Journal*, Vol. 1, No. 4, pp. 239-257.

Mayne, P. W. (2006b). "In-Situ Test Calibrations for Evaluating Soil Parameters". In *Characterisation and Engineering Properties of Natural Soils*, Vol. 3, CRC Press - Taylor & Francis Group.

Mayne, P. W. (2007). *Cone Penetration Testing*. National Cooperative Highway Research Program Synthesis 368, Transportation Research Board, Washington, DC.

Mayne, P. W. (2014). "Interpretation of Geotechnical Parameters from Seismic Piezocone Tests". Proceedings of CPT '14 - Third International Symposium on Cone Penetration Testing, pp. 47-73.

Mayne, P. W. (2017). "Stress History of Soils from Cone Penetration Tests". The 34th Manuel Rocha Lecture as published in *Soils and Rocks - An International Journal of Geotechnical and Geoenvironmental Engineering*, Vol. 40, No. 3, pp. 203-216.

Mayne, P. W. (2019a). "Effective Stress Friction Angle of Sands, Silts, and Clays from CPT". Presentation at the 22nd annual geotechnical seminar of the American Society of Civil Engineers, Connecticut Society of Civil Engineers Section, Geo-Institute/Connecticut Valley Chapter, Meriden, CT.

Mayne, P. W. (2019b). "Yield Stress of Soils from Cone Penetration Tests". Presentation at the 22nd annual geotechnical seminar of the American Society of Civil Engineers, Connecticut Society of Civil Engineers Section, Geo-Institute/Connecticut Valley Chapter, Meriden, CT.

Mayne, P. W. and Niazi, F. S. (2017). "Recent Developments and Applications in Geotechnical Field Investigations for Deep Foundations". Proceedings of the 3rd Bolivian International Conference on Deep Foundations, Santa Cruz de la Sierra, Bolivia, Vol. 1.

Mayne, P. W. and Sharp, j. (2019). "CPT Approach to Evaluating Flow Liquefaction Using Yield Stress Ratio". Proceedings of Tailings and Mine Waste 2019, Vancouver, BC, Canada, pp. 655-670.

Meyerhof, G. G. (1982). *The Bearing Capacity and Settlement of Foundations*. Tech-Press, Technical University of Nova Scotia, Halifax, NS, Canada.

Mosley, E. T. and Raamot, T. (1970). "Pile-Driving Formulas". *Highway Research Record 333*, Highway Research Board, Washington, DC, pp. 23-32.

Moss, C. J. (2015). "The Problem Posed by New York's "20-Foot Clay": Which is Wrong – Published Dates, Presumed Glacial Events, or Strata Origins?". Proceedings of the 22nd Annual Conference on Geology of Long Island and Metropolitan New York, Stony Brook University, Department of Geosciences, Long Island Geologists Program, Stony Brook, NY.

Moss, C. J. and Canale, T. D. (2017). "Mapping the Marine Wantagh Formation, Commonly Known as the 20-Foot Clay, in Queens and Brooklyn". Proceedings of the 24th Annual Conference on Geology of Long Island and Metropolitan New York, Stony Brook University, Department of Geosciences, Long Island Geologists Program, Stony Brook, NY.

NeSmith, W.M. and Siegel, T.C. (2009). "Shortcomings of the Davisson Offset Limit Applied to Axial Compressive Load Tests on Cast-in-Place Piles". Proceedings of the International Foundation Congress and Equipment Expo.

Niazi, F. S. (2013). "A Review of the Design Formulations for Static Axial Response of Deep Foundations from CPT Data". *DFI Journal*, Vol. 7, No.2, pp. 58-78.

Niazi, F. S. (2014). *Static Axial Pile Foundation Response Using Seismic Piezocone Data*. Dissertation submitted to the Georgia Institute of Technology, College of Engineering, School of Civil and Environmental Engineering, Atlanta, GA in partial fulfillment of the requirements for the degree of Doctor of Philosophy.

Niazi, F. S. and Mayne, P. W. (2015a). "Enhanced UniCone Expressions for Axial Pile Capacity Evaluation from Piezocone Tests". Proceedings of IFCEE 2015 - International Foundations Congress and Equipment Expo 2015, pp. 202-216.

Niazi, F. S. and Mayne, P. W. (2015b). "Operational Soil Stiffness from Back-Analysis of Pile Load Tests within Elastic Continuum Framework". *Geotechnical Engineering Journal of the SEAGS & AGSSEA*, Vol. 46, No. 2.

Niazi, F. S. and Mayne, P. W. (2015c). "Elastic Continuum Solution of Stacked Pile Model for Axial Load-Displacement Analysis". *Geotechnical Engineering Journal of the SEAGS & AGSSEA*, Vol. 46, No. 2.

Niazi, F. S. and Mayne, P. W. (2019). "Modeling of Shear Stiffness Reduction from Database of Axial Load Tests on Pile Foundations". Proceedings of IS-Glasgow 2019 - 7th International Symposium on Deformation Characteristics of Geomaterials.

Niazi, F. S., Mayne, P. W., and Woeller, D. J. (2010). "Case History of Axial Pile Capacity and Load-Settlement Response by SCPTu". Proceedings of CPT '10 - Second International Symposium on Cone Penetration Testing, Huntington Beach, CA.

Nordlund[211], R. L. (1963). "Bearing Capacity of Piles in Cohesionless Soils". *Journal of the Soil Mechanics and Foundations Division*, American Society of Civil Engineers, Vol. 89, No. SM3, pp. 1-35.

Nordlund, R. L. (1979). "Point Bearing and Shaft Friction of Piles in Sand". Presentation notes at the 5th Annual Fundamentals of Deep Foundation Design short course sponsored by the University of Missouri at Rolla, St. Louis, MO.

NYSDOT (2015). *Charts to Facilitate Computation of Skin Friction on Driven Non-Tapered Piles in Cohesionless Soil*. Revision No. 2 to Geotechnical Engineering Manual No. 11 (GEM-11), State of New York, Department of Transportation, Geotechnical Engineering Bureau.

Ouyang, Z. and Mayne, P. W. (2018). "Effective Friction Angle of Clays and Silts from Piezocone Penetration Tests". *Canadian Geotechnical Journal*, Vol. 55, pp. 1230–1247.

Paik, K., Lee, J., and Kim, D. (2011). "Axial Response and Bearing Capacity of Tapered Piles in Sandy Soil". *Geotechnical Testing Journal*, Vol. 34, No. 2, pp. 122-130.

---

[211] Appeared as "Norlund" in the heading of the original paper. Although this misspelling was corrected in the errata that appeared with Nordlund's closure in Vol. 90/No. SM4 (July 1964), this spelling error has, unfortunately, propagated through the literature over the years since the original paper was published.

Pando, M. A., Ealy, C. D., Filz, G. M., Lesko, J. J., and Hoppe, E. J. (2006). *A Laboratory and Field Study of Composite Piles for Bridge Substructures*. Report FHWA-HRT-04-043, U.S. Department of Transportation, Federal Highway Administration, Office of Infrastructure R&D, McLean, VA.

Peck, R. B. (1958). *A Study of the Comparative Behavior of Friction Piles*. Special Report 36, Highway Research Board, Washington, DC.

Peck, R. B., Hanson, W. E., and Thornburn, T. H. (1953). *Foundation Engineering*. Wiley.

Petroski, Henry (1992). *The Evolution of Useful Things*. Knopf.

Poulos, H. G. (1987). "Analysis of Residual Stress Effects in Piles". *Journal of Geotechnical Engineering*, Vol. 113, No. 3, pp. 216-229.

Poulos, H. G. and Davis, E. H. (1974). *Elastic Solutions for Soil and Rock Mechanics*. Wiley.

Poulos, H. G. and Davis, E. H. (1980). *Pile Foundation Analysis and Design*. Wiley.

Randolph, M. F. and Wroth, C. P. (1978). "Analysis of Deformation of Vertically Loaded Piles". *Journal of the Geotechnical Engineering Division*, American Society of Civil Engineers, Vol. 114, No. 12, pp. 1465–1488.

Rao, P., Cui, J., and Li, J. (2014). "Elastoplastic Solutions of Cylindrical Cavity Expansion Considering the $K_o \neq 1$". Proceedings of Geo-Shanghai 2014.

Rieke, R. D. and Crowser, J. C. (1987). "Interpretation of Pile Load Test Considering Residual Stresses". *Journal of Geotechnical Engineering*, Vol. 113, No. 4, pp. 320-334.

Robertson, P. K. (2012). "Interpretation of In-Situ Tests - Some Insights". Proceedings of the Fourth International Conference on Site Characterization, Recife, Brazil.

Robertson, P. K. and Cabal, K. L. (2010). "Estimating Soil Unit Weight from CPT". Proceedings of CPT '10 - Second International Symposium on Cone Penetration Testing.

Robertson, P. K. and Cabal, K. L. (2012). *Guide to Cone Penetration Testing for Penetration Testing*, 5th edition. Gregg Drilling & Testing, Inc., Signal Hill, CA.

Robertson, P. K. and Cabal, K. L. (2015). *Guide to Cone Penetration Testing for Penetration Testing*, 6th edition. Gregg Drilling & Testing, Inc., Signal Hill, CA.

Robinsky, E. I. (1963). *Effect of Shape and Volume Displacement on the Capacity of Piles in Sand*. Dissertation presented to the University of Toronto, Department of Civil Engineering, Toronto, ON, Canada in partial fulfillment of the requirements for the degree of Doctor of Philosophy.

Robinsky, E. I. and Morrison, C. F. (1964). "Sand Displacement and Compaction around Model Friction Piles". *Canadian Geotechnical Journal*, Vol. 1, No. 2, pp. 81-93.

Robinsky, E. I., Sagar, W. L., and Morrison, C. F. (1964). "Effect of Shape and Volume on the Capacity of Model Piles in Sand". *Canadian Geotechnical Journal*, Vol. 1, No. 4, pp. 189-204.

Rybnikov, A. M. (1990). "Experimental Investigations of Bearing Capacity of Bored-Cast-in-Place Tapered Piles". *Soil Mechanics and Foundation Engineering*, Vol. 27, No. 2, pp 48–52.

Sabry, M. (2005). *In Situ Stresses and Capacity of Driven Piles in Sand*. Dissertation submitted to Concordia University, Department of Building, Civil, and Environmental Engineering, Montreal, QC, Canada in partial fulfillment of requirements for the Degree of Doctor of Philosophy.

Sabry, M. and Hanna, A. (2009). "Earth Pressure Acting on Single Driven Piles in Sand". Proceedings of the 17th International Conference on Soil Mechanics and Geotechnical Engineering, pp. 1255-1258.

Sakr, M., and El Naggar, M.H. (2003). "Centrifuge Modeling of Tapered Piles in Sand". *Geotechnical Testing Journal*, Vol. 26, No. 1, pp. 22–35.

Sakr, M., El Naggar, M. H., and Nehdi, M. (2004a). "Load Transfer of Fibre-Reinforced Polymer (FRP) Composite Tapered Piles in Dense Sand". *Canadian Geotechnical Journal*, Vol. 41, No. 1, pp. 70-88.

Sakr, M., El Naggar, M. H., and Nehdi, M. (2004b). "Novel Toe Driving for Thin-Walled Piles and Performance of Fiberglass-Reinforced Polymer (FRP) Pile Segments". *Canadian Geotechnical Journal*, Vol. 41, No. 2, pp. 313-325.

Sakr, M., El Naggar, M. H., and Nehdi, M. (2005). "Uplift Performance of FRP Tapered Piles in Dense Sand". *International Journal of Physical Modelling in Geotechnics*, Vol. 2, pp. 1–16.

Sakr, M., El Naggar, M. H., and Nehdi, M. (2007). "Wave Equation Analyses of Tapered FRP-Concrete Piles in Dense Sand". *Soil Dynamics and Earthquake Engineering*, Vol. 27, No. 2, pp. 166–182.

Salgado, R. and Prezzi, M. (2007). "Computation of Cavity Expansion Pressure and Penetration Resistance in Sands". *International Journal of Geomechanics*, Vol. 7, No. 4, pp. 251-265.

Salgado, R., Mitchell, J. K., and Jamiolkowski, M. (1997). "Cavity Expansion and Penetration Resistance in Sand". *Journal of Geotechnical and Geoenvironmental Engineering*, Vol. 123, No. 4, pp. 344-354.

Sandven, R. (1990). *Strength and Deformation Properties Obtained from Piezocone Tests*. Dissertation submitted to the Norwegian University of Science & Technology, Trondheim, Norway in partial fulfillment of the requirements for the degree of Doctor of Philosophy.

Sandven, R., and Watn, A. (1995). "Interpretation of Test Results; Soil Classification and Parameter Evaluation from Piezocone Tests; Results from Oslo Airport". Theme lecture in the proceedings of the International Symposium on Cone Penetration Testing, Vol. 3, Swedish Geotechnical Society Report SGF 3:95, Linköping, Sweden, pp. 35–55.

Senneset, K., and Janbu, N. (1985). "Shear Strength Parameters Obtained from Static Cone Penetration Tests". In *Strength Testing of Marine Sediments*, Special Technical Publication No. 883, American Society for Testing and Materials, pp. 41–54.

Senneset, K., Sandven, R., and Janbu, N. (1989). "Evaluation of Soil Parameters from Piezocone Tests". *Transportation Research Record 1235*, Transportation Research Board, Washington, DC, pp. 24–37.

Siegel, T. and McGillivray, A. (2009). "Interpreted Residual Load in an Augered Cast-in-Place Pile". Proceedings of the Deep Foundations Institute 34th Annual Conference on Deep Foundations, Kansas City, MO.

Smith, E. A. L. (1954). "Impact and Longitudinal Wave Transmission". Presented at the 1954 Annual Meeting of the American Society of Mechanical Engineers and published in the August 1955 *Transactions of the American Society of Mechanical Engineers* with discussions and closure, pp. 963-973.

Smith, E. A. [sic] (1957). "What Happens When Hammer Hits Pile". *Engineering News-Record*, 5 September.

Smith, E.A.L. (1960). "Pile Driving Analysis by the Wave Equation". *Journal of the Soil Mechanics and Foundations Division*, American Society of Civil Engineers, Vol. 86, No. 4, pp. 35-64.

Smith, E.A.L. (1962). "Pile Driving Analysis by the Wave Equation" with discussions and closure. *Transactions*, American Society of Civil Engineers, Vol. 127, Part I, pp. 1145-1193.

Spillers, W. R. and Stoll, R. D. (1964). "Lateral Response of Piles". *Journal of the Soil Mechanics and Foundations Division*, American Society of Civil Engineers, Vol. 90, No. 6, pp. 1-10.

Stark, T. D., Arellano, D., Horvath, J. S., and Leshchinsky, D. (2004a). *Geofoam Applications in the Design and Construction of Highway Embankments*. National Cooperative Highway Research Project Web Document 65, Transportation Research Board, Washington, DC.

Stark, T. D., Arellano, D., Horvath, J. S., and Leshchinsky, D. (2004b). *Guideline and Recommended Standard for Geofoam Applications in Highway Embankments*. National Cooperative Highway Research Project Report 529, Transportation Research Board, Washington, DC.

Suits, L. D., Sheahan, T. C., Sakr, M., and El Naggar, M. H. (2003). "Centrifuge Modeling of Tapered Piles in Sand". *Geotechnical Testing Journal*, Vol. 26, No. 1.

Suits, L. D., Sheahan, T. C., Paik, K., Lee, J., and Kim, D. (2011). "Axial Response and Bearing Capacity of Tapered Piles in Sandy Soil". *Geotechnical Testing Journal*, Vol. 34, No. 2.

Tan, S. A. and Fellenius, B. H. (2016). "Negative Skin Friction Pile Concepts with Soil-Structure Interaction". *Institution of Civil Engineers Geotechnical Research Journal*.

Tavenas, F. and Audy, R. (1972). "Limitations of the Driving Formulas for Predicting the Bearing Capacities of Piles in Sand". *Canadian Geotechnical Journal*, Vol. 9, No. 1, pp. 47-62.

Terzaghi, K. (1942). Discussion of the "Progress Report of the Committee on the Bearing Value of Pile Foundations". *Proceedings of the American Society of Civil Engineers*, Vol. 68, pp. 311–323.

Terzaghi, K. (1943). *Theoretical Soil Mechanics*. Wiley.

Timber Pile Design & Construction Manual (2016). Southern Pressure Treaters' Association, 2002 edition with 2016 revisions.

Timoshenko, S. P. and Goodier, J. N. (1970). *Theory of Elasticity*, 3rd edition. McGraw-Hill.

Togliani, G. (2010). "Pile Capacity Prediction Using CPT - Case History". Proceedings of CPT '10 - Second International Symposium on Cone Penetration Testing, Huntington Beach, CA.

Vali, R., Khotbehsara, E. M., Saberian, M., Li, J., Mehrinejad, M., and Jahandari, S. (2019, published online in June 2017). "A Three-Dimensional Numerical Comparison of Bearing Capacity and Settlement of Tapered and Under-Reamed Piles". *International Journal of Geotechnical Engineering*, Vol. 13, No. 3, pp. 236-248.

Vardanega, P. J. and Bolton, M. D. (2011). "Practical Methods to Estimate the Non-Linear Shear Stiffness of Fine Grained Soils". Proceedings of the International Symposium on Deformation Characteristics of Geomaterials, Seoul, R.O.K., pp. 372-379.

Vesic, A. S. (1970). "Load Transfer in Pile-Soil Systems". Proceedings of the Conference on Design and Installation of Pile Foundations and Cellular Structures, Lehigh University, Bethlehem, PA, pp. 47-73.

Vesic, A. S. (1972). "Expansion of Cavities in Infinite Soil Mass". *Journal of the Soil Mechanics and Foundations Division*, American Society of Civil Engineers, Vol. 98, No. 3, pp. 265-290.

Vesic, A. S. (1975). *Principles of Pile Foundation Design*. Soil Mechanics Series No. 38, Duke University, NC.

Vesic, A. S. (1977). *Design of Pile Foundations*. NCHRP Synthesis of Highway Practice 42, Transportation Research Board, Washington, DC.

Wang, H. and Zhang, L. M. (2008). "Estimated Residual Stresses in Long Driven Piles in Weathered Soils". Proceedings of the 8th International Conference on the Application of Stress-Wave Theory to Piles, IOS Press BV, pp. 265-270.

Wei, J. Q. (1998). *Experimental Investigation of Tapered Piles*. Thesis submitted to The University of Western Ontario, Faculty of Graduate Studies, London, ON, Canada in partial fulfillment of the requirements for the degree of Master of Engineering Science.

Wei, J. Q. and El Naggar, M. H. (1998). "Experimental Study of Axial Behaviour of Tapered Piles". *Canadian Geotechnical Journal*, Vol. 35, No. 4, pp. 641-654.

Wolfe, R. W. (1989). "Allowable Stresses for the Upside-Down Timber Industry". Paper prepared for publication in the proceedings of the First International Conference on Wood Piles and Poles.

York, D. L., Brusey, W. G., Clemente, F. M., and Law, S. K. (1994). "Setup and Relaxation in Glacial Sand". *Journal of Geotechnical Engineering*, Vol. 120, No. 9, pp. 1498-1513.

York, D. L., Miller, V. G., and Ismael, N. F. (1974). "Long-Term Load Transfer in End-Bearing Pipe Piles". *Transportation Research Record 517*, Transportation Research Board, Washington, DC, pp. 48-60.

Yu, H. S. and Houlsby, G. T. (1991). "Finite Cavity Expansion in Dilatant Soils: Loading Analysis". *Géotechnique,* Vol. 41, No. 2, pp. 172-183.

Zelada-Tumialan, G., Konicki, W., Westover, P. and Vatovec, M. (2013). "Untreated Submerged Timber Pile Foundations; Part 1: Understanding Biodegradation and Compressive Strength". *STRUCTURE magazine*, December, pp. 9-11.

Zelada-Tumialan, G., Konicki, W., Westover, P. and Vatovec, M. (2014). "Untreated Submerged Timber Pile Foundations; Part 2: Estimating Remaining Service Life". *STRUCTURE magazine*, January, pp. 10-13.

Zhan, Y.-G., Wang, H., and Liu, F.-C. (2012). "Numerical Study on Load Capacity Behavior of Tapered Pile Foundations". *The Electronic Journal of Geotechnical Engineering*, Vol. 17, Bundle N, pp. 1969-1980.

Zhang, L. M. and Wang, H. (2007). "Development of Residual Forces in Long Driven Piles in Weathered Soils". *Journal of Geotechnical and Geoenvironmental Engineering*, Vol. 133, No. 10, pp. 1216-1228.

Zhou, H., Liu, H., and Kong, G. (2014). "Influence of Shear Stress on Cylindrical Cavity Expansion in Undrained Elastic–Perfectly Plastic Soil". *Géotechnique Letters*, Vol. 4, No. 3, pp. 203-210.

Zil'berberg, S. D. and Sherstnev, A. D. (1990). "Construction of Compaction Tapered Pile Foundations". Translated from *Osnovaniya, Fundamety I Mekhanika Gruntov*, Plenum Publishing Corporation, No. 3, pp. 7–10.

This page intentionally left blank.